EXPLORING THE UNKNOWN

ISBN 0-16-048899-0

9 780160 488993

90000

NASA SP-4407

EXPLORING THE UNKNOWN

Selected Documents in the History of the
U.S. Civilian Space Program
Volume II: External Relationships

John M. Logsdon, Editor
with Dwayne A. Day and Roger D. Launius

The NASA History Series

National Aeronautics and Space Administration
NASA History Office
Washington, D.C.   1996

Library of Congress Cataloguing-in-Publication Data

Exploring the Unknown: Selected Documents in the History of the U.S. Civil Space
Program/John M. Logsdon, editor ... [et al.]
p. cm.—(The NASA history series) (NASA SP: 4407)

    Includes bibliographical references and indexes.
    Contents: v. 1. Organizing for exploration
    1. Astronautics—United States—History.   I. Logsdon, John M., 1937–
II. Lear, Linda J., 1940–   III. Series.   IV. Series: NASA SP: 4407.
TL789.8.U5E87   1996   96-9066
387.8'0973–dc20   CIP

For sale by the U.S. Government Printing Office
Superintendent of Documents, Mail Stop: SSOP, Washington, DC 20402-9328
ISBN 0-16-048899-0

Contents

Chapter One

Essay: "The Development of International Space Cooperation," by John M. Logsdon . . 1

Documents

Chapter Two

Essay: "Invitation to Struggle: The History of Civilian-Military Relations in Space,"

Documents

Chapter Three

Essay: "The NASA-Industry-University Nexus: A Critical Alliance in the

Documents

Acknowledgments

This volume is the second in a series that had its origins almost a decade ago. The initial idea for creating a set of reference works that would include documents seminal to the evolution of the U.S. civilian space program came from then-NASA Chief Historian Sylvia K. Kraemer. She recognized that while there were substantial primary resources for future historians and others interested in the early years and evolution of the U.S. civilian space program available in many archives, and particularly in the NASA Historical Reference Collection of the History Office at NASA Headquarters in Washington, D.C., this material was widely scattered and contained a mixture of the significant and the routine. It was her sense that it was important to bring together the "best" of this documentary material in a widely accessible form. The several volumes of this collection, and any long-term value it may have, are the result of that vision. Once Dr. Kraemer left her position as NASA Chief Historian in 1990 to assume broader responsibilities within the agency, the project was guided with a gentle but firm hand by her successor, Roger D. Launius. His contributions have been so substantial that he fully deserves being listed as one of the primary collaborators on this volume.

Jannelle Warren-Findley, an independent intellectual/cultural historian, and Linda J. Lear, an adjunct professor of environmental history at George Washington University, approached the Space Policy Institute of George Washington University's Elliott School of International Affairs with the suggestion that it might serve as the institutional base for a proposal to NASA to undertake the documentary history project. This suggestion found a positive response. The Space Policy Institute was created in 1987 as a center of scholarly research and graduate education regarding space issues—and as a resource for those interested in a knowledgeable but independent perspective on past and current space activities. Having the kind of historical base that would have to be created to carry out the documentary history project would certainly enhance the Space Policy Institute's capabilities, and so the Director of the Institute, John M. Logsdon, joined with Warren-Findley, Lear, and Ray A. Williamson of the congressional Office of Technology Assessment in preparing a proposal to NASA. Much to our delight, we were awarded the contract for the project in late 1988, and the enterprise was officially under way in May 1989.

The undertaking proved more challenging than anyone had anticipated. The combination of getting ourselves started in the right direction, canvassing and selecting from the immense documentary resources available, commissioning essays to introduce the various sections of the work from external authors and writing several essays ourselves, and dealing with conflicting demands on the time of the four principals in the project has led to a delay in publishing the initial volume far beyond what we anticipated when first undertaking the project. The final pieces of the manuscript for Volume I were not delivered to NASA until the end of 1993, and the published volume itself did not appear until late 1995. By that time, both Jannelle Warren-Findley and Linda Lear had long ago moved on to the next steps in their careers, and Ray Williamson, who had taken a nine-month leave from the Office of Technology Assessment in 1990 to work on the project, had returned to his primary job. (However, after the Office of Technology Assessment in 1995 was closed by congressional decision, he returned to the staff of the Space Policy Institute, and he will be deeply involved in preparing subsequent volumes in this series.) This meant that Warren-Findley and Lear did not have the opportunity to make the kinds of contribution to the overall series that they had anticipated; nevertheless, without their initiative, the effort would not have been located at George Washington University, and they both made crucial contributions to conceptualizing and organizing the work in its early stages and to gathering the material from which the documents to be included in the collection have been selected. For all of that, they deserve high credit.

In its start-up phase, the project profited from the advice of a distinguished advisory panel that met twice formally; members of the panel were always available for individual consultation. Included on this panel were: Carroll W. Pursell, Jr., Case Western University (chair); Charlene Bickford, First Congress Project; Herbert Friedman, Naval Research Laboratory; Richard P. Hallion, Air Force Historian; John Hodge, NASA (retired); Sally Gregory Kohlstedt, University of Minnesota; W. Henry Lambright, Syracuse University; Sharon Thibodeau, National Archives and Records Administration; and John Townsend, NASA (retired). Certainly, none of these individuals bear responsibility for the final content or style of this series, but their advice along the way was invaluable.

We owe thanks to the individuals and organizations that have searched their files for potentially useful materials, as well as to the staffs at various archives and collections who have helped us locate documents. Without question, first among them is Lee D. Saegesser of the History Office at NASA Headquarters, who has helped compile the NASA Historical Reference Collection that contains many of the documents selected for inclusion in this work. All those who in the future will write on the history of the U.S. space program will owe a debt of thanks to Lee; those who have already worked in this area realize his tireless contributions.

Among those who have been particularly helpful in identifying documents for inclusion in this volume are: Lorenza Sebesta, European Space Agency History Project, European University Institute; R. Cargill Hall, Air Force History Office; Mark Erikson, Air Force Academy; Roy Houchin, Air Staff History Office; and Bill Burr, National Security Archive. L. Parker Temple III and Charles Cook deserve particular thanks for their work on an earlier overview of civil-military relations that served as a basis for the essay by Dwayne A. Day that appears in this volume.

Essential to the project was a system for archiving the documents collected. Charlene Bickford, on the basis of her experience with the First Congress Project, advised on our approach to archiving and to developing document headnotes. The archiving system was developed by graduate student John Morris, who also assisted with initial document collection. The documentary archive has been nurtured with fervor by Space Policy Institute research associate Dwayne Day; Dwayne has made many major contributions to all aspects of the project, including the essay on NASA-Department of Defense relations in this volume. Other students who worked on the project in its early years include Max Nelson, Jordan Katz, Stewart Money, Michelle Heskett, Robin Auger, and Heather Young. All were a great help.

Beginning with Linda Lear, a series of individuals struggled to bring editorial consistency to the essays and headnotes introducing the documents included in Volume I, thereby setting the standard for that and subsequent volumes. They included Erica Aungst, Kathie Pett Keel, and Kimberly Carter. Their contributions were essential to the lasting quality of the end product. Alita Black also helped set up the initial indexing system.

John M. Logsdon, George Washington University

****

There are numerous people at NASA associated with historical study, technical information, and the mechanics of publishing who helped in myriad ways in the preparation of this documentary history. J.D. Hunley, of the NASA History Office, edited and critiqued the text before he departed to take over the History Program at the Dryden Flight Research Center; his replacement, Stephen J. Garber, prepared the biographical appendix and helped in the final proofing of the work. Nadine Andreassen of the NASA History Office performed editorial and proofreading work on the project; the staffs of the NASA Headquarters Library, the Scientific and Technical Information Program, and the NASA Document Services Center provided assistance in locating and preparing for publication the documentary materials in this work. The NASA Headquarters Communications Management Division, under the leadership of Y. Diane Powell, developed the layout and handled printing. Specifically, we wish to acknowledge the work of Jane E. Penn, Patricia Lutkenhouse Talbert, Jonathan L. Friedman, and Kelly L. Rindfusz for their design and editorial work. In addition, Michael Crnkovic and Craig A. Larsen saw the book through the publication process. Thanks are due to all of them.

Roger D. Launius, NASA

Introduction

One of the most important developments of the twentieth century has been the movement of humanity into space with machines and people. The underpinnings of that movement—why it took the shape it did; which individuals and organizations were involved; what factors drove a particular choice of scientific objectives and technologies to be used; and the political, economic, managerial, and international contexts in which the events of the space age unfolded—are all important ingredients of this epoch transition from an Earth-bound to a spacefaring people. The desire to understand the development of spaceflight in the United States sparked this documentary history series.

The extension of human activity into outer space has been accompanied by a high degree of self-awareness of its historical significance. Few large-scale activities have been as extensively chronicled so closely to the time they actually occurred. Many of those who were directly involved were quite conscious that they were making history, and they kept full records of their activities. Because most of the activity in outer space was carried out under government sponsorship, it was accompanied by the documentary record required of public institutions, and there has been a spate of official and privately written histories of most major aspects of space achievement to date. When top leaders considered what course of action to pursue in space, their deliberations and decisions often were carefully put on the record. There is, accordingly, no lack of material for those who aspire to understand the origins and evolution of U.S. space policies and programs.

This reality forms the rationale for this series. Precisely because there is so much historical material available on space matters, the National Aeronautics and Space Administration (NASA) decided in 1988 that it would be extremely useful to have a selective collection of many of the seminal documents related to the evolution of the U.S. civilian space program that was easily available to scholars and the interested public. While recognizing that much of the space activity has taken place under the sponsorship of the Department of Defense and other national security organizations, for the U.S. private sector, and in other countries around the world, NASA felt that there would be lasting value in a collection of documentary material primarily focused on the evolution of the U.S. government's civilian space program. Most of this activity has been carried out under the NASA's auspices since its creation in 1958. As a result, the NASA History Office contracted with the Space Policy Institute of George Washington University's Elliott School of International Affairs to prepare such a collection. This is the second volume in the documentary history series; at least two additional ones detailing programmatic developments will follow.

The documents collected during this research project were assembled from a diverse number of both public and private sources. A major repository of primary source materials relative to the history of the civilian space program is the NASA Historical Reference Collection of the NASA History Office, located at the agency's headquarters in Washington, D.C. Project assistants combed this collection for the "cream" of the wealth of material housed there. Indeed, one purpose of this series from the start was to capture some of the highlights of the holdings at headquarters. Historical materials housed at the other NASA installations—as well as at institutions of higher learning, such as Rice University, Rensselaer Polytechnic Institute, and Virginia Polytechnic Institute and State University (Virginia Tech)—were also "mined" for their most significant materials. Other collections from which documents have been drawn include the Eisenhower, Kennedy, Johnson, and Carter Presidential Libraries; the papers of T. Keith Glennan, Thomas O.

Paine, James C. Fletcher, George M. Low, and John A. Simpson; and the archives of the National Academy of Sciences, the Rand Corporation, AT&T, the Communications Satellite Corporation, INTELSAT, the Jet Propulsion Laboratory of the California Institute of Technology, and the National Archives and Records Administration.

Copies of more than 2,000 documents in their original form collected during this project (not just the documents selected for inclusion), as well as a database that provides a guide to their contents, will be deposited in the NASA Historical Reference Collection. Another complete set of project materials is located at the Space Policy Institute at George Washington University. These materials in their original form are available for use by researchers seeking additional information about the evolution of the U.S. civilian space program or wishing to consult the documents reprinted herein in their original form.

The documents selected for inclusion in this volume are presented in three chapters, each covering a particular aspect of the evolution of U.S. space exploration. These chapters address (1) the relations between the civilian space program of the United States and the space activities of other countries, (2) the relations between the U.S. civilian space program and the space efforts of national security organizations and the military, and (3) NASA's relations with industry and academic institutions. Volume I of this series covered the antecedents to the U.S. space program, the origins and evolution of U.S. space policy, and NASA as an organizational institution. Future volumes will address space science activities, space application programs, human spaceflight, and space transportation activities.

Each chapter in this volume is introduced by an overview essay, prepared either by a member of the project team or by an individual particularly well-qualified to write on the topic. In the main, these essays are intended to introduce and complement the documents in the section and to place them in a chronological and substantive context. In certain instances, the essays go beyond this basic goal to reinterpret specific aspects of the history of the civilian space program and to offer historiographical commentary or inquiry about the space program. Each essay contains references to the documents in the section it introduces, and many also contain references to documents in other sections of the collection. These introductory essays were the responsibility of their individual authors, and the views and conclusions contained therein do not necessarily represent the opinions of either George Washington University or NASA.

The documents appended to each chapter were chosen by the essay writer in concert with the project team from the more than 2,000 assembled by the research staff for the overall project. The contents of this volume emphasize primary documents or long-out-of-print essays or articles and material from the private recollections of important actors in shaping space affairs. The contents of this volume thus do not comprise in themselves a comprehensive historical account; they must be supplemented by other sources, those both already available and to become available in the future. Indeed, a few of the documents included in this collection, particularly in the chapter on civilian-military relations, are not complete; some portions of them were still subject to security classification. As this collection was being prepared, the U.S. government was involved in declassifying and releasing to the public a number of formerly highly classified documents. As this declassification process continues, increasingly more information on the early history of NASA and the civilian space program will come to light.

The documents included in each chapter are for the most part arranged chronologically, although some thematic organization is used when appropriate. Each document is assigned its own number in terms of the chapter in which it is placed. As a result, the first document in the third chapter of this volume is designated "Document III-1." Each document is accompanied by a headnote setting out its context and providing a background narrative. These headnotes also provide specific information about the people and events discussed. We have avoided the inclusion of explanatory notes in the documents themselves and have confined such material to the headnotes.

The editorial method we adopted for dealing with these documents seeks to preserve spelling, grammar, paragraphing, and use of language as in the original. We have sometimes changed punctuation where it enhances readability. We have used ellipses to note sections of a document not included in this publication, and we have avoided including words and phrases that had been deleted in the original document unless they contribute to an understanding of what was going on in the mind of the writer in making the record. Marginal notations on the original documents are inserted into the text of the documents in brackets, each clearly marked as a marginal comment. When deletions to the original document have been made in the process of declassification, we have noted this with a parenthetical statement in brackets. Except insofar as illustrations and figures are necessary to understanding the text, those items have been omitted from this printed version. Page numbers in the original document are noted in brackets internal to the document text. Copies of all documents in their original form, however, are available for research by anyone interested at the NASA History Office or the Space Policy Institute of George Washington University.

We recognize that there are significant documents left out of this compilation. No two individuals would totally agree on all documents to be included from the more than 2,000 that we collected, and surely we have not been totally successful in locating all relevant records. As a result, this documentary history can raise an immediate question from its users: Why were some documents included while others of seemingly equal importance were omitted? There can never be a fully satisfactory answer to this question. Our own criteria for choosing particular documents and omitting others rested on three interrelated factors:

- Is the document the best available, most expressive, most representative reflection of a particular event or development important to the evolution of the space program?

- Is the document not easily accessible except in one or a few locations, or is it included (for example, in published compilations of presidential statements) in reference sources that are widely available and thus not a candidate for inclusion in this collection?

- Is the document protected by copyright, security classification, or some other form of proprietary right and thus unavailable for publication?

As editor of this volume, I was ultimately responsible for the decisions about which documents to include and for the accuracy of the headnotes accompanying them. It has been an occasionally frustrating but consistently exciting experience to be involved with this undertaking. My associates and I hope that those who consult it in the future will find our efforts worthwhile.

John M. Logsdon
Director
Space Policy Institute
Elliott School of International Affairs
George Washington University

Biographies of Volume II Essay Authors

Dwayne A. Day is a Guggenheim Fellow at the National Air and Space Museum of the Smithsonian Institution and a staff member of George Washington University's Space Policy Institute in Washington, D.C. He is the author of numerous articles on the development of space policy in the United States in such periodicals as *Space Policy, Spaceflight,* and *Quest: The Magazine of Spaceflight History.* He also was a co-editor of Volume I of *Exploring the Unknown.*

W. Henry Lambright is professor of political science and public administration at the Maxwell School at Syracuse University in Syracuse, New York. A premier scholar of the management of high technology in the federal government, he is the author of Governing Science and Technology (Oxford University Press, 1976), *Shooting Down the Nuclear Plane* (Bobbs-Merrill, 1976), *Technology Transfer to Cities* (Westview Press, 1979), *Presidential Management of Science and Technology: The Johnson Presidency* (University of Texas Press, 1985), and *Powering Apollo: James E. Webb of NASA* (Johns Hopkins University Press, 1995).

John M. Logsdon is Director of both the Center for International Science and Technology Policy and the Space Policy Institute of George Washington University's Elliott School of International Affairs, where he is also a professor of political science and international affairs. He holds a B.S. in physics from Xavier University and a Ph.D. in political science from New York University. He has been at George Washington University since 1970; he previously taught at The Catholic University of America. Dr. Logsdon's research interests include space policy, the history of the U.S. space program, the structure and process of government decision-making for research and development programs, and international science and technology policy. He is author of *The Decision to Go to the Moon: Project Apollo and the National Interest* (MIT Press, 1970) and has written numerous articles and reports on space policy and science and technology policy. In January 1992, Dr. Logsdon was appointed to Vice President Dan Quayle's Space Policy Advisory Board and served through January 1993. He is a member of the International Academy of Astronautics, of the Board of Advisors of The Planetary Society, of the Board of Directors of the National Space Society, and of the Aeronautics and Space Engineering Board of the National Research Council. In past years, he was a member of the National Academy of Sciences's National Academy of Engineering Committee on Space Policy and the National Research Council Committee on a Commercially Developed Space Facility, NASA's Space and Earth Science Advisory Committee and History Advisory Committee, and the Research Advisory Committee of the National Air and Space Museum. He also is a former chair of the Committee on Science and Public Policy of the American Association for the Advancement of Science (AAAS) and of the Education Committee of the Interna-tional Astronautical Federation. He is a fellow of the AAAS and the Explorers Club and an associate fellow of the American Institute of Aeronautics and Astronautics. In addition, he is North American editor for the journal *Space Policy.*

Glossary

AA . . . . .Associate Administrator
AACB . . . . .Aeronautics and Astronautics Coordinating Board
AAS . . . . .American Astronomical Society
ABA . . . . .American Bar Association
ABMA . . . . .Army Ballistic Missile Agency
ACD . . . . .Architectural Control Document
AD/DA . . . . .Deputy Administrator
ADCA . . . . .Arms Control and Disarmament Agency
ACJP . . . . .Air Corps Jet Propulsion
AEC . . . . .Atomic Energy Commission
AEDC . . . . .Arnold Engineering Development Center
AF . . . . .Air Force
AFB . . . . .Air Force Base
AFBMD . . . . .Air Force Ballistic Missile Division
AFSC . . . . .Air Force Space Center
AIAA . . . . .American Institute of Aeronautics and Astronautics
AM . . . . .amplitude modulation
AMC . . . . .Air Materiels Command
AMR . . . . .Atlantic Missile Range
AO . . . . .Announcement of Opportunity
AOMC . . . . .Army Ordnance Missile Command
APM . . . . .Attached Pressurized Module
ARDC . . . . .Air Research and Development Command
ARPA . . . . .Advanced Research Projects Agency
ASEE . . . . .American Society of Electrical Engineers
ASPA . . . . .Armed Services Procurement Act (of 1947)
ASPR . . . . .Armed Services Procurement Regulation
ASTP . . . . .Advanced Space Technology Program *or* Apollo-Soyuz Test
Project
ATS . . . . .Applications Technology Satellite
AU . . . . .astronomical unit
BCD . . . . .Baseline Configuration Document
Caltech . . . . .California Institute of Technology
CCIR . . . . .Comite Consultatif International des Radiocommunications
(International Radio Consultive Committee)
CCSDS . . . . .Consultative Committee for Space Data Systems
CCZ . . . . .Command and Control Zone
CG . . . . .commanding general
CIA . . . . .Central Intelligence Agency
CIT . . . . .California Institute of Technology
CNES . . . . .Centre Nationale des Etudes Spatiales (French Space Agency)
COBE . . . . .Cosmic Background Explorer
COP . . . . .Composite Operations Plan
COPUOS   Committee on the Peaceful Uses of Outer Space (United
Nations)
COSPAR . . . . .Committee on Space Research
COUP . . . . .Consolidated Operations and Utilization Plan
CRAF . . . . .Civil Reserve Air Fleet

```
CSM  . . . . . . . . . . . . . . .Command and Service Module
CSOC  . . . . . . . . . . . . . .Consolidated Space Operations Center
CSTI  . . . . . . . . . . . . . .Civil Space Technology Initiative
CUP . . . . . . . . . . . . . . .Composite Utilization Plan
CY . . . . . . . . . . . . . . . .calendar year
DD . . . . . . . . . . . . . . . .Defense Directive
DDE  . . . . . . . . . . . . . .Dwight D. Eisenhower
DDTE  . . . . . . . . . . . . .design, development, test, and evaluation
DEW  . . . . . . . . . . . . . .Defense Early Warning (Line)
DMSP  . . . . . . . . . . . . .Defense Meteorological Satellite Program
DOD/DoD  . . . . . . . . . .Department of Defense
DOT  . . . . . . . . . . . . . .Department of Transportation
DSOC  . . . . . . . . . . . . .Defense Space Operations Committee
ECS  . . . . . . . . . . . . . . .Environment Control System
EDRS . . . . . . . . . . . . . .ESA's Data Relay Satellite (system)
EDT . . . . . . . . . . . . . . .Eastern Daylight Time
ELDO  . . . . . . . . . . . . .European Launcher Development Organization
ELV . . . . . . . . . . . . . . .Expendable Launch Vehicle
EML  . . . . . . . . . . . . . .European Microgravity Laboratory
E.O./EO  . . . . . . . . . . .Executive Order
EOM  . . . . . . . . . . . . . .end of mission
EOS . . . . . . . . . . . . . . .earth orbital shuttle
ESA  . . . . . . . . . . . . . . .European Space Agency
ESC . . . . . . . . . . . . . . .European Space Conference
ESF . . . . . . . . . . . . . . .European Space Foundation
ESRO  . . . . . . . . . . . . .European Space Research Organization
ESTEC . . . . . . . . . . . . .European Space & Technology Centre
ET . . . . . . . . . . . . . . . .External Tank
ETR . . . . . . . . . . . . . . .Eastern Test Range
EVA . . . . . . . . . . . . . . .extravehicular activity
FCDA  . . . . . . . . . . . . .Federal Civil Defense Authority
FPR . . . . . . . . . . . . . . .Federal Procurement Regulations
F.R. . . . . . . . . . . . . . . .Federal Register
FSE  . . . . . . . . . . . . . . .flight support equipment
FTS  . . . . . . . . . . . . . . .Flight Telerobotic System
FY . . . . . . . . . . . . . . . .fiscal year
GALCIT  . . . . . . . . . . .Guggenheim Aeronautical Laboratory at the California
                            Institute of Technology
GAO  . . . . . . . . . . . . . .General Accounting Office
GEO  . . . . . . . . . . . . . .geosynchronous equatorial orbit
GLOW . . . . . . . . . . . . .gross liftoff weight
GMS  . . . . . . . . . . . . . .Geostationary Meteorological Satellite
GNP  . . . . . . . . . . . . . .gross national product
GOES  . . . . . . . . . . . . .Geostationary Operational Environmental Satellite
GOJ . . . . . . . . . . . . . . .Government of Japan
GPS . . . . . . . . . . . . . . .Global Positioning System
GRO  . . . . . . . . . . . . . .Gamma Ray Observatory
GSA . . . . . . . . . . . . . . .General Services Administration
GSE . . . . . . . . . . . . . . .ground support equipment
GSFC . . . . . . . . . . . . . .Goddard Space Flight Center
```

HEAO . . . . .High Energy Astronomy Observatory
H.R. . . . . .House Resolution
HQ . . . . .headquarters
HUD . . . .(Department of) Housing and Urban Development
ICBM . . . . .intercontinental ballistic missile
ICD . . . . .Interface Control Document
ICSU . . . . .International Council of Scientific Unions
IEWG . . . . .International Evolution Working Group
IGY . . . . .International Geophysical Year
IMP . . . . .Interplanetary Monitoring Platform
INMARSAT . . . . .International Mobil (formerly Maritime) Satellite (organization)
INTELSAT . . . . .International Telecommunications Satellite (consortium)
IOCWG . . . . .International Operational Concepts Working Group
IP . . . . .Increment Plan
IPL . . . . .integrated payload
IRAS . . . . .Infrared Astronomical Satellite
IRBM . . . . .intermediate range missile
ISAS . . . . .Institute of Space and Astronautical Science (Japan)
ISEE . . . . .International Society of Electrical Engineers
ISO . . . . .Infrared Space Observatory
ISPM . . . . .International Solar Polar Mission
ITL . . . . .Integrate-Transfer-Launch (complex)
ITU . . . . .International Telecommunications Union
IUCWG . . . . .International Utilization Coordination Working Group
IUE . . . . .International Ultraviolet Explorer
IUS . . . . .Interim *or* Inertial Upper Stage
IVA . . . . .intravehicular activity
JATO . . . . .jet-assisted takeoff
JCS . . . . .Joint Chiefs of Staff
JEM . . . . .Japanese Experiment Module
JPDRD . . . . .Joint Program Definition and Requirements Document
JPL . . . . .Jet Propulsion Laboratory
JPP . . . . .Joint Program Plan
JPRD . . . . .Joint Program Requirements Document
JSC . . . . .Johnson Space Center
JWG . . . . .joint working group
KSC . . . . .Kennedy Space Center
KW . . . . .kilowatt
LEM . . . . .Lunar Excursion Module
LH . . . . .liquid hydrogen
LLVPG . . . . .Large Launch Vehicle Planning Group
LM . . . . .Lunar Module
LOX . . . . .liquid oxygen
LPMB . . . . .Lunar and Planetary Missions Board
LST . . . . .Large Space Telescope
MCB . . . . .Multilateral Coordination Board
MCC . . . . .Mission Control Center
MGCO . . . . .Mars Geoscience-Climatology Orbiter
MIT . . . . .Massachusetts Institute of Technology
MMD . . . . .MSC Maintenance Depot

MOA . . . . .memorandum of agreement
MODS . . . . .Manned Orbital Development System
MOL . . . . .Manned Orbital Laboratory
MORL . . . . .Manned Orbital Research Laboratory
MOSST . . . . .Ministry of State for Science and Technology (Canada)
MOU . . . . .memorandum of understanding
MOUSE . . . . .Minimum Orbital Unmanned Satellite of the Earth
MRB . . . . .Material Review Board
MRS . . . . .Mobile Remote Servicer
MSC . . . . .Manned Spacecraft Center *or* Mobile Servicing Center
MSFC . . . . .Marshall Space Flight Center
MSFEB . . . . .Manned Space Flight Experiments Board
MSTS . . . . .Military Sea Transportation Service (Navy)
MTF . . . . .Mississippi Test Facility
MTFF . . . . .Man-Tended Free Flyer
NAA . . . . .North American Aviation
NACA . . . . .National Advisory Committee for Aeronautics
NAS . . . . .National Academy of Sciences
NASA . . . . .National Aeronautics and Space Administration
NASC . . . . .National Aeronautics and Space Council
NATO . . . . .North Atlantic Treaty Organization
nm . . . . .nautical mile
NIH . . . . .National Institutes of Health
NMI . . . . .NASA Management Instruction
NOAA . . . . .National Oceanic and Atmospheric Administration
NORAD . . . . .North American Air Defense
NOSS . . . . .National Oceanic Satellite System
NRC . . . . .National Research Council
NSC . . . . .National Security Council
NSD . . . . .National Security Directive
NSDD . . . . .National Security Decision Directive
NSDM . . . . .National Security Decision Memorandum
NSF . . . . .National Science Foundation
NSPD . . . . .National Space Policy Directive
NSSD . . . . .National Security Study Directive
NSSM . . . . .National Security Study Memorandum
OAO . . . . .Orbiting Astronomical Observatories
OART . . . . .Office of Aeronautical Research and Technology (NASA)
OMB . . . . .Office of Management and Budget
OMS . . . . .Orbital Maneuvering System
OMSF . . . . .Office of Manned Space Flight (NASA)
ONR . . . . .Office of Naval Research
OOS . . . . .orbit-to-orbit shuttle
OSD . . . . .Office of the Secretary of Defense
OSFP . . . . .Office of Space Flight Programs (NASA)
OSS . . . . .Office of Space Science (NASA)
OSSA . . . . .Office of Space Science and Applications (NASA)
OSTP . . . . .Office of Science and Technology Policy (White House)
PCC . . . . .Program Coordination Committee
PDRD . . . . .Program Definition and Requirements Document
PMR . . . . .Pacific Missile Range

POIC	Payload Operations Integration Center
PRD	Program Requirements Document
PSAC	President's Science Advisory Committee
PVO	Pioneer Venus Orbiter
R&A	Research and Applications (program)
R&D	research and development
R&E	(Defense) Research and Engineering
RAM	Research Application Module
RCA	Radio Corporation of America
RCS	Reaction Control System
RFP	request for proposals
RMS	Remote Manipulator System
ROBO	rocket bomber
RSA	Russian Space Agency
RTLS	return to launch site
S&ID	Space and Information Systems Division (North American Aviation)
SAC	Strategic Air Command
SAI	Space Astronomy Institute
SAMSO	Space and Missile Systems Organization
SCA	Shuttle Carrier Aircraft
SCF	Shuttle Carrier Flight
SET	Science, Engineering, and Technology (OMB division)
SIG	Senior Interagency Group (Space)
SIRTF	Shuttle Infrared Telescope Facility
SITE	Satellite Instructional Television Experiment
SL	Spacelab
SOHO	Solar and Heliospheric Observatory Satellite
SOP	System Operations Panel
SP	Special Publication
SPDM	Special Purpose Dexterous Manipulator
SRB	Solid Rocket Booster
SSCB	Space Station Control Board
SSCC	Space Station Control Center
SSE	Software Support Environment
SSIS	Space Station Information System
SSME	Space Shuttle Main Engine
SSUS	Spinning Solid Upper Stage
ST	Space Telescope
STA	Science and Technology Agency (Japan)
STOCC	Space Telescope Operations and Control Center
STOL	short takeoff and landing
STS	Space Transportation System
SYG	secretary general
T&DA	Training and Data Acquisition (NASA)
TAC	Tactical Air Command
TACAN	Tactical Air Navigation
TCP	Technological Capabilities Panel
TDRSS	Tracking and Data Relay Satellite System
TDY	tour of duty

TIROS . . . . .Television and Infrared Operational Satellite
TMIS . . . . .Technical Management Information System
TOP . . . . .Tactical Operations Plan
TPS . . . . .Thermal Protection System
TV . . . . .television
TVA . . . . .Tennessee Valley Authority
TWA . . . . .Trans World Airlines
UCLA . . . . .University of California at Los Angeles
UK/U.K. . . . . .United Kingdom
UN . . . . .United Nations
UOP . . . . .User Operations Panel
URA . . . . .Universities Research Association, Inc.
US/U.S. . . . . .United States
USA/U.S.A. . . . . .United States of America *or* United Space Alliance
USAF . . . . .U.S. Air Force
U.S.C. . . . . .U.S. Code
USRA . . . . .University Space Research Association
USSR/U.S.S.R. . . . . . . .Union of Soviet Socialist Republics
VAFB . . . . .Vandenberg Air Force Base
VHF . . . . .very high frequency
VOIR . . . . .Venus Orbiting Imaging Radar
VSIA . . . . .Vehicle System Integration Activity
VTOL . . . . .vertical takeoff and landing
WMO . . . . .World Meteorological Organization
WS . . . . .Weapons System
WSMR . . . . .White Sands Missile Range

Chapter One

The Development of International Space Cooperation

by John M. Logsdon

The National Aeronautics and Space Act of 1958, among its many provisions, indicated that NASA "under the foreign policy guidance of the President, may engage in a program of international cooperation in work done pursuant to this Act, and in the peaceful applications of the results thereof, pursuant to agreements made by the President with the advice and consent of the Senate."[1] The new space agency interpreted this provision as giving it authority to take the initiative in international space dealings. Therefore, within six months, NASA began to develop a program of international cooperation in space that over the following three decades has resulted in agreements with more than 100 countries and in major cooperative projects with both traditional U.S. allies and the country's erstwhile competitor in space, the Soviet Union. The engagement of other countries with the space activities of the United States has been a hallmark of the U.S. space program.

The language of the Space Act seemed to present unintentionally a formal obstacle to NASA in taking the lead in initiating such cooperative activities. The Space Act stated that the United States would enter into cooperative activities "pursuant to agreements entered into by the President with the advice and consent of the Senate." This seemed to require that the executive branch treat every cooperative space agreement as if it had the status of an international treaty, which certainly was not the intent of the congressional drafters of the final bill nor the desire of the White House. Thus, as he signed the document on July 29, 1958, President Dwight D. Eisenhower stated that he regarded this section of the Space Act "as recognizing that international treaties *may* [emphasis added] be made in this field, and as not precluding, in appropriate cases, less formal arrangements for cooperation."[2] [I-1][3] With that clarification, NASA felt free to begin exploring the possibilities of cooperative activity with other countries interested in space—and particularly in the new scientific opportunities made available by the ability to place instruments aboard orbiting satellites and into space beyond the near vicinity of Earth. Cooperation in space science (in addition to the creation of the international agreements needed to locate tracking and data reception sites in other countries) dominated the first decade of

---

1.   "National Aeronautics and Space Act of 1958," Public Law 85-568, 72 Stat., 426, Section 205. Signed by the president on July 29, 1958. This is document II-17 in John M. Logsdon, gen. ed., with Linda J. Lear, Jannelle Warren-Findley, Ray A. Williamson, and Dwayne A. Day, *Exploring the Unknown: Selected Documents in the History of the U.S. Civil Space Program, Volume I: Organizing for Exploration* (Washington, DC: NASA Special Publication (SP)-4407, 1995), 1: 334-45.

2.   Office of the Press Secretary, "Statement by the President," July 29, 1958, Presidential Files, NASA Historical Reference Collection, NASA History Office, NASA Headquarters, Washington, DC.

3.   Unlike most other chapters in *Exploring the Unknown*, the documents supporting this essay are listed in the order in which they appear, rather than in chronological order, because of the unique nature of the international effort in space and the importance of ensuring a regional continuity. This has been done at the expense of maintaining a chronological unity to the essay, but the overall understanding of this complex subject is enhanced as a result.

NASA's international activities, and this has continued as a centerpiece of U.S. cooperative activities to the present.

From its start, space cooperation was linked to broader U.S. foreign policy and national security objectives. The first comprehensive post-Sputnik statement of U.S. space policy, NSC 5814, suggested: "International cooperation in certain outer space activities appears highly desirable from a scientific, *political and psychological* [emphasis added] standpoint. . . . International cooperation agreements in which the United States participates could have the effect of . . . enhancing the position of the United States as a leader in advocating the uses of outer space for peaceful purposes. . . ."[4] The considerations of American leadership have been associated with the nation's approach to international cooperation from the beginning.

Early Space Science Cooperation

The initial NASA approach to space cooperation was crafted by individuals who had been involved in the U.S. activities related to the International Geophysical Year (IGY), which ended on December 31, 1958. These included Hugh Dryden, Deputy Administrator of NASA; Homer Newell, who came to NASA in October 1958 as its first head of space science; and Arnold Frutkin, who had worked on IGY matters with the National Academy of Sciences and then became NASA's second director of international affairs in September 1959 (a position he held for almost two decades).[5]

Under the coordinating umbrella of the International Council of Scientific Unions (ICSU), the nongovernmental scientific academies of participating states had carried out most IGY activities. NASA and National Academy of Sciences leaders hoped that ICSU could provide a venue for discussing, and perhaps coordinating, emerging cooperative activities in space, although some in the United States suggested that the North American Treaty Organization (NATO) would be a more appropriate body to perform this function. At the urging of the United States, ICSU created the Committee on Space Research (COSPAR) in October 1958. At the time of COSPAR's second meeting in March 1959, Richard Porter, the delegate from the National Academy of Sciences, after consultation with NASA, communicated to the president of COSPAR, H.C. van de Hulst of the Netherlands, a groundbreaking offer. The United States hoped that COSPAR "could serve as an avenue through which the capabilities of satellite launching nations and the scientific potential of other nations may be brought together." To facilitate such a development, the United States was willing to launch on U.S. boosters "suitable and worthy experiments proposed by scientists of other countries. This can be done by sending into space either single experiments as part of a larger payload or groups of experiments comprising complete payloads."[6] [I-2, I-3]

---

4.   National Security Council, NSC 5814, "U.S. Policy on Outer Space," June 20, 1958. This was published as Document II-18 in Logsdon, gen. ed., *Exploring the Unknown*, 1: 349.

5.   Arnold W. Frutkin's 1965 book, *International Cooperation in Space* (Englewood Cliffs, NJ: Prentice-Hall, 1965) provides an insider's view of the early years of cooperative space activity. Another source that describes this period is Homer E. Newell, *Beyond the Atmosphere: Early Years of Space Science* (Washington, DC: NASA SP-4211, 1980), Chap. 18.

6.   Richard W. Porter to Professor Dr. H.C. van de Hulst, President, Committee on Space Research (COSPAR), March 14, 1959, Space Policy Institute Documentary History Collection, George Washington University, Washington, DC.

It soon became clear that COSPAR was not well suited for the actual coordination of cooperative scientific missions; instead, for the most part, NASA would be cooperating with an appropriate government body in a partner country. The first country to respond to the U.S. invitation was the United Kingdom; even before the U.S. invitation to COSPAR, U.K. and U.S. scientists had been discussing possible cooperative projects. British Prime Minister Harold McMillan personally announced on May 12, 1959, that a delegation led by Professor Harrie S.W. Massey would visit the United States to discuss specific cooperative projects. Massey was chairman of the British National Space Committee, which had been formed by the Royal Society (the U.K. academy of science) in close consultation with the British government.[7] The British delegation met with NASA counterparts from June 25 through July 3, 1959, and reached agreement in principle on initial cooperative activities. This agreement was reflected in an exchange of correspondence between Massey and NASA Administrator T. Keith Glennan; although this was not a formal agreement, the exchange provided the basis for beginning NASA's first cooperative project. [I-4] Only in September 1961 did the U.S. and U.K. governments exchange diplomatic notes that put the cooperation on a formal basis.[8] The first of the cooperative U.S.-U.K. satellites, *Ariel 1*, was launched on April 26, 1962.[9]

Then, as Europe decided in the early 1960s to undertake most of its space science activities through a new multinational entity, the European Space Research Organization (ESRO), NASA quickly extended its cooperative offer to that new body.[10] [I-5]

These initial cooperative efforts and most others since were carried out within the framework of a set of guiding principles that were developed during the first year of U.S. space activity.[11] These principles were relatively conservative in character; they did not commit the United States to help pay for other countries' shares of cooperative projects. Rather, they provided some specific and rather limiting criteria that cooperative proposals would have to meet, as follows:

1. *Designation by each participating government of a central agency for the negotiation and supervision of joint efforts*
2. *Agreements on specific projects rather than generalized programs*
3. *Each country's acceptance of financial responsibility for its own contributions to joint projects*
4. *Projects of mutual scientific interest*
5. *General publication of scientific results*[12]

Added to this framework for cooperation in later years were the requirements that each cooperating partner assume technical as well as financial responsibility for its contributions and that there be simple technical interfaces between the contributions from different countries. This latter requirement was originally established to minimize

7.   NASA, "Statement by NASA," Release 59-193, July 29, 1959, Press Release Files, NASA Historical Reference Collection.

8.   Arnold S. Levine, *Ariel 1: An Experiment in International Cooperation*, Goddard Historical Note Number 4, NASA Goddard Space Flight Center, September 1967, NASA Historical Reference Collection.

9.   Frutkin, *International Cooperation*, pp. 42-43.

10.   See John Krige and Arturo Russo, *Europe in Space, 1960-1973* (Noordwijk, The Netherlands: European Space Agency, 1994), for a brief account of the origins of ESRO.

11.   See Newell, *Beyond the Atmosphere*, p. 306, and Frutkin, *International Cooperation*, pp. 32-36, for a discussion of the development of the NASA guidelines for cooperation.

12.   NASA, *International Programs*, 1962, NASA Historical Reference Collection.

the managerial complexity of cooperative projects, but in later years it also became an important safeguard to prevent unwanted technology transfer as a result of such projects.[13]

Operating under these guidelines over the years, NASA and its partners have been able to conduct numerous space science projects that have been scientifically productive, of increasing technical complexity, and in general free of rancor. On balance, the record of cooperation in space science is strongly positive, with both political and scientific benefits to all involved. This is not to say, however, that the path has been totally smooth. While Canada and most European countries worked closely with the United States in developing capabilities for performing space science, Japan chose to develop that capability on its own.[14] [I-6] Only after its Institute for Space and Astronautical Science in the 1970s developed an autonomous space science program, including its own launch vehicle, was Japan ready to enter into cooperative scientific projects with the United States.

Despite efforts from the late 1950s on to engage the Soviet Union in scientific cooperation (described in more detail below), such cooperation was slow to emerge, being constantly "held hostage" to the state of the overall political relationship between the two global superpowers. In the mid-1960s, the United States also initiated cooperative interactions with emerging spacefaring countries such as India and Brazil. [I-7] For many years, however, America's primary cooperative partner was Europe.

Even with Europe, there were difficulties.[15] On the U.S. side, scientists eager to have their instruments and experiments fly in space could not participate as principal investigators in the missions undertaken by ESRO and its successor, the European Space Agency (ESA), which was created in 1975. Europe gave priority to nurturing its own space scientists and did not want to have them compete with their U.S. counterparts for the limited payload space available on European missions. In contrast, European scientists from the beginning were afforded such an opportunity to participate in NASA missions. Meanwhile, U.S. scientists complained that scarce space on U.S. scientific missions was being allocated to non-U.S. scientists and instruments, both for political reasons and because the non-U.S. contributions did not cost NASA any money. Securing European participation in the kind of large science missions that became the NASA norm in the 1970s required delicate and lengthy negotiations.[16]

On the European side, there were reservations about the U.S. role as the dominant partner in almost every cooperative project. This often meant that NASA and U.S. scientists would define the objectives and content of a scientific mission and only then invite non-U.S. scientists to participate. In addition, the value of any international contribution to a U.S. mission depended on NASA's ability to sustain political and budgetary support for that mission.

These reservations peaked in the 1979-1981 period. First, Europe, frustrated by U.S. vacillation over whether to undertake a mission to Halley's comet during its 1986 visit to

---

13.   Personal communication from Richard Barnes, former Director of NASA's Office of International Affairs, to John M. Logsdon, December 11, 1995.

14.   Newell, *Beyond the Atmosphere*, pp. 309-11. See Joan Johnson-Freese, *Over the Pacific: Japanese Space Policy into the Twenty-First Century* (Dubuque, IA: Kendall/Hunt Publishing, 1993), for an overview of the Japanese space program.

15.   See Roger Bonnet and Vittorio Manno, *International Cooperation in Space: The Example of the European Space Agency* (Cambridge, MA: Harvard University Press, 1994).

16.   For an expansion of this point, see Noel Hinners, "Space Science and Humanistic Concerns," in Jerry Grey and Lawrence Levy, eds., *Global Implications of Space Activities* (New York: American Institute of Aeronautics and Astronautics, 1981), pp. 38-39; John M. Logsdon, "U.S.-European Cooperation in Space: A 25-Year Perspective," *Science* 223 (January 6, 1984): 11-16.

the inner Solar System, decided to undertake a Halley mission of its own. Europe did not want to wait for the United States to make up its mind whether it would initiate a Halley mission and then invite Europe to participate.[17] Then, in early 1981, the United States informed Europe that it was withdrawing its spacecraft from the cooperative two-spacecraft International Solar Polar Mission (ISPM). [I-8, I-9] This withdrawal was forced by the decision of the new administration of Ronald Reagan to reduce the federal budget. The White House required NASA to cancel one of its three major approved science missions (the Hubble Space Telescope, the Galileo mission to Jupiter, and ISPM), and the space agency chose the ISPM spacecraft for cancellation.[18] The European Space Agency and individual European countries protested loudly, but the decision was not reversed. Europe left this experience with a reminder of its dependence on U.S. political and budgetary decisions for achieving its own objectives in its many cooperative efforts with the United States. [I-10]

While the ISPM controversy cast a temporary chill on cooperation in space science, its effect did not linger (although the incident was frequently raised during subsequent negotiations concerning cooperative arrangements for other U.S.-European projects). The mutual benefits of cooperation in space science were evident to all. By 1983, for example, NASA and the European Space Agency had established regular consultations regarding areas for possible future cooperation. [I-11] Similar consultations between the United States and Japan and between the United States and Canada have provided the venue for discussions on when cooperative scientific projects were the best ways of achieving the objectives of the participating scientific communities, as well as on when those objectives were best served (in the case of infrared astronomy, for example) by mounting separate missions. In addition, the Inter-Agency Consultative Group, with the space agencies of the United States, the Soviet Union, Europe, and Japan as members, has proven a valuable forum for coordinating multilateral scientific undertakings.[19] From its beginning during the IGY to the present, space science remains the paradigm for successful international space cooperation.[20]

Cooperation in Human Spaceflight: Post-Apollo, the Spacelab, and the Space Station

The Apollo program was, by its very nature, competitive and unilateral in character. Its objective was to demonstrate to the world U.S. technological and managerial competence by being the first to and on the Moon. Although President Kennedy had considered turning Apollo into a cooperative U.S.-Soviet undertaking in 1963 (see below), as the first lunar landing grew near, it was clearly viewed as a symbol of U.S. power and capability.[21] [I-12]

---

17. See John M. Logsdon, "Missing Halley's Comet: The Politics of Big Science," *Isis* 80 (1989): 268-70.

18. See Joan Johnson-Freese, "Canceling the U.S. Solar-Polar Spacecraft: Implications for International Cooperation in Space," *Space Policy* 3 (February 1987): 24-37, for more details on this incident.

19. For a description and analysis of the Inter-Agency Consultative Group, see Joan Johnson-Freese, "A Model for Multinational Space Cooperation: The Inter-Agency Consultative Group," *Space Policy* 5 (November 1989): 288-300; Joan Johnson-Freese, "From Halley's Comet to Solar Terrestrial Science: The Evolution of the Inter-Agency Consultative Group," *Space Policy* 8 (August 1992): 245-55.

20. See Bonnet and Manno, *International Cooperation in Space*, for a European perspective on this.

21. Again, as mentioned previously in footnote 3, the documents associated with this chapter are arranged in the order in which they are discussed, rather than in strict chronological order. This organization was chosen to best represent the diverse character of NASA's international relationships.

The idea of expanded international space cooperation had been under discussion within the top levels of the U.S. government since the mid-1960s, and these discussions continued after Richard Nixon became president in January 1969. [I-13] With the Apollo 11 mission a success, NASA and the Nixon administration made a conscious decision to broaden the basis of international participation in U.S. post-Apollo efforts in space.[22] This broadening took two directions: (1) attempting to engage the Soviet Union in more substantial cooperative undertakings (discussed later in this essay) and (2) inviting U.S. allies to participate in the human spaceflight and technology development parts of the U.S. program (from which they previously had been largely excluded).

Such a broadening was one of the recommendations of the Space Task Group, which had been established by President Nixon in February 1969 to advise him on post-Apollo space plans. In its September 1969 report, the Space Task Group recommended "the use of our space capability not only to extend the benefits of space to the rest of the world, but also to increase direct participation of the world community in both manned and unmanned exploration and use of space." The group recognized that for other industrial countries "the form of cooperation most sought after . . . would be technical assistance to enable them to develop their own capabilities." The group also suggested that the United States should "move toward a liberalization of our policies affecting cooperation in space activities, should stand ready to provide launch services and share technology wherever possible, and should make arrangements to involve foreign experts in the detailed definition of future United States space programs. . . ."[23]

Armed with these recommendations and what he believed was a direct mandate from President Nixon to seek such expanded cooperation, during late 1969 and the early months of 1970, NASA Administrator Thomas O. Paine visited Europe, Canada, Japan, and Australia for initial discussions of cooperative opportunities in the post-Apollo period. [I-14, I-15, I-16] Paine believed that he could use the post-Apollo proposals spelled out in the September 1969 report of the Space Task Group as the basis for what the U.S. program during the 1970s would be—and thus for what cooperative possibilities might be open for U.S. allies. The reactions to Paine's overtures were varied. In addition, NASA found that some in the Nixon White House were far less enthusiastic about cooperation in large hardware programs than were Paine and the Space Task Group and that President Nixon had no intention of approving in toto the group's recommended program that had been the basis of Paine's briefings to foreign officials.

Early on, Australia indicated that space was not among its highest priorities and that it was not able to spend the considerable amount of money required to cooperate with the United States on a meaningful basis. The Japanese response was somewhat similar.[24] In the late 1960s, Japan had decided to accelerate its acquisition of an autonomous capability for space launch and satellite construction for missions other than space science. Japan asked the U.S. government to allow U.S. aerospace firms to license space technology to Japan to "jump start" that capability development process. Although NASA and the Department of

---

22.   See Arnold W. Frutkin, "International Cooperation in Space," *Science* 169 (July 24, 1970): 333-38, for an early discussion of this policy change. More recently, see Lorenza Sebesta, "The Politics of Technological Cooperation in Space: US-European Negotiations on the Post-Apollo Programme," *History and Technology: An International Journal* 11 (1994): 317-41.

23.   Space Task Group, *The Post-Apollo Space Program: Directions for the Future* (Washington, DC: U.S. Government Printing Office, September 1969), pp. 7, 16.

24.   For a discussion of the evolution of U.S.-Japanese space relations, see John M. Logsdon, "U.S.-Japanese Space Relations at a Crossroads," *Science* 255 (January 17, 1992): 294-300; Johnson-Freese, *Over the Pacific.*

Defense opposed such licensing, advocates of the diplomatic and strategic importance of the U.S.-Japanese alliance, particularly within the State Department, prevailed. [I-17]

On July 31, 1969, the United States and Japan signed an agreement that cleared the way for firm-to-firm cooperation between the two countries.[25] This agreement and its subsequent modifications in 1976 and 1980 facilitated Japanese acquisition of launch-related technology that was used and modified as the basis of Japanese booster capability for more than twenty years; by contrast, there was limited transfer of satellite-related technology. During most of the 1970s, cooperation between the United States and Japan was at the firm-to-firm, not the government agency-to-government agency, level. Although Japan eventually indicated limited interest in participating in post-Apollo efforts, clear priority was given to Japanese autonomy in space, and the Japanese interest did not lead to a post-Apollo cooperative agreement.

Canada, in contrast, indicated that it was interested, in the context of its modest space effort, in making a contribution to the U.S. post-Apollo program. [I-18] After several years of harmonious negotiations, it was agreed in 1975 that Canada would be responsible for developing the Remote Manipulator System for use aboard the Space Shuttle.

The most difficult post-Apollo interactions were between the United States and Europe. In addition to the uncertainties surrounding which of those systems proposed in the Space Task Group report the Nixon administration would approve, and then what contributions Europe could best make to such systems, there were two background issues that influenced the discussions. One was the question of whether the United States would guarantee to launch communication satellites for European regional use or whether Europe would have to develop its own launch vehicle to guarantee its access to, in particular, geosynchronous orbit. The second issue was the concern by some at the White House that space cooperation could be a means for European firms to gain access, at minimum cost, to advanced or sensitive U.S. technology. In the background of both of these issues was a concern on the part of some in the White House that NASA and the Department of State were advocating an approach to international cooperation that was not in the administration's interest. [I-19]

During this period, a goal of U.S. policy was to discourage Europe from developing its own launcher capable of lifting sizable payloads to orbit, thereby connecting much of Europe's future in space to cooperative projects with the United States launched on U.S. boosters, particularly the Space Shuttle. For example, NASA Administrator Thomas Paine, in his November 7, 1969, letter to President Nixon, indicated that his "fundamental objective was to stimulate Europeans to rethink their present limited space objectives, to avoid their wasting resources on obsolescent developments [a thinly veiled reference to European development of an expendable launch vehicle at the time that the United States was planning to develop a totally reusable Space Shuttle], and eventually to establish more considerable prospects for future international collaboration on major space projects."[26] One reason for this posture was to maintain the U.S. monopoly on access to space for such payloads and to create customers for the Space Shuttle once it became available. A second reason was the U.S. attempt to protect U.S. communications satellite manufacturers—companies that in 1970 had a "free-world" monopoly on the technology.

---

25.   Department of State, "Agreement between the United States of America and Japan, July 31, 1969," *Treaties and Other International Acts*, Series 6375.

26.   Thomas O. Paine, NASA Administrator, to the President, November 7, 1969, Administrators Files, NASA Historical Reference Collection.

Also, the United States and its partners in the International Telecommunications Satellite Consortium (INTELSAT) were participating in difficult negotiations over the future of the organization, and the United States wanted to block the emergence of regional competitors to INTELSAT in Europe and elsewhere.[27]

After lengthy discussions and negotiations within the U.S. government [I-20, I-21] and between the United States and the European Space Conference (a policy-level entity created to coordinate European responses to U.S. positions on space issues), the United States on September 1, 1971, set forth a policy on the availability of U.S. launchers for European satellites. The United States also made it clear that the issue of post-Apollo cooperation was, in U.S. thinking, not linked to the launcher issue.[28] [I-22]

The other obstacle—concern that the United States would be forced to transfer valuable technology to Europe to ensure that Europe could successfully complete its share of the post-Apollo program—proved more difficult to surmount. Assistant to the President Peter Flanigan, who had responsibility within the White House for space issues, raised concerns about technology flows related to U.S.-European cooperation. Flanigan suggested that President Nixon's 1969 mandate to NASA seeking expanded international cooperation was really focused on greater European participation as astronauts or in-orbit scientific investigators, not on Europe as a developer of hardware to be used by the United States. [I-23]

The debate over European involvement in the post-Apollo program had continued throughout 1971. As the Space Shuttle finally gained approval in January 1972 as the major post-Apollo development project, the question of European participation was still undecided.[29] Over the next several months, debates over how to proceed continued within the administration. [I-24]

Among those attempting to find a position satisfactory to both the advocates of cooperation within NASA and the State Department and the skeptics inside the White House was the executive secretary of the National Aeronautics and Space Council, former astronaut William A. Anders. In a March 17, 1972, memorandum to Flanigan, Anders suggested that Europe be invited only to develop a "sortie can"—a pressurized laboratory to fly inside the shuttle's payload bay—rather than be allowed to develop a technologically more challenging orbit-to-orbit "space tug" or parts of the shuttle orbiter itself.[30] [I-25] Responding to the Anders proposal on April 29, Secretary of State William Rogers argued against reversing the U.S. position that the space tug might be developed by Europe on a cooperative basis and against limiting European cooperation to developing the sortie can (also called the Research and Application Module).[31] [I-26] Commenting on Secretary Rogers's memorandum, NASA indicated that its "preferred objective is to obtain

27.   For some background on how the issue of launch guarantees interacted with European-U.S. negotiations on post-Apollo cooperation, see Douglas R. Lord, *Spacelab: An International Success Story* (Washington, DC: NASA SP-487, 1987), Chap. 1; Sebesta, "Politics of Technological Cooperation in Space."

28.   U. Alexis Johnson, Under Secretary of State, to Theo Lefevre, Chairman of European Space Conference, September 1, 1971, NASA Historical Reference Collection.

29.   James C. Fletcher, NASA Administrator, memorandum to George M. Low, NASA Deputy Administrator, "Summary of Peter Flanigan Meeting," March 3, 1972, Administrators Files, NASA Historical Reference Collection.

30.   William A. Anders, Executive Secretary, National Aeronautics and Space Council, to The Honorable Peter M. Flanigan, March 17, 1972, with attached: "Position Paper on European Participation in our Post Apollo Space Program," Space Policy Institute Documentary History Collection.

31.   William P. Rogers, Secretary of State, Memorandum for the President, "Post-Apollo Relationships With the Europeans," April 29, 1972, NASA Historical Reference Collection.

European agreement to develop a specific type of sortie module" and that European development of the space tug was "a distinctly second choice, and much less desirable."[32] [I-27]

The State Department argument did not prevail, and U.S. representatives informed their European colleagues in June 1972 that only the sortie can was an acceptable contribution to post-Apollo efforts. [I-28] European governments and industry were displeased by this outcome; substantial study funds had been invested in the tug, and the sortie can was technologically a much less interesting development.[33] Within a little more than a year, however, a U.S.-European agreement on the terms for the cooperative development of the sortie can (renamed the Spacelab) had been achieved [I-29], committing Europe to a cooperative project with the United States that was much more expensive and highly visible than previously had been the case.[34] Europe's agreement to develop the Spacelab came in the context of a comprehensive "package deal" that also committed European nations to develop their own launch vehicle (in spite of the 1971 U.S. assurance of access to U.S. launchers) and to begin work on a maritime communications satellite. In addition, Europe decided to develop a single space organization, to be called the European Space Agency (ESA), to manage these large projects and other European efforts in space. The European nations' experience in post-Apollo dealings with the United States was a major factor in convincing leading European countries to pool the better part of their future projects in a multilateral alliance for space.[35]

Despite the difficulties in reaching agreement on a mutually satisfactory undertaking, as well as substantial European dissatisfaction with the results of the cooperative effort, European-U.S. cooperation on the Spacelab created a precedent for contemplating—even expecting—similar close cooperation on any subsequent large-scale program that NASA might undertake. In fact, within a year after signing the Spacelab agreement, some at NASA were already thinking about international participation in a space station program. [I-30] When NASA Headquarters once again began active planning for a space station in 1982, the head of the Space Station Task Force, John Hodge, asked NASA Director of International Affairs Kenneth Pedersen—Arnold Frutkin had left NASA in 1979—for his ideas on what might be learned from the post-Apollo experience with respect to preparing for international participation in a space station. In reply, Pedersen prepared a lengthy memorandum containing his thoughts not only on the relevance of past experience but also on a strategy for obtaining international commitments to the emerging station program. [I-31]

Pedersen's ideas largely governed the NASA approach to international participation in the space station during 1982 and 1983. In addition, in August 1982 the Reagan administration adopted an overall policy statement with respect to international space relationships that provided a context for making the station an international project. [I-32]

When Ronald Reagan announced the approval of the space station program in his January 25, 1984, State of the Union address, he also stated that "NASA will invite other countries to participate so we can strengthen peace, build prosperity and expand freedom

32. James C. Fletcher, Administrator, NASA, to Honorable Henry A. Kissinger, Assistant to the President for National Security Affairs, May 5, 1972, with attached: "NASA's Comments on Secretary Rogers' Memorandum of April 29, 1972," Administrators Files, NASA Historical Reference Collection.

33. See, for example, "Europeans Delay Post-Apollo Meeting," *Aviation Week & Space Technology*, July 17, 1972, p. 19.

34. See Lord, *Spacelab*, for a participant's account on the international dimensions of the undertaking.

35. See Krige and Russo, *Europe in Space*; Michiel Schwarz, "European Policies on Space Science and Technology, 1960-1978," *Research Policy* 8 (1979): 205-44; Michiel Schwarz, *Europe's Future in Space* (London: Routledge & Kegan Paul, 1988), Chap. 4, for discussions of the early years of European space cooperation.

for all who share our goals."[36] NASA Administrator James Beggs, acting as Reagan's personal emissary, extended the president's invitation to join the space station program to leaders in Europe, Canada, and Japan during a series of rapid visits during March 1984. After returning from his visits to Europe and Japan, he reported the following to Secretary of State George Shultz:

The reaction so far to the President's call for international cooperation has been strongly positive and openly appreciative. It has been positive in the sense that our principal Allies are moving quickly, or have already moved, to take political decisions to participate. And their reactions clearly show appreciation for the major foreign policy benefits that will flow from open and collaborative cooperation on such a bold, visible, and imaginative project.[37] [I-33]

Beggs also wrote a letter to a senior official in each of the countries he had visited, summarizing his understandings, clarifying issues that had been raised, and laying out the next steps. He reiterated the basic U.S. position:

President Reagan has committed the U.S. to building an $8B fully functional Space Station to be operational by the early 1990s, but has also set the stage for working together to develop a more expansive international Space Station with even greater benefits and capabilities for all to use. Thus, we are inviting your Government to take a close look at our plans and concepts and then, based on your long-term interests and goals, share with us your ideas for cooperation that will expand the capabilities of the Space Station.[38] [I-34]

It would take more than four years of difficult negotiations to develop a framework for cooperation in the space station program that was acceptable to the United States and its partners—ten European countries acting through the European Space Agency, Japan, and Canada. In agreeing to that framework, the station partners launched what was the most expensive, longest duration in international scientific and technological cooperation ever undertaken. The station agreements included a multilateral intergovernmental agreement among the governments of the United States, Japan, Canada, and the nine European countries,[39] as well as three separate and lengthy memoranda of understanding (one between NASA and ESA, another between NASA and its counterpart agency in Canada, and the third between NASA and Japan's space agency.[40] [I-35] In its scope and complexity, international space cooperation had come a long way from the initial, quickly negotiated, informal, and straightforward 1959 agreement that, almost three decades earlier, had led to the U.S.-U.K. *Ariel* project.

---

36.   See John M. Logsdon, *Together in Orbit: The Origins of International Participation in Space Station Freedom* (Washington, DC: Space Policy Institute, George Washington University, December 1991), for an account of the process of internationalizing the U.S. space station program.

37.   James M. Beggs, NASA Administrator, to George P. Shultz, Secretary of State, March 16, 1984, Administrators Files, NASA Historical Reference Collection.

38.   James M. Beggs, NASA Administrator, to Kenneth Baker, MP, Minister of State for Industry with Special Responsibility for Space and Information Technology, United Kingdom, April 6, 1984, Administrators Files, NASA Historical Reference Collection.

39.   At the time this essay was written, the Department of State had not yet published the space station intergovernmental agreement. When it is published, it will appear in the Department of State series *United States Treaties and Other International Acts.* However, a copy of the agreement does appear in Hein's Microfiche Service, *United States Treaties and Other International Acts,* No. KAV 2383.

40.   Each memorandum of understanding was slightly different but covered essentially the same points. The one with Japan was signed by a representative of the Science and Technology Agency in the name of the government of Japan; the Japanese National Space Development Agency was not formally a government body. The Canadian signatory was the Ministry of State for Science and Technology, because the new Canadian Space Agency had not yet been formally established in September 1988.

Cooperation With the Soviet Union

From their inception, the space programs of the United States and the Soviet Union were closely linked to the global political and military rivalry between the two superpowers. Issues of U.S.-U.S.S.R. space cooperation have historically received separate treatment in U.S. policy making from those related to cooperation with U.S. allies. Although the IGY provided the context for the first scientific satellite launches, space cooperation was the most disappointing aspect of the IGY, primarily because the Soviet Union shared only very limited information about the substantive character of its satellite programs with other IGY participants. Also, the Soviet Union did not make the data obtained by those satellites available for analysis to scientists outside of its borders.[41]

Although the Soviet Union had refused to discuss the possibility of space cooperation with the United States during the Eisenhower administration, newly inaugurated President John F. Kennedy attempted to open the door to such discussions in his January 20, 1961, inaugural address, stating "let both sides seek to invoke the wonders of science instead of its terrors. Together let us explore the stars. . . ."[42] Kennedy added more detail to this call in his first State of the Union address ten days later:

I now invite all nations—including the Soviet Union—to join with us in developing a weather prediction program, in a new communications satellite program, and in preparation for probing the distant planets of Mars and Venus, probes which someday may unlock the deepest secrets of the universe.[43]

To back up his cooperative initiative, Kennedy in early February asked his science advisor, Jerome Wiesner, to establish a NASA-Department of State panel on international space cooperation. During its meetings over the following few weeks, the panel considered not only the cooperative possibilities mentioned by President Kennedy in the State of the Union address but also such bold initiatives as the creation of an international scientific outpost on the Moon. In its report, the panel listed more than twenty possibilities for U.S.-Soviet space cooperation. [I-36]

However, by the time the panel's final report was completed, its recommendations had been overtaken by events. The first orbital flight by Yuri Gagarin, on April 12, 1961, had stimulated John F. Kennedy to a competitive response.[44] With the announcement of the decision to go to the Moon a few weeks later, the chances for significant cooperation in space with the Soviet Union disappeared, at least for the time being.[45]

Tense U.S.-U.S.S.R. relations during 1961, symbolized by the Kennedy-Khrushchev confrontations at a June summit meeting in Vienna and the August erecting of the Berlin Wall, seemed to make any chance of cooperation in space remote. However, even after challenging the Soviet Union to a space race, President Kennedy never abandoned his

---

41. See Frutkin, *International Cooperation*, pp. 19-20, for this assessment.
42. Quoted in Dodd L. Harvey and Linda Ciccoritti, *U.S.-Soviet Cooperation in Space* (Miami, FL: University of Miami Center for Advanced International Studies, 1974), p. 65. The following account of cooperative initiatives in the first months of the Kennedy administration is taken from pages 66 through 74 of this study.
43. Ibid.
44. This story has been told in a series of documents contained in Volume I of this series. See Documents III-6 through III-12 in Logsdon, gen. ed., *Exploring the Unknown*, 1: 423-54.
45. See John M. Logsdon, *The Decision to Go to the Moon: Project Apollo and the National Interest* (Cambridge, MA: MIT Press, 1970).

hope of turning space into an arena for cooperation rather than competition. Apparently, Nikita Khrushchev was willing to change slightly the prior Soviet position linking the possibility of space cooperation with progress in the U.S.-U.S.S.R. disarmament talks. In late 1961, the Soviet Union joined with the United States in supporting United Nations resolution 1721 (XVI), which among other things called for strengthening international space cooperation. In his February 21, 1962, message to President Kennedy, which congratulated the United States on the orbital flight of John Glenn, Khrushchev suggested:

> If our countries pooled their efforts—scientific, technical, and material—to master the universe, this would be very beneficial for the advance of science and would be joyfully acclaimed by all peoples who would like to see scientific achievements benefit man and not be used for "cold war" purposes and the arms race.[46]

Quickly seizing what seemed to be an opening, the next day President Kennedy cabled Khrushchev. Kennedy stated that he was "instructing the appropriate officers of this Government to prepare new and concrete proposals for immediate projects of common action."[47] Those proposals were contained in a March 7 letter from Kennedy to Khrushchev. [I-37] Kennedy proposed specific cooperative undertakings in five areas: weather satellites, tracking services, mapping of the Earth's magnetic field, satellite communications, and space medicine.[48]

Khrushchev's reply came within two weeks. [I-38] It in effect accepted the notion of cooperation in most of the areas suggested by Kennedy, and it added other areas as candidates for cooperation. Most importantly, Khrushchev agreed to a meeting between U.S. and Soviet representatives to discuss how to implement the proposals that he and President Kennedy had made. However, Khrushchev also made it clear that the scope of U.S.-U.S.S.R. space cooperation was limited by broader considerations; he noted that "the scale of our cooperation in the peaceful conquest of space, as well as the choice of lines along which such cooperation would seem possible, is to a certain extent related to the solution of the disarmament problem."[49]

President Kennedy appointed NASA Deputy Administrator Hugh Dryden as his representative to the U.S.-U.S.S.R. talks, and Khrushchev appointed academician Anatoli Blagonravov. Both were experienced in international scientific diplomacy. Their first meeting took place on March 27 through 30, 1962 [I-39]; by June the two sides had agreed on three areas of initial cooperation.[50] [I-40] During October 1962 (in the midst of the Cuban missile crisis), an exchange of letters between NASA Administrator James Webb and the president of the Soviet Academy of Sciences, M.V. Keldysh, put the agreements into effect.[51]

---

46.   Quoted in Harvey and Ciccoritti, *U.S.-Soviet Cooperation in Space*, p. 86.
47.   Ibid.
48.   John F. Kennedy, "Text of Letter Dated March 7, 1962, From President Kennedy to Chairman Khrushchev Re Cooperation in Peaceful Uses of Outer Space," March 7, 1962, Space Policy Institute Documentary History Collection.
49.   Nikita Khrushchev, "Text of Letter From Chairman Khrushchev in Reply to President Kennedy's Letter of March 7, 1962," March 7, 1962, Space Policy Institute Documentary History Collection.
50.   For an account of the Dryden-Blagonravov negotiations, see Harvey and Ciccoritti, *U.S.-Soviet Cooperation*, pp. 92-102; Frutkin, *International Cooperation*, pp. 94-105.
51.   For the text of the letters, see U.S. Congress, Senate Committee on Aeronautical and Space Sciences, *Documents on International Aspects of the Exploration and Use of Outer Space, 1954-1962*, Staff Report, May 9, 1963, pp. 357-58.

Early results from this initial cooperative agreement were disappointing. By September 1963, however, the Kennedy administration was considering a bold initiative—turning Project Apollo from an exercise in U.S.-U.S.S.R. competition into a cooperative undertaking.[52] [I-41] The two countries had signed the Limited Test Ban Treaty in August 1963, and President Kennedy wanted to build on that agreement to move toward a broad détente between the superpowers. Joining together in sending people to the Moon would be a dramatic symbol of such détente, and on September 20, 1963, in a speech to the General Assembly of the United Nations, Kennedy asked:

Why, therefore, should man's first flight to the Moon be a matter of national competition? Why should the United States and the Soviet Union, in preparing for such expeditions, become involved in immense duplication of research, construction, and expenditure? Surely we should explore whether the scientists and astronauts of our two countries—indeed of all the world—cannot work together in the conquest of space, sending some day in this decade to the Moon not the representatives of a single nation, but the representatives of all our countries.[53]

Kennedy's proposal got a mixed reaction within the United States—and no response from the Soviet Union—but the president was not dissuaded. On November 12, 1963, Kennedy directed NASA Administrator James Webb "to assume personally the initiative and central responsibility within the government for the development of a program of substantive cooperation with the Soviet Union in the field of outer space." This program, said Kennedy, should include "cooperation in lunar landing proposals."[54] [I-42]

Ten days later, Kennedy was assassinated. While President Lyndon B. Johnson was also in favor of U.S.-U.S.S.R. space cooperation, the continued lack of a Soviet response to Kennedy's September 20 invitation and the absence of Kennedy's personal involvement led to the initiative gradually fading away. By the time NASA responded to the November directive on January 31, 1964, the focus had shifted to cooperation on the robotic predecessors to a human landing on the Moon. [I-43] Even that did not happen, and throughout the rest of the 1960s, U.S.-U.S.S.R. cooperation in space continued at a very modest level.[55]

As mentioned earlier, in the aftermath of the first lunar landing on July 20, 1969, the Nixon administration decided to broaden the basis of its cooperative space activities. One set of initiatives was directed at U.S. allies; other overtures were made to the Soviet Union. Rather than proposing cooperation across a broad range of space activities, NASA Administrator Thomas O. Paine, in a July 31, 1970, letter to Keldysh, proposed that cooperation focus on the issue of astronaut safety—and particularly on a program to achieve compatible rendezvous and docking systems between U.S. and Soviet spacecraft.[56] [I-44]

---

52.   This account of the consideration of making the Moon landing a cooperative undertaking is drawn from Harvey and Ciccoritti, *U.S.-Soviet Cooperation*, pp. 112-40; Frutkin, *International Cooperation*, pp. 105-19.

53.   Quoted in Harvey and Ciccoritti, *U.S.-Soviet Cooperation*, p. 123.

54.   National Security Action Memorandum No. 271, "Cooperation with the USSR on Outer Space Matters," November 12, 1963, Space Policy Institute Documentary History Collection.

55.   For an assessment of the record of U.S.-U.S.S.R. cooperation during the 1960s, in addition to the sources cited above, see U.S. Congress, Office of Technology Assessment, *U.S.-Soviet Cooperation in Space*, Technical Memorandum (Washington, DC: U.S. Government Printing Office, July 1985).

56.   Thomas O. Paine, NASA Administrator, to Academician M.V. Keldysh, President, Academy of Sciences of the USSR, July 31, 1970, Administrators Files, NASA Historical Reference Collection. For a full history of what eventually became the Apollo-Soyuz Test Project, see Edward Clinton Ezell and Linda Neuman Ezell, *The Partnership: A History of the Apollo-Soyuz Test Project* (Washington, DC: NASA SP-4209, 1978).

This proposal produced a positive response from the Soviet Union, and on October 26, 1970, a NASA delegation traveled to Moscow to discuss the feasibility of such a program with its Soviet counterparts. [I-45] This was the first official NASA delegation to visit the Soviet Union. These meetings went quite well, and they seemed to break a logjam in U.S.-U.S.S.R. cooperative relations. In January 1971, Acting NASA Administrator George Low and an accompanying delegation traveled to Moscow to meet with Keldysh and his colleagues, and they reached a preliminary agreement on renewed and expanded cooperation in a variety of areas. Following an exchange of letters between Low and Keldysh, a broad agreement on areas of U.S.-U.S.S.R. cooperation was announced on March 31, 1971. To implement the agreement, U.S.-U.S.S.R. joint working groups on meteorological satellites, meteorological rocket soundings, the natural environment, the exploration of near-Earth, the Moon and the planets, and space biology and medicine were established. These working groups and their successors have been the principal mechanisms for planning U.S.-U.S.S.R. space cooperation since 1971.

Meanwhile, discussions on the feasibility of a 1975 joint test flight involving the in-orbit rendezvous of U.S. and Soviet spacecraft had proceeded to the point where George Low, now back in his position as NASA's deputy administrator, journeyed to Moscow in April 1972 to determine whether the United States should commit to such a mission. This trip was made secretly at the request of the White House, because a formal agreement on such a mission was to be announced at a planned May 1972 summit meeting between President Richard Nixon and Soviet Premier Leonid Brezhnev.[57] [I-46]

Low's recommendation was to go ahead with the mission, which was accepted by the White House. [I-47] On May 24, 1972, President Richard Nixon and Chairman of the Soviet Council of Ministers Alexei Kosygin signed the government-to-government "Agreement Concerning Cooperation in the Exploration and Use of Outer Space for Peaceful Purposes."[58] Although this agreement incorporated all the areas of cooperation that had been agreed to in 1971, its centerpiece was the announcement of the Apollo-Soyuz Test Project (ASTP). The agreement had a five-year lifetime, with the possibility of an extension by mutual agreement.

Soviet and U.S. engineers, managers, and astronauts met frequently over the next three years. [I-48] This led to the successful ASTP mission, which took place from July 14 to 24, 1975. The mission itself was accomplished without major problems.[59]

Even as the launch dates for the ASTP mission approached, George Low and Keldysh began to discuss follow-on cooperation, particularly in human spaceflight and robotic missions to the far side of the Moon and to Mars. [I-49] These discussions continued over the next two years, and by May 1977, when the U.S.-U.S.S.R. space cooperation agreement was renewed for five more years, the two countries had also agreed to consider a joint mission between the U.S. Space Shuttle and the Soviet *Salyut* space station, as well as cooperation in the development of an "international space platform."[60] [I-50]

As it turned out, however, the ASTP mission marked the high point of U.S.-U.S.S.R. space cooperation for some time to come. That cooperation fell prey to a deterioration in the overall state of U.S.-U.S.S.R. relations during the presidency of Jimmy Carter and the first White House term of Ronald Reagan. The Carter White House by 1978 was ques-

---

57.   George M. Low, NASA Deputy Administrator, "Visit to Moscow, April 1972, to Discuss Compatible Docking Systems for US and USSR Manned Spacecraft," April 4-6, 1972, with attached: "Addendum, Moscow Trip, April 4-6, 1972," May 30, 1972, Deputy Administrators Files, NASA Historical Reference Collection.

58.   For more on the agreement, see Ezell and Ezell, *The Partnership*, pp. 192-93.

59.   See *ibid.* for details.

60.   Office of Technology Assessment, *U.S.-Soviet Cooperation*, p. 32.

tioning whether it was in the U.S. interest to be seen as a highly visible cooperative partner with a Soviet Union that it was accusing of human rights violations, and the shut-tle-*Salyut* project was set aside. As part of the U.S. reaction to Soviet involvement in the declaration of martial law in Poland in 1981, the U.S.-U.S.S.R. space cooperation agree-ment was allowed to lapse when it came up for renewal in 1982. With that development, any cooperation in space between the United States and the Soviet Union had to be on a scientist-to-scientist basis, with no formal government involvement or funding.

This situation lasted for several years. Then in 1986, the United States, in response to the reforms of new Soviet leader Mikhail Gorbachev and to increasing pressure from sci-entists and others in the United States who viewed space as an arena for demonstrating a new, post–Cold War superpower relationship, indicated its willingness to resume formal cooperative relations in space with the Soviet Union. Negotiations led to the April 1987 signing of a new government-to-government cooperative agreement that identified six-teen areas for initial cooperation.[61] [I-51]

The U.S.-U.S.S.R. space relationship, always a barometer of the overall state of rela-tions between the two countries, was once again on the upswing. Cooperation increased after the collapse of the Soviet Union and the emergence of Russia as the inheritor of most of the former Soviet Union's space capabilities. By the mid-1990s, U.S.-Russian coop-eration had become the centerpiece of NASA's international space activities, as the two countries in effect merged their programs of human spaceflight in the period leading to the creation of the International Space Station. Initial steps in this direction came in 1992, when Russia created a civilian space agency and when Russian President Boris Yeltsin and President George Bush agreed to broaden U.S.-Russian space interactions. [I-52] This accelerated under President Bill Clinton; the United States and Russia in effect merged most of their programs of human spaceflight. [I-53, I-54, I-55] Russia was invited to become a key participant in the International Space Station. [I-56]

After almost forty years, cooperation had replaced competition as the central focus of U.S.-Russian space relations. The January 1961 hopes of President John Kennedy finally neared realization.

---

61.   George P. Shultz and Eduard Shevardnadze, "Agreement Between the United States of America and the Union of Soviet Socialist Republics Concerning Cooperation in the Exploration and Use of Outer Space for Peaceful Purposes," April 15, 1987, with attached: "Agreed List of Cooperative Projects," Space Policy Institute Documentary History Collection.

Document I-1

Document title: Office of the Press Secretary, Statement by the President, July 29, 1958.

Source: NASA Historical Reference Collection, NASA History Office, NASA Headquarters, Washington, D.C.

The congressional drafters of changes to the Eisenhower administration's version of a bill that set out the goals and organizational features of the U.S. civilian space program were strong advocates of international cooperation in carrying out the new program. They inserted in the bill a provision, contained in Section 205, that appeared to require Senate approval for any cooperative agreement into which the executive branch or the new space agency might enter. This was intended to signal congressional interest in international space issues, but legal experts pointed out after the bill had passed Congress that such approval could be interpreted as trespassing on the power of the president and his appointees to enter into nontreaty agreements for cooperation. At the time he signed the National Aeronautics and Space Act of 1958, President Eisenhower placed this statement on the record to dispel the possibility of such an interpretation.

---

[1] IMMEDIATE RELEASE   July 29, 1958

James C. Hagerty, Press Secretary to the President

THE WHITE HOUSE

Statement by the President

I have today signed H. R. 12575, the National Aeronautics and Space Act of 1958.

The enactment of this legislation is an historic step, further equipping the United States for leadership in the space age. I wish to commend the Congress for the promptness with which it has created the organization and provided the authority needed for an effective national effort in the fields of aeronautics and space exploration.

The new Act contains one provision that requires comment. Section 205 authorizes cooperation with other nations and groups of nations in work done pursuant to the Act and in the peaceful application of the results of such work, pursuant to international agreements entered into by the President with the advice and consent of the Senate. I regard this section merely as recognizing that international treaties may be made in this field, and as not precluding, in appropriate cases, less formal arrangements for cooperation. To construe the section otherwise would raise substantial constitutional questions.

The present National Advisory Committee for Aeronautics (NACA), with its large and competent staff and well-equipped laboratories, will provide the nucleus for the NASA. The NACA has an established record of research performance and of cooperation with the Armed Services. The combination of space exploration responsibilities with the NACA's traditional aeronautical research functions is a natural evolution.

The enactment of the law establishing the NACA in 1915 proved a decisive step in the advancement of our civil and military aviation. The Aeronautics and Space Act of 1958 should have an even greater impact on our future.

Document I-2

Document title: Hugh L. Dryden, Deputy Administrator, NASA, to T. Keith Glennan, *et al.*, March 12, 1959.

Source: NASA Historical Reference Collection, NASA History Office, NASA Headquarters, Washington, D.C.

Document I-3

Document title: Richard W. Porter, to Professor Dr. H.C. van de Hulst, President, Committee on Space Research (COSPAR), March 14, 1959.

Source: Archives, National Academy of Sciences, Washington, D.C.

The National Academy of Sciences (NAS) had managed U.S. participation in the International Geophysical Year, including related U.S. space activities, and had suggested to the International Council of Scientific Unions that it establish a continuing Committee on Space Research (COSPAR). When the new space agency, NASA, was formed, its initial staff handling both space science and international affairs had close ties to the NAS. Once a decision had been made not to use the United Nations for carrying out international space programs, NASA worked closely with the NAS. They authorized the U.S. delegate to COSPAR, Richard Porter, to invite other members of COSPAR to plan experiments that would be launched on U.S. boosters and to cooperate with NASA in getting started in space. NASA's head scientist, Homer Newell, attended the COSPAR meeting and stayed in touch with Deputy Administrator Hugh Dryden as the U.S. offer took final shape. Dryden in turn informed others at NASA headquarters about the final changes to the U.S. offer; Administrator T. Keith Glennan wrote "OK—Good" on his copy of Dryden's memorandum, which is reproduced here. The final offer came in the form of Richard Porter's letter to COSPAR President H.C. van de Hulst. With this letter, the United States initiated a program of productive cooperation in space science that has continued to the present.

Document I-2

[1]   March 12, 1959

Memorandum for   Dr. T. Keith Glennan
                   Mr. Henry E. Billingsley
                   Mr. Franklyn W. Phillips
                   Mr. Harold R. Lawrence

1.   On March 11th Homer Newell telephoned me from Holland to ask further advice on the statements that could be made with respect to cooperation with COSPAR in satellite launchings. The Executive Committee of COSPAR had met during the day in the absence of the Russian member, Dr. Federov, who arrived after the meeting had adjourned.

2.   We had previously authorized to the U.S. delegation to offer cooperation in the provision of payload space or possibly a booster for a complete COSPAR payload. The U.S. delegation felt, from the nature of the discussion at the Executive Committee meeting, that it would be desirable to make this offer somewhat more specific. After discussion with

Newell, I authorized him to make the definite offer for a booster for a COSPAR payload, the booster not to be named although we have in mind either Scout or Thor Delta. The payload would be described as 150 to 500 pounds, depending on the specific booster; and the general nature of the available orbits would be described. The booster would be available in 1-1/2 to 2 years. It was agreed that the payload should be recommended by COSPAR, that it should be acceptable to NASA, and that it should pass environmental tests prescribed by NASA.

3.   The question was raised as to whether the foreign scientists would be permitted to be present at the launching. I suggested that this question be avoided for the present for we are not in a position to commit the Atlantic or Pacific Missile Range.

4.   The meeting of COSPAR itself takes place today.

Hugh L. Dryden
Deputy Administrator

Document I-3

[1]   March 14, 1959

Professor Dr. H. C. van de Hulst,
President,
Committee on Space Research (COSPAR),
Paleis Nooreinde 68,
THE HAGUE, The Netherlands.

Dear Mr. President,

COSPAR has a truly historic opportunity to become an effective force for international cooperation in space research. This cooperation will be most fruitful and meaningful if the maximum opportunity to participate in, and contribute to, all aspects of space research can be provided to the entire scientific community. In this regard, COSPAR can serve as an avenue through which the capabilities of satellite launching nations and the scientific potential of other nations may be brought together.

The United States will support COSPAR in this objective by undertaking the launching of suitable and worthy experiments proposed by scientists of other countries. This can be done by sending into space either single experiments as part of a larger payload or groups of experiments comprising complete payloads.

In the case of individual experiments to become part of a larger payload, the originator will be invited to work in a United States laboratory on the construction, calibration, and installation of the necessary equipment in a U.S. research vehicle. If this is impossible, a U.S. scientist may be designated to represent the originator, working on the project in consultation with him. Or, in the last report, the originator might prepare his experiment abroad, supplying the launching group with a final piece of equipment, or "black box," for installation. However, this last approach may not be practical in most cases.

In the case of complete payloads, the United States also will support COSPAR. As a first step, the delegate of the U.S. National Academy of Sciences is authorized to state that the U.S. National Aeronautics and Space Administration will undertake to launch an entire payload to be recommended by COSPAR. This payload may weigh from 100 to 300 pounds and can be placed in an orbit ranging from 200 to 2,000 miles altitude. It is expected that the choice of the experiments and the preparation of the payload may require a period of 1 1/2 to 2 years. NASA is prepared to advise on the feasibility of proposed experiments, the design and construction of the payload package, and the necessary pre-flight environmental testing. The U.S. delegate will be pleased to receive COSPAR's recommendations for the proposed payload when they can be readied.

In further support of COSPAR, the U.S. delegate would like to call attention to the availability of resident research associateships at the National Aeronautics and Space Administration in both theoretical and experimental space research. These provide for stipends of $8000 per annum and up.

Very truly yours,
Richard W. Porter

Document I-4

Document title: T. Keith Glennan, Administrator, NASA, to Professor H.S.W. Massey, United Kingdom Scientific Ministry, July 6, 1959, with attached: H.S.W. Massey, "U.K.-U.S. Technical Discussions on Space Research, June 25-July 3, 1959," July 3, 1959.

Source: NASA Historical Reference Collection, NASA History Office, NASA Headquarters, Washington, D.C.

This exchange of documents constituted what was, in effect, the first (albeit informal) cooperative agreement concluded by NASA and a counterpart space organization (in this case, the British National Space Committee of the nongovernmental Royal Society). Upon its return to the United Kingdom, the proposals for cooperation developed by Professor Massey and his delegation were quickly approved by the U.K. government, although a formal U.S.-U.K. agreement implementing the cooperative program was not signed until 1961. The first mission resulting from this agreement, Ariel 1, was launched in less than three years, on April 26, 1962.

---

[1] July 6, 1959

Professor H. S. W. Massey
United Kingdom Scientific Mission
1907 K Street, Northwest
Washington 6, D.C.

Dear Professor Massey:

I have examined your statement of July 3, 1959, of the discussions held here between your group and our people and find it accurate and acceptable, from my standpoint.

It is my understanding that this statement represents only an informal technical understanding between us and does not commit either of our governments to proceeding with this undertaking until further arrangements have been consummated. On both sides, it must be recognized that the exact content and pace of the proposed program is subject to the success we may have with supporting technological developments and the financial resources made available by our respective governments.

Because of the importance of this matter and its relationship to the improvement of international understanding between our two countries in this field, I hope we shall hear from you further after you have reported the results of our talks to your people in London.

I trust this understanding is satisfactory to you and that your trip here has been pleasant and satisfactory in all regards.

Sincerely,

T. Keith Glennan
Administrator

****

[Attachment page 1]

U.K.-U.S. Technical Discussions on Space Research
June 25-July 3, 1959

Discussions have been held between N.A.S.A. and the team of British scientists led by myself, to consider the offer made by N.A.S.A. to assist other countries in scientific work with satellites, in the interest of developing a programme of international cooperation in space research. Specifically, this paper records the outcome of meetings held in Washington in late June and early July. As far as the U.K. side is concerned the outcome of these discussions, as recorded below, will be reported as soon as possible to the U.K. Steering Group on Space Research:

1.   N.A.S.A. confirm that they would be ready to offer facilities to the U.K. for launching U.K. instrumented satellites in the U.S.A. by means of the SCOUT vehicle.

2.   The U.K. side consider that three SCOUT satellites should provide sufficient space for first flights of the instruments which would be required for a programme of experiments of the type and range provisionally listed in Document NCSP 41(59).

3.   On present plans, N.A.S.A. hope that the launching of three British-instrumented SCOUT satellites could be completed within 3 to 4 years from now—with the aim that the first launching would be in about 2 years from now.

4.   If it is decided to accept the N.A.S.A. offer, the U.K. side would hope to provide specific proposals for the instrument content of the first satellite in the near future. Consideration of the possible content of further satellites would proceed as soon as possible, but would obviously be affected by success or otherwise in launching.

[2] 5.   In the first instance, N.A.S.A. would provide the satellite shell and auxiliary facilities, including solar cells and batteries, temperature control and data transmission.

6.   During the whole process of planning and constructing the satellites, the U.K. would do everything possible to maintain the necessary direct scientific and technical cooperation between the experts.

7.   Both sides agreed that a detailed exchange of views was desirable about possible extension of worldwide tracking and telemetry reception stations.

8.   N.A.S.A. said that the financial basis of their offer was "no billing" on either side.

9.   In making this offer it is N.A.S.A.'s position that this should be a mutually cooperative programme, with benefits flowing to both parties, for the advancement of science. It is contemplated that the experiments, conducted with the instruments flown by the British scientists, would form an integral part of the total spectrum of scientific experiments in space, and mutually agreed upon as a reasonable and important group of experiments.

The U.K. side explained that they were not at this stage authorized to come to any agreement, as this was a matter for the authorities in London.

July 3, 1959   H.S.W. MASSEY

Document I-5

Document title: Hugh L. Dryden, NASA Deputy Administrator, to Sir Harrie Massey, European Preparatory Commission for Space Research, June 27, 1962.

Source: NASA Historical Reference Collection, NASA History Office, NASA Headquarters, Washington, D.C.

At the urging of many leading scientists in Europe, the European governments decided in 1962 to create a multilateral European Space Research Organization as the primary mechanism for carrying out a European space science effort. It was general U.S. policy to favor the development of multilateral institutions in Europe, and NASA wasted no time in extending to the new organization the same cooperative offer it had earlier extended to individual European countries.

[1]   June 27, 1962

Sir Harrie Massey, President
European Preparatory Commission for Space Research
36, rue La Perouse
Paris 16, France

Dear Sir Harrie:

On behalf of the National Aeronautics and Space Administration, I should like to extend felicitations upon the recent signing of the Convention for establishment of the European Preparatory Commission for Space Research.

NASA will welcome ESRO as a major new force contributing to the advancement of space science and technology in the context of peaceful cooperation among nations. It is our hope that we may participate with ESRO in cooperative efforts which may enhance our respective programs and our contributions toward this goal. There are many possibilities for specific projects in which we may join our efforts. It may serve a useful purpose to suggest a few of them:

1. We should be very happy to enter with ESRO into a joint program of the type proposed in COSPAR on our behalf in March 1959. In such a program, NASA would provide a suitable launching vehicle to place in orbit a satellite of scientific interest which might be prepared by ESRO.

2. As you know, NASA plans the launching of complex "Observatory" satellites, each bearing a variety of experiments. Some of these satellites are to be placed in polar orbits, others into highly eccentric orbits of lower inclination to the equator. We would propose to notify ESRO of opportunities for it to recommend the incorporation of specific European experiments in these satellites. Such recommendations would be given the same consideration as proposals from American scientists for the same satellites.

[2] 3. With the concurrence of the Department of State, certain satellite boosters manufactured in the United States would be available under reasonable conditions for
purchase by ESRO for scientific purposes.

4. NASA is prepared to accommodate in its own laboratories, in conjunction with
agreed programs of the above character, technicians and scientists sponsored and
supported by ESRO for such training and orientation as desired. In addition, fellowships in American university laboratories devoted to space theory and experimentation will continue to be available to ESRO nominees.

I hope that ESRO will soon be prepared to discuss these and other possibilities for useful cooperation. At such time, or at any stage in your preparation for such discussions, we
will be happy to welcome here scientific representatives of ESRO to discuss possible satellite or sounding rocket experiments.

Sincerely yours,

Hugh L. Dryden
Deputy Administrator

Document I-6

**Document title: James E. Webb, Administrator, NASA, to Dean Rusk, Secretary of State,
May 24, 1966.**

**Source: NASA Historical Reference Collection, NASA History Office, NASA
Headquarters, Washington, D.C.**

*Beginning in late 1965, the United States tried to expand the scope and intensity of its space science
cooperation with other countries. Initial initiatives were made with Europe, but the hope was to
involve Japan as well. In this letter, NASA Administrator James E. Webb suggests to Secretary of State
Dean Rusk why he was not optimistic about the potential for enhanced U.S.-Japanese space cooperation. At this point, Japan's space effort was concentrated at an institute of the University of Tokyo.*

---

[1] Honorable Dean Rusk
Secretary of State
Washington, D.C.

Dear Dean:

I share the views expressed in your letter of May 12 regarding the desirability of cooperation with Japan in the space field and would be happy to send to Japan a team similar
to that which visited Europe in connection with President Johnson's offer on the occasion
of Chancellor Erhard's December visit. In order to work out an appropriate approach to
the Japanese, I am asking Mr. Arnold Frutkin, Assistant Administrator for International
Affairs, to work with Mr. Pollack.

There are certain limitations with respect to an approach to the Japanese which I
think you will want to consider. As you know, the President's proposals for cooperation on
the European side were of a scale and character necessarily multilateral. Proposals of the
same dimensions could not well be made to the Japanese since Japan could not be expect-

ed to carry a burden of spacecraft development which would be appropriate to the combined resources of the leading European nations. Thus, it seems inevitable that the subject matter of any space cooperation with Japan would have to be scaled down to the bilateral level, unless Europe itself should invite Japan to participate in its broader multilateral project.

When the question of Japanese participation with Europe came up in the Advance Team discussions in Europe last February, the Team made it quite clear that the U.S. would welcome such an overture by the European countries. Unfortunately, the prospect of positive European action on the President's proposal is in itself uncertain. While the matter is probably best not further complicated at this particular time, it may be that the possibility of Japanese contributions might be reopened in Europe in terms of European interest.

[2] With regard to bilateral cooperation with the Japanese, the pertinent history is rather dubious. Over the years, Dr. Dryden and Mr. Frutkin both made strenuous efforts to interest Japan in a bilateral satellite program comparable with those which have been entered into with virtually all other advanced nations. Japan clearly made a conscious decision to proceed on its own without involvement with the United States. (At the COSPAR meeting in Vienna early in May, Japanese scientists for the first time officially reported plans to launch small satellites this year and next year in the course of a self-contained Japanese program.) In order to obtain this decision, the dominant figures in Japanese space activity, particularly Professor Itokawa at the University of Japan, have quite consciously distorted the conditions of cooperation with the United States. Professor Itokawa is understood to have a close family association with the Prime Minister. Professor Kaneshige, with whom we have maintained very close touch, who is the Japanese Chairman of the US-Japan Scientific Commission, and who has been the senior Japanese scientific adviser to the Prime Minister, only last month told me that his continuing efforts to promote space cooperation with the United States have failed. He frankly confirmed the policy of deliberate distortion of our program (a matter which we have sought to bring to Embassy attention).

It is our understanding that the reception given in Tokyo to the Vice President's remarks on cooperation was rather cool. In the circumstances, a team, even of the caliber sent to Europe in February, would almost certainly find its efforts contained within a Japanese policy of noncooperation at quite senior levels. You may wish then to consider how persistent the U.S. ought to be and through what channels in pressing an objective that has met quite calculated and entrenched opposition over a period of years.

Sincerely yours,

James E. Webb
Administrator

Document I-7

Document title: James E. Webb, NASA Administrator, to U. Alexis Johnson, Deputy Under Secretary of State for Political Affairs, May 19, 1966.

Source: Administrators Files, NASA Historical Reference Collection, NASA History Office, NASA Headquarters, Washington, D.C.

During late 1965 and 1966, the United States considered ways to increase its cooperative activities in space. Most of the focus was on increased cooperation with Europe and Japan. However, there was interest in involving some of the larger developing countries with space ambitions of their own, such as India and Brazil, in cooperative efforts with NASA. This letter outlined for the first time a possible initiative with India to use a NASA communications satellite to broadcast educational programming to Indian villages. Almost a decade after the idea was first broached, between August 1, 1975, and July 31, 1976, the United States and India cooperated on the Satellite Instructional Television Experiment (SITE). During this experiment, the NASA Applications Technology Satellite ATS-6 was moved to an orbital position over India, and educational programming was broadcast to more than 5,000 Indian villages.

---

[1]   May 19, 1966

U. Alexis Johnson
Deputy Under Secretary for Political Affairs
Department of State
Washington, D.C. 20520

Dear Alex:

In my letter of April 26, I forwarded to you some thoughts for extending international cooperation in space. I would now like to supplement those with an additional suggestion that could prove valuable in opening new avenues for US-Indian collaboration in the practical applications of space.

The proposal should be understood on several levels: (1) A technical experiment in direct broadcasting, (2) A pilot project in the social impact of direct broadcasting, (3) A stimulus to Indian industrial electronics, [and] (4) An attack upon the food and population problems of India. The only step proposed is a joint US-India study of feasibility—which would have political value in and of itself.

If they agree, India and the United States would jointly establish a study group to examine the feasibility, the advantages and disadvantages, and the trade-off considerations of alternate approaches as these factors relate to a continuing experiment in direct broadcasting. In this experiment, the US would build and position a synchronous satellite near India in such a way that broadcasts from it could be received over the major part of the Indian subcontinent. Most of the basic technology for this experiment exists already and it should prove possible to beam the broadcasts tightly enough and on such frequencies that no interference would be caused to adjacent nations.

India, for its part, would use its nascent electronics capability now focussed [sic] at the atomic energy center at Bombay to develop (with some design assistance from the US) improved television receivers to be established in perhaps a thousand rural population centers. India has [2] already demonstrated a significant capacity to contribute to such a task: Bombay is currently turning out analog computers, oscilloscopes, solid state components, and the like. In addition, there exists within India a considerable capacity for the production of radio receivers and other, simpler electronics devices.

The satellites would be turned over to India for its own use in broadcasting to its people news and other material of an informational and educational value. The US would probably want to limit its commitment to provide satellites to perhaps two with a total projected life expectancy of five years. Following this experimental period, India would be expected to arrange with the US or INTELSAT for continuing service if desired. The cost to the US would be that of placing two large synchronous satellites in orbit. The cost to India would be that of the receivers themselves. One thousand such receivers might cost about $1.6 million, much of it in rupees. Since there are over 100,000 villages that might benefit from this program it would be up to India to decide the extent of its commitment beyond the initial one thousand receivers. In any case, much could be done by moving receivers from village to village to maximize the size of audience.

I would not understate the problems we would be likely to encounter. The cost of the program might be more than either we or the Indian Government would care to bear. Television appears to possess a significant informational and educational potential, but there may be many pitfalls to its application on a scale such as this. We should certainly have to reach definite prior agreement with the Indians concerning the uses to which space broadcasting was put. We obviously could not tolerate its use to defame us or our actions or to embarrass us in our relations with nations such as Pakistan. We might also have to face the question of continuing subsidies after the experimental satellite had gone dead; India might not be able to finance new satellites solely from its own pocket. We should also have to be careful that the experiment remained clearly separate from commercial communications projects and did not prejudice relations with INTELSAT or the concept of a single global communications satellite system.

Nevertheless, there remain powerful arguments in favor of discussing such an experiment with the Indian Government. The discussions and the institution of a joint study group would be a convincing display to that Government of our intent to use the most advanced technologies in helping it to cope with its urgent human problems. The joint study itself would bring Indians and Americans into intimate technical collaboration. India could learn from the study new technological and management approaches to education and to the uses of informational media to weld together a nation-state. The US would, in turn, learn more about the Indians and their most pressing problems.

[3] Should the project come to fruition, then important additional benefits would flow. India would, on its own initiative and with its own resources, begin the accelerated development of a modern electronics industry. This "bootstrapping" operation would materially raise India's technological base and contribute thereby to the development of other, similar industries. Some Indian energies might also be diverted from concern with nuclear weapons development, the more so perhaps as the success of the experiment contributed to India's prestige in Asia. The posture of the US would also be improved through a generous demonstration of its willingness to share the benefits of advanced space technology with underdeveloped nations.

As I view the proposal, we would at no time be exposing ourselves by unconsidered commitments or precipitous action. Each step, from the initial, private discussions with the Indian Government, through the careful and detailed examination by the joint study group would provide renewed opportunities to reexamine initial premises and search for possible flaws in all facets of the proposal. Even should it prove infeasible in the end, both we and the Indians could not fail to have profited by the intimacy of our cooperation in a joint technological venture.

I look forward with interest to hearing your reaction to this proposal. I do want to add that we are already negotiating or entertaining a number of other prospective projects for the near or distant future ad so have excluded these possibilities from the category of suggestions for new cooperation. . . .

Sincerely yours,

James E. Webb
Administrator

Document I-8

Document title: Robert J. Allnutt, for A.M. Lovelace, Acting Administrator, NASA, via Margaret Finarelli, NASA Headquarters, to Erik Quistgaard, Director General, European Space Agency, Telegraphic Message, February 20, 1981.

Source: NASA Historical Reference Collection, NASA History Office, NASA Headquarters, Washington, D.C.

In February 1981, the Office of Management and Budget, under the new administration of President Ronald Reagan, ordered NASA to cancel one of its major science programs. This represented a way of reducing the NASA budget in future years. NASA chose to cancel the spacecraft it was committed to provide as part of the International Solar Polar Mission, a joint venture of NASA and the European Space Agency. Because of the secrecy surrounding budget decisions, NASA was unable to alert ESA of the possibility of such a cancellation until a telephone call on February 18, the same day that President Reagan announced the cuts he was making in the budget submitted by outgoing President Jimmy Carter. Formal notice of the cancellation was provided by this telegram, signed by Robert Allnutt, one of the top staff of NASA's Acting Administrator Alan Lovelace, and forwarded by Margaret Finarelli of the Office of International Affairs.

---

[1]

TO: MR. ERIK QUISTGAARD   INFO:
     DIRECTOR GENERAL   MR. JAMES MORRISON
     EUROPEAN SPACE AGENCY   NASA EUROPEAN REP.
     8-10 RUE MARIO NIKIS   RUFNPS AMEMBASSY
     75738 PARIS CEDEX 15   PARIS, FRANCE
     FRANCE

                                  MR. WILFRED MELLORS
                                  ESA WASHINGTON OFFICE
                                  (CODE LID-18 TO MAIL)

As I indicated to you in our telephone conversation yesterday, the administration's budget for FY82 requires severe cuts in the full range of NASA's programs. Because work on vital shuttle activities must continue, we have been forced to cancel or otherwise forego a number of major programs in the science and applications areas. We are foregoing starts for both VOIR and GRO; monies for the development of Spacelab payloads have been cut back even further: NOSS has been cancelled; and numerous other programs in the applications area such as our agricultural forecasting program have suffered deep cuts.

[2] We have endeavored, and will continue to endeavor, to honor our international commitments to the fullest extent possible. Nonetheless, the deep budget cuts have necessitated cancellation of part of the joint NASA/ESA ISPM Mission, namely the US Spacecraft which was to have participated in the Solar Polar mission. In view of the major scientific importance of Solar Polar research, we hope that ESA will continue with the mission which can now be launched in 1986 on a Shuttle/Centaur and that we will be able to maintain its cooperative nature.

As I indicated to you yesterday, the NASA budget will permit support of the remaining spacecraft, including the U.S. experiments previously planned for the ESA spacecraft.

[3] As I mentioned to you on the telephone, I propose that Dr. Stofan, NASA's Acting Associate Administrator for Space Science, and Dr. Trendlenberg, ESA's Director for Scientific and Meteorological Programs, meet promptly to review the current spacecraft situation, assess the impact of the budget reduction on the scientific value of the mission and determine the most effective way to proceed with the ESA spacecraft. When Stofan and Trendlenberg have concluded their discussions regarding the spacecraft, you and I should then discuss the mission further.

I want to assure you that cancellation of the US Spacecraft in the ISPM mission is taken with great reluctance and was unavoidable given the broad and deep cuts occurring throughout NASA and throughout the US Government budget. I would like to reiterate my deep appreciation for ESA's cooperation with NASA in the past and my continuing sense of commitment to working with ESA on programs of mutual interest.

[4] I share your view about the value of looking closely at our existing consultation procedures to see if, within the constraints on both sides, improvements can be made. I have asked Ken Pedersen to be sure this topic gets a high place on the agenda for our informal talk on March 9. I am looking forward to seeing you again.

> Robert J. Allnutt signed for
> A. M. Lovelace
> Acting Administrator

Document I-9

Document title: W.J. Mellors, Head, Washington Office, European Space Agency, "Aide Memoire, International Solar Polar Mission (ISPM)," February 24, 1981.

Source: NASA Historical Reference Collection, NASA History Office NASA Headquarters, Washington, D.C.

Not surprisingly, the European Space Agency (ESA) and its member states reacted with outrage to the U.S. cancellation of its spacecraft contribution to the International Solar Polar Mission. Diplomatic protests were lodged on a government-to-government basis, and Congress was made aware of Europe's unhappiness. These notes, prepared by Wilfred Mellors, the head of ESA's Washington office, were the basis of his initial formal presentation of the ESA response to the cancellation.

---

[1]   24th February, 1981

Aide Memoire
International Solar Polar Mission (ISPM)

1. Last week the Acting Administrator of NASA informed the Director General of the European Space Agency that the cuts announced by President Reagan in his speech on February 18th, 1981 included the cancellation of the US spacecraft and the associated U.S. experiments for the above mentioned mission. At a meeting held on February 23rd NASA confirmed this situation.

2. I am to say that:
    a) The cancellation of the NASA satellite, which was effected without consultation, is a unilateral breach of the ISPM MOU; this cancellation is totally

unacceptable and ESA requests full restoration of the programme to its original level.

b) If the cancellation were permitted to stand, there would be serious long term damage to European/United States cooperation in space.

c) Naturally, there has been a very unfavorable reaction in Europe. No less than seventeen European scientific institutes are involved in the United States spacecraft and would consequently be unable to fly. These experimenters have already committed more than 50% of the total cost of their experiments. Indeed, Europe has already made a major investment of the order of one hundred millions of US dollars, equivalent to the whole of ESA's annual budget for Space Science, in the ISPM programme.

3.   It is to be noted that at the time ESA decided to participate in ISPM, other candidate missions were considered, including some purely European projects with no American participation. However, ESA decided to collaborate with NASA because first, the ISPM mission—as it was with two spacecraft—was of outstanding value to the scientific community in Europe and in the United States as it permitted simultaneous measurements over the Northern and Southern hemispheres of the sun and, secondly, (of equal and even greater importance), it was believed there was great value in such transatlantic cooperation.

[2] 4.   I am further to say that this present cancellation of the US spacecraft is not the first instance of a unilateral action by the US in this project which has had serious consequences for ESA. I am to point out that in March of 1980 the revision of the NASA FY 81 budget resulted in a delay of two years in the launch date which cost ESA and the European scientific institutes supporting the mission at least an additional twenty millions of dollars; while in January of this year a decision was taken to change the upper stage from the IUS [Inertial Upper Stage] to the CENTAUR vehicle, the full consequences of which have not yet been established.

5.   In view of the above the Director General has today requested each Member State immediately to make through its Washington Embassy at Ambassador level, the strongest possible protest against the announced cancellation and to request a full restoration of the two spacecraft mission.

6.   Finally I am to point out that it is ESA's view that such unilateral actions as now taken by the United States could destroy the basis for collaborations of this nature in the future and that the impact could well go beyond the field of scientific space research.

W. J. Mellors

Head, Washington Office
European Space Agency

Document I-10

Document title: "Meeting of NASA Administrator and ESA Director General, 17 June 1982, ESA Head Office," with attached: "General Principles for NASA/ESA Cooperative Agreements."

Source: ESA Collection, European Community Archives, Florence, Italy.

In the months following the cancellation of the International Solar Polar Mission, the European Space Agency (ESA) and its member states conducted a comprehensive assessment of the desirability of continuing Europe's close cooperation with the United States. The conclusion of this assessment was that such cooperation remained desirable but the terms and conditions under which it would be carried out had to provide more protection to European interests. These terms and conditions were agreed to at a June 1982 meeting between the heads of NASA and ESA.

[1]

Meeting of NASA Administrator and ESA Director General
17 June 1982, ESA Head Office

FUTURE NASA/ESA COOPERATION

A. *Principles and terms of future agreements.*
   Having confirmed their intention to continue their long-standing cooperation, ESA and NASA agreed on the "General Principles for NASA/ESA Cooperative Agreements" attached to these minutes (Annex 1).
B. *Procedures for regular exchange of views on future programmes in space science and applications.*
   NASA proposed three measures to set up such procedures:
   – schedule regular discussions between the respective division directors responsible for astrophysics, environmental observations, and earth and planetary exploration programmes;
   – encourage exchange of information and ideas among US and European scientists who participate in the respective NASA and ESA programmes;
   – encourage regular meetings between the NASA Associate Administrator for Space Science and Applications and the ESA Directors of Scientific Programmes and Application Programmes.
   ESA gave a brief outline of ground rules recently approved by its Science Programme Committee for future scientific cooperation.
   These ground rules are attached (Annex 2).
[2] As to applications, ESA stated to be in favor of regular meetings at working level in the field of earth observation. In the telecommunications sector further ad hoc discussions on specific topics could be envisaged.
   ESA agreed to the three measures proposed by NASA and outlined above. . . .

****

General Principles for NASA/ESA
Cooperative Agreements

The National Aeronautics and Space Administration and the European Space Agency confirm their desire to continue conducting cooperative space projects. They recognize that in the past, cooperation has in general, been of great mutual interest to both parties.

NASA and ESA intend to continue formalizing such cooperation through either Memoranda of Understanding (MOUs) between the two agencies, the standard form of agreement for joint projects, or, for some specific major programmes, Intergovernmental Agreements between ESA Member States and the Government of the United States of America. Before a proposed MOU is concluded, it will be submitted by NASA to the U.S. Department of State, which will determine whether it constitutes an international agreement as defined by the provisions of Public law 92-403. This action is necessary because of Sec. 503 of Public law 92-426, which requires that the Secretary of State be informed and consulted before any agency of the United States Government takes any major action, primarily involving science or technology, with respect to any foreign government or international organization, and also because of Sec. 504 of that law which stipulates that the Secretary of State has primary responsibility for coordination and oversight with respect to all major science or science and technology agreements and activities between the United States and foreign countries or international organizations. NASA will notify ESA of the U.S. Department of State opinion before submission of the proposed MOU to the ESA Council.

As such international agreements have general limitations within the U.S. legal system and may have to contain, for legal or institutional reasons, specific limitations regarding their liabilities, the parties note that some degree of risk is involved in joint projects. Thus, NASA and ESA agree that, in order to alleviate the uncertainties and the risks, they will from now on apply the following guidelines:

1. In developing the type and degree of assurances to be included in a particular agreement, NASA and ESA will take into consideration the degree of risk and the importance of the project for each of the parties. The calculation of risk will include such factors as the possibility of one party losing all or a major part of its investment if the other party cannot further fulfill its commitment, the cost burden assumed by each party, the overall cost of the mission, and the time criticality of the mission. Both parties, in the process of negotiating an agreement, will undertake to provide within that agreement for a degree of assurances appropriate to the degree of risk resulting from the factors mentioned above.

[4] 2. NASA and ESA will endeavor to inform each other of any legislative or regulatory provisions existing, or coming into force subsequent to the signing of a[n] MOU, that might limit or prevent implementation of the agreement's provisions.

3. NASA and ESA recognize the importance of timely and full consultation to the effective implementation and completion of joint projects. Consultation is particularly important when one party experiences difficulties in meeting its obligations as stated in the project agreement of its annexes, if any. NASA and ESA will, to the fullest extent practicable in such cases, consult before a decision is taken.

4. NASA and ESA will consider whether a proposed project is suited to being implemented in discrete phases which could be the subject of separate agreements. The purpose of this approach would be to permit, after conclusion of each phase, each party to review its interest in continuing with the project. It is recognized that many projects will not be amenable to this approach.

5. In the course of negotiation [of future cooperative project agreements, NASA and ESA will examine possibilities for proving assurances relative to alternative flight opportunities or developed flight hardware in the event the agreement is not able to be executed in full.

Document I-11

Document title: Burton I. Edelson, Associate Administrator, Space Science Applications, NASA, and Roger M. Bonnet, Director, Scientific Programme, ESA, "NASA/ESA Space Science Planning Meeting, ESA Head Office—27th-29th June 1983," June 29, 1983.

Source: NASA Historical Reference Collection, NASA History Office, NASA Headquarters, Washington, D.C.

By 1983, the unhappiness of the European Space Agency (ESA) had not completely disappeared with respect to the 1981 cancellation of the U.S. spacecraft intended as part of the joint NASA-ESA International Solar Polar Mission. However, the two organizations still recognized the benefits of regular consultations on their future space science plans to identify productive synergies and opportunities for collaboration. For example, the following minutes of a June 1983 meeting show that together the two agencies identified areas, such as solar terrestrial research, in which enhanced cooperation would be fruitful. They also recognized other areas, such as infrared astronomy, in which each would pursue separate missions.

[1]

NASA/ESA Space Science Planning Meeting ESA Head Office—27th-29th June 1983

At their meeting on 27th, 28th, 29th June 1983 in Paris, ESA and NASA reaffirmed the great advantage of international cooperation in space science which they consider of particular importance from the point of view of scientific, technological, political and financial considerations.

The meeting was held with the aim of reaching an in-depth understanding of the other party's goals on fundamental as well as more immediate practical issues. Three areas of cooperation were discussed in more detail:
– Infrared Astronomy
– Solar Terrestrial Research
– Planetary Exploration
Each of these areas is treated in the following sections.

1. *General Framework of Cooperation between ESA and NASA*
It was agreed during the meeting that measures should be taken to improve in the future the framework of cooperation between the two parties. Two such measures have already been identified:

a) ESA and NASA agree to set up an international committee to advise the two par-
    ties on specific issues related to cooperation. The committee will be co-chaired by
    R. Bonnet from ESA and F. MacDonald from NASA and will include four
    American and four European senior scientists with experience in international
    cooperative programmes. The committee will in particular analyze the best ways
    of implementing the principle of reciprocity by which American participation in
    European programmes is applied in recognition of the US principle of opening
    their AOs [Announcements of Opportunity] to the non-US scientific community.
    The committee will report to ESA and NASA before the end of January 1984.
b) In order to widen their cooperation at the level of the younger scientists, the two
    parties agreed to formalize an exchange of fellowship programmes whereby a
    number of post-doctoral European fellows will each year be based in NASA cen-
    tres and the same number of American fellows based in the Space Science
    Department of ESA/ESTEC [European Space Technology Education Centre].
    The exact procedure for selection of the scientists and for reviewing their
    research programmes will be analyzed in detail before the next bilateral meeting.
Action: NASA, F. MacDonald; ESA, D.E. Page

2. *Infrared Astronomy*
    ESA and NASA note the technological and scientific success of IRAS [Infrared
Astronomical Satellite] and reaffirm their commitment to infrared astronomy in space.
[2] They agree to continue to explore further joint effort in infrared space astronomy.
    ESA notes NASA's revised plans to make SIRTF [Space Infrared Telescope Facility] a
long duration, reserviceable mission operating in a Shuttle/Space Station compatible
orbit, and NASA's strong interest in collaborating to develop a single major international
infrared Space telescope facility.
    NASA notes the firm commitment of ESA to the approved Infrared Space
Observatory, ISO, which is an Ariane launched mission with an elliptical orbit.
    The parties discussed the possibility for US participation in the ISO mission and
European participation in the SIRTF mission by furnishing focal plane instruments and
exchange of observing time. It is noted that the differences in orbit and launch vehicle
restrict any major hardware collaboration between these two missions as currently
defined.
    It is recognised that in the post-IRAS time frame, coordination in programme plan-
ning is desirable to optimize the overall scientific return. It is therefore agreed to set up a
joint study team to:
a) analyse the objectives and anticipated performances of ISO and SIRTF to identi-
    fy their complementarity;
b) identify characteristics of focal plane instruments in both facilities which could
    optimise the overall performance capability of these two missions;
c) identify elements in both programmes which could be considered as reciprocal
    contributions.
    The joint study team will be headed by Dr. Nancy Boggess of NASA and Dr. Henk
Olthof of ESA and will meet during the autumn with a preliminary report in time for the
next ESA/NASA Space Science Planning Meeting.

3. *Future Solar Terrestrial Research*
    The ESA/NASA representatives surveyed the large number of missions under study
in the USA, Europe and Japan in the area of Solar Terrestrial physics (DISCO, SDO,
SOHO, Cluster, OPEN, OPEN J, Plasma Turbulence Explorer) and agreed that a need
exists for an integrated look at all these missions.

There seems to be considerable merit in considering a joint NASA/ESA/ISAS [Institute of Space and Astronautical Science of Japan] mission which would cover major parts of solar heliospheric physics of DISCO, SDO and SOHO and at the same time cover magnetospheric and interplanetary physics and thereby replace IPL of OPEN in conjunction with the utilization of an enhanced OPEN J as the EML portion of OPEN.

[3] It was agreed that NASA and ESA will set up a preparatory meeting, to which ISAS will be invited, with two or three representatives from each Agency and one or two representatives from each of the projects mentioned above. The meeting will be organized by NASA in Washington DC in late September 1983. The goal of this meeting should be to look for joint missions which can satisfy the main scientific requirements in a cost-effective way.

Following reporting to the advisory committees of the Agencies in October/November, and a further round of meetings of the preparatory committee and advisory committees in January 1984, the aim is to define missions which can go into ESA phase A and NASA studies in approximately March 1984.

4. *Planetary Exploration*

The ESA/NASA representatives reviewed the status of the plans and studies of the two Agencies in the area of planetary exploration in order to identify mutually beneficial opportunities for cooperative missions.

a) *Saturn-Titan Probe Mission*

Pending the recommendation of ESA's advisory committees, NASA and ESA agree to undertake a joint assessment study in 1984 of a Saturn-Titan probe mission for launch around 1992. This mission would call for an FY 1989 NASA new start. The mission would be based on the ESA Cassini proposal and on the Titan probe mission identified by NASA's Solar System Exploration Committee, and would take into account the recommendations of the NAS/ESF Joint Working Group.

b) *Small Bodies Rendezvous Mission*

NASA and ESA plan to undertake a joint assessment study of a small bodies rendezvous mission using a European Solar Electric Propulsion System for launch in the 1990s. The mission would be based on the ESA Agora proposal and on the Multi-Mainbelt Asteroid mission identified by NASA's Solar System Exploration Committee, and would take into account the recommendation of the NAS/ESF Joint Working Group. The organization and timing of this study will be addressed at the next NASA/ESF Space Science Planning Meeting.

c) *Mars Missions*

The Announcement of Opportunity (AO) for the NASA Mars Geoscience-Climatology Orbiter (MGCO) mission is planned for release in 1985 and, as such, is well timed for coordination with an ESA AO for a 1992 Kepler mission, if this mission is approved for launch in that year.

[4] 5. *Next Meeting*

The next ESA/NASA Space Science Planning meeting is scheduled to take place in the US in December 1983/January 1984.

Burton I. Edelson
Associate Administrator
Space Science Applications
NASA

Roger M. Bonnet
Director
Scientific Programme
ESA

Paris, 29th June 1983

[5] *List of Participants*

NASA
Burton Edelson
Charles Pellerin
Shelby Tilford
Frank MacDonald
Geff Briggs
Richard Barnes
Lyn Wigbels

ESA
Roger Bonnet
Vittorio Manno
Edgar Page
Henk Olthof
George Haskell
Gordan Whitcomb
Roger Emery
Arne Pedersen
Brian Taylor
Valerie Hood

Document I-12

Document title: George M. Low, Manager, Apollo Spacecraft Program, to Director, Apollo Spacecraft Program, "Flag for Lunar Landing Mission," January 23, 1969.

Source: NASA Historical Reference Collection, NASA History Office, NASA Headquarters, Washington, D.C.

In his January 20, 1969, Inaugural Address, Richard Nixon had suggested that "as we explore the reaches of space, let us go to the new worlds together—not as new worlds to be conquered, but as a new adventure to be shared." Unsure of the intent behind the new president's words, NASA headquarters began to think of how best to make the first lunar landing appear to be more of an international accomplishment. To those working on the Apollo program who saw the enterprise primarily in nation-alistic terms, this was a troubling development. This memorandum from George Low, who had assumed personal direction of the Apollo spacecraft program after the Apollo 204 capsule fire, gives a sense of this reaction. George Hage, mentioned in the memorandum, was an official of the Apollo Program Office at NASA headquarters. The Apollo 11 mission eventually carried very lightweight flags of every country, which were returned to Earth and presented, along with a small lunar sample, to heads of state. A plaque saying "We Came in Peace for All Mankind" was attached to the lunar lander and left on the Moon.

---

[1]
AA /Director   January 23, 1969

In reply refer to:
PA-9-1-40

PA/Manager, Apollo Spacecraft Program

Flag for lunar landing mission

I received a call from George Hage indicating that, in light of Nixon's inaugural address, many questions are being raised in Headquarters as to how we might emphasize the international flavor of the Apollo lunar landing. Specifically, it was suggested that we might paint a United Nations flag on the LM [Lunar Module] descent stage instead of the United States flag. My response cannot be repeated here. I feel very strongly that planting the United States flag on the moon represents a most important aspect of all of our efforts; I indicated that, from a personal point of view, I would have no objection to carrying some small United Nations flags to the moon and bringing them back for subsequent presentation to the UN (provided, of course, that they don't weigh too much).

I thought that you should be aware of these discussions since the subject will probably come up again on several occasions.

George M. Low

Document I-13

Document title: Thomas O. Paine, NASA Acting Administrator, to President Richard Nixon, February 12, 1969.

Source: Thomas O. Paine Papers, Manuscript Division, Library of Congress, Washington, D.C.

As Richard M. Nixon assumed the presidency on January 20, 1969, interest in space was at a peak. The December 1968 Apollo 8 circumlunar mission, commanded by astronaut Frank Borman, had captured the imagination of the world and cleared the way for an initial lunar landing attempt. Thomas O. Paine, who had been NASA Deputy Administrator since March 1968 and had become Acting Administrator after James E. Webb retired in November 1968, stayed on during the presidential transition. In this letter, Paine provides to the president an assessment of the space situation in Europe and of U.S.-European space relations. The Dr. DuBridge mentioned in the letter is Lee A. DuBridge, the president's science advisor.

---

[1]   February 12, 1969

The President
The White House
Washington, D. C.

Dear Mr. President:

Dr. DuBridge has informed me of your desire for a summary of European space activities in connection with your forthcoming visit abroad and for advice on space-related matters that might be appropriate for you to discuss with the Europeans.

Frank Borman's visits are being enthusiastically received and may serve to generate more interest in space at the time of your trip than would otherwise be the case. The Borman family is now visiting the countries on your itinerary and we have offered your staff any assistance our people in Europe may be able to give in advance work and arrangements for your trip.

The following brief review covers national and regional space activities in Europe, European cooperative activities with NASA, and suggested positions which you might take on both European and cooperative space activities. This review has been coordinated with the State Department people concerned and accords with their views.

I have also considered two suggestions for additional ways in which you might express your personal interest in space cooperation while you are in Europe. One would be through your participation in a ceremony in Rome to confirm two pending agreements. The other would be to extend personal invitations to the Chiefs of State you meet to attend the historic Apollo 11 launch now scheduled to undertake a lunar landing this summer. The Department of State feels, however, that both suggestions could create problems that might outweigh the advantages, and we concur in their view that these proposals should not be pursued without further careful consideration by State.

I - General

Although much study and discussion has taken place, the European countries have not yet defined and agreed upon their individual and joint basic policies in space. They are making limited [2] investments in national programs at a level of about $300 million annually. They have pooled resources in two intergovernmental regional bodies: the European Space Research Organization (ESRO), and the European Launcher Development Organization (ELDO). These are maintained on a marginal basis only, however, with severe internal divisions as to purpose, structure, funding level, contract-sharing, and future direction and pace.

The countries you will visit all belong to the 65-nation communications satellite consortium, Intelsat, for which the US Comsat Corporation is manager and NASA is the launching agency on a reimbursable basis. Intelsat has made excellent progress toward a global satellite communications system, but certain quarters (particularly France) argue that the United States unduly dominates Intelsat through its technological advantages, large voting rights, designation of the US Comsat Corporation as manager, NASA's position as the only source of suitable launch vehicles, and by obtaining (competitively) the largest share of contracts for US industry. These basic issues will be the subjects of negotiating sessions beginning this month in Washington to arrive at definitive arrangements for Intelsat's future. Also at issue will be the place, if any, for domestic or regional communications satellite systems inside or outside of the Intelsat.

There has been and continues to be significant and productive cooperation between NASA on the one hand and the European national space agencies and ESRO on the other. This includes: a dozen European satellites launched or to be launched by NASA with full international data-sharing, some twenty European experiments contributed for flight on NASA satellites, dozens of joint scientific sounding rocket launchings, important support for meteorological and communications experiments, accommodation and operation of U.S. tracking and data acquisition facilities overseas, advanced information exchange programs, and joint fellowship and training programs.

Nevertheless, the European nations have still not determined whether they should rely ultimately on cooperation with the United States or should develop a completely independent capability for space operations. Near the heart of this issue is the specific question in Europe whether they should develop an independent launch capability for communications satellites, or should remain dependent upon U.S. boosters only, thereby submitting to the alleged American domination of Intelsat.

In the meantime European Space Agency-NASA cooperation proceeds very satisfactorily on the technical level and is proving most productive. It appears limited, however, to essentially small scientific satellites and one larger undertaking now nearing agreement with Germany until the above fundamental issue is resolved. Efforts on NASA's part to increase the scale of cooperation in the past several years have been [3] viewed in Europe against the background of the Intelsat issue. Thus, we have been suspected of attempting to divert European activities toward scientific pursuits and away from "high pay-off" projects in space communications, and our offers of boosters for their satellites have been interpreted in some quarters as calculated to undermine support for ELDO's development of a European booster. In general, however, you will find a positive view of American space cooperation, and a very enthusiastic view of NASA in the wake of Astronaut Borman's highly successful Presidential good-will tour.

II - National Situations

France is the third "space power" and the only country besides the U.S. and the USSR to have launched its own satellites with its own launchers. It possesses an excellent space laboratory at Bretigny and is developing a unique equatorial-polar launch site in Guiana. Cooperation between NASA and its French counterpart (CNES) has been professional, extensive, and scientifically rewarding. The first French scientific satellite, FR-1, was launched by NASA in December 1965. Another major French satellite, EOLE, is to be launched by NASA in 1970 to determine the feasibility of a satellite-balloon system for mapping global winds systems. Of five French scientific experiments accepted for flight on NASA satellites, four have already flown.

With respect to other space relationships, France has reflected Gaullist policies, has sought to dominate both ESRO and ELDO, has led the attempt to direct both organizations toward local communications satellite objectives, and had led the most severe criticism of alleged U.S. domination of Intelsat. France is now engaged with Germany in an experimental communications satellite, Symphonie. France is the only western nation to have reached a cooperative agreement with the Soviet Union for the actual development of a scientific satellite. This was delayed by French budget cuts and by French scientists' efforts to obtain access to Soviet launch sites necessary for validating their work, and is now reported abandoned.

The United Kingdom has contributed three scientific satellites for launching by NASA, agreement has been reached on a fourth, and a fifth is under consideration. British scientists have also contributed eleven outstanding experiments selected for flight on NASA satellites (more than any other country) and have made major contributions to ESRO satellites. This is significant, since the contribution of an individual experiment for a NASA satellite may cost the contributing country perhaps $300 thousand. Although Britain initiated ELDO in [4] 1962, it has led the current movement to scuttle the organization on grounds of excessive cost, poor reliability, and the ready availability of proven U.S. launch vehicles.

Germany was slow to initiate space activity but is now developing the largest space budget in Europe, over $100 million annually. Two small satellites are being prepared for launching in 1970 and 1972 on NASA launch vehicles. A space probe will be launched in 1970 by another NASA launch vehicle, and an ambitious solar probe, HELIOS, is in the final stages of joint definition. This will carry German and U.S. experiments closer to the sun than has yet been done, again using a U.S. launch vehicle. Germany usually aligns itself with France on European regional space issues and has joined with France in the Symphonie communications satellite project. These projects are straining Germany's project management capability to the utmost.

Italy has focussed [sic] mainly on cooperative satellite agreements with NASA (signed in Rome by then Vice-President Johnson in 1962). Under these agreements, Italy has developed an imaginative launch complex on towable platforms moored in the Indian Ocean off Kenya.

Here, the San Marco satellite was launched by Italians using a contributed NASA booster to make unique measurements of the density of the spatial medium. A jointly-instrumented satellite will be launched here in the next cooperative effort in 1970. NASA has a new agreement pending with Italy for the launching of two U.S. spacecraft from this complex on a reimbursable basis; the platform's location on the equator permits us to use smaller boosters than would otherwise be required to achieve equatorial orbits, thereby saving NASA $2-3 million per launch. Italy is the weakest supporter for ELDO and ESRO at the present time.

The Netherlands and Belgium maintain small but high-quality space science programs, primarily in selected university laboratories. The principal ESRO laboratory is located in Holland at Noordwijk. Dutch scientific groups have made contributions to ESRO and NASA satellites out of all proportion to their modest domestic support. Dutch scientific and industrial interests are pressing their government to propose the cooperative launching by NASA of a small but sophisticated astronomical satellite. Both Belgium and Holland possess excellent laboratory facilities in aeronautics as well as in space science. Both countries support the regional space institutions in Europe, although Belgium has tended to follow France's hostile lead with regard to Intelsat.

[5]   **III - Regional Organizations**

ESRO is a ten-member intergovernmental agency for the development and operation of spacecraft and sounding rockets for scientific purposes and practical applications. It spends about $50 million a year and has developed highly professional facilities at Noordwijk in Holland, other facilities elsewhere, and a small tracking and data acquisition network. NASA has, on a cooperative basis, launched ESRO's first two scientific satellites and, on a reimbursable basis, has launched a third. NASA and ESRO have developed a sophisticated integrated data exchanged system and conduct a jointly-funding training program. The ELDO crisis and financial and contract-sharing difficulties have strained ESRO and currently limit opportunities for enlarging the scale of U.S. cooperation.

ELDO is a seven-member intergovernmental organization, spending now about $90 million annually to develop a large European launch vehicle. England has developed the first rocket stage with U.S. technology, France the second stage, and Germany the third, while Belgium, the Netherlands and Italy are contributing ancillary systems. The Australian launch site at Woomera has been used for test launchings but the vehicle will ultimately be shifted to the French Guiana range. The three-stage ELDO launch vehicle falls between the U.S. Thor and Atlas rockets, but has yet to function successfully as a whole, though it probably will in time. Severe cost overruns and a decision by the UK to discontinue membership after 1971 have thrown ELDO into a serious crisis which jeopardizes its future as well as that of ESRO. ELDO has called on NASA only for minor assistance through visits or discussions relating to technical background and management systems. U.S. policy has supported both ESRO and ELDO as having European institutional values. Other U.S. policies, however, conflict to the extent that they restrict technical assistance which might conceivably be used to support European communications satellite capabilities inconsistent with Intelsat.

NATO. With regard to larger U.S. policy, efforts were made before the establishment of ELDO and ESRO to develop a European regional space activity based on NATO. European interests nevertheless insisted on: (1) projecting an uncompromising civilian posture in space, (2) making it possible for non-NATO nations like Switzerland and Sweden to join with others, and (3) preserving the option for an independent European space effort.

It is not yet clear whether Europe will be able to save and strengthen ELDO and ESRO, although efforts are in progress and the situation is very sensitive, particularly with regard to putative U.S. motivations [6] and European goals. European leaders have discussed an ultimate possibility that ESRO and ELDO might be merged into a single European "NASA" but plans for this purpose are not due for consideration until the end of this year.

IV - Suggested Positions During Your Trip

We anticipate that the Intelsat question would be the major space-related matter that might arise during your visit. This is a matter of central concern to the Department of State and other agencies. NASA is in complete agreement with the State Department's position that the United States should respond to questions and criticisms on Intelsat to the effect that these matters are negotiable in the Intelsat definitive negotiations beginning later this month. In particular, the French and German space commissions have jointly asked NASA whether we would launch their joint experimental communications satellite, Symphonie, on a reimbursable basis. With the guidance of the Department of State, we have responded positively. This was considered the best answer under the circumstances, though it was recognized that some Europeans would interpret this positive answer as designed to undercut ELDO's European launcher programs, just as they would have interpreted a negative answer as designed to monopolize satellite communication experiments by denying launching assistance in this area to European nations. We believe it important to continue to maintain as positive a posture on this point as possible.

Against this background, it would appear to us desirable if you could reassure Europeans, wherever space matters arise, that the U.S. is not seeking to impose its will on the direction of future West European space activities and that we recognize that European nations should determine their own courses based on their own assessments of where their interests lie. If U.S. cooperation can figure positively to our mutual advantage, it will indeed be available. There is a strong positive interest in NASA to further develop international cooperation in space in *both the science and applications areas*, on the basis of mutual interest.

Respectfully yours,

T. O. Paine
Acting Administrator

Document I-14

Document title: Thomas O. Paine, NASA Administrator, to the President, August 12, 1969.

Document I-15

Document title: Thomas O. Paine, NASA Administrator, to the President, November 7, 1969.

Document I-16

Document title: Thomas O. Paine, NASA Administrator, to the President, March 26, 1970.

Source: All in Thomas O. Paine Papers, Manuscript Division, Library of Congress, Washington, D.C.

These three letters record the initiatives that NASA, and particularly Administrator Thomas O. Paine, took in the aftermath of the Apollo 11 landing to increase international participation in the U.S. post-Apollo space program. Paine believed that he had a mandate directly from President Richard M. Nixon, delivered as they flew to the Apollo 11 landing (splashdown) in the Pacific Ocean, to actively seek enhanced international cooperation. Paine based his briefings to leading officials in other countries on the future plans laid out in the report of the Space Task Group, chaired by Vice President Spiro T. Agnew. As the Nixon administration made it clear in early 1970 that it did not intend to approve the program recommended by the Space Task Group and as the president's advisors raised concerns about the potential of technology transfer to other countries through cooperative space programs, the early enthusiasm about the possibility of major cooperative initiatives faded.

Document I-14

[1]   August 12, 1969

The President
The White House
Washington, D.C.

Dear Mr. President:

This is a brief status report on our current efforts and immediate plans to find new ways to increase international participation in space programs in the favorable situation generated by Apollo 11.

1.   On August 12, I met with Professor Herman Bondi, Director General of the European Space Research Organization (ESRO), briefed him fully on U.S. post-Apollo thinking and urged him to begin serious consideration of new approaches to achieve more substantial European participation in the manned and unmanned exploration and utilization of major space systems in the 1970's and 1980's. European thinking with respect to space activity has been relatively restricted heretofore [because] ESRO's current annual budget is slightly over $50 million and the European Launch Development Organization budget is slightly over $90 million. In addition, individual national efforts total over $160 million, for a total European space effort of something in excess of $300 million.

Professor Bondi agreed that a series of presentations should be made by top NASA personnel to senior space officials in Europe within the next few months to raise their sights to more advanced projects of greater mutual value.

2. To initiate these presentations and to conduct more direct and private discussions with officials in the best position to respond positively, I plan to brief senior (government) officials of the European Space Conference on future U.S. programs and the concrete opportunities they will [have] for rewarding participation. I will also talk with Ministers of Science in the three principal countries but especially with Minister Stoltenberg in West Germany, which is probably in the best position to consider substantial new participation. While we cannot achieve immediate commitments of a major character from these first discussions, we do hope to gain early agreement to an arrangement which could involve the Europeans ever more closely with us and place the benefits of participation constantly in their view. To this end, I plan to propose to the leading European space agencies that they associate their top space experts with us in phased program studies which we will be undertaking for important post-Apollo missions. The knowledge and interest which we jointly develop should then open the door to more substantial [2] participation in specific projects which flow out of these studies, and which would be suitable for European attention to the opportunities which would then develop to associate their own astronauts with us in future programs in the context of substantive joint contributions to space exploration and application. This could generate greater public interest and support abroad for participation with the United States in this venture.

3. Professor Bondi's mission to the U.S. was to obtain information needed to decide whether the European Launch Development Organization should continue the costly development of an already-outmoded medium launch vehicle, duplicating those we have had for years, or should halt this work and rely on reimbursable launch services from NASA. Europeans have heretofore feared that the U.S. would not provide launchings for regional communications satellites, which has motivated them toward small independent efforts rather than major joint ventures along the lines we will be proposing. A forthcoming response to Dr. Bondi has now been obtained from the Department of State and will, we hope, remove a long-standing negative element in the environment and facilitate our discussions looking to more significant cooperation. If Europe should now decide to abandon its trouble-plagued and obsolescent launch vehicle program in favor of purchasing U.S. launchings, European funds would be freed for more constructive cooperative purposes and a modest additional dollar market would be created for our vehicles and launch services.

4. Among other promising near-term prospects for significant cooperation with Europe are a prototype North Atlantic Air Traffic Control and Navigation Satellite Program, and a Synchronous Meteorological Satellite Program. NASA would develop the former in partnership with ESRO to meet requirements defined by the Department of Transportation (FAA) and its European counterparts. The latter would be developed with the French Space Commission as a contribution to the Global Atmospheric Research Program. We are pursuing both these prospects energetically.

5. We have recently significantly extended our data exchange arrangements with ESRO to the point where they now constitute, we believe, the most extensive and sophisticated international data system in existence. ESRO uses NASA computer software systems and formats to collect the European technical literature and feed it into their own and into NASA's computer banks making possible a totally integrated space publication and search system. ESRO has also introduced the NASA Recon (Remote Control) System to Europe. An international on-line computerized aerospace information network is thus enabling researchers at a number of scattered locations in Europe and in the U.S. to

retrieve from the NASA ESRO data bank in "real time," scientific and technical information for immediate use. This is the first international system of its kind and is being studied both in Europe and this country as a model for similiar [sic] systems.

6.   NASA welcomes and will participate enthusiastically in the review called for by Dr. Kissinger to consider U.S. policies on space and other technology exports. This is a timely opportunity to clear away unnecessary restrictions which could seriously obstruct the increased international activity which you have called for.

[3] 7.   With regard to potential cooperation with the Soviet Union, I have recently written top Soviet space authorities offering to discuss carrying Soviet scientists' experiments of future NASA planetary probes. I am now inviting Soviet scientists to attend a preparatory briefing next month for scientists from many other countries on our Viking Mars mission with a view to discussing possible participation in that mission and the achievement of some measure of cooperation between U.S. and Soviet planetary programs. Whether the Apollo 11 success will moderate past Soviet negativists in this area is not yet clear.

8.   Japan, Australia, and Canada are the principal remaining areas whose potential for greater participation will be carefully explored. I believe NASA has contributed to a reasonable formulation of the new agreement with Japan to initiate that country's purchase of certain space technology here and we will play a role in providing for the implementation of the agreement. Under your recent directive, we will provide Canada launch services for her planned communications satellite system; this action has clearly improved relationships in this area, and we are already discussing with Canadian officials their active interest in possible participation in our advanced earth resources technology satellite series. I discussed yesterday with our new Ambassador to Australia the great services that have been rendered through Australian operation of our large tracking and data acquisition complex there and our strong interest in further participation. I expect to visit these three countries at the earliest opportunity for greater international cooperation in those quarters.

I will, of course, report to you the results of my forthcoming visit to Europe immediately upon my return.

Respectfully yours,

T. O. Paine
Administrator

Document I-15

[1]   Nov. 7, 1989

The President
The White House
Washington, D.C.

Dear Mr. President:

This is to report to you the results of my recent three-day visit to Europe and related actions seeking to promote greater international participation in future U.S. space programs.

1.   On October 13, 14, and 15, I met with Ministers of Science and senior space program officials of the Federal Republic of Germany, France and the United Kingdom plus

the Committee of Senior Officials of the European Space Conference in Paris. I described for them the principal elements of our space program in the next decades—the reusable Space Shuttle, the multi-purpose space station, and the advanced nuclear stage—as recommended in the Space Task Group Report. I invited their careful study of these plans so that Europeans might assess the implications for their own planning and determine what interest they may have in constructive participation with us.

Our audiences were clearly impressed by the prospects for development of an economic, shuttle-based space transportation system and by the prospects for a space station as a platform for work in both practical applications and science. The Europeans appear to recognize that the shuttle and space station together clearly imply the gradual convergence of manned and unmanned flight programs and that this may well outmode their previous assumption that automated missions might suffice for Europe in the next decade.

Our fundamental objective was to stimulate Europeans to rethink their present limited space objectives, to [2] help them avoid wasting resources on obsolescent developments, and eventually to establish more considerable prospects for future international collaboration on major space projects. In these respects, I believe our visits were more successful than might have been expected in the present circumstances of very limited budgets and organizational difficulties in European space affairs. We were given to understand privately that the general reaction to our discussions was that current European space planning must indeed be thoroughly reassessed in the light of the opportunities inherent in the proposed U.S. programs. Chancellor Brandt's speech of last week called for increased cooperation in direct response to our suggestions.

2.   On October 16 and 17, NASA convened in Washington a conference of industrial firms to critique concepts for the Space Shuttle and to lay our design considerations for next steps in the program to develop the shuttle. At our invitation, some 43 foreign participants and observers attended from Germany, France, the UK, Netherlands, Canada, Sweden, and Italy, as well as the European Launcher Development Organization. This event interacted most favorably with my visit to Europe, lending credibility to my statements that the U.S. would welcome broader participation in our overall programs. In turn, the broad opportunities described during the European visit provided a meaningful framework for international participation in the Space Shuttle conference. We plan to continue this pattern of activity to the extent that substantive European interest permits.

3.   In the area of earth resources surveys by satellite, we have moved forward in several respects to follow through on your recent remarks to the United Nations General Assembly:

(a) An invitation was circulated to the entire UN membership to send observers to the 1969 International Symposium on Remote Sensing of the Environment, conducted at the University of Michigan last week. Some 41 foreign experts from 12 countries attended.

(b) If suitable arrangements can be made, we plan to invite the United Nations Outer Space Committee and representatives of the UN specialized agencies to inspect earth resources program work [3] and facilities at NASA's Manned Spacecraft Center in Houston at an early date.

(c) We are proceeding with several domestic universities to provide a number of graduate fellowships covering work in the earth resources disciplines; this fact will be reported to the Outer Space Affairs Group of the United Nations so that training possibilities will be generally known; and

(d) We are also moving forward with plans for an international workshop in 1970 to review the status of research and experimentation in the earth resources field for all interested nations.

4.   I believe you know already of the agreement signed by NASA and an Indian counterpart agency in mid-September to make available access to a NASA satellite for an experiment in instructional TV broadcasting to 5,000 remote Indian villages, beginning in 1973. Our ability to make available a share of the time of an advanced satellite in the course of an on-going program and to suggest a programmatic framework for the experiment stimulated India to a very considerable effort which will include the construction of augmented TV village receivers, the planning of instructional programs, and the logistics system required to coordinate and support all elements of the system. Such programs have the greatest implications for benefit to the developing world and for political value to the United States as a generous source of advanced technology able to serve the interests of the LDC's.

5.   I have in the past weeks written several times to President Keldysh of the Academy of Sciences of the Soviet Union. We invited him to send Soviet scientists to a briefing on our Viking Mars mission with a view to discussing possible participation in that mission as well as possibilities for coordination between American and Soviet planetary programs. Another letter assured Keldysh that NASA will welcome proposals from Soviet scientists for the analysis of lunar samples. Finally, I am forwarding to him copies of the Space Task Group [4] Report, suggesting that this may be an appropriate time for a meeting to discuss the possibilities of complementary or cooperative space programs. The exchange of astronaut/cosmonaut visits may indicate a greater receptivity on their part to such discussions.

Beyond this, I plan visits to Canada, Australia and Japan to provide the same sort of briefing and open the same opportunities to these nations as in my European visit. I shall continue to report to you as progress is made in any of the relevant areas and in particular to the extent that any substantial European interest develops.

Respectfully yours,

T. O. Paine
Administrator

Document I-16

[1]   March 26, 1970

The President
The White House

Dear Mr. President:

My recent talks in Australia and Japan completed the first round of foreign visits to discuss our space plans for the next decade and to stimulate consideration of new and more extensive international participation in the development and realization of those plans.

In Australia, I met with the Minister of Supply, his principal colleagues, and senior officials of the Department of Education and Science. A number of representatives of other agencies with interest in the practical applications of space technology participated in broader discussions. My impression is that our proposals for increased international participation in space activities will receive thoughtful consideration. Australian interest will probably focus on future application satellite programs and the possibilities for a role in operational aspects of space station/Space Shuttle activities.

In Japan, my discussions were with the leadership of the Science and Technics Agency, the Ministries of Education, International Trade, Transport, Posts and Telecommunications, Foreign Affairs, and with Japan's space agencies. I was encouraged by evidence of top-level industrial interest in our programs. Our meetings included a full afternoon session with major corporation executives who are members of the Keidanren, the federation of Japanese industry. Japan clearly construes its interests in participation in the proposed international program in hard and practical terms. As one deputy minister stated, Japan realizes that in the future it must go beyond quality initiative work and move on to undertake new, highly creative enterprises. Participation in the proposed major space development projects for the '70s may offer Japanese industry a unique opportunity for such technical creativity.

[2] Upon my return to Washington, NASA held an important meeting on March 13 attended by 40 space officials and representatives from 17 countries and from the three regional European space organizations: the European [Space] Research Organization, the European Launcher Development Organization, and the European Space Conference. These visitors participated in a quarterly review by NASA management of contractor design and definition studies for our space station and Space Shuttle programs. The principal discussion centered on the potential of these new systems for replacing many of the space systems which had previously been proposed in their development programs for the 1970's. It seems clear that our proposed space station/Space Shuttle systems would obsolete many of their proposed developments before they became fully operational. For this reason our proposals for international participation are receiving thoughtful attention.

The stakes are high and the issues complex here, so we should expect an extended period of up to a year during which foreign governments and their space agency officials will be increasing their grasp of the technical details and potentials of our new space systems for the '70s. European circles are now giving more serious and open consideration to the possibilities for their participation (an example is the attached item from today's Christian Science Monitor). The choices are, however, difficult ones. Many in Europe believe that they must choose either an independent European space effort of a limited and retrograde character or commit to a much bolder joint program that will be dominated by the United States. We are discussing with the Department of State the kinds of assurances of access to and use of the proposed jointly developed new systems that we should be prepared to give foreign collaborators in order to win their participation.

We will continue to involve foreign space interests in government and industry more closely with us, to stimulate their interest, and to begin to formulate for their consideration more specific proposals and institutional formats for joint development work.

Respectfully yours,

T. O. Paine
Administrator

Document I-17

Document title: Secretary of State, Telegram 93721 for U.S. Ambassador, Tokyo, "Space Cooperation with Japan," January 5, 1968.

Source: Lyndon B. Johnson Library, Austin, Texas.

This telegram transmitted policy guidance to U.S. Ambassador to Japan U. Alexis Johnson regarding new initiatives in U.S.-Japanese space cooperation. Johnson had pressured his colleagues in Washington to approve firm-to-firm licensing agreements that would help Japan develop launch a capability equivalent to an early model of the U.S. Delta booster, as well as application satellite capability. NASA and the Department of Defense opposed such an arrangement, but Johnson and the Department of State prevailed. The terms and conditions suggested in this telegram were incorporated in a July 31, 1969, exchange of diplomatic notes.

---

[1] R 050540Z Jan 68
FM SecState WashDC
To AmEmbassy Tokyo 0000
Info CINCPAC . . .

[2] For Ambassador

Subject: Space Cooperation with Japan

Reference: Tokyo 3837

 1. Agreement in November 15 communique between President Johnson and Prime Minister Sato (para 9) opens way for expanded space cooperation with Japan and we would like to initiate discussion with [the Government of Japan (GOJ)] to this end. We consider close cooperation with Japan in field of space very much in US interest. Such cooperation, first, entirely consistent with our basic relationship with Japan and national policy of closest possible partnership with Japan in both bilateral relations and joint actions to strengthen non-Communist position in east Asia. [Remainder of paragraph excised during declassification review]
 2. Therefore, on basis of discussions with you here, we have developed following policy regarding space cooperation with Japan:
  A. Under NSAM 338, we are prepared to cooperate in all aspects of communications satellite development and launch on the assumption that both governments will continue to act in sphere in conformity with their INTELSAT commitments.
[2] Therefore, we would approve technology transfers only after determining to our satisfaction that it would be used only in (i) purely experimental (as opposed to operational) systems or (ii) operationally domestic systems compatible with INTELSAT arrangements as they evolve. FYI—Neither Japanese nor we are in position to predict just what arrangements for satellite ownership and control other than by INTELSAT will be reflected in renegotiated INTELSAT agreements. However, President Johnson, in his message of August 14 (CA 1299, dated August 15), committed us to support continuation of INTELSAT and to avoid course of action which is incompatible with our support for a global system. Under these circumstances, it is not unreasonable to indicate to Japanese that we could not assist them if their policy is to contrary. End FYI.

    B.   [Paragraph excised during declassification review]

    C.   Only unclassified US technology is involved.

    D.   There must be Japanese [government] commitment on third country controls transferes [sic] of technology derived from US cooperation to Communist China and Soviet Union must be explicitly excluded. Sales or technical exchanges involving other third countries will require prior US [Government]/GOJ agreement, based on common policy for US and GOJ suppliers.

    E.   [Paragraph excised during declassification review]

[4] 3.   Guidelines for review of applications for licenses, export of equipment or technology will be:

    A.   [Paragraphs A and B excised during declassification review]

    C.   Licenses for export of equipment or technology will be granted if equipment or technology is unclassified and related to an identified Japanese peaceful space program or objective:

    and

    D.   We are satisfied that end-use of technology applied to communications satellites will be consistent with INTELSAT arrangements as they develop.

    4.   On basis above, we propose moving ahead with GOJ along following lines:

    A.   We anticipate space cooperation could be extended at two levels: (1) government-to-government and/or agency-level agreements in specific joint projects, including provision of reimbursable launch services and (2) industry-to-industry licensing arrangements requiring government approval under munitions control procedures and consistent with provisions of NSAM 338.

    B.   We are prepare [sic] to adopt positive position in all areas of peaceful space cooperation including technology, reimbursable launch services, and assistance in development of launch vehicles nessary [sic] for application satelites [sic].

    C.   The Japanese should understand that we take our commitment to INTELSAT seriously and would not act inconsistently with it. Therefore, we would approve technology transfers [5] only after determining to our satisfaction that it would be used only in (i) purely experimental (as opposed to operational) systems or (ii) operationally domestic systems compatible in INTELSAT arrangements as they evolve.

    D.   We would want an agreement with Japanese government (preferably through exchange of notes) covering two points:

    (1) Technology or equipment transferred under either government-to-government agreements or industry-to-industry arrangements will be for peaceful purposes except as may be otherwise mutually agreed; and,

    (2) Technology or equipment derived from US cooperation cannot be transferred under any circumstances to Communist China or the Soviet Union and can be transferred to other third countries only after mutual agreement based on common export policies.

    5.   We suggest you undertake appropriate discussions with GOJ. If, in your judgment, GOJ [is] sincere on end-use technology consistent with INTELSAT arrangements, we will be in position to move ahead vis-a-vis NSAM 338 on government, agency, and later on industry levels as appropriate. Action on proposals involving NSAM 294 would be undertaken following appropriate agreement as set forth [in] para 3 D above. When it becomes clear that such an arrangement is acceptable to the Japanese, we would want to undertake appropriate congressional consultation prior to formalizing agreement with the Japanese. GP-3 Rusk

Document I-18

Document Title: Arnold W. Frutkin, NASA Assistant Administrator for International Affairs, to Administrator, "Canadian Interest in Remote Manipulator Technology to be Used with the Space Shuttle," April 3, 1972.

Source: NASA Historical Reference Collection, NASA History Office, NASA Headquarters, Washington, D.C.

After President Nixon gave his go-ahead to the Space Shuttle program on January 5, 1972, it was time to decide what contributions, if any, other countries might make to the program. While U.S.-European negotiations on this question were rather acrimonious, the discussions on a Canadian contribution proceeded relatively smoothly. This memorandum summarizes the prospects for U.S.-Canadian post-Apollo cooperation as of early 1972; a final agreement that Canada would contribute a remote manipulator system (later named "Canadarm") to the Space Transportation System was reached in 1975. NASA Administrator James Fletcher, in a handwritten note to NASA staffer Donald Morris on the first page of this memorandum, stated: "O.K. to start discussions, but let's not get as far into it as we have on the shuttle (Post-Apollo) with the Europeans. I don't want any embarrassment if we decide not to go ahead."

---

[1]   APR 3, 1972

Memorandum

TO: A/Administrator

FROM: I/Assistant Administrator for International Affairs

SUBJECT: Canadian interest in remote manipulator technology to be used with the Space Shuttle

The only result of the NASA offer to the Canadians on post-Apollo participation has been interest in possible development of remote manipulator equipment which might be used in the Space Shuttle to service the Large Space Telescope [LST], and possibly other orbiting spacecraft. This offer stems from a specialized Canadian capability and technology resulting from the development of extensible booms for use in space, and the particular requirements of their nuclear power reactors—which are fueled without shutting down.

Two methods are now under consideration for servicing of the Large Space Telescope. The first involves a special RAM-telescope combination which would be serviced by technicians entering the RAM from the docked Space Shuttle for film and subsystem recovery and replacement. The second involves the use of an end-effector deployed by a technician-operator from the Shuttle, and designed to detach and replace equipment packages on the Large Space Telescope. Current study activity sponsored by [the Office of Manned Space Flight] is directed to a choice between these two options sometime next summer.

The Canadian Department of Trade, Industry and Commerce would like to have a Canadian industry team assist in a Goddard-conducted interface study of the Space

Shuttle and LST, which will explore the second option described above. Their goal is to indicate sincere Canadian interest in hardware development participation, to promote better understanding in Canada of the factors which would be involved in possible Canadian participation in any subsequent development effort, and to demonstrate to NASA what they have to offer.

[2] In considering possible Canadian association in this early study phase, we have examined the following factors relevant to subsequent hardware development:

1. Remote manipulator technology related to the LST would not affect the Space Shuttle development schedule. It would involve relatively simple interfaces with the Shuttle itself. The Shuttle Program would develop the basic manipulator and the Canadians would develop the end-effector to attach to the manipulator.

2. The development of remote manipulator end-effectors is not comparable to the kinds of "bits and pieces" of the Space Shuttle central to current discussions of post-Apollo participation. Manipulator end-effectors are related more to the payloads under consideration than to the development of the Shuttle itself.

3. The Canadians have a very special capability in this field which our people feel would be of benefit to the program.

4. Canadian association in the study effort and in a subsequent development effort would not require transfer of US technology to Canada. Interface and parametric data will be provided.

5. Although the Canadians would expect us to agree to procure a certain number of production units (in the same manner we have suggested to the Europeans we would be prepared to do in the case of Sortie modules or Tugs) the ratio of development cost to production costs is reasonable. Goddard estimates that development of a manipulator to work with the LST would cost the Canadians about $7-9 million while the cost per unit for production should be about $2.5 million.

Both [Associate Administrator for Manned Space Flight] Dale Myers and I believe this proposed activity lies more in the area of "separable items" than in "bit and pieces" of the Shuttle and that NASA would stand to benefit from Canadian participation in development of remote manipulators if this option is chosen by [the Office of Manned Space Flight]. Therefore, unless you feel [3] differently we would propose to respond positively to the Canadian request to work with us on the current interface study on a no-commitment basis along the lines of the attached draft letter. I would, of course, discuss this with State and John Walsh before proceeding.

Arnold W. Frutkin [initialed]

Document I-19

Document title: Peter M. Flanigan, "Memorandum for John Erlichman," February 16, 1971, with attached: Clay T. Whitehead, "Memorandum for Mr. Peter Flanigan," February 6, 1971.

Source: Nixon Project, National Archives and Records Administration, Washington, D.C.

Thomas Paine, supported by the Department of State, had taken a bullish approach to expanded international space cooperation. Paine, however, was frustrated by the Nixon administration's unwillingness to approve a large post-Apollo space program, and he resigned in September 1970. Deputy Administrator George Low, who had come to Washington from Houston after the Apollo 11 mission, became Acting Administrator. Within the White House staff, Assistant to the President Peter

Flanigan had responsibility for policy oversight of NASA. His assistant, Clay T. "Tom" Whitehead, worked with Flanigan on NASA and telecommunications policy issues, even after he became the director of the newly established Office of Telecommunications Policy. This memorandum is an early indication of the split within the executive branch over the approach to be taken with respect to European involvement in the post-Apollo program. President Nixon had made it clear that he wanted increased international cooperation, but just what that meant was a subject of some debate within top policy circles.

[1]   February 16, 1971

Memorandum for John Erlichman

FROM: PETER. M. FLANIGAN

Attached is a thoughtful memorandum which I asked Tom Whitehead to prepare on NASA. One obvious use of this memorandum is to give it to the new Administrator when he comes on board (I am expecting that Jim Fletcher will take the job in about four weeks).

You will particularly note the discussion starting in the middle of page two regarding international cooperation in space. I suggest that either you or I, or both of us, talk to the President about this before we get ourselves too deeply committed. If the President is not, as I suspect, committed to the current sharing program, then I think I should immediately get George Low in and discuss with him the kind of international cooperation that is desired.

****

[attachment page 1]   February 6, 1971

Memorandum for Mr. Peter Flanigan

This Administration has never really faced up to where we are going in Space. NASA, with some help from the Vice President, made a try in 1969 to get the President committed to an "ever-onward-and-upward" post-Apollo program with continued budget growth into the $6-10 billion range. We were successful in holding that off at least temporarily, but we have not developed any theme or consistency in policy. As a result, NASA is both drifting and lobbying for bigger things [parenthetical comment: "'the bigger the best' correct"]—without being forced to focus realistically on what it ought to be doing. They are playing the President's vaguely defined desire for international cooperation for all it's worth, and no one is effectively forcing them to put their cooperative schemes in any perspective of whether they are good or not so good, what are their side effects, and are they worth the candle. For the last two years, we have cut the NASA budget, but they manage each year to get a "compromise" of a few hundred million on their shuttle and space station plans. Is the President really going to ignore a billion or so of sunk costs and industry expectations when he gets hit for the really big money in a year or two?

I will try to be constructive by sketching out a few thoughts on the subject that might suggest what we should do about all this.

NASA is—or should be—making a transition from rapid razzle-dazzle growth and glamour to organizational maturity and more stable operation for the long term. Such a transition requires wise and agile management at the top if it is to be achieved successfully. NASA has not had that. (Tom Paine may have had the ability, but he lacked the inclination—preferring to aim for continued growth.) They have a tremendous overhead structure, far too large for any reasonable size space program, that will have to be reduced. There will be internal morale problems of obvious kinds. The bright young experts attracted by the Apollo adventure are leaving or becoming middle-aged bureaucrats with vested interests and narrow perspectives. (Remember when atomic power was a young glamour technology? Look at [the Atomic Energy Commission] now and you see what NASA could easily become.)

[2] There needs to be a sense of direction, both publicly and within NASA. The President's statement on the seventies in space laid the groundwork, but no one is following up. What do we expect of a space program? We need to define a balance of science, technology development, applications, defense, international prestige and the like; but someone will have to do that in a way that really controls the program rather than vice-versa. In particular, we need a new balance of manned and unmanned space activity, for that one dimension has big implications for everything else. We need a more sensible balance of overhead expenditures and money for actual hardware and operations; the aerospace industry could be getting a lot more business than they are, I suspect, with the same overall NASA budget if we could get into all that overhead.

NASA is aggressively pursuing European funding for their post-Apollo program. It superficially sounds like the "cooperation" the President wants, but is this what the President would really want if we really thought it through? We have not yet decided what we want our post-Apollo program to be or how fast it will go, but if NASA successfully gets a European commitment of $1 billion, the President and the Congress will have been locked into NASA's grand plans because the political cost of reneging would be too high. I assume the President wants space cooperation as a way of building good will and reducing international tensions. But it does not follow that all joint ventures will have that effect. INTELSAT, for example, is a fully cooperative space venture and less political than the post-Apollo effort now envisaged would be, but most would agree it has been more of a headache than a joy and has created new tensions and contentions rather than good will and constructive working relationships. [parenthetical comment: "yes!"] Finally, the U.S. trade advantage in the future will increasingly depend on our technological know-how. The kind of cooperation now being talked up will have the effect of giving away our space launch, space operations, and related know-how at 10 cents on the dollar. [parenthetical emphasis: "!!"] It does seem to me that taking space operations out of the political realm and putting it more nearly in the commercial area would diminish international bickering and give U.S. high technology industries the advantages and opportunities they deserve; this may or may not prove fully feasible but the point is, no one in this Administration is seriously trying to find out.

[3] The key thing missing, I think, is management attention to these issues. We need a new Administrator who will turn down NASA's empire-building fervor and turn his attention to (1) sensible straightening away of internal management and (2) working with [the Office of Management and Budget] and White House to show us what broad but concrete alternative the President has that meet[s] all his various objectives. [parenthetical comment: "implying Paine was not"] In short, we need someone who will work with us rather than against us, and will seek progress toward the President's stated goals, and will shape the program to reflect credit on the President rather than embarrassment. We need a generalist who can understand dedicated technical experts rather than the opposite. But we

also need someone in the Executive Office who has the time, inclination, and authority to coordinate policy aspects. Separate handling of political, budget, technical, and international aspects of NASA planning here means that we have no effective control over the course of events because all these aspects are interrelated.

We really ought to decide if we mean to muddle through on space policy for the rest of the President's term in office or want to get serious about it.

Clay T. Whitehead [signed "Tom"]

Document I-20

Document title: Memorandum from Edward E. David, Jr., Science Advisor, for Henry Kissinger and Peter Flanigan, "Post-Apollo Space Cooperation with the Europeans," July 23, 1971, with attached: "Technology Transfer in the Post-Apollo Program" and Henry A. Kissinger, National Security Advisor, to William P. Rogers, Secretary of State, no dates.

Source: NASA Historical Reference Collection, NASA History Office, NASA Headquarters, Washington, D.C.

Document I-21

Document title: George M. Low, NASA Deputy Administrator, to NASA Administrator, "Items of Interest," August 12, 1971.

Source: James C. Fletcher Papers, Special Collections, Marriott Library, University of Utah, Salt Lake City, Utah.

These two documents give a sense of the state of discussion on post-Apollo cooperation during the summer of 1971. The president's science advisor, Edward David, was one of those in the White House trying to find a position acceptable to those holding more nationalistic views, such as Peter Flanigan, and those in the State Department and NASA's international office taking a more internationalist perspective. James Fletcher had become NASA Administrator in May 1971, and George Low, who had been Acting Administrator from the time that Thomas Paine left in September 1970 until Fletcher was sworn in May 1971, had returned to his position as Deputy Administrator. Fletcher and Low worked closely together in dealings with the White House and the Executive Office of the President and let each other know what they were doing through frequent private memoranda.

Document I-20

[1]   July 23, 1971

Memorandum for
Henry Kissinger
Peter Flanigan

Subject: Post-Apollo Space Cooperation with the Europeans

Background
It was agreed at our meeting with Jim Fletcher on April 23, 1971, that NASA should prepare an evaluation of (1) the degree of technology transfer to the Europeans, which would take place if the proposed U.S.-European cooperation on development of a space transportation system (STS) were to materialize; and (2) alternative subjects for U.S.-European cooperation. I have now reviewed NASA's informal paper (summary attached) and discussed the subject with Jim Fletcher, who concurs with the course of action recommended in this memorandum.

Pending further consideration of the details of the NASA analysis, and additional discussions at the technical level between the U.S. and European space groups, I am not prepared to have the U.S. commit itself to this cooperative program of STS development. Although the NASA study (concurred in by Jim Fletcher) suggests that the technology transfer question as well as management complications are not of significant proportions, my personal concerns on these points have not yet been answered to my full satisfaction, nor can they be answered until there is a better understanding of the potential European contribution. Furthermore, U.S. shuttle planning is not sufficiently definitive at present to permit any agreement on the shuttle with the Europeans in the near future. Nonetheless, I do believe that a resumption of technical-level discussions with the Europeans would be in order at this time for the purpose of more clearly defining, without any precommitment, the potential interests and contributions of both sides.

[2] It is also apparent from recent telegrams from Europe that a reply to Minister Theo Lefevre's letter to Alex Johnson of March 3, requesting a statement of the U.S. position on post-Apollo space cooperation, cannot he delayed much longer. Europe's space officials must move ahead with their own planning for the future. I believe this matter can be resolved by separating the issue into two components and addressing each separately.

The urgent question before the Europeans is whether U.S. launchers will be available at a fair price and on a non- discriminatory basis for launching European satellites. If the answer is no, the Europeans will likely proceed to develop their own EUROPA-III launch vehicle, with little or no funds left for cooperation with the U.S. in any areas; if yes, they will most probably abandon their launcher development plans, freeing funds for increased cooperation with the U.S. and/or for other space developments of their own.

The first alternative would require European expenditures of almost a billion dollars to build a launch capability which has already existed in the United States for several years. In the process, it will doubtless engender some bitterness on the part of those countries who oppose this choice on practical grounds, but would feel constrained to support it on political grounds. However, this approach will by 1976-78 provide the Europeans with a capability to launch their own geosynchronous satellites independently of U.S. views or influence.

The second alternative would perpetuate European dependence on the U.S. for launch services, would generate sales for U.S. booster manufacturing firms, and would preserve the chance for a major European input to a cooperative program with the U.S. This alternative would seem more attractive than the first for longer-range U.S. interests.

Although the availability of U.S. launchers might also enable the Europeans to compete with U.S. firms for satellite construction contracts from other countries, both the U.S. aerospace industry and I believe that this would not be a significant commercial threat, in view of our vastly superiority [sic] satellite technology.

[3] *Recommendation*

Accordingly, I propose that we separate the two elements of launch assurances and space cooperation and that State be advised to proceed along the lines of the attached draft letter to Bill Rogers. If you are in agreement, I believe this course of action provides a satisfactory exit from the present impasse.

Edward E. David, Jr.
Science Advisor

Attachments

****

[Attachment page 1]
Technology Transfer in the Post-Apollo Program

As background for a decision on the course of action to be pursued in defining a mutually acceptable set of tasks for European participation in the post-Apollo space program, NASA was asked to examine the implications of cooperation in Space Shuttle development, particularly from the standpoint of technology transfer. The detailed report on this effort is attached.

One conclusion of the NASA study was that development of specific components of the shuttle, such as the vertical tail or elements of the attitude control system by the Europeans could provide technology benefits to both the United States and Europe, and that the transfer of critical technology to Europe would be a relatively small percentage of the program value. European development of the space tug might entail a broader range of technology transfer, but would he amenable to some controls. Other potential cooperative projects in the Post-Apollo Program such as payload modules would generally fall between these two cases. European cooperation in payload development could vary from zero transfer to modest transfer, depending on the policies we choose to follow in selecting and approving proposals.

In general, it has been understood that the major thrust of our international post-Apollo effort is to obtain foreign contributions primarily through the exercise of *foreign* capabilities and not through utilization of U.S. technology transferred abroad for that purpose. It is already widely understood abroad that NASA means to accept foreign participation only in those tasks for which Europe has an existing or potential capability and that this capability must be *validated* by joint teams from NASA, NASA's contractors, and the foreign governments concerned. If we could determine in some cases that it is in our own interest to provide certain elements of a task in order to make possible larger foreign contribution, we will still retain an option to provide those elements either *as technology* or, if they are particularly sensitive in character, on a "black-box" or end-product basis.

[2] It was judged that the transfer of management knowhow and systems engineering capability that would occur as a byproduct of European participation with the U.S. in a large-scale development such as the shuttle would be one of the principal objectives of such European participation. The significance of the transfer that might take place is open to question, and future implications are difficult to assess.

In the longer term, the impact of transfer of technology or management expertise will depend more upon the degree to which these elements can be transferred to other activities in the commercial sector, than upon their direct application to advanced space systems. It has been our experience that transfer of aerospace technical capabilities to other commercial areas has not been an easy or very successful process.

At the present time, direct commercial benefits from use of space systems have been restricted to communications satellites operated by the Intelsat Consortium. European contractors are playing an increasing role in supplying subsystems, satellites and ground elements of the Intelsat system. In the future, there may be additional space-based systems that provide income to the supplier from sale of satellites and services in areas such as navigation, traffic control, mobile communications, pollution monitoring, earth resources and crop surveys, and an increase in technical sophistication in European industry would enable a greater degree of competition with potential U.S. suppliers. In a meeting with U.S. aerospace industry managers, it was quite apparent that they are not concerned about being unable to compete for such contracts with European firms, as a result of cooperation on post-Apollo or the technology transfer which might ensue.

Furthermore, it is characteristic that the service provided by a space system is international in nature, requiring agreements and cooperation between nations if the potential benefits are to be realized. In the future, therefore, the U.S. is likely to depend upon the ties that can be developed with other nations in order to insure a role for U.S. industry and U.S. interests in service provided by space systems. It would be preferable to develop cooperative programs that foster these tips, rather than to force nations to develop capabilities that permit decisions independent of the U.S. Similarly, in a commercial sense, it is likely to be of greater value to involve many nations in cooperative systems with shared contracting than to see separate systems developed that may isolate the U.S.

[3] The previous record of major cooperative development programs, such as Skybolt, the Main Battle Tank, US/FRG fighter aircraft, and Concorde, have left some doubt whether such programs foster closer ties between the participants or act as an irritant which limits full development of cooperative relationships. The record of cooperative space projects conducted by NASA, on the other hand, has been excellent and provides an indication of those characteristics which produce favorable results from cooperative enterprises. They are in part:

(1) Mutual interest and mutual benefits.
(2) Financial contributions by both partners—usually no transfer of funds.
(3) Clearly defined interfaces and objectives.
(4) Capability for performance of agreed tasks can be assured.

While the scale of potential European participation in the shuttle program is much larger than previous programs, it appears possible to define tasks that meet these criteria.

One of the principal European contributions to the Space Shuttle program could be development of the Space Tug. This propulsion module would represent a major technological and economic challenge to Europe, and would fit the above criteria—particularly the ability to define interfaces and objectives clearly, since there would be minimum impact on design characteristics of the respective systems, as a result of changes within each element. The U.S. will be required to have some technical involvement in the development of the Tug, and in general, some technology support would be required for both the propulsion and avionics modules, as well as in total systems engineering. As the program develops, some additional technology may be required to alleviate unanticipated problems which arise.

In the future, it is expected that there will be sustained production of the Space Tug and that it will be used for both DOD and NASA missions. This would imply dependence upon a European supplier, or alternatively development of an independent U.S. production capability, perhaps on a license basis.

[4] A similar, highly separable component of the shuttle system that would be an attractive candidate for European development is a Research and Applications Module that would be used as the payload of a shuttle orbiter, providing a structure for observing instruments and other experiments, either manned or man-attended. In addition, European technology appears adequate to support design and construction of major structural elements of the shuttle such as wing and tail surfaces and the thrusters for the auxiliary propulsion and control system.

Alternatives to major participation in shuttle system development are limited, particularly in view of the unique scope, challenge, and economic implication of the shuttle program; the narrow focus of European space interests; the degree to which the U.S. is considered abroad to be committed to welcoming post-Apollo participation; and the wide range of existing international space programs and overtures by the U.S. None of the alternative cooperative ventures that have been developed appear to be acceptable, either individually or collectively, as replacements for shuttle participation. They may be pursued, however, on their own merits. If the U.S. should withdraw from the Shuttle program or decide to pursue it unilaterally, discussions of possible other projects would certainly continue.

A final question concerns launch assurance. It is generally understood that Europe desires assured access to U.S. launchers on a fee basis if she is to give up the development of her own launch vehicles so as to free funds for contribution to post-Apollo tasks. The response which the U.S. gave last September on this question was widely regarded as satisfactory in Europe but has since been reversed in part and become confused.

A restatement of the U.S. policy regarding provision of launch services, valid for all nations, is being developed through the interagency committee on space cooperation established under NSSM 72, and, if approved, should reduce European concerns about launch assurances and separate this issue from the question of post-Apollo cooperation. The proposed policy statement would have the effect of assuring availability of launch services for payloads that are for peaceful purposes and are consistent with international agreements.

****

[Attachment page 1]

Dear Bill:

Uncertainties in U.S. domestic shuttle planning and a need for additional review of the problems of technology transfer and management complications in undertaking a joint program of space transportation system (STS) development with the Europeans have delayed this reply to your letter to the President of March 23.

Although that review is not yet complete, the President feels it is now possible to develop a reply to Minister Lefevre and the European Space Conference (ESC) and to resume a dialogue with the Europeans; however, in a way that does not condition U.S. launch assurances for European payloads upon substantial European participation in a joint STS program, but treats each of these two matters separately.

A first priority would be to prepare a position for discussion with the Europeans, indicating U.S. willingness to provide launch assurances for foreign satellites of a peaceful nature. Language acceptable to the Europeans, but recognizing overall U.S. obligations to Intelsat, should be sought for such assurances.

[2] However, one possible formulation which would be acceptable to the President, if such a degree of assurance is necessary to avoid European charges that the U.S. seeks to retain a veto over their space plans, would provide for launch services by the U.S. of foreign systems approved under Article 14 of the definitive arrangements of Intelsat; and would permit sale of the necessary launch vehicle for "unapproved" systems, leaving to the launching nation the interpretation of its obligations under Article 14.

Renewed discussions with the ESC about post-Apollo cooperation should be undertaken at the technical working level. Their purpose would be to seek to define a possible cooperative relationship between Europe and the U.S. in a program of STS development, with full understanding that no commitment on either side is expected or assured until the results of these discussions have been referred to the involved governments for review and final decision. Although no cooperative programs have been discussed in the present context with the Europeans to compare in magnitude with STS development, it will be useful in the course of these talks to keep in mind the full range [3] of potential cooperative opportunities, in the eventuality that a satisfactory agreement is not reached on the STS program and assuming that the Europeans do respond to the offer of U.S. launch assurances by abandoning EUROPA-III.

The President hopes that this course of action will address the pressing European concern regarding launcher availability, will permit a continued dialogue with the Europeans directed toward mutually beneficial space cooperation with full protection of U.S. interests, and will avoid locking the U.S. prematurely into a commitment or schedule for the STS.

Sincerely,

Henry A. Kissinger

Honorable William P. Rogers
Secretary of State
Washington, D.C. 20520

Document I-21

[1]

August 12, 1971

Memorandum

TO: A/Administrator

FROM: AD/Deputy Administrator

SUBJECT: Items of Interest

Kissinger Meeting

Attendees were Ed David, Alex Johnson, Tom Whitehead, Herman Pollack, Arnold Frutkin, Kissinger's new staff man whose name is Michael Guhin, and Low.

On the subject of launch services, Johnson proposed that the United States would guarantee launch services if (a) there is a positive two-thirds finding by the INTELSAT body that the launch should proceed; or (b) in the absence of such a finding, if the United States itself is not opposed to the launch. Johnson believed that in the context of a post-Apollo participation (which presumably the Europeans want), they would be willing to accept this formula of launch assurances. David and Whitehead, on the other hand, believed that these assurances do not go far enough, that they constitute a "blatant U.S. veto," and that we should in addition offer to sell launch vehicles to the Europeans to launch from their own soil for whatever peaceful purposes they desire. Johnson indicated that this would be unacceptable to COMSAT and Senator Pastore. Low did not enter into the discussion in any major way, but did support Ed David's point of view.

On the subject of technology transfer Kissinger understands that this transfer would not be large and would be essentially controllable. He also understands [2] that what would be transferred and what is desired by the Europeans is systems engineering and systems management know-how.

On the subject of continuing the technical discussions with the Europeans, David and Whitehead felt that the shuttle should be de-emphasized in these discussions and that, instead, the "content" of the space program should be emphasized. After some debate on this subject, it was agreed that technical discussions on the Space Shuttle/space tug could be continued, but that they would be broadened to include payloads as well.

In the context of the technical discussions with the Europeans, I had an opportunity to mention the significance of our recent budgetary guidelines. Although I did this in a relatively low key way, Kissinger immediately reached a conclusion that we had been given a guideline that would essentially stop manned space flight for the United States, which was confirmed by Whitehead. David, on the other hand, stated that this was only a preliminary guideline to "force NASA to consider alternatives to the very expensive shuttle concept." Kissinger stated that stopping manned space flight in the United States is entirely unsatisfactory, and that he would do everything in his power to prevent this from happening.

The conclusion of the meeting was that Kissinger would notify Alex Johnson by the end of the week (by August 13th) of his decisions in all these matters. In the meantime, the State Department (with Arnold Frutkin's help) is drafting a response to Lefevre's February letter. In this response, the current United States position on launch assurances will be stated, and a continuation of the technical discussions on post-Apollo, plus pay-loads, will be urged. Frutkin understands that from NASA's point of view, we would like to delay the start of discussions until at least September 30th, so that we can have made up our own mind concerning the shuttle in the context of the Fiscal Year 1973 budget posture. . . .

Document I-22

Document title: Department of State Telegram, "Johnson Letter to Lefevre," September 7, 1971.

Source: James C. Fletcher Papers, Special Collections, Marriott Library, University of Utah, Salt Lake City, Utah.

This telegram communicated to concerned U.S. officials, including the U.S. embassy in Brussels, Belgium, the text of a letter from Under Secretary of State U. Alexis Johnson to Belgian Minister Theo Lefevre. As the chairman of the European Space Conference, Lefevre led the European delegation in negotiations with the United States concerning launch assurances and post-Apollo cooperation. This letter spelled out the conditions under which the United States would provide launch services for European satellites. It also discussed issues with respect to European participation in the U.S. post-Apollo program. The letter was the product of more than six months of debate within the executive branch in Washington. While the launch assurances were not totally acceptable to Europe, the letter did clear the path for additional discussions of post-Apollo cooperation.

---

[1] R 012307Z Sep 71
FM SecState WashDC
To AmEmbassy Bern . . .

[2] Subject: Johnson Letter to Lefevre

Refs: A) State CA-5237 October 9, 1970
       B) State 30947 Feb 24, 1971
       C) Brussels 774 Mar 6, 1971

    1. Under Secretary Johnson has written a letter to the Honorable Theo Lefevre, Chairman, European Space Conference, Brussels in response to Lefevre's letter of March 3, 1971, Ref (C). Instructions for action posts given in paras 7 and 8 below:
    2. Letter is dated September 1, 1971 and is marked "confidential" in view of US desire to avoid public discussion at this time. Text follows:
    Quote Dear Minister Lefevre: Para This letter is in response to yours of March 3, 1971, concerning possible European participation in post-Apollo space programs. It sets out our current views on the matters of consequence which were involved in our discussions this past February and in September, 1970. It overtakes my letter to you of October 2, 1970.
    Para I regret that it has not been possible to respond to you earlier. We felt that our mutual interests would be served best if we took sufficient time to review our position carefully in the light of your letter and of events since our discussions in February. As I stated during those discussions, our ultimate views on most of these matters remain contingent on choices yet to be made in Europe as to the measure and character of European participation and on further development of our own plans for post-Apollo programs.

Para Since we have understood that the [3] matter of greatest concern to the European Space Conference is the availability of launchers for European satellite projects, we have reviewed our position so as to meet the concerns expressed in your letter and during our earlier discussions. Our new position in this regard, described in the numbered paragraphs below, is not conditioned on European participation in post-Apollo programs. I believe it should provide a basis for confidence in Europe in the availability of US launch assistance.

Specifically:

Para (1.) We recognize the concern of the European Space Conference with regard to the availability of launch assistance for European payloads. In this respect US launch assistance will be available for those satellite projects which are for peaceful purposes and are consistent with obligations under relevant international agreements and arrangements, subject only to the following:

Subpara (A) With respect to satellites intended to provide international public telecommunications services, when the definitive arrangements for INTELSAT come into force the US will provide appropriate launch assistance for those satellite systems on which INTELSAT makes a favorable recommendation in accordance with Article XIV of its definitive arrangements. If launch assistance is requested in the absence of a favorable recommendation by INTELSAT, we expect that we would provide launch assistance for those systems which we had supported within INTELSAT so long as the country or international entity requesting the assistance considers in good faith that it has met its relevant obligations under Article XIV of the definitive arrangements. In those cases where requests for launch assistance are maintained in the absence of a favorable INTELSAT recommendation and the US had not supported the proposed system, the United States would reach a decision on such a request after [4] taking into account the degree to which the proposed system would be modified in the light of the factors which were the basis for the lack of support within INTELSAT.

Subpara (B) With respect to future operational satellite applications which do not have broad international acceptance, we would hope to be able to work with you in seeking such acceptance, and would favorably consider requests for launch assistance when broad international acceptance has been obtained.

Para (2.) Such launch assistance would be available, consistent with US laws, either from US launch sites (through the acquisition of US launch services on a cooperative or reimbursable basis) or from foreign launch sites (by purchase of an appropriate US launch vehicle). It would not be conditioned on participation in post-Apollo programs. In the case of launchings from foreign sites the US would require assurance that the launch vehicles would not be made available to third without prior agreement [with] the US.

Para (3.) With respect to European proposals for satellites intended to provide international public telecommunications services, we are prepared to consult with the European Space Conference in advance so as to advise the Conference whether we would support such proposals within INTELSAT. In this connection we have undertaken a preliminary analysis of the acceptability of European space segment facilities for international public telecommunications services separate from those of INTELSAT, in terms of the conditions established by Article XIV, and find that the "example of a possible operational system of European communication satellites," which was presented during our discussions in February, would appear to cause measurable, but not significant, economic harm to INTELSAT. Thus, if this specific proposal were submitted for our consideration, we would expect to support it in INTELSAT.

[5] Para (4.) With respect to the financial conditions for reimbursable launch services from US launch sites, European users would be charged on the same basis as comparable non-US Government domestic users.

Para (5.) With respect to the priority and scheduling for launching European payloads at US launch sites, we would deal with these launchings on the same basis as our own. Each launching would be treated in terms of its own requirements and as an individual case. When we know when a payload will become available and what its launch window requirements will be, we would schedule it for that time. We expect that conflicts would rarely arise, if at all. If there should be a conflict, we would consult with all interested parties in order to arrive at an equitable solution. On the basis of our experience in scheduling launchings, we would not expect any loss of time because of such a conflict to be significant.

(Note to posts: Remaining paragraphs of this letter are unnumbered.)

Para The United States is considering the timing and manner of public release of this position. Accordingly, it is requested that there be no public disclosure of this position without prior agreement with US.

Para With regard to post-Apollo cooperation, as you know, the United States had not yet taken final decisions with respect to its post-Apollo space programs, nor can we predict with assurance when such decisions will be taken.

Para With respect to the more detailed questions on post-Apollo collaboration posed in your letter of March 3, 1971 and in our earlier discussions in September 1970 and February 1971, our views [6] remain broadly as we put them to you in my letter of October 2, 1970 and in our meetings of last September and February. We would much prefer to continue the consideration of such questions in the context of specific possibilities for collaboration rather than in the abstract.

Para The relationship we are seeking with Europe with respect to post-Apollo space programs would, we believe, be well served if we can jointly consider the possibilities for collaboration in the context of a broader examination of the content and purposes of the space programs of the late 1970s and 1980s.

Para Accordingly, we suggest broadening your earlier suggestion for a joint expert group to conduct technical discussions. The purpose of these discussions will include the definition of possible cooperative relationships between Europe and the U.S. in a program of development of the space transportation system, but would be broadened to include an exchange of views regarding the content of space activities in which Europe might wish to participate in the post-Apollo era. The technical questions relevant to such participation, including the remaining questions raised in your letter of March 3, would be examined as well. the joint group would carry on its activities with no commitment on either side. the US representation would be Charles W. Mathews, Deputy Associate Administrator, Office of Manned Space Flight, NASA.

Para This group could most usefully commence its work after the end of September when the results of NASA's current technical studies of space transportation systems become available.

Para I trust, Mr. Minister, that this summary of our present views is a helpful response to the matters raised in your letter of March 3. I am pleased to confirm our continuing interest in [7] cooperating with interested European nations in the further exploration and use of space. Sincerely, U. Alexis Johnson, Unquote.

3. Comment for posts: It has become evident that the matter of greatest concern to the ESC is assured availability of launchers for European satellite projects, and it is our view that the new position set forth above achieves this goal. it important to note that launch assistance we are prepared to furnish (as given in the numbered paragraphs of the above letter to Lefevre) is not [repeat] not conditioned on European participation in post-Apollo programs.

4. Johnson letter also reiterates our offer made at February meeting with ESC representatives to consult with ESC in advance so as to advise them whether we would support within INTELSAT European proposals for satellites intended to provide international public communications at February meeting, Europeans presented a document entitled "example of a possible operational system of European communications satellites." Analysis of this example led to conclusion that we would expect to support such a proposal if it were submitted to INTELSAT.

5. The new position reserves to the US decisions with respect to "future operational satellite applications which do not yet have broad international acceptance." In maintaining this reservation we have in mind applications such as direct broadcasting satellites which do not yet have the broad international acceptance necessary to assure that this application will not be source of international tensions.

6. Letter to Lefevre also endorses Lefevre's suggestion that joint expert group be established to consider technical and scientific tasks which Europe might wish to perform as part of joint program.

Action requested:

7. For Brussels: Pass above text of Under Secretary Johnson's letter to Lefevre as soon as feasible. Word "confidential" should appear just above salutation. Call Lefevre's attention to paragraph of this letter requesting that there be no public disclosure of launch assistance position without prior agreement with US. Ask that his response be sent through diplomatic channel. Advise Department and other action addressees when delivery has been made. (Signed copy of letter pouched to Embassy today.)

8. For other action addressees: On the day after receiving Brussels's confirmation that Lefevre has received the letter, pass copies of text to foreign ministries and other space-related ministries at highest appropriate level and explain the importance of our new launch assurance position. Repeat caveat to Brussels (para 7) re: our [8] desire to avoid publicity at this time. We hope this new position will be widely accepted by the European nations as a satisfactory basis for confidence in the availability of US launch assistance. Rogers

Document I-23

Document title: "Memorandum for Peter Flanigan from the President," November 24, 1969.

Source: Nixon Project, National Archives and Records Administration, Washington, D.C.

Richard Nixon was intrigued by the possibility of flying non-U.S. astronauts aboard U.S. spacecraft. Astronaut Frank Borman first suggested this idea after his post-Apollo 8 overseas tour. NASA Administrator Thomas Paine had interpreted the president's mandate to him, while they traveled to the Apollo 11 splashdown, in terms of seeking increased international cooperation in space through cooperation in hardware development. Others believed, however, that the president was most interested in flying foreign astronauts and experiments. This memorandum was directed at getting more attention paid to the latter possibility.

November 24, 1969

Memorandum for
Peter Flanigan
from the President

Is there still no feasible way to get multi-national participation in some of our future space flights? I have raised this with Paine and Borman and I know there are some technical problems but it is a pet idea of mine and I would like to press it. Raise it with Borman and see whether we can jog the bureaucracy in that direction.

Document I-24

Document title: George M. Low, Excerpts from Personal Notes: No. 63, February 1, 1972; No. 67, March 26, 1972; No. 68, April 17, 1962; No. 69, undated; No. 71, June 3, 1972; No. 72, June 17, 1972.

Source: George M. Low Papers, Institute Archives and Special Collections, Rensselaer Polytechnic Institute, Troy, New York.

Each week, NASA Deputy Administrator George Low dictated his views on the preceding week's events. These notes comprise a fascinating first-hand account of personalities and decisions. The excerpts portray the confused character of the debate over post-Apollo cooperation during the first months of 1972. Among those mentioned are: Russell Drew, the staffer who handled space in the White House Office of Science and Technology; Herman Pollack, the top science and technology official in the Department of State; Phil Culbertson, a top NASA technical manager; and National Aeronautics and Space Council Executive Secretary Bill Anders.

---

[1]   **Personal Notes No. 63 . . .**

[2] *International Aspects of Post-Apollo Program*

Ever since our visit to San Clemente on January 5, 1972, Arnold Frutkin and the International people have been pushing for major activities with the Europeans in the Space Shuttle development. This work has been supported by the State Department, but has generally been opposed by Tom Whitehead, Bill Anders, and John Walsh, who's working for Kissinger. Walsh put together a group consisting of Russ Drew, Bill Anders, Arnold Frutkin, Phil Culbertson, Herman Pollack, Tom Whitehead, and perhaps others to review the situation. According to Frutkin, this group has now been converted to be in favor of post Apollo Shuttle development activities with the Europeans. According to Whitehead and Anders the group is still opposed and NASA would do well to get out of this activity. In an internal meeting within NASA, Fletcher and I felt that we should only undertake this work if it were really in the interest of the White House and the State Department to do so, and that we should not be pushing for it unless we were pushed into it. As a result of these views, we decided that Fletcher would visit with Henry Kissinger or Al Haig to tell him about NASA's concerns about full participation by the Europeans for the Shuttle development, and ask them whether they really want us to do this. We are still under the impression that we may be getting wrong signals from Frutkin (I have gotten wrong signals from Paine before that) and that we're really moving into an area that we don't want to get into and that the White House also does not want us to get into. At any rate, we need clarification, which Fletcher will seek before we proceed. . . .

****

[1] **Personal Notes No. 67 . . .**

[2] *Post-Apollo International Situation*

This matter is still terribly confused. The Flanigan/David Rice side of the White House feels that NASA pressured the White House into undertaking these international initiatives "in order to make it less likely to have the shuttle cancelled." NASA's position is that, at least since Tom Paine left, we have consistently stated we will do what the State Department and the White House want us to do, but that if we were on our own we would like to build the shuttle and its equipments all within this country. The State Department feels that we strung them along and we are now letting them down. Out of all this, Bill Anders was given the job by Flanigan to pull together a position that would be acceptable to the White House. The State Department, however, feels it should be its job to pull that position together, while at least John Rose in the White House is concerned that if the State Department were allowed to do this a position would be established that is not in the best interest of the United States. The underlying argument in the White House against having active participation in the post-Apollo programs is based upon a concern of too much technology transfer to Europe which is probably not a valid concern as well as a concern about being beholden to the Europeans for their piece of action in case they want to hold us up for it. The State Department concern on the other hand is that we have now gone so far that any backing down might cause serious international repercussions. I seem to be in the middle with Herman Pollack coming to visit me privately for a "nonmeeting" giving me his concerns and presumably the State Department concerns (Alex Johnson is still recuperating from a heart attack), while Bill Anders calls me and asks for help to consolidate the White House position.

In my meeting with Herman Pollack, I once again made it very clear that it was NASA's view that from the programmatic standpoint we would like to do the whole program [3] domestically; however, there are many options of doing things in Europe if it is in the United States' international interest to do so. I also took the same position in my recent discussions with Bill Anders. A memo for the record of my latest conversation with him is attached, as is a copy of a paper that he prepared for Peter Flanigan. . . .

****

[1] **Personal Notes No. 68**

These are the personal notes for the week ending April 1, 1972, as well as the week ending April 15, 1972. (I have already prepared special notes for the week ending April 8.) . . .

[2] *International Cooperation*

The situation in the post-Apollo international cooperation, primarily with the Europeans, is still very much up in the air. Bill Anders, in reaction to what he thought was

Peter Flanigan's request, had prepared a very one-sided document, indicating that at most we should let the Europeans develop the sortie module. In this document he did not air the two sides of the story so that it would have been impossible for the President to really pursue any alternatives. When Flanigan saw the document, he wisely stated that this was not at all what he wanted and, in effect, sent Bill back to the showers. The over-all situation is still that we have implied commitments to the Europeans that they could participate in the development of bits and pieces of the Shuttle itself, provided they also worked on either a sortie module or a tug. Our position now is that we don't want them to get involved in bits and pieces of the Shuttle because it makes for a very difficult technical and management problem, and at the same time we don't want them involved in the tug because we think this is too difficult a technical bite for them to take. We think the Europeans really are no longer deeply interested in working with us either and that, if we let matters stand as they are, they are going to die of their own weight. However, Bill Anders feels that this is not the proper way to proceed and that we should indeed take positive action to turn the Europeans off. Basically my own conclusion is that this is no longer a matter of substance because we have pretty well decided what to do and don't want the Europeans to do, and the entire situation merely becomes a matter of tactics.

My latest suggestion to Frutkin was that we should allow the Europeans to participate in an annual program review with us with the understanding that they would file a report within [3] 60 days, suggesting solutions to any problems that we might be facing. This they would do in addition to the development of a sortie module or, if they can demonstrate competence, the development of a tug. . . .

****

[1] **Personal Notes No. 69 . . .**

Post-Apollo European Cooperation

This subject is still nearly as confused as it has been for a long time.

State Department has now formally taken a position in a letter from Rogers to the President . . . that we should encourage the Europeans to participate through the development of a sortie can; that we should defer participation on the tug until they have conducted further studies; and that we should allow them to build bits and pieces of the shuttle, provided this is tied in with participation in the major elements such as the sortie module.

Peter Flanigan's position protecting the domestic economy is that he has no objection to Europe's participation in the development of the sortie can, but that we should absolutely not allow any participation in the tug or bits and pieces of the shuttle. Peter also believes that Kissinger will defer to him in this area and that his position is the one that will prevail with the President. Incidentally, Bill Anders has been asked by Flanigan to prepare a Presidential action paper reflecting Flanigan's views.

NASA's position is a fairly straightforward one. First, we state that given our own preference, in the absence of any international considerations, we would, of course, prefer to do everything in connection with the shuttle in the United States. Secondly, we state that given a strong Presidential directive that for international considerations Europe must participate in the shuttle, we would first [3] of all prefer their participation in the sortie can; secondly as a very poor second choice we would allow their participation in the tug, provided they can demonstrate through studies that they can indeed work on the tug and their technology is sufficiently advanced to do so; third, we would dislike their participation in bits and pieces of the shuttle because this would make our job much more difficult, but this too could be accomplished if we were directed to do so. Our views were expressed in a memorandum to Kissinger commenting on Secretary Rogers' memorandum.

At the time of this dictation, John Walsh on Kissinger's staff is preparing two action papers for the President, presumably they will be signed by both Flanigan and Kissinger. The first of these, which is the one that Flanigan prefers and the one that will go unless Kissinger feels strongly enough to get personally into the act, says in effect that the Europeans should be allowed to participate in the sortie can only, and we should turn off the tug and the bits and pieces of the shuttle immediately. The second paper also states that participation in the sortie can should be encouraged; that participation in the tug should be denied; but that participation in the bits and pieces of the shuttle should be "discouraged but not excluded." Presumably all of this will be steeled within the next several weeks.

In the meantime, the Europeans are champing at the bit because they are trying to meet a deadline that we have imposed on them to make up their mind by approximately the first of July. To do this, they need answers to some detailed questions that they posed to us two weeks ago; however, we, of course, are not in any position to answer these questions until we have settled our policy issues.

For the record, it might be worthwhile to review a bit of history here also. When I first came to NASA Headquarters, I learned from Tom Paine that the President was extremely interested in European participation in Space Shuttle development. I had no reason to disbelieve this and, as a matter of fact, I had seen letters from Paine to Kissinger reporting on his various visits to Europe, Japan, etc. and reports coming back from Kissinger saying in effect "keep [4] up the good work." When I became Acting Administrator, I continued what Paine had started and continued to send reports both to Kissinger and to Peter Flanigan. In all cases, I received replies from Kissinger and Flanigan (generally quite late) encouraging me to go ahead. However, it wasn't too long before I got views expressed by both Don Rice and Ed David that what I was doing might not really be what the President wanted us to do. I then tried for a long time to see Kissinger to get his personal views on this before continuing any further discussions; however, Kissinger was never able to see me on this subject. It wasn't until about the time that Fletcher came on board that we had the first meeting in Kissinger's office involving Kissinger, Whitehead for Flanigan, and several others where again the views expressed were inconclusive except that it became quite apparent that the domestic side of the White House was very much opposed to the kinds of things that Tom Paine had been doing. Ever since then, this had been tried to be brought to a resolution in the White House, and it may be that we are now close to this resolution. . . .

****

[1]   **Personal Notes No. 71 . . .**

Post-Apollo European Cooperation

I forget when I last discussed this situation, but, briefly, the facts are these. We have been trying to get the White House to resolve the difference of opinion that exists among NASA, State Department, and some of the White House staff. Basically, the State Department would like to cooperate to the greatest extent possible. The White House would like to have an isolationist policy and no cooperation whatsoever. NASA has taken a "hands off" approach indicating that we would like to undertake a minimum amount of cooperation in order to simplify our own problems, but that we could do almost anything that was necessary if the President so directed it because he felt they were overriding in a national consideration to do it. Out of all of this came a set of instructions signed by Kissinger which indicated that we should encourage the Europeans to develop one of several forms of Sortie modules; that we should discourage but permit European participation in the development of selected bits and pieces of the Shuttle; that we should under no conditions let Europe develop the tug; and that for future cooperative ventures we should concentrate on payloads and not launch systems hardware.

[3] This set of instructions is perfectly acceptable to us and it is now planned to meet with the Europeans within the next two weeks to respond to their long-standing questions concerning what we would be willing to do in cooperation with them on the post-Apollo projects.

Incidentally, Europe is quite confused concerning where the United States now stands with respect to all technological interchanges. I recently met Mr. Boelkow, President of Messerschmitt, who feels that our recent policies are so isolationist that we are going to hurt Europe and ourselves. He believes that we could reach agreements, particularly in aeronautics, that would help both countries both technically and economically. He wanted to discuss these with Magruder. However, I told him that Magruder was not the right person nor do we have a right person to do it. Nobody has really examined both the short-range and long-range economic effects of technology transfer. I subsequently discussed this with Fletcher and we decided to try to see Pete Peterson (Secretary of Commerce) to see whether we can't get him interested in the problem. . . .

****

[1]   **Personal Notes No. 72 . . .**

European Post-Apollo Cooperation

After more than a year of indecision, we finally received a memo from Kissinger concerning the extent of European post-Apollo cooperation. Specifically, the memorandum indicated that we should seek participation in the Sortie Module, should deny the tug to the Europeans, and should discourage [3] but allow essential participation in bits and pieces of the Shuttle. This is the package that Jim Fletcher and I had hoped for, and with the exception of the bits and pieces of the Shuttle, it is exactly how we would like to handle it. Meetings with the Europeans were held during the past week (at the sub-Ministerial level) and they very quickly got the message. It appears now that they might join us in a Sortie Module, and we have strongly encouraged them to do so. The real question is whether they can move quickly enough in pulling themselves together. At their informal request, we have set a deadline of this summer for them to make up their mind because the people who were here felt that if we did not do this, Europe would continue to argue about this for several years to come. . . .

Document I-25

Document title: William A. Anders, Executive Secretary, National Aeronautics and Space Council, to The Honorable Peter M. Flanigan, March 17, 1972, with attached: "Position Paper on European Participation in our Post-Apollo Space Program."

Source: George M. Low Papers, Institute Archives and Special Collections, Rensselaer Polytechnic Institute, Troy, New York.

Document I-26

Document title: William P. Rogers, Secretary of State, Memorandum for the President, "Post-Apollo Relationships With the Europeans," April 29, 1972.

Source: George M. Low Papers, Institute Archives and Special Collections, Rensselaer Polytechnic Institute, Troy, New York.

Document I-27

Document title: James C. Fletcher, NASA Administrator, to Honorable Henry A. Kissinger, Assistant to the President for National Security Affairs, May 5, 1972, with attached: "NASA's Comments on Secretary Rogers' Memorandum of April 29, 1972."

Source: NASA Historical Reference Collection, NASA History Office, NASA Headquarters, Washington, D.C.

These documents capture the character of the closing stages of the debate inside the U.S. government regarding international participation in the post-Apollo program. They demonstrate the concern with difficulties in completing large-scale technological projects. They also reflect the ever-present worry over technology transfer and the possibility of the United States losing its edge in a highly competitive international arena.

Document I-25

[1] March 17, 1972

Memorandum for
The Honorable Peter M. Flanigan

Pursuant to our conversation at lunch on March 3, I have summarized what I believe are the issues, objectives, and options for international participation in the post-Apollo space program. The outstanding problem is that in the past, NASA, interpreting a Presidential sanction, emphasized joint shuttle development with the Europeans, whereas our involvement would appear to have been greatly more in tune with the President's desire if it had been focused on joint manned operations and mutual utilization of space.

Joint European participation in our hardware programs has always seemed to me to have little national advantage and several drawbacks. However, as a country we have gone some distance down the pike with the Europeans, and an abrupt, visible change in policy will probably create a foreign relations problem of measurable but uncertain magnitude.

Possibly the problem can be reduced by a careful selection of options and tactics. Taking the factors I see bearing on the problem into account and weighing them as best I can, I have proposed a strawman cooperative program in this paper which, *if* it could be accepted by the Europeans, would be to the net advantage to the U.S. This program, consisting of payload cooperation and joint manned flight, plus European development of the Sortie can, is acceptable to NASA from their viewpoint as program managers. State will likely view this course of action as not responsive to Europe's expectations and as representing a significant change in previous policy. They can be expected to resist such a change or urge some intermediate concession by the U.S. A possible concession is discussed in the attached paper, whereby the U.S. prime contractor for the shuttle does a nominal amount of subcontracting in Europe; however, NASA would agree to this arrangement only if directed to as a concession to our foreign relations.

Please excuse the length of the paper, but there is a several year history of the development of this issue and a significant difference in motivations that are relevant to an understanding of our commitment and posture. Your [2] reaction to this paper and the strawman proposal, which has been coordinated with Jim Fletcher and John Walsh, of Kissinger's staff, and discussed with others, would be most timely if available by Tuesday a.m. The State Department has opened the post-Apollo policy for reexamination and will be meeting that afternoon. Since I will be attending, I could see that your views and whatever guidance you may have are put forward. Attention to and resolution of this messy issue should be soon since decision dates (e.g., NASA selection of prime contractors) are approaching inexorably and NASA needs a clear directive on how to proceed.

William A. Anders

Enclosure

****

[Enclosure page 1]

Position Paper on European Participation in our Post-Apollo Space Program

This paper examines our current position re European participation in our post-Apollo space program, how we got to this position, what are our commitments, and the options for decisions. A pragmatic program is proposed, and tactics for its implementation are discussed. Because of the technical content of the post-Apollo program and some semantic confusion, a definition of terms is desirable.

Definition of Terms

Post-Apollo literally encompasses all of the U.S. space program that comes after Apollo, starting in 1973. In the context of European cooperation, however, it has meant, at various times, the partnership development and utilization of the space station or Space Shuttle, then the shuttle alone, and now the shuttle, tug, or RAM. These elements of the post-Apollo system have the following characteristics:

The Space Station was a multi-manned, permanent orbital laboratory, which was dropped from NASA's plans on cost grounds, at least until the shuttle is completed and operational.

The Shuttle is a partially reusable launcher used to put a payload plus upper stage ("tug") into a 100 to 200 mile orbit, and to return them to earth. The shuttle can also be used to carry, support, and return a small manned space laboratory. The shuttle and later the tug will be used both in DOD and the civil space program. Development cost of the shuttle is projected to be $5.5B, unit cost will be $250M with an anticipated production of 5 units, and the operating cost is estimated between $10 to $12M per flight.

The Tug is a reusable upper stage, carried and returned in the shuttle payload bay, which moves payloads from the altitude of the shuttle orbit to higher altitudes, and returns payloads in the same fashion. Virtually all payloads above 200 n.m. will use the tug (or an expendable transfer stage), but owing to reuse, the production run for the tug will not be great—perhaps 25 altogether. Costs are estimated to be $1B for development, $20M per production unit, and $0.5M per flight for operations.

[2] *RAM* (Research and Applications Module) refers to a family of small manned (or unmanned) laboratories to be carried to orbit and supported there, internally or externally, by the shuttle, and then returned in the shuttle bay. (The first version has been referred to as a sortie module or sortie can.) In later versions, the laboratories may be left in orbit independently and recovered on a later shuttle flight. Because of distinctly different uses of the system, there will be several different versions of RAM, and each version can be developed and equipped independently. For each version the production run might be 10 units, development cost of $150 to $200M, and unit cost $15 to $20M, though a basic "stripped" version might be less.

Subcontracts. This term needs to be defined because of the confusion resulting from its dual usage in the post-Apollo negotiations.

European Contributed "Subcontracts" was until very recently the concept under discussion, wherein the European governments would pay for their industry to develop certain parts of the shuttle, which we would then use. This arrangement was necessitated by the NASA operating rule of no exchange of funds in foreign cooperative projects. A government-to-government agreement would cover the arrangement; this type of arrangement is felt to have a number of unattractive features which are discussed later in this paper under "Options." In February, the possibility of having more normal (company-to-company) commercial subcontracts was raised by the Europeans, and so now the intended definition of the term subcontract is further confused in dealing with the Europeans and among ourselves.

Normal Commercial Subcontracts. Subcontracts of this nature are undertaken between industries with no unusual government involvement. The prime contractor chooses certain parts of a system for outside development and production, selects the winner among bidders for the work (with NASA's concurrence in the case of the shuttle contract), and then has sole control of managing and paying the subcontractor. In this context, European industry would not be precluded from bidding on the shuttle subcontracts, and under normal economic pressures to use low bid from a qualified supplier, they·could conceivably win $10 to $100M of the subcontracts. However, because of the nature of [3] R&D contracts, such as for the shuttle, there is little inherent pressure on our industry to choose low bid subcontractors; rather the most important considerations are minimizing programmatic, schedule, and management risks, and thereby maximizing the possibility of receiving their incentive. Historically, Europe has won no subcontracts of significance on space systems. Relaxation of implied restrictions and guidance to our industry to be more receptive to qualified bidders in Europe could be employed by us as a bargaining tool in the post-Apollo negotiation; and if the dollar flow is considered a problem, it might be balanced through some reciprocal arrangements. These alternatives are discussed later under "Options."

U.S. Motivations and Objectives in Post-Apollo Cooperation

It has been U.S. policy and President Nixon's desire to promote international cooperation in space and to share the benefits (and burdens) of space with all mankind. It has also been U.S. policy to strengthen our allies and alliances, and to foster a sense of community among the Europeans and to encourage their joint undertakings. The desire to implement these policies and also to make a new program more attractive to Congress (and also less cancelable), led NASA to seek European partnership in the post-Apollo space program over two years ago. The prospect of a European financial contribution to our program was thought to be a further plus. There was, however, ambivalence in our understanding of how much of the Administration's desire for international participation in space focused on joint usage and how much on joint development of space hardware. In recent weeks there has been some clarification of Presidential preference; his interest is primarily in European involvement in the use of space, coming from the development of payloads and operations rather than from big joint engineering projects, and specifically to share in the use of our post-Apollo space systems for international manned operations.

Whatever cooperative program is devised, we seek maximum benefit for ourselves in terms of (1) creating togetherness and good will, or at least minimizing any ill will, (2) drawing their interest away from undertaking separate space systems (e.g., the Europa III booster, aerosat, or those competitive with Intelsat), and (3) gaining some technology from areas of European special qualification, and possibly obtaining some minor components at a lower cost. At the same time, we want to minimize (1) increased risk and management complexity of our development program, (2) technology/dollar/job outflow, and (3) foreign relations impairment resulting from disputes as the program progresses.

[4] European Motivations and Objectives in Post-Apollo Cooperation

A major European objective is to gain large systems management capability and some technology. Their government/industry technocrats were very impressed by our success with Apollo, and they believe that by participating with us in a major systems development, such as the shuttle, they can learn how to better manage and build their own big technical projects (Europa III being a possible example). Their willingness to pay for the development of part of our shuttle is, in their view, a ticket to participate in or at least get a front row seat to our management process. A second European objective is to have the use of the world's most advanced space system, the shuttle, to carry out more complex science and applications programs in space, and, in spite of no explicit European plans at this time, there may be awakening interest in sharing in the prestige and greater capabilities of manned flight. Finally, the science-technology ministers and the international space organizations are looking for big projects that their respective governments will support (bureaucratic empire building). Also, of course, the European aerospace industry, which is in a decline analogous to ours, wants to get some business, particularly if that business might have fallout that would improve their competitive posture in other high technology areas. The direct business prospect appears to them as twofold: the R&D money from European governments and then the sale of production items to both European and U.S. users.

The History of the U.S. Commitment to Post-Apollo Cooperation

It has been a U.S. attitude that space like Antarctica is inherently international, only to be explored for humanitarian reasons. Whatever benefits that derive from being in space can be benefits to all mankind, except, of course, where military utility is involved. The one challenge, thus far, to this viewpoint has been in the use of satellites for communications, where commercial exploitation exists for point-to-point communications and is in dispute for mobile usage (aerosat). Such challenges will become more common as the shuttle opens up the commercial utilization of space. All Presidents since the inception of the space program have called for international cooperation in space, many in Congress favor it, and the Space Act, which formed NASA, urges it. President Nixon publicly promoted it in his statement of March 7, 1970.

[5] NASA has had an international outlook and has engaged other nations in many useful joint science projects. Partially because of this international orientation and partially because of the desire to make the program more attractive (and less cancelable), Tom Paine in private discussions with President Nixon at the time of Apollo 11 raised the issue of seeking greater international participation in our space program after Apollo. Paine reported that the President concurred in the desirability of this course of action, though it was not made clear as to the relative preference between participation in hardware development or participation in manned flight and science payloads. Paine then went to Europe to test and stimulate the Europeans' interest, and at the same time he narrowed the candidates for cooperation to the joint development and use of the space station or shuttle, and then only to the latter when the space station was dropped from our plans due to funding reductions. NASA did report to the White House on its progress in obtaining European involvement, and these reports elicited acknowledgments which were possibly of a somewhat perfunctory nature. NASA, however, accepted these acknowledgments as direction to continue. Operating from the same background and with stimulation from NASA and in response to European overtures, the State Department conducted two minister-level exploratory talks with the Europeans on the basis of U.S. "desire for maximum partnership in the post-Apollo program consistent with mutual desires and capabilities." This came to mean to NASA, Europe, and the State Department, a partnership in the development and construction of the shuttle, with possible involvement in the tug or the sortie can version of RAM. It was also understood that the U.S. would guarantee to use the particular European product, if that product was completely satisfactory to us. Talks have continued between U.S. and European technical groups to define areas of possible cooperation, meanwhile the Europeans have spent roughly $5M studying the shuttle and tug in order to decide where their work might be concentrated. They are now expanding their tug studies and are also studying RAM (sortie can).

The initial U.S. stipulations to cooperation were that there be no exchange of funds and that the management/technology level of the European undertaking be in keeping with their current capability and not rely on technology infusion from the U.S. A later stipulation was that the Europeans would have to contribute a significant portion of the effort (10% of the program's cost). This stipulation was dropped, however, after the U.S. decided on separating the issues of post-Apollo and launch assurances. (The launch assurance issue involved Europe's concern about obtaining U.S. launches of their payloads, The U.S. has now agreed to launch any European payload having a peaceful [6] purpose, except where we believed the payload violated international agreements (e.g. , military systems or those competing with Intelsat). These launches would use our present boosters and the costs would be reimbursed.) Our most recent stipulation is that they would have to commit themselves to a "package deal" for the development of the tug or RAM before we would settle on their government-supported "subcontractual" involvement in the shuttle. An implied stipulation was that neither Europe nor we would try to recover our respective development costs through amortization in the unit or use prices.

There has been growing concern in the Executive Office and with top NASA management that we are getting ourselves involved in a situation that is not advantageous. A recent informal sounding of Presidential desire indicated that his interest would be almost fully served through joint use of space, and partnership construction of complex space hardware is not a strong motivation. In some response to these feelings, NASA has been directed to attempt to shift European interest away from the shuttle and onto the tug or RAM.

Present status is as follows: the Europeans are now trying to decide whether or not to develop a tug or RAM. If their decision is affirmative, they have been led to believe that they can, if they wish, develop a few prescribed, "simple" parts of the shuttle, with certain restrictions on funding control. The Europeans must make up their minds by early summer if they wish to avail themselves of this "package deal." The decision is very hard for them because they have not thoroughly studied what is involved in the development of the tug or RAM, and they are going to have to decide with major technical and cost uncertainties facing them. Meanwhile, our change in signals on aerosat has caused them additional concern as to our motives in space, and has produced some European "threats" against post-Apollo; apparently they believe us to be eager for their involvement.

Options for U.S.-European Involvement

The four main options, some having suboptions, that are open to the U.S. are listed below in increasing order of complexity as far as program management is concerned (except possibly for 4b).

1. *Complete Disengagement.* The most obvious option is to disengage and have no international participation in our space program, other than at the scientific level as we already have. This option guarantees no technology or dollar outflow, does not restrict our future political or programmatic decisions, and adds no technical and management complications to an [7] already complex program. This, in fact, may be the outcome anyway, since European interests may well not be sufficiently strong to underwrite an expensive program having a nebulous *quid pro quo.* But if we force this option, the Europeans will correctly view this as a major shift away from the commitment they accepted from U.S. officials as our government's policy. Foreign relations harm may result and, in fact, may have wider effects than space matters usually do because this would closely follow other unsatisfactory space negotiations in the European view and also may seem to show a quixotic approach to policy formulation in the U.S.

2. *International Cooperative Payloads.* This option is to indicate that our interest in international participation is focused on the usage of the shuttle for mutual benefit, including manned flight, and not on development of the hardware. This option probably should be emphasized whatever else we jointly undertake because it appears to be at the heart of the President's actual desire. However, the Europeans will probably not view this as a significant concession since we are talking about events eight years from now, and furthermore the Europeans may believe this already to be U.S. policy.

3. *European Development of an Element of the Post-Apollo Program Other than the Shuttle (Tug or Sortie Can Version of RAM).* A third option is to allow the present situation to continue to the extent that Europe is free to choose between the development of a tug or sortie can, with a U.S. guarantee to use the item if it meets our required specifications. Either would meet Europe's perception of the U.S. commitment. The possible advantages to us of their undertaking the tug is the savings of a substantial R&D cost and the availability of the system several years earlier than otherwise. A possible other advantage is that the diversion of European funds to the tug would preclude their development of Europa III, and thus limit the expansion of their independent launch capability. (Any lesser commitment of European funds to post-Apollo, such as doing a RAM-sortie can and/or parts of the shuttle, would leave open the possibility of doing Europa III. However, It is possible that the cost and difficulty of Europa III will discourage the Europeans from undertaking it regardless of their post-Apollo involvement; and if undertaken it is even more possible that it would not be completed, as greater realization of its relative inadequacy became more apparent.) Any advantages to the U.S. of a European tug project seem to be more than offset by several disadvantages: the probability of Europe producing an unacceptably low performance system, the likelihood of technology outflow, the enhancement of their own booster capability, the dollar outflow to buy production units (perhaps up to $500M), and the difficulty in accommodating DOD's unwillingness to rely on a foreign supplier.

[8] Concerning the other side of this option, the advantages of Europe developing the sortie can version of RAM is that the task clearly can be within their capabilities, has minimum risk of technology transfer, could contribute a useful element to the post-Apollo program, and has no military implication. The cost to the U.S. to buy units from Europe would depend on the degree of equipping but may be fairly nominal, in the range of $20 to $60M over a period of several years. This expenditure would be offset by European purchase of the other versions of RAM produced in the U.S.

Given that the tug is an unacceptable European project for several reasons, and that the sortie can would be acceptable, a difficult problem faces us in causing redirection of European interests. We could easily end up with the foreign relations disadvantages listed under 1 even though we are trying to take a conciliatory approach in offering a moderate program of participation. This problem is discussed further under "Tactical Considerations," but anticipating that discussion, no fully satisfactory tactic is evident.

4. *European Involvement in the Development of the Shuttle.* This option is in two parts: the first being a continuation of the current position and the second a possible fallback maneuver as a possible foreign relations concession.

a. European Government-Supported "Subcontracts." This option is also a continuation of the current situation, namely, to accept Europe as a limited partner in the development of the shuttle, with them building at their expense certain "simple" parts of the hardware. The advantage to us in this arrangement is that it further meets European understanding of our commitment. It had been a NASA position that sufficiently simple tasks had been identified to make this arrangement feasible, however, many now feel that the increased risks and technology/management outflow may well more than offset the dollar or good will value of a European government-supported contribution to the shuttle. There is also serious concern that the normal supplier problems in big and complicated development programs would, on occasion, be elevated into international disputes, thereby producing the reverse of the President's desire for good will. Furthermore, this arrangement amounts to a U.S. government guarantee to supply certain components to our prime contractor, thus removing some of our government's leverage and some of the contractors overall responsibility for the integration and management function. During the course of the program, the prime contractor could well use this as an excuse for schedule, cost, or design changes. Withdrawing this option, however, will have a negative effect on European attitude toward the U.S., and a possible concession to lessen this impact is suggested by the following option.

[9]   b.   Normal Commercial Subcontracts (A possible foreign relations concession to offset the negative impact of withdrawing shuttle participation as an option). If some European involvement in shuttle development was felt to be necessary as a foreign relations concession due to our past stimulation and commitments, a possible fallback from the above government arrangement would be for the prime contractor to do some nominal amount of normal commercial subcontracting with qualified bidders in Europe, once Europe has committed to a RAM or tug. This would partially satisfy their industry's desire to do some work on the shuttle, and would not have the serious disadvantage of involving their governments directly in the arrangements, nor of having European participation in the management of the overall system. Also, the U.S. might benefit by some minor technology flow in our direction. To mitigate outward dollar flow, some balancing amount of work might be subcontracted by Europe in the U.S. on their RAM or tug, though this may happen anyway depending upon the degree of assistance they need on their task, or balancing might be achieved through other offset arrangements to achieve no net exchange of funds. This alternative is not favored by NASA, but if directed to choose between 4a and 4b for foreign relations reasons, this latter alternative is less odious and is doable.

A Proposed Program

A program agreeable to NASA, and which attempts to maximize the net advantage to the U.S. and at the same time appears to be reasonably attentive to our commitment to Europe, has been selected from parts of the above options.

System Use: European operational involvement with us in some joint manned orbital missions, plus reimbursable use of our space transportation system to orbit their science and applications satellites, as a natural continuation of our present launch assurances.

System Development: If European interest continues to include working on hardware development, we should agree only to their building the sortie can version of the family of RAM's. We would agree to buy from them the basic components of this item, while other versions of RAM would be built by the U.S. and would be for sale to the Europeans.

[10] The second part of the above program, system development, has the most immediate impact and also major difficulties associated with it in a foreign relations sense. In visibly removing the tug and shuttle from the list of acceptable projects for participation, we will antagonize the Europeans, even if they were not going to opt for these projects. Coming on top of the bad deals they believe they have been dealt in aerosat and Intelsat, a narrowing of our post-Apollo policy in this fashion may well have serious repercussions in a broader context: we may be increasingly seen as unreliable partners and allies. For

this reason, some concession may be in order, and the views of [the National Security Council] and State would help to guide the policy in this regard. A concession could be made either re the tug or shuttle. However, because of the difficulty of developing a satisfactory tug and the potential for sizable technology and dollar outflow, and also because of DOD's concern in this area, we should preclude European development of this project. We would simply be trading off a short-term foreign relations problem for a longer-term one. In regard to shuttle involvement, the management and foreign relations problems associated with government-to-government subcontracting are unacceptable, but we might accept European subcontracting on a normal company-to-company basis. Though not to their liking, NASA could informally direct our U.S. shuttle contractors to select and use qualified, low-bid European subcontractors on tasks the prime contractors choose, perhaps up to the level of $50 to $100M out of a $3 to $4B shuttle contract. Dollar outflow could be balanced by our requirement that the Europeans subcontract at least a compensatory amount in the U.S. for their RAM development, if the two to three year delay in balancing is acceptable to us. Otherwise, balancing can be achieved through other offset arrangements. NASA would prefer not to make a foreign relations concession of this nature because of their long-standing adherence to an internal rule against exchange of funds and its potential political impact. If, however, State and [the National Security Council] urge this concession, NASA sees this arrangement as less odious than government-to-government subcontracting, and could implement it.

Strategy and Tactics for Implementing the Proposed Program

Two levels of action should be pursued: a longer-term (months) strategic move to gain European political appreciation of and accommodation to the differences in European and U.S. motivations re space, and a short-term (weeks) tactical move to decide on and offer to the Europeans a moderate program of participation in the post-Apollo development Phase, having net advantage to the U.S.

1. *Strategic Considerations.*
    Complicating our discussions on space cooperation with the Europeans are the differences in our respective backgrounds and orientations with respect to space. To those who ran the U.S. space program, particularly the Apollo program, and conducted our side of the talks with the Europeans, space has been a non-commercial venture encompassing exploration, science, and technology, and space's commercial value has played only an emerging role in their thinking. Commercial utilization has been handled by our [11] private sector; while in Europe both the exploration and utilization of space are government functions. European interest in post-Apollo is more in the vein of commerce than adventure. Obtaining a mutually satisfactory cooperative program has been difficult because the two sides have seen it as offering different payoffs. Therefore, our strategy must not simply be to bring a shift in emphasis on what piece of hardware Europe might supply, but should develop a basic accommodation through mutual understanding and acceptance of objectives.
    We must attempt, for example, to stimulate recognition in European science-minister/political leaders, and their staffs, of the political-prestige value of manned space flight. No significant effort has been made by the U.S. to determine the latent political interest in manned flight, nor has any coordinated attempt been made to guide them persuasively into the program. NASA seems to have taken the European view at face value, and all of our negotiations on cooperation have generally reflected our axiomatic acceptance of European disinterest in manned space flight. We also should try to obtain an understanding with Europe that the development of launchers duplicates skills and equip-

ment that already are well developed in the U.S., does not really enhance the direct derivation of benefits from space given the availability of launches, and does heighten U.S. concern because of technology flow and security considerations. There is some doubt that Europe can learn our management skills simply by sitting in on the shuttle management, but it is a risk to us for reasons of future competitive posture. We should attempt to make it clear that we expect them to join us in a cooperative space program primarily for non-commercial reasons, and they should disabuse themselves of the idea of making money from us or learning our technology and know-how. They may feel that it is their financial contribution to the program that motivates U.S. interest in cooperation, and hence they are entitled to get something significant and tangible out of the program. They are wrong on both counts, and we must clarify this matter to them. Discussion should begin informally and individually, not group-wise, recognizing, however, that the prospects of evangelizing are not great, a priori.

2. *Tactical Considerations.*

The most immediate problem is to persuade the European space technocrats that a RAM-sortie can is a challenging and important task, and that it opens the part of the post-Apollo program having the greatest direct benefit, [12] namely, payload development and use. The tug should be ruled out because of its difficulty and its high potential for technology and dollar outflow. If the Europeans insist on also participating in shuttle development, we can, on grounds of avoiding government involvement in contractor-subcontractor disputes, offer the possibility of their industry functioning as normal commercial subcontractors to our U.S. prime contractor at a moderate level ($50 to $100M). The Europeans have purportedly inquired about this possibility last month, and so a change in our position of this nature can be offered as acquiescence to their proposal. There would be an understanding with Europe that the dollar flow inherent in this arrangement would have some balance through European subcontracting in the U.S. for parts of its RAM, or through other offset arrangements.

The fact must be faced that the European technicians have been strongly motivated toward [a] tug; it is the biggest and most challenging post-Apollo project available to them, and has the greatest technology stimulation and spin-off to other high technology capabilities. Moreover, nothing the U.S. has said to the Europeans in almost two years would indicate anything other than our desire for them to undertake the tug. And at our encouragement they have spent $1 to $2M studying their capability for its development. Changing signals is therefore going to be difficult without irritating them (justifiably). Because it postpones the problem, there have been suggestions that we wait for Europe to come to its own understanding or demonstration of its inability to build an acceptable tug. The Europeans' anger and frustration would increase, though, in proportion to the amount of time and money they waste on a project we reject. It may be that the best course is to take the flak now and admit our concern over their abilities and over the technology/dollar outflow we envision, and withdraw the tug from consideration. In order to ease the foreign relations impact and some of the pressure their industry is applying to their governments to undertake development tasks that are unacceptable to us (tug or European-contributed shuttle work), we might allow them some normal subcontractor participation in the shuttle as qualified bidders.

The timing of these tactics is a major difficulty. We would have to get these messages across and obtain European agreement by July if European subcontractors are to be used on the shuttle; our prime contractor cannot wait past that period. If Europe only undertakes a RAM-sortie can, timing is no longer critical to us, but the Europeans themselves say they must decide by mid-summer because of the coupling with their decision on whether or not to go ahead with Europa III.

[13] The State Department is now reviewing the post-Apollo policy, and the receipt of directions to propose a modified program to the Europeans would be most timely. Some resistance within our government to an alteration in direction can be anticipated, if for no other reason than the psychological momentum of the people that have been involved in obtaining European participation. Considering the many factors involved, no more time should lapse before a decision is made and guidance given.

Document I-26

[1]

Memorandum for the President

Subject: Post-Apollo Relationships with the Europeans

I wish to bring to your attention my increasing concern about developing U.S. attitudes toward European participation in the development of the post-Apollo Space Transportation System and the need for prompt U.S. decisions in this matter, if we are to control the play of events.

Your name has been closely identified with U.S. efforts over the past several years to encourage European participation in the development of that System—the Shuttle, the Tug and associated research applications modules (RAMs). Tom Paine, alluding repeatedly to what he described as your views, visited each major European capital to invite such participation. In October of 1970 and again in February 1971, Alex Johnson and a sub-cabinet team met with the European space and science Ministers. These and other activities of responsible U.S. officials, including our Ambassadors, have provided the Europeans every reason to believe that the U.S. was seriously interested in having them participate in the development of certain parts of the Shuttle, in one or more of the RAMs, especially in the Tug. As an indication of their interest the European governments have already spent or committed a total of $11.5 million on preliminary technical studies.

The European space and science Ministers are scheduled to meet in three weeks (May 19th) to formulate their views with respect to participation, and again in early July to take a final position. We can expect a visit of a high-level European delegation shortly after the May meeting.

Within the last several months U.S. views that we should minimize European participation have begun to harden. These views hold that we should not permit European participation in development of the Shuttle because of domestic economic consideration and the difficulties of sharing such a task with foreign governments and subcontractors. With respect to the Tug they hold that the development task will be too [2] difficult technically to rely on European performance. European participation would thus be limited to development of one or more of the RAMs.

Were the European share of Shuttle development to be truly substantial, these economic and management considerations might well be overriding. However, the extent of their possible participation is now limited to a few specific projects totaling about $100 million out of the total Shuttle program costing $5.15 billion. The advantages of denying their participation at this level do not justify the loss of U.S. integrity abroad.

There is no need to reverse our position now on European development of the Tug, since it is a vary advanced project which will require several more years of design study. The Europeans are, as yet, not convinced that they should undertake it.

My basic worry is that we will buy more trouble with the Europeans than can be justified by the ephemeral domestic advantages that we may gain by denying their participation. To limit them now to development of only a RAM would be judged by them as a clear reversal of our previous policy. Your reputation as a consistent advocate of international cooperation in space and specifically with Europe on the post-Apollo program would inevitably suffer. Furthermore, we ought not to ignore altogether the very real political values that would result from European participation with us in the development as well as the use of the Space Transportation System.

Balancing all these considerations I suggest:

1. That we accept, but not encourage, European participation in the tasks in the development of the shuttle already identified by NASA conditioned on a prior commitment by the European Space Conference (ESC) that it will undertake the subsequent development of one or more RAMs.
2. That we bring the Europeans to agree that consideration of their undertaking the development of the Tug will be deferred pending further mutual study.
[3] 3. That we conduct negotiations on these matters so as to avoid indicating a major change in our policy toward European participation (i.e.: in the proposals which we have already made to the ESC).

I urge that you approve this course of action in principle and instruct me to reach agreement with the Europeans along these lines.

William P. Rogers

Document I-27

[1]   May 5, 1972

Honorable Henry A. Kissinger
Assistant to the President for National Security Affairs
The White House
Washington, DC 20500

Dear Henry:

Dr. Walsh of your staff has requested formal comments from NASA on Secretary Rogers' memorandum to the President of April 29 on Post-Apollo relations with the Europeans.

Our comments are attached.

Sincerely,

James C. Fletcher
Administrator

Enclosure

****

[Enclosure Page 1]

NASA's Comments on Secretary Rogers' Memorandum of April 29, 1972

NASA's comments on Secretary Rogers' memorandum of April 29 for the President, on the subject of post-Apollo relationships with the Europeans, follow:

Our preferred objective is to obtain European agreement to develop a specified type of sortie module for use with the shuttle, reserving other types for our own development. We regard this as a desirable contribution to the space transportation system and one which should present no undue problems technically or managerially.

We agree with the Secretary's letter that the tug requires further study. It is, therefore, a distinctly, second choice, and much less desirable. We believe that European study of the tug should be on element of an agreement only if it is clear that the US commitment to a European undertaking to develop the tug could be considered only after extensive European study and only on the basis that we might well decide not to pursue the tug after such study. The Europeans would have to understand, even before undertaking such a study, that the definition of the tug and European capability [2] to develop it are uncertain; even in the event we were both persuaded by studies and proposed management schemes that the project appeared feasible in Europe, we would still want to reserve the right to escape from an agreement if interim review indicated that the tug would be substantially delayed or fall short of agreed specifications. Of course, the Europeans might not wish to participate in the study on such a tenuous basis. Unless directed by the President, we would not anticipate NASA technical support of the European study. For all of these reasons, we do not recommend European involvement in the tug.

With regard to specific shuttle tasks that Europe might perform, we continue to feel such European participation is highly undesirable and that it would complicate our shuttle management problems. However, if it is considered by the President, on the basis of international factors, that Europe's participation in the shuttle itself is of overriding importance, we believe that we could accept such participation if suitable management terms cam be established. In essence, acceptable management terms would call for US prime contractor selection and direction of European subcontractors, with the [3] European side responsible for both estimated costs and overruns and the US side responsible only for those out-of-scope changes imposed by us. As stated in the Secretary's letter, the European performance of shuttle items would be conditioned upon European development of a sortie module.

[4]   **INTERNATIONAL PARTICIPATION**
                          IN THE SHUTTLE

A. NASA wants:
    1. U.S. Shuttle (now)
    2. U.S. Sortie
    3. U.S. Tug (2-3 years later)
B. NASA can do (if required):
    1. A European Sortie Can (no significant management problem)
    2. European "bits and pieces" of shuttle (now)
    3. Maybe European Tug (later) (Technology and Management Problem)

C. Europe Wants (?)
   1. European Tug
   2. "Bits and pieces" of shuttle
      a. or involvement some way in shuttle
   3. Sortie (?)
D. U.S. should (1) offer and (2) accept
   1. (1) Sortie Can
   2. (2) Shuttle Items or C 2 a. Plus D 1
   3. (2) Europe study of Tug in full knowledge of questions re definition, European capability, performance, minimal number of procurements, interim review and escape procedure for NASA
[5] E. NASA Position
   1. Sortie OK
   2. Tug only after detailed study, poor second choice, et cetera
   3. We are directable to do shuttle items plus 1 (if we can control management method)

Document I-28

Document title: European Space Conference, Committee of Alternates, "Report of the ESC Delegation on discussions held with the U.S. Delegation on European participation in the Post-Apollo Program, Washington, 14-16 June 1972," CSE/CS (72) 15, June 22, 1972, excerpts.

Source: ESA Collection, European Community Archives, Florence, Italy.

On June 1, 1972, Assistant to the President for National Security Affairs Henry Kissinger communicated to the Department of State and NASA, among others, a presidential decision that removed a reusable space "tug" as a candidate for the European contribution to the Space Transportation System. President Nixon was also discouraging the idea of European firms building portions of the space shuttle itself. These preferences were the basis of the U.S. position announced in a meeting with a European Space Conference delegation on June 14 through June 16. After the meeting, Europe decided over the next year that its post-Apollo contribution would be a research and applications module (subsequently renamed Spacelab).

[1]   Neuilly, 22nd June 1972

Report of the ESC Delegation on discussions held with the U.S. Delegation on European participation in the Post-Apollo Program Washington, 14-16 June 1972

1. According to instructions given by ESC Ministers in an informal meeting of 19 May 1972, held in Paris, a European Space Conference Delegation met a U.S. Delegation in Washington, D.C., 14-15 June 1972.
    A list of the European and U.S. delegates is attached to this report (Annexes I and II).

General considerations

2.   At the beginning of the discussions, Mr. Herman Pollack, head of the U.S. Delegation, made an opening statement (see Annex III) in which he recalled the developments which had occurred since the last meeting between President Lefevre and Under Secretary of State Johnson in early 1971 and provided a brief overview of the current U.S. attitude towards cooperation with Europe in the post-Apollo Program:

> (1) The concept of European participation in the Shuttle development has changed considerably and would be subject to such stringent conditions that it may become almost unattractive for Europeans;
>
> (2) The U.S. has concluded that it is not prudent to continue discussions on the possibility of tug development by Europe;
>
> (3) The U.S. encourages Europe to undertake the development of one or more of the Research and Application Modules which in its opinion would constitute a desirable form of cooperation;
>
> [2] (4) The U.S. also urges Europe to anticipate and make extensive use of the Space Shuttle system when it becomes operational and to participate in payload development, both manned and unmanned. It was mentioned in that respect that participation of European astronauts in shuttle flights would be welcomed.

3.   In the course of the discussion which followed the statement, it was made clear to the European Delegation that the U.S. attitude was defined at top governmental level and that no change in it could be expected; it was also stressed within the limits so described Europe could submit any proposal of participation.

4.   In his concluding remarks (see Annex IV) Mr. Herman Pollack drew the attention of the European Delegation to the fact that the U.S. feels that the "potential of outer space" which would become possible through the post Apollo program is so far-reaching that it can no longer be the subject solely of national decision. This is the reason why the U.S. is seeking ways to make it possible for other qualified and interested nations to participate with it in the development and utilization of this new capability.

The enduring nature of the ties that bind the U.S. and Europe motivated the U.S. in it search for European participation in the post-Apollo program. The motivations were purely political and commercial or technical factors had practically no influence. . . .

[1]   ANNEX III

Opening Remarks by Mr. Herman Pollack
Meeting with ESC Delegation on Post-Apollo Cooperation
June 14, 1972

Welcome.

Many of us sat in this room for the second of the two meetings between Minister Lefevre and Under Secretary Johnson and their delegations 16 months ago in early 1971.

A good deal has occurred during those 16 months to enable us all to have a clearer definition of the post-Apollo program and a somewhat better understanding of each others' readiness and interest in cooperating in that program. In retrospect perhaps the most significant of these developments have been:

1. The development by the U.S. of a launch assurance policy, which stands independent of European participation in the development of the reusable Space Transportation System or its use. I refer to the launch assurances conveyed in Under Secretary Johnson's letter to Minister Lefevre of September 1, 1971.
2. The discussions held between NASA and technical representatives of the European Space Conference.
3. The decision of our President to proceed with the development of the Space Shuttle System, and the development timetable which follows from that decision.
4. The preparations under way in Europe for Ministerial decisions, prospectively this summer, on a broad range of matters affecting European space activities.
5. Considerable changes in the economic perceptions and budgetary circumstances in the U.S. I imagine the same is true in Europe.

We meet now, at your request, specifically to discuss the questions which you have raised in the agenda before us.

It is our understanding that these discussions are not negotiations. Obviously we will not reach decisions here. Rather, we anticipate informal and frank exchange of views in which we seek to understand more precisely each others' preferences and interests on the matters which you have raised.

In the absence of a clear indication of the measure of European interest in possible participation, we shall do our best to make the U.S. views regarding the questions you have raised as helpful as we can. Were it possible during the early part of our discussions to obtain a clearer understanding of the measure of European interest, and possible participation, our views could possibly be more responsive and useful to you.

[2] Now, if I may, I should like to present a brief overview of U.S. attitudes toward cooperation with Europe in the post-Apollo program.

1. We urge Europe to anticipate and make extensive use of the Space Shuttle System when it becomes operational, and to participate in payload development, both manned and unmanned.
2. We have concluded that from our point of view, as well as yours as we understand it, that the development by Europe of one or more of the Research Applications Modules would constitute a desirable form of cooperation, and we encourage you to undertake such a task.
3. With the passage of time the concept of European participation in the development of the Shuttle itself has changed considerably. We are now strongly impressed by the potential difficulties that might ensue from an intergovernmental effort to produce a relatively small number of components of a massive piece of highly complex hardware whose timetable is pressing and in whose success the political and economic stakes are so high. Cooperation in some of the Shuttle items is not precluded. However, it will be necessary for Europe to undertake to meet rather stringent conditions designed to satisfy fully U.S. concerns. In candor I must report that the conditions the United States finds necessary may diminish the attractiveness to Europe of participating in the Shuttle items.
4. Since the definition of the Tug is still uncertain and the decision by the United States to proceed with its development has not yet been made, and there are no hard predictions as to when it will be made, the United States has concluded that it is not prudent to continue discussions of the possibility of cooperation on this task.

As I indicated earlier I have presented this overview in the interests of making our discussion here today more constructive and to help illuminate the responses we shall make to the questions you have raised.

I have, as you know, participated in these discussions from their outset. If words alone were all that were required to get cooperation under way we would be in full orbit by now. I want to assure you that European cooperation in this program, while evolving in form with passing time and changing circumstances, continues to be an objective of the United States. Let me say, however, that this is not essentially a commercial transaction we are discussing. Above all, it is a political act. In the absence of mutual political will to achieve a state of cooperation the real and apparent hazards and pitfalls will assume inordinate proportions and I fear that this venture will founder. It is my hope that our discussion today, and any that may subsequently follow, will be strongly motivated by a mutual desire to find a basis for agreement.

That concludes my opening remarks.

[1]   ANNEX IV

Concluding Remarks by Mr. Herman Pollack
Meeting with ESC Delegation on Post-Apollo Cooperation
June 16, 1972

In this meeting we have tried to be entirely forthcoming, realizing fully the difficulty and the importance of the decisions that are to be made in Europe and the value to you of the clearest possible understanding of what the United States has in mind. It is our hope that we have provided the facts you are seeking and that they will enable your Governments to arrive at affirmative decisions when your Ministers meet in July. Some of the facts, however, which I think are relevant to the decisions of your Governments cannot be expressed with mathematical precision but are nevertheless important, and perhaps fundamentally of greater importance than some of the hard information we have provided you with during this meeting.

For example, it is important that both sides keep in mind the basic, enduring nature of the ties that bind the United States and Europe. These are well understood on both sides of the Atlantic and need not be elaborated here. But, it is this compelling and fundamental fact of life that above all else has motivated the United States in its search for European participation in the Post-Apollo program.

Another major but somewhat ineffable motivation arises out of the awe which United States leaders viewed the potential of outer space which would become possible once capability such as that contemplated in the post-Apollo program became a part of mankind's competence. We felt then and continue to feel now that this potential is too great, its implications to mankind too far reaching to be properly the subject solely of national decision. We therefore began to seek ways to make it possible for other qualified and interested nations to participate with us in the development and utilization of this new capability.

I repeat my statement made on the first day that commercial or technical factors have practically no influence in motivating our desire for European participation in a post-Apollo program. Rather, the considerations I mentioned above have generated this objective and keep it alive and strong today.

When we began our discussions with Europe we ourselves did not fully understand the nature of the system whose construction we shall embark on this summer.

Furthermore, it is clear in retrospect, that we approached these opportunities in prospect of a considerable interest abroad in participating in the development and use of a new Space Transportation System.

[2] You have participated with us in the preliminary definition of that System and, indeed, have made significant contributions to our changing perspectives and deepening understanding of it. Positions which originated several years ago relied heavily on predictions—indeed speculation—both as to the System itself and your interest in it. These position have been altered and modified as our mutual comprehension grew.

Thus we have arrived at a point in time at which your participation in the development of the Shuttle on a significant scale, as originally conceived, has been overtaken by time and, for the reasons we have enumerated during our discussions, can no longer be encouraged by us even on the limited scale we are still discussing. Consideration of mutual development of the Tug has of necessity been set aside. The opportunity to develop Sortie modules and to plan together for the use of the over-all Space Shuttle System and actually to make use of it, nonetheless constitute a major challenge and would be a significant response to our earlier expectations. We hope we have made it clear that we would warmly welcome your participation in these two areas.

Finally, let me repeat that for over two years we have sought European participation in this program and let me emphasize that we continue to do so. It is my hope that for your own reasons as well as for those which move us, we shall be able to come this summer to an agreement to move forward together on this historic project.

Document I-29

Document title: Arnold W. Frutkin, Assistant Administrator for International Affairs, NASA, Memorandum to Administrator, "Government Level Negotiations on Sortie Lab," May 9, 1973.

Source: NASA Historical Reference Collection, NASA History Office, NASA Headquarters, Washington, D.C.

This memorandum describes the final issues that had to be resolved before the United States and Europe could sign a memorandum of understanding regarding European development of the Spacelab (SL). The French L3S vehicle mentioned in the memorandum was later renamed Ariane. AEROSAT was a proposed cooperative satellite system for air traffic control, which was controversial at the time this memorandum was written; the United States later withdrew from discussions regarding the development of this system.

[1]   May 9, 1973

Memorandum

TO: A/Administrator
FROM: I/Asst. Administrator for International Affairs
SUBJECT: Government Level Negotiations on Sortie Lab

Formal but secondary level efforts to arrive at a consolidated US/European text between the European side and the US side were carried out on May 2-3 in Washington. Pollack chaired the US side (myself, Elliott, Rattinger) and Trella chaired the European side (with representation from seven countries).

A large number of essentially cosmetic changes were made in the existing US draft to accommodate European interests. In the end the European side cited three "reservations." It was agreed the European side would attempt to respond in about two weeks to indicate whether the reservations stood so as to require further discussion or were removed.

Among the many cosmetic changes were the following:

1.   The words "United States" were dropped where they preceded space shuttle system in order to parallel our removal of the word "European" before Spacelab. We do not want to emphasize ownership of the Spacelab by Europe and feel there is no need, in view of all the facts of life, to emphasize US ownership of the Shuttle.

2.   Under obligations of the US, "assistance" was removed to come under a requirement for mutual agreement and relevant US law and regulation.

3.   We agreed that the US right to provide assistance as hardware rather than know-how could be exercised "in exceptional cases." In fact, we do not now know of any requirements for such reservations.

4.   We agreed to state the legal situation in response to the European request for technology beyond that necessary to execute the SL program, namely, that the US will consider [2] such requests on a case-by-case basis.

5.   We made the same arrangement for the use of required technology for additional purposes outside the SL.

6.   We agreed that cooperative (non-reimbursable) European proposals would be given preference over third countries if at least equal in merit, but that cost reimbursable proposals by Europe will get such preference in the event of payload limitations or scheduling conflicts. This was agreed by our side to be only what would be the case under the President's launch assurance policy in practice.

7.   It is specified, again pursuant to the President's launch policy, that commercial use of Shuttles and SLs will be non-discriminatory.

8.   On the first SL we get "full control . . . including the right to make final determination as to its use . . ." and, except for joint *planning* of the *first* flight on a cooperative basis and encouragement of cooperative use of the first SL *unit* throughout its life, we get "unrestricted use of the first SL free of cost." It is made clear that we may charge Europe for use of the first SL.

9.   The term of agreement—five years after the first operational flight of the Shuttle—is restated as lasting until January 1, 1985 but at least for five years following the first flight.

The European chairman ended with three reservations:

1.   The group was not convinced that the European risk or contribution was adequately balanced by European benefits. He was specifically referring to their hope that we would agree to *try* to balance the European procurements in the US with US procurements in Europe. We said this is out of the question.

2.   The Europeans feel that the agreement ought to extend to 1988 in order to draw out our obligation to buy SLs. We feel that a term beyond 1985 is totally unrealistic in the light of our present knowledge and future uncertainties.

3.   The European side would still hope to improve the US launch assurance policy. We've made clear that it is totally impractical to think of working out changes in the US Government on this policy at this time.

[3] My own assessment is that the three reservations are tactical, to keep things open, while the final arrangements in Europe fall into place on such matters as funding of the French L3S launch vehicle. I believe that the European leaders are essentially committed to participation with us and that only unexpected rebuffs by us, congressional reversal, or

a serious falling out in Europe can change this. The AEROSAT problem is one such threat on the horizon.

Arnold W. Frutkin

Document I-30

Document title: Arnold W. Frutkin, Assistant Administrator for International Affairs, NASA, Memorandum to Deputy Administrator, "International Space Station Approach," June 7, 1974.

Source: George M. Low Papers, Institute Archives and Special Collections, Rensselaer Polytechnic Institute, Troy, New York.

This memorandum, prepared by longtime NASA Assistant Administrator for International Affairs Arnold Frutkin, contains some of the earliest thinking within top NASA circles with respect to what was expected to be the major "post-shuttle" program—a space station.

---

[1]   June 7, 1974

TO: AD/Deputy Administrator

FROM: I/Asst. Administrator for International Affairs

SUBJECT: International Space Station Approach

REF: Your Oral Request of May 15

Note: I have assumed that it would be easier to get domestic clearance to explore a space station internationally than to get domestic approval for a space station per se before inviting international participation. Therefore, we propose an approach here in which all elements of the project would be attacked on an international basis: justification, definition, design, construction, operation and use.

1. *Prospective Partners*
     The three plausible partners for an international space station effort would be the US, the Soviet Union, and the European Space Agency. All have space experience, will have had manned flight experience, and have the necessary resources. (Canada is extremely limited in resources and Japan has shown no disposition to contribute to a non-national space purpose. The possible participation on a secondary level of these and other countries will be discussed later.)
     There are a number of options as to how to approach participation by the senior three. We could approach either the USSR or Europe bilaterally, but I believe that each would be reluctant to enter into a strictly bilateral arrangement in the foreseeable future—Europe because of conservative space funding views and current space commitments, the Soviet Union because of political and security considerations. I do not think we should put the USSR forward as the senior partner since Europe would be quite offended (in view of the Spacelab agreement). Moreover, Europe might really be a better partner operationally, technically, financially and politically.

[2] Our best bet would be to approach both Europe and the USSR simultaneously, holding over each the possibility that we might be going ahead with the other. On this basis, we may be able to motivate *both* to work with us on a tripartite basis. This would give us the strongest basis for a large undertaking—economically and politically. Our approach (below) is calculated to make the USSR and Europe feel they have very little to lose and perhaps something to gain by entering into the particular procedure we are proposing.

2.   *Domestic Clearance*
The multi-step approach outlined below would, of course, have to be cleared domestically with the usual offices. The prospects for such clearance would be greatly improved for the following reasons:
(a) No significant funds would be required until about the fourth year of the relationship.
(b) The procedure would advance us vary cautiously, step by step, into the project, with a long-deferred final commitment, and ample and specific opportunity to decide not to proceed beyond any given step.

3.   *Initial International Approach*
After obtaining clearance, NASA would approach the Soviet Academy (or Aerospace Ministry) and the ESA (with backup visit at least to the principal ESA members—Germany, UK, France and Italy) on the following basis:
"The next major step in space, following the Shuttle/Spacelab, could well be a space station—'permanent,' resuppliable, recycling, etc. Such a space station would represent a facility of very wide interest and potential value to nations. As a very considerable and beneficial undertaking, it would require and deserve the pooling of resources. Accordingly, we feel that the question of such a space station should be explored as a *possible* international undertaking.
"We recognize that it is entirely too early for any nation to consider any commitment whatever to such a project. At the same time, in view of the protracted study and development which would be required before a space station could become a reality, it is out too early to undertake the very first preliminary inquiries regarding the purposes and character of such a facility.
[3] "We propose a very cautiously and conservatively structure[d] approach which would allow the three principal space power centers to examine into the question. They would begin independently, coordinate their next steps and, assuming that progress is satisfactory, proceed on an increasingly integrated basis.
"Thus, we are asking you, the USSR and Europe to agree to explore with us, according to a very highly protective procedure, your possible interest in proceeding into the design, development and operation of a truly international space station."

4.   *Detailed International Procedure*
This procedure would proceed on two separate tracks. The first track would move from study and definition to planning the design, construction and establishment of the space station. The second track would aim at setting up the relationships and arrangements among the participants for undertaking, managing, and operating the space station.
The two-track approach is designed to separate the technical from the political problems in the early phases of the project. The second track, on management and operation, begins only at a specific point, when it should be clear whether the first track is making satisfactory progress.

This arrangement will give us the best chance to see if we can agree on what it is that we might want to do together. We could then turn to how we want to do it. Our entire experience, especially the Spacelab negotiations, demonstrates that if you get into how before what, you argue about abstractions and principles and problems which will never arise instead of arguing about a specific job to be done.

Farther down the road, the two tracks must be brought together so that we end up with a coherent plan and program.

(a) *First Track*

i. The three participants would initiate independent conceptual (pre-phase A) studies of a space station in order to develop their own very preliminary notions of its possible objectives, benefits, character and use, configuration and approach to placing and maintaining the station on orbit. (3/75)

ii. The participants would interchange their independent studies and consider them. (9/75)

[4] iii. The participants would form a (technical) joint study group to produce a single "strawman" concept. This would have no necessary relationship to the prior independent studies nor would the participants be bound by the views of their technical representatives. (12/75)

iv. The participants would independently review the "strawman" study. (9/76)

v. Technical representatives of the three participants would convene to formulate plans for a formal Phase A conceptual study to be conducted on an integrated (joint) basis. These plans would provide for integrated management of the study with a joint project team supported by "contractors," public or private, in the three countries. (12/76)

vi. Implementation of the Phase A study per above. (3/76-3/78)

vii. Those participants prepared to proceed, on the basis of Phase A results, to formulate a Phase B plan would do so. (They would do whatever might be appropriate to preserve the participation of any member with reservations at any step.) (6/78)

viii. Implementation of Phase II Program. (6/78-6/80)

ix. Review of Phase B program in manner similar to the review of Phase A above. (6/80-9/80)

x. Formulation of Phase C/D program. (9/80-3/81)

xi. Review of formulation and commitment to Phase C/D. (3/81-6/81)

(b) *Second Track*

i. At approximately the time when the formal Phase A study is agreed in Track One, the parties would convene a separate Joint Implementation Working Group to initiate an implementation plan for managing a possible Phase B and beyond.

ii. The Joint Implementation Working Group would address the financial arrangements, the form and location of management, method of decision making, provision for a systems integration mechanism, the division of labor among the participants, the degree to which management [5] authority should reflect the relative responsibility of the parties, questions of mutual access, and ultimately the operation and use of the system. The target for final agreement on an implementation plan would be the conclusion of the formal study of Phase A.

5. *Technology Transfer*

Each participant will expect the ultimate space station to be fully available for his use, alone or in concert with the other participants. Therefore, each will want essentially a total knowledge of the system. In all likelihood, narrow commercial processes could be held back as proprietary. I believe we should face this prospect squarely—in approaching domestic clearances and later in reaching understandings with the other participants.

Thus, for the present, we would not hold out any tenuous, complex or artificial prospects for restricting technology in the course of the program. I believe we could support such a policy on the basis that the space station itself would be a valuable consideration and that the technology entering into its structure would be unlikely to have significant commercial application. I propose, however, to keep this question under review to see if an alternative is feasible.

6.   *Additional Accessions*

The question of participation by additional countries might be handled in the following way:

(a) In the design and development phase, any one of the three major participants could absorb personnel from additional countries within its own participation without in any way reducing its own responsibilities. It could not sub-contract its responsibilities to foreign companies or the equivalent without the specific knowledge and consent of the other major participants.

(b) In the use phase, other nations might apply for use of the system to a combined use-control board composed of the major participants according to provisions which would have to be worked out by the Joint Implementation Working Group.

7.   *Assurances*

The participants would have to exchange government-level assurances relating to peaceful purposes, the openness of scientific results, [and] descriptive information of technical [6] activities (allowing for reservation of proprietary rights to industrial processes tested or employed in space, etc.).

Arnold W. Frutkin

Document I-31

Document title: Kenneth S. Pedersen, Director of International Affairs, NASA, to Director, Space Station Task Force, "Strategy for International Cooperation in Space Station Planning," undated [August 1982].

Source: NASA Historical Reference Collection, NASA History Office, NASA Headquarters, Washington, D.C.

This memorandum by NASA Director of International Affairs Kenneth Pedersen, who had joined the space agency in 1979, was prepared in response to a query from John Hodge, Director of NASA's Space Station Task Force, about the elements of a strategy for international cooperation in a possible station program. The Space Station Task Force had been established earlier in 1982 as a focal point for developing a NASA proposal for such an undertaking, which new Administrator James Beggs and Deputy Administrator Hans Mark had made a top priority in their approach to the space agency's future.

[1]

TO: MFA-l3/Director, Space Station Task Force

FROM: LI-15/Director of International Affairs

SUBJECT: Strategy for International Cooperation in Space Station Planning

In your July 30 memo, you raised some interesting questions concerning international cooperation strategies. I would like to set down some thoughts on the matter, beginning with a quick look back to how we proceeded during Post-Apollo planning.

LESSONS LEARNED.

The most important lesson is to avoid making premature commitments or promises. Along with this, we must be careful to avoid broad statements that can be misconstrued.

Based on long-standing U.S. policy to cooperate internationally in space (that was reconfirmed in a 1969 White House Space Task Force report) and to encourage a European community, President Nixon told NASA that the U.S. should have European participation in Post-Apollo activities. NASA, as a result, immediately began to seek European cooperation in its Shuttle and Space Station activities. In late 1971, NASA and the European Space Conference (ESC) agreed that Europe would study a Tug and RAM (a Spacelab-like module that could either be operated within the Shuttle bay or on a space station), and that European companies would formally team with U.S. companies during Phase A and look at specific parts of the Shuttle (i.e., tail, payload bay doors). Also, NASA and European labs engaged in studying technologies needed for Post-Apollo activities. All in all, Europe spent about $20M on these Phase A (and in some cases, Phase B) studies. In the end, Europe's main interest was to develop the Tug.

Meanwhile, the U.S. position with respect to the level and kind of European participation it wanted crystallized. First, the Administration's interest in cooperating was later interpreted as an interest in European involvement in the use of space; i.e., the development of payloads and international manned operations rather than joint engineering projects. Second, [2] NASA, through an extensive review of European industry, found that European industry lagged approximately five to ten years behind U.S. industry. Therefore, NASA dropped the idea of joint development of technology, speculating that the U.S. might stand to lose more than it would gain. Third, NASA also decided that it did not want to depend on foreign countries for critical items on the Shuttle, so that the Shuttle could fly independent of foreign activities. Fourth, NASA decided that, for safety reasons, it did not want to fly a Tug using liquid propellants, the only type Europe was studying. Moreover, there was real concern that Europe did not have all the technology to develop a Tug.

A smaller lesson learned was the undesirability of formal teaming in the study phase. While this teaming was for joint development of [the] Shuttle, which was eventually dropped for the reason stated above, it did reveal the possibility of losing flexibility in subsequent development. NASA might prefer certain European companies while not wanting to choose the U.S. companies with which they were teamed, or vice versa. Most important, individual European companies could be denied participation in the program if they had prematurely teamed with a U.S. company which did not end up winning the bid.

The U.S. Government thus found itself in the position of having to walk back from the European perception of the cooperative possibilities in Post-Apollo that were encouraged by the way the U.S. and Europe had proceeded to define that cooperation. In some quarters in Europe, these misperceptions still exist, particularly as they concern the reasons why we rejected European development of the Tug.

Therefore, it must be decided whether certain systems and subsystems are going to be off-limits before we enter Phase B, so that we can avoid not only dashed expectations but also the possibility of missused [sic] foreign funding. Looking at it from the positive side, we should seek to identify, as soon as possible after we understand the basic design of the space station and before Phase B, what systems are realistic possibilities for foreign cooperation.

Further along in our program as we begin our negotiations with potential partners on an MOU, we need also to avoid some of the features which have proved troublesome in the Spacelab agreement. Neither NASA nor potential foreign partners will want an arrangement where one piece of hardware is contributed, and NASA is obligated to buy the additional units. Instead, our foreign partners will probably want preferential or free access to the Space Station as the quid pro quo. As an internal exercise, they will probably want to assure themselves there will be industrial spin-offs for their industry. And while NASA may indeed want to buy additional units from the foreign source, it is not desirable for NASA to be either [3] obligated to do so or restricted from developing similar types of hardware.

MISSION REQUIREMENTS STUDIES . . . HOW WE'VE PROCEEDED TO DATE.

The first key step was to involve foreigners early on in the process. This is responsive to their longtime requests for earlier participation in major NASA projects. It creates some ambiguity such as schedules, false starts, etc., but not enough so that it outweighs the benefits of beginning this way. Therefore, I think we are on the right track.

Each space agency that is undertaking a parallel mission requirements study has made the mission requirements aspect the first effort of its study, so that the results of its study can be factored into NASA's similar efforts. This results from the numerous times NASA has emphasized in its discussions with foreign officials the importance of identifying the potential uses of a space station, and from the realization by foreign officials that key U.S. players need to be apprised of the requirements that justify developing a space station.

Nonetheless, these same space agencies are also studying possible hardware contributions to a NASA space station. These efforts result from their own political realities at home; i.e., they have to justify spending their resources in a space station not only on potential space station *utilization*, but on potential industrial return as well. It also derives from the fact that if NASA is successful in its attempts to get Phase B approval for FY84, then they are going to have to move quickly to get big bucks from their governments to fund their Phase B activities. Thus, they are preparing the information necessary for their governments to determine if they are interested.

While this is acceptable, we must not let the emphasis on requirements get lost in the next several months. We can accomplish this by immediately addressing the question of how we plan to exchange results of the studies, thus reenforcing [sic] in the foreign space officials' minds that we are most interested in this aspect. I believe we will accomplish this at the September 13 meeting. I have outlined a proposal for this in the next section.

HOW WE SHOULD PROCEED FROM HERE.

Phase A

We have already laid the groundwork for this Phase. We saw an intense interest in our Space Station plans, and we have effectively translated that interest into several foreign mission requirements studies which are useful to both NASA and our potential partners. The complementary studies are designed to determine how much foreign interest exists in contributing to and using a U.S. Space Station. The results of these [4] studies will help both NASA and the potential partners decide if there is mutual interest in continuing cooperative activities in Phase B.

The next eight months—the duration of the mission requirements studies—are a very important time. It is incumbent upon NASA to maintain the emphasis on these require- ments studies. I believe we have to continue to demonstrate that these studies are impor- tant for all the reasons we discussed at the beginning.

Thus, the scenario I would like to see happen for the rest of Phase A is as follows. Since the mid-term review of the U.S. contractors will be done individually with each con- tractor, it is not appropriate to invite our potential partners to observe this review. However, immediately following these reviews, I propose we invite the agencies which are undertaking parallel studies to a NASA summary of the mid-term reviews, and in turn, request each potential partner to brief us on their mid-term results. The final review of the U.S. contractors, I understand, will be an *open* review with all contractors reporting at the same forum. Thus, it is appropriate to invite our foreign agency study managers to attend this review, and present the final results of their studies. Finally, just as the U.S. con- tractors will prepare a final report on their studies, we should request copies of the reports of the foreign studies. We should also offer to provide to them the unclassified portion of the U.S. contractors' reports.

In addition to the above, we should use any of our NASA trips abroad in the next year as an opportunity to pulse the progress of these foreign studies. Likewise, we should wel- come any requests to meet with us at NASA Headquarters to discuss the status of our Space Station planning activities. Given that the current space station activities are being run out of NASA Headquarters, I believe we should request foreign visitors to meet with Headquarters rather than Center personnel. Our foreign visitors should not be needless- ly exposed to the Center politics now going on, which could only arouse further confu- sion as to NASA's objectives at the present time. In addition, if opportunities for NASA personnel to address conferences arise that are attractive and useful, we should accept them, and use the conferences as additional opportunities to meet offline with foreign officials. In particular, we should attempt to find conferences that include potential for- eign users of a Space Station: scientists, business groups, and applications-oriented groups. This office will be on the lookout for all such possibilities. Finally, this office will ensure that all foreign visitors to NASA are apprised of our Space Station planning activi- ties and, where relevant, set up meetings with Space Station Task Force personnel. We may be able to identify *new* potential users through this process.

[5] As we planned for our Phase A activities, we discussed with our potential foreign part- ners the best way for NASA to work with them and with foreign industry. Based on these and our own internal discussions, we decided that NASA would work directly with the for- eign agencies, which in turn would keep their respective industries informed. In this way, NASA maintains its ties with its foreign space agency partners. And, since we do not want any formal industry teaming during Phase A, this strategy best suits our objectives. I believe we should maintain this strategy throughout Phase A. However, if a foreign part- ner invites NASA to address a meeting it has convened for its foreign industry, it might be

beneficial for us to attend. But if we decide to accept one, we must be ready to respond positively to *all* such foreign requests. Finally, we should encourage foreign space agencies to invite foreign industry to the planned January NASA industrial symposium that the [American Institute of Aeronautics and Astronautics] is arranging for us.

Phase B

The following discussion on Phase B presupposes that we get the go-ahead for Phase B in FY84. If not, then NASA would in effect be winding down Phase A activities or stretching them out another year. Our potential foreign partners will only proceed beyond these requirement studies when NASA gets Phase B approval.

Planning for and discussions with our potential partners on Phase B should rightly take place as Phase A winds down and as we have a better fix on the conduct of Phase B. I believe it is too early now to begin publicly speculating in great detail about how Phase B might look. This would only result in the danger of appearing to be overcommitting, and may, in fact, raise expectations in our potential foreign partners that we should not now raise. Although we must begin developing our views on how collaboration in Phase B would look, it would be premature to talk at length with our potential foreign partners at this stage.

The way our potential partners interact with NASA in Phase B is largely dependent on the management scheme we choose. I understand that one of the Task Force's working groups is now looking at various Phase B management schemes and that choice will probably be made around January. However, I would like to set down a few thoughts that I have concerning the way the international aspect of this Phase should proceed, regardless of the management approach chosen.

By the time we reach Phase B, special foreign interests in system and subsystem areas will probably have developed. These can be explicitly recognized within an agreement and used to focus respective studies, but no commitments to hardware development should be made. In a sense, this second phase would be the time when NASA and the potential partners would be [6] trying each other out, to see if cooperation really makes sense, both in the hardware and in the policy sense. But neither side will be quite ready to make a formal commitment.

While the main focus of Phase B is on designing hardware concepts, I believe each party should continue to refine requirements. Mission requirements analysis efforts should be an ongoing activity throughout the life of the Space Station program. In Phase B, we will be farther along in our design of the Space Station, which should help us discover additional uses. In addition, results from past Shuttle flights and other space endeavors should reveal new uses.

Whatever Phase B management scheme we decide upon, it may be desirable to avoid formal teaming of U.S. and foreign industry. The Shuttle experience suggests that it can reduce NASA's flexibility to choose certain foreign proposals because they are so tied to U.S. companies that eventually lose out in the development Phase. We all realize, however, how hard discouraging formal teaming will be, given our recent experience with Phase A.

If foreign space agencies fund Phase B activities, then they are half way there in seeing the merits of cooperating on the development phase. However, they will need to be convinced that a) the piece they eventually build is of significant value to the total system and, b) the returns to them are worth the costs of building it. Thus, regular discussions with our potential partners *throughout* Phase B is important to ensure that we mutually determine the best possible combination of cooperative possibilities that satisfy all our needs. Thus, in addition to meeting with our potential partners on the specifics of the Space Station activities, we must also use these opportunities to begin this type of dialogue.

Phase C/D

While MOU negotiations are at least 1 1/2 to two years away, over the next year, International Affairs will be re-examining past cooperative agreements (in particular Spacelab) to determine what features NASA ought to retain or avoid. In addition, we will be holding informal discussions with our potential partners to determine the things they will be looking for in their MOU's. As we proceed, we will continue to consult with you.

GUIDELINES FOR COOPERATIVE PROPOSALS.

First, we must determine if each specific proposal is beneficial to the U.S.; i.e., contributes to the overall objectives of the Space Station Program, to NASA's scientific and technological goals, and in a broader sense to overall U.S. foreign policy objectives. From our experience, a proposal can be beneficial by:

[7] –   encouraging foreign STS and space station use on both a cooperative and reimbursable basis, thus tying other countries' programs to ours;
–   sharing the cost of U.S. programs by stimulating contributions from abroad;
–   extending ties among scientific and national communities;
–   enlarging the potential for the development of the state-of-the-art;
–   supporting U.S. foreign relations and foreign policy.

Second, we should be confident that the industrial and technological infrastructure exists within a foreign country in order to handle the tasks proposed. This point is crucial, because it is one of the most important ways we assure ourselves that little or no technology will be transferred. During the Post-Apollo discussions, we were not that familiar with the European aerospace industry and subsequently toured that industry to make an assessment. We are much more fortunate today in that we have now worked with both European and Canadian industry on STS, and with Japanese industry on several scientific and applications projects. Thus, our analysis will probably be done much more quickly this time around.

We should also assure ourselves that the proposals are realistic in terms of the projected costs involved. ESA, for example, sets ceilings on the amount of money a program will cost. Yet, many times, unforeseen design changes or launch slips will push that cost up. We would want to make sure that the proposals' cost projections include an adequate contingency to hopefully avoid the potential problem of foreign attempts to have NASA pay for these charges and slips.

Finally, while it might be attractive if the potential partner proposed hardware that matched its utilization requirements, I believe that we should be satisfied that the partner has utilization requirements for the Space Station system as a whole, and that its specific proposal contributes to that system.

OTHER U.S. GOVERNMENT INVOLVEMENT.

The decision to involve international partners in NASA's Post-Apollo Program was made before that program was either defined or approved. The White House Space Task Force Group headed by Vice President Agnew confirmed that there would continue to be international involvement in NASA's Post-Apollo programs, and President Nixon reinforced this to Administrator Tom Paine. Afterwards, there was a long, intensive interagency review to determine just what President Nixon meant when he [8] said he wanted international involvement in the Post-Apollo Program. As the program was defined, NASA, too, determined what it thought the optimal international involvement should be. As the development evolved, some conclusions were confirmed, and others were identified as things to avoid in future undertakings.

The 1958 Space Act gave NASA the statutory responsibility to seek international cooperation in its space activities. This policy was interpreted by the Nixon Administration as applying to NASA's Post-Apollo activities. President Reagan's July 4 space policy statement reconfirmed that policy with respect to present and future NASA activities. Therefore, NASA should proceed with pursuing the best possible international involvement in a space station that is beneficial to U.S. interests.

Given this, NASA is responsible for making sure that all U.S. Government agencies or portions thereof that have foreign policy responsibilities are kept informed of our activities. Furthermore, informing them early on in the planning process gives us a much better opportunity to have them onboard as potential supporters for this program.

We started by briefing the interested offices within the State Department. The Space Station Task Force has kept the relevant DOD offices informed of NASA's international activities. NASA is briefing the export control community since U.S. companies are now seeking approval for information exchange agreements during the Phase A mission requirements study. Other agencies such as OSTP, OMB, DOD, NSC and ACDA are probably interested in the international aspects as well as the programmatic ones. We should consider augmenting the briefings the Task Force is giving to these organizations. Further, we typically prepare briefing materials for White House and other U.S. Government agency personnel as they attend foreign and international S&T conferences, summits, etc.; we will include in these materials information on the Space Station activities being undertaken by foreign space agencies.

As we proceed towards designing the Space Station, we will be much more aware of the level and type of DOD involvement expected. It is possible that DOD may express concerns that might drive an interagency review of the international component of a Space Station similar to the type experienced in Post-Apollo planning. Otherwise, I expect the international aspect will be considered within the context of the overall decision on the Space Station program. NASA's best strategy in such a policy review would have to be determined once we saw how the arguments were shaping up.

In terms of the normal State Department review of NASA's international agreements, it will only review the final MOUs.

[9] SPACE STATION APPROVAL—CAN FOREIGN INVOLVEMENT HELP?

From the onset, we must be fully aware that a Space Station will be built by the U.S. Government because the U.S. needs and wants it. However, after having said that, I believe there are several ways our potential partners can help NASA gain approval to proceed with building it.

The first is already underway. If the foreign requirements studies reveal that the potential foreign utilization rate is large or moderate, then this can help bolster NASA's contention that it is timely to develop a Space Station. Foreign industrial support can help expand the overall industrial interest in a Space Station and willingness to fund space R&D that can contribute to Space Station utilization. Thus, a larger corps of domestic industry (besides aerospace) may visibly support the Space Station.

Second, foreign contributions will reduce the cost to the U.S. of the Space Station program, something that can help us in our budget deliberations with both OMB and the Congress.

On a different level, the fact that foreign governments are willing to put funds into a U.S. program again shows additional support for the Space Station concept.

Our development of foreign cooperative relationships must be consistent with U.S. foreign policy objectives. While making the argument that this is politically feasible and desirable will never be a sole justification for the program, it is important to recognize that it could help NASA bring in members of the foreign policy community as supporters, and help produce a willing Presidential ear.

Congress has consistently been an ardent supporter of international space programs. Foreign involvement in a U.S. Space Station will be kindly received there. Moreover, it may help allay Congressional fears that the civil space program is being unduly influenced by the military. Thus, we could see active Congressional support both before and after Presidential approval of the program.

FOREIGN REACTION TO MILITARY INVOLVEMENT.

This is an important issue, since the interest and debate over the militarization of space is at an all-time high—much more intense than during the Post-Apollo planning activities. Foreign reaction to military involvement in a U.S. Space Station will largely rest on three factors: 1) the nature of the military involvement and the architecture of the Space Station; 2) the manner in which these countries already interact with the U.S. military; and 3) the tradeoff these countries perceive between: (a) the benefits from participating and (b) the domestic and foreign reaction to such participation.

The first depends on the final Space Station design and how the U.S. structures military involvement. If military operational weapons systems are to be part of one U.S. space station, other countries would probably be reluctant to join since doing so would constitute tacit acceptance of weapons activities in space. They might also be concerned that the station could be considered an attractive military target. If DOD use of a single space station were restricted to peaceful military purposes (i.e., reconnaissance and communications), the reluctance would be greatly reduced since both of the above concerns would be lessened. If there are two space stations (one military and one civil), foreign participation on the civil unit should pose no problem to anyone.

In that case, we must then look at our potential partners' current activities with and attitudes toward the U.S. military. [10]

• Most member nations of ESA are also members of NATO (exceptions being Austria, Sweden and Switzerland; France, while still a NATO member, has withdrawn from all NATO military activities). Therefore, most of the ESA member states have a long-standing involvement and NATO commitment to work with the U.S. militarily. Thus, while ESA is a civilian space agency, there is solid foundation among a majority of its member states to cooperate in a program that may have some military aspects. In fact, ESA did make that decision a decade ago when it decided to cooperate with NASA on STS and contribute Spacelab, despite the fact that the U.S. military *would use* STS and possibly Spacelab. In fact, ESA wanted the Spacelab MOU and Intergovernmental Agreement to state that Spacelab would be used by the *U.S. Government* (not just NASA) for peaceful purposes. However, it is important to point out that, on the one hand, Sweden chose not to participate in Spacelab because it did not want to contribute to any system which would be used by the U.S. military; on the other hand, neutral Switzerland has participated. Thus, we can foresee a situation where ESA might sign on, while some of its individual member countries might choose not to participate for political reasons. France has raised an additional concern; that is, the possibility that military involvement would mean that international users could be bumped. In fact, this is a current Ariane claim with respect to STS reliability to provide launch services to domestic and international commercial users. The question of how military involvement would infringe on access rights to the station is a vital issue—probably in the end the single most important factor influencing foreign participation.

•   The Science Minister of the ruling LDP Party in Japan has recently stated that Japan's participation in cooperative projects such as space station would be "unavoidably narrowed" if the U.S. plans to use them for largely military purposes. This statement is not unexpected. Since World War II, Japan has been consistent in not wanting for political and economic reason to divert national resources for military reasons, even [11] if defensive in nature. The LDP is extremely sensitive to opposition party charges that its policies are tied too much to what the U.S. wants. For mainly economic reasons, Japan needs to be highly sensitive to Third World attitudes, including the current focus on "the militarization of space." Japan is critically dependent on the Third World suppliers for virtually all of its energy needs and raw materials. Given Japanese interest in the Third World as both a supplier and a consumer, Japan could thus be expected to be very cautious about participating with the U.S. in a space station perceived as largely military. However, the above has to be balanced with Japan's strong ties to the U.S. for defense. Therefore, Japan's participation will largely depend on the tradeoff between the benefits it sees from a Space Station and potential domestic and foreign negative reaction if the station has obvious military roles. Japan's assessment of involvement in a Space Station will also be driven by a frank eagerness to join the U.S. and other developed countries in the next major step, since Japan feels it missed a key opportunity to participate in the Shuttle.

•   Canada is probably the country that would least object to any military involvement in a space station. Canada is also a member of NATO, but even more than that, is part of NORAD and has several defense sharing agreements with the U.S. The line between civil and military for Canada is probably slightly fuzzier than ours. Furthermore, Canada strongly supports DOD use of [the Remote Manipulator System (RMS)], and would work hard to ensure that DOD did not use an alternative.

When making an analysis like this, we must keep in mind that this is the situation as we perceive it today. Who knows what the political situation might be like a few years from now when we are ready to make a commitment to cooperate? In the interim, these countries will stay in the game because they do not want to be the only developed country to miss out and because they want to make sure they are ready to participate when the time comes to sign on the dotted line. It is at that point that each country will weigh the pros and cons of their participation.

From NASA's perspective, I believe it is important to be fairly straightforward at all times on the probability and level of DOD involvement expected. Since NASA wants to maintain and even strengthen the civil role of space activities in the next few years, it is to our advantage to actively seek and encourage international civil involvement in our next major step. We should be working to accommodate both civil and military uses within the basic design of the space station, so that one does not make the other impossible.

[12] *TECHNOLOGY TRANSFER.*

The greatest source of technology transfer, in my mind, is through industry to industry relationships. NASA's cooperative programs have been structured carefully to avoid technology transfer. Historically, our partners have agreed to provide a discrete piece of the overall project, and have then been fully responsible for the R&D on that piece. Only the minimum amount of technical information necessary to achieve a successful interface among the various elements of a project has been exchanged.

Secondly, while it might have been true ten years ago that U.S. industry was several years ahead of foreign industry overall, I do not think the same claim can be made today. During the past decade, we have seen measurable growth in foreign space budgets and capabilities. European, Japanese and Canadian industries are challenging U.S. industry in several fields: communications, remote sensing and launch vehicle development. We see increasing evidence that foreign governments are adopting sophisticated strategies to enhance their aerospace industries' competitive positions. Many foreign governments support their space industry not only through direct R&D funding (which often is targeted to areas with demonstrable commercial payoff), but also by price subsidization and financing assistance, development of attractive package deals, and creation of quasi-governmental marketing organizations. As a result, the U.S. probably stands to gain as much as our potential partners.

I want to reemphasize what I said earlier, that one of the more important criteria we should use in evaluating specific proposals for cooperation is that the cooperating country has the necessary industrial and technological infrastructure to successfully complete the job. If we carefully choose the cooperative arrangements—for example, we might make sure that they are discrete hardware pieces with minimal interfaces—we can minimize the potential for technology transfer in the normal conduct of the project.

However, even if we at NASA are satisfied we have structured a program which minimizes the opportunity for technology transfer, we must be sensitive to the growing interest in this topic throughout the government. Evidence of closer application of existing export guidelines and review of appropriate future steps in staunching the flow of advanced technology is readily apparent. In a long-term, multi-faceted program of this type, we must maintain close and continuing contacts with the export control community. Thus, we must keep the export control community continually informed on our activities and our efforts to protect against technology transfer. As I mentioned above, this process has been initiated.

[13] *POTENTIAL FOREIGN CONTRIBUTIONS TO SPACE STATION.*

An assessment of potential foreign contributions to a U.S. Space Station can only be a speculative one. Foreign decisions and commitments on participation will be reached during the end of Phase B, at least two years away. Impacting each country's decision will be the domestic and international economic situation at that time.

However, it is possible to make some assumptions based on the size of foreign space budgets and the level of contributions already made to NASA STS-related programs. Ultimately, the size of the contributions will be related to the potential benefits perceived by the contributors and the terms of cooperation proposed by NASA.

ESA's current annual budget is approximately $750 million a year. In addition, the combined space budgets of the ESA Member States is approximately $1.5 billion, apportioned between ESA contributions and individual space programs.

Canada recently increased its space budget by one third to almost $500 million for the next four years. Japan's annual space budgets for recent years have been on the order of $500 million and could be expected to remain at least at that level.

In sum, our potential partners now have moderate-sized space budgets that have greatly increased over the past decade, reflecting a realization by these nations of the importance and benefits from space activities.

Our STS partners contributed roughly 11-12% of the cost of the development program. ESA contributed a $1 billion Spacelab and Canada contributed a $100 million RMS. Italy currently plans to contribute $30 million to Tether. I believe it is reasonable to expect similar percentage contributions from these countries to Space Station, if they choose to cooperate. Japan's GNP is roughly half that of the ESA Member States. Thus, it is not unrealistic to expect them to contribute half the European contribution. Furthermore, Japanese space industry has advocated doubling its space budget in the near future. Therefore, a 20% increase (approximately $100 million/year) for Space Station activities would not be unrealistic, given a strong Japanese industrial interest in a Space Station.

We should keep in mind that other space activities and comparable competing concepts for these funds exist. The Canadian Minister for Science and Technology recently told Mr. Beggs that while Canada is interested in cooperating on a Space Station, Canada is already planning several communications and remote sensing missions. Its economy would have to improve before it could take on new space projects. ESA will be considering additional Ariane upgrades at the same time it will [14] consider participating in a NASA Space Station. France has been studying its own robotics-space station, Solaris.

Ultimately, the willingness of these countries to contribute will depend on both prevailing economic conditions and the perceived benefits. Foreign partners will be willing to consider large investments only if they will lead to direct quid pro quos which are highly attractive, such as preferred or free access to the station, and also to spin off benefits which magnify the returns to their industry.

I would be happy to discuss with you any of these topics in greater detail.

Kenneth S. Pedersen

Document I-32

Document title: NASA Fact Sheet, "Space Assistance and Cooperation Policy," August 6, 1982.

Source: NASA Historical Reference Collection, NASA History Office, NASA Headquarters, Washington, D.C.

This statement of U.S. policy concerning launch assistance and international cooperation was an update and revision of a similar policy approved by the Nixon administration on August 30, 1972, and contained in National Security Decision Memorandum 187. The earlier policy statement formalized the modified U.S. approach to international space issues adopted in the post-Apollo period, and the 1982 revision made few changes in the basic principles set out a decade earlier.

---

[1]   August 6, 1982

Space Assistance and Cooperation Policy

I.   INTRODUCTION
The fundamental aspects of National Security Decision Memorandum (NSDM) 187 of August 30, 1972, as they apply to today's international space activity have been reviewed. This review highlighted the substantial lead the U.S. enjoys in a wide variety of technological and space related areas—a lead which should be maintained when considering and implementing any international activity or transfer governed by the following directive. Based upon this review, this directive which replaces NSDM 187 is approved and provides general guidance for U.S. space launch assistance; space hardware, software and related technologies assistance; and international space cooperation. Specific implementing guidelines are being issued by the Assistant to the President for National Security Affairs.

II. *POLICY GOVERNING SPACE LAUNCH ASSISTANCE*
   In dealing with requests from foreign governments, international organizations or foreign business entities for assistance in launching foreign spacecraft, the following general policy guidance is provided.
   [Paragraph excised in declassification review]

[2] III. *POLICY GOVERNING SPACE HARDWARE AND RELATED TECHNOLOGIES ASSISTANCE*
   In dealing with requests for the transfer of, or other assistance in the field of space hardware, software and related technologies, the following general policy guidance is provided.
   Sales of unclassified U.S. space hardware, software, and related technologies for use in foreign space projects will be for peaceful purposes; will be consistent with relevant international agreements and arrangements and relevant bilateral agreements and arrangements; [phrase excised in declassification review] will contain restrictions on third country transfers; will favor transfers of hardware over transfers of technology; will not adversely affect U.S. national security, foreign policy, or trade interests through diffusion of technology in which the U.S. has international leadership; and will continue to be subject to the export control process. A special interagency coordinating group chaired by the Department of State will be established to consider special bilateral agreements covering the transfer of space hardware, software, and related technologies.

IV. *OBJECTIVES OF INTERNATIONAL COOPERATION IN SPACE ACTIVITIES*
   The broad objectives of the United States in international cooperation in space activities are to protect national security; promote foreign policy considerations; advance national science and technology; and maximize national economic benefits, including domestic considerations. The suitability of each cooperative space activity must be judged within the framework of all of these objectives.

[Attachment page 1]

Implementing Guidelines to the Space Assistance and Cooperation Policy

A. *Policy Governing Space Launch Assistance*
   1. Space launch assistance will be available, consistent with U.S. laws, either from U.S. launch sites through the acquisition of U.S. launch services on a cooperative or reimbursable basis or from foreign launch sites by purchase of an appropriate U.S. launch vehicle (see policy guidance under Section B). In the case of launchings from foreign sites, the U.S. will require assurance that the launch vehicles will be used solely for peaceful purposes and will not be made available to third parties without prior agreement of the U.S.
   2. Although due consideration is to be given to Intelsat definitive arrangements, the absence of a favorable Intelsat recommendation regarding such arrangements should not necessarily preclude U.S. launching of public domestic or international telecommunications satellites when such launching is determined to be in the best interests of the U.S.
   3. With respect to the financial conditions for reimbursable launch services from U.S. launch sites, foreign users (including international organizations) will be charged on the same basis as comparable non-U.S. Government domestic users.

4. With respect to the priority and scheduling for launching foreign payloads at U.S. launch sites, such launchings will be dealt with on the same basis as U.S. launchings. Each launching will be treated in terms of its own requirements and as an individual case. Once a payload is scheduled for launch, the launching agency will use its best efforts to meet the scheduling commitments. Should events arise which require rescheduling, such as national security missions, the U.S. will consult with all affected users in an attempt to meet the needs of the users in an equitable manner.

5. Interface drawings and hardware (i.e., spacecraft attach fittings, etc.) provided in connection with the launch assistance provisions of this policy shall be exempt from the provisions of Section B.

B. *Space Hardware, and Related Technologies Assistance*
    1. For the purpose of this policy, the following distinctions are recognized:
[2]   a. Hardware, software, and related technical information include:
        (1) Equipment in the form of launch vehicle components and spacecraft including subsystems and components thereof, associated production and support equipment.
        (2) General physical and performance specifications, and operating and maintenance information on the above equipment.
    b. Technical assistance technology, data and know-how necessary for design, development and production of space hardware and software, including pertinent laboratory and test equipment or performance of functions and/or the conveyance of oral, visual or documentary information involving the disclosure of information relating to:
        (1) Development and testing activities, detailed design drawings and specifications, managerial and engineering know-how and problem solving techniques.
        (2) Production activities in the form of licenses, detailed production drawings, process specifications, and identification of requirements for production equipment.
    2. [Sentence excised in declassification review] This does not mean that transfer of certain "technical assistance" under appropriate safeguards should not be considered on a case-by-case basis. In those cases in which "technical assistance" is provided, it should be done under safeguards which ensure protection of U.S. national security and foreign policy interests. Thus, whether the sale involves "hardware, software and related technical information," or "technical assistance," or some combination, adequate assurances to control replication and retransfer and ensure peaceful use must be provided in advance of the transfer through bilateral agreements, export licensing procedures or other mechanisms. [Sentence excised in declassification review]
[3] 3. All requests for the export or exchange of either space "hardware, software and related technical information" or "technical assistance" as defined above must specify the end use for which it is sought.
    4. All such requests shall be examined on a case-by-case basis in accordance with applicable U.S. laws and regulations to determine the net advantage to the U.S. The determination shall take into account relevant international agreements and arrangements, relevant bilateral agreements and arrangements, and our objectives for international cooperation in space activities (see Section C).
    5. U.S. space "hardware, software and related technical information" or "technical assistance" as defined above shall be made available solely for peaceful purposes. No U.S. space "hardware, software and related technical information" or "technical assistance" as defined above shall be made available by a recipient to a third party without the express prior agreement of the U.S. This includes any cases where U.S. space hardware is launched from a foreign site.

6. U.S. space "hardware, software and related technical information" or "technical assistance" as defined above, or any hardware, software, or technical information and processes derived from such transfers, will not be used to contribute to or assist in the development of any foreign weapon delivery system. Further, any officially promulgated national security policy directive is overriding with respect to the transfer of military-related missile hardware, information or technology within its purview.

7. In view of the sensitivity of space technology, the following distinctions shall be applied in reaching decisions as to its export. These distinctions shall apply both to transfer abroad by federal agencies and to commercial export.

    a. Proposals or requests for the export of space "hardware, software and related technical information" should be met, when in the interests of the U.S., through the provision of "hardware, software and related technical information" rather than "technical assistance" as defined above, whenever possible and reasonable to do so.

    b. "Technical assistance" as defined above shall be exported only under adequate safeguards providing for its use and protection.

[4] 8. In instances where space "hardware, software and related technical information" and "technical assistance" are intended specifically for use in operational communication satellite projects to provide public domestic or international telecommunications services, its export shall be governed as specified in Section III of the Space Assistance and Cooperation Policy and Section A, paragraph 2 above.

9. Recognizing distinct U.S. national interests, special bilateral agreements covering the transfer of space launch vehicle "hardware, software and related technical information" or "technical assistance" may be considered under the following guidance:

    a. The Department of State will convene and chair a special interagency coordinating group consisting of representatives from DOD, ACDA, NASA, NSC, OSTP, DCI, and other interested agencies as appropriate to recommend policy and to decide upon, formulate, negotiate, and provide general guidance on implementation oversight activities regarding bilateral agreements covering transfer to selected foreign governments and international organizations.

    b. Such agreements with selected foreign governments and international organizations will contain provisions for peaceful use assurances, restrictions on third country transfers and other appropriate safeguards as may be deemed necessary and mutually agreed.

    c. Any agreements that would result in funding demands on the U.S. Government must be approved through the budgetary process prior to any commitment with a foreign entity.

    d. Transfer of specific space "hardware, software and related technical information" and "technical assistance" under such agreements will continue to be subject to the export control review process.

10. The U.S. should encourage other supplier nations of space "hardware and related technical information" and "technical assistance" to establish controls on their exports which are comparable to those set forth in this policy.

C. *Objectives of International Cooperation in Space Activities National Security Objectives*
    [Paragraph excised in declassification review]

[5] *Foreign Policy Objectives*
    a. To gain other countries' support for the U.S. in general by promoting the U.S. national interest through bilateral and multilateral cooperation.
    b. To assist in the achievement of foreign policy objectives through:

(1) Strengthening our allies and improving our working relationships with them.
(2) Promoting multilateral cooperation with, and among, other nations similar to on-going U.S. cooperation with the European Space Agency through suitable cooperation with their programs, on a commercial or joint program basis, in the event they desire such cooperation.

c.   To encourage other countries to associate their interests with our space program.

d.   To enhance U.S. prestige and ensure the U.S. position as the world's leader in science and technology.

e.   [Paragraph excised during declassification review]

f.   To demonstrate that the U.S. is a reliable partner in international ventures.

Scientific and Technological Objectives

a.   To foster cooperation in basic scientific research.

b.   To develop precedents and experience in substantial cooperative undertakings which will lend themselves to other international scientific and technological activities.

[6]   c.   To obtain support and assistance in the development of our national program through (1) acquisition of scientific and technical contributions from areas of excellence abroad and (2) use of facilities abroad that are necessary for mission support—tracking stations, overflights, contingency recovery, etc.

Economic Objectives

a.   To maximize economic benefit by appropriately weighing:
(1) Implications of releasing technology which involves commercial "know-how";
(2) [Paragraph excised during declassification review]
(3) ensuring a reasonable return on the American investment in space technology; and
(4) promoting positive effects on domestic employment and our balance of payments.

b.   [Paragraph excised during declassification review]

c.   To seek opportunities to enhance our overall competitive position in space technology.

d.   To seek more productive aggregate use of American and foreign resources and skills.

e.   [Paragraph excised during declassification review]

f.   To enhance the cost-effectiveness of space systems through increased and more effective use.

D.   Effective immediately, National Security Decision Memorandum 187 is rescinded.

[Attachment page 1]

Fact Sheet
Space Assistance and Cooperation Policy

Introduction
On August 6, 1982, the President signed a directive which establishes U.S. national space assistance and cooperation. This policy directive highlights the substantial lead the U.S. enjoys in a wide variety of technological and space related areas—a lead which should be maintained when considering and implementing any international activity or transfer. This directive provides general guidance for U.S. space launch assistance; space hardware, software and related technologies assistance; and international space cooperation.

Policy Governing Space Launch Assistance
In dealing with requests from foreign governments, international organizations or foreign business entities for those space projects which are for peaceful purposes and are consistent with U.S. laws and obligations under relevant international agreements and arrangements (such as Intelsat) as determined by the U.S. Government.

Policy Governing Space Hardware, and Related Technologies Assistance
In dealing with requests for the transfer of, or other assistance in the field of space hardware, software and related technologies, the following general policy guidance is provided.

Sales of unclassified U.S. space hardware, software, and related technologies for use in foreign space projects will be for peaceful purposes; will be consistent with relevant bilateral and international agreements and arrangements; will serve U.S. objectives for international cooperation in space activities (see the following section); will contain restrictions of third country transfer; will favor transfers of hardware over transfers of technology; will not adversely affect U.S. national security, foreign policy, or trade interests through diffusion of technology in which the U.S. has international leadership; and will continue to be subject to the export control process. The Department of State will chair an interagency coordinating group when it becomes necessary to consider bilateral agreements which cover the transfer of space hardware, software, and related technologies.
[2] *Objectives of International Cooperation in Space Activities*
The broad objectives of the United States in international cooperation in space activities are to protect national security; promote foreign policy consideration; advance national science and technology; and maximize national economic benefits, including domestic considerations. The suitability of each cooperative space activity must be judged within the framework of all these objectives.

Document I-33

Document title: James M. Beggs, NASA Administrator, to Honorable George P. Shultz, Secretary of State, March 16, 1984.

Source: NASA Historical Reference Collection, NASA History Office, NASA Headquarters, Washington, D.C.

From March 3 to 13, 1984, a NASA delegation led by Administrator James Beggs visited London, Rome, Bonn, Paris, and Tokyo to extend personally President Ronald Reagan's invitation to U.S. "friends and allies" to participate in the U.S. space station program that Reagan had announced in his January 25, 1984, State of the Union address. (The group visited Ottawa later in March.) In this letter to the Secretary of State, Beggs reported the results of his trip.

[1]   MAR 16, 1984

Honorable George P. Shultz
Secretary of State
Washington, DC 20520

Dear George:

As you recall, the President recently asked me to visit certain of our Allies to invite international participation in our Space Station program. This followed up, of course, on the President's announcement of this initiative during January's State of the Union address. I've just come back from Europe and Japan. Before heading off to Ottawa next week, I wanted to fill you in on the first stage of our consultations.

The reaction so far to the President's call for international cooperation has been both strongly positive and openly appreciative. It has been positive in the sense that our principal Allies are moving quickly, or have already moved, to take political decisions to participate. And their reactions clearly show appreciation for the major foreign policy benefits that will flow from open and collaborative cooperation on such a bold, imaginative and visible project. I heard nothing but praise and admiration for the President's foresight and leadership in making this decision. Prime Minister Nakasome and other Japanese officials, while still cautious in public, made it obvious that Japan will participate in a significant way. The Japanese believe they made a mistake in not joining us on the Shuttle and are determined not to be left behind again. In Europe, Italian Prime Minister Craxi was openly ebullient about the prospect of cooperation and strong Italian participation is assured. Mitterrand and Cheysson were both well informed and prepared to move ahead. Mitterrand, in particular, has obviously thought deeply about the need to press ahead with the exploration and exploitation of space. The French will be tough bargainers, and obviously intend to pursue their own independent space programs, but I am confident that we can agree on mutually beneficial terms for cooperation. By the way, you will be interested that Mitterrand observed to me that his recent proposal for a European military space station fell on deaf ears in Europe.

[2] Chancellor Kohl was in Washington during my stop in Bonn, but the relevant Ministers were quite clear that a major German contribution will be forthcoming. The British were more cautious, and, while I believe they will participate, it will probably be on the same terms that have marked their recent space-related activities—relatively small scale projects done on a multilateral basis.

While in Paris, I also met with the executive leadership of the European Space Agency (ESA) and with delegates from the Agency's eleven member countries—encompassing essentially all of our friends and allies in Western Europe. Here, too, the reception was warm and positive. ESA will almost certainly play a key role in managing Europe's Space Station participation, just as it did in the highly successful Spacelab project.

As businessmen, we both understand the importance of protecting intellectual property if we're to motivate private sector investment in this program. Not surprisingly, the Europeans and the Japanese are as concerned about this—from their point of view—as we are. The whole technology transfer question will obviously be an area where I will look to you, and other relevant agencies, for advice as discussions on the details of cooperation get more specific in the months ahead. I also explained our policy on the possibility of military use of the Space Station. I was pleased to find, even in Japan where the need for caution is clear, general acceptance of our announced position: that while no military use is contemplated, the Space Station will be a national facility open to any paying customers—including DOD—for peaceful uses.

As a final item, I raised the President's desire to have the London Economic Summit endorse the principle that members will cooperate in developing an international Space Station. Germany, Italy, France and Japan were all supportive. Again, the British were more cautious and will need more convincing. The next step here—as laid out by Bud McFarlane—is for NASA, with State's help, to prepare a report on approaches to international collaboration before the Summit. I plan to present that report to the President and also to report to him on my trip. I hope you will join me in that meeting.

I'd like to thank you for the excellent support provided by the Department and by our Embassies at every step of the way. I especially want to express my appreciation and gratitude for the fine work done by Mark Platt and Mike Michalak who accompanied me on the trip. They are true professionals whose [3] involvement was instrumental in helping to produce the positive reception the President's initiative received. I look forward to continuing to work with you and your staff in the months ahead in the same productive and cooperative spirit.

Sincerely,

James M. Beggs
Administrator

Document I-34

Document title: James M. Beggs, Administrator, NASA, to Kenneth Baker, MP, Minister of State for Industry with Special Responsibility for Space and Information Technology, April 6, 1984.

Source: NASA Historical Reference Collection, NASA History Office, NASA Headquarters, Washington, D.C.

After his initial round of visits in Europe, Japan, and Canada to extend President Reagan's invitation to participate in the U.S. space station program, NASA Administrator Beggs wrote essentially identical letters to the most senior official with whom he had met in each country visited. The following is the letter to the minister in charge of space matters for the United Kingdom (U.K.), Kenneth Baker. In his letters, Beggs spelled out what he believed were the results of his visits, and he restated the basic U.S. policy with respect to the station program and international participation in it. He also outlined the next steps in the process of developing international station partnerships.

[1] Office of the Administrator   APR 6, 1984

Mr. Kenneth Baker, MP
Minister of State for Industry with Special Responsibility
  for Space and Information Technology
Department of Industry
Ashdown House
123 Victoria Street
London SW1E 6RB
UNITED KINGDOM

Dear Mr. Minister:

Having recently returned from my visit to Europe, Japan and Canada, I wish to take this opportunity to summarize my impressions of the trip and to express my appreciation for your generous hospitality. Overall, I was extremely pleased by the reactions I received to President Reagan's Space Station initiative. Government and industry leaders at each of my stops exhibited great interest in the possibilities for an international Space Station. I believe this reflects the successful legacy of cooperation already established among us in the space age, as well as the groundwork we have laid together over the past two years for embracing this challenge. I hope you feel as I do that our discussions were quite useful for getting the dialogue started for our next step in the planning process. I am quite optimistic about the prospects for international cooperation on the Space Station project, and will soon be sharing these views with the President.

As we discussed, the President believes that international participation in the manned Space Station program can provide a highly positive centerpiece for demonstrating Free World unity, goodwill and technological progress. He has proposed that the international Space Station be discussed at the London Economic Summit with an eye towards agreement in principle that Summit partners will participate in the development of the station. A Summit declaration will serve us all well by establishing the political underpinnings for this joint technological venture. With this firm basis for our collaboration, we will be able to arrange mechanisms that will allow us to interact more closely during the planning phase of the Space Station project.

I believe that our working closely together over the next year is extremely important. This will ensure that our respective planning activities and definition studies are [2] complementary. During the next two years, NASA will conduct an extended definition phase study of the Space Station in order to design the Station best capable of meeting requirements, facilitating management and providing flexibility for growth. As time goes on, there will be less and less flexibility in the Space Station design to accommodate the interests and needs of potential partners. Early participation in the planning process, either directly or through ESA, is therefore essential. I believe insight into this planning process will allow participants to hone their ideas for participation; it will also allow them to speak directly to their proposals so that the final Space Station design can accommodate them.

As I mentioned, NASA will hold frequent international workshops over the next two years to permit this cross-fertilization to occur. We will hold the first such workshop in June at which time we can all review our activities. For our part, we will brief you on our preparation of the domestic U.S. "Request for Proposals" for Phase B. These RFP's will cover the $8B fully functional Space Station that the U.S. will provide. As I described to you, President Reagan has committed the U.S. to building an $8B fully functional Space Station to be operational by the early 1990's, but has also set the stage for working together to develop a more expansive international Space Station with even greater benefits and capabilities for us all to use. Thus, we are inviting your Government to take a close look at our plans and concepts and then, based on your long-term interests and goals, share with us your ideas for cooperation that will expand the capabilities of the Space Station.

Also during the June meeting, we will discuss additional mechanisms for working together over the next two years. In the course of my trip, I heard many proposals for such mechanisms which we are currently evaluating. Mr. Kenneth Pedersen, NASA Director of International Affairs, will be contacting you in the near future with the necessary details on the June meeting.

During the past 18 months, we have worked hard to make sure that our Space Station concepts are compatible with and responsive to user communities. We will continue this emphasis in the next two years of planning as well as throughout the lifetime of the Space Station program. As I mentioned, the U.S. is committed to maintaining a strong space science and applications program. I have received a commitment from the President that the NASA budget will grow 1% per year in real terms in order to maintain a balanced space program. Indeed, this year, the President requested Congress to authorize two new starts in space science along with the Space Station.

Because I understand that the relationship between scientific objectives and the Space Station program is [3] important to you, I would be pleased for you to designate an observer to our Space Station Science Advisory Committee, chaired by Professor Peter Banks, which was recently established to assist NASA in scientific planning for the Space Station. As you know, ESA has already designated two observers, so you may wish to work through them. The first meeting of the Banks Committee will take place at NASA Headquarters on April 25 and 26. A second meeting is planned in June. There will also be a week-long workshop held later this summer. One of the key early objectives of the Committee is to influence the Space Station Phase B RFP so that the Space Station is designed to optimize space science and application uses. In addition, an Industrial Committee similar to the Peter Banks Committee will be established to ensure that the Space Station maximizes the commercial opportunities of space, another important objective that we all share. We will welcome observers on that Committee, as well. Once we agree more formally on our respective activities during the planning phase, then we would look forward to having our partners as permanent members on both Committees.

Another topic which we discussed is the importance of protecting against the unwarranted transfer of technology. Technology transfer has been an increasing concern on all our parts in the past few years, and we will need to work together to make sure we are protecting our respective technology bases in this partnership. Major international partners in the Space Station will receive assured access to the Station. Therefore, protection of intellectual property is a prime requirement if we are to stimulate private sector investment and involvement in this program over the long term.

During my trip I was also asked frequently about the extent of U.S. military involvement in the U.S. Space Station. The U.S. Space Station program is a civil program which will be funded entirely out of NASA's budget, with no national security funds to be used. While the Defense Department worked with NASA in the early planning for [the] Space Station by reviewing their near- and long-term requirements for space, they concluded they had no requirements for a manned Space Station. NASA, therefore, constructed its proposal to the President on the basis of civil and commercial requirements. The Space Station that the President directed NASA to build is a civil Space Station. Of course, like the Shuttle, the Space Station will be available for users. If there are any national security users, like national and international users, they will be able to pay to use the facility. As provided in the Outer Space Treaty, however, all activity on the Space Station will be limited to peaceful, non-aggressive functions.

Finally, on behalf of the U.S. Delegation, I would like to thank you for your gracious hospitality during our visit. I especially appreciated your giving me the opportunity to meet [4] numerous U.K. Government and industrial representatives at the fine luncheon you hosted. It was a pleasure seeing you, and I am looking forward to seeing you again soon in the near future.

With warm personal regards.

Sincerely,

James M. Beggs
Administrator

Document I-35

Document title: Dale D. Myers, NASA, and Reimar Leust, European Space Agency (ESA), "Memorandum of Understanding Between the United States National Aeronautics and Space Administration and the European Space Agency on Cooperation in the Detailed Design, Development, Operation and Utilization of the Permanently Manned Civil Space Station," September 29, 1988.

Source: NASA Historical Reference Collection, NASA History Office, NASA Headquarters, Washington, D.C.

This NASA-ESA memorandum of understanding (MOU) and two similar documents—one between NASA and the Science and Technology Agency of Japan and the other between NASA and the Canadian Ministry of State for Science and Technology—contained the detailed agreements that would guide the international partners during the lifetime of the space station program. The MOUs were the end product of lengthy and contentious negotiations between NASA and its potential station partners. These MOUs operated within a policy and legal framework established by a multilateral intergovernmental agreement signed at the same time by representatives of the governments (rather than of the respective space agencies). The intergovernmental agreement and the three MOUs established the most ambitious experiment in international technological cooperation ever undertaken.

Memorandum of Understanding Between the United States National Aeronautics and Space Administration and the European Space Agency on Cooperation in the Detailed Design, Development, Operation and Utilization of the Permanently Manned Civil Space Station

[1] The National Aeronautics and Space Administration (hereinafter "NASA") and the European Space Agency (hereinafter "ESA"),

Recalling that in his State of the Union Address of January 25, 1984, the President of the United States directed NASA to develop and place into orbit within a decade a permanently manned Space Station and invited friends and allies of the United States to participate in its development and use and to share in the benefits thereof, in order to promote peace, prosperity and freedom,

Recalling the terms of Resolution Number 2 adopted on 31 January 1985 by the ESA Council meeting at ministerial level on participation in the Space Station program,

Recalling the terms of Resolution Number 2 adopted on 10 November 1987 by the ESA Council meeting at ministerial level on participation in the Space Station program,

Recalling the NASA Administrator's letter of April 6, 1984, to the ESA Director General,

Having successfully implemented the Memorandum of Understanding between NASA and ESA for the Conduct of Parallel Detailed Definition and Preliminary Design Studies (Phase B) Leading toward Further Cooperation in the Development, Operation and Utilization of a Permanently Manned Space Station, which entered into force on June 3, 1985,

Considering the Agreement among the Government of the United States of America, Governments of Member States of the European Space Agency, the Government of Japan and the Government of Canada on Cooperation in the Detailed Design, Development, Operation and Utilization of the Permanently Manned Civil Space Station (hereinafter "the Intergovernmental Agreement") and particularly Article 4 thereof,

[2] Considering the Memorandum of Understanding between NASA and the Science and Technology Agency of Japan (STA) for the Cooperative Program Concerning Detailed Definition and Preliminary Design Activities of a Permanently Manned Space Station, which entered into force on May 9, 1985, and the Memorandum of Understanding between NASA and the Ministry of State for Science and Technology of Canada (MOSST), for a Cooperative Program Concerning Detailed Definition and Preliminary Design (Phase B) of a Permanently Manned Space Station, which entered into force on April 16, 1985,

Considering the Memorandum of Understanding between NASA and the Government of Japan (the GOJ) on Cooperation in the Detailed Design, Development, Operation and Utilization of the Permanently Manned Civil Space Station and recognizing that the GOJ has designated STA in that Memorandum of Understanding as its Cooperating Agency, as provided for in Article 4 of the Intergovernmental Agreement,

Considering also the Memorandum of Understanding between NASA and MOSST on Cooperation in the Detailed Design, Development, Operation and Utilization of the Permanently Manned Civil Space Station,

Convinced that this cooperation among NASA, ESA, STA and MOSST implementing the provisions established in the Intergovernmental Agreement will further expand cooperation through the establishment of a long-term and mutually beneficial relationship and will further promote cooperation in the exploration and peaceful use of outer space,

Have agreed as follows:

Article 1 - Purpose and Objectives

1.1.   The purpose of this Memorandum of Understanding (MOU) is, pursuant to Article 4 of the Intergovernmental Agreement and on the basis of genuine partnership, to establish arrangements between NASA and ESA (hereinafter "the Parties") implementing the provisions of the Intergovernmental Agreement concerning the detailed design, development, operation and utilization of the permanently manned civil Space Station for peaceful purposes, in accordance with international law. In drafting this MOU, the Parties intended it to be consistent with the provisions of the Intergovernmental Agreement. This MOU will be subject to the provisions of the Intergovernmental Agreement. It defines [3] the nature of the genuine partnership, including the respective rights and obligations of the Parties to this MOU.

1.2.   The specific objectives of this MOU are:
-   to detail the roles and responsibilities of NASA, ESA, STA and MOSST (hereinafter the "partners") in the detailed design, development, operation and utilization of the Space Station and also to record the commitments of NASA and ESA to each other and to STA and MOSST;
-   to establish the management structure and interfaces necessary to ensure effective planning and coordination in the conduct of the detailed design, development, operation and utilization of the Space Station;
-   to provide a framework that maximizes the total capability of the Space Station to accommodate user needs and that ensures that the Space Station is operated in a manner that is safe, efficient and effective for both Space Station users and Space Station operators; and
-   to provide a general description of the Space Station and the elements comprising it.

1.3.   Relevant definitions and explanations are to be found in Article 22.

Article 2 - General Description of the Space Station

2.1.   NASA has a Space Station program which will produce a core U.S. Space Station. ESA has a Columbus program, and STA and MOSST also have space programs to produce significant elements which, together with the core U.S. Space Station, will create an international Space Station complex with greater capabilities that will enhance the use of space for the benefit of all participating nations and humanity. MOSST's contribution will be an essential part of the infrastructure of the permanently manned civil international Space Station complex (hereinafter "the Space Station").

2.2.   The Space Station will be a unique, multi-use facility in low-Earth orbit, comprising both manned and unmanned elements: a permanently manned base comprising elements provided by all the partners; unmanned platforms in near-polar orbit; a mantended free-flying laboratory to be serviced at the manned base; and Space Station-unique ground elements to support the operation and utilization of the elements on orbit.

[4] 2.3. The Space Station will enable its users to take advantage of human ingenuity in connection with its low-gravity environment, the near-perfect vacuum of space and the vantage point for observing the Earth and the rest of the Universe. Specifically, the Space Station and its evolutionary additions could provide for a variety of capabilities, for example:
-   a laboratory in space, for the conduct of science and applications and the development of new technologies;
-   a permanent observatory, with elements in low inclination and near-polar orbits, from which to observe Earth, the Solar System and the rest of the Universe;
-   a transportation node where payloads and vehicles are stationed, assembled, processed and deployed to their destination;
-   a servicing capability from which payloads and vehicles are maintained, repaired, replenished and refurbished;
-   an assembly capability from which large space structures and systems are assembled and verified;
-   a research and manufacturing capability in space, where the unique space environment enhances commercial opportunities;
-   an infrastructure to encourage commercial investment in space;
-   a storage depot for consumables, payloads and spares; and

– a staging base for possible future missions, such as a permanent lunar base, a manned mission to Mars, unmanned planetary probes, a manned survey of the asteroids, and a manned scientific and communications facility in geosynchronous orbit.

Article 3 - Space Station Elements

3.1. The Space Station will consist of elements comprising both flight elements and Space Station-unique ground elements. The elements are summarized in the Annex to the Intergovernmental Agreement and are further elaborated in this Article. Their requirements are defined and controlled in appropriate program documentation as provided for in Article 7.

[5] 3.2. NASA Space Station Flight Elements: NASA will design, develop and provide the following flight elements including subsystems, the Extra Vehicular Activity (EVA) system, the Space Station Information System, flight software and spares as required:
– one permanently attached Habitation Module with complete basic functional outfitting to support habitation for a crew of up to eight, including primary storage of crew provisions
– one permanently attached multipurpose Laboratory Module, located so as to contain the center of gravity of the manned base, with complete basic functional outfitting and including provisions for storage of NASA spares, secondary storage of crew provisions, and storage for safe haven capability
– two sets of Attached Payload Accommodation Equipment for accommodation of payloads externally attached to the Space Station Truss Assembly
– four Resource Nodes which provide pressurized volume for crew and equipment, connections between manned base pressurized elements and support of pressurized attached payloads
– Truss Assembly which is the manned base structural framework
– Solar Photovoltaic Power Modules which serve as the manned base electrical power source, providing 75kw of total power
– Propulsion Assembly
– at least three sets of Logistics Elements (pressurized and unpressurized Integrated Logistics System carriers) which provide systems operation support and user ground-to-orbit and return logistics and on-orbit supply for extended periods
– Airlock/Hyperbaric Airlock for purposes of crew and equipment transfer
– one Flight Telerobotic System (FTS)
– one Mobile Transporter which will serve to provide translation capability for the Mobile Servicing Center
– one Polar Platform to work together with the ESA-provided Polar Platform
[6] 3.3. ESA Space Station Flight Elements: ESA will design, develop and provide the following flight elements including subsystems, flight software and spares as required:
– one Attached Pressurized Module (APM), with volume equivalent to that of four Spacelab segments, permanently attached to the manned base, with complete basic functional outfitting and including provisions for storage of ESA spares, secondary storage of crew provisions, and storage for safe haven capability
– one Polar Platform to work together with the NASA-provided Polar Platform
– one Man-Tended Free Flyer (MTFF), including a pressurized module, with volume equivalent to that of two Spacelab segments, capable of autonomous operational periods of six months or longer

3.4.   STA and MOSST Space Station Flight Elements: As reflected in the MOU between NASA and the GOJ and in the MOU between NASA and MOSST:

3.4.a.   STA Space Station Flight Elements: STA will design, develop and provide the following flight elements including subsystems, flight software and spares as required:

    –   one Japanese Experiment Module (JEM), a permanently attached multipurpose research and development laboratory, consisting of a pressurized module and an Exposed Facility, at least two Experiment Logistic Modules, and including a scientific equipment airlock, the JEM remote manipulator and IVA control/monitoring of the JEM Remote Manipulator System (JEM-RMS), with complete basic functional outfitting, including provisions for storage of STA spares, secondary storage of crew provisions, and storage for safe haven capability

3.4.b.   MOSST Space Station Flight Elements: Canadian elements will be developed to play the predominant role in satisfying the following functions for the Space Station:

    –   attached payload servicing (external)
    –   Space Station assembly
    –   Space Station maintenance (external)
    –   transportation on Space Station
    –   deployment and retrieval functions
    –   EVA support

3.4.b.1. MOSST will design, develop and provide the following flight elements, including subsystems, flight software and spares as required:

[7]   –   one Mobile Servicing Center (MSC) which comprises a Mobile Remote Servicer (MRS) and the NASA-provided Mobile Transporter

    –   one MSC Maintenance Depot (MMD), primarily for maintenance of the MSC, including external storage of MOSST element spares. (Necessary internal storage of MOSST element spares will be provided in the NASA-provided elements.)

    –   one Special Purpose Dexterous Manipulator (SPDM)

3.5.   Space Station-unique ground elements will be provided by NASA, ESA and the other partners. These elements will be adequate to support the design and development (including assembly and verification), the continuing operation and the full international utilization of each partner's flight elements listed above. The requirements for these elements will be defined and controlled in appropriate program documentation as provided for in Article 7.

3.5.a.   NASA will provide the following Space Station-unique ground elements to support the flight elements listed in Article 3.2: equipment required for specialized or unique integration or launch; ground support equipment (GSE) and flight support equipment (FSE) including necessary logistics; engineering support centers and user support centers; a polar platform control center; and test equipment, mock-ups, simulators, crew training equipment, software and any facilities necessary to house these items. To support the Space Station as a whole, NASA will provide Space Station-unique ground elements including the Space Station Control Center (SSCC), the Payload Operations Integration Center (POIC), subsystem testbeds and elements related to logistics support and to software development including the Software Support Environment.

3.5.b.   As will be agreed and documented in the program documentation as provided for in Article 7, ESA will provide, at defined locations, a defined capacity of the following Space Station-unique ground elements to support the ESA flight elements listed in Article 3.3: equipment required for specialized or unique integration or, as the case may be, for launch or return to Earth; GSE and FSE including necessary logistics; operations control centers, engineering support centers and user support centers; and test equipment, mock-ups, simulators, crew training equipment, software and any facilities necessary to house these items.

3.5.c.   As reflected in the MOU between NASA end the GOJ and in the MOU between NASA and MOSST, STA and MOSST will provide, at defined locations, a defined capacity of the following Space [8] Station-unique ground elements to support their flight elements listed in Article 3.4: equipment required for specialized or unique integration or, as the case may be, for Launch or return to Earth; GSE and FSE including necessary logistics; engineering support centers and user support centers; and test equipment, mockups, simulators, crew training equipment, software and any facilities necessary to house these items.

Article 4 - Access to and Use of the Space Station

4.1.   NASA and ESA will each assure access to and use of their Space Station flight elements listed in Article 3, in accordance with allocation commitments detailed in Articles 8.3.a, 8.3.b, and 8.3.c. Beyond these allocation commitments, the capabilities of the Space Station will be made available to the partners subject to specific arrangement between the relevant partners.

4.2.   The partners' utilization of flight elements listed in Article 3 will be equitable, as provided in the allocation commitments set forth in Article 8 of this MOU and of the corresponding MOU's between NASA and the GOJ and between NASA and MOSST.

4.3.   In accordance with the procedures in Article 8, NASA and ESA will each assure access to and use of their Space Station-unique ground elements referred to in Article 3.5 by each other and the other partners in order to support fully the utilization of the flight elements in accordance with the Consolidated Operations and Utilization Plan provided for in Article 8.1.c. As provided in Article 8, NASA and ESA will each also assure access to and use of their Space Station-unique ground elements by each other and the other partners for system operations support.

4.4.   As requested by ESA for its design and development activities, access to and use of the Space Station-unique ground elements provided by NASA to support the Space Station as a whole will be provided for in appropriate program documentation as provided for in Article 7. Access by ESA and NASA to each other's remaining Space Station-unique ground elements for design and development activities will be subject to specific arrangements on a space-available basis.

[9]   ### Article 5 - Major Program Milestones

5.1.   The Space Station program of NASA and the Columbus program of ESA each include detailed design and development. The NASA and ESA programs also include Space Station operation and utilization. Because of the extended period required to assemble the Space Station, the design and development activities will overlap the operation and utilization activities. After the completion of detailed design and development which includes assembly of the Space Station and one year of initial operational verification (Phase C/D), mature operations and utilization (Phase E) will begin.

5.2.   Major target milestones for the Space Station are as follows:

–	Initiation of NASA's Phase C/D	Dec 1987
–	Initiation of ESA's Phase C/D	Feb 1988
–	NASA-provided Polar Platform Preliminary Design Review	Jan 1989
–	First Space Station Element Launch	Jan 1994
–	NASA-provided Laboratory Module Launch	Jan 1995

–	Permanently Manned Capability	Oct 1995
–	NASA-provided Polar Platform Launch	Oct 1995
–	ESA-provided APM Launch	Oct 1996
–	Completion of Manned Base Assembly	Nov 1996
–	ESA-provided Polar Platform Launch	Mar 1997
–	Completion of NASA's Phase C/D; Initiation of Phase E	Nov 1997
–	First Station Servicing of MTFF	Jun 1998

5.3.   NASA and ESA will develop, maintain and exchange coordinated implementation schedules. These schedules, including the dates for the above milestones, the delivery dates for the ESA-provided elements and the assembly sequence for all elements of the Space Station, will be updated as necessary and formally controlled in appropriate program documentation as provided for in Article 7.

Article 6 - Respective Responsibilities

6.1.a.   While undertaking the detailed design and development of the Space Station elements described in Articles 3.2 and 3.5.a, and within the scope of the Parties' responsibilities established elsewhere in this MOU, NASA will:

[10] 1.   provide overall program coordination and direction;
2.   perform overall system engineering and integration and perform system engineering and integration for NASA-provided elements consistent with these responsibilities;
3.   establish, in consultation with the other partners, overall verification, safety, reliability, quality assurance and maintainability requirements and plans and develop verification, safety, reliability, quality assurance and maintainability requirements and plans for the NASA-provided elements that meet or exceed these overall requirements and plans, which address the elements in Articles 3.2 and 3.5.5;
4.   confirm that the ESA verification, safety, reliability, quality assurance and maintainability requirements and plans for the APM, for the MTFF insofar as it has effects on the manned base associated with its servicing at the manned base, and for the ESA-provided Polar Platform insofar as it has effects on the NASA Space Transportation System (STS) associated with its servicing by the STS, developed by ESA in accordance with Article 6.2.a.5, meet or exceed the overall Space Station verification, safety, reliability, quality assurance and maintainability requirements and plans;
5.   provide regular progress and status information on NASA Space Station program activities and plans;
6.   provide, as applicable, program information, systems requirements information and technical interface information necessary for the integration of ESA-provided elements described in Article 3.3 into the Space Station and/or the coordinated operation and utilization of ESA-provided elements;
7.   develop, with ESA, the agreed joint documentation described in Article 7.1;
8.   perform ground integration tests as necessary to assure on-orbit compatibility and perform verification and acceptance tests for the flight elements in Article 3.2 and accommodate ESA representation at such tests as necessary for NASA and ESA to fulfill their respective responsibilities under this MOU;
9.   conduct overall Space Station preliminary design reviews, critical design reviews, design certification [11] reviews, safety, reliability and quality assurance reviews, operations readiness reviews and flight readiness reviews in order for NASA to certify, following

the certifications at element level provided by NASA and the other partners, that all Space Station elements to be launched on the STS, including the ESA-provided APM, are acceptable for launch, on-orbit assembly and orbital operations; that the ESA-provided Polar Platform, to be launched on Ariane-5, is acceptable for servicing by STS; and that the ESA-provided MTFF, to be launched by Ariane-5, is acceptable for servicing at the manned base; and accommodate ESA representation as necessary for NASA and ESA to fulfill their respective responsibilities under this MOU;

10. conduct for the elements it provides preliminary design reviews, critical design reviews, design certification reviews, and safety, reliability and quality assurance reviews; and accommodate ESA representatives as necessary for NASA and ESA to fulfill their respective responsibilities under this MOU;

11. support, as appropriate, and provide information necessary for ESA to conduct the reviews identified in Article 6.2.a.11;

12. deliver on-orbit the ESA-provided APM and its initial outfitting in accordance with Article 12 and the assembly sequence controlled in appropriate program documentation as provided for in Article 7; [and] assemble on-orbit and verify interfaces of Space Station flight elements, including the flight elements that ESA will provide, with assistance from ESA, in accordance with agreed assembly, activation and verification plans;

13. assist in the on-orbit activation and performance verification of the APM provided by ESA in accordance with agreed assembly, activation and verification plans;

14. for each NASA-provided flight element, provide necessary ground and flight support equipment and initial spares; and perform qualification and acceptance tests of this equipment according to Space Station program requirements and interfaces as set forth in the documents described in Article 7.1;

15. establish in Europe and accommodate in the U.S. agreed liaison personnel as provided in Article 7.2;

[12] 16. participate with ESA and the other partners in Space Station management mechanisms as provided in Articles 7 and 8, including the development of the Operations Management Plan and the Utilization Management Plan;

17. work with ESA and the other partners to ensure that the Space Station Composite Utilization Plan described in Article 8.3.f can be accommodated by the elements provided by NASA, ESA and the other partners—in particular, work with ESA and the other partners to establish standard interfaces between the elements and user-provided hardware and software; provide standard and special user integration and user operations support as described in Articles 8.3.e, 8.3.h, and 8.3.l to users of the other partners or the other partners as users who are to use the NASA-provided flight elements; perform rack-level physical integration on the ground of NASA users of the APM; plan and conduct user operations; and make available Space Station-unique ground elements to support the Space Station Composite Utilization Plan. In addition, NASA will work with ESA in order that NASA and MOSST, respectively, may establish the capabilities to distribute data to NASA and MOSST users of the APM directly from the NASA Tracking and Data Relay Satellite System (TDRSS) space network and to process NASA and MOSST user commands to the APM through the TDRSS space network;

18. establish in consultation with ESA and the other partners, information format and communication standards for a technical and management information system, and establish and maintain a computerized technical and management information system. This system is to work in conjunction with a compatible ESA computerized information system in accordance with the documents described in Article 7.1;

19. develop a Space Station Information System (SSIS) architecture for the end-to-end data transmission between the Space Station data source and the data user; [and]

establish and maintain a Software Support Environment (SSE), including necessary hardware and Space Station software standards to be established by NASA in consultation with ESA and the other partners, to work in conjunction with an ESA software development facility, in accordance with the documents described in Article 7.1;

20. develop and maintain flight and ground software related to elements it provides in accordance with Space Station software standards described in Article 6.1.a.19;

[13] 21. develop an Integrated Logistics System for the manned base in accordance with the documents described in Article 7.1;

22. provide spares for the NASA-provided elements as required to support assembly and initial operational verification;

23. provide operations support and logistics support for the NASA-provided flight elements; and

24. develop and provide to the System Operations Panel described in Article 8 baseline operations plans and maintenance plans for the NASA-provided elements describing routine systems capabilities and defining maintenance requirements, including logistics requirements, necessary for sustaining their functional performance.

6.1.b.   Beginning upon the initiation of Space Station operations and utilization, and within the scope of the Parties' responsibilities established elsewhere in this MOU, NASA will:

1.   participate in Space Station management mechanisms and development of documentation as provided in Articles 7 and 8, and in the sharing of Space Station operations costs as provided in Article 9;

2.   provide sustaining engineering, spares, operations support and logistics support for the Space Station elements it provides;

3.   maintain overall systems engineering, integration and operations support capability for Space Station operations and utilization;

4.   provide resupply and logistics management/integration support for Space Station operations;

5.   work with ESA and the other partners to prepare and implement plans for the integration and operation of user activities in the Space Station Consolidated Operations and Utilization Plan described in Article 8.1.c. In order to accomplish this, provide standard and special user integration and user operations support as described in Articles 8.3.e, 8.3.h, and 8.3.l; perform rack-level physical integration on the ground of NASA users of the APM; make available its Space Station-unique ground elements to support this Consolidated Plan; support planning for future utilization activities; and, using the capabilities provided for in Article 6.1.a.17, NASA and MOSST, respectively, may distribute data to NASA and MOSST [14] users of the APM directly from the TDRSS space network and process NASA and MOSST user commands to the APM through the TDRSS space network;

6.   provide logistics flights for the NASA-provided elements in accordance with Articles 9 and 12, and provide logistics flights for the ESA-provided elements in accordance with Articles 9 and 12;

7.   provide the Space Station Control Center and the Payload Operations Integration Center for manned base operations control; a polar platform control center for the NASA-provided Polar Platform; and engineering support centers for the NASA-provided elements as provided in Article 8;

8.   maintain the Software Support Environment including hardware and software standards for the support of Space Station operations;

9. maintain its flight and ground software in accordance with the Space Station software standards described in Article 6.1.a.19;

10. upon completion of manned base assembly plus a one-year operational verification period, provide docking, access and servicing for the MTFF at the manned base as required by ESA, however, no more frequently than once every six months; and

11. if appropriate STS capability exists, provide for STS servicing of the NASA-provided Polar Platform and, if ESA selects to use this STS capability and with details to be agreed by NASA and ESA, provide STS servicing of the ESA-provided Polar Platform in accordance with Articles 9 and 12.

6.2.a.   While undertaking the detailed design and development of the Space Station elements described in Articles 3.3 and 3.5.b, and within the scope of the Parties' responsibilities established elsewhere in this MOU, ESA will:

1. perform system engineering and integration for the APM consistent with NASA's overall system engineering and integration responsibilities;

2. design the APM to be compatible with the STS and with the Space Station Information System which includes use of TDRSS;

3. design and develop the ESA-provided MTFF; insofar as the MTFF has effects on the manned base associated with its [15] servicing at the manned base, the design and development of the MTFF will comply with otherwise established manned base requirements, capabilities and interfaces, including safety; the MTFF will be capable of autonomous operational periods of six months or longer;

4. design and develop the ESA-provided Polar Platform; insofar as the ESA-provided Polar Platform has effects on the STS associated with its servicing by the STS, its design and development will comply with the operational and safety requirements of the STS;

5. develop, in consultation with NASA, verification, safety, reliability, quality assurance and maintainability requirements and plans for the APM, for the MTFF insofar as it has effects on the manned base associated with its servicing at the manned base, and for the ESA-provided Polar Platform insofar as it has effects on the STS associated with its servicing by the STS that meet or exceed the overall Space Station verification, safety, reliability, quality assurance and maintainability requirements and plans established in Article 6.1.a.3, which address the elements in Articles 3.3 and 3.5.b;

6. provide regular progress and status information on Columbus Program activities and plans;

7. provide, as applicable, program information, systems requirements information and technical interface information necessary to understand the impact of the ESA-provided flight elements on the Space Station configuration and/or on the coordinated operation and utilization of the Space Station, and necessary to integrate those flight elements into the Space Station;

8. develop, with NASA, the agreed joint documentation described in Article 7.1;

9. perform interface verification tests as necessary to assure on-orbit compatibility and perform verification and acceptance tests for the flight elements in Article 3.3, and accommodate NASA representation at such tests as necessary for NASA and ESA to fulfill their respective responsibilities under this MOU;

10. maintain, and provide to NASA on request, ground and on-orbit verification test procedures and results as necessary to assess that the ESA-provided APM complies with overall Space Station program requirements and interface requirements, and, insofar as they have effects on the STS and the manned base, that the [16] ESA-provided Polar Platform and MTFF comply with the operational and safety requirements associated with servicing

of these ESA-provided elements by the STS and at the manned base, respectively, as set forth in the documents described in Article 7.1;

11. conduct for the elements it provides preliminary design reviews, critical design reviews and other reviews as set forth in the documents described in Article 7.1 which will include review of safety, reliability and quality assurance, and accommodate NASA representation as necessary for NASA and ESA to fulfill their respective responsibilities under this MOU;

12. support as appropriate, and provide information necessary for NASA to conduct, the reviews identified in Article 6.1.a.9;

13. support, as appropriate, and provide information necessary for NASA to conduct the reviews identified in Article 6.1.a.10;

14. following design and development of the APM, arrange for the on-orbit delivery of the APM and its initial outfitting in accordance with Article 12 and in accordance with the assembly sequence controlled by appropriate program documentation as described in Article 7;

15. launch and operate the MTFF so that its first servicing at the manned base will be no earlier than the completion of the one-year manned base operational verification period, and launch and operate the ESA-provided Polar Platform;

16. assist in the on-orbit assembly and interface verification of the ESA-provided APM in accordance with agreed assembly, activation and verification plans;

17. activate on-orbit and verify performance of the ESA-provided APM, with assistance from NASA, in accordance with agreed assembly, activation and verification plans; activate on-orbit and verify performance of the ESA-provided MTFF, in accordance with the appropriate program documentation as described in Article 7 which addresses the MTFF insofar as it has effects on the manned base associated with its servicing at the manned base; and activate on-orbit and verify performance of the ESA-provided Polar Platform;

18. for each ESA-provided flight element, provide necessary ground and flight support equipment and initial spares; [17] and perform qualification and acceptance tests of this equipment according to Space Station program requirements and interfaces as set forth in the documents described in Article 7.1;

19. establish in the United States and accommodate in Europe agreed liaison personnel as provided in Article 7.2;

20. participate with NASA and the other partners in Space Station management mechanisms as provided in Articles 7 and 8, including the development of the Operations Management Plan and the Utilization Management Plan;

21. work with NASA and the other partners to ensure that the Space Station Composite Utilization Plan described in Article 8.3.f can be accommodated by the elements provided by NASA, ESA and the other partners—in particular, work with NASA and the other partners to establish standard interfaces between the elements and user-provided hardware and software; provide standard and special user integration and user operations support as described in Articles 8.3.e, 8.3.h, and 8.3.l to users of the other partners or the other partners as users who are to use the ESA-provided flight elements; support and provide information necessary for NASA and MOSST to perform rack-level physical integration on the ground of NASA and MOSST users of the APM; plan and conduct user operations; make available Space Station-unique ground elements to support the Space Station Composite Utilization Plan; and support and provide information necessary for NASA and MOSST, respectively, to establish the capabilities to distribute data to NASA and MOSST users of the APM directly from the TDRSS space network and to process NASA and MOSST user commands to the APM through the TDRSS space network;

22. establish and maintain, in accordance with the documents described in Article 7.1, a compatible computerized technical and management information system to work in conjunction with the NASA computerized information system referred to in Article 6.1.a.18; ESA will be responsible for the provision of necessary hardware and software based on information format and communication standards established by NASA, in consultation with ESA and the other partners;

23. establish and maintain the necessary hardware and software for software production to work in conjunction with the Software Support Environment;

[18] 24. develop and maintain flight and ground software related to elements it provides; for the ESA-provided APM, the development and maintenance of this software will be in accordance with Space Station software standards described in Article 6.1.a.19;

25. provide spares for the ESA-provided elements as required to support initial operational verifications, including assembly for the APM;

26. provide operations support and logistics support for the ESA-provided flight elements; and

27. develop and provide to the System Operations Panel described in Article 8 baseline operations plans and maintenance plans describing routine systems capabilities and defining maintenance requirements, including logistics requirements, necessary for sustaining the functional performance of the ESA-provided APM, for the MTFF insofar as it has effects on the manned base associated with its servicing at the manned base and for the ESA-provided Polar Platform insofar as it has effects on the STS associated with its servicing by the STS.

6.2.b.   Beginning upon the initiation of Space Station operations and utilization, and within the scope of the Parties' responsibilities established elsewhere in this MOU, ESA will:

1. participate in Space Station management mechanisms and development of documentation as provided in Articles 7 and 8, and in the sharing of Space Station operations costs as provided in Article 9;

2. provide sustaining engineering, spares, operations support and logistics support for the Space Station elements it provides;

3. work with NASA and the other partners to prepare and implement plans for the integration and operation of user activities in the Space Station Consolidated Operations and Utilization Plan described in Article 8.1.c. In order to accomplish this, provide standard and special user integration and user operations support as described in Articles 8.3.e, 8.3.h, and 8.3.l; support and provide information necessary for NASA and MOSST to perform rack-level physical integration on the ground of NASA and MOSST users of the APM; make available its Space Station-unique ground elements to support this Consolidated Plan; support planning for future utilization activities; and support and provide information necessary for NASA and MOSST, respectively, to distribute data to NASA and MOSST users of the APM [19] directly from the TDRSS space network and to process NASA and MOSST user commands to the APM through the TDRSS space network;

4. arrange for logistics flights related to the ESA-provided elements in accordance with Articles 9 and 12;

5. provide operations control centers and engineering support centers for the ESA-provided APM, Polar Platform and MTFF, as provided in Article 8; and

6. maintain its flight and ground software for the elements it provides; for the ESA-provided APM, the maintenance of this software will be in accordance with Space Station software standards described in Article 6.1.a.19.

Article 7 - Management Aspects of the Space Station Program
Primarily Related to Detailed Design and Development

7.1.   Management/Reviews

7.1.a.   NASA and ESA are each responsible for the management of their respective Space Station Phase C/D activities consistent with the provisions of this MOU. This Article establishes the management mechanisms to coordinate the respective Space Station design and development (including assembly and verification) activities of NASA and ESA, to establish applicable requirements, to assure safe operations, to establish the interfaces between the Space Station elements, to review decisions, to establish schedules, to review the status of activities, to report progress and to resolve issues and technical problems as they arise.

7.1.b.   The NASA/ESA Program Coordination Committee (PCC), co-chaired by the NASA Associate Administrator for Space Station and the ESA Director of Space Station and Platforms, will meet periodically throughout the lifetime of the program or promptly at the request of either Party to review the Parties' respective design and development activities. The Co-Chairmen will together take those decisions necessary to assure implementation of the cooperative design and development activities related to Space Station flight elements and to Space Station-unique ground elements provided by the Parties, including, as appropriate, to design changes of the Parties' flight elements during Phase E. In taking decisions regarding design and development, the PCC will consider operation and utilization impacts, and will [20] also consider design and development recommendations from the Multilateral Coordination Board described in Article 8.1.b. However, decisions regarding operation and utilization activities will be taken in accordance with Article 8. The Co-Chairmen will each designate their respective members and will decide on the location of meetings. If the Co-Chairmen agree that a specific design and development issue or decision requires consideration by another partner at the PCC level, the NASA/ESA PCC may meet jointly with the NASA/STA PCC and/or the NASA/MOSST PCC.

7.1.c.   Multilateral Program Reviews will be organized by NASA and will meet as necessary at the request of any partner so that the Parties to this MOU and the other partners can report progress and discuss the status of their Phase C/D program activities.

7.1.d.   The manned base and NASA-provided Polar Platform requirements, configuration, housekeeping resource allocations for design purposes, and element interfaces will be controlled by the Space Station Control Board (SSCB) chaired by NASA. The SSCB will also control Space Station activities through the completion of assembly and initial operational verification, and other Space Station configuration control activities related to the manned base, related to the MTFF insofar as it has effects on the manned base associated with its servicing at the manned base, and related to the ESA-provided Polar Platform insofar as it has effects on the STS associated with its servicing by the STS. ESA will be a member of the SSCB, and of such subordinate boards thereof as may be agreed, attending and participating when these boards consider items which affect the APM, interfaces between the NASA-provided and the ESA-provided elements, interfaces between the ESA-provided elements and the STS, interfaces between the ESA-provided elements and other partner-provided elements, or the accommodation on the manned base of the Composite Utilization Plan and the Composite Operations Plan described in Article 8. Decisions by the SSCB Chairman may be appealed to the PCC, although it is the duty of the SSCB Chairman to make every effort to reach consensus with ESA rather than have issues referred to the PCC. Such appeals will be made and processed expeditiously. Pending resolution of appeals, ESA need not proceed with the implementation of an SSCB decision as far as its provided elements are concerned; NASA may, however, proceed with an SSCB decision as far as its provided elements are concerned. NASA will be a member of the Columbus Control Board chaired by ESA, and of such subordinate boards thereof as may be agreed, attending and participating as appropriate. As far as the elements separated from the manned base are concerned, NASA will assume management responsibility for the design and development of the NASA-provided Polar Platform, including meeting requirements related to polar [21] platform user interfaces and polar platform STS servicing; ESA will assume management responsibility for the design and development of the ESA-provided Polar Platform, including meeting requirements related to polar platform user interfaces and polar platform STS servicing; [and] ESA will also assume management responsibility for the design and development of the MTFF and for meeting requirements related to its effects on the manned base associated with its servicing at the manned base.

7.1.e.   NASA will develop an overall Program Plan for Space Station design and development based on information provided by all the partners detailing overall program content, implementation approach and schedules. ESA will develop a Columbus Program Plan for design and development detailing ESA program content, implementation approach and schedules. A Joint Program Plan [JPP] for design and development, signed by the NASA Associate Administrator for Space Station and the ESA Director of Space Station and Platforms, will cover the interrelationship between the ESA program and the overall program. Any modification or any addition to the JPP will be approved by the PCC.

7.1.f.   NASA will develop a Program Requirements Document (PRD) based on information provided by all the partners providing the programmatic basis for the overall conduct of Phase C/D. A Joint PRD (JPRD), signed by the NASA Associate Administrator for Space Station and the ESA Director of Space Station and Platforms, will represent the top-level requirements related to the APM, the MTFF insofar as it has effects on the manned base associated with its servicing at the manned base and the ESA-provided Polar Platform insofar as it has effects on the STS associated with its servicing by the STS. The JPRD will identify the applicability to the ESA program of all paragraphs in the PRD, including any which are added or modified. Any modification or any addition to the JPRD will be approved by the PCC.

7.1.g.   NASA has developed an overall Program Definition and Requirements Document (PDRD) based on information provided by all the partners which contains requirements for Space Station flight element hardware and software and provides the technical basis for the overall conduct of Phase C/D. A Joint PDRD (JPDRD), signed by the NASA Program Director and the ESA Program Manager, contains the detailed requirements related to the APM, the MTFF insofar as it has effects on the manned base associated with its servicing at the manned base and the ESA-provided Polar Platform insofar as it has effects on the STS associated with its servicing by the STS. The JPDRD identifies the applicability to the ESA program of all paragraphs in the PDRD including any which are added or modified. Any modification to the PDRD will be approved by the SSCB. Any modification or any addition to the co-signed JPDRD will be mutually agreed and [22] jointly signed by the NASA Program Director and the ESA Program Manager.

7.1.h.   NASA will develop Architectural Control Documents (ACD's) which define and control the end-to-end architecture of the manned base distributed systems and control the interfaces of these systems with each other and with the flight elements. In addition, NASA will develop, in consultation with the appropriate partners, Interface Control Documents (ICD's) which control interfaces between: the flight elements comprising infrastructural elements and the flight elements comprising accommodations elements as defined in Article 8.1.d; between the flight elements comprising infrastructural elements; and, as appropriate, between any other flight elements, between flight and ground or among ground elements. NASA will also develop a Baseline Configuration Document (BCD) based on information provided by all the partners which controls the configuration of the manned base and of the NASA-provided Polar Platform. The ACD's and the BCD will be developed by the start of NASA's Phase C/D; the ICD's will be developed early in Phase C/D. Any modification or any addition to the ACD's, the BCD and the ICD's will be approved by the SSCB. Joint interface documentation, which identifies the applicability to the ESA-provided APM of all interfaces in the ACD's, BCD and ICD's, including any which are modified, will be developed by NASA and ESA. This joint interface documentation will be mutually agreed and jointly signed by the NASA Program Director and the ESA Program Manager. Any modification or any addition to this joint interface documentation will be mutually agreed and jointly signed by the NASA Program Director and the ESA Program Manager. NASA and ESA will jointly develop an ICD which will govern the interfaces between the ESA-provided MTFF and the manned base in connection with the docking, access and servicing of the MTFF at the manned base, in accordance with Article 6.2.a.3. NASA and ESA will also jointly develop an ICD in which they will agree on standard user interfaces for the polar platforms they provide; this ICD will also govern the interfaces between the ESA-provided Polar Platform and the STS. The MTFF ICD will be developed early in Phase C/D; the Polar Platform ICD will be established no later than the Preliminary Design Review for the NASA-provided Polar Platform. The MTFF and Polar Platform ICD's will be mutually agreed and jointly signed by the NASA Program Director and the ESA Program Manager. Any modification or addition to these documents will be mutually agreed and jointly signed by the NASA Program Director and the ESA Program Manager.

7.1.i.   Program Management Reviews will be held as necessary at which the NASA Program Director and the Program Managers representing ESA and the other partners will report on the status of their respective design and development activities, including schedule, element performance parameters and element [23] interface requirements. These formal Program Management Reviews will be held at least quarterly and will be chaired by NASA. Less formal status reviews will be held monthly; representatives of the partners' Program Managers will attend these reviews.

7.1.j.   ESA will participate in selected NASA reviews on Space Station requirements, architecture and interfaces as defined in the JPP. Similarly, NASA will participate in selected ESA reviews as defined in the JPP; the other partners will participate as appropriate.

7.1.k.   Through participation in the above management mechanisms, NASA and ESA agree to achieve commonality on the manned base as required by the overall Space Station safety requirements as defined pursuant to Article 10. NASA and ESA also agree to provide standard interfaces for Space Station users both in the permanently attached pressurized laboratories and on the polar platforms. Exceptions to these requirements for commonality may be agreed on a case-by-case basis between NASA and ESA. In addition, NASA and ESA will work through the above management mechanisms to seek agreement on a case-by-case basis regarding the use of interchangeable hardware and software in order to promote efficient and effective Space Station operations, including reducing the burden on the Space Station logistics system.

7.2.   Liaison. The NASA Office of Space Station and ESA Space Station and Platforms Directorate are responsible for NASA/ESA liaison activities. ESA may provide representative(s) to NASA Headquarters in Washington, D.C., and NASA may provide representative(s) to ESA Headquarters in Paris. In order to facilitate the working relationships between the NASA Program Director and the ESA Program Manager, ESA will provide and NASA will accommodate ESA liaison to the NASA Space Station Program Office. Similarly, NASA will provide and ESA will accommodate NASA liaison to the ESA Space

Station Program Office. In addition, by mutual agreement, ESA may provide and NASA will accommodate ESA liaison to NASA Centers involved in the Space Station program, and NASA may provide and ESA will accommodate liaison to ESA Centers involved in the ESA Space Station program. Arrangements specifying all conditions relating to the liaison relationships will be agreed and co-signed by the Co-Chairmen of the PCC.

Article 8 - Management Aspects of the Space Station Program
Primarily Related to Operations and Utilization

8.1.   General

8.1.a.   NASA and ESA each have responsibilities regarding the management of their respective operations and utilization [24] activities and the overall Space Station operations and utilization activities, in accordance with the provisions of this MOU. NASA will have the responsibility for the overall planning for and direction of the operation of the manned base (including all elements within the operational Command and Control Zone (CCZ) of the manned base as defined in the program documentation provided for in Article 7) and the NASA-provided Polar Platform. ESA will have the responsibility for the planning for and direction of the operation of the elements it provides which are separated from the manned base (specifically, the MTFF when outside the operational CCZ of the manned base and the ESA-provided Polar Platform when outside the operational CCZ of the STS, as defined in the program documentation provided for in Article 7). Operations and utilization activities will comprise long-range planning and top-level direction and coordination, which will be performed by the strategic-level organizations; detailed planning and support to the strategic-level organizations which will be performed by the tactical-level organizations; and implementation of these plans which will be performed by the execution-level organizations.

8.1.b.   A Multilateral Coordination Board (MCB) will be established as soon as possible after the start of NASA's Phase C/D and will meet periodically over the lifetime of the program or promptly at the request of any partner with the task to ensure coordination of the activities of the partners related to the operation and utilization of the Space Station. The Parties to this MOU and the other partners will plan and coordinate activities affecting the safe, efficient and effective operation and utilization of the Space Station through the MCB, except as otherwise specifically provided in this MOU. The MCB will comprise the NASA Associate Administrator for Space Station; the ESA Director of Space Station and Platforms; the MOSST Deputy Secretary, Space Policy Sector; and the STA Director-General of the Research and Development Bureau. The NASA Associate Administrator for Space Station will chair he MCB. The Parties agree that all MCB decisions should be made by consensus. However, where consensus cannot be achieved on any specific issue within the purview of the MCB within the time required, the Chairman is authorized to take decisions. The Parties agree that, in order to protect the interests of all partners in the program, the operation and utilization of the Space Station will be most successful when consensus is reached and when the affected partners' interests are taken into account. MCB decisions will not modify rights of the partners specifically provided in this MOU. Decisions regarding the operation and utilization of the ESA-provided elements which are separated from the manned base and which do not have effects on the manned base associated with servicing at the manned base or have effects on the STS associated with servicing by the STS will be taken by ESA, except as otherwise specifically provided in Article 8.3.

[25] 8.1.c.   The MCB will establish Panels which will be responsible for the long-range strategic coordination of the operation and utilization of the Space Station, to be called

the System Operations Panel and the User Operations Panel respectively, described in detail below. The MCB will develop a charter that will define the organizational relationships and responsibilities of these Panels, and the organizational relationships of these Panels with the tactical- and execution-level organizations described below. Any modifications to the charter will be approved by the MCB. The MCB will approve, on an annual basis, a Consolidated Operations and Utilization Plan (COUP) for the Space Station based on the annual Composite Operations Plan and the annual Composite Utilization Plan developed by the Panels and described below. In doing so, the MCB will be responsible for resolving any conflicts between the Composite Operations Plan and the Composite Utilization Plan which cannot be resolved by the Panels. The COUP will be prepared by the User Operations Panel and agreed to by the System Operations Panel. The charter for these Panels will also delineate the Panels' delegated responsibilities with respect to adjustment of the COUP. The COUP will be implemented by the appropriate tactical- and execution-level organizations.

8.1.d.   Manned Base Hardware. The following is provided to explain the relationships between the different types of elements on the manned base which are allocated for use by the partners. The Space Station manned base includes:

–   accommodations elements; and
–   infrastructural elements.

The accommodations elements are the NASA-provided Laboratory Module, the ESA-provided APM, the STA-provided JEM including the Exposed Facility and the Experiment Logistics Modules, and the NASA-provided Attached Payload Accommodation Equipment. The infrastructural elements comprise all other manned base elements, including servicing elements and other elements that produce resources which permit all manned base elements to be operated and used.

8.1.d.1. Housekeeping. Both accommodations elements and infrastructural elements will be used for assembly, for verification and for maintenance of the manned base in an operational status, and also for the storage of element spares, crew provisions and safe haven capability, with secondary storage of crew provisions to be distributed equally among the three laboratories. In such use, they are referred to, respectively, as providing:

[26]   –   housekeeping accommodations; and
       –   housekeeping resources.

During Phase C/D, these housekeeping accommodations and housekeeping resources will be controlled in appropriate program documentation as provided for in Article 7. During Phase E, these housekeeping accommodations and housekeeping resources will be controlled according to the mechanisms in Article 8.2.d.

8.1.d.2. Utilization. The accommodations and resources not required to maintain the manned base in an operational status will be used in connection with Space Station utilization, and are referred to, respectively, as:

–   user accommodations; and
–   utilization resources.

Details regarding the allocation of the Space Station user accommodations and utilization resources are provided in Article 8.3. NASA and ESA agree to seek to minimize the demands for housekeeping accommodations and housekeeping resources in order to maximize those available for utilization.

8.1.e.   Platforms and MTFF. Because of the different character of the platforms and the MTFF, differentiation between accommodations and resources is not required. Mechanisms governing the operation of these elements are to be found in Article 8.2 and mechanisms governing the utilization of these elements are to be found in Article 8.3.

8.2. Operations

8.2.a. It is the goal of the Parties to this MOU to operate the Space Station in a manner that is safe, efficient and effective for both Space Station users and Space Station operators. To accomplish this, the MCB will establish, within three months of its establishment, a System Operations Panel (SOP) to coordinate strategic-level operations activities and operations planning activities as provided for in Article 8.1.c.

8.2.b. The SOP will comprise one member each from NASA, ESA and the other partners. Members may send designated alternates to SOP meetings. In addition, each partner may call upon relevant expertise as necessary to support SOP activities. The SOP will take decisions by consensus; in the event of failure to reach consensus on any issue, the issue will be forwarded to the MCB for resolution. In the interest of efficient management, NASA and ESA recognize that the SOP should take the responsibility routinely to resolve all operations issues as expeditiously as possible rather than refer such issues to the MCB.

[27] 8.2.c. The SOP will develop, approve and maintain an Operations Management Plan for the operation, maintenance and refurbishment of and logistics for the manned base, the NASA-provided Polar Platform and the ESA-provided Polar Platform insofar as these platforms have effects on the STS associated with their servicing by the STS, and the MTFF insofar as it has effects on the manned base associated with its servicing at the manned base during Phase E. This Plan will describe relationships among the strategic, tactical and execution levels of operations management, where the strategic level is coordinated by the SOP; the tactical level, by the tactical operations organization referred to in Article 8.2.e; and the execution level, by implementing organizations and field centers. Consistent with the other provisions of this Article, the Operations Management Plan will also address operational requirements for the manned base, the NASA-provided Polar Platform and the ESA-provided Polar Platform insofar as these platforms have effects on the STS associated with their servicing by the STS, [and] the MTFF insofar as it has effects on the manned base associated with its servicing at the manned base and Space Station-unique ground elements. The Operations Management Plan will provide the procedures for preparation of the baseline operations plans and maintenance plans provided for in Articles 6.1.a.24 and 6.2.a.27, annual refinements to these baseline plans, and the Composite Operations Plan described in Article 8.2.d, including procedures for adjustment of these plans as further information becomes available.

8.2.d. On an annual basis, NASA and ESA will each provide to the SOP any significant refinements to their baseline operations plans and maintenance plans five years in advance. Using the operations and maintenance plans and these refinements provided by all of the partners, including requirements for use of Space Station-unique ground elements, the SOP will develop and approve an annual Space Station Composite Operations Plan (COP) consistent with the annual Space Station Composite Utilization Plan described in Article 8.3.f. The COP will also identify the housekeeping accommodations and housekeeping resources required for maintenance of the manned base in an operational status. Compatibility of the COP and the Composite Utilization Plan must be assured through coordination between the SOP and the User Operations Panel, described in Article 8.3.d, during the preparation and approval process.

8.2.e. NASA, with the participation of all the partners, will be responsible for integrated tactical-level activities for Space Station manned base operations. To this end, NASA will establish an integrated tactical operations organization and the other partners will participate in discharging the responsibilities of this organization. ESA and the other partners will provide personnel to the integrated tactical operations organization who will bring expertise on the elements [28] each provides and will participate in overall integrated tactical operations activities. NASA and ESA will consult and agree regarding the responsibilities to be discharged by the ESA personnel. NASA and ESA will also consult and agree regarding the number of ESA personnel and all administrative conditions related to these personnel. In conjunction with the integrated activities, NASA, ESA and the other partners will each perform distributed tactical-level activities related to the elements each provides, such as decentralized system operations support planning, user support planning, logistics planning, and the accommodations assessments described in Article 8.3.h. Tactical-level activities will include planning for system operations and for user support activities across all manned base elements. Tactical-level activities for elements separated from the manned base when outside the operational CCZ of the STS or the manned base, as defined in the program documentation provided for in Article 7, will be performed by the element provider. However, where the same services, such as transportation, logistics and communications, are required by both the manned base and elements which are operating separated from the manned base, planning for these services will be performed by the integrated tactical operations organization.

8.2.f.   Tactical Operations Plans (TOP's) for the manned base and for the MTFF insofar as it has effects on the manned base associated with its servicing at the manned base will be developed by the tactical operations organization described in Article 8.2.e to implement the COUP. Each TOP will include Increment Plans (IP's) for a period of two years prior to launch of the STS to the manned base for a specific increment. (An increment is normally the interval between visits of the STS for the purpose of resupply in support of manned base operations and utilization as approved in the COUP.) Each IP will describe the detailed manifest of user payloads, systems support equipment and supplies needed to support the increment. Each IP will also describe changes to the complement of hardware and software to be flown during that increment and the payload and system support activities needed to carry out the activities approved in the COUP for that increment. The IP will identify the crew complement and define logistics requirements including STS interface requirements, changes to housekeeping resource requirements, changes to housekeeping accommodation requirements and communication requirements, including TDRSS use and requirements for distribution of data, to support the subject increment.

8.2.g.   NASA, with the participation of all the partners, will be responsible for integrated execution-level planning for and execution of the day-to-day operation of the manned base. ESA and the other partners will participate in discharging the responsibilities of the Space Station Control Center (SSCC), [29] established and managed by NASA, which will conduct execution-level activities and support tactical planning. ESA and the other partners will provide personnel to the SSCC. These personnel will bring expertise on the elements that partner provides, will participate in overall SSCC-based activities, and will support real-time on-orbit activities with emphasis on the elements each provides. NASA and ESA will consult and agree regarding the responsibilities to be discharged by the ESA personnel. NASA and ESA will also consult and agree regarding the number of ESA personnel and all administrative conditions related to these personnel. In conjunction with the integrated activities, NASA, ESA and the other partners will each perform distributed execution-level activities related to the elements each provides, such as monitoring and support of real-time systems operations. NASA, ESA and the other partners will provide engineering support centers to perform detailed engineering assessments and real-time operations support to the SSCC required for the operational control of the manned base elements they provide. Execution-level activities for elements separated from the manned base when outside the operational CCZ of the STS or the manned base, as defined in the program documentation provided for in Article 7, will be the responsibility of the element provider. The partners may also participate in and provide personnel to other execution-level activities at other sites as agreed.

8.2.h.   The International Operational Concepts Working Group (IOCWG), established by the Space Station Phase B MOU's, will continue to advise the Parties to this MOU in planning for the establishment of the SOP. Once the SOP is established, the activities of the IOCWG will end.

8.3.   Utilization

8.3.a.   Manned Base

8.3.a.1. NASA and MOSST will provide Space Station manned base infrastructural elements to assemble, maintain, operate and service the manned base; NASA and MOSST will also provide resources derived from these infrastructural elements to the other partners as provided in Article 8.3.a.2. ESA will retain the use of 41% of the user accommodations on its APM; NASA will retain the use of 97% of the user accommodations on its accommodations elements; NASA and ESA will each provide MOSST 3% of the user accommodations on their accommodations elements; and ESA will provide NASA the remaining user accommodations on its APM. NASA, ESA and MOSST will each control the selection of users for their allocations of user accommodations; such NASA, ESA and MOSST control of the selection of users for their allocation of user accommodations will he exercised in accordance with the procedures in this MOU and in the NASA-MOSST MOU for developing the Composite Utilization Plan.

[30] 8.3.a.2. Allocation of manned base resources among the partners will be in accordance with the following approach. Housekeeping resources required by all elements, and provided as noted in Article 8.1.d.1, will be set aside. The utilization resources will be allocated as follows: 20% of utilization resources will be allocated to NASA because of its Attached Payload Accommodation Equipment; 3% of utilization resources will be allocated to MOSST; [and] the remaining utilization resources will be apportioned equally among the three laboratory modules. ESA will be allocated 50% of the utilization resources apportioned to the ESA-provided APM and STA will be allocated 50% of the utilization resources apportioned to the ESA-provided JEM. NASA will be allocated 100% of the utilization resources apportioned to the NASA-provided Laboratory Module, the remaining 50% of the utilization resources apportioned to the ESA-provided APM and the remaining 50% of the utilization resources apportioned to the ESA-provided JEM. The above allocation of utilization resources is to the partner, not to the elements, and may be used by the partner on any Space Station element consistent with the COP and the Composite Utilization Plan. More than this allocation of any utilization resource may be gained by each partner through barter or purchase from other partners.

8.3.a.3. ESA's allocation of user accommodations and utilization resources will begin once the APM is verified following assembly to the manned base.

8.3.a.4. Manned base utilization resources are power, user servicing capacity, heat rejection capacity, data handling capacity, total crew time and EVA capacity. The initial list of manned base utilization resources to be allocated is power, user servicing capacity and total crew time. All other manned base utilization resources may be used without allocation. To support the operation and full international utilization of the Space Station manned base as defined in Article 3, NASA plans to provide the number of STS flights per year baselined by the SSCB during Phase C/D. From the total Space Station user payload capacity available on STS flights actually flown to and from the manned base each year, each partner will have the right to purchase STS launch and return services for its Space Station utilization activities, up to its allocated percentage of utilization resources. (The foregoing does not apply to STS launch and return capacity provided to and from the manned base in connection with Space Station evolutionary additions.) Similarly, the partners will have the right to purchase, up to their allocated percentage of utilization resources, TDRSS data transmission capacity available to the manned base. The User Operations Panel, defined in Article 8.3.d, will update the lists of utilization resources and allocated utilization resources as necessary as NASA and the other partners gain experience.

[31] 8.3.b. Platforms

8.3.b.1. In recognition of the fact that platforms are separate elements that do not require extensive support from the infrastructural elements of the manned base, platforms are treated separately from the manned base.

8.3.b.2. NASA and ESA will share the use of each other's polar platforms on a balanced reciprocal basis, recognizing that the two platforms may have different capabilities and that the user community may propose specific splits based on actual payloads; such proposals must be agreed to by NASA and ESA, and by MOSST with respect to its 3% utilization of the polar platforms provided for in Article 8.3.b.3, and processed by the User Operations Panel as part of the development of the Composite Utilization Plan provided in Article 8.3.f.2. NASA and ESA will also provide associated user integration and user operations support to each other and each other's users.

8.3.b.3. MOSST will be provided 3% utilization of both the NASA and ESA polar platforms together with the associated user integration and user operations support. STA may purchase, barter or enter into other arrangements for platform utilization.

8.3.c.   Man-Tended Free Flyer

8.3.c.1. ESA will retain the total use of the MTFF it provides.

8.3.c.2. Notwithstanding Article 8.3.c.1, each year, NASA will have an option to use up to 25% of MTFF utilization capacity by purchase at prices ESA routinely charges comparable customers or by barter such as for an amount of utilization resources and/or user accommodations. The conditions of such purchase or barter will be agreed between NASA and ESA.

8.3.c.3. In case of total use of the MTFF by ESA, all accommodations and resources required to service the MTFF at the manned base will come out of the user accommodations and utilization resources available to ESA as provided in Article 8.3.a.

8.3.d.   It is the goal of the Parties to use the Space Station in a safe, efficient and effective manner. To accomplish this, the MCB will establish, within three months of its establishment, a User Operations Panel (UOP), to assure the compatibility of utilization activities of the manned base, the polar platforms, and use by the MTFF of manned base utilization resources and user accommodations. The UOP will comprise one member each from NASA, ESA and the other partners. Members may send designated alternates to UOP meetings. In addition, each partner may call upon relevant expertise as necessary to support [32] UOP activities. The UOP will take decisions by consensus; except as noted in Article 8.3.f.2, in the event of failure to reach consensus on any issue, the issue will be forwarded to the MCB for resolution. In the interest of efficient management, NASA and ESA recognize that the UOP should take the responsibility to routinely resolve all utilization issues as expeditiously as possible rather than refer such issues to the MCB.

8.3.e.   The UOP will develop, approve and maintain a Utilization Management Plan which will describe relationships among the strategic, tactical and execution levels of utilization management, where the strategic level is coordinated by the UOP; the tactical level, by the integrated tactical operations organization described in Article 8.2.e; and the execution level, by implementing organizations and field centers. The Plan will also establish processes for utilization of the Space Station elements, including the user support centers and other Space Station-unique ground elements provided by all the partners, consistent with Article 8.3.d; define standard user integration and user operations support; and describe the approach to distributed user integration and operations. The Plan will provide procedures for preparation of the partners' Utilization Plans and Composite Utilization Plan described in Article 8.3.f, including procedures for adjustment of these Plans as further information becomes available.

8.3.f. Utilization Plan for the Manned Base and the Polar Platforms

8.3.f.1. On an annual basis, five years in advance, NASA and ESA each will develop a Utilization Plan for all proposed uses of its allocation of manned base user accommodations and utilization resources, for all proposed uses of unallocated manned base utilization resources and Space Station-unique ground elements, and for all uses of the polar platforms. Each partner will satisfy the requirements of its users for storage within the user accommodations available to that partner, with the exception of temporary on-orbit storage in the Integrated Logistics System carriers in which user equipment, including MTFF equipment, is launched or returned to Earth as specified in the applicable Increment Plan. As regards the MTFF, the ESA Utilization Plan will include all uses of manned base user accommodations and utilization resources required to service the MTFF at the manned base, information necessary to determine whether any planned utilization of the MTFF would have effects on the manned base associated with its servicing at the manned base, and information related to Article 9.8(e) of the Intergovernmental Agreement. NASA and ESA each will prioritize and propose appropriate schedules for the user activities in its Utilization Plan, including the use of user support centers and other Space Station-unique ground elements to support the [33] utilization of the flight elements. These individual Utilization Plans will take into consideration all factors necessary to assure successful implementation of the user activities, including any relevant information regarding crew skills and special requirements associated with the proposed payloads.

8.3.f.2. NASA and ESA each will forward its Utilization Plan to the UOP. Using the Utilization Plans of NASA, ESA and the other partners, the UOP will develop the Composite Utilization Plan (CUP), covering the use of both flight and Space Station-unique ground elements, based on all relevant factors, including each element-provider's recommendations regarding resolution of technical and operational incompatibilities among the users proposed for its elements. In its use of the Space Station, each partner will seek, through the mechanisms established in this MOU, to avoid causing serious adverse effects on the use of the Space Station by the other partners. In the event of failure of the UOP to reach consensus on the utilization of the manned base and/or related Space Station-unique ground elements, the issue will be forwarded to the MCB for resolution. In the event of failure of the UOP to reach consensus on the utilization of the ESA-provided Polar Platform, ESA will take the decision, and in the event of failure of the UOP to reach consensus on the utilization of the NASA-provided Polar Platform, NASA will take the decision; however, in either event, NASA and ESA will respect the utilization rights of Canada and of each other in any such decisions.

8.3.f.3. Utilization Plans proposed by NASA, ESA and the other partners which fall completely within their respective allocations and do not conflict operationally or technically with one another's Utilization Plans will be automatically approved. However, Articles 9.8(a), 9.8(b) and 9.11 of the Intergovernmental Agreement will apply.

8.3.g. Utilization Plan for the MTFF

8.3.g.1. The MTFF Utilization Plan will be developed and approved by ESA. As appropriate, MTFF utilization will be consistent with Articles 8.3.c.2 and 8.3.f.1.

8.3.h. Each partner will participate in integrated tactical-level planning of user activities. To this end, each partner will provide personnel to the operations organization described in Article 8.2.e. These personnel will participate in integrated tactical-level planning of user activities; they will also support the strategic-level planning of user activities. NASA and ESA will consult and agree regarding the responsibilities to be discharged by the ESA personnel. NASA and ESA will also consult and agree regarding the number of ESA personnel and all administrative conditions related to these personnel. In addition, partners providing user accommodations [34] will be responsible for providing standard user integration and user operations support to users of other partners or other partners as users, including conducting assessments of the flow of payload integration activities for all payloads manifested in the user accommodations they provide. Accommodation assessments for individual payloads manifested in a laboratory module covering engineering, operations and software compatibility will also be performed by the partner providing that laboratory module in support of the preparation and execution of Tactical Operations Plans and Increment Plans. Similarly, MOSST will be responsible for providing standard user integration and user operations support for users of the other partners or other partners as users of the flight elements provided by MOSST; and NASA will be responsible for providing standard user integration and user operations support for users of the other partners or other partners as users of the manned base systems/subsystems provided by NASA.

8.3.i.   Each partner will participate in discharging the responsibilities of the Payload Operations Integration Center (POIC) established and managed by NASA which will be responsible for assistance to manned base users in planning and executing user activities on the manned base, for overall direction of the execution of user activities on the manned base, and for interaction with the SSCC in order to coordinate user activities with systems operations activities. Each partner will provide personnel to the POIC. NASA and ESA will consult and agree regarding the responsibilities to be discharged by the ESA personnel. NASA and ESA will also consult and agree regarding the number of ESA personnel and all administrative conditions related to these personnel. The interaction between the POIC and SSCC will be described in the Operations Management Plan. Both NASA and ESA will provide user support centers which will function within the framework of NASA's responsibilities for the POIC. The interactions between the user support centers and the POIC will be described in the Utilization Management Plan. NASA and ESA will each be responsible, relative to the elements they provide which are separated from the manned base, for assistance to users in planning and executing user activities, for direction of the execution of user activities and for interaction with the MTFF and polar platform control centers to coordinate user and element operations activities.

8.3.j.   In working out problems which may arise after the development of the COUP, in the case of a technical or operational incompatibility between users, the partner(s) providing the element(s) in which the users have accommodations, as well as other impacted partners, will provide appropriate analyses and recommendations to the appropriate strategic-, tactical- or execution-level organization for resolution of conflicts. However, if such conflict only has impacts within a single manned base element and only impacts users of the [35] provider of that element, the partner providing that manned base element will be responsible for resolving such conflicts in accordance with the content of the COUP; conflicts related to proposed polar platform utilization will be resolved as provided in Article 8.3.f.2.

8.3.k.   NASA, ESA and the other partners may at any time barter for, sell to one another or enter into other arrangements for any portion of their Space Station allocations, and are free to market the use of their allocations individually or collectively, according to the procedures established in the Utilization Management Plan. The terms and conditions of any barter or sale will be determined on a case-by-case basis by the parties to the transaction. The partner providing allocations will ensure that the obligations it has undertaken under this MOU are met. NASA, ESA and the other partners each may retain the revenues they derive from such marketing.

8.3.l.   NASA and ESA will make their Space Station-unique ground elements, including user support centers, available for use by each other and the other partners in order to support fully both the standard and special user integration and operations support approved in the CUP and the requirements in the COP. Any special user integration or user operations support provided by a partner to users of the other partners or other partners as users will be provided on a reimbursable basis at prices routinely charged comparable users for similar services.

8.3.m.   The International Utilization Coordination Working Group (IUCWG), established by the Space Station Phase B MOU's, will continue to advise the Parties to this MOU in planning for the establishment of the UOP. Once the UOP is established, the activities of the IUCWG will end.

8.4.   In order to protect the intellectual property of Space Station users, procedures covering all personnel, including Space Station crew, who have access to data will be developed by the MCB.

8.5.   The partners will seek to outfit the NASA-provided Laboratory Module, the ESA-provided APM and the STA-provided JEM to equivalent levels by the end of Space Station assembly in Phase C/D.

Article 9 - Operations Costs Responsibilities

9.1.   The Parties will seek to minimize operations costs for the Space Station. The Parties will also seek to minimize the exchange of funds, for example, through the performance of specific operations activities.

[36] 9.1.a.   The costs associated with ESA's providing personnel to undertake integrated tactical- and execution-level activities as provided for in Articles 8.2.e, 8.2.g, 8.3.h, and 8.3.i will be agreed between NASA and ESA and will be a contribution towards the satisfaction of ESA's common system operations costs responsibilities established below.

9.2.   Element operations costs

9.2.a.   NASA and ESA will each have operational responsibilities for the elements it provides as detailed in Article 8. Such operational responsibilities mean that NASA and ESA will each be financially responsible for element operations costs, that is, costs attributed to operating and to sustaining the functional performance of the flight elements that it provides, such as ground-based maintenance, sustaining engineering, provision of spares, launch and return costs for spares, launch and return costs of the fraction of the Integrated Logistics System carriers provided for in Article 3.2 that is attributable to spares, and also costs attributed to the maintenance and operation of element-unique ground centers.

9.3.   Common system operations costs

9.3.a.   Manned Base. Other than the element operations costs covered in Article 9.2.a, NASA, ESA and the other partners will equitably share the common system operations costs; that is, the costs attributed to the operation of the manned base as a whole. The categories comprising common system operations costs are: integrated tactical planning activities performed by the integrated tactical operations organization provided for in Article 8.2.e, including user integration planning and maintenance of common documentation; space systems operations (SSCC-based operations, SSCC maintenance and common elements of the Software Support Environment); POIC-based operations and POIC maintenance; Integrated Logistics System operations, including consumables and common inventory management activities; prelaunch/post landing processing of logistics carriers; launch to orbit and return of consumables, crew and crew logistics, and launch and return of the fraction of the Integrated Logistics System carriers provided for in

Article 3.2 that is attributable to consumables and crew logistics; and transmission of housekeeping data between the manned base and the ground (SSCC, POIC and launch and landing sites). Each partner will be responsible for a percentage of common system operations costs equal to the percentage of Space Station utilization resources allocated to it in Article 8.3.a.2. ESA's responsibility for sharing common system operations costs will begin following the assembly and verification of the APM.

[37] 9.3.b.   Platforms. NASA and ESA will each be responsible for the common system operations costs for the platforms which they provide.

9.3.c.   Man-Tended Free Flyer. ESA will be responsible for the common system operations costs for the MTFF it provides.

9.3.d.   Any changes to the list of common system operations costs in this Article will be made by agreement among the partners.

9.4.   The Parties to this MOU and the other partners will work through the SOP to identify the detailed contents to be included in each common system operations cost category. The partners will also, each year, report to the SOP on their forecasts for future years for all costs included in the common system operations costs of the manned base and on their identified actual annual common system operations costs. The SOP will develop detailed procedures for implementing this Article. If possible, after the partners have gained experience in the operation of the Space Station, the SOP will endeavor to establish a fixed value for the annual common system operations costs.

9.5.   Costs of user activities such as payload/experiment design, development, test and evaluation (DDT&E); payload ground processing; provision of payload/experiment spares and associated equipment; launch and return of payloads/experiments, spares and associated equipment; launch and return of the fraction of the Integrated Logistics System carriers provided for in Article 3.2 that is attributable to user payloads/experiments, spares and associated equipment; and any special user integration or user operations support, including specialized crew training, will be the responsibility of Space Station users of the partners or of individual partners as users. Such costs will not be shared among NASA, ESA and the other partners, nor will such costs contribute toward the satisfaction of common system operations costs responsibilities. In addition, the DDT&E and operations costs of the users' support centers will not be shared among NASA, ESA and the other partners.

9.6.   NASA, ESA and the other partners will not recoup their DDT&E costs for their elements from one another in the operation and utilization of the Space Station.

9.7.   In case of failure of any partner to perform its operations responsibilities or to provide for its share of common system operations costs, the partners will meet to discuss what action should be taken. Such action could result in, for example, an appropriate reduction of the failing partner's rights to its allocations.

[38]   *Article 10 - Safety*

10.1.   In order to assure safety, NASA has the responsibility, working with the other partners, to establish overall Space Station safety requirements and plans covering Phase C/D and Phase E. Such requirements and plans for Phase C/D have been established, and development of further safety requirements and plans for Phase C/D and Phase E and changes to safety requirements and plans will be processed, according to the procedures in Articles 7 and 8. As far as the elements separated from the manned base and their payloads are concerned, NASA has the responsibility to establish and implement overall safety requirements and plans governing the NASA-provided Polar Platform, and ESA has the responsibility to establish and implement overall safety requirements and plans governing the ESA-provided Polar Platform and the MTFF. The overall Space Station safety requirements and plans will be applicable to the MTFF insofar as it has effects on the manned base associated with its servicing at the manned base. STS safety requirements will be applicable to the ESA-provided Polar Platform insofar as it has effects on the STS associated with its servicing by the STS.

10.2.   Each partner will develop detailed safety requirements and plans, using its own standards where practicable, for its manned base hardware and software that meet or exceed the overall Space Station safety requirements and plans. Each partner will have the responsibility to implement applicable overall and detailed Space Station safety requirements and plans throughout the lifetime of the program, and to certify that such safety requirements and plans have been met with respect to the Space Station manned base elements and payloads it provides. ESA will have the responsibility to certify that the MTFF and ESA-provided Polar Platform and their payloads are safe. However, NASA will have the overall responsibility to certify that all Space Station manned base elements and payloads are safe, including the MTFF and its payloads insofar as they have effects on the manned base associated with their servicing at the manned base. NASA will also have the responsibility to certify that the ESA-provided Polar Platform and its payloads are safe insofar as they have effects on the STS associated with their servicing by the STS.

10.3.   NASA will conduct system safety reviews which ESA will support. NASA, ESA and the other partners will also conduct safety reviews of the elements and payloads they provide; NASA will participate in and support such reviews by the other partners. MOSST will also participate in and support safety reviews by the other partners as appropriate related to the MOSST-provided elements and MOSST payloads. NASA and MOSST [39] support to such safety reviews will include provision of necessary safety-related information to enable the other partners to conduct their reviews. Furthermore, status reports on safety requirements and plans will be a standard agenda item at the Program Management Reviews provided for in Article 7.1.i. The partners will participate as appropriate in any Space Station safety review boards established by NASA.

10.4.   NASA will have the responsibility for taking any decision necessary to protect the safety of the manned base, including all elements operating in conjunction with the manned base, or its crew in an emergency.

Article 11 - Space Station Crew

11.1.   ESA has the right to provide personnel to serve as Space Station crew from the time that ESA begins to share common system operations costs as provided in Article 9.3.5. NASA will provide flight opportunities for ESA Space Station crew satisfying the percentage of the total crew requirement equal to the percentage of manned base utilization resources allocated to ESA in Article 8.3.a.2. Flight of ESA Space Station crew will be satisfied over time, not necessarily on each specific crew rotation cycle. The SOP will review the implementation of this paragraph on a biennial basis.

11.2.   During assembly and verification, a fully trained ESA crew member will participate in the on-orbit assembly and system verification of the ESA-provided APM and other assigned flight element assembly and system verification tasks planned during that on-orbit period as provided in the verification plan described in Articles 6.1.a.4. and 6.2.a.3. Further, during the first two servicings of the MTFF at the manned base, a fully trained ESA crew member will participate in the relevant activities.

11.3.   Space Station crew will meet medical standards and security and suitability requirements developed by NASA in consultation with ESA and the other partners regarding Space Station crew qualifications for long-term manned space flight. NASA and ESA will jointly certify that these standards and requirements have been met by the ESA Space Station crew. Furthermore, the MCB may establish additional criteria for Space Station crew. Following certification, all Space Station crew will enter into an appropriate training cycle in order to acquire the skills necessary to conduct Space Station operations and utilization. Such training will be conducted in groups, subject to the requirements of different functional specializations. The training will include integrated manned systems operations training conducted primarily at NASA centers [40] and element-specific operations training conducted primarily by the partner providing the element at appropriate centers of all of the partners. In full consultation with ESA regarding the flight assignments of ESA crew members, NASA will designate, from among the certified Space Station crew, specific crew complements, which include the Space Station Commander, for specific crew rotation cycles, consistent with Article 11.1. NASA will designate specific crew complements to support payload requirements identified in the COUP. A specific crew complement will be trained as a team in preparation for a specific crew rotation cycle, subject to requirements of different functional specializations.

11.4.   NASA and ESA will be financially responsible for all compensation, medical expenses, subsistence costs on Earth, and training for Space Station crew which they provide. Full training for all assigned duties will be required.

11.5.   The Code of Conduct for the Space Station will be developed by NASA, with the full involvement of ESA, MOSST and the GOJ, and approved for the Space Station program in accordance with the principles for reaching decisions established in Article 8.1.b. It will, inter alia: establish a clear chain of command; set forth standards for work and activities in space, and, as appropriate, on the ground; establish responsibilities with respect to elements and equipment; set forth disciplinary regulations; establish physical and information security guidelines; and provide the Space Station Commander appropriate authority and responsibility, on behalf of all the partners, to enforce safety procedures and physical and information security procedures in or on the Space Station.

11.6.   ESA crew selected for operating the MTFF outside the operational CCZ of the manned base are not considered Space Station crew, pursuant to this Article, for the purposes of that activity.

Article 12 - Transportation, Communications and Other Non-Space Station Facilities

12.1.   Transportation

12.1.a.   For purposes of design of Space Station elements and payloads, NASA's STS is the baseline launch and return transportation system for the Space Station manned base and for the NASA-provided Polar Platform. ESA's Space Transportation System is the baseline launch transportation system for the MTFF and the ESA-provided Polar Platform.

12.1.b.   NASA will provide reimbursable STS launch services to ESA in connection with the assembly of the ESA-provided APM to the manned base and its initial outfitting in accordance with the program documentation described in Article 7.1. NASA will [41] also provide reimbursable launch and return services in connection with the logistics requirements of manned base elements. NASA will also provide reimbursable launch and return services in connection with the MTFF when it is serviced at the manned base and in connection with manned base users; availability of STS services for such purposes is as provided in Articles 8.3.a.4 and 8.3.c. NASA will also provide reimbursable launch services in connection with servicing of the ESA-provided Polar Platform, with details to be agreed by NASA and ESA, if appropriate STS capability exists and if ESA selects to use this capability. Reimbursement for such launch services may be in cash or agreed kind. All reimbursable STS services will be provided under launch services agreements. NASA will also provide launch and return services in connection with manned base common system operations logistics; costs for such services will be shared among the partners as provided in Article 9.3. ESA will provide the initial launch of the MTFF and the ESA-provided Polar Platform. ESA will also provide launch and return services in connection with the logistics requirements of the MTFF when it is not serviced at the manned base.

12.1.c. Other government or private sector space transportation systems of partners may be used in connection with the Space Station if they are compatible with the Space Station. Specifically, ESA will have the right of access to the Space Station manned base using the ESA Space Transportation System, including Ariane and Hermes. Recognizing that the responsibility for developing these systems and for making them technically and operationally compatible with the manned base rests with ESA, NASA will provide to ESA that information necessary for ESA to make them compatible. Technical, operational and safety requirements for access to the manned base will be controlled in appropriate program documentation as provided for in Articles 7 and 8.

12.1.d. With respect to financial conditions, NASA and ESA will provide reimbursable launch and return services to each other, to the other partners and to each other's and the other partners' users at prices they routinely charge comparable users. Launch and return services related to manned base common system operations logistics will also be made available by NASA on the same basis.

12.1.e. Both NASA and ESA will use their best efforts to accommodate additional launch and return requirements in relation to the Space Station, as well as proposed requirements and flight schedules related to the Space Station activities described above.

12.1.f. Each partner will respect the proprietary rights in and confidentiality of appropriately marked data and goods to be transported on its space transportation system.

[42] 12.2.   Communications

12.2.a. Space Station communications will involve space-to-ground, ground-to-space, ground-to-ground and space-to-space data transmission. The TDRSS space network is the baseline communication system for the manned base elements and payloads, as well as for the NASA-provided Polar Platform and its payloads. ESA's Data Relay Satellite system (EDRS) is the baseline communication system for the ESA-provided Polar Platform and the MTFF and their payloads. ESA will be responsible for ensuring communications compatibility of the MTFF with the manned base for proximity operations, docking and servicing and of the ESA-provided Polar Platform with the STS for servicing as applicable. On a reimbursable basis, NASA and ESA will use their best efforts to accommodate, with their respective communication systems, specific Space Station-related requirements of each other and the other partners. With respect to financial conditions, NASA and ESA will provide such communication services at prices no higher than those they routinely charge comparable customers. Other communication systems may be used on the manned base by ESA, the other partners or Space Station users if such communication systems are compatible with the manned base and manned base use of TDRSS. Technical and operational requirements related to Space Station communications will be controlled in appropriate program documentation as provided for in Articles 7 and 8.

12.2.b. NASA and ESA will consult regarding the possible future addition of manned base capability to accommodate ESA-provided facilities permitting manned base use of EDRS, if compatible with the manned base and with manned base use of TDRSS.

12.2.c. Unless otherwise agreed by NASA and ESA, ground-to-ground transmission of polar platform data from one partner to the other partners or the other partners' users will conform to the communications transportation formats, protocols and standards agreed to by the Consultative Committee for Space Data Systems (CCSDS).

12.2.d. Partners and users of the partners may implement measures to ensure confidentiality of their utilization data passing through the Space Station Information System and other communication systems being used in connection with the Space Station. (Notwithstanding the foregoing, data which are necessary to assure safe operations will be made available according to procedures in the Utilization Management Plan and their use will be restricted to safety purposes only.) Each partner will respect the proprietary rights in, and the confidentiality of, the utilization data passing through its communication systems, including its ground network and the communication systems of its contractors, when providing communication services to another partner.

[43]12.3.   Other Non-Space Station Facilities

12.3.a. Should ESA desire to use the Space Shuttle, Spacelab, or other NASA facilities on a cooperative or reimbursable basis to support the development of its Space Station Utilization Plan or to support its Space Station detailed design or development activities, NASA will use its best efforts to accommodate ESA's proposed requirements and schedules. Likewise, should NASA desire to use Ariane, Hermes or other ESA facilities on a cooperative or reimbursable basis to support the development of its Space Station Utilization Plan or to support its Space Station detailed design or development activities, ESA will use its best efforts to accommodate NASA's proposed requirements and schedules.

12.3.b. If NASA and ESA agree that it is appropriate and necessary for the conduct of the cooperative program, NASA and ESA will use their good offices in connection with attempting to arrange for the use of U.S. and European Governments' or contractors' facilities by the Parties and/or their contractors. Such use will be subject to separate arrangements between the user and the owner of the facilities.

Article 13 - Advanced Development Program

13.1.   NASA and ESA each are conducting Space Station advanced development programs in support of their respective detailed design and development activities. Cooperation in such advanced development activities will be considered on a case-by-case basis and entered into where it is advantageous to both sides and where there are reciprocal opportunities.

13.2.   ESA proposals to use NASA advanced development test beds or other NASA facilities in support of ESA's Space Station advanced development program will be considered on a case-by-case basis either on a cooperative or reimbursable basis. Likewise, NASA proposals to use ESA's facilities in support of NASA's Space Station advanced development program will be considered on a case-by-case basis either on a cooperative or reimbursable basis.

13.3.   Should ESA desire to use the Space Shuttle or Spacelab on a cooperative or reimbursable basis to support ESA Space Station advanced development activities, NASA will use its best efforts to accommodate ESA's proposed requirements and flight schedules. Likewise, should NASA desire to use ESA launch vehicles on a cooperative or reimbursable basis to support NASA Space Station advanced development activities, ESA will use its best efforts to accommodate NASA's proposed requirements and flight schedules.

[44]   *Article 14 - Space Station Evolution*

14.1.   The partners intend that the Space Station will evolve through the addition of capability and will strive to maximize the likelihood that such evolution will be effected through contributions from all the partners. To this end, it will be the object of the Parties to provide, where appropriate, the opportunity to the other partners to cooperate in their respective proposals for additions of evolutionary capability. The Space Station together with its additions of evolutionary capability will remain a civil station, and its operation and utilization will be for peaceful purposes, in accordance with international law.

14.2.   This MOU sets forth rights and obligations concerning only the elements listed in Article 3, except that this Article and Article 16 of the Intergovernmental Agreement will apply to any additions of evolutionary capability. As such, this MOU does not commit either Party to participate in, or grant either Party rights in, the addition of evolutionary capability.

14.3.   NASA and ESA agree to study evolution concepts for the Space Station during Phase C/D and Phase E. NASA will be responsible for development of overall manned base evolution concepts, in consultation with ESA and the other partners, and for integrating ESA's and the other partners' evolution concepts into an overall manned base evolution plan. ESA will be responsible for development and decision on subsequent implementation of evolution concepts for the ESA-provided Polar Platform and for the MTFF insofar as they have no technical or operational impacts on the STS or the manned base, in accordance with Articles 14.6 and 14.7.

14.4.   NASA, ESA, and the other partners will participate in an International Evolution Working Group (IEWG) to coordinate their respective evolution studies and to consider overall Space Station evolution concepts and planning activities.

14.5.   The MCB will review specific evolutionary capabilities proposed by any partner, assess the impacts of those plans on the other partners' elements and on the manned base, and review recommendation for minimizing potential impacts on Space Station activity during the addition of evolutionary capabilities.

14.6.   Following the review and assessment provided for in Article 14.5, and consistent with the provisions of the Intergovernmental Agreement, cooperation between or among partners regarding the sharing of addition(s) of evolutionary capability will require either amendment of the relevant NASA-ESA, NASA-GOJ and NASA-MOSST MOU's or a separate agreement to which, to the extent that such addition is on the manned base or has a technical or operational impact on the STS or the manned base, NASA is a party to ensure that such addition is [45] consistent with NASA's overall programmatic responsibilities as detailed in this MOU.

14.7.   Following the review and assessment provided for in Article 14.5, and consistent with the provisions of the Intergovernmental Agreement, the addition of evolutionary capability by one partner will require prior notification of the other partners, and, to the extent that such addition is on the manned base or has a technical or operational impact on the STS or the manned base, an agreement with NASA to ensure that such addition is consistent with NASA's overall programmatic responsibilities as detailed in this MOU.

14.8.   The addition of evolutionary capability will in no event alter the rights and obligations of either Party to this MOU concerning the elements listed in Article 3, unless otherwise agreed by the affected Party.

Article 15 - Cross-Waiver of Liability: Exchange of Data and Goods;
Treatment of Data and Goods in Transit; Customs and Immigration;
Intellectual Property; Criminal Jurisdiction

The Parties note that, with respect to the cross-waiver of liability, exchange of data and goods, treatment of data and goods in transit, customs and immigration, intellectual property and criminal jurisdiction, the relevant provisions of the Intergovernmental Agreement apply.

Article 16 - Financial Arrangements

16.1.   Each Party will bear the costs of fulfilling its responsibilities, including but not limited to costs of compensation, travel and subsistence of its own personnel and transportation of all equipment and other items for which it is responsible under this MOU. However, as provided in Article 9.3, the partners will equitably share common system operations costs.

16.2.   The ability of each Party to carry out its obligations is subject to its funding procedures and the availability of appropriated funds.

16.3.   In the event that funding problems are arising that may affect a partner's ability to fulfill its responsibilities under this MOU, that partner will promptly notify and consult with the other partners. Further, the Parties undertake to grant high priority to their Space Station programs in developing their budgetary plans.
[46]

16.4.   The Parties will seek to minimize the exchange of funds while carrying out their respective responsibilities in this cooperative program, including, if they agree, through the use of barter, that is, the provision of goods or services.

Article 17 - Public Information

17.1.   NASA and ESA will be responsible for the development of an agreed Public Affairs Plan that will specify guidelines for NASA/ESA cooperative public affairs activities during the detailed design, development, operation and utilization of the Space Station.

17.2.   Within the Public Affairs Plan guidelines, both NASA and ESA will retain the right to release public information on their respective portions of the program. NASA and ESA will undertake to coordinate with each other, and, as appropriate, with the other partners, in advance concerning public information activities which relate to each other's responsibilities or performance in the Space Station program.

Article 18 - Consultation and Settlement of Disputes

18.1.   The Parties agree to consult with each other and with the other partners promptly when events occur or matters arise which may occasion a question of interpretation or implementation of the terms of this MOU.

18.2.   In the case of a question of interpretation or implementation of the terms of this MOU, such question will be first referred to the NASA Associate Administrator for Space Station and the ESA Director of Space Station and Platforms for settlement. The Parties recognize that in the case of a question concerning the commitments made in this MOU to STA and/or MOSST, the consultations will be broadened so as to include the STA Director General of the Research and Development Bureau and/or the MOSST Deputy Secretary, Space Policy Sector.

18.3.   Any question of interpretation or implementation of the terms of this MOU which has not been settled in accordance with Article 18.2 will be referred to the NASA Administrator and the ESA Director General for settlement. The Parties recognize that in case of a question concerning the commitments made in this MOU to STA and/or MOSST, the matter will also be referred to the Minister of State for Science and Technology of Japan and/or the Secretary of MOSST.

18.4.   Any issues arising out of this MOU not satisfactorily settled through consultation, pursuant to this Article may be [47] pursued in accordance with the relevant provisions of the Intergovernmental Agreement.

18.5.   Unless otherwise agreed between NASA and ESA, implementation of decisions made pursuant to mechanisms provided for in this MOU will not be held in abeyance pending settlement of issues under this Article.

Article 19 - Entry into Force

19.1.   Pursuant to the Arrangement Concerning Application of the Space Station Intergovernmental Agreement Pending its Entry into Force, which became effective on September 29, 1988, this MOU will enter into force after signature of both the NASA Administrator or his designee and the ESA Director General or his designee, upon written notification by each Party to the other that all procedures necessary for its entry into force have been completed.

19.2.   Pending the entry into force of the Intergovernmental Agreement between the United States and the European Partner in accordance with Article 25 of that Agreement, the Parties agree to abide by the relevant terms of that Agreement.

19.3.   If the United States or the European Partner withdraws from the Arrangement Concerning Application of the Space Station Intergovernmental Agreement Pending its Entry into Force, the corresponding Cooperating Agency will be deemed to have withdrawn from this MOU effective from the same date.

19.4.   If, by December 31, 1992, the Intergovernmental Agreement has not yet entered into force between the United States and the European Partner in accordance with Article 25 of that Agreement, the Parties will consider what steps are necessary and appropriate to take account of that circumstance.

19.5.   If the United States or the European Partner gives notice of withdrawal from the Intergovernmental Agreement in accordance with Article 21 of that Agreement, the corresponding Cooperating Agency will be deemed to have withdrawn from this MOU effective from the same date.

Article 20 - MOU Amendments

This MOU may be amended at any time by written agreement of the Parties. Any amendment must be consistent with the Intergovernmental Agreement. To the extent that a provision of this MOU creates specific rights or obligations accepted by another partner, that provision may be amended only with the written consent of that partner.

[48]   ### Article 21 - Review

Upon the request of either Party, the Parties will meet for the purpose of reviewing and promoting cooperation in the Space Station. In the process of this review, the Parties may consider amendments to this MOU.

Article 22 - Definitions and Explanations

22.1.   In addition to the definitions specified in the Intergovernmental Agreement, the following definitions will apply to this MOU:

"international Space Station complex," also "Space Station," means the collection of elements listed in Article 3;

"manned base" means Space Station flight elements excluding the polar platforms and the MTFF;

"Parties" means NASA and ESA;

"partners" means NASA, ESA, STA and MOSST.

22.2. Explanation of the following terms may be found in this MOU in the Articles noted:

"Accommodations" - Article 8.1.d
"Command and Control Zone (CCZ)" - Article 8.1.a
"Common system operations costs" - Article 9.3
"Composite Operations Plan (COP)" - Article 8.2.d
"Composite Utilization Plan (CUP)" - Article 8.3.f
"Consolidated Operations and Utilization Plan (COUP)" - Article 8.1.c
"Flight elements" - Article 3
"Increment Plan (IP)" - Article 8.2.f
"Infrastructure" - Article 8.1.b
"Multilateral Coordination Board (MCB)" - Article 8.1.b
"Payload Operations Integration Center (POIC)" - Article 8.3.i
"Program Coordination Committee (PCC)" - Article 7.1.b
"Resources" - Article 8.1.d and Article 8.3.a.4
"Space Station Control Board (SSCB)" - Article 7.1.d
"Space Station Control Center (SSCC)" - Article 8.2.g
"Space Station-unique ground elements" - Article 3
"System Operations Panel (SOP)" - Article 8.2.a and Article 8.2.b
"Tactical Operations Plan (TOP)" - Article 8.2.f
"User Operations Panel (UOP)" - Article 8.3.d

[49] DONE at Washington, this 29th day of September, 1988, in two originals in the English, French, German and Italian languages, each version being equally authentic.

[50] FOR THE UNITED STATES NATIONAL AERONAUTICS AND SPACE ADMINISTRATION:

FOR THE EUROPEAN SPACE AGENCY:

POUR L'ADMINISTRATION NATIONALE DE L'AERONAUTIQUE ET DE L'ESPACE DES ETATS UNIS:

POUR L'AGENCE SPATIALE EUROPEENNE:

FÜR DEI NATIONALE LUFT UND RAUMFAHRTORGANISATION DER VEREINIGTEN STAATEN:

FÜR DEI EUROPAISE WELTRAUMORGANISATION:

PER L'AMMINISTRAZIONE NAZIONALE PER L'AERONAUTICA STATI UNITI:

PER L'AGENZIA SPAZIALE EUROPEA:

signed by Dale D. Myers

signed by Reimar Leust

Document I-36

Document title: "Draft Proposals for US-USSR Space Cooperation," April 4, 1961.

Source: NASA Historical Reference Collection, NASA History Office, NASA Headquarters, Washington, D.C.

President John F. Kennedy called for U.S.-Soviet space cooperation in his January 20, 1961, inaugural address and his first State of the Union address a few days later. To examine the possibilities for such cooperation, presidential science advisor Jerome Wiesner set up both an external advisory group and an internal government study group. A number of drafts of a white paper on the topic were prepared. As the white paper was nearing completion, the Soviet Union launched Yuri Gagarin into orbit on April 12, 1961. A few days later, President Kennedy decided that he had to compete—not cooperate—in space, and the white paper was temporarily set aside.

[1]   April 4, 1961

Draft Proposals for US-USSR Space Cooperation

OBJECTIVES

The objectives are to confirm concretely the U.S. preference for a cooperative rather than competitive approach to space exploration, to contribute to reduction of cold war tensions by demonstrating the possibility of cooperative enterprise between the U.S. and the USSR in a field of major public concern, and to achieve the substantive advantages of cooperation that in major projects would impose more of a strain on economic and manpower resources if carried out unilaterally.

GUIDELINES

The proposals seek to (a) maximize acceptability by the USSR, and (b) minimize the potential for misunderstanding and obstructionism which must be recognized to exist in any joint program with the Soviet Union. The proposals therefore have, in general, the following character:
  (1) Valid scientific objectives.
  (2) Comparable contributions by U.S. and USSR.
  (3) Technical and economic feasibility for U.S. portion.
  (4) Minimal interference with on-going U.S. programs.
  (5) Minimal grounds for Soviet suspicions of U.S. motives (success, surveillance, etc.)
  (6) Opportunities for third-nation participation at appropriate time.
The proposals fall into three categories:
  (a) The employment of existing or easily attainable ground facilities for exchange of information and services in support of orbiting experiments.
  (b) The coordination of independently-launched satellite experiments so as to achieve simultaneous but complementary coverage of agreed phenomena.
[2] (c) Coordination of or cooperation in ambitious projects for the manned exploration of the moon and the unmanned exploration of the planets.
The three categories of proposals are advanced in order to offer the Soviet Union a wide range of choice and avoid the appearance of "pushing" a pre-selected objective. While the costs are estimated by NASA to range from relatively insignificant levels in Category (a) to $15-20 million in Category (b) and, very roughly, $10 billion in Category

(c), it may be assumed that the Soviet Union as well as ourselves is likely to pursue the more costly programs in any event.

Such cooperation as is discussed here should be proposed and carried out on the basis of an expanding U.S. program of space science and exploration, and without prejudice to continuing joint enterprise with and assistance to the free world.

PROCEDURE

Overtures should be made at Governmental levels, inviting the USSR to engage in cooperative enterprise such as the proposals below. Soviet counter-suggestions of areas of cooperation would also be invited. The initial discussions would seek a go-ahead for exploratory technical talks preliminary to agreements in principle. Privacy in all such discussions would appear to enhance the chances of success. Technical advice should be available at all times.

[3] *PROPOSALS*
  Category (a)
These proposals for the most part call for the use of ground facilities for mutual service:

(i) The U.S. and the USSR might agree to provide ground-based support on a reciprocal basis for space experiments, e.g.,

– When either nation launches a satellite or probe carrying a magnetometer experiment, the other would collect rapid-run magnetograms at its ground observatories. (A Soviet scientist has recently promised to do this in connection with the U.S. P-14 probe, following a private request.)

– When either nation launches a meteorological satellite, the other would carry out routine and special (airborne, balloon-borne, all-sky camera) weather observations synchronized with the passes of the satellite, analyze the data from both sources, and participate in scientific exchanges of the results.

– Similar arrangements would be useful in connection with ionospheric, auroral, and other geophysical researches.

(ii) The U.S. and the USSR could agree to record telemetry from each other's satellites, exchanging the resulting tapes as requested. Each would furnish the necessary orbital information and telemetry calibrations to the other. This would be of particular value in sun-related experiments and could extend to the exchange of command signals to permit the best-situated nation to energize a given experiment under certain conditions of solar activity.

(iii) In the communications field, the USSR may wish to employ a ground facility for long-distance experimental transmission of voice or TV signals by means of communications satellites to be launched by NASA after mid-1962 (Projects Relay/Rebound). Such facilities are being prepared also by the U.K. and France. Transmissions may be effected between the latter and the USSR (by means of a U.S. satellite) as usefully as between the U.S. and the USSR. (If *supplementary* equipment peculiar to such experimental testing in this case is required by the USSR, NASA could provide it at costs ranging up to $2 million.)

[4] The exchanges proposals in (a) have been sought, almost with complete unsuccess, at government agency and scientific society levels since the beginning of the IGY. They are included because of their inherent desirability and because a somewhat greater chance of acceptance may follow if initiated at higher levels. (The programs in Categories (b) and (c) have not yet been proposed to the Soviet Union.)

The proposals made in Category (a) are for *coordinated* rather than *interdependent* efforts and thus would avoid difficulties which may be associated with the latter type of cooperation with the USSR.

[5] *Category (b)*

(i) Weather satellites promise broad near-future benefits to the peoples of the world. Equal participation by the U.S. and the USSR in coordinated launching of experimental satellites capable of providing typhoon warnings, etc., would have great impact.*

One specific proposal is that the U.S. and the USSR each place in polar orbit a meteorological satellite to record cloud-cover and radiation-balance data, such that

- The two satellites have reasonably overlapping lifetimes (at least three months).
- The satellites orbit in planes at right angles to each other, providing at least six-hour coverage of the earth.
- The data characteristics permit reception and analysis interchangeably, if possible.
- Each country may receive telemetry from the other's satellite through continuous readout if power sources permit or by command if otherwise.
- Camera resolutions are appropriate only for the objective—photographs of cloud cover.
- The results are to be made available to the scientific community (World Data Centers and WMO).

(ii) Coordinated programs including experimental or research satellite launchings in other fields than meteorology (e.g., communications) could also be of value. In the field of geophysics, for example, there are possibilities for the useful coordination of the orbits of contemporaneous satellites so as to obtain measurements under contrasting or complementary conditions.

(iii) Simultaneous and coordinated rocket launchings from a number of stations covering a wide range of latitudes and longitudes would for the first time provide a global picture of the properties of the atmosphere at a given instant of time, if conducted on a scale greater than now done during International Rocket Weeks.

[6] The first proposal in Category (b) above falls in the meteorological field, in which the U.S. appears to lead. While the USSR has not yet done anything in this field, it has on one occasion indicated at the highest scientific level that space meteorology is favorably viewed as an area for cooperation. A generous time-scale (or offer to provide instrumentation) might moderate the negative factor.

The proposals made in Category (b) are, like those in Category (a), for *coordinated* rather than *interdependent* efforts and thus would avoid difficulties which may be associated with the latter type of cooperation with the USSR.

[7] *Category (c)*

These proposals related to the exploration of celestial bodies.

(i) Mars or Venus Programs.

Planetary investigations are immensely difficult undertakings requiring protracted programs of great complexity and variety, progressing through fly-bys, orbiters, hard and soft landings, and surface prospecting. The U.S. and the USSR could coordinate their independent programs so as to provide for a useful sequencing and, perhaps, sharing of experimental missions, with scientific benefits and economics. Full data exchange, guaranteed by provision of telemetry calibrations, should be provided. If cooperation is interrupted, no less is sustained and the programs may proceed independently.

The U.S. and USSR could, alternatively, enter into a joint program that would mean more intimate involvement; such a program would include cooperative development of

---

* Broader cooperation in meteorology is possible and desirable. A specific proposal for a major worldwide cooperative meteorological program, in which satellites would be a part, is being developed separately.

equipment and sharing of experimental missions, and would point toward eventual joint launching of probes.

(ii) Manned Exploration of Moon.

The presence of man will immeasurably enhance the scientific investigation of the Moon—so critical for understanding the origin of the solar system—by providing the resourcefulness, flexibility and opportunity for improvisation available only with man.

As a first step in non-limited cooperative effort, the U.S. and the USSR would each undertake to place a small party (about 3) of men on the moon for scientific purposes and return them to earth.

As in planetary programs, a more extensive cooperative program could also be envisaged in which the U.S. and USSR enter into a joint manned lunar program, including cooperative development, planning, and international exploration.

The proposals made in Category (c), in the lunar and planetary fields, suggest programs for which the USSR has demonstrably greater existing capability. Inclusion of both categories in proposals to the USSR may therefore be effective.

[8] No significant Mars probe capability now exits in the U.S. By 1964, Centaur should permit significant fly-bys only, while Saturn C-1 would put about 300 pound payloads in orbit after 1964.

The Mars/Venus program is a long-range one whose cost varies widely with numbers of launchings, nature of payloads, and extent of back-up. A balanced program (unmanned), including some 15 Venus shots and 8 Mars shots in the next decade, may cost in the order of $1 billion.

Neither country now possesses a capability for a manned lunar project. It will require boosters of the order of Saturn C-2 using orbital rendezvous and refueling techniques (still to be attempted and perfected) for the upper stages. At least six Saturn C-2's would be required for a single mission, plus appropriate back-up. The time-scale is probably a decade, during which some 70-80 Saturns would be required for developmental purposes, and the cost is roughly of the order of $10 billion. During the decade, alternative vehicle systems may conceivably become available, obviating the difficult rendezvous requirement.

In the suggestions for cooperation given above, it can be seen that the degree of involvement between the U.S. and the USSR can in principle be varied from coordination of national programs to full cooperation on joint endeavors.

It is possible to *restrict* proposals which may be made to the Soviet Union to the level of coordination of essentially independent programs. Benefits would derive from joint planning and organization of such coordinated efforts. This might have the advantages of greater acceptability in the U.S. and in the Soviet Union (where suspicions of U.S. motivations would be present in any case). It may also be more realistic in terms of the technical exchange and access which may be feasible.

On the other hand, it would be possible to indicate a *range* of possible relationships to the Soviet Union, extending to interdependent programs and leaving it to them to select the starting level.

As we contemplate programs that involve greater degrees of cooperation, we must also anticipate certain increased difficulties. These would include the risk that the whole program would be lost if one or the other participant withdrew because of political or other reasons: the fact that we would have to be prepared to admit Russians to installations such as Cape Canaveral and to show them details of our booster and payload systems (of course, the Russians [9] would have to do the same if they agreed to intimate cooperation), and the possibility that Congressional, scientific and public support might also be more difficult because of the very high costs involved, coupled with the potential damage to our program if the Soviets became obstructive or withdraw. Positive factors must also be considered, of course, such as the impact on U.S./USSR relations growing out of intimate cooperation on large and meaningful projects, and the advantages occurring to both countries in carrying out space programs utilizing the best of what each has to offer without unnecessary time pressures.

At any level of relationships, proposals for cooperation in Category (a) have the greatest potential for matching the President's theme that "Both nations would help themselves as well as other nations by removing those endeavors from the bitter and wasteful competition of the Cold War." The United States considers exploration of the celestial bodies, particularly manned space exploration, to be perhaps the most challenging adventure of this century. This venture should be conducted on behalf of the human race and the earth as a whole, not on behalf of any single nation. The vigorous and accelerating United States space exploration program is proceeding in this spirit. If the Soviet Union shares this conception, then planning should be undertaken promptly for cooperative manned exploration of the moon and unmanned exploration of Mars and Venus. These projects should of course be open to the participation of all interested countries [and might come under the auspices of the United Nations]. They could, however, be undertaken most constructively only if the United States and the Soviet Union agree on objectives and on coordination of their efforts for the most rapid progress and the most efficient use of human and natural resources.

Document I-37

Document title: John F. Kennedy, to Soviet Union Chairman Nikita Khrushchev, March 7, 1962.

Source: NASA Historical Reference Collection, NASA History Office, NASA Headquarters, Washington, D.C.

From the day he was inaugurated, President John F. Kennedy had hoped that the Soviet Union would be willing to cooperate with the United States in space exploration and exploitation. Kennedy decided in 1961 that he had to compete with the Soviet Union in dramatic space achievements, but he still hoped that other areas of space could serve as arenas for cooperation. Nikita Khrushchev seemed to open the door to such cooperation in his February 21, 1962, message to Kennedy, which congratulated the United States on its first human orbital flight, the Freedom 7 *Mercury mission of John Glenn. Kennedy replied immediately, telling the Soviet premier that the United States would soon forward specific proposals for cooperation. After a rapid review of cooperative possibilities within the U.S. government, Kennedy forwarded this letter on March 7, proposing specific cooperative initiatives to the Soviet Union. This letter marked the beginning of substantive cooperation between the two space superpowers.*

---

[1] Dear Mr. Chairman:

On February twenty-second last I wrote you that I was instructing appropriate officers of this Government to prepare concrete proposals for immediate projects of common action in the exploration of space. I now present such proposals to you.

The exploration of space is a broad and varied activity and the possibilities for cooperation are many. In suggesting the possible first steps which are set out below, I do not intend to limit our mutual consideration of desirable cooperative activities. On the contrary, I will welcome your concrete suggestions along these or other lines.

1.   Perhaps we could render no greater service to mankind through our space programs than by the joint establishment of an early operational weather satellite system. Such a system would be designed to provide global weather data for prompt use by any nation. To initiate this service, I propose that the United States and the Soviet Union each launch a satellite to photograph cloud cover and provide other agreed meteorological services for all nations. The two satellites would be placed in near-polar orbits in planes approximately perpendicular to each other, thus providing regular coverage of all areas. This immensely valuable data would then be disseminated through normal international meteorological channels and would make a significant contribution to the research and service programs now under study by the World Meteorological Organization in response to Resolution 1721 (XVI) adopted by the United Nations General Assembly on December 20, 1961.

2.   It would be of great interest to those responsible for the conduct of our respective space programs if they could obtain operational tracking services from each other's territories. Accordingly, I propose that each of our countries establish and operate a radio tracking station to provide tracking services to the other, utilizing equipment which we would each provide to the other. Thus, the United States would provide the technical equipment for a tracking station to be established in the Soviet Union and to be operated by Soviet technicians. The United States would in turn establish and operate a radio tracking station utilizing Soviet equipment. Each country would train the other's technicians in the operation of its equipment, would utilize the station located on its territory to provide tracking services to the other, and would afford such access as may be necessary to accommodate modification and maintenance of equipment from time to time.

[2] 3.   In the field of the earth sciences, the precise character of the earth's magnetic field is central to many scientific problems. I propose therefore that we cooperate in mapping the earth's magnetic field in space by utilizing two satellites, one in a near-earth orbit and the second in a more distant orbit. The United States would launch one of these satellites while the Soviet Union would launch the other. The data would be exchanged throughout the world scientific community, and opportunity for correlation of supporting data obtained on the ground would be arranged.

4.   In the field of experimental communications by satellite, the United States has already undertaken arrangements to test and demonstrate the feasibility of intercontinental transmissions. A number of countries are constructing equipment suitable for participation in such testing. I would welcome the Soviet Union's joining in this cooperative effort which will be a step toward meeting the objective, contained in United Nations General Assembly Resolution 1721 (XVI), that communications by means of satellites should be available to the nations of the world as soon as practicable on a global and non-discriminatory basis. I note also that Secretary Rusk has broached the subject of cooperation in this field with Minister Gromyko and that Mr. Gromyko has expressed some interest. Our technical representatives might now discuss specific possibilities in this field.

5.   Given our common interest in manned space flights and in insuring [sic] man's ability to survive in space and return safely, I propose that we pool our efforts and exchange our knowledge in the field of space medicine, where future research can be pursued in cooperation with scientists from various countries.

Beyond these specific projects we are prepared now to discuss broader cooperation in the still more challenging projects which must be undertaken in the exploration of outer space. The tasks are so challenging, the costs so great, and the risk to the brave men who engage in space exploration so grave, that we must in all good conscience try every possibility of sharing these tasks and costs and of minimizing these risks. Leaders of the United

States space program have developed detailed plans for an orderly sequence of manned and unmanned flights for exploration of space and the planets. Out of discussion of these plans, and of our own, for undertaking the tasks of this decade would undoubtedly emerge possibilities for substantive scientific and technical cooperation in manned and unmanned space investigation. Some possibilities are not yet precisely identifiable, but should become clear as the space programs of our two countries proceed.

[3] In the case of others it may be possible to start planning together now. For example, we might cooperate in unmanned exploration of the lunar surface, or we might commence now the mutual definition of steps to be taken in sequence for an exhaustive scientific investigation of the planet Mars or Venus, including consideration of the possible utility of manned flight in such programs. When a proper sequence for experiments has been determined, we might share responsibility for the necessary projects. All data would be made freely available.

I believe it is both appropriate and desirable that we take full cognizance of the scientific and other contributions which other states the world over might be able to make in such programs. As agreements are reached between us on any parts of these or similar programs, I propose that we report them to the United Nations Committee on the Peaceful Uses of Outer Space. The Committee offers a variety of additional opportunities for joint cooperative efforts within the framework of its mandate as sets forth in General Assembly Resolutions 1472 (XIV) and 1721 (XVI).

I am designating technical representatives who will be prepared to meet and discuss with your representatives our ideas and yours in a spirit of practical cooperation. In order to accomplish this at an early date I suggest that the representatives of our countries, who will be coming to New York to take part in the United Nations Outer Space Committee, meet privately to discuss the proposals set forth in this letter.

Sincerely,

John F. Kennedy

Document I-38

Document title: Nikita Khrushchev, to President John F. Kennedy, March 20, 1962.

Source: NASA Historical Reference Collection, NASA History Office, NASA Headquarters, Washington, D.C.

Nikita Khrushchev replied to President Kennedy's March 7 letter within two weeks. With his acceptance in principle of the concept of U.S.-U.S.S.R. space cooperation, discussions could begin between NASA and its Soviet counterparts regarding specific cooperative undertakings. While the need for progress on disarmament was mentioned in the Khrushchev letter, it was not made a precondition for cooperation.

---

[1] Dear Mr. President:

Having carefully familiarized myself with your message of March 7 of this year, I note with satisfaction that my communication to you of February 21 containing the proposal that our two countries unite their efforts for the conquest of space has met with the necessary understanding on the part of the Government of the United States.

In advancing this proposal, we proceeded from the fact that all peoples and all mankind are interested in achieving the objective of exploration and peaceful use of outer space, and that the enormous scale of this task, as well as the enormous difficulties which must be overcome, urgently demand broad unification of the scientific, technical, and material capabilities and resources of nations. Now, at a time when the space age is just dawning, it is already evident how much men will be called upon to accomplish. If today the genius of man has created space ships capable of reaching the surface of the moon with great accuracy and of launching the first cosmonauts into orbit around the earth, then tomorrow manned spacecraft will be able to race to Mars and Venus, and the farther they travel the wider and more immense the prospects will become, for man's penetration into the depths of the universe.

The greater the number of countries making their contribution to this truly compli-cated endeavor, which involves great expense, the more swiftly will the conquest of space in the interests of all humanity proceed. And this means that equal opportunities should be made available for all countries to participate in international cooperation in this field. It is precisely this kind of international cooperation that the Soviet Union unswervingly advocates, true to its policy of developing and strengthening friendship between peoples. As far back as the beginning of 1958 the Soviet Government proposed the conclusion of a broad international agreement on cooperation in the field of the study and peaceful use of outer space and took the initiative in raising this question for examination by the United Nations. In 1961, immediately after the first space flight by man had been achieved in the Soviet Union, we reaffirmed our readiness to cooperate and unite our efforts with those of other countries, and most of all with your country, which was then making prepa-rations for similar flights. My message to you of February 21, 1962 was dictated by these same aspirations and directed toward this same purpose.

[2] The Soviet Government considers and has always considered the successes of our country in the field of space exploration as achievements not only of the Soviet people but of all mankind. The Soviet Union is taking practical steps to the end that the fruits of the labor of Soviet scientists shall become the property of all countries. We widely publish notification of all launchings of satellites, space ships and space rockets, reporting all data pertaining to the orbit of flight, weight of space devices launched, radio frequencies, etc.

Soviet scientists have established fruitful professional contact with their foreign col-leagues, including scientists of your country, in such international organizations as the Committee of Outer Space Research and the International Astronautical Federation.

It seems to me, Mr. President, that the necessity is now generally recognized for fur-ther practical steps in the noble cause of developing international cooperation in space research for peaceful purposes. Your message shows that the direction of your thoughts does not differ in essence from what we conceive to be practical measures in the field of such cooperation. What, then, should be our starting point?

In this connection I should like to name several problems of research and peaceful use of space, for whose solution it would in our opinion be important to unite the efforts of nations. Some of them, which are encompassed by the recent U.N. General Assembly res-olution adopted at the initiative of our two countries, are also mentioned in your message.

1.   Scientists consider that the use of artificial earth satellites for the creation of international systems of long-distance communication is entirely realistic at the present stage of space research. Realization of such projects can lead to a significant improvement in the means of communication and television all over the globe. People would be pro-vided with a reliable means of communication and hitherto unknown opportunities for broadening contacts between nations would be opened. So let us begin by specifying the definite opportunities for cooperation in solving this problem. As I understood from your message, the U.S.A. is also prepared to do this.

2. It is difficult to overestimate the advantage that people would derive from the organization of a world-wide weather observation service using artificial earth satellites. Precise and timely weather prediction would be still another important step on the path to man's subjugation of the forces of nature; it would permit him to combat more successfully the [3] calamities of the elements and would give new prospects for advancing the well-being of mankind. Let us also cooperate in this field.

3. It seems to us that it would be expedient to agree upon organizing the observation of objects launched in the direction of the moon, Mars, Venus, and other planets of the solar system, by radio-technical and optical means, through a joint program.

As our scientists see it, undoubted advantage would be gained by uniting the efforts of nations for the purpose of hastening scientific progress in the study of the physics of interplanetary space and heaven[ly] bodies.

4. At the present stage of man's penetration into space, it would be most desirable to draw up and conclude an international agreement providing for aid in searching for and rescuing space ships, satellites and capsules that have accidentally fallen. Such an agreement appears all the more necessary, since it might involve saving the lives of cosmonauts, those courageous explorers of the far reaches of the universe.

5. Your message contains proposals for cooperation between our countries in compiling charts of the earth's magnetic field in outer space by means of satellites, and also for exchanging knowledge in the field of space medicine. I can say that Soviet scientists are prepared to cooperate in this and to exchange data regarding such questions with scientists of other countries.

6. I think, Mr. President, that the time has also come for our two countries, which have advanced further than others in space research, to try to find a common approach to the solution of the important legal problem with which life itself has confronted the nations in the space age. In this connection I find it a positive fact that at the UN General Assembly's 16th session the Soviet Union and the United States were able to agree upon a proposal on the first principles of space law which was then unanimously approved by the members of the UN: a proposal on the applicability of international law, including the UN charter, in outer space and on heavenly bodies; on the accessibility of outer space and heavenly bodies for research and use by all nations in accordance with international law; and on the fact that space is not subject to appropriation by nations.

Now, in our opinion, it is necessary to go further.

[4] Expansion of space research being carried out by nations definitely makes it necessary to agree also that in conducting experiments in outer space no one should create obstacles for space study and research for peaceful purposes by other nations. Perhaps it should be stipulated that those experiments in space that might complicate space research by other countries should be the subject of preliminary discussion and agreement on an appropriate international basis.

I have named, Mr. President, only some of the questions whose solution has, in our view, now become urgent and requires cooperation between our countries. In the future, international cooperation in the conquest of space will undoubtedly extend to ever newer fields of space exploration if we can now lay a firm foundation for it. We hope that scientists of the USSR and the U.S.A. will be able to engage in working out and realizing the many projects for the conquest of outer space hand in hand, and together with scientists of other countries.

Representatives of the USSR on the UN Space Committee will be given instructions to meet with representatives of the United States in order to discuss concrete questions of cooperation in research and peaceful use of outer space that are of interest to our countries.

Thus, Mr. President, do we conceive of—shall we say—heavenly matters. We sincerely desire that the establishment of cooperation in the field of peaceful use of outer space facilitate the improvement of relations between our countries, the easing of international tension and the creation of a favorable situation for the peaceful settlement of urgent problems here on our own earth.

At the same time it appears obvious to me that the scale of our cooperation in the peaceful conquest of space, as well as the choice of the lines along which such cooperation would seem possible is to a certain extent related to the solution of the disarmament problem. Until an agreement in general and complete disarmament is achieved, both our countries will, nevertheless, be limited in their abilities to cooperate in the field of peaceful use of outer space. It is no secret that rockets for military purposes and spacecraft launched for peaceful purposes are based on common scientific and technical achievements. It is true that there are some distinctions here; space rockets require more powerful engines, since by this means they carry greater payloads and attain a higher altitude, while military rockets in general do not require such powerful engines—engines already in existence can carry warheads of great destructive force and assure their arrival at any point, on the globe.

[5] However, both you and we know, Mr. President, that the principles for designing and producing military rockets and space rockets are the same.

I am expressing these considerations for the simple reason that it would be better if we saw all sides of the question realistically. We should try to overcome any obstacles which may arise in the path of international cooperation in the peaceful conquest of space. It is possible that we shall succeed in doing this, and that will be useful. Considerably broader prospects for cooperation and uniting our scientific-technological achievements, up to and including joint construction of spacecraft for reaching other planets—the moon, Venus, Mars—will arise when agreement on disarmament has been achieved.

We hope that agreement on general and complete disarmament will be achieved; we are exerting and will continue to exert every effort toward this end. I should like to believe that you also, Mr. President, will spare no effort in acting along these lines.

Yours respectfully,

N. Khrushchev

Moscow, March 20, 1962

Document I-39

Document title: "Record of the US-USSR Talks on Space Cooperation," March 27, 28, and 30, 1962, with attached: Arnold W. Frutkin, Director, Office of International Programs, NASA, "Topical Summary of Bilateral Discussions With Soviet Union," May 1, 1962.

Source: NASA Historical Reference Collection, NASA History Office, NASA Headquarters, Washington, D.C.

Following up on the exchange of letters between John Kennedy and Nikita Khrushchev, both the United States and the Soviet Union appointed delegations to begin discussions on space cooperation possibilities. NASA Deputy Administrator Hugh L. Dryden headed the U.S. delegation, while Professor Anatoli A. Blagonravov of the Soviet Academy of Science led the Soviet delegation. This document records their first three days of meetings, which laid the foundation for more formal negotiations a few months later.

[1]
Record of the US-USSR Talks on Space Cooperation

Held in New York City
on March 27, 28 and 30, 1962

First Meeting - March 27

Participants:

United States	USSR
Dr. Hugh Dryden	Prof. A. A. Blagonravov
Dr. John W. Townsend	Mr. Y. A. Barinov
Dr. Donald F. Hornig	Mr. Roland H. Timerbaev
Mr. Lewis Bowden	Mr. Valentin A. Zaitzev
Mr. Peter Thacher	Mr. G. S. Strashevsky
	(Interpreter)

The first in a series of bilateral conversations between the US and the USSR was held March 27 at USUN [United Nations]. It was agreed at the outset by Dryden and Blagonravov that these were preliminary, informal talks designed to prepare the basis for further, formal negotiations between US and Soviet experts to discuss specific areas of practical cooperation in outer space as suggested in the exchange of correspondence between President Kennedy and Mr. Khrushchev. Blagonravov stressed the need for initial cooperation in practical fields, such as weather satellites and communication systems, which would be meaningful to the man in the street. They agreed to take up the subject of meteorological satellites at the outset.

Meteorological Satellites. Dryden suggested that the US and the USSR put up meteorological satellites in complimentary [sic] orbits. The US TIROS satellite was a relatively crude, experimental craft, and we had in mind making NIMBUS the basis of our contribution to an operational system. The first launching of NIMBUS would be within a year. It would be stabilized as to scan the earth continuously from a polar orbit, and we had in mind equipment which would permit transmittal of data direct to any nation's ground station, including pictures of overhead cloud cover. We would in addition, of course, transmit information to WMO [the World Meteorological Organization].

[2] Blagonravov made what appeared to be a general statement to the effect that cooperation must develop stage by stage; he noted that launch systems were closely related to other aspects (military); therefore the achievement of broadest cooperation will be related to progress in disarmament. Conversely, progress in cooperation will aid the development of mutual trust between nations.

Turning to the meteorological project, the USSR will transmit to the US all data they receive from NIMBUS. They expect to launch their own meteorological satellite and are prepared to come to an agreement on coordination of orbits. They will transmit all meteorological data from their own system to other countries. He noted that speedy transmittal of data is essential.

Dryden commented that he had in mind the problem of access to launch sites and therefore was proposing only coordination; in any case, we will not seek information the Soviets do not wish to give. He noted that the recent Soviet launch, which was first in a new series, was said to include devices for measurement of cloud coverage.

Blagonravov said that they intend to launch meteorological satellites on a national basis and to exchange data. He had in mind that WMO would insure [sic] the proper transmittal of information to other countries.

Dryden questioned what the next step might be. Should there be a meeting of both sides' experts at the time of the coming COSPAR [Committee on Space Research] Conference in Washington, or should the problem be left to WMO? Blagonravov said the best way would be to continue through WMO. He drew attention to an April 23 symposium scheduled for Washington. Dryden asked if there should be private meetings between US and Soviet experts at that time. Blagonravov said this particular symposium will not attract experts in the field of satellite weather forecasting, but nonetheless the experts present could explore the problem in a preliminary way.

Dryden noted that the SYG of WMO has obtained the presence of two US and Soviet meteorological experts in Geneva to help with the preparation of WMO's report on this subject. Blagonravov said he had no information about the details of their discussion and was unable to judge the results.

[3] Dr. Hornig commented that success in this field will depend in large part on the compatibility of information sought and obtained. He asked if we could discuss this aspect with a view to standardizing equipment in satellites. Blagonravov preferred to leave the job of determining technical requirements to WMO experts.

Communications Satellites. Blagonravov said they are ready to take part in studies of principles and of design plans for a system which should be organized through ITU [the International Telecommunications Union]. They are ready to take part in experimental projects, and they are ready to supply information to the US on radio signals bounced off ECHO. He thought the time had come to make a "symbolic start" in this field.

Dryden noted that ECHO has become smaller, and the surface is considerably wrinkled; it is therefore less satisfactory for radio relay purposes. We plan to launch within a year a large, 140-foot sphere which will be more rigid and therefore more suitable. Blagonravov indicated that they were agreeable to using the larger sphere.

Dryden suggested the USSR might wish to join experiments with active relays. Blagonravov said they lack experience. Dryden said we also lack experience but noted that several European states are building ground stations for this purpose and suggested that the USSR might also. Blagonravov noted that active relays require extensive equipment somewhat like the enormous receivers that the USSR is now building for deep space probes, such as to Venus. Dryden suggested it might be possible to modify some of these large dishes so that they could receive signals from active-relay satellites. Blagonravov said they would prefer to leave it up to ITU experts to organize cooperation in this field. Townsend noted this would be difficult for ITU because the problem is one essentially of equipment, a subject ITU does not normally handle.

Dryden felt the subject needed further bilateral discussions between experts. He noted that CCIR has recently been discussing the problem of the sharing of frequencies between satellite and ground-based microwave systems. He thought that both countries might cooperate in studying this possible source of interference, a subject also suitable for [4] bilateral discussion. Blagonravov noted the problem is already under review by ITU. Dryden said we do not presently have any active communications satellites; they are at present only in a research and development phase which will include one low-altitude launch later this year. Blagonravov commented this first experiment may help to clarify the situation.

Geomagnetic Research. Dryden noted the desirability of coordinating data gathering in this field. Blagonravov said he could not yet say when the USSR will be prepared to launch research vehicles to measure the earth's magnetic field. Their first interest will be the measurement of field components; later they will seek to measure the dimension of the total field. Nonetheless, the time is now right to organize an exchange of data on geomagnetic measurements. Dryden thought this was already on the agenda of COSPAR and wondered if Soviet experts would be present at the COSPAR meeting. It was agreed that Dryden and Blagonravov would meet with their experts during COSPAR. Townsend asked if the USSR had decided which orbit, high or low, they would undertake. Blagonravov replied that it does not make much difference; they could do it at any altitude. He noted that at a previous COSPAR meeting US experts had suggested that the Soviet Union take the high orbits but no decision had been reached. Dryden commented that this suggestion had been in recognition of greater Soviet thrust capabilities. Although not exciting for the man in the street, Dryden felt this is a field of great interest to scientists.

Space Medicine. Dryden announced that the US will publish on April 6 a detailed report containing all medical information resulting from the Glenn flight. He said Blagonravov and other Soviet scientists would be welcome at the time. On the US side many ideas for cooperation in this field are being discussed, such as the establishment of an international laboratory, and possible coordination in manned-flight experiments, but he suspected that the Soviets might prefer an exchange of information. Blagonravov expressed preference for a broad exchange of information. Dr. Hornig noted that much background other than from manned-flight space is available; he hoped that the exchange would include ground laboratory and animal data. Blagonravov agreed. Dryden asked if Blagonravov had considered visits to laboratory facilities. Blagonravov said he had not discussed this subject with appropriate Soviet experts before leaving Moscow and, therefore, could not answer.

[5] *Salvage and Rescue.* Blagonravov raised the problem of insuring [sic] the return of astronauts and vehicles from other states. Dryden noted this was largely a legal and political problem, but worth exploring here. Blagonravov said their ideas had not advanced beyond general terms. Dryden said it was no question but that the US would use its facilities to aid a Soviet astronaut in difficulty and he hoped the same would be true for Americans. Blagonravov stated this would, of course, be so. Dryden asked if the Soviets favor some form of international agreement or treaty, of the sort, for example, which govern civil aircraft. Blagonravov felt some means should be found to assure that all UN members agree to the return of capsules. Dryden called on Thacher who suggested that it might be appropriate for the UN Outer Space Committee to recommend an appropriate resolution for adoption at the next session of the General Assembly. Timerbaev felt he and Thacher should discuss this bilaterally in the context of the committee.

It was agreed that there should be no announcement made to the press concerning these talks and that the next meeting would take place on March 28 at the Soviet Mission.

Second Meeting - March 28

Participants:	*United States*	*USSR*
	Dr. Dryden	Prof. Blagonravov
	Dr. Townsend	Mr. Barinov
	Dr. Hornig	Mr. Timerbaev
	Mr. Frutkin	Mr. Zaitzev
	Mr. Bowden	Mr. Aldoshin
	Dr. Porter	Mr. Strashevsky
	Mr. Thacher	(Interpreter)

The second in a series of US-Soviet bilateral discussions about possible cooperation in outer space matters was held at the Soviet Mission to the UN during the morning of March 28.

[6] *Contamination of Space.* Blagonravov believed the problem of pollution deserves studying. He had in mind radioactive contamination, bacteriological contamination, interference with radio transmission from the earth to satellites and from satellites to the earth, and possible physical interference of the sort many feared would result from Project WESTFORD. He commented that there were grounds for fear of interference by the needles on two counts: radio astronomy, and physical damage to other satellites, particularly optical equipment. He did not feel it necessary to exclude this type of experiment, rather he felt there should be some procedure for preliminary discussion which would analyze all possible harmful effects and thereby dispel the fears of interested scientists. Dryden noted that in his letter, Khrushchev had placed the subjects of radioactive and bacteriological contaminations primarily in a legal context. He noted that there is broad consultation by the US with interested scientists and, on the international level, through COSPAR which is a useful means of bringing about understanding of the scientific aspects of experiments. Blagonravov and Dryden agreed that nuclear engines would be needed for long distance probes and presented a number of technical problems. Dryden commented these were primarily problems relating to contamination of the surface of the moon and planets, not of intervening space. Porter noted that in Florence he and Blagonravov had agreed on three principles: (1) radioactive components should be so packaged as to prevent dispersal in case of impact; (2) radioactive materials should be chosen with short half-life times; and (3) radioactive materials should be chosen which did not occur in nature. Dryden felt there was not much left to discuss about radioactive and bacteriological contamination and asked if we should consider the problem of frequency allocation. Thacher noted that the UN Committee would probably consider the problem of contamination in its technical subcommittee. Porter suggested it might be wise to delegate to COSPAR the task of studying this problem. Dryden felt that the ultimate decision must rest with launching states. As to the problem of terminating satellite transmissions, he found it hard to make a general rule where so much depends on the precise nature of the experiment. For example, it did not seem desirable arbitrarily to exclude experiments from which continual transmission could be expected. Blagonravov agreed and said the Soviet idea is mainly to stress the importance of preliminary [7] exchanges which can dispel apprehensions whenever they seem likely. Although thankful for information given him by Porter about the problem of physical interference by WESTFORD, apprehension nonetheless arose and Blagonravov felt it might be necessary to have meetings between scientists. Porter invited Blagonravov to express these apprehensions as soon as they arise to the National Academy of Sciences.

Tracking for Deep Space Probes. After Blagonravov appeared to have completed his remarks, Dryden noted that in the course of conversation all but one of the general topics suggested in the letters of President Kennedy and Chairman Khrushchev had been touched upon. The one remaining was Khrushchev's suggestion which appeared to relate to tracking facilities for deep space probes. He turned to Blagonravov.

Blagonravov said he would prefer to hear Dryden first. Dryden said that President Kennedy had suggested an exchange of "tracking stations" but that our interest was more in the field of telemetry data rather than in observation of satellite orbits. This is particularly true with regard to those scientific satellites, such as Van Allen's which broadcast continuously and do not store data. He noted that emphasis was placed in Khrushchev's letter on the need for observation and contact with deep space shots. It seemed logical that useful exchange could be found. Dryden commented that to a large extent the technical problem in following a deep space probe relates to the transmitter. Our stations in Southern California, Australia and South Africa, for example, are equipped to handle only certain frequencies which cannot easily be altered. Therefore, if our receivers were to be of help, the satellite transmitter should be at appropriate frequencies. He wondered if this presented any technical problems for the Soviets.

Blagonravov replied there is no problem in tracking US satellites over the USSR, and if the US supplied frequencies the USSR will devote the necessary facilities to track them and receive them, and will supply resulting data. Conversely, when the Soviets are interested in receiving data from us they will supply the frequencies and the codes to us.

[8] Dryden commented this could be done in two ways, either by recording telemetry signals on tape and sending the tape to the launchers, or by supplying the code, in which case the recipient could reduce the data for the launchers. Blagonravov asked which Dryden preferred.

Dryden commented we found our own scientists prefer to work out their own results. He cited as an example the case of Van Allen and the Japanese scientist who had been given the code but whose results were out of line because he had not realized that one of the channels was malfunctioning. Blagonravov felt that both ways were possible and that the decision would depend on the specifics in each case. Townsend suggested this would be a good area for progress.

Next Steps. Dryden asked where we were to go from here. Blagonravov replied our approach may vary from problem to problem. Some, as had been suggested, may be appropriate for COSPAR; the general subject of frequency allocation is appropriate for ITU; others are appropriate for WMO experts.

Dryden agreed that discussions ultimately should take place with other states in an appropriate international forum. But we felt it more useful to start bilateral talks at the time of the COSPAR meeting. For example, it might then be useful to start discussions on meteorological satellites and geomagnetic research. So far as the meteorological satellite is concerned, he felt it would be wise to distinguish between the research and development phase, and the operational system. He thought talk should start without delay about the experimental stage; as a result the two sides could come to an agreement on the type of information to be sought. Continuing with the general outline, Dryden suggested there might be later discussion in Moscow about such matters as coordination of planetary exploration. In the meantime, he thought it would be helpful if we could follow up the general discussion of the past days with specific discussions on certain subjects. He asked if it would be possible to agree to try to arrange discussions at the time of the COSPAR meeting on meteorological satellites and geomagnetic research.

[9] Blagonravov drew attention to his inability to consult with appropriate experts but said he would be prepared to get in touch with Dryden. It would be helpful if Dryden could list his ideas as to the priority of subjects which he could take with him to Moscow, and later he, Blagonravov, could respond with proposals. Dryden suggested that he could prepare plans and meet with Blagonravov again next week. Blagonravov said he would be leaving for Moscow this weekend, and it would be desirable to receive a list before that time. Dryden suggested we could select a few steps, although we are to respond to all, and suggested meeting again on March 30.

Blagonravov made an evasive reply. Dryden said we therefore would prepare and give to Blagonravov on Friday specific proposals for later discussion by the experts during the COSPAR meeting in Washington. These proposals would involve meteorological satellites, geomagnetic research and the general area of telemetry. At the same time, we would be prepared for later discussions, perhaps in Moscow, regarding communication satellites, space medicine, and inter-planetary exploration.

It was agreed that the next meeting would take on "neutral ground" (at the Soviet's insistence), i.e. at the UN. It was agreed that Thacher and Timerbaev should prepare a joint press release for issuance after the following meeting which would respond to the desires of both countries for forward movement in the area of US-Soviet cooperation.

Military Reconnaissance. During the course of general conversation which followed, Blagonravov commented that the climate for cooperation would be greatly improved if both sides would issue a declaration to the effect that neither would use satellites for the purpose of military reconnaissance. Blagonravov expressed himself as certain that Dryden was not in a position to comment on this aspect of outer space. Nonetheless, Blagonravov hoped that the US Government was as attentive to the opinion of its scientists as was his, the Soviet, Government. Before coming to New York, his colleagues had asked him to urge his American colleagues to persuade the US Government to issue such a declaration. (The translator failed to make clear, as had Blagonravov in Russian, that he had been instructed by his Government to raise this matter.)

[10]   *Third Meeting - March 30*

Participants: *United States*   USSR
                  Dr. Dryden   Prof. Blagonravov
                  Dr. Townsend   Mr. Barinov
                  Dr. Hornig   Mr. Timerbaev
                  Mr. Frutkin   Mr. Zaitzev
                  Mr. Bowden   Mr. Strashevsky
                  Dr. Porter   (Interpreter)
                  Mr. Thacher

The third in a series of US-Soviet bilateral discussions about possible cooperation in outer space matters was held at the United Nations Headquarters during the afternoon of March 30.

Dryden presented to Blagonravov the three tentative proposals worked out by the American side on collaboration in the fields of weather satellites, geomagnetic survey, and telemetry. Dryden pointed out that these proposals were being handed to the Soviets in order that they might study them and be prepared to discuss them in a concrete fashion at the April COSPAR meeting and in other forums. Blagonravov, with the aid of his interpreter, scanned the tentative proposals quickly and said that he would take them back to Moscow and discuss them with the relevant Soviet specialists.

While Blagonravov was reading our proposals, Thacher of USUN and Timerbaev of the Soviet UN Mission attempted to come to agreement on the wording of the joint statement to be made upon the conclusion of the talks that day. The US draft had proposed listing the three topics mentioned above since these were the fields in which further concrete talks were planned. Timerbaev insisted that if topics touched on in the three days of talks were listed they would necessarily have to include mention of military intelligence reconnaissance satellites. Thacher and Timerbaev did not reach an accord on the matter, and it was placed before Dryden and Blagonravov. The latter reiterated Timerbaev's stand and Dryden demurred, pointing out that the subject of military reconnaissance satellites did not fall within the frame of reference agreed to for the talks. Agreement was finally reached that the statement issued would simply say that the items mentioned in the Kennedy and Khrushchev letters respectively had been discussed, as well as additional topics. The problem of listing the specific items touched on was, therefore, obviated.

Dryden and Blagonravov then met with the press and released their statement.*

****

[Attachment page 1]

Topical Summary of Bilateral Discussions With Soviet Union

(Note: For negotiations of March 27, 28, 30, 1962)

1. Meteorological Satellites

Blagonravov indicated that a series of scientific satellites which had just begun with the launching of COSMOS I would seek meteorological data, although this was not necessarily true of the first launching in the series. The Soviet Union intends to launch meteorological satellites to photograph cloud cover and would be agreeable to coordinating their orbits and other details with the US and to exchange the data. Like the US, the USSR would wish to relate any such program to WMO activities and sponsorship. Blagonravov said that meteorological satellites should be launched on a national basis with data coordination through WMO. (Subsequent private discussions suggest that the USSR nevertheless recognizes the fundamental necessity of bilateral coordination in flight programs.) Dr. Dryden suggested the possibility of using Nimbus as a basis for the joint program in about a year.

2. Communications Satellites

Khrushchev had given priority to cooperation in the field of communications satellites, and Blagonravov indicated that the Soviet Union would desire to take part in experimental projects. Nevertheless, he was not yet ready to identify suitable modes of cooperation in this field. He said that (Soviet) experience was lacking on active repeaters and seemed to feel that a position must be worked out on communications systems in the ITU before such could be done. The Soviet interest in communications satellites cooperation actually appeared directed primarily toward operational matters rather than experimental. As a "symbolic" gesture, however, Blagonravov made a point of expressing readiness to utilize the US ECHO satellite for a communication demonstration between the US and the Soviet Union. (It should be noted that Blagonravov later stated that he did not mean, by the use of the word symbolic, that the cooperative use of ECHO would not have real value.) The US delegation considered that ECHO had deteriorated too much to permit a satisfactory demonstration, and the two sides agreed to look toward the ECHO follow-on program for such a demonstration. There was some indication that the Russians would wish to utilize a new deep space probe dish, or dishes, which they are now constructing for communications experiments.

[2] 3. Magnetic Field Survey

The Soviets would be willing to coordinate with the US in an effort in which each country placed a satellite in orbit to measure the Earth's field. Blagonravov said that the Soviets could place a satellite at either of the higher or lower altitudes required for this project and could measure the field components as well as strength. It is still undecided whether the USSR would devote a special satellite for the program or join the experiment

---

* See "Preliminary Summary Report" of these conversations prepared by Dr. Dryden.

with others in a multi-purpose satellite. There vas some indication that the Soviets would measure the field components as early as, or earlier than, the scalar values. When it was suggested that the standards established for the World Magnetic Survey for measurement of vectors were quite stringent, Blagonravov said he was not personally familiar with them but would make sure that they were brought to the attention of the proper scientists of the USSR.

4. Data Acquisition

Blagonravov made clear that the Soviet Union was not ready to exchange tracking and data acquisition station equipment. Instead, he said that the Soviet Union would make available on its own territory equipment to American specifications to provide desired services. Blagonravov did not exclude the possibility of equipment exchange at a later stage. Soviet interests were clearly directed more toward deep space tracking and data acquisition than toward the acquisition of telemetry for scientific satellites as desired by the US. There was some appearance of the possibility of an agreement for appropriate trade-offs here. The question of exchanging telemetry codes along with the exchange of telemetry tapes was discussed, and it was agreed that such exchanges would have to be worked cut on a case-by-case basis. Dr. Dryden pointed out that there was some difficulty in providing calibrations for telemetry, both because of the sensitivities of prime experimenters and the empirical requirements for calibration adjustment in the period after satellite launch. Dr. Dryden pointed out, in addition, that public errors had been made, as by the Japanese, in using calibrations not fully understood by them. With regard to deep space probe tracking and telemetry, it appeared that some activity is going on in the Soviet Union to strengthen its capabilities. It was also understood that both sides would be launching deep space probes at approximately the same periods due to the "window" situation and that therefore each country might be limited in providing services to the other.

[3] 5.   Deep Space Activities

With regard to cooperation in lunar and planetary activities per se, Blagonravov stated that current programs were too far along to permit coordination at this date. The coordination of future progress with respect to physical quantities measured by the probes launched by the US and USSR seemed possible.

6. Space Medicine

There was relatively little discussion of space medicine. Dr. Dryden suggested this might be an appropriate area for broad exchanges. He indicated that some people in this country feel it may be useful to exchange laboratory visits. Blagonravov appeared to believe that laboratory visits would not be easy to arrange at this stage but rested on a lack of information as to the situation in his own country. The matter was left for further definition.

7. "Pollution" of Space

Blagonravov expressed concern about several types of possible interference in the space activities of one country by reason of the activities of another. In this category, he included biological contamination, radio nuisances and interference, and the dispersion of particles as in Project WESTFORD. It is not clear whether he had in mind legal prohibitions. Specifically, with regard to Project WESTFORD, Blagonravov indicated fear of damage or interference with optical experiments in satellites. It was clear that he did not argue to prohibit WESTFORD but rather to provide for preliminary discussion to avoid harmful effects. Dr. Dryden reviewed the procedures followed in the US to assure that the scientific community has no substantial concerns regarding any proposed experiment, referred to the descriptions by a [Jet Propulsion Laboratory] representative of our contamination procedures at a recent meeting in the Soviet Union, to continuing consideration by COSPAR of this subject, and to the coordination of radio frequency uses by the

ITU. The US delegation indicated its belief that the technical aspects were largely for the Technical Subcommittee. Dr. Dryden concluded with the observation that each launching country would undoubtedly expect to retain the final judgment over action to be taken in any given case of possible or alleged pollution. Blagonravov seemed to be in complete agreement. Dr. Dryden explored Soviet attitude toward the use of nuclear power or propulsion sources. Blagonravov agreed that there was no objection to these per se, assuming general safeguards. His response to this was so prompt as to reflect current Soviet consideration of nuclear propulsion or power sources.

[4] 8.   Spy Satellites

During the session held at the Soviet UN Mission, Blagonravov brought up, almost apologetically, a proposal which he stated he had been "instructed" to raise. He said that it would be desirable if the scientists of the US would join with those of the Soviet Union in a pledge to reserve space for peaceful purpose and to prohibit the use of satellites for surveillance purposes. Blagonravov suggested that Dr. Dryden might not be prepared to comment on this. Dr. Dryden replied that the subject was outside the scope of the present technical discussions. There was some further discussion in a rather bantering vein about this subject with Blagonravov expressing the belief that scientists in his country could not devote themselves to non-peaceful purposes in space research. Dr. Dryden observed that this was an interesting remark to come from an old artillery observer. The subject was raised again by Blagonravov at the end of a subsequent session as an item to be included in the joint press release at the end of the first round of discussions. It was offered as a counterproposal to the US desire to specify the three subjects of greatest [discussion, with plans for future talks,] identified in the negotiations. The Soviets wished then to include all other subjects, plus the spy-in-the-sky pledge, or, in the alternative, remove the specific references to subjects discussed. The US side held to the position that the press release should not go beyond those matters discussed in the letters. The implication left in the press release was that the current, as well as future, discussions would be based upon the matters identified in the Kennedy-Khrushchev correspondence.

9.   Balloons

When the question relating to balloon-borne experiments arose, Blagonravov made clear the Soviet dislike for the use of balloons.

10.   Procedures

It was agreed that the first round of discussions constituted informal exploratory talks prior to formal negotiations. Dr. Dryden's official summary of the opening sessions and the text of the press release which terminated them are attached. These indicate that formal negotiations will begin either at the time of the COSPAR meeting in Washington at the end of April or at the time of the meeting of the Technical and Legal Subcommittees of the UN Outer Space Committee in Geneva at the end of May. Continuity between the two separate sessions was assured by (1) leading the Soviet delegation to agree privately to the identification of three subjects as most promising for early and [5] more detailed investigation, and (2) providing to the USSR somewhat expanded papers on each of these three subjects for their study and future comments. It was agreed that Soviet scientists would consider these papers and Blagonravov indicated that his side would provide similar papers. It was agreed that the working papers would not be published. It should be noted, however, that none of the subjects indicated in the Kennedy-Khrushchev correspondence is excluded from further investigation, although Dr. Dryden indicated that certain aspects might be more appropriate for the Legal Subcommittee of the UN or other forms.

Dr. Dryden specifically asked Blagonravov whether the Soviet view would permit agreement on individual cooperative projects as agreement could be reached upon them or whether the Soviets felt it was necessary to achieve a total package before any agreement could be reached. Blagonravov strongly indicated his belief that the first procedure should be followed. This may be interpreted as a hopeful sign, particularly in view of the fact that this discussion followed immediately upon the heels of Blagonravov's efforts to write a spy-in-the-sky pledge into a final joint press release. The sequence would suggest that the Soviets do not at this time mean to impose political preconditions upon cooperative projects of the character discussed in the Kennedy-Khrushchev correspondence.

With regard to locations of meetings, Dr. Dryden indicated readiness to hold a future meeting in Moscow, after the Washington or Geneva meeting. The Soviets welcomed this since they appear to attach some value to rotating meetings at among western, neutral and Soviet sites.

11. General

While political considerations were trotted out by Blagonravov at various times during the course of the discussions, they did not appear ever to be raised with the purpose of obstructing conversation. Blagonravov repeated Khrushchev's statement that more ambitious cooperative efforts would have to wait upon disarmament. The spy-in-the-sky pledge discussed above was raised with good humor and appeared to have been fitted in outside the central framework of the negotiations. In the formulation of the press release, the Soviet political offices did ask to reverse the priority of the references to Kennedy and Khrushchev, presumably on the basis of Khrushchev's initiation of their correspondence. [6] On several occasions, the junior member of the Soviet delegation Barinov indicated he was considerably impressed by the scope and size of the NASA program as reflected in the briefings given during the week for the Outer Space Committee at the US Mission.

At one point, Dr. Dryden described in detail the US working relations with the UK on their joint satellite program. Blagonravov stated that he hoped for similar relationships between the US and the USSR.

Arnold W. Frutkin, Director
Office of International Programs
National Aeronautics and Space Administration   May 1, 1962

Document I-40

Document title: McGeorge Bundy, Memorandum for the President, July 13, 1962, with attached: George Ball, Under Secretary of State, Memorandum for the President, "Bilateral Talks Concerning US-USSR Cooperation in Outer Space Activities," July 5, 1962.

Source: NASA Historical Reference Collection, NASA History Office, NASA Headquarters, Washington, D.C.

The White House monitored closely the initial U.S.-U.S.S.R. talks on space cooperation to make sure that any agreements reached did not go beyond the bounds of political feasibility in the United States. McGeorge Bundy was President John F. Kennedy's Assistant for National Security Affairs. With this memorandum, he forwarded to Kennedy the Department of State's report on the initial talks and agreements between NASA's Deputy Administrator Hugh L. Dryden and Soviet representative Anatoli Blagonravov.

[1] July 13, 1962

Memorandum for the President

Here for your approval is a memorandum from George Ball on the results of the Dryden and Blagonravov outer space negotiations. At pages three and four it gives recommended procedure from here on out.

I know you have been concerned lest Dryden make agreements that might come under political attack. I believe these three specific projects are quite safe. They have been reviewed with a beady eye by CIA and Defense, and they have been reported in detail to determined and watchful Congressmen like Tiger Teague, with no criticism. In essence they provide for the kind of cooperation in which we get as much as we give, and in which neither our advanced techniques nor our cognate reconnaissance capabilities will he compromised.

McG. B.

****

[Attachment page 1] July 5, 1962

Memorandum for the President

Subject: Bilateral Talks Concerning US-USSR Cooperation in Outer Space Activities

On May 15 the Secretary wrote to you describing the developments in this matter prior to the recent talks in Geneva between Dr. Dryden and Professor Blagonravov. These talks commenced on May 29 and continued concurrently with meetings of the subcommittees of the UN Outer Space Committee. As a result, technical arrangements for three specific cooperative projects were agreed ad referendum to the US and Soviet Governments in a joint memorandum signed by Dr. Dryden and Professor Blagonravov on June 8. (See Enclosure 1.) On the same day, Dr. Dryden and Professor Blagonravov issued a joint Press Communique summarizing briefly the results of these discussions. (See Enclosure 2.)

The three projects involve (1) exchange of weather data from satellites and the eventual coordinated launching of meteorological satellites, (2) a joint effort to map the magnetic field of the earth by means of coordinated launchings of geomagnetic satellites and related ground observations, and (3) cooperation in the experimental relay of communications via the ECHO satellite. It was also agreed that there should be further discussion of the possibility of broader cooperation in experiments using active communications satellites to be launched in the future. These arrangements are quite limited in [2] scope and have been drawn carefully to assure reciprocal benefit. They have been developed in the context of multilateral programs (e.g., the program of the World Meteorological Organization for the acquisition and world-wide distribution of weather data, and the program being planned by the International Union of Geodesy and Geophysics for a world geomagnetic survey). The Soviets appeared quite anxious to achieve these agreements.

The arrangements proposed in the joint Dryden-Blagonravov memorandum represent a sound way of proceeding so long as they are adhered to by the Soviet Government and are developed in such a way as not to foster an impression abroad that they represent a more significant step toward US-Soviet cooperation than they actually do or that US-USSR cooperation will in any way preempt the cooperation already being developed with other countries.

There remain three other specific projects which were suggested in your exchange of correspondence with Chairman Khrushchev last March, but on which no specific conclusions or proposals have been reached during the technical discussions so far, i.e.: (1) the acquisition of data obtained through tracking facilities located in each other's countries but operated by the host governments, (2) joint observation of solar and interplanetary probes, and (3) space medicine. Although it seems clear that the Soviets are not interested in cooperating in tracking and it appears doubtful that they are really interested in joint observation of space probes, it would be well to afford them the opportunity to discuss all these projects further.

Upon Dr. Dryden's return from Geneva, Under Secretary McGhee, who is coordinating this matter for the Department, [3] convened a meeting of the interested agencies of government in which Dr. Dryden, Dr. Welsh, Dr. Reichelderfer, and representatives of Dr. Wiesner, Mr. Bundy, the Defense Department, the Air Force and CIA participated. A review of the recent discussions in Geneva and of the specific proposals contained in the joint Dryden-Blagonravov memorandum resulted in agreement to proceed as follows:

1. After a reasonable interval and if no serious objections have been raised by any of the interested agencies, Dr. Dryden will inform Professor Blagonravov that we have no changes to suggest in their joint memorandum. (The memorandum provided for a two-month waiting period during which either party could propose changes.)

2. Upon notification from Professor Blagonravov that the Soviets do not desire changes which would be unacceptable to us (or at the conclusion of the two-month waiting period), we will, assuming the Soviets still wish to proceed, exchange notes with the Soviet Government to confirm government-level agreement to these proposals.

3. It was suggested that when that agreement has been obtained, you may wish to write to Chairman Khrushchev noting both the agreement to proceed with the specific arrangements at hand and the prospects of further technical discussions on additional topics. A draft of such a letter will be submitted for your approval.

4. Meanwhile, Under Secretary McGhee and Dr. Dryden will report these developments to members of Congress who have a specific interest and responsibility in this field, and the Department will prepare a report to be sent to the Secretary General of the United Nations when formal agreement has been reached with the Soviets.

5. Dr. Dryden will, in cooperation with the interested agencies, proceed now to arrange nominations for US membership in the joint US-Soviet working groups which are to [4] develop the detailed implementation of the meteorological and geomagnetic proposals. These working groups will not, however, be activated until formal agreement has been reached with the Soviet Government.

6. The joint Dryden-Blagonravov memorandum will be treated as CONFIDENTIAL, pending government-level agreement by the Soviets or earlier Soviet public release.

7. After formal agreement has been obtained, Dr. Dryden will arrange directly with Professor Blagonravov for further technical discussions, possibly in Moscow this fall, concerning broader cooperation in communication via satellites and the possibility of cooperation in such of the remaining topics dealt with in your exchange of letters with Chairman Khrushchev as may seem worthwhile to pursue further.

It is our feeling that the present low key, step-by-step approach through informal talks by scientific representatives continues to be the preferable means of moving toward further cooperation and that we should plan to proceed on this basis after government-level agreement has been reached on the specific arrangements already proposed.

George Ball

Document I-41

Document title: McGeorge Bundy, Memorandum for the President, "Your 11 a.m. appointment with Jim Webb," September 18, 1963.

Source: NASA Historical Reference Collection, NASA History Office, NASA Headquarters, Washington, D.C.

The possibility of turning Project Apollo into a cooperative undertaking with the Soviet Union was under active consideration in NASA and the White House as President John F. Kennedy met with NASA Administrator James E. Webb on September 15, 1963. The political climate was much different than it had been in 1961, when the President had decided to race the Soviet Union to the Moon; the high levels of spending for Apollo were coming under criticism in the United States. Kennedy's National Security Advisor McGeorge Bundy suggested to the President that a cooperative mission was desirable, if technically, institutionally, and politically feasible. Two days later, in an address to the United Nations General Assembly, Kennedy suggested that the United States and the Soviet Union take the lead in making the first human voyages to the Moon an undertaking of all countries.

[1]   September 18, 1963

Memorandum for the President

SUBJECT: *Your 11 a.m. appointment with Jim Webb*

Webb called me yesterday to comment on three interconnected aspects of the space problem that he thinks may be of importance in his talk with you:

1. *Money*. The space authorization is passed at $5.350 billion, and he expects the appropriation to come out at about $5.150 billion. While the estimates are not complete, his current guess is that in early '64 he will require a supplemental of $400 million ($200 million requiring authorization and $200 million appropriation only) in order to keep our commitment to a lunar landing in the 1960's.

2. *The Soviets*. He reports more forthcoming noises about cooperation from Blagonravov in the UN, and I am trying to run down a report in today's *Times* (attached) that we have rebuffed the Soviets on this. Webb himself is quite open to an exploration of possible cooperation with the Soviets and thinks that they might wish to use our big rocket, and offer in exchange the advanced technology which they are likely to get in the immediate future. (For example, Webb expects a Soviet landing of instruments on the moon to establish moon-earth communications almost any time.)

The obvious choice is whether to press for cooperation or to continue to use the Soviet space effort as a spur to our own. The *Times* story suggests that there is already low-level disagreement on exactly this point.

3. *The Military Role*. Webb reports that the discontent of the military with their limited role in space damaged the bill on the Hill this year, with no corresponding advantage to the military. He thinks this point can and should be made to the Air Force, and he believes that the thing to do is to offer the military an increased role somehow. He has already had private exploratory talks with Ros Gilpatric for this purpose.

[2] Webb thinks the best place for a military effort in space would be in the design and manning of a space craft in which gravity could be simulated, in preparation for later explorations. He thinks such a space craft may be the next logical step after Gemini. On

the other hand, he is quite cool about the use of Titan III and Dinosoar [sic] and would be glad to see them both cancelled. You will recall that McNamara has just come out on the other side on Titan III.

My own hasty judgment is that the central question here is whether to compete or to cooperate with the Soviets in a manned lunar landing:

1.   *If we compete,* we should do everything we can to unify all agencies of the United States Government in a combined space program which comes as near to our existing pledges as possible.

2.   *If we cooperate,* the pressure comes off, and we can easily argue that it was our crash effort on '61 and '62 which made the Soviets ready to cooperate.

I am for cooperation if it is possible, and I think we need to make a really major effort inside and outside the government to find out whether in fact it can be done. Conceivably this is a better job for Harriman than East-West trade, which might almost as well be given to George Ball.

McG. B.

Document I-42

Document title: National Security Action Memorandum No. 271, "Cooperation with the USSR on Outer Space Matters," November 12, 1963, with attached: Charles E. Johnson, Memorandum for Mr. Bundy, December 16, 1963.

Source: NASA Historical Reference Collection, NASA History Office, NASA Headquarters, Washington, D.C.

Ten days before he was assassinated, President John F. Kennedy signed this memorandum giving the NASA Administrator the lead within the Executive Branch in developing substantive proposals for enhanced U.S.-U.S.S.R. space cooperation. This action was a followup to Kennedy's September 20 speech before the United Nations. Note that the attached memorandum from Charles Johnson to McGeorge Bundy has Anatoli Blagonravov's last name misspelled twice.

<hr />

[1]   November 12, 1963

National Security Action Memorandum No. 271

**Memorandum for the Administrator,
National Aeronautics and Space Administration**

SUBJECT: Cooperation with the USSR on Outer Space Matters

I would like you to assume personally the initiative and central responsibility within the Government for the development of a program of substantive cooperation with the Soviet Union in the field of outer space, including the development of specific technical proposals. I assume that you will work closely with the Department of State and other agencies as appropriate.

These proposals should be developed with a view to their possible discussion with the Soviet Union as a direct outcome of my September 20 proposal for broader cooperation between the United States and the USSR in outer space, including cooperation in lunar landing programs. All proposals or suggestions originating within the Government relating to this general subject will be referred to you for your consideration and evaluation.

In addition to developing substantive proposals, I expect that you will assist the Secretary of State in exploring problems of procedure and timing connected with holding discussions with the Soviet Union and in proposing for my consideration the channels which would be most desirable from our point of view. In this connection the channel of contact developed [2] by Dr. Dryden between NASA and the Soviet Academy of Sciences has been quite effective, and I believe that we should continue to utilize it as appropriate as a means of continuing the dialogue between the scientists of both countries.

I would like an interim report on the progress of our planning by December 15.

Information copies to:

Chairman, National Aeronautics and Space Council
Secretary of State
Secretary of Defense
Director of Central Intelligence
Chairman, Atomic Energy Commission
Director, National Science Foundation
Special Assistant to the President
for Science and Technology
Director, Bureau of the Budget
Director, U.S. Information Agency

****

[Attachment page 1]   December 16, 1963

Memorandum for Mr. Bundy

Mac—

The attached interim report to the President from NASA in response to NSAM 271 follows the line I suggested to NASA. It is intended to show that work is actively progressing on the development of a concrete approach to the Soviets following on the Kennedy-Johnson initiatives. I am following the progress of this project and will try to ensure that it stays on the timetable described by Dryden.

There has been an additional development since the preparation of the interim report. Our Embassy Moscow reports the receipt of a letter from Blaganravov to Dryden, the cable is attached. This is the first communication from Blaganravov in eight months. NASA still has its institutional fingers crossed as to whether this represents a substantive response on the part of the Soviets. They are awaiting the final text (being pouched) before reacting to the letter.

Charles E. Johnson [initialed]

Document I-43

Document title: James E. Webb, Administrator, NASA, to the President, January 31, 1964, with attached: "US-USSR Cooperation in Space Research Programs."

Source: NASA Historical Reference Collection, NASA History Office, NASA Headquarters, Washington, D.C.

This letter from NASA Administrator Webb transmitted to President Lyndon Johnson the report on possible U.S.-Soviet cooperative initiatives related to the lunar landing program that had been requested by President Kennedy in November 1963. It also summarized the contents of the lengthy report. None of the suggested cooperative initiatives was ever implemented because the Soviet Union decided in 1964 to carry out its own lunar landing program on a crash basis.

---

[1]   JAN 31, 1964

The President
The White House

Dear Mr. President:

The attached report on possible projects for substantive co-operation with the Soviet Union in the field of outer space is provided to you in accordance with National Security Action Memorandum 271, dated November 12, 1963, and my interim report to you of December 13, 1963. It has been coordinated with the Department of State, the Department of Defense, the Executive Secretary of the Space Council, the Central Intelligence Agency, the Office of the Science Adviser, and White House staff.

Since space technology is closely related to and in some measure interchangeable with technology of military interest, careful examination of the attached report is desirable in connection with further initiative in this field.

1. An appendix to the report reviews the status of agreements already reached between NASA and the Soviet Academy of Sciences for cooperation in three areas: (1) coordinated meteorological satellite program; (2) passive communications satellite experiments with the ECHO II satellite launched this month; and (3) geomagnetic satellite data exchange. The appendix also reviews Soviet rejection of numerous specific offers of space cooperation made in the past by the US. At this writing, the Soviet Academy, while in communication with NASA in regard to the agreements between us, has failed to meet time limits on most agreed action items but has conducted optical observations of the ECHO II satellite as agreed and apparently intends to proceed with communications experiments between the USSR and the Jodrell Bank Observatory. Other tests of Soviet intentions under these agreements will materialize shortly.

2. The report focuses upon possible cooperation in manned and related unmanned lunar programs. (Possibilities for cooperation in other space programs have been and will continue to be advanced in the channel between NASA and the Soviet Academy.)

[2] 3. The report recommends these guidelines to govern foreseeable negotiations with the Soviet Union in the space field: substantive rather than propaganda objectives alone; well-defined and comparable obligations for both sides; freedom to take independent action; protection of national and military security interests; opportunity for participation by friendly nations; and open dissemination of scientific results.

4. The report recognizes that cooperation with the Soviet Union must ultimately rest on specific projects. However, the advantages and disadvantages of specific proposals are not absolute. They may vary significantly, depending upon Soviet objectives, techniques, procedures, and schedules relative to ours. Lacking sufficient information of these factors, we remain uncertain of the security and tactical aspects of specific proposals which might be advanced to the Soviets.

5. Accordingly, the report outlines a preferred structured approach calculated to determine a level of confidence in any Soviet response, to gain information on basic elements of the Soviet program, and to merit confidence and support by the public and the Congress.

Briefly, this approach provides for maximum exchange of past results (generally subject to verification from other US sources), proceeds then to sufficient disclosure of the future planning of both sides to identify areas favorable for cooperation, and concludes with the joint definition of specific projects. Examples of specific projects would be put forward in the initial presentation of this approach to lend credibility and substance to it.

6. The report recognizes that the Soviet Union is unlikely to be amenable to such an approach. In that case, it would be possible to proceed directly to specific proposals. Some 15 examples of possible projects are described in the report and evaluated in such terms as our current knowledge of the Soviet program permits.

However, limitations (described in the report) attach to virtually all these proposals. These limitations reflect the general climate of US-Soviet relations and are therefore subject to change—which might bring *any* of the proposals within the range of realistic negotiation. At present, a change in sentiment appears necessary even for small steps in cooperation; for example, in the exchange of purely scientific data relating to solar radiation and micrometeorites, the Soviet Union has within the past year declined to provide details of instrumentation and calibration required for their understanding. Given a change in sentiment, however, such [3] exchanges would be useful and some cooperation might be proposed and developed in several areas including those listed below and, in addition, mutual tracking support and the recovery and return of manned capsules after their return to earth.

7. On balance, the most realistic and constructive group of proposals which might be advanced to the Soviet Union, with due regard for the uncertainties and limitations discussed above and detailed in the report, relates to a joint program of unmanned flight projects to support a manned lunar landing. These projects should be linked so far as possible to a step-by-step approach, ranging from exchange of data already obtained to joint planning of future flight missions. They include projects for the determination of:

(a) Micrometeoroid density in space between earth and moon.
(b) The radiation and energetic particle environment between earth and moon.
(c) The character of the lunar surface.
(d) The selection of lunar landing sites.

8. I believe this affords flexibility for positive action, utilizing either a variant of the structured approach (paragraph 5) or, with necessarily greater caution, selected specific proposals without reference to the structured approach (paragraph 7).

9. With regard to the timing and form of further US initiatives toward the Soviet Union, the report recommends the following:

(a) Continuing interest should be expressed through the existing NASA-Soviet Academy channel, in a positive Soviet response to the proposals for cooperation already made by President Kennedy and by you.

(b) No new high-level US initiative is recommended until the Soviet Union has had a further opportunity (possibly three months) to discharge its current obligations under the existing NASA-USSR Academy agreement, or, in the alternative, until the Soviets respond affirmatively to the proposal you have already made in the UN.

(c) If Soviet performance under the existing agreement is unsatisfactory, a high-level initiative on a non- [4] public basis would seem desirable to prod the Soviet Union to better performance; additional public steps might be considered if this proves unavailing.

(d) If Soviet performance under the existing agreement proves satisfactory, personal initiative by you would still be required to extend this success to cooperation in manned lunar programs. Because the scope of initiative by Soviet Academy representatives seems limited, Mr. Khrushchev's personal interest and support would also seem to be required for any significant extension of joint activity. It is believed that your initiative will be more effective if taken privately in the first instance.

(e) A US initiative should establish our interest in the preferred structured approach described above. If it then becomes feasible to proceed with technical negotiations, the NASA-Soviet Academy channel should continue to be the vehicle used; as in the past, technical proposals to be considered in such negotiations should be made available for prior interdepartmental comment. (It may become appropriate to consider an effort to induce the Soviet Union to make personnel available who are closer to their technical program.)

(f) Agreements reached in technical negotiations should be embodied in memoranda of understanding, explicitly subject to review and confirmation by governments.

(g) To demonstrate the serious intentions of the US with regard to international cooperation in space and to maintain some pressure upon the Soviet Union to follow suit, we should continue to expand our current and successful joint projects with other nations to the degree possible.

This report will be kept under continuing review in NASA in concert with other interested offices and agencies, and we shall keep you advised of our progress with the Soviet Academy under the current agreement between us. I believe we are well prepared to support whatever initiative you determine to be appropriate in light of this report and stand ready to provide such additional information and judgment as you may require.

Respectfully yours,

James E. Webb
Administrator

Enclosure

****

[Enclosure page 1]

US-USSR Cooperation in Space Research Programs

President Kennedy and President Johnson have affirmed and reaffirmed the desirability of exploring further joint efforts with the Soviet Union and other countries in cooperative space activities, including manned lunar programs. (See Appendix I.) In support of these initiatives and in anticipation of possible discussions with the Soviet Union, this report examines technical proposals which might be put forward by the United States, as well as other considerations appropriate to such discussions.

For two reasons, this report concentrates upon possible cooperation in lunar programs: (1) cooperation in lunar programs was the focus of President Kennedy's September 1963 initiative and of President Johnson's confirmation of that initiative and, in particular, of his State-of-the-Union reference to the subject; (2) cooperation in other areas of space research and exploration was covered in the Kennedy-Khrushchev correspondence of February-March 1962 in both specific and general terms, has progressed to the point of firm agreement on three projects, and is the subject of an apparently continuing relationship pursuant to that correspondence and agreement. At issue now is an extension of this relationship to the only major field effectively excluded from it, i.e., manned lunar programs and related unmanned efforts. (A brief review of the current relationship [2] appears in Appendix II.)

This report necessarily assumes that the Soviet Union is engaged to some degree in a program looking toward eventual manned lunar landings. Soviet statements on this point have been ambiguous as to timing and status but clearly positive on balance. If there is *not* a Soviet program, the Soviet Union will probably confuse the issue for an indefinite period. (In that case, it has been suggested that US pressure for cooperation might even induce the Soviet to undertake manned lunar efforts not now planned. Viewed positively, this could divert Soviet resources from less desirable preoccupations; seen negatively, it could lead the Soviet Union into new technology. We believe that the safest assumption is that the Soviet Union does not exclude a manned lunar program and that no significant danger to us is involved if this assumption is incorrect.)

I.

Guidelines which have been applied in the preparation of this report follow:

(1) The central objective is to bring about continuing cooperation with the Soviet Union, rather than to achieve propaganda gains as such. (In his September 20 speech at the UN, President Kennedy stated, ". . . we must not put forward proposals merely for propaganda purposes;").

(2) In order to achieve real gains, we should press for [3] substantive rather than token cooperation.

(3) Cooperation with the Soviet Union should be well defined and the obligations of both sides made clear and comparable. (This will facilitate implementation as well as clarify responsibility in the event of failure and withdrawal.)

(4) In the present state of US-Soviet relations, we should undertake no project or other arrangement which night make us dependent upon Soviet performance, thereby impairing or limiting our independent capability in space.

(5) National security interests and military potential must be fully protected. No exchanges impinging upon security should be considered in the absence of certain, comparable, and verifiable information from the Soviet side.

(6) Opportunity for participation by other countries should be preserved and all results made available to them.

II.

Ultimately, any program of substantive cooperation with the Soviet Union must rest upon positive proposals of specific character. Such specific proposals can be defined almost without limit, and numerous examples of different modes of cooperation with the Soviet Union are provided in this report. *However, the advantages and disadvantages of specific proposals are not fixed by the terms of those proposals in an absolute sense.* The positive and negative [4] values to us may vary markedly, depending upon Soviet objectives, techniques, procedures, and schedules relative to ours. It is therefore most desirable that we seek information on these aspects of the Soviet program so that we can evaluate and shape our own proposals effectively and prudently. Lacking such information, we would inevitably remain uncertain in matters of security, tactics, and bona fides.

Accordingly, we should define, and attempt to hold to, an approach to the Soviet Union which is calculated to (1) *determine the level of confidence* which we can place in the Soviet Union in this subject area, (2) *provide information* of the basic elements of the Soviet program, and (3) merit the confidence and support of the public and the Congress.

An approach structured to achieve these ends is spelled out in the next section of this report. *If such a structured approach is not acceptable in whole or in part to the Soviet Union, the President and the Department of State may, nevertheless, depending upon the circumstances and apparent attitude of the Soviet Union, determine that technical negotiators should proceed to the direct presentation of specific proposals. Such flexibility is desirable—but with clear recognition that different considerations will apply to the same proposals, depending upon whether they are offered with [5] or without some confidence and knowledge of Soviet plans.*

III.

The preferred approach to negotiations with the Soviet Union entails the discharge of outstanding obligations, followed by an escalating series of exchanges which are, in the initial stages, subject to verification. It is thus calculated to build a level of confidence upon which progressively significant cooperative activities may be based.

Since negotiation on manned lunar programs necessarily presages significant new relationships with the Soviet Union, requiring evidences of good faith, the first steps should be directed to clearing the slate as much as possible.

A most desirable first step would be material progress on both sides to implement the existing bilateral (Dryden-Blagonravov) space agreement in which the Soviets remain, at this writing, delinquent (although they have resumed communication).

A second step more directly following upon the US overtures in the UN would be the detailed exchange of data and information of the two countries' manned space programs *to date.* (This should include past flight, biomedical, and training data and could extend to early spacecraft technology.) The virtues of this step would be that it would represent a clean start, requiring from us little new information yet obliging the Soviet Union to present [6] considerable information not previously made available publicly. Since elements of the USSR contribution at this stage would be subject to verification through independent sources, a practical and useful test of Soviet intentions would be available at the earliest contact, and a first confidence level could be established.

If this step should prove a significant obstacle to further progress, it might, in the interests of flexibility, be downgraded, as it were, and subsumed quite naturally under the third step (below). It should, in any event, be tested since other means of determining the degree of Soviet good faith are not readily apparent. Opportunities for establishing a confidence level for dealing further with the Soviets would be diminished in proportion to de-emphasis of this second step.

The third step would be the exchange of gross descriptions of our respective manned lunar programs. Again, this step would not place an undue burden upon us because of the publicity already given to our own intentions, but it would for the first time require the Soviet Union to describe its conceptual approach to the lunar landing problem. This step appears virtually indispensable for it is hardly possible to proceed intelligently or safely to coordinated, cooperative, or joint effort without some over-view of the proposed Soviet program.

The fourth step would seek, through more precise descriptions of our respective lunar programs, to isolate [7] elements of conflict or duplication and to discover opportunities for trade-off, complementary procedure, or joint action. Significant security considerations do not arise until this step is reached.

* * *

Examples of cooperative relationships that might develop at various stages of the above procedure follow:

– Conflict between the two programs could arise, as a crude illustration, through plans to use the same "window" for independent lunar missions on the same radio frequencies. It would be of mutual interest to eliminate any such conflicts.

– Unnecessary duplication, illustrated by independent but adequate programs for exploration of the lunar surface, would offer opportunities for thinning out or otherwise adjusting our respective programs so as to provide, together, only required information— the exact degree of thinning out depending upon the confidence level established at the time.

– In other cases, a desirable redundancy of effort might be recognized and specific provisions for data exchange made to increase reliability and confidence.

– Discovery that both sides planned to apply limited resources to the same facet of a broader problem (e.g., examination of the lunar surface in a relatively narrow region) would permit a reordering of efforts to cover additional facets of the problem on a shared-effort basis, with subsequent exchange of the results.

[8] – Some tradeoffs can be visualized, arising from differentials in schedules and capability in the two programs; e.g., the possibility that the Soviet Union might acquire a sample of the lunar surface before the United States, taken together with our twenty-four hour deep space tracking capability, suggests a trade-off between the two; medical data obtained in the Vostok flights might be ·traded for radiation or micrometeorite data obtained in our scientific program.

If an improved confidence level is achieved through the modest but meaningful arrangements suggested above, progress toward more advanced, integrated relationships could be made.

IV.

At various steps in the above procedure, specific projects should be put forward as appropriate to lend concrete substance to the negotiations. A relatively detailed description of such projects follows:

(Negative or uncertain values reflected in this description follow from our current lack of knowledge of Soviet plans; a more positive evaluation should be possible in each case if serious intentions on the part of the Soviet Union motivate a sufficient exchange of the necessary background information. A negative assessment of Soviet interest or desire in a given case does not necessarily mean that the proposal should not be put forward; it is intended [9] solely to reflect realistically the *present* prospects for a substantive advance of our purpose. These apparent prospects may well change in light of any information forthcoming from the Soviet side relative to their program and interests. Close examination of the comments provided in each case will show that the framing of proposals with positive appeal to both sides requires knowledge of the *objectives, modes* of attack, and relevant *schedules* of both sides. The same knowledge is necessary to determine what critical *tactical or security advantages* may be conferred or lost in a given project. These defects grow in direct proportion to the significance of the proposal contemplated.)

A. Data Exchange

1. *On Micrometeoroid Flux*—Both the US and the USSR could profit from a full exchange of information on the temporal and spatial distribution, mass penetration characteristics, and shielding of micrometeorites in earth-to-moon space. The security aspects are minimal, and present indications are that information obtained will not present radical problems of an unexpected nature. However, as recently as June 1963, Soviet scientists, in precisely such an exchange relating to their Mars and our Venus flights, declined to give us instrumentation and programming information necessary for meaningful interpretation of their data. Also, the USSR must be expected to be quite [10] reluctant to provide data on shielding materials and results.

2. *On Radiation and Solar Events*—Both sides seek greater knowledge of radiation end particle fluxes in cislunar space, particularly that associated with solar proton events. Such information is necessary to improve the predictability of proton showers so as to fix manned flight schedules safely and permit the design of optimum shielding. This is likely to be a long-range program requiring constant monitoring and predisposes both sides to welcome an exchange of information. We could advance a proposal to define a project of investigation and exchange on this subject to be carried forward by a joint working group consisting of designated representatives of both sides. There is some question, however, whether the Soviets are yet on a par with us in this work. Also, we anticipate that the USSR will continue [to be] reluctant to discuss the detailed interrelationships of data, instrumentation, and programming in adequate depth. Nor could we be sanguine about exchange relating to shielding or other countermeasures.

3. *Lunar Surface Characteristics*—Both sides require information on the characteristics of the lunar surface for final design of spacecraft to land on the moon. Whether there is the basis for an exchange relationship depends in part on the relative schedules of the two programs; if the Soviets are ahead of us, as is possible at this early stage, they will have acquired intelligence [11] of the lunar surface before we do and have little interest in any contribution we can make on this point. On the other hand, if we are on similar schedules and the lunar surface is discovered to have radical characteristics not anticipated, such information could become critical to equipment design and even mission success. It could thus become an important element in the space race itself, with critical tactical and even security value. Either side might well wish to withhold knowledge of this kind.

4. *Selection of Lunar Landing Sites*—The same considerations discussed immediately above apply to exchange of information in the survey and selection of lunar landing sites. Assuming a Soviet lunar landing program, both sides are faced with the same gross requirement, and thus there should be in principle a basis for cooperation. However, the actual degree of interest and potential for cooperation would depend in good part upon technical requirements and relative time schedules; if the latter are not close, the leading side could be expected to be relatively disinterested, whereas if they are close, information on a suitable site could become critical in a closely competitive situation.

5. *Astronaut Training and Experience*—Each side must be assumed in principle to have interest in the other's astronaut training techniques, flight experience, space medicine results, and spacecraft technology. The US has already been quite open in publishing its material along [12] these lines, and has not yet had comparable periods in orbit. The Soviet Union must therefore be presumed to have less interest than we. Indeed, a Soviet representative to a very recent International Academy of Astronautics meeting declined to participate in a second conference on manned flight, asserting that there was little new to be expected from the American program in the next year or so. (No additional manned flights can be expected in the US program for upwards of a year.) In sum, it would appear that we cannot offer mutuality for a considerable time in flight results and

space medicine. Indeed, we would appear to be leading from weakness if we pushed for exchanges in these fields. Exchanges in the related areas of astronaut training and spacecraft technology would, if they were to be meaningful, impinge upon flight systems, security considerations, and simulator techniques, and must be regarded as most difficult to approach in the initial instance with the Soviet Union.

 B. Operational Cooperation

  1. *Mutual Tracking Support*—Several modes of cooperation in tracking and data acquisition have been explored from time to time with the Soviet Union: the USSR was offered the support of the Mercury network for any manned flight of their own, with no strings attached (Glennan); it was asked to consider an exchange of tracking stations, each side to place a station in the other country, each to operate its own station (Kennedy); and the USSR itself [13] suggested cooperation in the tracking of deep space probes (Khrushchev), but later retracted this offer, privately implying security considerations. Despite seeming Soviet disinterest in this area and the fact that lunar missions are conducted at particular times (windows) when both sides may launch missions of their own, it seems probable that both could gain from mutual tracking arrangements. Since windows are a function of launch site and tracking station locations, mission profile and objectives, and payload capabilities, the two sides would probably utilize somewhat different windows. We might then provide twenty-four hour ground coverage (lacked by the USSR) in exchange for greater flexibility afforded by use of their land and ship-based nets.

  2. *Capsule Recovery (earth)*—Both sides face the possibility of spacecraft returns to earth in areas not planned. Accordingly, they might both have an interest in exchanging the signals and recovery procedures to be utilized in emergency recoveries. Either side could then proceed to the rescue of astronauts in areas under their control. The exchange of such signals could in principle also permit either side, somewhat more readily than now, to interfere with recovery operations by the other. However, this appears a very small risk and one which might very well be taken. Such a project would appear to have few negative aspects, little prospect for wide implementation, but possibly considerable public value.

[14] 3. *Capsule Recovery (space)*—It is possible to frame a proposal that both sides agree upon common docking hardware so as to permit either to "rescue" the spacecraft of the other in distress. In fact, it is not known whether hardware common to the two competing systems would be feasible, but assuming it is, rescue operations of this kind, given current limits to spacecraft maneuverability, would require compatible trajectories and orbits, compatible oxygen supply arrangements, an agreed communications, rendezvous, and docking procedure, common training, and possibly compatible aerodynamic configurations for re-entry purposes. At a minimum, guidance systems, docking hardware, and rendezvous and docking techniques, capabilities and limitations would all appear, at early stages, to be of security concern. A proposal of this sort would, therefore, not be attractive to either side.

  4. *Lunar Logistics*—Following the first manned lunar landings, it would appear possible to define a proposal for sharing logistic support for more ambitious lunar exploration. Such a proposal could be shaped in terms of a division of the logistic responsibilities or a division of responsibility as between logistics and personnel. A proposal of this type would have some appeal if the two sides were on roughly similar schedules and shared ambitious plans for lunar stations or exploration, something not known to be planned in either case. If one were well ahead [15] of the other or had no current plans for ambitious follow-on lunar projects, it would have relatively little appeal. A proposal of this type would have the disadvantage of subjecting us to reliance on the honorable and

competent discharge by the USSR of its responsibilities over a period of years. In any case, the proposal would not appear to promise early realization and should be deferred for subsequent consideration in the course of a progressive and satisfactory development of more immediate projects.

5.   *Trade-Offs*—Where mutual benefits cannot be established in symmetrical projects, it may be possible to relate dissimilar activities to a single balanced cooperative effort. For example, we could offer the Soviets the support of our twenty-four hour deep space tracking capability (in periods when it is not directed to our own use) in exchange for data (or samples) of the lunar surface, which the Soviets might acquire before the US.

C.   Integrated Projects

Substantial integration of major elements of flight configurations is circumscribed by two factors: (1) virtually all major contracts for accomplishment of project APOLLO have already bean placed, establishing a heavy and costly commitment in design and development; (2) the placement of responsibility in the Soviet Union for integral elements of our own program would enable the Soviets to obstruct our progress while proceeding [16] clandestinely on their own. Nevertheless, certain cooperative projects requiring close integration are widely entertained and some comment is appropriate. More important, there may be some integrated effort which is, nevertheless, possible at a relatively early stage. At least one proposal of this type is noted below.

1.   *USSR Booster/US Spacecraft*—It has been widely proposed that we suggest to the Soviet Union a manned lunar effort based upon the use of their greater boosting capability and the most advanced spacecraft of the US. The Soviet Union is *not* now known to possess a booster capable of manned lunar landing and return although they are developing engines which, if clustered, could provide this capability. The US *is* building such a booster. It is not consistent with the US objective of achieving a leading space capability to delegate the development of an adequate booster to the Soviet Union. A reversal of the proposal would not appear to be in the national interest since it would employ an advanced US capability to place a Soviet spacecraft first on the moon. It would also entail Soviet access to US launching sites and techniques without the possibility of access to USSR sites under comparable circumstances.

The heart of the problem posed by a proposal of this type lies in the very extensive exchange of technology required to integrate the spacecraft of one side with the booster of the other. Such an exchange applies to all [17] significant characteristics of the booster system in design and performance, including guidance, and requires the launching authority to have full information of the spacecraft system. A continuing and extensive mutual interplay on technical terms is known (through experience in domestic as well as international satellite programs with friendly nations) to be required for spacecraft-booster integration if success and avoidance of recrimination are to be achieved. Extensive access would be required by both sides to the launch site, and, by reason of the unsymmetrical basis for the project, such access would be one-sided. The experience with the Soviet Union in areas with (or, indeed, without) military implications suggests that even a small fraction of the interchange required would be forthcoming from them.

2.   *Turner Proposal*—A Republic Aviation engineer, Thomas Turner, has proposed in *Life* (October 11, 1963) a cooperative effort to circumvent (some of) the difficulties noted immediately above. According to his proposal, the US would forego the development of a large booster and concentrate simply on placing its lunar excursion module (LEM) in earth orbit. The Soviet Union would at the same time place a very large and powerful spacecraft in earth orbit. The two would rendezvous, then utilize the Soviet's spacecraft propulsion to transfer to a lunar orbit, at which time the LEM would separate and descend to the lunar surface with both a Soviet and an American aboard. It would

then [18] return to lunar orbit, the occupants would transfer to the Soviet spacecraft, abandoning the LEM, and return to earth. According to Turner, the sole requirements are common docking hardware and a communications agreement. The proposal is an ingenious one but implies that neither side would develop the total resources to conduct a manned lunar program by itself. We regard this, at this time and in the present context, as an unacceptable interdependence, prejudicing seriously our ability to proceed with our own program in the event that the Soviets do not live up to their agreement over the extended period of years required to implement it. The US requires a major booster for its own posture and broad national interest. Thus, no real saving would be effected by the Turner proposal. The notion that the necessary lunar orbit docking could be conducted without common training and practice procedures on earth is not tenable. In addition, this raises most of the questions which are specified in item B.(3) above. Our conclusion is that the Turner proposal is neither practicable nor desirable at this stage in US/USSR relationships. It could be held in abeyance until a progressive improvement in the discharge of cooperative obligations by the USSR warrants its consideration at a later date.

3.   *Interchange of Astronauts*—The US could propose a reciprocal arrangement under which astronauts of each side are accepted by the other for extended periods [19] of training leading to participation in flight missions. It is apparent that such an exchange would entail long-term and extensive access to training facilities and programs, flight hardware and systems, launching sites, and so forth, as well as language preparation; however, reciprocity might be assured through synchronized phasing of the program in both countries. The US would have far more to gain than to lose from such reciprocity in view of the relative secrecy of the Soviet program to date. The prospect is particularly attractive because of its implications for opening up Soviet operations. We are informed, however, that it may be politically premature.

As always in dealing with the Soviet Union, it may be feared that comparable access, information, and training will not be afforded the American astronaut(s) exchanged with the Soviet Union. The concept of synchronized phasing of the training of the two would go a long way to correct this, since the two astronauts would move from one phase to another of the two countries' programs on a par and we could withdraw our man if we were dissatisfied. The prospects of such dissatisfaction must be regarded as rather high, given experience with exchange programs with the Soviets in the past. It may be, therefore, that greater success could be had with this same project if, again, it were developed in the course of a progressively improving relationship with Soviet space authorities. It remains, in any case, one of the more attractive possibilities. [20] In fact, early instruction of selected astronauts in the Russian language has been suggested to remove at least one obstacle to its realization.

V.

Questions of initiative, timing, and procedure for negotiations with the Soviet Union have been considered. (The pertinent background and status of past negotiations with the USSR is briefly summarized in Appendix II.)

1.   As contacts continue at the agency (Dryden-Blagonravov) level, we should clearly express our continuing interest in a response from the Soviet Union on the question of extending cooperation to lunar programming and other subjects.

2.   No new top-level action (by the President, Secretary of State, or Ambassador) is recommended until—

(a) the Soviet Union is given a further opportunity to evidence the discharge of its obligations under the existing NASA-USSR Academy space agreement, or

(b) the Soviets respond to US initiatives already taken in the UN.

3.	After the Soviet Union has had a further opportunity to deliver or default on the existing agreement, a further top level initiative would seem appropriate.

The nature of such a US initiative might be along the following lines:

(a) In the event of continued failure of the Soviet Union to discharge existing obligations in the Dryden- [21] Blagonravov agreement, a top level US/USSR initiative would seem desirable, privately in the first instance. If Soviet intransigence persists, it may then become appropriate to tax the Soviet Union publicly with their failure in matters of cooperation.

(b) If the prospects for an extension of existing agreements to the manned lunar landing area become promising—either because of performance in the existing agreement or because of a response from the Soviet Union to our UN initiative—a further top level US action should be taken, privately in the first instance. For example, the President may wish to inform Khrushchev that we propose an orderly, structured approach toward a developing cooperation, beginning with the maximum exchange of past results, proceeding to sufficient description of future planning to permit identification of possible areas of cooperation, and concluding with the definition of specific projects. (Examples of possible projects would be included in the presentation of this structured approach to lend it credibility.) Again, if the Soviets are intransigent, consideration might be given to stating our position publicly in order to increase pressure on the Soviet Union. In such a public statement, the US approach could be openly described to domestic and foreign advantage.

[22] 4. Whether a further US initiative is taken or a specific Soviet response to the President's UN offer received, in either case making negotiations possible, it is then our considered view that our action should be for the express purpose of preparing the way for technical discussions. The NASA-Soviet Academy channel, which has been successfully opened by Dr. Dryden, should continue to be the vehicle for technical exploration and negotiation of the possibilities for cooperation with the Soviet Union. (If it should prove technically desirable or necessary, consideration should be given to requesting the Soviets to assign to the negotiations personnel closer to the technology of their program.) As in the past, proposals to be considered in such negotiations should be made available for prior inter-departmental consideration.

5.	Any agreements reached at this technical level should be embodied in memoranda of understanding, explicitly subject to review and confirmation by governments.

6.	As a tactical device, calculated to put pressure upon the Soviet Union, demonstrate our serious intentions, and gain good will from certain nations, consideration should be given to means by which "other countries" than the Soviet Union might be further identified with our lunar programs. (See Appendix III.)

****

[1]

Appendix I

(A)	President Kennedy made the following statement regarding United States-Soviet cooperation in outer space in his address before the United Nations General Assembly on September 20, 1963:

"Finally, in a field where the United States and the Soviet Union have a special capacity—the field of space—there is room for new cooperation, for further joint efforts in the regulation and exploration of space. I include among these possibilities a joint expedition to the moon.

"Space offers no problem of sovereignty; by resolution of this Assembly, the members of the United Nations have foresworn [sic] any claims to territorial rights in outer space or on celestial bodies, and declared that international law and the U. N. charter will apply. Why should the United States and the Soviet Union, in preparing for such expeditions, become involved in immense duplications of research, construction and expenditure? Surely we should explore whether the scientists and astronauts of our two countries—indeed, of all the world—cannot work together in the conquest of space, sending some day in this decade to the moon, not the representatives of a single nation, but the representatives of all humanity."

[2] (B) President Johnson reaffirmed the above statement through Ambassador Adlai E. Stevenson who made the following remarks in Committee I of the United Nations General Assembly during debate on international cooperation on outer space, on December 2, 1963:

"As you also know, President Kennedy proposed before the General Assembly last September to explore with the Soviet Union opportunities for working together in the conquest of space, including the sending of men to the moon as representatives of all our countries. President Johnson has instructed me to reaffirm that offer today. If giant strides cannot be taken at once, we hope that shorter steps can. We believe there are areas of work—short of integrating the two national programs—from which all could benefit. We should explore the opportunities for practical cooperation, beginning with small steps and hopefully leading to larger ones.

"In any event, our policy of engaging in mutually beneficial and mutually supporting cooperation in outer space—with the Soviet Union as with all nations—does not begin or end with a manned moon landing. There is plenty of work yet to come before that—and there will be even more afterward."

[3] (C) In his State-of-the-Union address to the Congress on January 8, 1964, President Johnson said,

"Fourth, we must assure our preeminence in the peaceful exploration of outer space, focusing on an expedition to the moon in this decade—in cooperation with other powers if possible, alone if necessary."

****

[1]
Appendix II

The background of experience in negotiations with the USSR is briefly summarized: Progress at all levels has almost invariably required US initiative. It appears that new initiatives are successful only if the way is paved at the very highest levels. Negotiations are seriously hampered by the fact that Soviet representatives are drawn from the Academy complex which seems to be once removed from the actual conduct of the Soviet space program. (Soviet scientists do not often appear well informed of flight conditions or hardware.) Soviet reaction time to US initiatives and correspondence has been extremely slow. The USSR is currently delinquent on most action items scheduled in the Dryden-Blagonravov agreements; however, correspondence has been resumed by Blagonravov after more than three months of silence and agreed optical observations of the ECHO II satellite have now been performed by the Soviet Union.

The basic Soviet line for the past four years has been that significant cooperation cannot precede major improvements in the political atmosphere, including disarmament.

(The US proposals which led to the Dryden-Blagonravov agreement were apparently regarded as sufficiently modest to permit some departure from this line—though at least one of the agreed projects could lead to a joint global meteorological satellite system.)

[2] At various times the Soviet Union has rejected US offers of tracking support for manned flights, an interchange of overseas tracking stations for earth satellites or deep space probes, formal participation with NASA and other countries in experimental communication satellite tests, exchanges on standards and techniques to preclude contamination of the lunar and Martian environments, and repeated open-end offers to explore any items of interest to the Soviet Union.

With regard to Soviet plans for a manned lunar program, Khrushchev has said little more than that the USSR will not proceed until they are ready and that they are working on the problem, but it is not known whether they are developing a large enough booster although engines suitable for clustering for that purpose are reportedly under development. Khrushchev has spoken only ambiguously about cooperation and has actually seemed to accept competition as desirable.

On the other hand, some softening of the Soviet line may be indicated, not only by the Dryden-Blagonravov agreement, but also by the recent willingness of the Soviet Union to reach agreement on legal principles to apply to space activity and on radio frequencies to be used in space communications and research. The requirements for these agreements, however, are far from comparable to those applicable to cooperation in manned lunar programs.

A brief summary and evaluation of the status and content of the Dryden-Blagonravov agreement follows:

[3] A first US-USSR Bilateral Space Agreement was reached on June 6, 1962 and was then supplemented by an implementing Memorandum of Understanding which became effective August 1, 1963. Together, these agreements set forth the technical details and arrangements for cooperation in three areas:

    1.   Coordinated Meteorological Satellite Program
- Exchange of cloud cover photographs and weather situation analyses gained from each country's experimental meteorological satellites;
- Establishment of a full-time, conventional, facsimile quality communications link between Washington and Moscow for two-way transmission of these data;
- Coordinated launchings of future experimental weather satellites, and ultimately, of operational weather satellites.

    2.   Communications Satellite-Experiments
- Experimental transmissions at 162 mc/s between the USSR and the Jodrell Bank Observatory in England using the US passive reflector satellite ECHO II;
- USSR to consider experiments at higher frequencies;
- USSR to consider radar and optical observation of ECHO II;
- Future negotiations on possible joint experiments with active communications satellites.

    3.   Geomagnetic Satellite Data
- Launching by each country of a satellite equipped to measure the earth's magnetic field as part of research planned for the International Year of the Quiet Sun in 1965;
- Exchange of results of satellite measurements;
- Exchange of data from magnetic surveys of other types.

[4]

Dr. Dryden wrote Blagonravov in mid-August listing action items requiring early completion if the agreed deadlines for joint action were to be met, and conveying the United

States position on each. This communication went unanswered until December when Blagonravov acknowledged the letter, apologized for delay, indicated substantive replies were being prepared, and asked for the launch date for ECHO II. Dr. Dryden replied immediately by cable, giving the launch window and nominal orbital elements for the ECHO II satellite, and reiterating NASA's request for Soviet radar cross-section and optical observation of the satellite during the inflation stage (which occurs in part over the USSR on the first orbit). [5] This cable was immediately acknowledged by Blagonravov; as of this writing, he has provided a statement of intention to discharge at least the minimum requirements upon the Soviet Union for observation of ECHO II and communications tests with that satellite. He remains delinquent in other outstanding matters.

Although all joint action has slipped several months because of Soviet dilatoriness, this need not affect any of the proposed cooperative efforts substantively but may only delay their implementation. At this time, it seems likely that Soviet performance will continue [to be] ragged, with little regard for deadlines. The remoteness of the relationship maintained by the USSR detracts in some degree from the positive value of the cooperative association established; nevertheless, satisfactory completion of *any* of the steps prescribed in the agreements should provide the best basis for improved relationships and further progress.

****

[1]

Appendix III

Besides inviting the Soviet Union to cooperate in the lunar program in his recent UN speech, the President expressed a desire to bring other countries in as well. The possibilities include the following:

1. Tracking and data acquisition—We already enjoy the cooperation of a number of countries in the accommodation and operation of manned flight tracking and data acquisition stations and should publicize this fact along with our interest in extending the present level of participation.

2. Scientific experiments—We now give foreign scientists a chance to compete for space for their experiments in our observatory satellites. We should consider extending this practice to Gemini and Apollo, noting that these opportunities may be very limited even for our own scientists. (In addition to space and weight limitations, there could be difficulties growing out of Air Force participation in Gemini).

3. Contracts—If they materialize in sufficient number, publicity can be given to certain subcontracts entered into with foreign contractors (e.g., Canadian companies are developing and providing extensible antennae for the Gemini and Apollo missions, including the antenna to be used for rendezvous missions.) In addition, consideration [2] could be given to offering foreign governments the opportunity to take on the development and production of subsystems and parts, *on a cooperative basis* (i.e., at their own expense), to meet our design, standard, and schedule requirements. The technical and contracting limitations would, however, be severe and the takers few.

4. Astronaut orientation—A program might be organized under which foreign high performance pilots might be brought together for observation of, and limited participation in, NASA astronaut training (only) programs as a familiarization and orientation effort on a continuing basis (e.g., successive three-month classes).

5. Astronaut training and flight—The numerous and valid objections heretofore raised against including foreign pilots in our astronaut program are recognized.

The negative aspects are these: rivalry among interested foreign nations; further pressure upon our limited flight opportunities; resentment by current US astronauts; difficulties in application of commercial benefits to astronauts; security questions; pressures for flight priorities; feminist and congressional criticism; absence of practical application abroad for the training given here.

The positive aspects are these: few other single actions could more dramatically express the President's deep desire for cooperation; few other single actions could equal [3] the boost given by this one to US relations with Latin America or Asia, if pilots from those regions (many already trained here) were chosen; few other actions could do more in the next few years to eclipse Soviet propaganda in this area—or protect us more effectively against a similar Soviet move.

On balance, technical and political considerations suggest a negative conclusion on an offer of this kind and preference for the proposal reflected in item 4 above.

* * *

Perhaps the most acceptable position to meet the issue of the third country participation is represented by the recent statement of Senator Clinton P. Anderson before the AIAA, January 1964:

"... we can give validity to this nation's policy to internationalize space by asserting that the United States will accept offers of support from any nation which can contribute to the space program."

Such contributions should continue to be organized and implemented within the policies already applicable to existing (and uniformly successful) international programs of NASA.

Document I-44

Document title: Thomas O. Paine, NASA Administrator, to Academician M.V. Keldysh, President, Academy of Sciences of the USSR, July 31, 1970.

Source: Thomas O. Paine Papers, Manuscript Division, Library of Congress, Washington, D.C.

In this letter, NASA Administrator Thomas O. Paine, the successor to James E. Webb in 1968, followed up on earlier, more general overtures to the Soviet Union for enhanced post-Apollo cooperation. This was a specific proposal for cooperation in compatible docking arrangements between U.S. and Soviet spacecraft. Paine had also just announced his intention to resign as NASA Administrator in September, and he wanted to assure Keldysh that the desire for enhanced cooperation was a U.S. government position, not just his own preference. The Soviet reply to this letter was positive, and the two countries began discussions in October 1970 that led to the 1975 Apollo-Soyuz Test Project.

[1]   JUL 31, 1970

Academician M.V. Keldysh
President, Academy of Sciences of the USSR
Leninsky Prospect 14
Moscow, U-71, USSR

Dear Academician Keldysh:

We were encouraged to learn of the inquiries by your Embassy to the National Academy of Sciences regarding possible discussions of compatible docking arrangements in space. I had mentioned the subject to Dr. Handler as a possible item for consideration by NASA and your Academy prior to his recent trip to the Soviet Union.

As the government agency responsible for civil space activities, NASA has direct responsibility for any discussions with Soviet officials regarding actions we might take together to assure compatible docking systems in our respective manned space flight programs. If you agree that this subject should be discussed between us in the meeting which we have had in view for some time, we will be glad, in order to facilitate adequate mutual preparation, to receive two Soviet engineers at our Manned Spacecraft Center in Houston in the very nearest future to examine NASA's current spacecraft designs for docking purposes and to discuss future docking techniques. In the next step we would propose to proceed with the responsible Soviet officials to discuss our respective views with regard to the achievement of compatible docking configurations and techniques. If we can indeed agree on common systems, and I foresee no particular technical difficulty, we will have made an important step toward increased safety and additional cooperative activities in future space operations. This is particularly timely in my view as we proceed toward the initial experiments leading to the orbiting space station.

[2] You may already know that I have submitted my resignation as Administrator of NASA to the President for personal reasons. This, of course, will not change the policies and interests of NASA with respect to international cooperation in space. Thus, you should understand our past and current correspondence on [an] official rather than personal [basis], although this matter has my wholehearted personal support. I regret very much that I will not have the opportunity to carry through personally our discussions with you to fruition, but am optimistic that much can be accomplished and hope that we can continue to make progress in the next month.

Sincerely,

T.O. Paine
Administrator

Document I-45

Document title: Glynn S. Lunney, "Trip Report—Delegation to Moscow to Discuss Possible Compatibility in Docking," November 5, 1970.

Source: NASA Historical Reference Collection, NASA History Office, NASA Headquarters, Washington, D.C.

Glynn Lunney, who headed the Flight Director's Office at NASA's Manned Spacecraft Center in Houston—renamed the Lyndon B. Johnson Space Center in 1973—was part of the first-ever delegation of NASA engineers and other officials who traveled to Moscow, October 26 to 28, 1970.

The purpose of the trip was for the initial discussions of the feasibility of a U.S.-Soviet cooperative project regarding compatible docking between the two countries' spacecraft. This trip report captures both the human and the technical aspects of the NASA team's three days in Moscow. The figures referred to in the text have been omitted.

---

[1]

MEMORANDUM TO: See attached list

FROM: FC/Glynn S. Lunney

SUBJECT: Trip Report—Delegation to Moscow to Discuss
                                     Possible Compatibility in Docking

Before I discuss our technical meetings, so many people have asked me about personal observations that I have included some of these at the beginning. In general, everything was done to make our visit pleasant and productive. General comments are as follows.

1. Our time was scheduled very well, and we kept a busy schedule.
2. Transportation and a guide/interpreter were always at the ready.
3. Weather was mostly overcast and occasional drizzle, but just "raincoat" cold.
4. We stayed in a very large, modern hotel (the Russiya—4000 rooms) and the quarters were very adequate.
5. Breakfast was a buffet arrangement in the hotel.
6. Lunch in the middle of the afternoon and dinner in the late evening were scheduled each day at various places. The food was delicious, the Russian vodka is an excellent drink; the caviar is worth eating also.
7. There is a fair amount of apartment building going on in Moscow. From what we saw, there were essentially no single-family dwellings in the city; the 7 million population apparently lives in the apartment buildings. We were in only one apartment building which is provided specifically for the foreign embassy people. The rooms were comfortable and about the size of Houston apartments.
8. We did not see any downtown or remote shopping center areas. Mostly, there were shops of different merchandise in some of the first floors of buildings we passed.
[2] 9. There is a very extensive subway system we did not see, and there never was any real traffic problems although it slowed a little around quitting time. Their car, the Volga, is about the size of a 4-door Mustang if you can imagine that.
10. The people generally seemed to me to be more serious or somber than you might find in our country (outside New York), but that is really hard to justify on very limited contact in a large city.
11. The Bolshoi Theater is a beautiful place. For the talks, we met with the same Soviet delegation on all three days. These same men also accompanied us on most of our unofficial stops.

General comments are as follows.
1. The official people we visited were friendly and openly discussed various aspects of their program. They presented and answered questions on their technical areas.
2. They were also very interested in bringing our first talks to a productive conclusion and to provide the framework for future discussions. In attempting to summarize the

technical discussions, I will include the impressions from our visit to Star City where their cosmonauts live and train. We visited there on Sunday and were greeted by the Commandant, General K. _____, General Beregovoy, who was in the U.S. last year, and Colonel Shatalov, the rendezvous pilot on Soyoz [sic] 4 and 5 flights, [who] were our principal escorts. Star City is 40 minutes out of central Moscow in pleasant woods country. There were 3 or 4 apartment buildings (about 8 stories) and another one being finished (probably a sign of continued progress in manned space). We visited their exhibit area, a memorial area for Yuri Gagarin, saw a Gagarin film and, the highlight for me, visited two [3] different simulators. The first was a general purpose simulator for all bases including docking. With reference to Figure 1, this simulator has the command module below and the orbital module attached above (with a hatch to pass through).

In the order I visited them, the orbital module was a sphere about 7-8 feet in diameter. A sketch of it is shown in Figure 2, and the inflight films and stills would indicate the flight vehicle being very similar to these simulators. The walls were covered with a light-colored, felt-like material much like the ceiling covering in some of our earlier cars. The flight atmosphere is an air mixture, slightly greater then one atmosphere, I believe. In the sketch, you can see the central trench area with the hatch in the floor. From this view, the left compartment has a hinged lid for stowage. (I imagine their space suits are stowed in there.) The right compartment is a work area, with a top like a desk and a slightly-inclined-from-the-vertical control panel. I believe there is some access to the volume underneath from the side of the central trench. There was also a manual handle in this area for water condensation removal. (Sounded like a manually operated squeezer, but I could not tell if that was their primary mode—I doubt it.) There were 4 portholes (approximately 10" diameter) 90° apart and an ECS inlet and CO_2 scrubber against the wall on the opposite side of the central trench from the EVA hatch. The side (?) view in Figure 2 attempts to show that. Based on the answers given, they do not use replaceable cartridges but add other inlets and scrubber units, dependent on the flight. I am still a little surprized [sic] by that, but maybe we lost something in the translation.

The overall impression of this module is one of simplicity, and I will try to convey that by a discussion of what I will call the control area [4] on the right-hand compartment. This is shown in Figure 2 from memory, and we did not hear what all the buttons were for. From the front, the central panel has a stowage compartment on the left side with a food warmer mounted on the wall behind it. The control panel has buttons, a C&W panel (about 6x6 lights), a speaker for A/G voice and one of the very few gauges in the ship. The gauge has three readouts—command module pressure, orbital module pressure, and ECS pressure (source or regulated—I am not sure). An identical gauge exists in the command module.

The hatch can be operated remotely from the control panel by several buttons—for depress, open and close hatch (although I did not hear this, these functions are probably repeated in the command module). The depress and hatch actuation can also be done with manual valves, handles on the hatch itself. The hatch opened into the cabin. Several other buttons on the control panel were labeled for use with the TV camera mounted on the far right of the control panel and with a long length of power cord stowed in the compartment underneath. On the tabletop in front of the control panels were several switches and a small electrical package, apparently for experiments.

From their flight films, the orbital module is the living, sleeping, and experiment area where the crew spends most of their time. It, of course, is also an airlock. I think it is worth repeating how it impressed us—a roomy area with very simple controls and instruments, probably all of which were devoted to airlock, experiment, living and sleeping functions (as opposed to attitude control, etc.).

[5] Next, George Hardy and I went down to the command module with Colonel Shatalov. For three men this is a small volume, but is only used during takeoff, landing, rendezvous, and periodically in orbit (see Figure 3); and, since they wear flight coveralls for these portions, it is adequate. Also, the couches are essentially against the floor for most flight phases. Dr. Gilruth found out that the couches are raised toward the control panel for attenuation travel at some point in the deorbit-landing phase.

Again, the very strong impression is one of simplicity—no circuit breaker panels, no large number of switches, not many displays. The couches are not exactly parallel to the display panel. As a matter of fact, we almost sat on the horizontal couches to view the panel with the upper hatch to the orbital module overhead. In Figure 3, I have sketched the control/display panel as I remember it. From the left top, the G&N area had a 6-8" diameter globe of the earth which obviously rotated with the orbital position. There were several digital readouts in this area like latitude, longitude, altitude, period, and maybe one or two others. I could not determine exactly how these readouts were driven. From the rest of what I learned, I would guess they were set up manually (probably from ground instructions) and then are driven in some approximate way to provide the pilots with general navigation data. There was a round-face clock with, I believe, a couple of controls below that. There was a C&W panel (maybe about 6x8) with red, yellow, and blue lights from the top down which is also used in some fashion in periodic systems checkout. The TV screen showing the target vehicles was approximately 5-6" square and driven from either of two TV cameras up in the nose of the ship, around the probe or drogue mechanisms. Below this screen was a range/range rate meter. On [6] the upper right were 6 digital readouts with controls for setting them. I had the impression these were digital inputs to the control system for attitude and/or translation maneuvers, but I may not have that quite right. I believe there was something in the lower right which I cannot recall—perhaps radio controls. Below the panel and above the center couch right knee was the periscope view of the target. (We were concentrating on the rendezvous and docking aspects, but I gathered that they use it for earth observation also.) The identical pressure gauge was on top of the panel, and there was a "sun lamp" above that. George Hardy was questioning about that and I did not hear the conversation. My guess is that it was a device for the pilot to see how close the vehicle alignment was for solar inertial holds for their solar panels. As in ours, the left-hand T-handle was for translation, although some switches had to be used for fore-aft braking. The right hand T-handle device was for rotation.

On the right *and* left of the main display panel was the control device which I figured to be the heart of the ship control. There were about 12 buttons down the left side of the device which seemed to be used for operating a given phase. As a phase was selected; e.g., manual docking, the pilot would punch one of these buttons. Then, next to the buttons, a set of display windows with labels on them would be mechanically rotated into view. Some of these windows would be lit, some blue, and some not lit. Although it was difficult to get a clear understanding, this device seemed to be used for whatever configuring would be necessary (perhaps deadbands, for example) and for displaying and executing any sequential functions. These could be automatic or backed up manually. There were two columns of [7] buttons to the right of these windows which seem to be this manual function. I asked if a different phase; e.g., landing, would be selected on this panel and got an affirmative. I kind of concluded that this device, then, was used to select the flight phase—some automatic configuring is probably done according to the phase; and there can be auto or manual sequential functions performed. So, it seemed to be a combination of a sequence controller and a vehicle configurer according to the flight phase. (Admittedly, this is some extrapolation on my part.) The flight films we saw showed the pilots using this device in their periodic systems checkout. So I would also guess that the light patters there and on the C&W panel represented a checkout, monitoring tool. Also, one row of buttons across horizontally were red, indicating special precaution.

All in all, this was a fairly simple cockpit and we watched a docking exercise from about several hundred feet out into docking. Much as you would expect, the pilot monitored the TV, the periscope, and the range/range rate meter and brought the ship in to docking. Roll is easy with the displays, and at docking the periscope cross hairs were lined up on a flashing light on one of the other ships' booms. Again, no circuit breaker panels, few displays, and control switches, no attitude reference display (except periscope and perhaps sun lamp).

The second set of simulators were two command module elements—one active and one passive. There were 2 parallel tracks per module on which a model of the Soyoz [sic] spacecraft was brought towards the simulator, from about 150' away. One track was watched by the TV, the other through the periscope. The images were magnified to proper scale on the cockpit instruments. Inside the cockpit, all was very similar to the general [8] simulator and I concluded that this was a more accurate docking trainer, with a greater separation distance simulated than the general-purpose simulator. Roll control only needs to be within 15° but the cosmonauts always try for and generally make approximately 1 degree.

This simulator work was a great help in the following days of discussion. It was easy to watch and understand what was happening, but, in real specifics, it was more difficult to understand that sequencer, for example, with the time we had and the need to translate everything.

For our technical discussions with the Soviet delegation, two days were planned for a mutual exchange of experience and to outline a framework for future activities. On the third day, Wednesday, October 28, it was planned to formalize our discussions by approving a document containing the framework and schedule for future work. The members of two delegations were:

Dr. Robert Gilruth	B. N. Petrov (Academician—National Academy of Sciences)
Arnold W. Frutkin	K. P. Feoktistov (Deputy Director—Manned Space Program)
George K. Hardy	V. S. Syromyatnikov (Docking Assembly)
Caldwell C. Johnson	V. V. Suslennikov (Radio Guidance Equipment)
Glynn S. Lunney	V. A. Lavrow (Foreign Affairs)

On the first day, the U.S. side presented two discussions:

1. Rendezvous experience and techniques. (General vehicle capabilities, rendezvous techniques.)

[9] 2. Docking assemblies. (Gemini, Apollo designs, future possibilities.) The Soviet side presented two discussions, essentially parallel to our presentations, but with no reference to any future programs. Dr. Feoktistov presented the rendezvous discussion; Dr. Syromyatnikov presented the docking hardware discussion. Papers were given to us on

each of these subjects and on the radio guidance system presentation on the next day. We are in the translation process now, and these papers will be available.

With reference to rendezvous, the Soviet approach is to build a system for both unmanned and manned use. They view the rendezvous process in three distinct phases.

1.   Delivery of the active vehicle to the vicinity of the target. (Done in either direct ascent fashion, or a re-rendezvous vectored from the ground.)

2.   The zone of automatic rendezvous to station keeping. (The limits of this zone were not specifically identified, but the range was on the order of tens of kilometers and tens of meters per second.)

3.   Station keeping from about 300 meters to docking. (Relative velocity is very low during this phase.) The system discussion primarily centered around phases 2 and 3, and I understood the second phase discussion best. Phase 3 is easy manually, but I did not fully understand the implementation for the automatic option. The automatic rendezvous is started when the two vehicles acquire each other and orient nose-to-nose. This is done with 2 acquisition-type antennas, giving spherical coverage. The radio guidance radar heads are then locked to each other. The active ship has a gimballing head and [10] the passive ship has a fixed head with vehicle orientation to keep the nose pointed at the active ship. Range, range rate and the relative angular motion is measured by the active ship. The relative angular motion is then used to continually establish the plane in which the guidance system solves the problem. The mechanization of the guidance scheme is to establish and maintain a range/range rate corridor and to keep the relative angular motion within some deadband. This is done by firing the main engine (of which there are two [?] of about 800# thrust) in the direction required to satisfy the range corridor or the line of site motion deadband. This is an iterative, driving technique to bring the vehicles within a few hundred meters.

Once in this zone of docking, small thrusters of 20# are used and relative roll control is established for docking assembly. This can be done either automatically or manually, and, I believe, signal strengths to mutually aligned antennas are used in the auto mode although I am not real positive of that. The manual mode we watched in the simulators was a very reasonable one and the bright flashing lights can be used on the lit side of the earth.

Dr. Syromyatnikov presented a discussion of the docking assembly—a probe and drogue device very similar to ours, with a few exceptions.

1.   It was not designed to be removed for a tunnel transfer. (They use an EVA transfer)

2.   They use an electric motor for retracting which permits unlimited reuse.

3.   The docking interface automatically includes the mating of four electrical umbilicals with on the order of 20-30 pins apiece.

Once the head of the probe is engaged, there are mechanical guide pins (6" long, 1" diameter, approximately) for further alignment and then grooves to get down to a 1-minute accuracy. This must be required for [11] only the electrical umbilicals and I get a little fuzzy here. I believe that the umbilicals alone are controlled to 1 minute and the rest of the mechanism is a 1-degree fit, but I pass to Dr. Johnson who understood this portion very well. Their alignment and velocity tolerances seemed to be about the same as Apollo.

On the second day of discussions George Hardy presented a discussion of the Skylab program, and I think the long term aspects of this flight intrigued the Soviets, especially after the Soyoz [sic] 9 18-day flight. After this discussion, Dr. Suslennikov presented a more detailed paper on the radio guidance equipment used in the automatic rendezvous. This paper did not add much to my understanding of the rendezvous but did discuss some of the functional elements within the radio guidance equipment—modulators, doppler shift extraction, etc. After this discussion, the Soviets requested similar kind of information on our system which we agreed to do. The kind of information is in the Russian text and is available in many of our block diagrams.

After these exchange discussions, we entertained the subject of what areas to study for compatibility and how to proceed. We had previously discussed the subjects which would require attention and the Soviet delegation had essentially the same ones. We grouped these subjects into logical groups such that three working groups could handle the range of subject matter. There would be some overlap between groups, and the three groups suggested are:

1. Group to assure compatibility of *overall* methods and means for rendezvous, docking, and life support.

[12] 2. Group to assure compatibility of radio, optical guidance systems and communications.

3. Group to assure compatibility of docking assembly and tunnel.

The groups, a more detailed definition of the work required, and the proposed schedule is contained in a summary of results signed by both delegations. This summary is being presented to Dr. Low of NASA and Academician Keldysh of the National Academy of Sciences. Once agreed to by these two parties, I envision the work proceeding along the lines expressed therein. It is my belief that this effort will involve a rigorous, full-time effort by a relatively small number of personnel, but with the support of many other elements. This effort will be similar to early mission and techniques planning combined with ICD tradeoffs and definition and, finally, preliminary system design to assure compatibility.

Glynn S. Lunney

Document I-46

Document title: George M. Low, "Visit to Moscow, April 1972, to Discuss Compatible Docking Systems for US and USSR Manned Spacecraft," April 4-6, 1972, with attached: "Addendum, Moscow Trip, April 4-6, 1972," May 30, 1972.

Source: George M. Low Papers, Institute Archives and Space Collections, Rensselaer Polytechnic Institute, Troy, New York.

NASA Deputy Administrator George M. Low led a three-person NASA delegation on an April 1972 trip to Moscow to make a final technical determination of whether the United States and the Soviet Union should agree to a joint test flight. This would involve an in-orbit docking of a U.S. Apollo spacecraft and a Soviet Soyuz spacecraft. Low concluded that such a test flight was indeed desirable and feasible, and NASA recommended to the White House that the United States agree to it. The U.S.-Soviet agreement to carry out the Apollo-Soyuz Test Project was announced just as an overall agreement on space cooperation was signed at a U.S.-U.S.S.R. summit meeting in May 1972.

[1]   April 4-6, 1972

Visit to Moscow, April 1972,
to Discuss Compatible Docking Systems
for US and USSR Manned Spacecraft

Summary

In early April 1972, Arnold Frutkin, Glynn Lunney and I went to Moscow to meet with representatives of the Soviet Academy of Sciences on the subject of compatible docking systems for US and USSR manned spacecraft. The specific purpose of the trip was to determine whether the US side was ready to make a commitment to a joint test flight in 1975 involving a rendezvous and docking of US and USSR spacecraft in earth orbit. Such a commitment could be made in the forthcoming summit talks at the end of May 1972.

As a result of three days of meetings, we reached agreement on technical matters, as well as on the principles of managing and scheduling and conducting a 1975 joint test flight. Both sides affirmed the desirability of such a test flight and are ready to proceed with preparations for the flight on the basis of a prospective government-to-government agreement.

Background

Initial discussions concerning compatible docking systems, for future manned spacecraft took place in October 1970. Following those discussions, Bob Gilruth, who headed the US team to Moscow in October 1970, recommended that an early test flight using Apollo and Soyuz hardware would be highly desirable. After discussions with Henry Kissinger in San Clemente early in January 1971, I proposed such a joint test flight to Keldysh in Moscow when I was there to negotiate the Low/Keldysh agreement. During the next set of talks on the compatible docking systems in Houston in June 1971, the Soviet side agreed that an early test flight would be highly desirable, but suggested that the Salyut space station (which was then on its first and only flight) be used instead of the Soyuz spacecraft. Detailed work on an Apollo/Salyut mission for the 1975 time period continued into the Fall of 1971, and during meetings in Moscow in November/December 1971, the US and USSR agreed that such a mission would be technically feasible and desirable. [2] In the Fall of 1971, NASA also recommended to the White House that a final agreement on a test mission might be included in the agenda for the May 1972 summit meeting. As a result of several discussions on the subject, we were asked to make a firm recommendation by April 15, 1972, concerning the feasibility of conducting such a mission.

Lunney recommended that in order to assure this feasibility, we should get agreement in principle at least on three basic documents: a project technical proposal document, an organization plan, and a project schedules document. Draft versions of these documents had been prepared by MSC [the Manned Spacecraft Center] and had been transmitted to Moscow in late March 1972. At the same time, we asked for a meeting with Keldysh to explain the purpose of the documents and to establish a firm basis for discussing them. It turned out, however, that Keldysh had just entered the hospital and would not be available until early April.

We therefore decided that Frutkin, Lunney and I would go to Moscow during the week of April 2nd to discuss the documents, to reach agreement on the most important points, and especially to determine whether the Soviets really understood what we were talking about.

We decided that we would not publicize this trip, [Handwritten footnote: "This was at the request of the White House, because we were to discuss a possible agenda item for the following summit meeting." (footnote added 1-10-76)] this was at the request of the White House, because we were to discern a possible agenda item for the forthcoming summit meeting, and it took pains to make sure that only the smallest possible number of people would know that we had gone to Moscow. For example, insofar as MSC was concerned, Lunney was visiting Washington. In my own case I was on leave "to take care of family business." Then, on the day we left the United States, the New York Times carried a front-page story of an interview between John Noble Wilford and Petrov. In this interview, Petrov stated that there would be meetings in Moscow during the coming week on the compatible docking systems. Fortunately, however, at least at the time of this writing, nobody has yet asked whether anyone had indeed gone to Moscow or who had gone.

Chronology of Events

We left Washington via TWA on Easter Sunday, April 2, 1972, and arrived in Paris early the following morning. From Paris to Moscow, we were on Aeroflot (an Iluyshian 62) and arrived [3] in Moscow approximately 5:30 Monday evening, Moscow time. There we were met by Petrov, Vereshchetin, and Bushuyev. On the way to Moscow, Petrov told me that Keldysh was still in the hospital but that I would meet with the Acting President of the Academy, Academician Kotelnikov; however, Kotelnikov would not be available until Tuesday noon, and our meetings would start at that time.

Tuesday morning we had a brief meeting with Ambassador Beam, during the course of which he invited us to a luncheon on Thursday. I later found out that one of the invited guests was Bob Kaiser, the Washington Post correspondent in Moscow. I went back to see the Ambassador and told him in view of the White House and State Department desire not to publicize our trip, I felt this was a bad idea. The Ambassador assured me that this would be a purely social occasion, that he would take personal responsibility, and that Bob Kaiser would not know the purpose of our trip nor would he say anything about it. Although I was extremely skeptical about this, I had no way of avoiding the invitation.

Tuesday from noon to approximately 2 o'clock, we met with Kotelnikov, Petrov, Bushuyev, Rumyantsev, Vereshchetin, with Zonov as their interpreter. (We had also brought along our own interpreter, Cyril Murumcev.) From that session, we went to a typical Moscow luncheon at the Club of Scientists, which, I guess, is Moscow's Cosmos Club. After lunch we continued the discussions, with Petrov taking charge on the Soviet side and without Kotelnikov. We adjourned at close to 7 p.m. that evening.

We reconvened at 9:30 Wednesday morning, held discussions until approximately 2 o'clock, at which time we adjourned for lunch. The American party went to the U.S. Embassy for a quick lunch in their snack bar, as well as a complete reworking of our final document. The afternoon session started at 4 p.m. and lasted only until about 6. However, as a result of the document we had prepared during lunch, and as a result of the basic understandings reached in previous discussions, we were able to conclude the substance of our talks at that time.

[4] On Thursday morning, Frutkin and Vereshchetin worked on the editing of the final document, with the help of Jack Tech, who is the Science Attache at the American Embassy. Lunney and I continued our discussions until about 1 o'clock. This was followed

by luncheon at the American Embassy Residence (Spaso House) while the English version of the summary of the results of the talks was being typed at the Embassy.

Following lunch, we returned to the Presidium of the National Academy of Sciences (where all of the discussions had been held) in order to sign the documents. This was the usual signing ceremony in which each of us signed two English and two Russian texts. Incidentally, this signing ceremony took place in Kotelnikov's office, which he claims Napoleon used as his bedroom during his last night in Moscow on the way back to France. I also learned that the large table that I used in signing the Keldysh/Low agreement had been a desk used by Napoleon.

Thursday evening we had a farewell dinner with Kotelnikov, Petrov and the rest of the Russian delegation. There were the usual toasts, as there had been at the luncheon on Tuesday afternoon. (At the Tuesday luncheon, I had made a toast, stating that we here had an opportunity to make history and that the results of what we were trying to accomplish would probably be much more far reaching than any of us could at that time even imagine. During the Thursday evening dinner, Kotelnikov said in a toast that the true importance of what we were doing was that this could be an important step in bringing peace to men everywhere.)

Early Friday morning we left Moscow via Aeroflot to London, Pan Am to New York, and then back to Washington.

Highlights of the Talks

Tuesday Noon. This was the meeting with Kotelnikov, Petrov, Bushuyev, and Rumyantsev. After a brief welcome by Kotelnikov, I gave a brief opening statement in which I reviewed the history of 18 months of technical discussions and that the possibility now existed to reach a government-to-government agreement, perhaps during the forthcoming summit talks. I went on to say that before such an agreement can be reached, it is essential that we both understand that [5] this mission can indeed be carried out and that my specific assignment in these talks was to determine whether we are now ready to proceed. I pointed out that we had high confidence in understanding each other on technical matters, but that I was still less sure of a complete understanding on matters of schedule and organization. I concluded by stating that it was my hope that in these talks we could gain a common understanding of the basic principles for organizing, developing, scheduling and conducting a test mission so that I can advise the White House that we are indeed ready to commit to such a mission.

Kotelnikov, in his opening statement, said that they had reached a very important conclusion that they would like to lay on the table at this time. The conclusion was that they would use the Soyuz spacecraft instead of the Salyut space station for their rendezvousing vehicle.

This, of course, came as a major surprise, and we had a long discussion on the subject. The reasons for the switch, they said, were "technical and economic." They explained that the Salyut space station only had one docking port and that it would have to be redesigned completely to accept a second docked vehicle. This was a major redesign that would be extremely costly. They then took a close look at the Soyuz and found that it could be modified with all of the modifications that had already been discussed for the Salyut, and that they were prepared to do so. They were quite strong in stating that there would be no difference in any of the things that had already been agreed to. (My own assessment is that there are three possible reasons for the switch. These are: (1) the actual reason given by them; (2) major difficulties with Salyut identified during its first flight; and (3) the "political reason" that since we will not have a Skylab available for a future flight, they are unwilling to commit a Salyut to such a mission. My inclination is to believe that the reason they gave is the actual one.) I stated that barring any technical difficulties, Lunney would have to certify that the switch from Salyut to Soyuz would be acceptable to the United States and, in fact, reminded the Soviets that this was the vehicle that we had recommended in the first instance in January 1971. From the technical point of view, Lunney was unable to identify any difficulties with this mission and, in fact, [6] pointed out that operationally this could present a simpler problem, since it would involve only two coordinated launches (Apollo and Soyuz) and not three (Apollo, Salyut and Soyuz). I also tried to think through any "political" implications and found none. It would still be possible to exchange crews, which will have the major public impact of this mission. And having a Soyuz, instead of a Salyut, will have the added benefit of not calling attention to the fact that they have a space station flying at the time when we do not.

After we had settled this issue, I stated that I wanted to bring up another matter; namely, that of the lack of the Soviet responsiveness to our proposals concerning direct voice communications between the two project managers on a regular basis. (For background, this item had been proposed by us during the November/December 1971 talks and was supposed to be confirmed by the Soviets when the agreement of those talks was confirmed. This was not done, and I sent a telegram to Keldysh asking for confirmation. As of now, we have not received a response to that telegram.) I mentioned that I was not only interested in the substance of the issue but also concerned about the lack of responsiveness on their part which, if indicative of future relationships, would make it difficult to conduct the joint mission. Kotelnikov quickly understood why I attached importance to the issue and said we should settle it right then, which we did after considerable debate and discussion.

Finally, during the first session, we determined the agenda for the remaining stay in Moscow. Specifically, we agreed that we would attempt to reach an agreement on the basic principles of the "organizational plan"; the level of detail to be included in the schedules; and any technical matters that might have come about as a result of the switch from Salyut to Soyuz. Both sides also agreed that with the exception of any new technical problems that might have resulted from the switch, we knew of no other outstanding difficulties.

Tuesday Afternoon. The discussion proceeded after lunch, with the same participants with the exception of Kotelnikov. Lunney had prepared a document entitled, Apollo/Salyut Test Mission Consideration, dated March 23, 1972, a [7] copy of which is attached to these notes. This document essentially is a summary of the organizational plan, and we had hoped to agree to this plan in detail to make it part of our agreement of these Moscow talks. At this point, however, things got to be quite confusing, and we started spending an inordinate amount of time quibbling over the exact wording of each sentence. We quickly saw that we would be in Moscow for weeks rather than days were we to proceed in this way.

We had also brought along a "Summary of Results" which was to be the basic document of agreement concerning these talks. At this point in our proceedings, we, therefore, called for a quick recess to discuss our strategy for the meeting and to show the Soviets that what we really intended to sign was something like the Summary of Results. Further, we indicated that the document which I previously discussed we had hoped to make part of this summary and to include it as an appendix. Finally I pointed out that it would be most important to reach agreement and a full understanding of the "twelve principles governing mission conduct" which were an enclosure to the Apollo/Salyut Test Mission Consideration document, and that I felt it would be best if we started discussing those. The Soviet side agreed with this recommendation.

We had no problems in reaching a very quick understanding and agreement on the first six of the principles, which concern command, control, and communications. By that time, however, it was getting late, and we decided to review the remaining six principles only very quickly for subsequent discussion in tomorrow's meeting, In this quick review, however, we determined that we might have major problems on item seven concerning astronaut training and item 12 concerning public information release.

Wednesday Morning. On Wednesday morning, we continued the discussions of Tuesday afternoon, starting out with a detailed discussion of astronaut familiarization and training. After an in depth discussion, we did agree that it would be essential to identify candidate crews one to two years before the flight and that these crews would have to be trained in the other country on the other country's normal training equipment. The discussions continued then [8] with a relatively quick understanding on the need to transmit television downlinks from one control center to the other; the need to gain participation by flight operational personnel in the talks; and the need to have the flight crews understand the other country's language. We did have some difficulty in the discussion concerning the desire to locate a small team of flight-oriented personnel from each country in the other country's control center during the flight, but, on our side, decided this was not essential and, therefore, did not pursue the point but rather left it for further discussion by the project managers. Finally, on the point of public releases we again held a rather lengthy discussion. The Soviets agreed that everything during a normal flight should be released immediately and also pointed out that during a major disaster they would be willing to have speedy releases just as they did in the case of the deaths of the Soyuz 11 cosmonauts. Their main concern seems to be with minor abnormalities during a flight, which, in their words, might be misunderstood by the general public. They indicated, however, that in all areas of public information, they were loosening up and cited the recent announcement of the intended objective of the Venera 8 as an example. I, in turn, pointed out absolute need for us to continue to disclose publicly all information that is available at the American control center and received at American tracking stations. At the conclusion of the discussions, we agreed that we would develop a public information plan which would take into account the obligations and practices of both sides.

After we finished discussing the 12 basic principles, it became time to start thinking about the wording in the summary of the results of the talks. In the meantime, the Soviets had translated our draft summary and had made a number of changes in it, and then retranslated it back into English. This was to be the basis for our joint document. However, we quickly found that the document had been weakened to the point where it really said nothing of substance. To be a little more charitable, it said that we understood each other, but it didn't say that we had agreed to anything. After a long discussion on this point, I said that the document as written by the Russians was totally unacceptable to us and that unless we could come out of this meeting with a firm agreement on at least basic princi- [9] zation, as well as on the need to firm documentation and schedules, I would be in no position to recommend that we are ready to proceed with a test mission, and, in fact, would make a negative report when I returned to the United States. I further stated that I was prepared to stay in Moscow until we had hammered out the necessary words; that I believed that we did understand each other and it was now time to put all of this down on paper. Thereupon we adjourned for lunch.

Wednesday Lunch. We had a quick bite to eat in the Embassy snack bar, and then Frutkin, Lunney and I each took a piece of the summary of results that we had prepared before we left Washington and modified it to include all of the 12 basic principles, together with any changes that we had made in these principles during our previous discussions in Moscow. All of this, of course, had to be done in a great hurry, and the document was retyped before we returned to the Presidium at 4 o'clock for the afternoon session.

Wednesday Afternoon. When we returned with our new document, this came as a complete surprise to the Soviet side. It was just unthinkable for them that anybody could have recast the entire document so quickly. After a quick verbal translation by Zonov, the Soviets called for a recess of half an hour. During the course of that recess, they studied the document in detail, and when they returned, told us that the document was completely acceptable to them with the exception of some minor editorial changes. We then adjourned for the evening and agreed that Frutkin and Vereshchetin would form an editorial committee of two that would meet in the morning to go over the final document.

Thursday Morning. While Frutkin and Vereshchetin were editing the document, Lunney and I continued the discussions with Petrov, Bushuyev, and Rumyantsev. First, Bushuyev responded to the schedules document and gave an excellent discussion of his views of the need to control schedules. During the course of the scheduling discussion, we also discussed design reviews, which were understood and agreed to by both sides; joint testing, which was also understood and agreed to; and finally, the Soviet side stated that they agreed in principle to the entire organizational plan.

[10] Next I raised a question concerning the Soviet organization to do this mission. I pointed out that they knew clearly where each of us fit into our organization and what our responsibilities were. I asked if it would be possible to get the same kind of understanding of their organization. Petrov responded in some detail, but really said nothing. He said that Keldysh, as President of the Academy of Sciences, reported to the Council of Ministers, and had been charged with being responsible for the US/USSR cooperation in space. Petrov, in turn, reported directly to Keldysh, and Bushuyev to Petrov. I asked whether the same organization would be in force during the hardware and flight operational phase, and the answer was in the affirmative. Petrov indicated that they would bring additional people into the organization at that time, but that these people would still report to Bushuyev.

By this time, Vereshchetin and Frutkin had finished editing the "Summary of Results" and had prepared identical documents in English and in Russian. We reviewed these documents, had a few questions but no major hangups. Both sides agreed with the documents as they had been prepared.

Finally, Thursday morning Bushuyev discussed technically the Soyuz system and gave Lunney a document describing those systems. For the test mission in 1975, the Soyuz would fly only two men for a five-day period, plus one day in reserve. They proposed also that the Apollo spacecraft should be launched first and that the Apollo would be active in the rendezvous and docking maneuver. (In subsequent discussions with Lunney, I told him that from a policy point of view, I would actually prefer to have the Apollo launched first as the Soviets now recommended and that unless there is a good technical reason not to do so, we should accept this recommendation.)

Thursday Afternoon. After lunch at the American Embassy Residence, we returned to the Presidium to sign the Summary of Results. After the signing ceremony and after making the usual speeches, I discussed with Kotelnikov and the group the public posture relative to the meetings we had just completed. I mentioned, first, that we intended no public release of the meetings at all; second, that we do not intend to mention the fact that we were now discussing Soyuz instead of [11] Salyut; third, I indicated that if pressed and if we had to admit that meetings took place in Moscow during this week, we would say that we were preparing the agenda for the July meeting but that we could not discuss the content of the agenda; fourth, that if we were to take any different action from the above, we would so notify Petrov; and, fifth, that we would intend to remain in this posture until after the summit meeting. Kotelnikov completely agreed with this proposal, and with this we ended our formal discussions in Moscow.

Conclusions

A copy of the Summary of Results that was signed in Moscow is attached. From this, and particularly from the discussions that went along with the agreements that were reached and documented, I have reached the conclusion that we are ready to undertake this test mission. Insofar as hardware matters are concerned, we have reached an understanding and agreement on all issues which have been identified so far, and, furthermore, don't see any issues that we will be unable to agree on. On the management side, we have reached agreement on such matters as regular and direct contact through frequent telephone and telex communications, as well as visits; the requirement for and control of formal documentation; joint reviews of designs and hardware at various stages of development; the requirement for joint tests of interconnecting systems; early participation by flight operations specialists; the development of crew training plans; and the training in each country of the other country's flight crew and operations personnel. We also reached agreement on the requirement for and the level of detail of project schedules. Finally, in the area of flight operations, we reached agreement on the principles of communications command and control of the flight; the requirement for flight plans and mission rules for both normal and contingency situations; the immediate transmission of flight television received in one country to the other country's control center; the language problem; and the need to develop a public information plan, taking into account the obligations and practices of both sides.

Based on all of these agreements, it was my recommendation that the United States is ready to execute a government-to-government agreement and should now do so.

****

[Attachment page 1]   May 30, 1972

Addendum
Moscow Trip, April 4-6, 1972

This is an Epilogue to the special notes I prepared after my trip to the Soviet Union on April 4-6, 1972.

During the course of that visit to Moscow we reached an agreement (signed by Kotelnikov, the then Acting President of the USSR Academy of Sciences and myself) on matters concerning the technical details, the organization, management, operational details, and scheduling of a possible joint docking mission involving the United States Apollo spacecraft and the Soviet Union's Soyuz spacecraft. Upon my return from Moscow we recommended to the White House (Henry Kissinger) that, from NASA's point of view, we were prepared to proceed with such a mission in the 1975 time period, that no further NASA/USSR Academy meetings would be required, and that the form of the agreement between the United States and the Soviet Union could be a relatively simple and straightforward one. A copy of our proposed wording for that agreement is attached.

Between the middle of April and the middle of May (the summit meeting started on May 22), there was a great deal of interest by the press in the possibility of having a joint docking mission on the summit agenda, and a large number of interviews with NASA people was held. In all of these interviews, there was a great deal of speculation about the possibility of an agreement on the docking mission at the summit, but there was never any hint of the April 4-6 meeting, nor was there ever any hint that during that meeting the

Soyuz spacecraft was substituted by the Russians by the Salyut. In other words, from NASA's side we were able to avoid any discussions of NASA's preparation for the summit meeting or of the form that any agreement might take. This was possible only because such a very small number of NASA people had been involved in the activities leading up to the summit.

[2] It was only during the week before the summit meeting that the State Department worked on the specific wording of the agreement and made only minor changes in our previously submitted wording. Apparently State and the White House started coordinating the words with the Soviet Union only on the 18th or 19th of May (we have no idea in NASA why this was undertaken only at this late date). On May 20, the USSR responded to our proposed wording with a much lengthier document, which among other things, included the Keldysh-Low agreement of January 21, 1971, in addition to the docking agreement. Furthermore, with respect to the docking agreement, the Soviet words did not include by reference our previous meetings and, instead, some rather cumbersome wording was substituted.

Apparently when the Soviet response was received by our State Department, it was immediately discussed with Kissinger and Rogers, who were at the time over the Atlantic on their way to Salzburg, a stop on the way to Moscow. Kissinger asked that we prepare an appropriate response but that insofar as possible, we should not change the wording in the Soviet text. All of this was done in a meeting at State Department starting at 2:30 Saturday afternoon, the 20th, and ending in the middle of the night. During that time we straightened out the wording in the Preamble but kept by and large the Soviet meaning. With respect to the Keldysh-Low Agreement, we did not make any significant changes, with one exception. The Soviet document had incorporated words concerning communications satellites which had not been part of the January 21, 1971, agreement, and we therefore deleted these words. Finally, with respect to the docking agreement, we selected words similar to those that we had proposed in April in our memorandum to Kissinger and especially incorporated in that article the April 4-6 agreement by reference. This document, together with the clarifying document, was forwarded to the White House/Salzburg late that night. In the clarifying document we stated that NASA had no objection to the inclusion of the Keldysh-Low Agreement in the government-to-government agreement, but [3] pointed out that this was not necessary, nor had it been the intent. State Department on the other hand felt that it should not be included because it would make our relationships with the Europeans even more difficult in light of our recent lack of enthusiasm for space cooperation with the Europeans. With respect to the April 6 agreement, we stated in the clarifying telegram that NASA insisted that it be included by reference.

Following the Saturday meeting we had no additional information except persistent signals that the space agreement was scheduled to be signed in Moscow on Wednesday, the 24th. On the 23rd, I left for the West Coast for a talk in San Diego on the evening of the 23rd, and then a visit to JPL [the Jet Propulsion Laboratory] on the 24th. During the course of the evening in San Diego (after dinner and during the preliminaries leading up to my talk), I received a telephone call, through the State Department Operations Center, involving Arnold Frutkin, somebody in State Department, and myself. State had just received a final text as it had been agreed to tentatively in Moscow. In this text the Keldysh-Low Agreement was still included and there were words acceptable to us with respect to the docking mission. The April 6th agreement was specifically included. I accepted the words as they had been read to me just in time to get back into the ballroom (I had taken the telephone call at a hallway outside) to hear myself introduced as the evening's main speaker. It is interesting to note that by this time it was 6 a.m. in Moscow on the day that the agreement was actually signed.

On the next day, May 24, I went to JPL and soon learned that the agreement actually had been signed in Moscow at apparently 11 o'clock a.m. EDT. At 2:25 p.m. EDT, the Vice President introduced Jim Fletcher, Jim McDivitt, and Glynn Lunney, who held a press conference at the Executive Office Building. Sometime thereafter, Fletcher held another press conference at NASA Headquarters, and simultaneously, I held one at JPL.

[4] There has been no adverse criticism in this country concerning the space agreement in general, or the Apollo/Soyuz test project in particular, and, in fact, there has been a great deal of overwhelmingly favorable editorial comment. . . .

Document I-47

Document title: Henry A. Kissinger, to the President, "US-Soviet Space Cooperation," May 17, 1972, with attached: "Draft Agreement."

Source: Nixon Project, National Archives and Records Administration, Washington, D.C.

The final decision to proceed with what became known as the Apollo-Soyuz Test Project was not made until shortly before the May 1972 U.S.-U.S.S.R. summit meeting in Moscow. This decision memorandum, when approved by President Nixon, was the basis for project approval.

---

[1]   May 17, 1972

Memorandum for the President

FROM: Henry A. Kissinger

SUBJECT: US-Soviet Space Cooperation

In NSDM 153, you directed NASA and State to explore with the Soviets the possibility of a US-Soviet agreement on the desirability of a joint, manned space docking mission, so as to provide you with the option of announcing this agreement during the Moscow visit.

NASA's Deputy Administrator, Dr. George Low, held detailed talks on the possible joint mission with representatives of the Soviet Academy of Sciences in Moscow from April 4-6. Both sides had earlier agreed that such a mission was technically feasible and desirable. NASA Administrator Fletcher now reports that Dr. Low's April mission was successful; he was able to reach agreement on the principles of managing, scheduling and conducting a joint space docking mission.

The Soviets have informed NASA that they would like to reach formal agreement on space cooperation, including the joint manned mission, during the Moscow Summit. Programmatically, the US is ready to execute such an agreement, and NASA recommends that this be done.

The costs involved with the joint manned mission, which would be scheduled for 1975, are now estimated by NASA to be approximately $250 million. This estimate has been developed in coordination with the Office of Management and Budget.

Clark MacGregor has taken soundings to determine the likely Congressional reaction to the proposed joint mission. These soundings indicate that the proposal would gain acceptance by a 3-1 or 4-1 margin.

The text of the proposed space agreement could be quite brief, along the lines of the draft at Tab A. I recommend that you approve the proposed [2] US-Soviet space agreement, permitting the necessary steps to be taken prior to your Moscow visit to provide you with the option of announcing the agreement at the Summit.

Approve _____ Disapprove _____

With your approval, I will forward a copy of this memorandum to the Director, Office of Management and Budget, with the request that he arrange to take such budgetary steps as may be necessary to provide for implementation of the agreement.

Approve _____ Disapprove _____

****

[Attachment page 1]

Draft Agreement

The United States of America and the Union of Soviet Socialist Republics agree to a program of joint activities designed to enhance the safety of manned flight in space and provide a basis for possible cooperative space projects of mutual benefit.

Toward these goals, it is agreed that rendezvous and docking systems of future generations of manned spacecraft of both countries will be compatible, to permit rendezvous, docking, rescue, and possible joint experiments in space. It is further agreed that the first flight to test these future systems will be carried out in 1975, using specially modified Apollo-type and Soyuz-type spacecraft. In this flight the two spacecraft will rendezvous and dock in space, and cosmonauts and astronauts will visit in each other's spacecraft. This joint project will be conducted in accordance with the Summary of Results of the Meeting Between Representatives of the US National Aeronautics and Space Administration and the USSR Academy of Sciences held in Moscow on April 4 to 6, 1972.

Document I-48

Document title: George M. Low, Memorandum for the Record, "Visit to Moscow, October 14-19, 1973," November 1, 1973.

Source: George M. Low Papers, Institute Archives and Special Collections, Rensselaer Polytechnic Institute, Troy, New York.

Once the United States and the Soviet Union had agreed to carry out the Apollo-Soyuz Test Project, there were frequent interchanges of personnel between NASA and its counterparts in the U.S.S.R. NASA Deputy Administrator George Low made an October 1973 visit to Moscow for a top-level mid-term project review. This detailed memorandum for the record contains Low's observations on his time in Moscow.

[1]   November 1, 1973

Memorandum for the Record

SUBJECT: Visit to Moscow, October 14-19, 1973

BACKGROUND

On August 14, I had written to Academician H. V. Keldysh, President of the USSR Academy of Sciences, suggesting a mid-term review of the Apollo-Soyuz Test Project. A copy of my letter to Keldysh is attached. In the letter I also stated that in addition to reviewing the current status of the project, I would like to discuss in detail four specific subjects: system failures; participation in and observation of the test activity and flight preparation; project milestones; and the preparation of documentation. I further asked if it would be possible to visit some Soviet space facilities during the course of my visit. Keldysh responded favorably on August 30. (A copy of his letter is also attached.) Then, about a week before my visit, I received a telephone call from Chet Lee, who was already in Moscow, indicating that Keldysh was ill and would be unable to see me. He added, however, that the Soviet side clearly wanted me to come ahead and urged him to convey to me that this is not a "diplomatic illness" and that my visit would be most worthwhile. In order to further make it desirable for me to come, they promised that they would take me to the Soviet Mission Control Center near Moscow. The telephone call was followed by an official telegram from Keldysh and after discussions with Arnold Frutkin we decided that I should go ahead with the visit as planned. (Both Arnold and I asked about Keldysh's health on many occasions after we arrived in Moscow. The response we both received was that Keldysh is not really ill in the true sense of the word but is extremely tired and run-down. He had not taken a vacation after his major operation earlier this year and had worked extremely hard ever since then. He was therefore "ordered" by his physicians to take a rest and not to participate in any of the meetings with me. During the course of my visit, his office was always dark, his secretary was nowhere in sight, and it was quite clear that he was completely away from the office during this week.)

[2]   SUMMARY OF VISIT

Sunday, October 14
Arrived in Moscow with Frutkin early in the evening. Met at airport by Boris Petrov, Vereshchetin, Jack Tech from the U.S. Embassy, and one or two others. Rode to Hotel Rossia in Petrov's car and, as we had requested, did not participate in any official functions that evening.
Monday, October 15, 7:00 a.m.
Executive Session at Hotel Rossia with Lunney and his Working Group chairmen. According to Lunney, the two weeks of preparatory meetings had gone extremely well and much had been accomplished. The "Summery of Results" of their meetings had been prepared and a copy of this is attached. In addition, Donnelly had negotiated a first-phase (pre-flight) Public Affairs Plan which was to be ratified by Petrov and me. We discussed some of the technical results of the meeting but I will cover these later as I discuss each specific item.

Monday, October 15, 9:00 a.m.

We met at the Presidium of the Soviet Academy of Sciences for the Apollo-Soyuz "Midterm Review." Participating on our side were Low, Frutkin, Lee, Lunney, Cernan, Stafford, Smylie, Dietz, and Frank. Soviet attendees included Petrov, Bushuyev, Vereshchetin, Rumyantsev, Abduyevski (the Deputy Director of the Control Center), Cosmonaut Yeliseyev (the Flight Director), Cosmonaut Leonov (the Soyuz Commander); Tulin, Tsorev, and Kozorev of Intercosmos; Working Group Chairmen Timchenko, Legostaev, Syromyatnikov, Nikitin, Galin, and Lavrov; and their interpreter Zonov. During the course of the meetings, Bushuyev, Lunney and alternate Working Group Chairmen gave a technical review using a notebook of "Vu-graphs." Notebooks had been prepared in both languages so that all of us could follow the review.

Monday, October 15, lunch time

Frutkin, Lee, Lunney, and I joined Petrov, Bushuyev, and Vereshchetin for a small luncheon at the "Club of Scientists." [3] Even though this was very informal and there were not too many toasts, it was nevertheless a Soviet-size dinner, with five or six courses, which consumed the better part of two hours.

Monday, October 15, 3:00 p.m.

We returned to the Presidium for another session involving all participants. This was a relatively brief session with only a few questions asked by our side and responses given by their side. At the conclusion of the session, both Petrov and I agreed that good progress had been made in ASTP, that there were no open questions other than those raised by the technical Project Directors in their Summary, and that we had high confidence in meeting our launch date of July 15, 1975.

Monday, October 15, 4:00 p.m.

I had asked for an Executive Session to discuss some of the points raised in my letter to Keldysh which were not brought out during the technical meeting. Participating on our side were Low, Frutkin, Lee, and Lunney, and on their side Petrov, Bushuyev, Vereshchetin, Rumyantsev, Tulin, Tsorev, and Kozorev. During the course of this meeting, I brought up the subjects of systems failures, participation in factory installation of U.S. equipment, documentation, Stafford's desire to see actual spacecraft hardware and not only mock-ups, and the desirability of a press conference before our departure from Moscow. This was a very frank and forceful discussion with our side politely but firmly insisting on responsiveness by the Soviet side.

Monday, October 15, 7:00 p.m.

The Charge d'Affaires at the U.S. Embassy in Moscow had invited the two delegations for a small reception at the Embassy. This was quite informal and friendly with no detailed discussions about the business at hand. There was great interest in Skylab and the well-being of the Skylab's three astronauts on the part of a number of the Soviet delegation and they appeared to be amazed how well Bean and his crew had done after 59 1/2 days in space. I also picked up the following incidental piece of information from Petrov: It is the Soviet's view that TU-144 [4] accident was caused by a small French aircraft which flew into the TU-144's flight path. The TU-144 had to veer off and thus flew into the ground.

Monday, October 15, 8:00 p.m.

I met in my hotel room with Donnelly, Shafer, Frutkin, and Lee to discuss the Public Affairs Plan. Donnelly and Shafer appeared to be quite disturbed by some of the things that had happened while they were in Moscow but we agreed not to discuss this any further until we returned to Washington. We then discussed the substance of the Public Affairs Plan and agreed that it was not yet ready for ratification without further clarification.

Tuesday, October 16, 9:00 a.m.

I paid a brief call on Academician Kotelnikof, the Acting President of the Academy of Sciences. This was only a courtesy visit, with some small talk but no substance.

Tuesday, October 16, 10:00 a.m.

Visited the Institute of Geochemistry and Analytical Chemistry of the Academy of Sciences. Vinogradov was to have been our host, but we were told that he suffered a bad cold and we therefore met with his Deputy, whose name I believe is Sorkhov.

Tuesday, October 16, 11:00 a.m.

Next we visited the Institute of Space Research of the Academy of Sciences and met its new head, Prof. R. S. Sagdeyev. Sagdeyev speaks good English, is friendly and open, and looks like the sort of person with whom we ought to be able to develop good relationships.

Tuesday, October 16, 3:00 p.m.

Visited Academician V. A. Kirillin, the Deputy Chairmen of the Council of Ministers and Chairman of the State Committee for [5] Science and Technology. I had asked for this courtesy visit prior to my arrival in Moscow and as soon as I arrived there were many questions as to why I wanted to see Kirillin. I assured everybody that this was really only a courtesy visit.

Tuesday, October 16, 7:00 p.m.

Went to the ballet in the Kremlin and saw "Don Quixote" for the second time during one of my Moscow visits. For one who doesn't like ballet, this should be considered to be above and beyond the call of duty.

Wednesday, October 17, 8:45 a.m.

Left the hotel to visit the cosmonauts' training center at Star City. At Star City we were met by General Beregovoy since General Shatalov, who is now in charge, was visiting in Japan. We also met the Soyuz 12 cosmonauts, Lazarev and Makarov, as well as ASTP cosmonauts Leonov, Kubasev, and Filipchenko. Petrov and Bushuyev were with us, and we were also joined by Feoktistov, whom I had not seen since my January 1971 visit. The reason for this became apparent later. Feoktistov was there to show us through the Salyut mock-up. He knew Salyut as well as I had at one time known Apollo, and obviously is either the Chief Engineer or Program Manager on Salyut.

At Star City we had a sit-down briefing, a visit to the Soyuz simulators and docking trainers, a discussion of the ASTP version of Soyuz, and then a very detailed description of Salyut, with a tour of its high fidelity mock-up. We were also shown the Soyuz 12 space suit. We then had a quick tour of the museum and the usual seven- or eight-course dinner with the usual number (15 or 20) of toasts. I was a lot smarter this time, though, then I had been on the last visit to Star City. I did not participate in any of the "bottoms up" toasts and merely sipped my vodka politely each time.

Wednesday, October 17, 4:00 p.m.

I had asked for discussion on the ASTP Public Affairs Plan and Petrov and I decided to have this meeting while we were at [6] Star City. Participating in this meeting were the same ones who participated in the Executive Session on Monday afternoon. At the completion of this meeting we left for Moscow.

Wednesday, October 17, evening

The evening was free but Arnold Frutkin and I met in our hotel room for further discussions on the Public Affairs Plan. Here we wrote some words which we hoped would clarify the Plan, for additional discussions the next morning.

Thursday, October 18, 9:00 a.m.

Frutkin and I met with Petrov, Vereshchetin, and Rumyantsev on the ASTP Public Affairs Plan. During the course of this discussion, we reached a complete understanding of all points but did not reach agreement on them. Unfortunately, Donnelly had already left Moscow so he was unable to participate with us.

Thursday, October 18, 10:15 a.m.

We left the hotel for the visit to the Soviet Mission Control Center. This was a first for any Western visitors and, of course, of great interest to us. We arrived there approximately 45 minutes later and had a very detailed tour of the Center. Following the tour, at 2:00 p.m., we had lunch at the Control Center, complete with eight different wine, vodka, and brandy glasses in front of us, and served by waiters in dinner jackets. It was again a dinner with many, many courses and many, many toasts. Chris Kraft's cafeteria in the Houston Mission Control Center was really put to shame.

Thursday, October 18, 3:30 p.m.

We visited the Cosmos Pavilion of the USSR Exhibition of Achievements in National Economy. This is the USSR Space Museum, which I had seen once before. I, therefore, looked at only the new exhibits, which included Mars 3, Lunokhod, and several other lesser exhibits. We also were shown a countdown and launch [7] demonstration using a complete working model of the Baikonur launch complex.

Thursday, October 15, 5:15 p.m.

We were back at the Presidium for the "signing ceremony." Here we signed the Summary of Results of our meeting which, in this case, was very brief since the detailed Summary had been signed by Lunney and Bushuyev. The Summary, as well as the press release, had been worked out by Frutkin and Vereshchetin and had been previously approved by Petrov and me during our meeting at Star City. (Copies attached.)

Thursday, October 18, 5:30 p.m.

Petrov and I, in the company of Lunney and Bushuyev, held a press conference at the Presidium. Petrov preferred to call this a "meeting" with the press because he did not invite the foreign press corps (other than U.S.) nor many of the Soviet press corps. We had, however, insisted that the entire American press corps would be invited. After a brief introduction by Petrov, I gave an opening statement summarizing our entire visit. We then opened it up to questions. Unfortunately, the American press wasn't smart enough to ask some of the more difficult questions like "Where is the Mission Control Center?" or "What did you learn about the Soyuz II failure?" We were prepared on both of these questions. However, Lunney did talk to some of the American press after the press conference and did at that time get into the record that we had indeed been given a detailed report on the Soyuz II failure.

Thursday, October 18, 7:00 p.m.

The Soviet delegation had a dinner and reception in our honor at the "Hall of Mirrors" of the Hotel Prague. This was another formal sit-down dinner with many more toasts and, I might add, the second big dinner of the day. Somehow we all survived.

Friday, October 19, 8:00 a.m.

We left Moscow Airport on an Aeroflot flight for London and from there back to the U.S.

[8]   *GENERAL OBSERVATIONS*

Moscow

Moscow seemed to be a friendlier place this time than I remembered it from my previous visits. There were more cars, more lights, people appeared to be livelier, and even the hotel staff appeared to be less dour. Either there has been a change or perhaps we have become accustomed to their way of life. The fact that I could understand their language this time, at least at times, and the fact that I could speak it well enough to order breakfast, get my room key, and leave a wake-up call, may also have had something to do with the apparent change in attitude.

Relations with Academy of Sciences and ASTP Personnel

In general, both sides seemed to get to the point quicker and easier and appeared to reach a fuller understanding of each issue. Discussions were more direct and more open and frank. Each side made a special effort to make sure that there would be no misunderstandings in the agreements which were reached. (The single exception appeared to be in the negotiation of the Public Affairs Plan where our people have less experience in working with the Soviets.)

NASA Contingent

The NASA contingent under Glynn Lunney is doing an outstanding job. They are diplomatic but firm in all their dealings with their Soviet counterparts. They excel not only during the course of technical discussions but also at social functions.

USSR Reaction

The general reaction to us and to our work still appears to be one of inferiority, but at the same time one that seeks parity. After each visit we were asked, "How did you like it?" "What did you think?" "How does it compare with yours?"

International Situation

We were in Moscow at the height of the Middle East conflict and at a time when Handler and Keldysh were exchanging rather firm [9] letters on the Sakharov affair. Yet neither one of these subjects came up at any time during our visit and the situation appeared to be perfectly normal. (From our side, of course, we missed getting any news about the Middle East situation.) As a matter of fact, the *New York Times* concluded "The warm treatment of Mr. Low and a team of American specialists, working with their Soviet counterparts to prepare for the Apollo-Soyuz mission, was read as a deliberate gesture by Moscow to emphasize its interest in Soviet-American cooperation and the detente despite the frictions of the Middle East conflict."

Personal Reaction

I had learned a great deal about how to "survive" for a week in Moscow since my first visit and, therefore, this visit was very much easier than previous ones had been. Generally, I had only one meal per day, that is lunch, which, as I have mentioned previously, was always a full dinner. (On Thursday, however, we had two of these dinners.) I always had only a very minimal breakfast of tea, bread, and butter at the hotel "cafeteria" and more often than not no evening meal at all. I also learned that I could coax a single vodka through many toasts.

TECHNICAL STATUS OF ASTP

During the course of the status review, Bushuyev gave a basic introduction which was followed by status reports on internal preparations in the U.S. and USSR given by Lunney and Bushuyev, respectively. Next, each of the Working Group chairmen (either a Russian or an American) gave a progress report for their respective groups: mission model, operations plans, experiments, and spacecraft integration; guidance and control, and docking aids; mechanical design; communications and tracking; and life support and crew transfer. Each group gave a detailed schedule and report of progress against that schedule. By and large, all milestones were met and when they were not being met workarounds were available.

Agreements have been reached on five joint experiments; on reciprocal participation of specialists as observers during life support system tests of Apollo and Soyuz; participation in joint seal tests; on a number of safety assessment reports and others [10] that yet had to be written; on studies for the need of electro-magnetic compatibility tests of the cable communications system; and on the participation by U.S. specialists at the Soviet launch site during the pre-flight checkout of the VHF AM equipment. In addition, drawings had been exchanged on the Soyuz orbital module and the Apollo docking module. The problem of mixed crew descent had been discussed and it was decided that this would be considered an "unexamined contingency situation." Another area open for further discussion is additional dockings subsequent to the first undocking.

At the conclusion of the meeting, four potential problem areas were described. These were: documentation; the desirability of U.S. access to the factory in the event of problems during the installation of the VHF equipment; the launch window; and the need for continuing timely exchange of ground and flight test data on ASTP-type Soyuz and Apollo vehicles and systems.

The subject of documentation was discussed during the main meeting as well as during the executive session. I also brought it up privately with Petrov. It seems that a great deal of progress has been made by the Soviets in recent weeks in catching up in all areas where they were behind on documentation. Nevertheless, Lunney is concerned that as time grows shorter they will once again fall behind and we may stub our toe on the entire project. The Soviet solution to the problem is a better forecast of documentation requirements. We agree with this point of view but we say that this is not the complete solution because we can't possibly foresee all problem areas. I believe that Petrov finally understood what we were getting at and promised to personally keep an eye on the situation.

On the subject of access of U.S. specialists during the installation of the U.S. provided VHF equipment, it is quite clear that they do *not* want our people in their factory but have no objection to their presence at the launch site. We told them that we accepted their view on this but that they should consider now what they would do in the event they were to run into trouble and then really required our presence at the factory. I later told Petrov during the executive session that we understood that this might present difficulties and that he would be wise to work these out now for the *contingency* situation which might require our presence.

[11] Insofar as the launch window is concerned, it now closes on September 22 as a result of lighting constraints in the recovery area. Both sides agreed to work on this to see whether it cannot be extended into December.

The last point concerning the timely exchange of ground and flight test data is closely related to the documentation question which I have already discussed.

VISITS TO USSR FACILITIES

The present Soviet decision is that Star City, the Control Center, and the launch site will be open to our technical people. The Soyuz factory will not. Although we reached agreement only on pre-flight activities insofar as the launch site is concerned, Petrov let it be known during the press conference that there would be no problem with our specialists staying there during the time of the launch. Insofar as access for the news media is concerned, the present decision seems to be that Star City, or at least parts thereof, will be open to the news media but the Control Center and the launch site will not.

Tom Stafford had also voiced a concern to me about the fact that he would only see Soyuz simulators and never actual Soyuz flight hardware. I discussed this concern during the executive session. We were told that simulators really were exactly like the flight hardware but nevertheless I said that Stafford was looking for subtle differences and that it was quite important to him to see the actual flight hardware. I suggested perhaps that this too would be possible at the launch site since their spacecraft arrived there some four to six months before the launch. During the course of the executive session, Petrov agreed to look into this and later told Stafford that he, thought this would be possible.

SOVIET FAILURES

During the course of the technical visits preceding my review, the Soviets had made a detailed presentation of the Soyuz II failure and had given us a copy of their failure report. They had not discussed any other failures. In the failure report, they also stated that Cosmos 496 and Cosmos 573 were both [12] unmanned test flights of the changes made after the Soyuz II failure and prior to the Soyuz 12 flight. During the course of the technical review they also stated that there will be two or three more manned Soyuz flights in 1974 and prior to the ASTP flight. Soyuz 12, by the way, did not incorporate a docking system while the 1974 flights will incorporate the ASTP-type docking system.

During the course of the executive session, I told Petrov that we greatly appreciated their report on the Soyuz II failure but that we were also concerned about additional failures reported in the American press during the summer of 1973. I specifically mentioned Salyut 2, which the press had reported as a failure, and Cosmos 557, which some American press reports had also called a Salyut-type vehicle.

Petrov was obviously prepared for the Salyut 2 question, but not for the Cosmos 557 question. On Salyut 2, he said that this bore no relation to the Soyuz which we will use in our joint mission. He stated that Salyut 2 was an improved modernized version of the Salyut. Because of the significant changes, the Salyut 2 flight had been planned from the beginning as an automated flight and was never intended to be manned. We were told that many of the changes were in the automatic control system and these changes clearly required an unmanned flight. To add emphasis, this point was repeated many times. Petrov went on to say that Salyut 2 should be considered a flight for the development of future space stations, that the Salyut is completely independent of the Soyuz, and, finally, that it was not important where it returned to the earth, merely that it returned some place in the open sea.

In summary, it was never clearly said whether Salyut 2 was a failure or a success, but only that whatever it was did not concern us because it did not relate to Soyuz.

I again brought up the subject of Cosmos 557 since there was no response on this question. Petrov did not respond, but another in the group—I believe it was Tsorev—did. He said that Cosmos 557 bore no relation to a manned flight and was neither related to Salyut nor to Soyuz. He said the reports in our press obviously were mistaken.

[13]   STAR CITY

I saw more of Star City this time than I had during my previous visit. Of major significance is the amount of new construction underway at the present time. A new training building is being put up especially for ASTP training. It is a 4-story building which will include classrooms, lecture halls, display rooms for our spacecraft subsystems, etc. In addition, they are building a new hotel and dispensary for the United States team. I think both of these projects are underway so that astronaut treatment at Star City won't appear to be shabby in comparison to cosmonaut treatment in Houston. In addition, two or three other large buildings for training or to house simulators are under construction, as well as a large centrifuge with a capability of up to 20 g's at an onset rate of 2 g's per second for personnel or 4 g's per second for equipment. Both the ASTP classroom and the ASTP hotel buildings were started after the ASTP agreement had been reached, and neither will be quite ready at the time of the November visit but should be ready for the second visit of our astronauts.

Soyuz Simulator and Docking Simulator

I had seen both of these on my previous visit to Star City in January 1971. Leonov conducted the briefings on both. The basic change in the Soyuz reentry module is that it is equipped for only two cosmonauts now while it had room for three during my previous visit. There are also provisions to connect pressure suits and the new pressure relief and shut-off valves which were installed subsequent to the Soyuz II failure are very evident. We were told that the simulator was currently in the Soyuz 12 configuration. This configuration did not include a docking hatch. In the orbital module, we were shown the potassium superoxide air regeneration system and during the course of the discussion there was much talk about condensation removal. This must at one time have been a problem. On the way to the orbital module simulators, one passes through the room in which the optical systems for the displays are mounted. These included both Soyuz and Salyut models.

The docking trainer also showed no difference from 1971 except that the visual targets for docking now included both the Soyuz and the Salyut, whereas only the Soyuz was included in 1971.

[14] *Mock-up Area*

We next went to the mock-up area where Bushuyev went over the Paris Air Show display of the Soyuz with the new docking system, as well as an "external mock-up" of each of the two Soyuz modules. I put the words "external mock-up" in quotes because for all I know this might have been flight hardware. Of interest on this external mock-up was the external insulation, which is a fabric blanket, and the fact that the orbital module had an old style docking system, and it too was said to be in the Soyuz 12 configuration. Again we were told that the ASTP docking system will not be flown until 1974. Bushuyev also indicated that in the Soyuz 12 configuration, Soyuz is a 4-day vehicle if flown alone and a 60-day vehicle if flown with Salyut.

Space Suit

This was modeled by a technician and described by Cosmonaut Kubasev. It is a fairly lightweight garment which, according to Leonov, takes five minutes to don. It will be the type of garment used in ASTP. It is expected to be worn only for about two hours at any one time and, therefore, has no provisions for sanitation. The outer garment provides the strength. The inner garment is a thin rubber bladder, which is sealed by gathering up a bunch of rubber, twisting it, and then tying it with a large rubber band. This sealed garment is then tucked underneath the folds of the external garment which is laced shut. The suit is worn for launch, docking, undocking, and reentry.

Salyut

In the same mock-up building with the Soyuz Paris Air Show exhibit is also the Salyut mock-up. Incidentally, this is a fairly new building in which the ASTP training will also be conducted. It has a glass partition and we were told that the news media will be able to watch from behind that partition when our crews are there. (Even though the building is fairly new, somehow they managed to make the bathrooms look as though they were twenty years old.) Feoktistov was our guide around and through Salyut. (He had already met with Lunney earlier during the visit because Lunney had asked why we never see him anymore. At that time, [15] Lunney asked him when he would again visit the U.S. Feoktistov responded that he had many very serious problems and thought that he would not be able to visit for a long time to come.) Externally, the Salyut we saw differed from the pictures I had previously seen in that it had three solar panels mounted on the main part of the body. Two were mounted horizontally like wings on an airplane and the third vertically but in the same section as the horizontal ones. The horizontal ones could be pivoted to get a better exposure to the sun even while the Salyut was flying at an angle. (I don't recall whether the vertical one could also be pivoted.) Feoktistov told us that Salyut could fly in any attitude for an indefinite period of time without thermal problems.

We entered Salyut through a hatch on the side of what in Skylab would be the multiple docking adapter. I forgot to ask, however, whether it was possible to dock with more than one spacecraft at a time. I don't believe it is. We then went into the main section and first looked at the instrument panel which is very similar to that of the Soyuz. In fact, many of the instruments are identical, as are many of the subsystems. The propulsion system, for example, we were told is exactly like the Soyuz system, and the ECS is a version of the Soyuz system. In response to my question, Feoktistov said that Salyut nominally had a 60-day lifetime but that this could easily be extended to four months by trading on-board consumables for propellants. He also mentioned that food, water, and the air generation system could be resupplied but the propellant could not be resupplied. However, if the Salyut is in a sufficiently high orbit the amount of propellant used for attitude stabilization is minimal. There are no control moment gyros. We saw two rather primitive fire extinguishers, a bungie cord exerciser, including a treadmill, and a wall chart indicating the exercises to be taken. Sleep stations are tucked away around a 10-meter focal length solar telescope. There were a number of other scientific instruments—spectrometers, cameras, star sensors, sun sensors, etc.—all of which were explained in detail by Feoktistov. There is also a refrigerator and a food warmer. Finally, the bathroom is at the very tail end of the station and does not appear to be as complete as the Skylab bathroom. Also at the tail end of the station are two trash air-locks, both used for dumping garbage in bags to the outside. They are at approximately ± 45° from the vertical and appear to be of inordinately heavy construction.

[16] Incidentally, Lunney told me that he inferred from some discussions that there might be some heavy flight activity in the March-April time period next year since many of the specialists with whom he normally deals will then not be available.

SOVIET MISSION CONTROL CENTER

The drive to the Mission Control Center from the hotel took approximately 45 minutes. We headed out of town in a northerly direction, passed the Exhibition of Achievements in the National Economy (Space Museum), then the Moscow city limits, and then drove for another five minutes or so. The Center is located in the village of Kaliningrad. (After leaving the Center and on the way to the press conference, I asked Petrov how I should respond to a question concerning the Control Center's location. At first he stated that I should merely say that it is at the outskirts of Moscow, but apparently he checked this out after we reached the Academy of Sciences again and then told me that I could state, if asked, that it is in Kaliningrad. I was not asked.)

The Center is located within a large complex of buildings surrounded by a security wall. The way we entered and left the area it was difficult to see much of the other buildings. They all are several stories high and could house all sorts of equipment. There were no antennas in evidence. Some new construction is also going on. Within the Control Center building, all of the curtains on the street side were open but all of the curtains facing the rest of the complex were conspicuously drawn. The Control Center building is approximately three or four years old. It had been used in the past for the control of unmanned flights but the first manned flight under control of this Center was Soyuz 12. We were told that it would be used for all future manned flights, Soyuz as well as Salyut, but that not all Salyut flights would be controlled from there. Apparently, there will be some unmanned Salyut flights to be controlled from somewhere else. The building itself is well-constructed and well-appointed. (I will later describe the Institute of Space Research, which is very poorly constructed. By contrast, a lot more money was spent on the physical building of the Control Center than on the Institute of Space Research.) We were first taken into the conference room on the second floor where we were greeted by Abduyevski (the Deputy Director of the Control Center). Abduyevski was with us all of the time but answered few, if any, [17] questions. I have the feeling that he is relatively new in the Control Center and does not know a great deal about it yet. In fact, he may be there solely for the purpose of dealing with NASA. Next we were briefed by Yeliseyev, the Flight Director. He used three charts which had been prepared in English as well as in Russian. These charts depicted how the Control Center fits within the overall operations (launch, network, communications, control, etc.); the flow of information within the Control Center; and the organization of flight controllers within the mission operations control room. In the first order, there is no difference in any of these areas from the way we operate in Houston. It is possible, however, that some of the functions that are performed at Goddard for manned flight control in the U.S. are actually performed within this Control Center in the USSR.

Data flow from the tracking stations apparently without any preprocessing at the stations. They are then manipulated and formatted within various parts of the Control Center and finally displayed in digital form on TV displays in the Mission Operations Control Room. Voice transmissions to the spacecraft flow in the opposite direction. There are no electronic commands generated within the Control Center. Command decisions are made at the Control Center, of course, but the electronic command generation takes place at the tracking stations.

We left the conference room through a second door and found ourselves in the viewing room of the Mission Operations Control Room. This is on a balcony overlooking the main floor of the Control Room. I don't know exactly what I expected to see when I entered the Control Room, but somehow I was surprised and had the feeling that I had wound up in the midst of a Hollywood set. The Control Room is extremely well-appointed and well-outfitted. It is not very different in appearance from our Control Room in Houston. On the front wall there are a number of large screens for either optical or television displays. Television displays are handled with an eidophor just as they are in our case.

As we entered the Control Room, a playback of the Soyuz 12 final countdown was in progress. Across the top of the front wall are a number of clocks showing Moscow time, elapsed time, station acquisition time, and station loss-of-signal time. On the left hand screen were displayed a number of trajectory parameters—apogee, perigee, period, etc. The top of the center screen was [18] a world map with a lighted dot indicating the spacecraft location. The bottom part of the screen was a piece of flight plan concentrating on the "dynamic mode" which refers to the type of control of the spacecraft, as well as a display concerning the type of data being displayed (real time, playback, etc.). On the right hand screen the top half was a television display of the booster at the launch site (later on it switched to onboard television), while the bottom half of the right hand screen contained additional flight planning parameters. (We saw later that there was access to at least this screen from a typewriter at the back of the Control Room, and they were able to type the message "Welcome American colleagues" on that screen.

On the floor were four rows of consoles. The very back row, which is out of sight from the balcony, is for the people who set up the communications and data flow within the Control Center. Also the Project Director (Bushuyev) will sit in this back row. The Flight Director is in the next row from the back and is the focal point for all activity in the Control Center. To his left and right, and in the two rows of consoles in front of him, are the various support functions, which are pretty much the same as the functions within our own Control Center, except that there is no launch vehicle console. Each console has a number of television screens, and the Flight Controller at that console calls up all sorts of displays, either out of the computers or from any one of a number of hard copy projectors. Real time data apparently are only a few seconds behind the actual event. They are also able to generate within the computer a display which merely indicates whether all parameters on a given subsystem are normal or abnormal. If they are normal, that's the end of it. If they are abnormal, the Flight Controller can then go to another display to find out which function is specifically abnormal. There are no warning tones with any of the displays. The communications system allows the Flight Director to talk to any or all of the other consoles as well as to the back rooms. We learned that the Control Center takes over after the spacecraft has been separated from the launch vehicle in orbit. Until that time, the flight is under full control of the Launch Center. The reason for this was explained to us as follows:

First, there are no booster functions that can be performed by the astronauts themselves. Second, spacecraft functions must also be read out at the Launch Center for checkout purposes, and spacecraft experts are at the Launch Center for checkout purposes. [19] For both of these reasons it was more convenient then to handle all abort control at the launch site and not at the Mission Control Center. These facts were further borne out when we saw the onboard TV of the Soyuz 12 launch. The cosmonauts were lying in their couches with their hands folded in their laps. They are obviously just passengers during the launch phase.

In the Mission Operations Control Room Yeliseyev answered all questions concerning flight control. He has obviously been there before and has obviously worked in the Control Center on at least some simulations if not on Soyuz 12. The questions concerning the Control Center itself were answered by the "Deputy Flight Director for Measurements." I believe his name was Miltsin, but I am not sure of this. At any rate, he obviously knew the Control Center well and was able to answer every question which we asked. There was no holding back.

We left the Control Room floor and went behind the large screen where we saw the display projectors. From there we looked into a large number of rooms housing, first, communications equipment, and then computing equipment. We also went to one of the staff support rooms, which was located quite a distance from the Control Room floor, but was equipped with consoles similar or identical to those in the Control Room. Communications gear included a large number of teletype machines as well as all sorts of terminals, recorders, strip charts, and the kind of gear you see in any communications center.

We also saw rooms where all of the onboard tapes were being processed, but none for photographic processing. All of the computing equipment appeared to be made in the Soviet Union. There are three large digital computers, and my guess would be that they are of the generation we used for Mercury and Gemini and not of the Apollo generation. The external memory is a drum memory with 16 drums, each storing 32,000 48-bit words for each of the computers. I don't recall the numbers for the internal memories. In addition to the main computers, there are quite a few peripheral computers used for special tasks. The computers are used for trajectory as well as telemetry work.

As I said earlier, every one of our questions was answered in detail, and if there is anything we don't know it is only because we didn't have enough time or didn't know to ask the [20] right questions. Lunney and Frank, both of whom are very familiar with our own Control Center, should, of course, have a much better view of the real significance of what we saw. It was also of interest that the Control Center was obviously not controlling a flight while we were there. There was very little activity, although one or two people were in evidence in each or the rooms where we opened a door.

During one of the toasts at lunch, Abduyevski said that frankly they had been quite concerned about our visit because they knew of our wonderful technology and hoped that they compared favorably. Many of the private questions we were asked afterwards also concerned our views of their Center. They are obviously very proud of it.

VISIT TO INSTITUTE OF GEOCHEMISTRY

This is Vinogradov's institute where lunar samples are being analyzed. The area of sample handling and preliminary analysis is extremely primitive. Samples from Luna 16 and 20 and from Apollos 11 through 17 were all in storage. The various tools for sample analysis throughout the institute also appeared to be extremely primitive and mostly foreign made. We were shown equipment for spectrographic analysis, a scanning electron microscope, and equipment to measure magnetic spin resonance. I was impressed by neither the people nor the equipment.

INSTITUTE OF SPACE RESEARCH

This institute is in a brand new building which is not yet fully in operation. Apparently the building was constructed by a military labor battalion. It is the shoddiest construction I have ever seen.

We were taken to various laboratories in the Institute and saw flight instrumentation used in gamma ray astronomy, X-ray astronomy, particles and fields measurements, and ionospheric measurements. We also saw some of the instruments which are now on their way to Mars. Incidentally, I asked Sagdeyev whether the newspaper reports to the effect that no life sensing instruments were on the present Mars spacecraft were indeed true, and he said yes, they were not yet ready to send any instruments [21] that were capable of searching for life. He implied, however, that they were working on such instruments for the next Mars opportunity. He also asked how long it had taken us to develop the instruments we intend to fly on Viking. There was some additional discussion about the present flights to Mars and apparently one of the four spacecraft is having telemetry difficulties which have not yet been resolved.

The X-ray type instrumentation we saw apparently has already been flown and some results have been published. By their own admission, however, these results are not as good as those obtained with Uhuru. They indicated that since their satellite was not in an equatorial orbit and was only in orbit for a short period of time, they could not match Uhuru's results. The gamma ray instrumentation we saw had not yet flown on a satellite. Insofar as ionospheric measurements are concerned, they apparently have a very active program, both with sounding rockets and with satellites.

In summary, we saw instruments of the type flown in our physics and astronomy and planetary programs. Although earth resources work is also going on in the same institute, this was not discussed nor were we shown any of the work. Our guess is that they just don't have anything worth seeing.

The remaining time at the Institute of Space Research was spent on a discussion of the results of the Venus 8 spacecraft. (Sagdeyev pointed out that this was done especially at the request of Keldysh since we had discussed our Mars results with Keldysh.) The briefing was given by Abduyevski, who, as I mentioned earlier, is now the Deputy Director of the Control Center. Whereas he was a novice at the Control Center business, he knew all about the engineering of the Venus 8 spacecraft as well as the details of the scientific results. My guess is that he was deeply involved in the Venus 8 flight.

The Venus 8 spacecraft was designed to withstand the Venus surface temperatures for a short period of time (approximately 1 hour). This was achieved with good insulation and through precooling the spacecraft for several days before it arrived at Venus. Abduyevski made a major point of the fact that the insulating properties of the insulation change drastically with increasing pressures of the kind encountered at the surface of Venus (90 atmospheres), and that new materials with lower "filtration constants" had to be designed.

[22] The most interesting result was the measurement of surface lighting in an area near the Venus terminater. The conclusion is that there is adequate lighting on the surface of Venus for television, even near the terminater.

VISIT WITH KIRILLIN

As I mentioned before, this was a courtesy visit made at my request. After a few words of welcome by Kirillin, I opened the discussion by reviewing the status of ASTP and other joint projects.

Kirillin then asked my views concerning the practical results of the exploration of space. I spoke of the usual things—communications, weather, and earth resources—as well as the potential long-range results of some of the scientific efforts in space. Kirillin came back to the point that the future of space must be practical and added one subject which I had left out of my discussions of earth resources, and that is geology. He felt that major contributions to geology can be made from space.

I then asked Kirillin where he thought our future cooperation in space might go. My purpose in asking this question was to find out whether he had given the matter any thought. Apparently he had not and gave only a very vague answer.

Finally, I brought up the subject of aeronautics, reminding Kirillin that NASA, of course, has a major effort in aeronautics research and asking whether he had ever considered any cooperation in this area. His eyes immediately lit up and he started talking about some of the commercial discussions now underway with Boeing, General Dynamics, and McDonnell Douglas, but he wondered what I had in mind and how NASA might fit in. I told him that I had really nothing specific in mind when I brought up the subject but that any cooperative efforts with NASA would have to be in the areas of aeronautical research as opposed to in the commercial areas. Both of us agreed to think about future possibilities in possible cooperation in aeronautics and said that we might pursue this at a later time.

[23]   PUBLIC AFFAIRS PLAN

Donnelly had negotiated the first phase of a Public Affairs Plan covering preflight activities. This plan had been signed by Lunney and Bushuyev; it was to be confirmed by Petrov and me. When I met with Donnelly to review the plan he was concerned that the definition of news media in the plan was not clear and that it was quite likely that the Soviet side would not permit television cameramen to accompany television correspondents. Instead, he felt that they would want to impose on us the usual practice of having the Soviets take all television film and of selling that film through Novesty news agency. Donnelly, therefore, suggested that we should not confirm the plan until this issue had been settled. (Since this was an open issue, it is still not clear to me why he asked Lunney to approve the plan in the first place.)

In subsequent discussions with Petrov, it became clear to me that the plan as signed lacked in two other respects: first, it would be quite possible that the Soviet side would admit its own news media to a joint function without at the same time admitting U.S. news media; and secondly, Donnelly indicated that he had verbal agreements that our astronauts could be accompanied by their own documentary photographer. This was not written down in the plan.

In my first meeting with Petrov (the meeting at Star City), he appeared to understand all the points that needed to be covered, and also appeared to be in agreement with them.

We adjourned our meeting at Star City, and Frutkin and I wrote additions to the Public Affairs Plan in the area of the three points mentioned; that is, the definition of news media, the participation of news media from both sides in joint activities, and the possibility of bringing along a documentary photographer. When we met again the next morning, Petrov was not as willing to include these new additions as he had implied the night before. Obviously, he must have checked into this with somebody better versed in the ways of the press in the Soviet Union. He threw up a smoke screen about things like the copyright agreement and the lighting required whenever TV cameramen were present. I told him that I wanted him to understand that there is only one serious issue in the definition of news media and that concerns television [24] cameramen. Will U.S. cameramen be allowed in the Soviet Union or not? The meeting broke up without reaching any conclusion. Subsequently, Frutkin had additional discussions with Vereshchetin, and I had additional discussions with Petrov. Vereshchetin assured Frutkin before we left Moscow that they agreed in principle with all of our points, but they were not sure whether they could agree exactly with our language. They promised that they would send, at an early date, a new version of the Public Affairs Plan, incorporating the substance of our additions. We could then either confirm the plan or, if we still did not like it, we would have to have further negotiations.

MISCELLANEOUS ITEMS

Comet Kohoutek

I gave Petrov several reprints of the Kohoutek article which appeared in the October issue of *Aeronautics and Astronautics*, and asked whether the USSR would have any interest in participating in the planned observations. On the following day Petrov informed me that they would ordinarily be quite interested in participating, thanked me for the invitation, but told me that during the time of the Comet the weather would be so bad in the Soviet Union that it was unlikely that any of their ground observatories would be able to see it. I took this as a polite way of saying "no."

Reaffirmation of the Low-Keldysh Agreement

Frutkin informed me that he believed that the Low-Keldysh Agreement needed to be reaffirmed three years after it was approved, or in the spring of 1974. Although I was not quite sure that this was the case, I did bring up the subject with Petrov. He implied that the spring of 1974 would be a bad time because this will be the 250th Anniversary of the Soviet Academy of Sciences, and Keldysh is expected to be very busy. However, he suggested that we might get together in the summer or fall of 1974. Although he assumed that we would get together in the Soviet Union, I issued an invitation to do this in the United States. However, I am not sure how necessary it is to do anything other than to exchange letters on the subject.

George M. Low

Document I-49

Document title: George M. Low, Deputy Administrator, NASA, to Academician M.V. Keldysh, President, Academy of Sciences of the USSR, March 24, 1975.

Source: George M. Low Papers, Institute Archives and Special Collections, Rensselaer Polytechnic Institute, Troy, New York.

The United States viewed the Apollo-Soyuz Test Project as only the first step in an ambitious program of U.S.-Soviet space cooperation. As indicated in this letter, the United States was eager to begin discussing next cooperative steps with the Soviet Union even before the Apollo-Soyuz mission was completed.

---

[1]   March 24, 1975

Academician M. V. Keldysh
President
Academy of Sciences of the USSR
Leninsky Prospect 14
Moscow, V-71, USSR

Dear Academician Keldysh:

I understand that the ASTP Technical Directors have now agreed on the schedule of activities for the May meetings in the Soviet Union. Accordingly, I plan to arrive in Moscow on May 17 and to join in the visit to the launch area scheduled for May 19. I would return with the Technical Directors to Moscow and remain for the Flight Readiness Review on May 23.

This schedule would make May 21 and 22 available for other business. I understand from Academician Petrov your wish to defer the meeting of full delegations for detailed discussions of future cooperation because of the demands of ASTP on the time of your specialists and because of the demands of the Academy elections on your own time. We, of course, will accede to your wishes in this respect. At the same time, I believe it would be most desirable for us to take advantage of this opportunity to meet briefly.

To assure that your concerns are met, our meeting could be entirely informal in character, with no written record. I would plan to be accompanied only by Mr. Frutkin and our interpreter. I would expect to outline the status of our thinking here with regard to future possibilities for cooperation. You would, of course, be free to comment or to indicate Soviet thinking in the degree you wish. It would be understood that no commitments of any kind were implied by either side.

Our own present thinking, which I would expand on in our meeting, is along three lines:

1.   Projects in the area of manned space flight—We would be prepared to consider cooperative exploratory [2] *studies* of future space stations, with a view to pursuing such studies to further steps, if warranted. We are prepared also to consider such possible interim steps as a Space Shuttle/Salyut mission, as well as Soviet use of the Shuttle in cooperative projects of mutual value.

2.   Projects in the area of unmanned scientific missions—We have in mind the possibility of a lunar farside sample return mission, and we continue to find the long-term goal of a future Mars surface sample return mission attractive.

3.   Projects in the area of space applications—Here we have in mind such possibilities as coordinated environmental monitoring missions and the exchange of data relating to radiation balance, stratospheric ozone monitoring, and search and rescue.

In the informal conversation which I suggest, we might also refer to a subject which Dr. Lunney has already taken up with Prof. Bushuyev in a preliminary way. If the first NASA Space Shuttle mission is to have rendezvous and docking capability compatible with Soviet spacecraft of the 1979-80 time period, we would need, for development purposes, to have agreement by January 1976 on such parameters as diameter of the passageway, load factors, communications interface, and atmospheric pressures. To this purpose, we would want to put discussion of such parameters by our specialists on a schedule consistent with design and development requirements.

I hope it would be possible to use the occasion of my presence in Moscow for such an informal constructive conversation so that we can preserve the momentum which has been generated by our cooperation in ASTP.

Sincerely,

George M. Low
Deputy Administrator

Document I-50

Document title: A.P. Aleksandrov, USSR Academy of Sciences, and A.M. Lovelace, NASA, "Agreement Between the USSR Academy of Sciences and the National Aeronautics and Space Administration of the USA on Cooperation in the Area of Manned Space Flight," May 11, 1977.

Source: NASA Historical Reference Collection, NASA History Office, NASA Headquarters, Washington, D.C.

This agreement was the result of almost two years of discussions between the United States and the Soviet Union. It was signed at the time that the renewal of the U.S.-Soviet Space Cooperation Agreement for a second five-year term was announced. The agreement was never implemented. Carter administration displeasure with the Soviet record on human rights and then with Soviet involvement in Afghanistan led to low priority being given to U.S.-U.S.S.R. space cooperation overall.

[1]

Agreement Between the USSR Academy of Sciences and the National Aeronautics and Space Administration of the USA on Cooperation in the Area of Manned Space Flight

In accordance with the Agreement on Cooperation in the Exploration and Use of Outer Space for Peaceful Purposes between the USSR and the USA, dated May 24, 1972, and taking into account the results of discussions held in Washington, October 19-22, 1976, between the delegation of the USSR Academy of Sciences, headed by the Chairman of the Intercosmos Council of the USSR Academy of Sciences, Academician B. N. Petrov, and the delegation of the National Aeronautics and Space Administration of the USA, headed by the NASA Deputy Administrator, Dr. A. M. Lovelace, the Academy of Sciences and NASA agree to undertake the following steps for further development of cooperation between the USSR and USA in the exploration and use of outer space for peaceful purposes.

I. *Study of the Objectives, Feasibility and Means of Accomplishing Joint Experimental Flights of a Long-Duration Station of the Salyut-Type and a Reusable "Shuttle" Spacecraft (Salyut-Shuttle Program)*

In view of the fact that the long orbital stay-time of the Salyut-type station and the capabilities of the Shuttle spacecraft commend their use for joint scientific and applied experiments and for further development of means for rendezvous and docking of spacecraft and stations of both [2] nations, the two sides agree to establish two joint working groups (JWGs) of specialists, charging them with studying the objectives, feasibility and means of carrying out a joint experimental program using the Soyuz/Salyut and Shuttle spacecraft:

– a JWG for basic and applied scientific experiments.
– a JWG for operations.

Within 30 days after the Agreement becomes effective, the sides will inform each other of the initial leaders and composition of these JWGs. The work of both Joint Working Groups should begin simultaneously. The composition of the JWGs can be changed or enlarged at any time as necessary. Appropriate sub-groups can be formed.

In their studies, the JWGs should proceed on the assumption that the first flight would occur in 1981. The final date would be set in the course of the joint work.

First Phase of the Joint Working Groups' Activity

The following preliminary project documents should be prepared within 6-12 months after the agreement comes into effect:

– preliminary proposals for scientific experiments;
– preliminary technical proposals for carrying out the program;
– preliminary schedules for implementing the program.

[3] *Second Phase of the Joint Working Groups' Activity*

The JWGs should prepare the following definitive documents within one year of joint work in the second phase:

– a technical description of the joint program and its realization;
– a scientific program for the joint flight;
– a schedule for conducting the joint work;
– an organizational basis for implementing the program;
– a list of additional joint technical documentation which may be required.

The sides will make the final decision on implementing the program at the end of the second phase of the JWGs' activity.

The working period of the JWGs in the first and second phases of their activities can be shortened.

Each side will consider the accommodation on its spacecraft of payloads proposed by the other side for flight in the Shuttle-Salyut program. Such accommodation will be undertaken where both sides agree that the payloads concerned are of mutual value and interest.

II. *Consideration of the Feasibility of Developing an International Space Platform in the Future (International Space Platform Program)*

Both sides recognize that no commitments are made at [4] this stage concerning the realization of any project for creating an international space platform.

The sides agree to establish a Joint Working Group of specialists for preliminary consideration of the feasibility of developing an International Space Platform on a bilateral or multilateral basis in the future.

The JWG will carry out its work on the basis of studies conducted by each side independently and also by the two sides jointly, proceeding from each of the following stages to the next as may be mutually agreed:

    –   define at the first stage the scientific and technical objectives which would warrant the use of such a space platform.

    –   consider possible configurations appropriate to the objectives identified.

    –   formulate proposals on the feasibility and character of further joint work which may be desirable in this field.

At the first stage of its activity, the group will work in close coordination and contact with the JWGs set up to consider ways to realize the Salyut-Shuttle program.

The sides will appoint the initial leaders and members of the JWG for this program within two months after the Agreement goes into effect. This JWG should formulate preliminary proposals on possible scientific-technical objectives which could be achieved by an international station one year after beginning its work.

[5] This Agreement comes into force at the moment it is signed by both sides.

For the USSR Academy of Sciences	For the National Aeronautics and Space Administration of the USA
A. P. Aleksandrov	A. M. Lovelace

Document I-51

Document title: George P. Shultz and Eduard Shevardnadze, "Agreement Between the United States of America and the Union of Soviet Socialist Republics Concerning Cooperation in the Exploration and Use of Outer Space for Peaceful Purposes," April 15, 1987, with attached: "Agreed List of Cooperative Projects."

Source: NASA Historical Reference Collection, NASA History Office, NASA Headquarters, Washington, D.C.

As part of its overall hostile stance toward the Soviet Union, the administration of President Ronald Reagan allowed the basic U.S.-Soviet Space Cooperation Agreement, signed in 1972 and renewed in 1977, to lapse when it came up for renewal in 1982. U.S. policy toward the U.S.S.R. became much more friendly after Mikhail Gorbachev came into power in 1985, and by 1987 the two countries had agreed to restart formal cooperative activities in space. This agreement, signed by the U.S. secretary of state and the Soviet foreign minister, provided the framework for such cooperation, and an attached list identifies an initial sixteen areas of possible cooperation.

[1]

Agreement Between the United States of America and the Union of Soviet Socialist Republics Concerning Cooperation in the Exploration and Use of Outer Space for Peaceful Purposes

The United States of America and the Union of Soviet Socialist Republics, hereinafter referred to as the Parties;

Considering the role of the two States in the exploration and use of outer space for peaceful purposes;

Desiring to make the results of the exploration and use of outer space available for the benefit of the peoples of the two States and of all peoples of the world;

Taking into consideration the provisions of the Treaty on Principles Governing the Activities of States in the Exploration and Use of Outer Space, including the Moon and Other Celestial Bodies, and other multilateral agreements regarding the exploration and use of outer space to which both States are Parties;

Noting the General Agreement Between the Government of the United States of America and the Government of the Union of Soviet Socialist Republics on Contacts, Exchanges, and Cooperation in Scientific, Technical, Educational, Cultural, and other fields, signed on November 21, 1985;

Have agreed as follows:

[2]   *ARTICLE 1*

The Parties shall carry out cooperation in such fields of space science as solar system exploration, space astronomy and astrophysics, earth sciences, solar-terrestrial physics, and space biology and medicine.

The initial agreed list of cooperative projects is attached as an Annex.

ARTICLE 2

The Parties shall carry out cooperation by means of mutual exchanges of scientific information and delegations, meetings of scientists and specialists and in such other ways as may be mutually agreed, including exchange of scientific equipment where appropriate. The Parties, acting through their designated cooperating agencies, shall form joint working groups for the implementation of cooperation in each of the fields listed in Article 1. The recommendations of the joint working groups shall be subject to the approval of each Party in accordance with its appropriate national procedures prior to implementation. The designated cooperating agencies shall notify each other of the action taken by the parties on the recommendations within three months of their adoption by the joint working groups.

[3]   *ARTICLE 3*

The joint working groups shall begin their work with the projects listed in the Annex to this Agreement. Revisions to the list of projects in the Annex, which may include the identification of other projects in which cooperation would be of mutual benefit, may be effected by written agreement between the Parties through a procedure to be determined by them.

ARTICLE 4

Cooperative activities under this Agreement, including exchanges of technical information, equipment and data, shall be conducted in accordance with international law as well as the international obligations, national laws, and regulations of each Party, and within the limits of available funds.

ARTICLE 5

This Agreement shall be without prejudice to the cooperation of either Party with other States and international organizations.

ARTICLE 6

The Parties shall encourage international cooperation in the study of legal questions of mutual interest which may arise in the exploration and use of outer space for peaceful purposes.

[4]   *ARTICLE 7*

This Agreement will enter into force on the date of signature by the Parties and will remain in force for five years. It may be extended for further five-year periods by an exchange of notes between the Parties. Either Party may notify the other in writing of its intent to terminate this Agreement at any time effective six months after receipt of such notices by the other Party.

IN WITNESS WHEREOF the undersigned, being duly authorized by their respective Governments, have signed this Agreement.

DONE at Moscow, in duplicate, this 15th day of April, 1987, in the English and Russian languages, both texts being equally authentic.

[signed George P. Shultz]   [signed Eduard Shevardnadze}
FOR THE UNITED STATES OF   FOR THE UNION OF SOVIET
AMERICA: SOCIALIST REPUBLICS:

****

[Attachment page 1]
Agreed List of Cooperative Projects

1.   Coordination of the Phobos, Vesta, and Mars Observer missions and the exchange of scientific data resulting from them.

2.   Utilization of the U.S. Deep Space Network for position tracking of the Phobos and Vesta landers and subsequent exchange of scientific data.

3.   Invitation, by mutual agreement, of co-investigators and/or interdisciplinary scientists' participation in the Mars Observer and the Phobos and Vesta missions.

4.   Joint studies to identify the most promising landing sites on Mars.

5.   Exchange of scientific data on the exploration of the Venusian surface.

6.   Exchange of scientific data on cosmic dust, meteorites and lunar materials.

7.   Exchange of scientific data in the field of radio astronomy.

8.   Exchange of scientific data in the fields of cosmic gamma-ray x-ray, and sub-millimeter astronomy.

9.   Exchange of scientific data and coordination of programs and investigations relative to studies of gamma-ray burst data.

10. Coordination of observations from solar terrestrial physics missions and the subsequent exchange of appropriate scientific data.

11. Coordination of activities in the study of global changes of the natural environment.

12. Cooperation in the Cosmos biosatellite program.

13. Exchange of appropriate biomedical data from U.S. and U.S.S.R. manned space flights.

14. Exchange of data arising from studies of space flight-induced changes of metabolism, including the metabolism of calcium, from both space flight and ground experiments.

15. Exploration of the feasibility of joint fundamental and applied biomedical experiments on the ground and in various types of spacecraft, including exobiology.

16. Preparation and publication of a second amplified edition of the joint study "Fundamentals of Space Biology and Medicine."

Document I-52

Document title: Office of the Press Secretary, The White House, "Joint Statement on Cooperation in Space," June 17, 1992.

Source: NASA Historical Reference Collection, NASA History Office, NASA Headquarters, Washington, D.C.

The 1991 collapse of the Soviet Union and the emergence of the Russian Federation as its primary successor opened new prospects for space cooperation. The Russian Federation created a civilian space agency, the Russian Space Agency, in April 1992; its head was Yuri Koptev, formerly an official of the Soviet Ministry of General Machine Building. On April 1, 1992, a new NASA Administrator, Daniel S. Goldin, took office. The two agency heads met for the first time in June 1992 and quickly agreed that there were many opportunities for enhanced cooperation, particularly in the area of human spaceflight. During a summit meeting between Russian President Boris Yeltsin and U.S. President George Bush a few days later, the two countries announced their intention to broaden cooperative relations in space.

June 17, 1992

Joint Statement on Cooperation in Space

— The United States and the Russian Federation have agreed on steps to broaden cooperation in the use and exploration of outer space:
   – *Space Agreement:* A new space agreement has been signed today that puts space cooperation between the two countries on a new footing, reflecting their new relationship.
   – The new agreement provides a broad framework for NASA and the Russian Space Agency to map out new projects in a full range of fields; space science, space exploration, space applications and the use of space technology.
   – Cooperation may include human and robotic spaceflight projects, ground-based operations and experiments and other important activities, such as monitoring the global environment from space, Mir Space station and Space Shuttle missions involving the participation of U.S. astronauts and Russian cosmonauts, safety of spaceflight activities, and space biology and medicine.
   – Pursuant to the agreement, the two governments will give consideration to the following:
      • flights of Russian cosmonauts aboard a Space Shuttle mission (STS 60), and U.S. astronauts aboard the Mir Space Station in 1993; and
      • a rendezvous docking mission between the Mir and the Space Shuttle in 1994 or 1995.
   – An important part of the agreement involves annual subcabinet consultations led at the Under Secretary of State/Deputy Foreign Minister level, a new mechanism for high level government review of the bilateral civil space relationship between the two countries.
— *Joint Study of Space Technology:* The two governments are also announcing detailed technical studies of the possible use of space technology.
   – NASA is awarding a contract to the Russian firm NPO Energiya; the principal area being examined in the Russian *Soyuz-TM* spacecraft as an interim crew return vehicle for Space Station Freedom.
[2] – Other important areas to be studied are the suitability of the Russian developed *Automated Rendezvous and Docking System* in support of NASA spaceflight activities, the use of the *Mir Space Station* for long–lead time medical experiments, and other applications by NASA of Russian hardware.
— *Space Commerce:* Both governments also agreed on steps to encourage private companies to expand their search for new commercial space business.
   – The United States has accepted an invitation from the Russian Federation for American businessmen to visit Russia. The Department of Commerce will lead a delegation of U.S. aerospace firms to Russia in the near future on a space technology assessment mission.
   – The Russian Federation has accepted an invitation from the United States to send a delegation of business leaders to the United states to meet with their counterparts in the American aerospace private sector.
— *Space Launch:* Reflecting its support for economic reform in Russia, the United States has decided to consider favorably a decision expected by the INMARSAT Organization in July 1992 to launch one of the INMARSAT 3 satellites from Russia.

- The INMARSAT 3 satellite is manufactured primarily in the United States. If approved by INMARSAT, this would mark the first time that a U.S. manufactured commercial satellite would be launched from Russia.
- The United States and Russia have agreed to negotiate a bilateral agreement on technology safeguards for the INMARSAT 3 satellite to enable issuance of a U.S. export license.
- The United States and the Russian Federation support the application of market principles to international competition in the provision of launch services, including avoidance of unfair trade practices.
- Recognizing Russia's current transition to a market economy, and in order to allow consideration of future proposals involving Russian launch of U.S. satellites, the Russian Federation and the United States have agreed to enter into international negotiations on an expeditious basis to develop international guidelines concerning competition in the launch of commercial satellites.
- In the case of INMARSAT, the Russian Federation has also assured the United States that the terms and conditions of the Russian proposal, including pricing, are consistent with those that would normally be offered in the international market.

Document I-53

Document title: "Implementing Agreement Between the National Aeronautics and Space Administration of the United States of America and the Russian Space Agency of the Russian Federation on Human Space Flight Cooperation," October 5, 1992.

Document I-54

Document title: Office of the Vice President, The White House, "United States-Russian Joint Commission on Energy and Space—Joint Statement on Cooperation in Space," September 2, 1993.

Document I-55

Document title: "Protocol to the Implementing Agreement Between the National Aeronautics and Space Administration of the United States of America and the Russian Space Agency of the Russian Federation on Human Space Flight Cooperation of October 5, 1992," December 16, 1993.

Source: All in NASA Historical Reference Collection, NASA History Office, NASA Headquarters, Washington, D.C.

As a result of the U.S.-Russian dialogue on expanded space cooperation initiated in June 1992, NASA and the Russian Space Agency signed an agreement in October 1992 to exchange cosmonauts and astronauts on each others' human spaceflight missions and to dock the Space Shuttle with the Russian space station Mir. During its first year in office, the administration of President Bill Clinton and Vice President Al Gore moved to expand substantially existing U.S.-Russian cooperation in human spaceflight, in effect merging large portions of the efforts of the only two countries with the capability of sending people into space; such a move was announced in September 1993. The political decision to undertake this expansion was linked to broader U.S.-Russian foreign policy concerns, such as stemming the proliferation of missile technology capability and providing job opportunities for the Russian aerospace sector. The United States agreed to provide funding for various Russian activities and hardware associated with the expanded cooperation; this transfer broke with the long-standing NASA tradition that its cooperative programs did not involve an exchange of funds. After a few more months of discussion, NASA and the Russian Space Agency decided to increase the intensity of their interactions, particularly with respect to flights of the U.S. Space Shuttle to dock with the Russian space station Mir. On December 16, 1993, the heads of the two agencies signed a protocol to the October 1992 agreement that reflected this new level of activity.

Document I-53

[1]

Implementing Agreement Between the National Aeronautics and Space Administration of the United States of America and the Russian Space Agency of the Russian Federation on Human Space Flight Cooperation

PREAMBLE

The National Aeronautics and Space Administration (hereafter referred to as "NASA") and the Russian Space Agency (hereafter referred to as "RSA"), jointly referred to as "The Parties," have agreed to cooperate in the area of human space flight. This cooperative program consists of three inter-related projects: the flight of Russian cosmonauts on the U.S. Space Shuttle; the flight of U.S. astronauts on the Mir Space Station; and a joint mission involving the rendezvous and docking of the U.S. Space Shuttle with the Mir Space Station. These will be jointly referred to in the future as the "Shuttle-Mir Program."

The Parties have agreed as follows:

ARTICLE I: DESCRIPTION OF COOPERATION

1. The cooperation set forth in this Implementing Agreement will be undertaken in accordance with the Agreement Between the United States of America and the Russian Federation concerning Cooperation in the Exploration and Use of Outer Space for Peaceful Purposes, of June 17, 1992 (hereinafter the June 17, 1992 Agreement).

2. An experienced cosmonaut will fly aboard the Space Shuttle on the STS-60 mission, which is currently scheduled for November 1993. The cosmonaut will be an integral member of the orbiter crew, and will be trained as a Mission Specialist on Shuttle systems, flight operations, and manifested payload procedures following existing Shuttle practices.

3. The RSA will nominate two cosmonauts, for approval by NASA as candidates for the STS-60 Space Shuttle mission. In accordance with Article IV, one of the two cosmonauts will be designated the Primary Russian-sponsored crewmember, with the other being designated as a backup crewmember. Both crewmembers will receive [2] Mission Specialist Astronaut training, until the time that the STS-60 crew begins dedicated mission training. From that point, the backup crewmember will receive as much training as practical. The two cosmonauts will be scheduled for arrival at the Johnson Space Center in Houston, Texas, in October, 1992. Their names, experience and personal history will be provided to NASA by the RSA prior to the initiation of training.

4.   An experienced NASA astronaut will fly on the Mir Space Station as an integral long-duration crewmember (e.g., longer than 90 days) participating as an integral member of the crew in a variety of operations and experiments. The timing of this flight will be consistent with a Shuttle docking flight in 1994 or 1995. The astronaut will be flown to the Mir on a Soyuz transportation system. Special emphasis will be placed on science, particularly life science, as well as engineering and operational objectives. Astronaut and cosmonaut participation before, during and after the long-duration flight will be emphasized to accomplish all flight objectives.

5.   NASA will nominate two astronauts for approval by RSA as candidates for a long-duration Mir mission (e.g., longer than 90 days) to occur in conjunction with the rendezvous and docking of the Space Shuttle with Mir. In accordance with Article IV, one of the two astronauts will be designated as the primary U.S.-sponsored crewmember, with the other being designated as the backup crewmember. Both crewmembers will receive full cosmonaut training with their cosmonaut crew.

The two astronauts will be scheduled to begin training no later than 12 months prior to the agreed upon flight date. They will be U.S. citizens, and their names, experience and personal history will be provided to RSA by NASA no later than one month prior to the initiation of training.

6.   The Space Shuttle will rendezvous and dock with Mir in conjunction with the flight of the NASA astronaut aboard Mir. NASA will transport two Russian cosmonauts in the Shuttle to replace the two cosmonauts on board Mir. Training for these cosmonauts will be in accordance with Article V of this Implementing Agreement. Life sciences experiments involving the NASA astronaut and the two cosmonauts who have been on board the Mir for 90 days or more will be conducted while the Shuttle is docked to the Mir. The NASA astronaut and the two cosmonauts who have been on the Mir for 90 days or more will be returned in the Shuttle for continued postflight life sciences experiments.

7.   As part of the technical discussions leading up to the Mir rendezvous, joint implementation teams will explore the use of the Androgynous Peripheral Docking Assembly developed by NPO Energiya, consistent with the June 17, 1992 Agreement and this Implementing Agreement. (If such used appears technically [3] feasible, NPO Energiya will enter into a separate contract with an American company to provide, modify or integrate this device or its derivatives with the Shuttle.)

8.   Joint implementation teams will also consider exchange of Mir crewmembers, transportation of experimental and logistic equipment, and Extra Vehicular Activity (EVA), and will define the respective responsibilities of the Parties, consistent with the June 17, 1992 Agreement and this Implementing Agreement. The implementation teams will jointly develop a contingency plan which will cover procedures for investigation, consultation, and exchange of data in the event of a mishap which causes damage to equipment or injury to personnel during the conduct of the Shuttle-Mir Program.

9.   Consistent with the June 17, 1992 Agreement, each Party will be responsible for funding its respective responsibilities, consistent with its domestic laws and regulations, and subject to the availability of appropriated funds. All training, in-country travel and living arrangements, flight and other associated posts for each Party's crew members and dependents will be borne by the host country, in a manner it deems appropriate, at a standard afforded its own flight crews.

ARTICLE II: DESIGNATION OF REPRESENTATIVES AND ORGANIZATIONS

Designated Points of Contact for the implementation of the activities described herein are contained in Annex 1 to this Implementing Agreement. Annex 1 may be modified by either Party upon notification to the other Party. NPO Energiya and the Yuri Gagarin

Cosmonaut Training Facility will be the lead technical implementors of the Shuttle-Mir Program in Russia.

ARTICLE III: JOINT IMPLEMENTATION TEAMS

The Parties agree to establish joint implementation teams to coordinate and implement the activities described herein. Designated team members will be identified by each side within 30 days of the entry into force of this Implementing Agreement. Each Party may modify the membership of its joint implementation teams at its discretion. The joint implementation teams will develop a plan for implementation of the activities described herein on the basis of equality, reciprocity and mutual benefit, consistent with the June 17, 1992 agreement.

ARTICLE IV: SELECTION OF CANDIDATES

1. Selection of flight candidates will be based on mutual agreement prior to any announcement. Candidates selected will be [4] current, active members of each side's astronaut or cosmonaut corps.
2. Flight candidates selected will have previous space flight experience. The cosmonauts selected for training shall have sufficient knowledge in verbal and written English. The NASA astronauts selected for training shall have sufficient knowledge in verbal and written Russian. Information that each side's candidates meet the criteria in this Article shall be exchanged prior to any announcement on crew selections.

ARTICLE V: TRAINING

1. Throughout their training programs, the Russian cosmonauts will be based at the Johnson Space Center in Houston, Texas, and will be assigned to the Astronaut Office in the Flight Crew Operations Directorate. The NASA astronauts will be based at Yuri Gagarin Cosmonaut Training Facility ("Star City") in the Moscow Region.
2. At the beginning of the training programs, each Party will require its candidates to enter into a Standards of Conduct Agreement with the other Party, which will include, inter alia, installation safety and security matters, provisions related to prohibitions on use of position for private gain, authority of the Mission Commander, and limitations on use of information received during training and flight. Each Party will ensure that its candidates comply with the provisions of such an agreement.
3. The candidates will have completed all aspects of the required training to the full and final satisfaction of the host Party prior to certification for flight.
4. By mutual agreement, the Parties will identify any support personnel required for the flight candidates selected.

ARTICLE VI: SCIENCE

1. The Parties will establish a Scientific Working Group to coordinate appropriate scientific experiments and activities to be conducted by each side on the respective missions. Designated working group members will be identified by each side within 30 days of the entry into force of this Implementing Agreement. Each Party may modify the membership of its Scientific Working Group at its discretion.
2. Results of the scientific experiments conducted by each Party under this Implementing Agreement will be made available to the scientific community in general

through publication in appropriate journals of other established channels. In the event [5] such reports or publications are copyrighted, NASA and RSA shall have a royalty-free right under the copyright to reproduce, distribute and use such copyrighted work for their own purposes.

ARTICLE VII: LIABILITY

1.   A comprehensive cross-waiver of liability between the two Parties and their related entities (e.g., contractors, subcontractors, and other participating entities associated with the Parties including any state from which RSA procures a launch to carry out its obligations under this agreement) shall apply to the activities under this agreement. The cross-waiver of liability shall be broadly construed. The terms of the waiver are set out in Annex 2.

2.   Except as provided in Annex 2, the Government of the United States and the Government of the Russian Federation will remain liable in accordance with the Convention on International Liability for Damage Caused by Space Objects (the "Liability Convention") of March 29, 1972. In the event of a claim arising out of the Liability Convention, the governments will consult promptly on any potential liability, on any apportionment of such liability, and on the defense of such claim.

ARTICLE VIII: INVENTION AND PATENT RIGHTS

1.   With the exception of the intellectual property rights referred to in Article X, Exchange of Technical Data and Goods, and subject to national laws and regulations, provisions for the protection and allocation of intellectual property rights created during the course of cooperation under this Implementing Agreement are set forth in Annex 1 of the June 17, 1992 Agreement.

2.   Except as set forth in paragraph 1, nothing in this Implementing Agreement shall be construed as granting or implying any rights to, or interest in, patents or inventions of the Parties or their contractors and subcontractors.

ARTICLE IX: PUBLIC INFORMATION

Release of public information regarding these joint activities may be made by the appropriate agency for its own portion of the program as desired and, insofar as participation of the other is involved, after suitable consultation.

[6] ARTICLE X: EXCHANGE OF TECHNICAL DATA AND GOODS

Each Party is obligated to transfer to the other Party only those technical data and goods which both Parties agree are necessary to fulfill the responsibilities of the transferring Party under this Implementing Agreement, subject to the following:

1.   Interface, integration, training and safety data (excluding detailed design, manufacturing, and processing data, and associated software) will be exchanged by the Parties without restrictions as to use or disclosure, except as otherwise restricted by national laws or regulations relating to export controls.

2.   In the event a Party finds it necessary to transfer technical data other than that specified in paragraph 1 above, in carrying out its responsibilities under this Implementing Agreement that are proprietary, and for which protection is to be maintained, such technical data will be marked with a notice indicating that it shall be used and

disclosed by the receiving Party and its contractors and subcontractors only for the purposes of fulfilling the receiving Party's responsibilities under this Implementing Agreement, and that the technical data shall not be disclosed or retransferred to any other entity without prior written permission of the furnishing Party. The receiving Party agrees to abide by the terms of the notice, and to protect any such marked technical data from unauthorized use and disclosure.

3. In the event a Party finds it necessary to transfer technical data and goods in carrying out its responsibilities under this Implementing Agreement that are export-controlled, and for which protection is desired, the furnishing Party will mark such technical data with a notice and identify such goods. The notice or identification will indicate that such technical data and goods will be used and such technical data will be disclosed by the receiving Party and its contractors and subcontractors only for the purposes of fulfilling the receiving Party's responsibilities under this Implementing Agreement. The notice or identification will also provide that such technical data will not be disclosed, and such technical data and goods will not be retransferred, to any other entity without prior written permission of the furnishing Party. The Parties will abide by the terms of the notice or identification and will protect any such marked technical data and identified goods.

4. The Parties are under no obligation to protect any unmarked technical data or unidentified goods.

[7] ARTICLE XI: CUSTOMS AND IMMIGRATION

1. Each Party will facilitate the movement of persons and goods necessary to implement this Implementing Agreement into and out of its territory, subject to its laws and regulations. The RSA will take steps to expedite such movement of persons and goods to launch facilities it will utilize to fulfill its obligations under this Implementing Agreement.

2. Subject to its laws and regulations, each Party will facilitate provision of the appropriate entry and residence documentation for the other Party's nationals and families of nationals who enter, exit, or reside within its territory in order to carry out the activities under this implementing Agreement. The RSA will take steps to arrange for such provision for such activities at launch facilities it will utilize to fulfill its obligations under this Implementing Agreement.

3. The Parties agree to arrange for free customs clearance for entrances to, and exits from, their respective countries for equipment required for implementation of the activities described herein. The RSA will take steps to arrange for such clearances to and from launch facilities it will utilize to fulfill its obligations under this Implementing Agreement.

ARTICLE XII: SETTLEMENT OF DISPUTES

1. The Parties will consult promptly with each other on all issues involving interpretation or implementation of this Implementing Agreement. In the case of a continuing dispute, such matters will first be referred to the Points of Contact identified in Annex 1.

2. Any matter which has not been settled in accordance with the above paragraph will be referred to the NASA Associate Administrator for Space Flight and the First Deputy of the General Director of the RSA, or their designees, for resolution. Issues not resolved at this level will be referred to the NASA Administrator and the RSA General Director.

ARTICLE XIII: DURATION OF IMPLEMENTING AGREEMENT

1. This Implementing Agreement will terminate five (5) years following its entry into force or upon completion of all activities covered by this Implementing Agreement, whichever comes first. This Implementing Agreement may be extended or amended by written agreement of the Parties.

[8] 2. Either Party may terminate this Implementing Agreement upon six months written notice to the other Party. Termination of this Implementing Agreement shall not affect the Parties' continuing obligations under Articles VII, VIII and X, unless otherwise agreed to by the Parties.

ARTICLE XIV: ENTRY INTO FORCE

This Implementing Agreement will enter into force upon an exchange of diplomatic notes between the Governments of the United States of America and the Russian Federation confirming acceptance of its terms and that all necessary legal requirements for entry into force have been fulfilled.

IN WITNESS WHEREOF the undersigned, being duly authorized by their respective Governments, have signed this Implementing Agreement.

Done at Moscow, in duplicate, this 5th day of October, 1992, in Russian and English languages, both texts being equally authentic.

Daniel S. Goldin
FOR THE NATIONAL AERONAUTICS
AND SPACE ADMINISTRATION
OF THE UNITED STATES OF AMERICA

Yuri Koptev
FOR THE RUSSIAN SPACE AGENCY
OF THE RUSSIAN FEDERATION

Document I-54

[1]   THE WHITE HOUSE

Office of the Vice President

September 2, 1993

United States-Russian Joint Commission on Energy and Space

Joint Statement on Cooperation in Space

Having reviewed the status of the agreement between the United States of America and the Russian Federation Concerning Cooperation in the Exploration and Use of Outer Space for Peaceful Purposes dated June 17, 1992, the Parties note with satisfaction past agreement on the following: the flight of a Russian cosmonaut on the Space Shuttle System in 1993 and 1994, and American astronauts on the MIR station, the docking and a joint flight of these two space complexes in 1995. These activities are consistent with the national space programs of both countries and the overall development of a spirit of trust, partnership, and long-term political and scientific and technological cooperation between Russia and the United States.

Based on the agreement reached at a meeting of the U.S. and Russian Presidents in Vancouver on April 3-4, 1993 and June 17, 1992, the Parties see great promise and mutual benefit through cooperation in space science and exploration activities.

Given the particular importance for Russia and the U.S. of their respective efforts in developing a new generation of orbital stations for scientific and technological progress and human activities in space, the Parties regard further cooperation in this area as most important, and consistent with the interests of both Russia and the U.S., as well as the broader international community.

With this in mind it is the intent of the U.S. and Russia to undertake a cooperative human space flight program. Interim investigation has already indicated potential advantages of joint cooperative activities in a truly international space station program. The Parties intend to pursue such cooperation in accordance with the following principles:

-   joining on a mutually beneficial basis the resources and the scientific, technological, and industrial potentials of Russia and the U.S. in space activities to carry out a large-scale program of scientific, technical, and technological research;
-   working with each of our current partners, and in accordance with earlier international obligations assumed by each of the Parties under the Freedom and MIR projects;
[2] -   operating in an orbit which is accessible by both U.S. and Russian resources;
-   utilizing compatible service systems, enhancing reliability of the station and increasing the flexibility of transportation and technical maintenance;
-   performing activities under cooperative programs on mutually beneficial terms, and including on a contract basis the procurement of individual systems and units or the provision of services.

The first phase of our joint programs begins immediately and is designed to form a basis for resolution of engineering and technical problems. This initial phase encompasses an expansion of our bilateral program involving the U.S. Space Shuttle and the Russian MIR Space Station. The MIR will be made available for U.S. experiments for up to two years of total U.S. astronaut stay time. The number of Space Shuttle flights and the length of crew stay time will depend upon the details of the experiments to be defined by November 1, 1993. During phase one, the use of the Russian modules "Priroda" and "Spektr," equipped with U.S. experiments, could undertake a wide-scale research program. These missions will provide valuable in-orbit experience in rendezvous, docking, and joint space-based research in life sciences, microgravity, and Earth resources. It will bring to reality performance of large-scale space operations in the future. The Parties consider it is reasonable to initiate in 1993 the joint development of a solar dynamic power system with a test flight on the Space Shuttle and MIR in 1996, the joint development of environmental control and life support systems, and the joint development of a common space suit.

Subsequent joint efforts on the second phase will be directed to the use of a Russian MIR module of the next generation, in conjunction with a U.S. laboratory module and the U.S. Space Shuttle. This facility would provide an interim human-tended space science capability where significant scientific experimentation can take place in a microgravity environment and also provide practical experience gained out of the use of different transportation systems (including the U.S. Space Shuttle and the Russian Proton), performance of complex construction and assembly efforts and command and control process of orbital structure of considerable complexity. Successful implementation of this phase could constitute a key element of a truly international space station.

It is envisioned that the U.S. will provide compensation to Russia for services to be provided during phase one in the amount of $100 million dollars in FY 1994. Additional funding of $300 million dollars, for compensation of phase one and for mutually [3] agreed upon phase two activities, will be provided through 1991. This funding and appropriate agreements will be confirmed and signed by no later than November 1, 1993. Other forms of mutual cooperation and compensation will be considered as appropriate.

All the above programs are mutually connected and are considered as a single package, the main goal of which is to create an effective scientific research complex earlier and with less cost than if done separately. The Parties are convinced that a unified Space Station can offer significant advantages to all concerned, including current U.S. partners, Canada, Europe, and Japan.

The precise planning process and organization of drafted phases of joint activity will give the opportunity to benefit both countries through expanded cooperative efforts on the space station project.

The Parties hereby instruct NASA and RSA, in pursuance of this Joint Statement, to develop by November 1, 1993, a detailed plan of activities for an international space station. This will serve as the basis for early review and decision within each government and as the basis for consultations with the international partners. Upon conclusion of the process of government approval and consultation, appropriate implementing agreements will be signed. NASA and RSA will include within the plan overall configuration, volumes, and forms of contributions and mutual compensation for Russian and U.S. activities.

Document I-55

[1]

Protocol to the Implementing Agreement Between the National Aeronautics and Space Administration of the United States of America and the Russian Space Agency of the Russian Federation on Space Flight Cooperation of October 5, 1992

PREAMBLE

The National Aeronautics and Space Administration (hereafter referred to as "NASA") and the Russian Space Agency (hereafter referred to as "RSA"), jointly referred to as "the Parties";

Consistent with the Joint Statement on Cooperation in Space issued by Vice President Gore and Prime Minister Chernomyrdin on September 2, 1993; desiring to broaden the scope of the Implementing Agreement of October 5, 1992, on Human Space Flight Cooperation (hereinafter the October 5, 1992 Agreement) to encompass an expanded program of activities for cooperation involving the Russian Mir-1 Space Station and the U.S. Space Shuttle Program;

Having decided that the enhanced cooperative program will consist of a number of inter-related projects in two phases;

Having determined that Phase One will include those activities described in the October 5, 1992, Agreement and known as the Shuttle-Mir Program, including the exchange of the Russian Mir-1 crew and crew member participation in joint mission science, as well as additional astronaut flights, Space Shuttle dockings with Mir-1, and other activities;

Having further determined that Phase Two of the enhanced cooperative program will involve use of a Russian Mir module of the next generation mated with a U.S. laboratory module operated on a human-tended basis in conjunction with the Space Shuttle, operating in a 51.6 degree orbit which is accessible by both U.S. and Russian resources, to perform precursor activities for future space station-related activities of each Party, with launch to occur in 1997; and

Intending that activities in Phase Two would be effected through subsequent specific agreements between the Parties.

Have agreed as follows:

[2] ARTICLE I: DESCRIPTION OF ADDITIONAL ACTIVITIES

1. This Protocol forms an integral part of the October 5, 1992 Agreement.

2. An additional Russian cosmonaut flight on the Space Shuttle will take place in 1995. The back-up cosmonaut currently in training at NASA's Johnson Space Center will be the primary cosmonaut for that flight, with the STS-60 primary cosmonaut acting as back-up. During this mission, the Shuttle will perform a rendezvous with the Mir-1 Space Station and will approach to a safe distance, as determined by the Flight Operations and Systems Integration Joint Working Group established pursuant to the October 5, 1992 Agreement.

3. The Space Shuttle will rendezvous and dock with Mir-1 in October-November 1995, and, if necessary, the crew will include Russian cosmonauts. Mir-1 equipment, including power supply and life support system elements, will also be carried. The crew will return on the same Space Shuttle mission. This mission will include activities on Mir-1 and possible extravehicular activities to upgrade solar arrays. The extravehicular activities may involve astronauts of other international partners of the Parties.

4. NASA-designated astronauts will fly on the Mir-1 space station for an additional 21 months for a Phase One total of two years. This will include at least four astronaut flights. Additional flights will be by mutual agreement.

5. The Space Shuttle will dock with Mir-1 up to ten times. The Shuttle flights will be used for crew exchange, technological experiments, logistics or sample return. Some of those flights will be dedicated to resources and equipment necessary for life extension of Mir-1. For schedule adjustments of less than two weeks, both sides agree to attempt to accommodate such adjustments without impacting the overall schedule of flights. Schedule adjustments of greater than two weeks will be resolved on a case-by-case basis through consultations between NASA and RSA.

6. A specific program of technological and scientific research, including utilization of the Mir-1 Spektr and Priroda modules, equipped with U.S. experiments, to undertake a wide-scale research program, will be developed by the Mission Science Joint Working Group established pursuant to the October 5, 1992 Agreement. The activities carried out in this program will expand ongoing research in biotechnology, materials sciences, biomedical sciences, Earth observations and technology.

7. Technology and engineering demonstrations applicable to future space station activities will be defined. Potential areas include but are not limited to: automated rendezvous and docking, electrical power systems, life support, command and [3] control,

microgravity isolation system, and data management and collection. Joint crew operations will be examined as well.

8. The Parties consider it reasonable to initiate in 1993 the joint development of a solar dynamic power system with a test flight on the Space Shuttle and Mir in 1996, the joint development of spacecraft environmental control and life support systems, and the joint development of a common space suit.

9. The Parties will initiate a joint crew medical support program for the benefit of both sides' crew members, including the development of common standards, requirements, procedures, databases, and countermeasures. Supporting ground systems may also be jointly operated, including telemedicine links and other activities.

10. The Space Shuttle will support the above activities, including launch and return transportation of hardware, material, and crew members. The Shuttle may also support extravehicular and other space activities.

11. Consistent with U.S. law, and subject to the availability of appropriated funds, NASA will provide both compensation to the RSA for services to be provided during Phase One in the amount of US $100 million in FY 1994, and additional funding of US $300 million for compensation of Phase One and for mutually-agreed upon Phase Two activities will be provided through 1997. This funding will take place through subsequent NASA-RSA and/or through industry-to-industry arrangements. Reimbursable activities covered by the above arrangements and described in paragraphs 3-8 will proceed after these arrangements are in place and after this Protocol enters into force in accordance with Article III. Specific Phase One activities, schedules and financial plans will be included in separate documents.

12. Implementation decisions on each part of this program will be based on the cost of each part of the program, relative benefits to each Party, and relationship to future space station activities of the Parties.

13. The additional activities will not interfere with or otherwise affect any existing, independent obligations either Party may have to other international partners.

ARTICLE II: JOINT IMPLEMENTATION TEAMS

The coordination and implementation of the activities described herein will be conducted through the Joint Working Groups established pursuant to the October 5, 1992 Agreement or such other joint bodies as may be established by mutual agreement.

[4] ARTICLE III: ENTRY INTO FORCE

This Protocol will enter into force upon an exchange of diplomatic notes between the Governments of the United States of America and the Russian Federation confirming acceptance of its terms and that all necessary legal requirements for entry into force have been fulfilled.

IN WITNESS WHEREOF the undersigned, being duly authorized by their respective Governments, have signed this Protocol. Done at Moscow, in duplicate, this sixteenth day of December, 1993, in the English and Russian languages, both texts being equally authentic.

FOR THE NATIONAL AERONAUTICS
AND SPACE ADMINISTRATION
OF THE UNITED STATES OF AMERICA:
Daniel S. Goldin

FOR THE RUSSIAN SPACE
AGENCY
OF THE RUSSIAN FEDERATION:
Yuri Koptev

Chapter Two

Invitation to Struggle:
The History of Civilian-Military
Relations in Space

by Dwayne A. Day

The history of American civilian and military cooperation in space is one of competing interests, priorities, and justifications at the upper policy levels, combined with a remarkable degree of cooperation and coordination at virtually all operational levels. It is a history of the evolution of responsibility for space exploration. Both the Eisenhower and Kennedy administrations gradually decided which organization should be responsible for which activities, eventually establishing these responsibilities as fact. This process did not result in a smooth transition; first the Army and then the Air Force saw its hopes for assuming the predominant role in space exploration subsumed to larger national priorities. It proved to be most painful for the Air Force, which had the biggest dreams for space and saw them dashed as NASA achieved all of the glory during the Cold War space race.

This history can be separated into two broad eras—cooperation prior to NASA's creation and cooperation between NASA and the Department of Defense (DOD), with a transition period in between. This transition is an aspect that is frequently overlooked in discussions of the subject, for civil-military cooperation in space did not begin with the establishment of NASA—it *changed* with the creation of NASA, and it did so dramatically. Prior to NASA's establishment, the military had had the upper hand in determining *all* space priorities, and civilian interests, when considered at all, were clearly secondary. There were also multiple military space actors—primarily the Air Force and the Army—and it was not clear which would emerge dominant. After NASA was created, the Army space program largely disappeared—being subsumed by NASA. The Air Force became the dominant military space actor and often found itself playing a secondary, supporting role to the civilian program.

This history is also the history of the evolution of an idea—that space exploration, particularly human exploration, should be a civilian pursuit. Throughout history there is ample precedent for both civilians and the military undertaking exploratory missions with government support, but early American plans for human space exploration centered on military missions. Wernher von Braun's wheeled space station and planned trips to the Moon all involved the use of military crews in what were envisioned as essentially military missions. The popular culture of the day echoed this vision, as in B-grade science fiction films such as *Project Moonbase* and *The Conquest of Space*. Also, science fiction and pseudo-news articles depicted a military space force dedicated to conquering the heavens. Human space exploration seemed, at least in much of the popular consciousness, to be a logical evolution of existing military missions and an extension of the idea of military pacification of the frontier. Certainly, this was the view of the uniformed leadership of the Air Force immediately after Sputnik.

Reality was to prove to be more complex and more nuanced than the popular vision, however, in large part because of the desire to make the American space program stand as a positive, peaceful beacon for Western-style democracy. The U.S. Air Force strove to find a military mission for humans in space. It could not. Once the two main reasons to place humans in space—science and prestige—became civilian pursuits, the Air Force, after more than a decade trying, could find no cost-effective reason to place humans in orbit.

The idea that there was no role for military officers in space found resistance within the Air Force, which tried unsuccessfully to portray space as merely an expansion of its current operating realm. Prior to Sputnik, there was only limited enthusiasm within the Air Force for space programs and expenditures. There was a core group of space enthusiasts within the Air Force, but they lacked both authority and resources. After Sputnik, the top brass—particularly the Air Staff—embraced space, with a strong emphasis on human spaceflight. But it did so at precisely the same time that the political wind was shifting, and human spaceflight was determined to be better as a civilian, rather than a military, mission.

This essay also highlights the difference between the civilian and uniformed leadership of the military—particularly in the Air Force. Throughout the 1945-1988 period, both the civilian and uniformed leaders of the Air Force made major decisions concerning space, but most of the major policy decisions were made by the civilian leadership, not those in uniform, who had different priorities, biases, and interests.

Yet one of the important differences to note is that the uniformed officers represent the institutional memory of a military service. Secretaries of DOD, service secretaries, and undersecretaries come and go, making decisions during their reign of which they usually do not have to bear the consequences later. But military officers—particularly mid-ranking officers hoping to make general officer rank—often see the decisions get made, are responsible for implementing them, and then have to live with the consequences as they rise up through the ranks. The result is that uniformed officers may eventually resent decisions made by civilian officials long before their time; this can color their outlook as they rise to leadership positions. There is no better example of this than the Space Shuttle experience, which continues to shape NASA-DOD relations to this day.

Finally, this is a history of the attention to, and ignorance of, the issue of duplication by the civilian and military space programs. Virtually every presidential administration has referred to the "national space program" as if the separate civilian, military, and intelligence space programs were part of a unified whole. This was certainly the intent of the Eisenhower administration. But the creation of NASA itself duplicated missions that were already being addressed by DOD. Other policy decisions, such as giving NASA its own rocket development capability, created further redundancy.

This issue really came to the fore during the Kennedy administration. Secretary of Defense Robert S. McNamara sought to eliminate duplication among the parts of the "national space program," but with only limited success—killing the Dyna-Soar space plane while attempting to reduce duplication between DOD and civilian organizations, such as NASA and the Central Intelligence Agency (CIA). However, while he attempted to reduce duplication in certain aspects, McNamara allowed further divergence on rocket development. Finally, perhaps the biggest attempt to eliminate the duplication of functions—the Space Shuttle—failed spectacularly at that task and made the future convergence of military and civilian functions all the more difficult.

The First Era—Pre-NASA

The true genesis of the U.S. military space program predates Sputnik and even predates the well-known V-2 rocket research at White Sands at the end of World War II. American military rocket research began at the Guggenheim Aeronautical Laboratory at the California Institute of Technology (GALCIT), under Frank Malina, Hsue-shen Tsien, and others in the late 1930s and early 1940s.[1] Malina and Tsien speculated about the possibilities of ballistic missiles at GALCIT, an Army laboratory renamed the Jet Propulsion Laboratory (JPL) in 1943. But the U.S. military chose not to follow the German path of investing heavily in an immature technology with only limited immediate payoff. Instead, the military focused research on the development of a much more promising weapon, the atomic bomb.[2] As a result, U.S. rocket research during the war centered on more immediate and practical, if rather mundane, applications, such as short-range rocket projectiles and the misnamed jet-assisted takeoff (JATO) rockets for heavily laden aircraft.

In the immediate post-war years, the U.S. military conducted extensive research with captured German rocket technology. It was during this time that a precedent was established that would have a significant impact a decade later. Colonel Holger Toftoy, chief of the Army Ordnance Enemy Equipment Intelligence Section, had acquired the parts and documentation to assemble more than 100 captured V-2 rockets. Toftoy invited scientists from various organizations to participate in V-2 launches by providing test payloads and instrumentation for everything from upper atmosphere research to radio and radar propagation experiments.[3] The field of rocketry was so new that basic research was a high priority and the involvement of scientific groups was only natural. Out of this emerged the precedent for civilian government scientists to provide scientific payloads for military rockets, and indeed this was the genesis of a U.S. space science community.

Close military-civilian cooperation in basic research in many fields was a result of World War II, and a number of government-university research centers evolved. In the aviation field, the military already had a long track record of working with the civilian National Advisory Committee for Aeronautics (NACA). The military—primarily the U.S. Air Force—conducted a large number of aeronautics test and development projects with NACA throughout the 1950s.

It was from this early cooperation on space and aeronautics-related research that the NASA-military relationship was to expand and evolve. But early American proposals for the development of satellites and rockets were entirely military in nature.

---

1.   An early GALCIT report can be found as Document I-12 in Volume I of this series. See John M. Logsdon, gen. ed., with Linda J. Lear, Jannelle Warren-Findley, Ray A. Williamson, and Dwayne A. Day, *Exploring the Unknown: Selected Documents in the History of the U.S. Civil Space Program, Volume I: Organizing for Exploration* (Washington, DC: NASA Special Publication (SP)-4407, 1995), 1: 153-76.

2.   For a discussion of the limited military utility and tremendous drain on German resources of the V-2, see Michael J. Neufeld, *The Rocket and the Reich: Peenemünde and the Coming of the Ballistic Missile Era* (New York: Free Press, 1995).

3.   David H. DeVorkin, *Science With a Vengeance: How the U.S. Military Created Space Sciences After World War II* (New York: Springer-Verlag, 1992), pp. 59-61. See also Homer E. Newell, *Beyond the Atmosphere: The Early Years of Space Science* (Washington, DC: NASA SP-4211, 1980).

The Air Force and Army Space Studies

In May 1945, German rocket expert Wernher von Braun, who was brought to the United States after the war, prepared a report for the U.S. Army discussing the potential of Earth-orbiting satellites. In October, the U.S. Navy proposed its own satellite. In November, Army Air Force General H. H. "Hap" Arnold declared that a spaceship was entirely "practicable today."[4]

On April 9, 1946, the Army-Navy Aeronautical Board discussed the subject and decided to reconsider it a month later on May 14. Immediately after the first meeting, Major General Curtis E. LeMay, Director of Research and Development of the Army Air Forces, decided to commission an independent study of the issue. It was to be a three-week crash effort to return a report before the second Aeronautical Board meeting, apparently with the intention of securing this new field for the Army Air Forces.

Project RAND, a division of Douglas Aircraft Company's Santa Monica research laboratories, which had been established to serve as a "think tank" for the Army Air Forces, was given the responsibility for the satellite study. The result was the report titled "Preliminary Design for an Experimental World Circling Spaceship," issued on May 2, 1946. This was RAND's first study. In 324 pages, it concluded that it was entirely possible, using existing technology, to develop a satellite system, although the payload would be limited to less than 2,000 pounds. The satellite could be used to gather scientific information, as well as to conduct weather reconnaissance, weapons delivery, attack assessment, communications, and "observation." The report further noted that "the satellite offers an observation aircraft which cannot be brought down by an enemy who has not mastered similar techniques."[5]

If LeMay's concern had been to maneuver the Navy out of the satellite business, his tactic apparently worked, for Navy efforts soon disappeared. However, while the first study had concluded that a satellite vehicle was practical, it failed to create any great enthusiasm for it in the Army Air Forces, which did not want to ignore the possibilities of satellites—particularly for satellite reconnaissance—but was unwilling to pursue it in any meaningful way. The Army Air Forces ordered a second study, and RAND produced a series of documents on the subject during the winter of 1946-1947. One document noted that a satellite in polar orbit would be ideal for scanning the oceans for ships. Another noted that a satellite equipped with television equipment and one or more cameras could be used for reconnaissance. In September 1947, the Air Staff of the newly formed Air Force ordered the Air Materiel Command to evaluate RAND's studies. The Air Materiel Command returned a cautious report noting that the practicality of such systems was questionable and recommended a further study to establish Air Force requirements.[6]

In January 1948, General Hoyt S. Vandenberg, Vice Chief of Staff of the newly created U.S. Air Force, signed a "Statement of Policy for a Satellite Vehicle." This statement declared that the Air Force "as the Service dealing primarily with air weapons—especially strategic—has logical responsibility for the Satellite." The document also stated that the technology was immature and that a development decision lay some time in the future.

---

4.   R. Cargill Hall, "Early U.S. Satellite Proposals," *Technology and Culture* 4 (Fall 1963): 410-34. See also R. Cargill Hall, "Earth Satellites: A First Look by the United States Navy," in R. Cargill Hall, ed., *History of Rocketry and Astronautics: Proceedings of the Third through the Sixth History Symposia of the International Academy of Astronautics* (San Diego, CA: Univelt, Inc., 1986), AAS History Series, Vol. 7, Part II, pp. 253-78.

5.   Document II-2 in Logsdon, gen. ed., *Exploring the Unknown*, 1: 236-45.

6.   Merton E. Davies and William R. Harris, *RAND's Role in the Evolution of Balloon and Satellite Observation Systems and Related U.S. Space Technology* (Santa Monica, CA: The RAND Corporation, 1988), p. 15.

Until that time, the issue would be studied "with a view to keeping an optimum design abreast of the art, to determine the military worth of the vehicle—considering its utility and probable cost—to insure [sic] development in critical components, if indicated, and to recommend initiation of the development phases of the project at the proper time." [II-1]

With a very clearly stated position on the matter, the Air Force asked RAND in February 1948 to conduct further studies on the satellite. RAND contracted with several other organizations, including North American Aviation, the Radio Corporation of America (RCA), the Ohio University Research Foundation, and Boston University. This was a classic early Cold War research effort, uniting government, industry, and academia. By 1950, RAND's research was bearing fruit; in November, the Air Force Directorate of Intelligence recommended that further research and development was justified.[7]

The primary use envisioned for a satellite was reconnaissance. In February 1951, Colonel Bernard A. Schriever organized a conference during which he established several criteria for a satellite reconnaissance system. Early the next month, tests were conducted using television cameras to establish further baselines for these criteria. In April 1951, RAND released two further reports. The first, *Feasibility of Weather Reconnaissance from a Satellite Vehicle*, examined the requirements and value of weather forecasting from space. In particular, such a system enabled weather reconnaissance behind enemy lines, something crucial to strategic bombing campaigns. The second study was *Utility of a Satellite Vehicle for Reconnaissance*.[8]

This study led to yet another study, which eventually became known as Project Feed Back; it was presented to the Air Force in 1954. The report demonstrated that a space reconnaissance satellite was feasible, and it outlined the steps to develop it. In December 1948, the "first report" of the Secretary of Defense stated:

> *The Earth Satellite Vehicle Program, which was being carried out independently by each military service, was assigned to the Committee on Guided Missiles for coordination. To provide an integrated program with resultant elimination of duplication, the committee recommended that current efforts in this field be limited to studies and component designs; well-defined areas of such research have been allocated to each of the three military departments.*[9]

This statement seems to have been an anomaly, because the three services continued their individual studies on their own. Why it was written remains unknown. The Air Force's clearly stated claim on the satellite mission may have prompted it. But after the publication of the report, nothing changed—there was no centralization of the satellite mission, and the services continued their separate low-level studies. The report apparently was completely overlooked.

In the meantime, others in the civilian world had been working on different satellite ideas. During a spring 1950 meeting at scientist James A. Van Allen's home, the prospect of an International Geophysical Year (IGY) was discussed. S. Fred Singer, a physicist at the University of Maryland, proposed building a satellite for the IGY. Singer later proposed a Minimum Orbital Unmanned Satellite of the Earth (MOUSE) at the fourth Congress of the International Astronautical Federation in Zurich, Switzerland, in 1953.[10] Singer's

---

7.   *Ibid.*, pp. 17-19.
8.   *Ibid.*, pp. 23-30.
9.   Office of the Secretary of Defense, *First Report of the Department of Defense* (Washington, DC: Office of the Secretary of Defense, December 1948), p. 129.
10.   Document II-11 in Logsdon, gen. ed., *Exploring the Unknown*, 1: 314-24.

paper was based on a study prepared two years earlier by members of the British Interplanetary Society.

On June 23, 1954, Frederick C. Durant III, former president of the American Rocket Society and then president of the International Astronautical Federation, called Wernher von Braun at the Redstone Arsenal and invited him to a meeting two days later in Washington, D.C., at the Office of Naval Research, which had been involved in the earlier V-2 upper atmosphere experiments. At this meeting, plans were discussed for developing a satellite program using already existing rocket components. Further meetings followed at which the Army gave tentative approval, provided that the cost was not too great and the plan did not interfere with missile development. Von Braun's secret report, *A Minimum Satellite Vehicle: Based on Components available from missile developments of the Army Ordnance Corps*, was submitted to the Army.[11] It summarized what he had said at earlier meetings. The Air Force's declaration six years before that it was responsible for satellite development was either unknown or ignored by the Army.[12]

Sometime in 1952, President Truman discussed the satellite issue with his personal physician, Brigadier General Wallace Graham. Graham persuaded Truman to commission a study from Aristid Grosse, a chemical engineer who had worked on some military projects. Grosse conducted extensive discussions with Wernher von Braun. He delivered his rather slim report not to Truman, but to the Eisenhower administration.[13] Despite years of research on the subject, the space issue never reached the upper levels of the Truman White House.[14] There was no Truman space policy, and space issues remained largely the realm of a small group of engineers and analysts.

However, to say that the Grosse report had no effect is to overlook one key fact: although not delivered to the administration for which it was intended, it was delivered to the new Assistant Secretary of Defense for Research and Development, Donald A. Quarles. In the Eisenhower administration, Quarles was to play a major role in establishing the American space program.

The Killian Report

In September 1954, the Science Advisory Committee of the Office of Defense Mobilization, under orders from President Eisenhower, began a study of the problem of surprise attack.[15] One of the major reasons behind this study was the surprises the Soviet Union had achieved in regard to atomic weapon development. The main task of the committee was "obtaining before it is launched more adequate foreknowledge of a surprise attack, should one be planned, obtaining better knowledge of enemy capabilities."

This special group was headed by Massachusetts Institute of Technology (MIT) President James R. Killian, who later became Eisenhower's science advisor. The group became known as the Technological Capabilities Panel, and it issued its report, titled "Meeting the Threat of Surprise Attack," on February 14, 1955. Eisenhower and others often referred to this document as the "Killian Report."

---

11.   Document II-7 in *ibid.*, 1: 274-81.

12.   The earlier Air Force declaration was also apparently more of an internal document intended to authorize further Air Force studies of the issue rather than an external statement of policy; *ibid.*

13.   Document II-5 in *ibid.*, 1: 266-69.

14.   Rip Bulkeley, *The Sputniks Crisis and Early United States Space Policy* (Bloomington: Indiana University Press, 1991), p. 83.

15.   J.R. Killian, Jr., to General Curtis E. LeMay, September 2, 1954, Papers of Curtis LeMay, Box 205, Folder B-39356, Manuscript Division, Library of Congress, Washington, DC.

During the course of deliberations, the intelligence panel, headed by Polaroid's Din Land, became aware of two advanced proposals for intelligence collection. One was the nuclear-powered reconnaissance satellite using a television camera outlined in the Project Feed Back study. The other idea was for a high-flying strategic reconnaissance aircraft then under consideration by the Air Force. While investigating the latter, Land's panel became aware of a proposal by the Lockheed Skunk Works for its own high-flying strategic reconnaissance aircraft known as the CL-282. They brought this to the attention of President Eisenhower. Unlike the Air Force program, the CL-282 would be configured for strategic reconnaissance prior to hostilities—what was referred to as "pre–D-Day reconnaissance." This was a mission that the Strategic Air Command had previously rejected.

Eisenhower approved the CL-282 in the fall of 1954, and he placed it under the charge of the CIA. It eventually became known as the U-2, and Richard Bissell, a newcomer to the CIA was to manage the program. When the report was issued in the spring of 1955, it apparently never mentioned the aircraft, which was, however, detailed in a classified annex to the report. This was most likely for the "eyes only" of President Eisenhower, and he probably destroyed it along with another classified annex on submarine-launched ballistic missiles.[16]

It was obvious to those involved in the issue that overflight of another nation's territory by such an aircraft would constitute a clear violation of international law and could also be viewed as a hostile act. In fact, such issues were not abstract, because American aircraft flying on the periphery of the Soviet Union were being fired on and even occasionally shot down.

However, the other advanced reconnaissance proposal—a satellite—would fly much higher and would not necessarily violate international law because no clear definition existed of where "airspace" ended and "space" began. Realizing this, Land and the others on his panel decided to attempt to strongly influence the evolution of international law. They proposed that the United States first launch a scientific satellite to establish "Freedom of Space." By doing so, later military and intelligence satellites would be able to overfly Soviet territory following the precedent established by the earlier civilian satellite. The report's recommendation 9.b read:

Freedom of Space. The present possibility of launching a small artificial satellite into an orbit about the earth presents an early opportunity to establish a precedent for distinguishing between "national air" and "international space," a distinction which could be to our advantage at some future date when we might employ larger satellites for intelligence purposes. [II-2]

Land and others considered the reconnaissance satellite to be technologically unrealistic in the near future, but that should not prevent the United States from helping to establish the right to overfly other nations in space. This was best done with a satellite that was nonmilitary in nature.

---

16. Although the intelligence section of the Technological Capabilities Panel report remains classified, awaiting review as of mid-1996, the index has been declassified. It includes the word "satellites," but apparently in the context of satellite countries of the Soviet Union. Those who have seen the report confirm that it mentioned balloon and satellite programs, but it apparently did not mention the U-2 aircraft, except in a separate appendix that Eisenhower most likely destroyed. The information about the separate "eyes only" reports given to Eisenhower is contained in an interview with Killian. Other documents concerning the recommendations of the intelligence committee have also been released. "The Report to the President by the Technological Capabilities Panel of the Science Advisory Committee," February 14, 1955, Office of the Staff Secretary: Records of Paul T. Carroll, Andrew J. Goodpaster, L. Arthur Minnich, and Christopher H. Russell, 1952-61, Subject Series, Alphabetical Subseries, Box 16, "Killian Report—Technological Capabilities Panel (2)," Dwight D. Eisenhower Library, Abilene, Kansas.

The Scientific Satellite Program

In August and September of 1954, Wernher von Braun and his colleagues at the Army Ballistic Missile Agency (ABMA) in Huntsville, Alabama, teamed up with the Office of Naval Research to propose a satellite called *Orbiter*. This was essentially a slight re-work of von Braun's Minimum Unmanned Satellite Vehicle. *Orbiter* was to be a scientific satellite only, essentially mirroring the earlier upper atmosphere research conducted with the V-2 rockets at White Sands. Later in the year, the American Rocket Society prepared a detailed survey of possible scientific and other uses of a satellite and proposed it to the U.S. National Committee for the IGY, a group under the National Academy of Sciences (NAS).[17]

As it was, 1954 proved to be a very important year for the generation of significant ideas concerning scientific and intelligence collection systems. In addition to both the Project Feed Back and the Lockheed CL-282 ideas, the NAS was now considering a scientific satellite as well. These projects were inextricably linked politically.

While the Project Feed Back study and the Killian Report were both highly secret, *Orbiter* was not. The CL-282, in particular, was known to only a handful of people. One person who did know of all three projects, as well as the Technological Capabilities Panel report, was the Assistant Secretary of Defense for Research and Development, Donald Quarles. He was in charge of virtually all defense research projects.

On the same day as the release of the Technological Capabilities Panel report, the U.S. National Committee for the IGY presented a recommendation to National Science Foundation Director Alan T. Waterman at the NAS. The committee recommended that a scientific satellite be launched as part of the IGY.[18] Quarles lobbied Waterman to suggest this idea to the National Security Council (NSC), and four days later, Waterman sent a letter to Deputy Under Secretary of State Robert Murphy, proposing that the United States conduct such a scientific mission.[19]

Four days later, Murphy met with Waterman, NAS President Detlev Bronk, and Lloyd Berkner (who at the time was a member of the U.S. National Committee for the IGY) to discuss the issue. In a letter one month later, Murphy stated that such a proposal would "as a matter of fact, undoubtedly add to the scientific prestige of the United States, and it would have a considerable propaganda value in the cold war."[20] Having gained the concurrence of the Department of State, Waterman then discussed the issue once again with Quarles, who suggested that he consult CIA Director Allen Dulles on how to proceed. Waterman did so and gained Dulles's support for the program. He also spoke with Bureau of the Budget Director Percival Brundage to gain his cooperation when needed. Thus, the proposal now had the support of the Departments of State and Defense, the CIA, and the Bureau of the Budget. Waterman also agreed to formally propose the full program to an executive session of the National Science Board on May 20, and he notified Quarles of these events on May 13, 1955.[21]

---

17. Constance McLaughlin Green and Milton Lomask, *Vanguard: A History* (Washington, DC: Smithsonian Institution Press, 1971), pp. 22-23.

18. Joseph Kaplan, Chairman, United States National Committee, International Geophysical Year 1957-58, National Academy of Sciences, to Dr. A.T. Waterman, Director, National Science Foundation, March 14, 1955, Space Policy Institute Documentary History Collection, Washington, DC.

19. Alan T. Waterman, Director, Memorandum for Mr. Robert Murphy, Deputy Under Secretary of State, 18 March, 1955, Space Policy Institute Documentary History Collection.

20. Robert Murphy, "Memorandum for Dr. Alan T. Waterman, Director, National Science Foundation," April 27, 1955, Space Policy Institute Documentary History Collection.

21. Alan T. Waterman, Director, to Donald A. Quarles, Assistant Secretary of Defense (Research and Development), May 13, 1955, Space Policy Institute Documentary History Collection.

On May 20, 1955, the NSC approved a top-level policy document known as NSC 5520, "Draft Statement of Policy on U.S. Scientific Satellite Program," which stated that the United States should develop a small scientific satellite weighing 5 to 10 pounds.[22] Paragraph number 2 of the document stated (the newly released part is in roman type):

The report of the Technological Capabilities Panel of the President's Science Advisory Committee recommended that intelligence applications warrant *an immediate program leading to a very small satellite in orbit around the earth, and that re-examination should be made of the principles or practices of international law with regard to "Freedom of Space" from the standpoint of recent advances in weapon technology.*

The other major declassified portion of the document (paragraph number 5) stated:

From a military standpoint, the Joint Chiefs of Staff have stated their belief that intelligence applications strongly warrant the construction of a large surveillance satellite. While a small scientific satellite cannot carry surveillance equipment and therefore will have no direct intelligence potential, it does represent a technological step toward the achievement of the large surveillance satellite, and will be helpful to this end so long as the small scientific satellite program does not impede development of the large surveillance satellite.

NSC 5520 also stated (starting at the end of paragraph number 6):

Furthermore, a small scientific satellite will provide a test of the principle of "Freedom of Space." The implications of this principle are being studied within the Executive Branch. However, preliminary studies indicate that there is no obstacle under international law to the launching of such a satellite.
7. It should be emphasized that a satellite would constitute no active military offensive threat to any country over which it might pass. Although a large satellite might conceivably serve to launch a guided missile at a ground target, it will always be a poor choice for the purpose. A bomb could not be dropped from a satellite on a target below, because anything dropped from a satellite would simply continue alongside in the orbit.[23]

Although the document correctly noted the limited utility of satellites as active military offensive threats, this was not the purpose of the program. Also included in NSC 5520 was the clear stipulation that the program was not to interfere in any way with the ballistic missile programs.

Establishing a right of overflight was important, but developing the intercontinental ballistic missile (ICBM) and the intermediate range ballistic missile (IRBM) was considered even more important. Both considerations later established a framework for conducting the program—the U.S. scientific satellite, although developed by the U.S. military, would be handled in such a way as to both seem as disassociated from ballistic missiles as possible and interfere in their development as little as possible.

---

22.   Document II-10 in Logsdon, gen. ed., *Exploring the Unknown,* 1: 308-14.
23.   NSC 5520, May 20, 1955, Record Group 59, General Records of the Department of State: Records Relating to State Department Participation in the Operations Coordinating Board and the National Security Council, 1947-1963, Box 112, "NSC 5520," National Archives and Records Administration,Washington, DC.

Quarles oversaw the selection process that followed. It involved the creation of the Committee on Special Capabilities, headed by Homer Stewart. This committee evaluated the various proposals and rejected the Army's Jupiter rocket for reasons that included its obvious military ties.[24]

It was determined that the scientific satellite program should look as nonmilitary as possible—a rocket vehicle that was not the direct development of a ballistic missile was considered the best way to do this. The result was the selection of the Navy's Vanguard rocket, which had its genesis in a pure research program and would be developed virtually from the ground up as a space vehicle.[25]

At the time, there was no clear distinction made between military and civilian space exploration. The military was to bear responsibility for launching all U.S. payloads. The payloads could be either civilian, such as the NAS satellite, or military, such as the Project Feed Back satellite, but all would fly on military rockets. Meanwhile, a distinction was made among different degrees of what can only be labeled "militaristic" involvement. The Vanguard rocket, although developed by the Navy, had no direct connections to a weapons system. It was therefore a better choice politically to peacefully establish the right to overfly foreign territory. Fundamental to Eisenhower's philosophy at this time was not to inflame the superpower rivalry unnecessarily. Keeping the rocket program as far away from weapons development was an outgrowth of this attitude.

Lukewarm Military Enthusiasm for Space

The Air Force had made only a half-hearted effort at submitting a proposal for the scientific satellite program. At the time, the program was apparently too uninteresting to garner top-level Air Force support. General Bernard A. Schriever, commander of the Western Development Division and head of U.S. ICBM development, thought that the Air Force should concentrate on the military satellite instead. On March 16, 1955, the Air Force issued General Operational Requirement No. 80. Up until this time, the approval for

---

24.   Charles A. Lindbergh wrote in the foreword to the most detailed book on the Vanguard program (Green and Lomask, *Vanguard: A History*), which was written before the recent revelations on the origins of the program, stated on page vi:

Why was the Redstone-von Braun satellite project not supported? Answers vary with the person talked to: The Navy's brilliant developments in satellite instrumentation had tipped the choice to Vanguard, and budgetary restrictions had prevented a paralleling project. The name Redstone was too closely associated with military missiles. Vanguard offered lower costs, more growth potential, longer duration of orbiting. We would eventually gain more scientific information through Vanguard than through Redstone. To these observations, I can add from my own experience that inter-service rivalry exerted strong influence; also, that any conclusion drawn would be incomplete without taking into account the antagonism still existing toward von Braun and his co-workers because of their service on the German side of World War II.

25.   Eisenhower's staff secretary, then-Colonel Andrew Goodpaster, provided some insight into the Huntsville Germans' views and their lobbying on behalf of their work in a memorandum written in the summer of 1956:

On May 28th Secretary [Deputy Secretary of State] Hoover called me over to mention a report he had received from a former associate in the engineering and development field regarding the earth satellite project. The best estimate is that the present project would not be ready until the end of '57 at the earliest, and probably well into '58. Redstone had a project well advanced when the new one was set up. At minimal expense ($2-$5 million) they could have a satellite ready for firing by the end of 1956 or January 1957. The Redstone project is one essentially of German scientists and it is American envy of them that has led to a duplicative project.

I spoke to the President about this to see what would be the best way to act on the matter. He asked me to talk to Secretary Wilson. In the latter's absence, I talked to Secretary Robertson today and he said he would go into the matter fully and carefully to try to ascertain the facts. In order to establish the substance of this report, I told him it came through Mr. Hoover (Mr. Hoover had said I might do so if I felt it necessary).

Quoted in Colonel A.J. Goodpaster, "Memorandum for Record," June 7, 1956, White House Office, Office of the Staff Secretary: Records, 1952-1961, Box 6, "Missiles and Satellites," Eisenhower Library.

further satellite studies had been from a low level of the Air Force bureaucracy; now the go-ahead came from the top. On April 2, 1956, Schriever and General Thomas Power, the commander of the Air Research and Development Command, approved a full-scale development plan for what was called "Weapons System 117L" (WS-117L), a reconnaissance satellite program. It would utilize an Atlas launch vehicle and was to be fully operational by 1963. Air Force headquarters approved the plan on July 24, 1956, and allocated $3 million. This proved to be a major disappointment to all involved, because it was less than 10 percent as much as was needed to go to full-scale development.[26]

The Air Force, as a young organization that owed its very existence to modern technology, was also the most logical of the services to embrace new technology such as satellites and long-range rockets. But at the same time, the Air Force was also dominated by the culture of the manned strategic bomber, and any new missions often had to serve this culture. Thus, the concept of strategic rocketry was not one that was adopted readily or without resistance by the Air Force.[27]

The Air Force's strategic bombing emphasis had been one of the main reasons that the Western Development Division had been set up on the west coast instead of the pre-existing development operation at Wright Field in Dayton, Ohio. The satellite program was also more likely to receive the support it needed there than at Wright Field. But Donald Quarles, who had been promoted to Secretary of the Air Force in July 1955, apparently felt that reconnaissance satellites, although a very promising idea, were still a long way from being practical, and he did not provide the money for full-scale development.

In 1956, the Air Force also directed Bell Aircraft Company to conduct a study of a manned boost-glide reconnaissance system known as "Brass Bell." An earlier study, known as BOMI, evolved into a concept known as ROBO, for "rocket bomber." The Air Research and Development Command also issued a system requirement for a hypersonic research and development vehicle known as "Hywards." But the Air Force did not allocate any money to manned space operations in fiscal year 1957.[28]

Similarly, the early RAND studies about the possibilities of space did not receive an enthusiastic response from top leaders of the Air Force. Space was still an expensive and dubious proposition for the Air Force, which was more interested in spending its money on strategic bombers and, to a much lesser extent, the Atlas ICBM. As long as neither the Navy nor the Army was developing a military satellite system, the Air Force did not show much enthusiasm for the various military satellite systems—human and robotic—that it was evaluating. The WS-117L proceeded, and the Air Force even selected Lockheed as the prime contractor for the vehicle. One of the losers in the competition, RCA, then looked elsewhere for an agency to pay it to build a television-equipped satellite. It found the receptive ear of Wernher von Braun. In April 1957, the Army produced the Janus report, which was essentially the RCA bid for the WS-117L.[29]

---

26.   Davies and Harris, *RAND's Role*, pp. 73-74.

27.   See, for instance, Carl Builder, *The Icarus Syndrome* (New Brunswick, NJ: Transaction Books, 1994). To be fair, the manned strategic bomber was a proven technology, whereas ICBMs were not. Furthermore, manned bombers were more flexible, reliable, and accurate than ICBMs would be for a long time. They also were recallable, compared to the "push button" ICBM. As one Air Force historian has noted: "The Air Force's institutional penchant for equating the necessity for a manned bomber to fulfill its primary mission of strategic bombardment, and ensure its continued independence, hindered the incorporation of missile technology. The majority of Air Force leaders believed ballistic missiles should undergo a step by step development, followed by integration into the weapons inventory." Roy F. Houchin, "The Dyna-Soar Program: Why the Air Force Proposed the Dyna-Soar X-20 Program," *Quest: The History of Spaceflight Magazine* 3 (Winter 1994): 10.

28.   *Ibid.*

29.   "Briefing on Army Satellite Program," November 19, 1957, White House Office, Office of the Special Assistant for Science and Technology, Records (James R. Killian and George B. Kistiakowsky, 1957-61), Box 15, "Space (2)," Eisenhower Library.

Thus, various space programs within the military services received support, but primarily only for continued study, not for substantial development. These programs also produced core groups of enthusiasts. Schriever and his people at the Western Development Division in California were the Air Force space enthusiasts. Von Braun and his ABMA team in Huntsville were the Army space enthusiasts. But in the case of the Air Force, the program lacked support from both the top-level career military officers in the Air Staff and the civilian leadership. Schriever had mentioned satellites in a speech in February 1957. The Office of the Secretary of Defense told him not to mention "space" again—this was not a military priority for the administration, and Eisenhower did not want anyone to think it was, particularly when the White House was concerned about peacefully establishing "Freedom of Space."[30] In the case of the Army, the ABMA was specifically forbidden by the White House to develop satellites. The only satellite program that had all the money it needed was the Navy's Vanguard program, and it quickly ran way over its early estimated budget.

Even though von Braun and the Army were officially precluded from developing a satellite, he and his rocket team lacked faith in the Vanguard Project. In the spring of 1956, they lobbied for a reconsideration to allow the Army to attempt to launch a satellite atop a Jupiter-C missile. This proposal was rejected in the summer of 1956.[31]

In late 1956, after the Vanguard Project was well under way and running into cost overruns, one of von Braun's closest associates, Ernst Stuhlinger, made contact with James Van Allen, who in 1950 had shown an interest in a scientific satellite. Stuhlinger informed Van Allen that, although von Braun had been ordered not to place a satellite in orbit with the Jupiter-C, the team had grave doubts about the officially sanctioned Vanguard Project. Stuhlinger discussed possible scientific payloads capable of being carried atop a Jupiter-C. On November 23, 1956, he sent a letter to Van Allen thanking him for the meeting and proposing that Van Allen visit the ABMA to view their operations. Van Allen apparently did.[32]

Van Allen responded on February 13, 1957, with a list of possible scientific payloads. This letter was sent to William Pickering, Director of the Army's Jet Propulsion Laboratory (JPL) of the California Institute of Technology (the renamed GALCIT operation).[33] It was JPL that had begun to build the Explorer I satellite—work that was both clandestine and forbidden at the time. Also, at some point during this period, von Braun's team entertained RCA, reviewing its failed bid for the WS-117L program.

Meanwhile, establishing "Freedom of Space" continued to be an active concern in policy planning circles in Washington; the legal ramifications were being worked out in the State Department and elsewhere. [II-3] Furthermore, Vanguard ran severely over budget. The initial estimate had been $15 to 20 million for the program. By late 1957, the cost was estimated at ten times that amount. Money had to be found in various budgets to pay for it. Budget Director Percival Brundage said: "Apparently, both the Department of Defense and the National Science Foundation are very reluctant to continue to finance

30.   Neufeld, *The Rocket and the Reich*, p. 181.

31.   Dwayne A. Day, "New Revelations About the American Satellite Programme Before Sputnik," *Spaceflight*, November 1994, pp. 372-73.

32.   Ernst Stuhlinger, Director, Research Projects Office, Development Operations Division, Army Ballistic Missile Agency, U.S. Army Ordnance Corps, to Dr. Van Allen, Department of Physics, Iowa State University, November 23, 1956, James A. Van Allen Papers, Special Collections, University of Iowa Library, Iowa City.

33.   J.A. Van Allen, Head, Department of Physics, State University of Iowa, to Dr. Ernst Stuhlinger, Army Ballistic Missile Agency, February 13, 1957, James A. Van Allen Papers.

this project to completion. But each is quite prepared to have the other do so." The two had supplied some supplementary funds to the program, and, surprisingly, even the CIA contributed $2.5 million in funds. [II-4]

Why this was done is unknown. CIA Director Richard Bissell was kept abreast of the developments and may have realized the importance of "Freedom of Space" to future reconnaissance efforts. It is also true that he had a substantial discretionary fund to spend on unforeseen problems. This fund contained around $100 million and was often used to address pressing national security needs. Completing the Vanguard mission of shaping international law was considered a national security issue, and this may have been why CIA money funded part of the U.S. satellite for the IGY. What it certainly does illustrate, however, is the confluence of both civilian and national security interests in the early space program.

By the end of September 1957, the framework of the American space program was pretty much in place. The military was responsible for launching and supporting all satellites. Scientific satellites would be developed and manufactured by civilian scientists, most likely under the auspices of the NAS or at universities. The Army was not officially involved in any space programs. It was, however, actively studying large rocket proposals and also conducting numerous studies of possible satellite payloads.

The Air Force had the WS-117L under way but was underfunding it. In the summer of 1957, a proposal for a faster, interim reconnaissance satellite using film-return techniques was not received enthusiastically by the Air Force. The Air Force was also undertaking the ROBO, Hywards, and Brass Bell studies, but not at a significant level. Overall, the service's commitment to military space programs was weak—both in the Air Staff and in the civilian Office of the Secretary of the Air Force. At the same time, although military space programs had not received much high-level support in either the Air Force or the Army, within each service core groups of officers and scientists had formed—space enthusiasts who constantly advocated for bigger programs.

Under the restrictions of both NSC 5520 and President Eisenhower's conservative spending priorities, space seemed unlikely to become a major enterprise. Even after the scientific satellite had flown and established "Freedom of Space," it was unlikely that things would change substantially for either the Air Force or the Army. Both would have to face the continued fiscal conservatism of the president and the civilian and military leadership at the Pentagon. Sputnik changed all of that.

Turbulent Transition

On October 4, 1957, the Soviet Union launched Sputnik. The launch itself was not a great surprise to U.S. intelligence, which had ample warning that the Soviets were capable of launching a satellite.[34] The public reaction to the launch was greater than the administration expected, despite plenty of warning in various top-level policy documents.[35]

Eisenhower had failed to realize the degree to which a Soviet first in space could undercut his domestic priorities. He attempted to downplay the significance of Sputnik so

---

34. Document II-13 in Logsdon, gen. ed., *Exploring the Unknown*, 1: 329.

35. Document II-5 in *ibid.*, 1: 266-69. General Andrew Goodpaster, Eisenhower's staff secretary, stated that Eisenhower had been warned plenty of times of the propaganda effects of such a satellite launch but had always dismissed them. He also stated that the attitude in the White House was generally dismissive of the Sputnik launch for about 24 hours, before the public and scientific reaction of the country became known. After that, everybody's attitude changed. Interview with General Andrew Goodpaster, March 19, 1996, Washington, DC.

that he could "head off a stampede on the Treasury."[36] But if the public reaction was bad enough after Sputnik, it would soon get much worse. On November 3, the Soviets orbited Sputnik II, which weighed 1,121 pounds and carried the dog Laika. The sophistication and size of this satellite (partly because the upper stage remained attached to the payload) left no doubt in the minds of many that the Soviet Union possessed tremendous superiority in space launchers. The public uproar, and Khrushchev's gloating, took on even bigger dimensions when the Vanguard TV-3 launch—billed as a fully operational vehicle and broadcast on national television at the White House's urgings (and the muted protests of the engineers)—blew up on the launch pad on December 6.

The reaction to Sputnik within the military services was swift and startling—and alarmed even Eisenhower. On October 10, the Air Force rolled its three human space-flight proposals into one and labeled it "Dyna-Soar," for "dynamic soaring." In mid-October, someone leaked information to *Aviation Week* magazine about the WS-117L—including the involvement of Lockheed.[37] On October 26, the Army made a presentation to the Committee on Special Capabilities (which had rejected the Army's earlier scientific satellite proposal), recommending the development of its Janus reconnaissance satellite that would use a television system to photograph the Soviet Union. On December 10, the Air Force created in the Office of the Deputy Chief of Staff/Development a new department called the Directorate of Astronautics.

This enthusiastic response, particularly within the Air Force, came from the career military officers and not the civilian leadership, who shared Eisenhower's skepticism. After objections from Deputy Secretary of Defense Quarles and others, the order establishing the Directorate of Astronautics in the Air Force was revoked only three days after it was issued.[38]

Eisenhower clearly liked none of this. Soon after Sputnik, he admonished his officials not to comment on the issue of whether the United States could have "beaten" the Soviets into space. The reason was that talk about whether or not the Army could have launched a satellite sooner tended to make the matter look like a race, which was exactly what he wanted to avoid.[39]

By sheer coincidence, soon-to-be Secretary of Defense Neil McElroy was having dinner with ABMA Director General John Medaris and Wernher von Braun when the announcement of the Sputnik launch was made. Von Braun immediately pressured McElroy to let the ABMA team launch a satellite into orbit; they received permission on November 8. The ABMA's military leaders apparently had their own satellite in mind for the mission. But von Braun and JPL's leadership had their own, and this was initially named "Deal-1." As in a game of poker, if you are dealt a bad hand—as the country had been dealt with both Sputniks I and II—you fold and tell the person to deal you another. JPL Director William Pickering was able to convince ABMA Director Medaris that their satellite was the better choice.[40] Deal-1 was soon renamed Explorer I, and it was launched into orbit on January 31, 1958.

---

36.   Walter A. McDougall, . . . *The Heavens and the Earth: A Political History of the Space Age* (New York: Basic Books, 1985), p. 146.

37.   "USAF Pushes Pied Piper Space Vehicle," *Aviation Week*, October 14, 1957, p. 26.

38.   Lee Bowen, *The Threshold of Space: The Air Force in the National Space Program, 1945-1959*, USAF Historical Division Liaison Office, September 1960, p. 20.

39.   Brigadier General A.J. Goodpaster, Memorandum of Conference with the President, Office of the Staff Secretary: Records of Carroll, Goodpaster, Minnich, and Russell, 1952-61, Subject Series, Department of Defense Subseries, Box 6, "Missiles and Satellites, Vol. I (3) [September-December 1957]," Eisenhower Library.

40.   William E. Burrows, *Exploring Space* (New York: Random House, 1990), p. 76. The information on the story of the name Deal-1 comes from Dr. Jonathan McDowell at the Harvard-Smithsonian Center for Astrophysics.

The New Military Space Agency

In November, newly appointed Secretary of Defense McElroy proposed centralizing control of the various American space projects then under way, such as Vanguard and the WS-117L, along with advanced ballistic missile development. They would be placed into a Defense Special Projects Agency, which would be responsible for whatever projects the secretary would assign to it. The idea for this agency apparently arose from the President's Science Advisory Council in mid-October, just days after both Sputnik and McElroy's nomination.[41] Eisenhower himself expressed the opinion that a fourth service should be established to handle the "missiles activity."[42] McElroy said that he was thinking about a "Manhattan Project" for anti-ballistic missiles. The president thought that a separate organization might be a good idea for this problem.[43] In testimony before Congress, Quarles, who might easily have been regarded as an Air Force partisan, stated that long-range, surface-to-surface missiles had been assigned to the Air Force because it possessed the targeting and reconnaissance capabilities to use them, not because it was uniquely an Air Force mission.[44] Space could conceivably be treated in the same way.

Killian and the Science Advisory Committee of the Office of Defense Mobilization found McElroy more receptive than his predecessor.[45] On November 7, in a national television address, Eisenhower announced that he was elevating Killian to the position of Special Assistant for Science and Technology and head of the President's Science Advisory Committee. The press quickly labeled Killian the "Missile Czar." By this time, Killian was probably pushing the idea of a separate agency for space as well.[46]

---

41. Goodpaster interview, March 19, 1996.

42. Eisenhower's comments on this subject appear in numerous documents. For instance, in October 1957, Goodpaster reported: "The President went on to say he sometimes wondered whether there should not be a fourth service established to handle the whole missiles activity." Brigadier General A.J. Goodpaster, "Memorandum of Conference with the President, October 11, 1957, 8:30 AM," October 11, 1957, Ann Whitman File, DDE Diary Series, Box 67, "Oct. 57 Staff Notes (2)," Eisenhower Library. In January 1958, Goodpaster reported: "In the course of the discussion the President indicated strongly that he thinks future missiles should be brought into a central organization." Brigadier General A.J. Goodpaster, "Memorandum of Conference with the President, January 21, 1958," January 22, 1958, Office of the Staff Secretary: Records of Carroll, Goodpaster, Minnich, and Russell, 1952-61, Subject Series, Department of Defense Subseries, Box 6, "Missiles and Satellites, Vol. II (1) [January-February 1958]," Eisenhower Library. In February 1958, Goodpaster reported: "The President said that he has come to regret deeply that the missile program was not set up in [the Office of the Secretary of Defense] rather than in any of the services." Brigadier General A.J. Goodpaster, "Memorandum of Conference with the President, February 4, 1958 (following Legislative Leaders meeting)," February 6, 1958, Office of the Staff Secretary: Records of Carroll, Goodpaster, Minnich, and Russell, 1952-61, Subject Series, Department of Defense Subseries, Box 6, "Missiles and Satellites, Vol. II (1) [January-February 1958]," Eisenhower Library.

43. Brigadier General A.J. Goodpaster, "Memorandum of Conference with the President, October 11, 1957," Ann Whitman File, DDE Diary Series, Box 67, "Oct. 57 Staff Notes (2)," Eisenhower Library. In February, another memorandum states: "The President said that he has come to regret deeply that the missile program was not set up in [the Office of the Secretary of Defense] rather than in any of the services. Personal feelings are now so intense that changes are extremely difficult." Brigadier General A.J. Goodpaster, "Memorandum of Conference with the President, February 4, 1958 (Following Legislative Leaders meeting)," February 6, 1958, Office of the Staff Secretary: Records of Carroll, Goodpaster, Minnich, and Russell, 1952-61, Subject Series, Department of Defense Subseries, Box 8, "Missiles and Satellites, A National Integrated Missile and Space Vehicle Development Program, December 10, 1957," Eisenhower Library.

44. Robert Frank Futrell, *Ideas, Concepts, Doctrine, Vol. 1: A History of Basic Thinking in the United States Air Force* (Maxwell AFB, AL: Air University Press, 1989) p. 589. The comments on Quarles's partisanship come from the Goodpaster interview, March 19, 1996.

45. Richard Vernon Damms, "Scientists and Statesmen: President Eisenhower's Science Advisors and National Security Policy, 1953-1961," Ph.D. Diss., Ohio State University, 1993, p. 297.

46. Goodpaster interview, March 19, 1996.

The Defense Special Projects Agency would act as a central authority for all U.S. space programs and would essentially contract out missions to the separate services, civilian government agencies, and even universities and private industry. "Above the level of the three military services," McElroy said, "having its own budget, it would be able to concentrate on the new and the unknown without involvement in immediate requirements and inter service rivalries." McElroy also stated in front of Congress that "the vast weapons systems of the future in our judgment need to be the responsibility of a separate part of the Defense Department."[47] This proposal was placed in a DOD reorganization bill. At this point, it was still assumed that the entire American space program would remain under military control, although at the level of the secretary of defense, in an office specially created to manage it.

On December 6, McElroy received a letter from the Joint Chiefs of Staff stating their opposition to the creation of the Defense Special Projects Agency. They felt that line authority for space programs should remain within the services themselves. Schriever also objected. He wanted an authority that would be able to set policy, but not one that would actually manage programs for astronautics. This, he felt, would duplicate capabilities already within his own organization.[48] McElroy—and, more importantly, Eisenhower—did not agree. This was to be a constant source of contention for the next year and a half.

All of these events apparently were having a cumulative effect on Eisenhower, who was concerned that the military services were less focused on their missions and more interested in grabbing this newly opening frontier as their own turf. For Eisenhower, this was a constant worry. He had always been concerned about the parochialism and turf-building impulses of the military and became convinced that he was seeing it again. A separate military space agency seemed to be the way to avoid it.

A Separate Civilian Agency

At the end of December 1957, Killian drafted a "Memorandum on Organizational Alternatives for Space Research and Development." In it, he argued that the Defense Special Projects Agency was a good idea and should house the DOD space program. In addition, much space-related research and development properly belonged in such an agency. At the same time, however, the scientific community was arguing that purely scientific and nonmilitary aspects of space research should not be under the control of the military. There were two options for addressing this. The first option was to establish a central space laboratory within DOD with a broad charter that included basic space research. The second option was to establish a new civilian space agency formed around NACA.

Although Killian did not specifically recommend one option over the other, he concluded:

The overall plan, then, must keep steadily in view the need for those means and programs which will command the interest and participation of our best scientists. We must have far more than a program which appeals to the "space cadets." It must invoke, in the deepest sense, the attention of our best scientific minds if we as a nation are to become a leader in this field. If we do not achieve this, then other nations will continue to hold the leadership.[49]

---

47. *Organization and Management of Missile Programs*, Hearings before a Subcommittee of the Committee on Government Operations, U.S. House of Representatives, 86th Cong., 2d sess. (Washington, DC: U.S. Government Printing Office, 1959), p. 133.
48. Futrell, *Ideas, Concepts, Doctrine*, 1: 590.
49. Document IV-1 in Logsdon, gen. ed., *Exploring the Unknown*, 1: 628-31.

In January 1958, the Senate began a series of public hearings on the country's space program. They were ostensibly intended to investigate the status of the U.S. missile and space programs and to determine why the United States was apparently so far behind the Soviet Union in space. But Senate Majority Leader Lyndon Johnson also wanted to use them to publicly embarrass Eisenhower.

Before the hearings began, on January 7, 1958, McElroy requested that all three services list their proposed space projects. The ABMA, under von Braun, had an extensive list, such as reconnaissance, meteorology, basic science, and extensive rocket development for space missions, including the delivery of supplies to paratroopers in enemy territory.[50] The Navy was already responsible for the one satellite program that was actually building hardware and was not itself adverse to expanding its slice of the pie.

The Air Force expected to be lead agency in the new space program. The Air Staff by now had ambitious space plans that included reconnaissance, early warning, and hypersonic space planes. It also had expanded its wish list to include nuclear rockets to service lunar bases and soon added a proposal for placing an American in space sooner than the Dyna-Soar schedule would allow. The uniformed Air Force interpreted this request as an indication that not only was it being named lead agency for space, but that its grandiose program was about to be approved.[51] This propensity of the Air Force for thinking big was well known in the White House, and members of the President's Science Advisory Committee felt they had an obligation to "ridicule the occasional wild-blue-yonder proposals by a few Air Force officers for the exploitation of space for military purposes."[52]

At the same time, the Air Force signed several agreements with NACA concerning the Dyna-Soar program (also known as Weapons System 464L). [II-5, II-6, II-7] The Air Force was interested in the strategic bombardment aspects of the program, while NACA was interested in the possible civil applications of such a vehicle. What differentiated these agreements from earlier space cooperation was that both the military and civilian agencies were to cooperate on the development of a space payload, not simply focus individually on the payload or the launch vehicle. The precedent for this cooperation came from the previous Air Force-NACA work on the X-plane series, particularly the challenging X-15 program.

Discussion of the Defense Special Projects Agency continued within the administration. Its name was changed to the Advanced Research Projects Agency (ARPA), and Eisenhower sent a message to Congress on January 7, 1958, requesting supplemental appropriations for the agency.[53] In early January, the newly created President's Science Advisory Committee addressed the issue of ARPA. Other than opposing the placement of advanced ICBM research into a separate agency instead of keeping it with the current ICBM programs, the committee had no objection to ARPA.

On February 4, 1958, during a White House meeting between Eisenhower and key Senate Republicans to discuss legislation currently before Congress, the issue of space came up again. Eisenhower felt that all of the nation's space programs could be adequately housed within DOD, presumably with ARPA in charge. Eisenhower wanted to

50.   "Proposal: A National Integrated Missile and Space Vehicle Development Program," December 10, 1957, Report No. D-R-37, Office of the Staff Secretary: Records of Carroll, Goodpaster, Minnich, and Russell, 1952-61, Subject Series, Department of Defense Subseries, Box 8, "Missiles and Satellites, A National Integrated Missile and Space Vehicle Development Program, December 10, 1957," Eisenhower Library.

51.   Bowen, *The Threshold of Space*, p. 22.

52.   James R. Killian, Jr., *Sputnik, Scientists, and Eisenhower: A Memoir of the First Special Assistant to the President for Science and Technology* (Cambridge, MA: MIT Press, 1977), pp. 296-97.

53.   *Organization and Management of Missile Programs*, Hearings, U.S. House of Representatives, 86th Cong., 2d sess., p. 133.

avoid duplication of effort, and because military space programs were of paramount importance, he saw no need for creating a civilian space agency outside DOD.[54]

Killian expressed some reservations at having the military run the U.S. space program. The interests of civilian scientists were unlikely to be represented in such an organization, and Killian was, after all, a scientist himself. But it was Vice President Richard Nixon who stated that it was important for the United States to have a civilian space program entirely separate from the military. This, Nixon argued, would advance the American position in the world the most.[55]

On February 7, 1958, James Killian and Din Land, who was also a member of the President's Board of Consultants on Foreign Intelligence Activities, met with Eisenhower and his staff secretary, General Andrew Goodpaster. They briefed him on the potential of both a recoverable space capsule and a supersonic reconnaissance aircraft program, suggesting that to speed up the development of a reconnaissance satellite, the United States should pursue the recoverable capsule idea as an "interim" solution. Eisenhower apparently accepted this recommendation at that time.

An equally important result of this first meeting was the decision to finalize Secretary of Defense McElroy's proposal and create ARPA to house highly technical defense research programs. General Electric executive Roy W. Johnson was to serve as its director. Eisenhower decided to give ARPA control of all military space programs. The military "man-in-space" program, meteorological programs, and the WS-117L would all be turned over to ARPA.

During a second conference on February 8 concerning the recoverable satellite program, Eisenhower said "emphatically that he believed the project should be centered in the new Defense space agency, doing what CIA wanted them to do."[56] This was a major shift in the development of the reconnaissance satellite program; not only did it give it top-level approval, it also removed responsibility for the film-return satellite from the Air Force and granted it to the CIA, mirroring the earlier U-2 decision.

The importance of these meetings in early February cannot be overemphasized. In the course of only a few days, Eisenhower had not only taken the entire military space program, particularly the Air Force's ambitious plans, and given it to a newly created DOD agency, but he had also taken a key project in that program and given it to the CIA. Both decisions later had profound effects on the shape of the American military and civilian space programs. In addition, the president had begun to address the issue of creating a separate civilian space agency. This was being heavily discussed in Congress and the press, but until the February 4 meeting, Eisenhower apparently thought that the issue of duplication of effort justified keeping all space research located in DOD, centralized at a level above any of the rival armed services.

A month later, on the same day that the Air Force proposed the approval of a "man-in-space" program, Eisenhower announced his decision to create a separate civilian space agency, with NACA as its core. This was to forever change the nature of civil-military cooperation in the American space program.

---

54.   Eisenhower may have also been swayed by public opinion at the time, which was generally in favor of a separate civilian space program. See Robert Hotz, "NACA, The Logical Space Agency," *Aviation Week,* February 3, 1958, p. 21.

55.   Document IV-2 in Logsdon, gen. ed., *Exploring the Unknown,* 1: 631-32.

56.   "Memorandum of Conferences on 7 and 8 February, 1958," cited in "CORONA Program Profile," Lockheed Press Release, May 1995. This document is apparently the only written record of these meetings and the decision to proceed with CORONA. It is therefore one of the most important documents on the reconnaissance satellite program. Unfortunately, it remains classified in the Eisenhower Library.

Sputnik brought space to the attention of the top military and civilian Air Force leadership. It was suddenly a highly visible and exciting endeavor and one in which top Air Force officers naturally felt that the service should lead. As a result, the dreams of the service's space enthusiasts suddenly received high-level attention. Chief among these was the plan to place a human—an Air Force pilot, no less—in orbit around the Earth.

Sputnik also re-focused attention on Wernher von Braun's rocket team at the ABMA—a highly capable team of engineers who dramatically enhanced their reputation by launching Explorer I. The Army hoped that the ABMA would be the flag-carrier for a significant Army role in space.

However, the ambitious plans of both the Air Force and Army ran headlong into reality—and the civilian leadership of DOD. In February 1958, ARPA was formally created, and the interim reconnaissance satellite program (later called CORONA) was placed under CIA control. ARPA assumed control of the manned ballistic capsule project as well. One by one, the Air Force's other plans were gradually stripped away. The Army's programs did not receive serious support; despite its impressive capabilities, the "ground service" was not considered particularly well-suited to lead the country into space.

Thus, in the immediate post-Sputnik period, the Air Force saw its plans for becoming an "aerospace force" emerge and then quickly vanish—one by one lost to other agencies. In many of the programs that it had conceived and pioneered, it was thus reduced to a support role—almost the same as a contractor. Over the next few months, it became obvious that the projects it did not lose to ARPA would be lost anyway to the new civilian agency.

The one program of which the Air Force did maintain exclusive control was the Dyna-Soar project. This was not simply a consolation prize; it was, in fact, the most important mission to many within the Air Force space community. It had everything that an Air Force space program was expected to have—wings and a human in the cockpit. What it lacked was a clearly defined mission.

The Transition

The National Aeronautics and Space Act of 1958 established a purposely blurry line between NASA and the military space programs. Under the "Declaration of Policy and Purpose," the Space Act states:

The Congress declares that the general welfare and security of the United States require that adequate provision be made for aeronautical and space activities. The Congress further declares that such activities shall be the responsibility of, and shall be directed by, a civilian agency exercising control over aeronautical and space activities sponsored by the United States, except that activities peculiar to or primarily associated with the development of weapons systems, military operations, or the defense of the United States (including the research and development necessary to make effective provision for the defense of the United States) shall be the responsibility of, and shall be directed by, the Department of Defense; and that determination as to which such agency has responsibility for and direction of any such activity shall be made by the President in conformity with section 201 (e).[57]

This was not terribly clear policy guidance, particularly as the entire nature of space exploration and exploitation was still vague and under development. It was also not very clear considering that the entire issue of which organization—ARPA or NASA—

---

57.   Document II-17 in Logsdon, gen. ed., *Exploring the Unknown,* 1: 334-45.

would be responsible for human spaceflight was unresolved. For the time being, the military space program was under the control of ARPA in the Office of the Secretary of Defense. This was not popular with the military services, but it did serve to mitigate turf disputes over the proper location of space programs. Such decisions were made at the national level, and the services on their own were incapable of making significant movement on space programs with ARPA in control of initiating and budgeting programs. The Space Act made it clear that it was up to the president to decide which programs belonged where.

More importantly, the establishment of NASA to conduct scientific experiments in space undercut much of the Air Force's emerging justification for human spaceflight. The Air Force had proposed human spaceflight less for mission reasons than as an extension of aeronautical medicine—to study the reaction of the human body to spaceflight. This was now a mission that NASA was more appropriately suited to accomplish. Furthermore, if people were to be placed in space for prestige reasons, the civilian program was more suitable for this from a propaganda standpoint. The Air Force was thus largely left with the search for a practical reason to put people in space. As robotic systems improved, this practical justification became more and more elusive. Finally, in August 1958, Eisenhower formally assigned the role of human spaceflight to NASA.[58]

Over time, the issue of where to conduct human spaceflight began to be resolved by top officials. For instance, by November 1958, only two months after NASA officially came into being, NASA Administrator T. Keith Glennan and ARPA Director Roy Johnson signed a memorandum of understanding concerning a "Program for a Manned Orbital Vehicle." [II-8] This was to supplement the Dyna-Soar vehicle development (whose exact status had not been clearly defined, although it stayed within the Air Force and did not come under the control of ARPA). Eventually, the ballistic capsule concept totally migrated over to NASA. As long as the Air Force continued to have its own human spaceflight program, top Air Force officials did not complain too much about losing the less interesting ballistic capsule vehicle.

Other areas proved more contentious, however. NASA had acquired the three NACA research centers and their heavy emphasis on aeronautics research. But the new space agency lacked expertise in other areas, particularly the key ones of satellite and rocket development. It became obvious that NASA would have to acquire these as well. In the meantime, the Army was launching lunar and scientific probes on behalf of NASA, including Pioneer III, which traveled 63,580 miles toward the Moon, and Explorer IV, which took radiation measurements in space.

The obvious choice was for NASA to acquire the Army's JPL, which had technical expertise in the areas of guidance, communications, telemetry, rocket propellants, and satellites. JPL was primarily a research center, and the Army could continue to benefit from its research no matter who operated it. On December 3, 1958, the Army transferred JPL to NASA, along with its Explorer satellite program.[59] [II-9]

The other obvious entity to turn over to NASA was the ABMA in Huntsville, Alabama, which had produced the Jupiter and Redstone rockets. Jupiter was an IRBM and fulfilled the same role as the Air Force's Thor. Its days as a weapons system were limited. The ABMA had other rocket programs in the works. In October 1958, with the concurrence

58.   Loyd. S. Swenson, Jr., James M. Grimwood, and Charles C. Alexander, *This New Ocean: A History of Project Mercury* (Washington, DC: NASA SP-4201, 1966), pp. 101-102.
59.   Eddie Mitchell, *Apogee, Perigee, and Recovery: Chronology of Army Exploitation of Space*, N-3103-A (Santa Monica, CA: The RAND Corporation, 1991), p. 24.

of ARPA, the ABMA had initiated an effort known as Juno V, which was soon to be renamed Saturn. Juno V was a space rocket, not a missile, and the ABMA's other work was not in the IRBM or ICBM field (the latter being the exclusive domain of the Air Force).

However, the ABMA represented the Army's last vestige of long-range missile work, a concept that it had pioneered in the post-war years. Unlike JPL, it was also a major development command and, as such, represented a significant amount of money. The Army was therefore reluctant to give it up, especially if the money would no longer appear in the Army budget as well. There was even the appalling (for the Army) possibility that the Saturn rocket could be turned over to the Air Force.

Rather than turning the center over to NASA immediately, the Army negotiated to do this gradually. Eisenhower disagreed with this strategy, but he was willing to let NASA Administrator T. Keith Glennan work it out. [II-10] The Redstone program was transferred to NASA on December 3, 1958, and then the Saturn program was transferred in November 1959. Finally, from March through July 1960, the Army transferred the ABMA Development Operations Division, which included the 150 German scientists and engineers, 3,900 ABMA personnel, and 2,500 missile and satellite technicians. [II-11] The Army was officially out of the space business.

While NASA was busy acquiring facilities and personnel from the Army, it was also using the services of the Air Force and forging various agreements with that military service, particularly for the use of its powerful missile, the Atlas, as well as its ground stations. Paying for these systems became an issue; NASA and DOD signed an agreement in November 1959 for the reimbursement of costs. [II-12]

The move of the ABMA to NASA was the second important step in the creation of duplicative tasks for the civilian and military space programs. But it seems to have aroused little concern within the Eisenhower administration.

Although the core of NASA consisted of NACA, as the organization grew, it took on aspects of both the Army and the Air Force approaches to ballistic missile development. The Army approach centered on the arsenal system, which involved heavy in-house development of weapons using both uniformed personnel as well as civilian Army employees, but relatively few outside contractors. The Air Force adopted a more open, contractor-oriented approach; direction remained within the military, but civilian contractors did a large amount of the research and development work. NASA adopted both of these practices over time. As it rapidly acquired former Army laboratories, it developed a strong in-house technical capability for the development of hardware. But key NASA managers also came to the agency from the Air Force and brought with them both their experience and expertise of working with aerospace contractors, as well as long-standing close relationships with such contractors.[60]

A Rocky Road to Cooperation

The Space Act included provisions for a "Civilian-Military Liaison Committee," in which NASA and DOD were expected to "advise and consult with each other on all matters within their respective jurisdictions relating to aeronautical and space activities and shall keep each other fully and currently informed with respect to such activities." But almost from the beginning, this committee did not work very well.

---

60. For a discussion of the evolution of NASA as an institution, see Chapter IV in Logsdon, gen. ed., *Exploring the Unknown,* 1: 611-29.

In a December 15, 1958, interagency meeting on U.S. launch vehicles, representatives of the Air Force Ballistic Missile Division (AFBMD), speaking for ARPA, had discussed their upper stage vehicles with NASA. However, they failed to mention the Agena
B vehicle, which at the time was being considered for the CORONA and SAMOS reconnaissance satellites, as well as other payloads. NASA representatives discussed their Atlas-
Vega vehicle. Vega was to be a two-stage addition to the Atlas. The second stage would be
powered by a 33,000-pound thrust, liquid oxygen-kerosene engine. The third stage was to
be a restartable 6,000-pound thrust, storable-propellant engine developed by JPL.

On January 16, 1959, the AFBMD ordered Lockheed to initiate a study and a test program for a restartable booster. This occurred only a day after Convair submitted a proposal for a medium-energy upper stage for the Atlas-Vega. A week and a half later, on
January 27, NASA listed the Atlas-Vega as the first in a series of upper stage vehicles for
use in the national space program.[61] NASA signed contracts for the Atlas-Vega in March
and May of that year. In April and June, the AFBMD had worked out details for the Agena
B with Lockheed and authorized formal development work—without notifying NASA.[62]

Gradually, word of the Agena B reached NASA officials, and by December 1959, NASA
canceled the Vega as redundant. This duplication of effort had cost the country $16 mil-
lion. A Government Accounting Office review of the program placed most of the blame
on the Air Force for not informing NASA of its ongoing program.[63] The Civilian-Military
Liaison Committee had been intended to preclude just such a duplication of effort, and
it had failed because the Air Force decided to keep part of its program secret from another government agency. A year later, in September 1960, the Civilian-Military Liaison
Committee was eliminated, and NASA and DOD signed an agreement creating an
Aeronautics and Astronautics Coordinating Board. [II-13] Over the years, the importance
of the board has varied, depending on the issue and the personnel participating in it.

Taking the Military Space Program Away From ARPA

ARPA was never very popular with the military services. It removed a number of
key space programs from service control and placed it within DOD itself. Although the
services bowed to this reality, it became increasingly irksome to them as time went on. In
March 1958, soon after its creation, Director Roy Johnson informed the service secretaries
that he would bypass the service chiefs and deal with the heads of the commands directly.[64] Soon thereafter, the services began losing each of their programs.

When the structure of ARPA came up for review a year later, Air Force Brigadier
General James F. Whisenand, Special Assistant to the Chairman of the Joint Chiefs, stated
in a February 1959 memorandum to General Nathan Twining (the Chairman): "From the
military viewpoint, we would hope that ARPA would be phased out eventually and that
[the Office of the Secretary of Defense] could get back solely to policy direction."[65]

There was also concern that the Air Force would predominate once ARPA was
eliminated. A Department of the Army space policy in February clearly stated that the
Army would have a subordinate role in the national space program. But it also stated that
in its view, "Space is a new largely unknown medium which transcends the exclusive inter-

---

61.   "A National Space Vehicle Program," NASA, January 27, 1959, NASA Historical Reference
Collection, NASA History Office, NASA Headquarters, Washington, DC.
62.   Paul Means, "How the Two Programs Progressed," *Missiles and Rockets,* June 20, 1960, p. 20.
63.   Paul Means, "Vega-Agena-B Mix-Up Cost Millions," *Missiles and Rockets,* June 20, 1960, p. 19.
64.   Bowen, *The Threshold of Space,* p. 24.
65.   Brigadier General James F. Whisenand, Special Assistant to Chairman, Memorandum for General
Twining, "DOD Charter for ARPA," February 16, 1959, Record Group 218, Records of the U.S. Joint Chiefs of
Staff, Chairman's File, General Twining, 1957-1960, Box 34, "471.94 (1959)." National Archives.

est of any service. . . . No military department should be assigned sole responsibility for space activities."[66]

This situation also was unacceptable for the Navy. In April 1959, the chief of naval operations urged the Joint Chiefs of Staff to create a single military space agency. The Army, rapidly losing its space program to NASA, agreed. The Air Force chief of staff objected that this would remove the weapons systems from the unified commands. By July 1959, White House and DOD officials began evaluating this separate military space agency. It would report directly to the Joint Chiefs of Staff, and command would rotate among the services. It was tentatively called the Defense Astronautical Agency.[67] [II-14]

In September 1959, Defense Secretary Neil McElroy rejected the proposal for a separate military space agency. Furthermore, he removed military space from ARPA and gave it back to the separate services. Booster development was transferred to the Air Force, and payload development went to the Army, Navy, and Air Force based on competence and primary interest. Under this plan, the Saturn rocket was expected to be turned over to the Air Force. This ultimately did not happen, however, as administration leaders recognized that there was no military need for such a large booster; a month later, Saturn was turned over to NASA.[68]

During the first two years after Sputnik, there was a considerable philosophical change in the Eisenhower administration's approach toward space. Eisenhower had initially opposed the creation of a separate civilian space agency, which he thought would duplicate capabilities already at DOD. Yet he had been convinced to create NASA. His top officials, such as Killian, had also initially opposed the idea of giving NASA programs that duplicated those in the military services. However, first the ballistic space capsule and then Saturn and ABMA's rocket development facilities were given to NASA.

These later moves, in particular, were a much more dramatic shift. Giving NASA its own rocket development capability directly duplicated capabilities that *could* have been left solely with the Air Force, but they were not. This split—and the establishment of separate civilian and military rocket production facilities—was to have a profound effect on the relationship between NASA and the Air Force for years to come. In military terms, it created separate "stovepipes" that duplicated missions and capabilities. The creation of the National Reconnaissance Office only a few years later added a third stovepipe to the national space program, adding even more duplication. Gradually, by accretion and usually without much second thought, the separate programs grew beyond what Eisenhower had originally wanted when he created ARPA in early 1958.

The New Era

By the end of 1959, the Air Force had regained from ARPA control over most of its space program. Furthermore, it had been made lead authority for developing large military boosters. With the Army out of the picture, the Air Force was now clearly the premier military space agency.

The Air Force also had not abandoned some of the expansive dreaming that had begun in the immediate post-Sputnik period. In April 1960, the AFBMD produced a secret report for a "Military Lunar Base Program or S.R. 183 Lunar Observatory Study."

---

66. Office of the Adjutant General, Headquarters, Department of the Army, *Department of the Army's Interest, Capability, and Role in Space*, February 25, 1959, Record Group 218, Records of the U.S. Joint Chiefs of Staff, Chairman's File, General Twining, 1957-1960, Box 34, "471.94 (1959)," National Archives.

67. Bowen, *The Threshold of Space*, pp. 30-32.

68. *Ibid.*, p. 33.

[II-15] The base was billed as a "manned intelligence observatory" that could be developed into a "Lunar Based Earth Bombardment System." According to the report, the decision to place strategic weapons on the Moon could be deferred for a few years. "However, the program to establish a lunar base must not be delayed and the initial base design must meet military requirements. For example, the base should be designed as a permanent installation, it should be underground, it should strive to be completely self-supporting, and it should provide suitable accommodations to support extended tours of duty." The report recommended that "[t]he program for establishing a military lunar base be recognized as an Air Force requirement."[69]

The Air Force clearly still had its own designs on a large human spaceflight program. Within this atmosphere, on April 14, 1960, Air Force Chief of Staff Thomas D. White sent a letter to his staff, stating:

I am convinced that one of the major long range elements of the Air Force future lies in space. It is also obvious that NASA will play a large part in the national effort in this direction and, moreover, inevitably will be closely associated, if not eventually combined with the military. It is perfectly clear to me that particularly in these formative years the Air Force must, for its own good as well as for national interest, cooperate to the maximum extent with NASA, to include the furnishing of key personnel even at the expense of some Air Force dilution of technical talent. [II-16]

Unfortunately for White and the Air Force, the memorandum was leaked to Congressman Overton Brooks, the chair of the House Committee on Science and Astronautics. As Brooks characterized it, the statement indicated that White thought "that the military would ultimately take over NASA."[70] There was also much speculation within the press about the possible consolidation of the military and civilian space programs.

---

69.   The ideas of military bases on the Moon and orbital weapons were not new. One of the first mentions of orbital bombardment weapons appeared in *Forbes* magazine in 1946 (see Document II-1 in Logsdon, gen. ed., *Exploring the Unknown*, 1: 230-36). Apparently the first mention of a lunar-based bombardment system appeared in *Collier's* magazine in 1948 (see Robert S. Richardson, "Rocket Blitz From the Moon," *Collier's*, October 23, 1948, pp. 24-25; 44-46). Noted science fiction author Robert A. Heinlein used the idea of space bombardment in a short story called "The Long Watch" in *American Legion Magazine* in December 1949—and again in his popular novel *Space Cadet*. The same week that the creation of ARPA was being finalized, Brigadier General Homer A. Boushey, Air Force Deputy Director for Research and Development, wrote an article that advocated a lunar base as the ultimate deterrent (see Brig. Gen. Homer A. Boushey, "Who Controls the Moon Controls the Earth," *U.S. News & World Report*, February 7, 1958, p. 54). See also Lt. Col. S.E. Singer, "The Military Potential of the Moon," *Air University Review* 11 (1959), pp. 31-53. But by far the most noteworthy study was conducted by von Braun and his team at the ABMA, known as Project Horizon. It was presented in June 1959, and one of the justifications was the basing of weapons on the Moon to provide "International Law Enforcement" (*Project Horizon, Phase I Report*, Volume I, June 8, 1959, Space Policy Institute Documentary History Collection). Rather surprisingly, the Army was still discussing lunar bases long after the Apollo program was under way (see, for instance, *Space Information Briefing*, March 30, 1966, Future Weapons Office, R&D Directorate, U.S. Army Weapons Command, Space Policy Institute Documentary History Collection).

70.   *Defense Space Interests*, Hearings Before the Committee on Science and Astronautics, U.S. House of Representatives, 87th Cong., 1st sess. (Washington, DC: U.S. Government Printing Office, 1961), p. 91.

Robert S. McNamara and the "National Space Program"

Soon after the Kennedy administration took office on January 20, 1961, newly appointed Secretary of Defense Robert McNamara quickly put his own imprint on the military space program. On March 6, he issued a directive to the secretaries of the military services stating: "I have decided to assign space development programs and projects to the Department of the Air Force, except under unusual circumstances."[71] Such assignment, McNamara stated, was not to predetermine the assignment of operational responsibilities for the space systems. In addition, preliminary research could still be conducted by the individual services, but it would eventually have to be transferred to the newly created director of defense research and engineering for evaluation before proceeding to development. In light of that, "[r]esearch, development, test, and engineering of Department of Defense space development programs or projects, which are approved hereafter, will be the responsibility of the Department of the Air Force." [II-17]

Taken together, both memoranda made outside observers believe that the Air Force was about to attempt to take control of the majority of the civilian space program. In March 1961, Overton Brooks called hearings to discuss the issue. He was also concerned about the report of President Kennedy's transition group for space, which indicated that NASA was to be responsible for scientific research, while the military would play the predominant role in developing space systems. Shortly before the hearings began, he sent a letter to Kennedy asking for clarification on the matter. [II-18]

During the course of the hearings, General Thomas D. White declared that the leaked memorandum, which had caused such consternation in the press and the committee, was only a general marching order to his staff to *improve* its cooperation with NASA; it did not indicate any planning to take over NASA. General Bernard Schriever, then commander of the Air Research and Development Command, admitted that he was mostly to blame for White's memorandum, because he had resisted the transfer of Air Force personnel to NASA. White was trying to indicate to Schriever that he was not happy with this lack of cooperation. However, given the Air Force's secrecy over the Agena B, and its continuing expansive space plans, it was conceivable that the service's top officials had at least some designs on NASA's turf.[72]

The result of the hearings, and of Brooks's letter to Kennedy, came in Kennedy's reply on March 23, the final day of the hearings. Kennedy stated:

---

71. Robert S. McNamara, Secretary of Defense, Memorandum for the Secretaries of the Military Departments, *et al.*, March 6, 1961, reprinted in *ibid.*, p. 2.

72. Certainly, the Air Force was interested in expanding its missions and power. In the course of the hearings, White also denied that he wanted to gain control of all strategic nuclear forces, stating that "[t]he Air Force has no designs whatsoever, on the Polaris weapon system." *Ibid.*, p. 98. This question came up because sixteen months before White had sent a letter to the chair of the Joint Chiefs of Staff requesting that he recommend to the secretary of defense that he "assign control of the Polaris weapon system to the Strategic Air Command in view of its strategic capabilities." General Thomas S. Power, Commander in Chief, USAF, to General Nathan G. Twining, Chairman, Joint Chiefs of Staff, October, 1959, Record Group 218, Records of the U.S. Joint Chiefs of Staff, Chairman's File, General Twining, 1957-1960, Box 34, "471.94 (1959)," National Archives. After a protest from the Navy, the Air Force backed away from this request. White's actions, as well as those of others in the Air Force, indicate that the service was obviously interested in empire-building, but the uniformed leadership was being less successful at it than they had hoped. For a further discussion, see David Alan Rosenberg, "The Origins of Overkill: Nuclear Weapons and American Strategy, 1945-60," *International Security* 7 (Spring 1983): 3-71.

It is not now, nor has it ever been, my intention to subordinate the activities in space of the National Aeronautics and Space Administration to those of the Department of Defense. I believe, as you do, that there are legitimate missions in space for which the military services should assume responsibility, but that there are major missions, such as the scientific unmanned and manned explo- ration of space and the application of space technology to the conduct of peaceful activities, which should be carried forward by our civilian space agency. [II-19]

Kennedy's letter thus made it clear to the Air Force that NASA would have primary responsibility for both human spaceflight and the development of space technology in general. At the same time, he acknowledged a clear military role in space. This attitude would become clearer less than two months later with a joint memorandum to the presi- dent from NASA Administrator James E. Webb and Secretary of Defense Robert McNamara. The "Webb-McNamara Memo," as it became known, stated that space projects could be undertaken for one of four possible reasons. The first was scientific knowledge, the second was commercial/civilian value, and the third was military missions. The final reason was for purposes of national prestige. Such missions were "part of the battle along the fluid front of the cold war."[73]

This was in stark contrast to the position of President Eisenhower, who had explicitly rejected national prestige as a reason for space exploration and attempted to restrict both NASA and the military to strict utilitarian missions. By embracing their own view, and by calling explicitly for an "integrated" space program, Webb and McNamara also indicated that large, "prestige" missions were best carried out within NASA. They essentially applied a "strict scrutiny" approach to military space programs. If the programs did not serve clear military needs, then they should be either turned over to NASA or abandoned altogether.

Blue Gemini

On May 25, 1961, President Kennedy committed the United States to a major new undertaking in space, expressly for the purposes of national prestige.[74] Project Apollo resulted in a dramatic infusion of funds to NASA, along with the decision to ensure that the United States was ahead in every area of space technology. NASA was selected as the primary—and most visible—instrument for accomplishing this. As NASA's leadership planned out its program for reaching the Moon, it became obvious that certain tech- nologies and capabilities would have to be developed. Foremost among these was ren- dezvous in orbit. NASA quickly decided to develop a more advanced space vehicle than the Mercury to develop these new techniques and technologies. This first "operational" spacecraft was soon named Gemini.

As NASA increased in size and assumed a predominant role, its interests also tended to diverge at key points from those of DOD. On July 7, 1961, NASA Associate Administrator Robert Seamans proposed a joint study to determine mission models and requirements affecting the selection of large launch vehicles. NASA's Nicholas Golovin directed the study. As this study progressed, the different requirements and institutional interests of NASA and DOD became clear. Both agencies distanced themselves from the contents of the report. By the time the report was released on September 24, 1962, almost

---

73.   Document III-11 in Logsdon, gen. ed., *Exploring the Unknown*, 1: 439-52.
74.   Document III-12 in *ibid.*, 1: 453-54.

a year later, it had been obvious for some time that there would be very little cooperation between NASA and DOD on large launch vehicles. [II-20] The result was a further solidification of entirely separate and redundant rocket development programs in the civil and military spheres.

In February 1962, during congressional hearings on the Air Force space plan, Air Force officials first broached the idea of an Air Force version of the Gemini spacecraft. The idea became firmer in June when the Air Force's Space Systems Division began looking at the use of Gemini hardware for a preliminary Air Force space station known as MODS (Manned Orbital Development System). The Space Systems Division had been given the task of acting as a contractor to NASA for providing launch and target vehicles for Gemini. In August, those at the Space Systems Division started referring to the Air Force plan as "Blue Gemini."[75]

Although not officially sanctioned at the top levels of the Air Force, Blue Gemini became more appealing as other Air Force programs were cut back or slipped in schedule. A planned satellite interceptor was cut in the fall of 1962, and Dyna-Soar was still a long way from its first flight. The possibility of acquiring a simpler vehicle than Dyna-Soar to accomplish the rendezvous and reconnaissance agendas for the other two programs became very appealing at many levels of the Air Force.[76]

Many at NASA did not oppose the possibility of the Air Force taking a bigger role in the development of Gemini; they thought that DOD money flowing into the program could only help its development. In November 1962, the NASA Gemini program team met with representatives of the Air Force's Space Systems Division to discuss the coordination between the agencies. Soon after, NASA Administrator James Webb and Associate Administrator Robert Seamans visited the Pentagon to discuss increased DOD participation in Gemini with Deputy Secretary of Defense Roswell L. Gilpatric. However, Secretary of Defense McNamara was also there, and he surprised all of them by proposing the merging of the NASA Gemini program office with the Air Force office and moving it all to DOD.[77]

Retired Admiral W. Fred Boone became NASA Deputy Associate Administrator for Defense Affairs on December 1, 1962. Boone soon began working in earnest to build support against such a move. In early January 1963, NASA officials met with Pentagon officials and convinced them that taking over Gemini was a bad idea. McNamara and Gilpatric backed away from the takeover idea, but McNamara pushed for a joint management board for Gemini.[78]

In January 1963, Webb wrote Secretary of Defense McNamara and stated unequivocally his opposition to the joint management board for Gemini. [II-21, II-22, II-23] Webb had a major argument on his side; Gemini was vital to achieving the lunar goal, and DOD could not interfere with that mission. For DOD, Gemini was intended to be used to explore the utility of human spaceflight for the military—it was a much more open-ended and ambiguous mission. At the same time, there were those in the Air Force who were opposed to taking over Gemini because it would increase the chance of Dyna-Soar being killed. McNamara had to back away from the Gemini takeover attempt and ultimately accepted the creation of a Gemini Program Planning Board, which did not significantly alter the relationship between the actors.[79]

---

75.   Barton C. Hacker and James M. Grimwood, *On the Shoulders of Titans: A History of Project Gemini* (Washington, DC: NASA SP-4203, 1977), p. 118.

76.   *Ibid.*

77.   *Ibid.*, p. 119.

78.   *Ibid.*, p. 120.

79.   *Ibid.*, pp. 121-22.

In this context, and as Dyna-Soar moved toward the construction of hardware, that program became increasingly difficult for the Air Force to justify convincingly. Its proponents were forced to grasp at whatever justification they could find. Dyna-Soar was to be a reconnaissance craft. It was to be an offensive weapon, capable of striking the Soviet Union from virtually any direction, dropping up to two nuclear warheads. It was also to be an anti-satellite weapon, capable of destroying Soviet reconnaissance satellites. Some of these missions, however, could be accomplished more cheaply and more immediately with robotic spacecraft. Others, such as the bombing mission, were not really needed. Furthermore, as long as the fundamental utility of human spacecraft for military missions was in doubt, it made no sense to rely on a technologically challenging program to prove their worth. Gemini was perfect at the time for demonstrating the military value of human spaceflight because it was cheaper and easier than Dyna-Soar. The Air Force still remained wedded to the image of flying Air Force pilots in space, but this was an image that was more emotional than logical.

In April 1963, President Kennedy asked Vice President Johnson to conduct, in his role as chair of the National Aeronautics and Space Council, an overall review of the "national space program." [II-24] McNamara was asked to report to Johnson on this issue and did so, commenting that he and NASA Administrator Webb had worked hard to eliminate duplication between the civilian and military space programs. [II-25] The idea of a "national space program" was not McNamara's alone; indeed, the term had been used during the prior administration. But McNamara, with his dedication to efficiency, was the person most concerned about eliminating duplicative and wasteful programs.[80] McNamara was expansive in his view of his mission as well, and he was willing to reach beyond the DOD budget and programs to attempt to acquire or even to eliminate programs in other organizations that he did not see as worthwhile. Striving for McNamara's definition of "efficiency" was not always easy, but this was a central factor in DOD-NASA relations during much of the first decade of the space program.[81]

---

80.   The strive for efficiency was felt in other areas as well, including a major fight in late 1962 and early 1963 over whether or not the CIA's presence in the ultra-secret National Reconnaissance Office was still necessary. McNamara thought it was duplicative and wasteful and felt that the Air Force should run all satellite reconnaissance. He lost this fight. See Albert D. Wheelon, "Lifting the Veil on CORONA," *Space Policy* 11 (November 1995): 252-53.

81.   There was, however, an example of duplication in space programs that proved in the end to be positive. This centered on the meteorological satellite programs. NASA inherited its Television and Infrared Operational Satellite (TIROS) system from the Army. The first TIROS flew in April 1960. NASA and DOD began negotiating on the development of an operational system in October, but by December, Air Force Commander in Chief Thomas S. Power expressed an interest in the Air Force controlling the operational system to provide weather data to its forces and also to be used for reconnaissance satellite flights. Negotiations continued for several months before the Air Force withdrew and began its own program, within the secrecy of the National Reconnaissance Office (it was later apparently turned over to the Strategic Air Command). NASA and the National Weather Bureau signed a joint agreement to cooperate on the development of an operational satellite in January 1962, but the program did not proceed very well because of conflicts between the partners, and the Weather Bureau withdrew in September 1963. Then the Weather Bureau approached the Air Force for access to its system. TIROS IX was launched in 1966 and was based on the Air Force's satellite design. The majority of this story still remains classified, but it provides an interesting counter to McNamara's arguments for efficiency in the national space program. See Janice Hill, *Weather From Above* (Washington, DC: Smithsonian Institution Press, 1991), pp. 22-26. See also General Thomas S. Power, Commander in Chief, Strategic Air Command, to General Thomas D. White, December 1, 1960, Box 34, "2-15 SAC," Papers of Thomas D. White, Library of Congress, Washington, DC.

In March 1963, McNamara still had not made up his mind about the desirability of Dyna-Soar. He felt that the Air Force had not concentrated enough on exactly what it was to do in orbit, focusing solely on its flying characteristics. He suggested several missions that should be evaluated, including inspection and kill, reconnaissance, the vulnerability of space vehicles, and orbital weapons. But he was also interested in the test bed possibilities of any spacecraft and voiced this in a meeting with Boeing and NASA officials. One NASA official stated that according to the Space Act, such joint use might create a conflict, because regulations dictated that NASA was not to be involved in weapons development. McNamara responded to this with scorn, stating that he was willing to change the law if necessary.[82] His view of his authority and mission was quite expansive indeed.

During the summer of 1963, the Air Force began to seriously consider an orbital space station. It received authorization from the director of defense research and engineering to study the issue. The space station was not to be an end in itself; rather, it was to be used to "demonstrate and assess qualitatively the utility of man for military purposes in space."[83] The Air Force's initial study was completed by November, and it assessed a number of options, including the use of Gemini and Apollo spacecraft to service the military space station.

Dyna-Soar was an arguably duplicative program and also one that was becoming increasingly expensive as it moved away from purely theoretical research and into the development phase. In addition, Kennedy had been elected to some degree on the propaganda scare of a nonexistent "missile gap," from which he and McNamara later had to retreat. Kennedy's actions after the Cuban missile crisis of 1962 and the Nuclear Test Ban Treaty of 1963 also symbolized a movement away from boisterous displays of nuclear capabilities. In light of these events, as well as ongoing public and congressional concerns about "the militarization of space," the image of a piloted space bomber swooping in from orbit to obliterate Moscow became distinctly unattractive to the administration.

Another problem with Dyna-Soar was that the basic utility of humans for military space missions was in doubt. It was to be proven or disproven with the military space station, which was itself an experimental vehicle. Identifying the utility issue did not require an experimental vehicle, and using an experimental spacecraft to service an experimental space station seemed to be too risky and too expensive.

By 1963, the Kennedy administration was very aware of the value of satellite reconnaissance. It had even evaluated the possibility of sharing U.S. reconnaissance data with other nations. Satellite reconnaissance was viewed as a valuable national asset, not merely a military war-fighting tool. But the Air Force apparently continued to view reconnaissance solely in terms of military capabilities and thus sought a way of neutralizing Soviet reconnaissance satellites—doing so in a highly visible manner.

In short, Dyna-Soar would militarize space in all the ways that the administration did not want to see it militarized. It was largely unjustified and duplicative of missions that NASA was already conducting. It also now stood in the way of identifying clear military space missions for humans. Thus, by late 1963, Dyna-Soar was in clear trouble with Defense Secretary McNamara. The response from the Air Staff was a letter to the secretary of the Air Force outlining several space station missions, all involving Dyna-Soar. If money was a problem for the national space program, suggested the assistant to the vice

82.   Brockway McMillan, Assistant Secretary for Research and Development, "Memorandum for Secretary Zuckert," March 15, 1963, Space Policy Institute Documentary History Collection.
83.   Harold Brown, Director of Defense Research and Engineering, to Secretary of the Air Force, "Military Orbiting Space Station," August 30, 1963, Space Policy Institute Documentary History Collection.

chief of staff, then it was always possible to cancel Gemini (its role in the Apollo program was ignored).[84] This last ditch, vindictive effort to save Dyna-Soar failed.

On December 10, 1963, McNamara canceled the Dyna-Soar program. As consolation to the Air Force, DOD authorized money for a Manned Orbital Laboratory program utilizing the Gemini spacecraft. This laboratory program would continue for another five years, serving as the Air Force's hope for flying its own pilots in space. The laboratory was to serve as an occupied, real-time reconnaissance spacecraft with multiple cameras, demonstrating various reconnaissance and surveillance technologies. However, at the beginning of its life, the Manned Orbital Laboratory, similar to Dyna-Soar, was amorphous, with no clear, overriding purpose other than technology development and the ever-persistent Air Force desire to fly its own astronauts in space.

At the same time, NASA was investigating the possibility of developing a space station, and cooperation with DOD on this matter was only natural. [II-26, II-27] The two organizations even signed an agreement for the creation of a Manned Space Flight Experiments Board. [II-28] The agreement established the principle of reciprocity and the sharing of flight opportunities between NASA and DOD for both Apollo and the Manned Orbital Laboratory.

By 1968, the Manned Orbital Laboratory had solidified significantly and was to include a massive camera system with a ground resolution of four inches. The officers aboard it were to provide near real-time reconnaissance of the Earth. This had been an early goal of the Air Force's WS-117L and SAMOS programs, but it had proven a difficult one to achieve because of the technological challenges. The CIA had successfully developed its CORONA reconnaissance system, which, by the late 1960s, had already flown more than 100 missions and proved an astounding success. The Air Force had chosen another route, developing "close-look" systems for the technical assessment of Soviet weapons, but the service had never abandoned its desire for real-time reconnaissance. CORONA photographs could take more than a day to reach Washington and photo-interpreters. The Air Force wanted to reduce this to hours or less; such a quick turn-around would enable the photographs to be used in battlefield operations. This coincided well with the Air Force's dream of flying Air Force officers in space—hence a major impetus behind Dyna-Soar and, later, the Manned Orbital Laboratory.

With the Vietnam War waging, the DOD budget was under extreme pressure. The Manned Orbital Laboratory was the largest single item in the DOD budget and therefore an obvious target for being cut. In 1968, the laboratory was doomed, but it survived for one more year and the election of another president (Richard Nixon). Then it was killed. Once again, the Air Force's attempt to fly military officers in space had been thwarted.[85]

---

84.   Major General J.K. Hester, Assistant, Vice Chief of Staff, HQ, Air Force, Memorandum to the Secretary of the Air Force, "Approaches to Manned Military Space Programs," December 4, 1963, Space Policy Institute Documentary History Collection.

85.   The Manned Orbital Laboratory had also run afoul of other developments and a shortage of funds, this time not because of NASA, but from a different source entirely—the CIA. By the late 1960s, the CIA was beginning to develop its follow-on to the highly successful CORONA series of wide-area surveillance satellites. The CORONA follow-on and the Manned Orbital Laboratory, which were intended to perform entirely different missions, were competing with each other for funding within the highly secret world of the National Reconnaissance Office. Furthermore, the laboratory ran into some of the same problems as its linear predecessor, Dyna-Soar; it was far too visible for its own good, especially for a reconnaissance system. In a contentious meeting at the National Photographic Interpretation Center in 1968, Vice President Hubert Humphrey, chair of the National Aeronautics and Space Council, repeatedly complained that the Manned Orbital Laboratory would not be able to fly without "Walter Cronkite looking over your shoulder." While the nation's other reconnaissance satellite programs had remained remarkably secret, the laboratory had attracted much attention within the press and Congress. It had become a political football in Congress, where angry Florida senators and representatives wanted to know why the space station had to fly out of Vandenberg Air Force Base, California.

Human spaceflight was one of the key issues of military-civilian cooperation. During the 1960s, NASA had clear justifications for flying humans in space—medical research and prestige. The Air Force did not have these clear justifications, and its human space-flight program was thus focused first on demonstrating the utility of astronauts for mili-tary space missions. In the end, the Air Force *failed even to justify flying astronauts simply to perform this evaluation*, let alone to serve practical purposes in space. Robotic spacecraft as well as NASA experiments undercut the tenuous justifications the Air Force had advanced even for experimental missions. The costs were simply too high and the benefits viewed as too elusive. The experience with both Dyna-Soar and the Manned Orbital Laboratory apparently taught the senior uniformed leadership at the Air Force a lesson, and they were forever after very skeptical of human spaceflight.

The Military and the Space Shuttle

In early 1969, President-elect Richard Nixon appointed a Space Task Group to address the issue of the post-Apollo space program. Vice President-elect Spiro Agnew was appointed chair of the group, and its other members were NASA Administrator Thomas O. Paine and Secretary of the Air Force Robert C. Seamans (who had been deputy administrator at NASA). On March 22, 1969, the Space Task Group met to discuss the joint development of a Space Transportation System (STS). Less than two weeks later, on April 4, Paine asked Seamans to approve a joint NASA-Air Force study of an STS.[86]

The conclusion of the Space Task Group was that the country should undertake an ambitious space exploration program involving landing humans on Mars and developing a lunar base and space station. These missions would be serviced by a reusable Space Shuttle, intended to reduce the costs of transportation. President Nixon, however, did not accept this report and only gave his initial approval to the space station and shuttle options, postponing the former and tentatively agreeing to the latter.[87]

NASA and the Air Force had diverged on the issue of large launch vehicle develop-ment seven years before. While NASA developed the Saturn IB and the much larger Saturn V, the Air Force developed its Titan series of boosters. Versions of the Titan were used for ICBMs and various reconnaissance missions, and even larger versions were devel-oped first for Dyna-Soar and later the Manned Orbital Laboratory and CORONA follow-on. By early 1970, NASA officials such as Paine had recognized that DOD support would likely be essential for obtaining White House approval for the Space Shuttle program. NASA and Air Force officials met a number of times to discuss the design of the Space Shuttle and to establish terms of reference for such a system. [II-29]

In February 1970, NASA and the Air Force signed a joint agreement to cooperate by establishing a NASA-Air Force Space Transportation System Committee (STS Committee). They agreed that the program would be unclassified and would also involve international cooperation. Furthermore, both NASA and DOD would make substantial contributions to shuttle development and operations—which later became important for the establishment of shuttle pricing agreements. [II-30, II-31, II-32] The STS Committee was the mechanism through which the Air Force informed NASA of its requirements for

---

86.   DOD prepared a massive report for the Space Task Group, which served as a basis for transporta-tion mission models for the space shuttle. This document was essentially a listing of all possible DOD missions over the next several years. "DOD Space Programs, Options, Recommendations," August 7, 1969, Space Policy Institute Documentary History Collection.
87.   Documents III-25 and III-26 in Logsdon, gen. ed., *Exploring the Unknown*, 1: 522-46.

the Space Shuttle. During its first year of operation, the STS Committee laid considerable groundwork for the shuttle's design.

NASA initially wanted a smaller shuttle with only limited cross-range (that is, the ability of the shuttle to travel to either side of its ground track during landing). Low cross-range meant relatively small, straight wings, while high cross-range meant larger, delta-shaped wings for more maneuvering. Smaller payload size and smaller wings would presumably result in a smaller, easier (to build), and, hopefully, cheaper shuttle.

The Air Force, however, had two primary requirements. One was the ability to launch the largest payload in its inventory, by then the CORONA follow-on satellite (which the CIA had eventually turned over to the Air Force for development), with a little extra room and weight for growth. The second was the ability to launch polar-orbiting reconnaissance satellites. Polar orbit could not be reached from Cape Canaveral without overflying inhabited areas, and such launches therefore flew out of Vandenberg Air Force Base in California, heading south. For the shuttle, this proved problematic, for if there was an abort during liftoff, the shuttle had to be capable of returning to California to avoid landing with a highly classified payload in the Soviet Union. The rotation of the Earth would cause California to move during that time period, and the shuttle needed to catch up with it. It therefore needed a high cross-range capability—1,100 miles—in addition to the large payload capability.

NASA's initial proposal was for a shuttle with a 14-foot by 45-foot payload bay, which would eventually be expanded to 15 feet by 60 feet at a future date. The Air Force strongly objected to this, because it could not use a payload bay smaller than 15 feet by 60 feet for key missions. The Air Force stated that of the 149 military payloads forecast to be flown between 1981 and 1990, 71 would not fit in the smaller payload bay. Without the larger bay, these missions would have to fly on Titan III boosters instead, undercutting the justification for the Space Shuttle.[88]

To gain the Air Force's support for the development of the shuttle, NASA agreed to both the payload and cross-range design requirements.[89] [II-33] In addition, to place large payloads in high-Earth orbit, a "space tug" was needed. NASA and DOD began negotiating on the development of this vehicle as well. [II-34]

According to NASA's early cost models for the shuttle's development, virtually all American payloads had to be shifted to the shuttle for the vehicle to be cost-effective. This meant, in effect, that other launch vehicle production had to be eliminated, but the Air Force had not explicitly agreed to this. In 1973, Malcolm R. Currie, Director of Defense Research and Engineering, wrote to the secretary of the Air Force stating, that uncertainties about the operational availability of the shuttle dictated the maintenance of a back-up launch capability using expendable launch vehicles.[90] With congressional pressure mounting on NASA because of rising shuttle costs, NASA Administrator James Fletcher wrote to Secretary of Defense James Schlesinger, asking for his continued support of the shuttle, as well as continued dialogue with NASA on the issue. [II-35] Schlesinger, along with Deputy Secretary of Defense William Clements, met with Fletcher in August 1976 to discuss the shuttle issue.

---

88.   For a fuller discussion of the technical tradeoffs involved in the shuttle design, see M. Scott Pace, "Engineering Design and Political Choice: the Space Shuttle 1969-1972," M.A. Thesis, Massachusetts Institute of Technology, 1982.

89.   Documents III-28 and III-32 in Logsdon, gen. ed., *Exploring the Unknown*, 1: 546-59.

90.   Malcolm Currie, Director of Defense Research and Engineering, to Dr. Robert C. Seamans, Secretary of the Air Force, "DOD Space Shuttle Planning," August 7, 1973, Space Policy Institute Documentary History Collection.

In a letter to Fletcher, Clements stated for the first time: "Once the Shuttle's capabil-ities and low operating cost are demonstrated we expect to launch essentially all of our military space payloads on this new vehicle and phase out of inventory our current expendable launch vehicles." [II-36] This letter, although not a specific policy directive, is apparently the first clear statement of DOD intent to rely exclusively on the shuttle for access to space. This policy was not quickly or easily accepted within the Air Force, and even two and a half years later, a joint memorandum of understanding on the manage-ment and operation of the shuttle notably did not state that the shuttle would be the exclusive means for access to space. [II-37]

Two months later, John J. Martin, Assistant Secretary of the Air Force (Research and Development), and John F. Yardley, NASA Associate Administrator for Space Flight, signed an agreement that determined what DOD would pay for shuttle launch services. For the first six years of operation, DOD would pay NASA what amounted to the incre-mental costs of materials and services. [II-38] This later led to charges in Congress and the press that NASA was giving the Air Force a preferential deal on shuttle flights to main-tain its continued support. However, the Air Force had already agreed to significant costs of its own for using the shuttle.

The effects of the Air Force decision to cooperate with NASA on the shuttle were not felt for some time. There were gradual indications that this had been a mistake. The cost of developing a separate launch and landing facility at Vandenberg Air Force Base was increasing. It was planned that the shuttle use Space Launch Complex-6 (known as "Slick Six") at Vandenberg, which had originally been intended for Dyna-Soar, was then modi-fied for the Manned Orbital Laboratory, and had never launched a single rocket despite the expenditure of billions of dollars. The modification of "Slick Six" was expected to cost even more money than planned.

In addition, the Air Force was looking at the possible procurement of its own orbiters, but as the development cost rose, this became less attractive. Finally, the decision to coop-erate on the shuttle did not necessarily constitute an Air Force decision to make exclusive use of the shuttle for launching all payloads. However, the cost of supporting both the shuttle and the Air Force fleet of expendable boosters was also becoming apparent. By 1974, Secretary of Defense James Schlesinger and Secretary of the Air Force Malcolm Currie were becoming increasingly concerned about all of these costs.

NASA-Air Force relations during this time were not always cordial. As the shuttle design matured, NASA managers frequently made changes without including the Air Force in the decisions, only informing the service after the fact. Furthermore, the initial launch rate for the shuttle was set at 60 flights per year, with 40 from Kennedy Space Center and 20 from Vandenberg. NASA soon determined that this flight rate was unachievable without a five-orbiter fleet; in 1976, the space agency began calling for a fifth orbiter, expecting the Air Force to pay for it. The DOD leadership refused to acknowledge that its mission model dictated the need for the fifth orbiter, which it feared it would have to procure on its own. [II-39]

In 1977, Hans Mark became the new under secretary of the Air Force and the direc-tor of the National Reconnaissance Office. Mark previously had directed NASA's Ames Research Center and felt that the shuttle was in the best interests of the country. He entered office at a time when the shuttle was coming under increasing pressure from the new administration of President Jimmy Carter over cost increases and schedule delays.[91]

---

91.   Document III-33 in Logsdon, gen. ed., *Exploring the Unknown,* 1: 559-74.

Mark was an ardent shuttle supporter and argued that the vehicle itself was an important contributor to national defense.[92] To further justify the shuttle, Mark chose to eliminate the option of "dual-compatibility" and shift key national security payloads to a "shuttle-compatible" only policy. According to a report at the time, this meant "a payload design compatible with shuttle launch: it may or may not be compatible with [expendable launch vehicle] launch. The term 'Shuttle optimized' implies a payload designed to exploit the unique capabilities of the shuttle—i.e., retrieval, on-orbit service, large weight and volume, etc. The 'Shuttle optimized' payload is not likely to be compatible with existing [expendable launch vehicle] launch capability."[93] In anticipation of using the shuttle's unique capabilities, the procurement rate of national security satellites was reduced during the 1970s until the shuttle became operational. The result of this decision was a "bow wave" of unfunded requirements that drove up DOD space spending in the 1980s.[94]

In 1981, President Ronald Reagan, despite the objections of the uniformed Air Force, directed the transition of all U.S. government payloads to the Space Shuttle as expeditiously as possible, once "the capabilities of the STS are sufficient to meet its needs and obligations."[95] As a result, a number of national security payloads were modified so that they could only fly on the Space Shuttle. This was to have a profound effect on the military and intelligence space programs later in the 1980s.

The Death of Military Human Spaceflight

By the time the Space Shuttle became operational in the early 1980s, it had changed considerably from what the Air Force had originally anticipated. The Air Force faced launch costs totaling nearly $300 million per flight. In August 1982, Air Force Systems Command Commander General Robert T. Marsh, who had responsibility for Air Force participation in the STS, informed Air Force Chief of Staff General Charles Gabriel of rising shuttle costs. [II-40] The shuttle did not fare well when compared to the Air Force's other heavy booster, the Titan III. Not only had shuttle costs risen, but when added to the Air Force's internal costs for personnel, hardware, mission control, and so on, the overall cost to the Air Force was much higher than expected. It was becoming obvious to many within the Air Force that the shuttle posed a *major* budgetary burden. In addition, the shuttle program was also considerably behind schedule and was unlikely to meet anticipated flight rates.

---

92.   See Hans Mark, *The Space Station: A Personal Journey* (Durham, NC: Duke University Press, 1987).

93.   *Space Shuttle Appropriations for Fiscal Year 1979*, Committee on Appropriations, Hearings before Subcommittees of the Committee on Appropriations, U.S. House of Representatives, 95th Cong., 2d sess. (Washington, DC: U.S. Government Printing Office, 1980), p. 363.

94.   From 1973 to 1976, DOD strove to design satellites for "dual-compatibility" with both the shuttle and expendable launch vehicles. This meant that the satellites had to be designed and tested in the different acoustic and dynamic environments of the shuttle and expendable launch vehicles. Such a design, however, proved to be more difficult in practice than in theory, because of widely different operating environments and other conditions. As a result, satellite program managers tended to defer changes until the next "block change" in the satellite, when the costs could be folded into other necessary design changes as well. Vice President's Space Policy Advisory Board, *A Post Cold War Assessment of U.S. Space Policy* (Washington, DC: U.S. Government Printing Office, December 1992), p. 6.

95.   National Security Decision Directive 8, "Space Transportation System," November 13, 1981, and National Security Decision Directive 42, "National Space Policy," July 4, 1982, Space Policy Institute Documentary History Collection.

In March 1983, Lieutenant General Richard C. Henry, Commander of the Air Force Space Division, wrote a letter to General Marsh at the Systems Command. Henry expressed growing concern that carrying humans aboard a vehicle designed merely to deliver payloads to orbit created an unnecessary expense. After the initial ground-processing delays of the shuttle *Challenger,* Henry wrote:

A four orbiter-only fleet, experiencing problems similar to those of Challenger, would develop a backlog of launches that would take months to years to work off. This represents a considerable threat to the continued vitality of the national space program and in particular, could impact national security through inadequate launch support of priority DOD spacecraft.

Henry's letter outlined for the first time the idea of a "mixed fleet" of launch vehicles and also mentioned the possibility of commercializing launch vehicles, such as the Delta and the Atlas. [II-41] This was at a time when the Air Force was rapidly preparing to close down its expendable launch vehicle production lines.

DOD continued to support the shuttle despite strong reservations, particularly among top Air Force officers. In early 1984, however, Secretary of Defense Caspar Weinberger issued a directive that established a need for a complementary expendable launch vehicle to supplement the Space Shuttle. [II-42] This move was not popular with top NASA officials, who viewed it, correctly, as a lack of faith in the Space Shuttle, but they could not address the problem because it was an Air Force policy issue. In the Air Force's view, the Space Shuttle was nowhere near reaching its definition of "operational status," even more than three years since the first launch. [II-43] DOD initially ordered ten complementary expendable launch vehicles, based on a modified Titan 34D design. This eventually became known as the Titan IV.

A year after the complementary expendable launch vehicle decision, Undersecretary of the Air Force Edward C. Aldridge, who was also the director of the National Reconnaissance Office, discussed with NASA Administrator James Beggs the possibility of preserving other expendable launch vehicle lines in addition to the Titan. Having completed a competition to select the complementary expendable launch vehicle, Aldridge needed NASA to concur with the decision. He reached an agreement with Beggs, and this was taken to the National Security Council for the president's signature. It became National Security Decision Directive 164 (NSDD 164), "National Security Launch Strategy," signed on February 25, 1985, which stated that the shuttle would continue to be the primary space launch system for both the military and civilian space programs. This directive authorized DOD to develop the complementary expendable launch vehicle; it also stated that the two organizations should begin developing a second-generation STS. [II-44]

After the *Challenger* accident, however, the military was placed in a tremendous bind. Although DOD had already begun shifting some of its payloads away from the shuttle, it had also designed a number of them so that they could be carried *only* by the shuttle. With the primary launch vehicle for many of these payloads out of service for an indeterminate amount of time, the depth of the shuttle cooperation mistake became apparent to virtually everyone in the Air Force and DOD. Classified satellites that could only fly on the shuttle began to pile up at various "clean rooms" around the country, creating a backlog of payloads that needed to be in orbit. Furthermore, several other expendable launch vehicle failures at the same time left the United States grounded and resulted in the destruction of several valuable reconnaissance payloads. Finally, the on-orbit constellation of reconnaissance, early warning, communications, and other satellites continued to age.

For a period of several years, the United States was left with only one reconnaissance satellite in orbit, a situation that was totally unacceptable from a national security point of view.[96]

The Shuttle Legacy for NASA-DOD Relations

Air Force involvement in the shuttle came largely at the urging of the civilian leadership of the service, not the general officers or the Air Staff. This is not terribly surprising because the shuttle was a NASA-initiated program, and NASA officials had negotiated with their civilian counterparts in the Office of the Secretary of the Air Force. Assistant Secretary of the Air Force Grant Hansen was one of the principal contacts with NASA during early negotiations, as was Secretary Seamans. Later in the 1970s, Undersecretary of the Air Force Hans Mark further entwined the Air Force's fate with the performance of the Space Shuttle.

At the same time, support for involvement with the shuttle received only lukewarm response from uniformed personnel. This represented a decided shift from the previous major military space initiatives in the Air Force, where the uniformed officers had been pushing the programs and the civilian leadership—both at the secretary level and in the Office of the Secretary of Defense—had opposed them. This characteristic had begun with the WS-117L reconnaissance program, which had been underfunded by Secretary of the Air Force Quarles. It was also seen in such instances as General Schriever being warned by the Office of the Secretary of Defense not to use the word "space" in speeches. It was certainly common in the immediate post-Sputnik era, when the Air Staff had lobbied extensively for a number of new space missions, only to see its authority stripped by Secretary of Defense McElroy with the creation of ARPA. And it was in evidence under McNamara, when the Air Staff had bold plans for Dyna-Soar, which met opposition among the civilian leadership. It even applied to areas that were well within the Air Force's space mission, such as the development of the MIDAS early warning satellite, which McNamara refused to approve for operational development over the objections of Schriever and others.[97] By the time that the shuttle decision was made, however, the Air Staff had apparently lost much of its enthusiasm for space, particularly for human spaceflight missions. Why this is so is not clear. At the very least, solely military "man-in-space" missions were apparently out of the question, and cooperative missions with NASA were not particularly attractive to the uniformed military.

---

96.   A report by the Air Force's Scientific Advisory Board in June 1983 further symbolized the uniformed Air Force's move away from the dream of a military "man-in-space" program. A special Ad Hoc Committee on the Potential Military Utility of a Manned National Space Station concluded that the most valuable use to the military of a space station was the ability to conduct research and test new technology with human crews in attendance. However, the committee did not feel that this mission justified major involvement or funding; DOD could be a potential customer of the planned NASA space station once it was operational without being an active participant in designing, managing, or funding the station. This time, the Air Force, rather than striving to develop its own program for human spaceflight or even cooperating with NASA as it did with the shuttle, would be content to serve merely as a customer. This later caused some controversy when Secretary of Defense Weinberger insisted that no agreement be signed with an international partner that prevented the United States from conducting military experiments on the station.

97.   General B.A. Schriever, Commander, Air Force Systems Command, to Eugene M. Zuckert, Secretary of the Air Force, "DOD Program Change (4.4.040) on MIDAS (239A)," August 13, 1962, Box B167, Curtis E. LeMay Papers, Library of Congress.

This is not to say that the civilian leadership of DOD in general, or the Air Force in particular, rushed enthusiastically into a major development project with NASA. Certain important members of DOD required much convincing before signing the agreements that increased cooperation with NASA on the Space Shuttle. Later on, in the 1980s, particularly under the leadership of first Undersecretary and then Secretary of the Air Force Edward Aldridge, the civilian leadership at the Air Force became particularly suspicious and distrustful of the total reliance on the Space Shuttle. It is also true that by the 1980s, the military space program had clear priority within the White House. Even the policy-making apparatus for space decisions, centered as it was in the National Security Council, was biased in favor of DOD over NASA.

The *Challenger* accident did not create the problems for DOD in general and the Air Force in particular in terms of cooperation with NASA. However, it did throw them into harsh relief; it confirmed the grumblings and second-thoughts of much of the uniformed military. All of this is important to recognize with respect to what happened later to Air Force-NASA relations—*Challenger* was not the cause, merely the most blatant symptom of a long-standing tension.

Civilian DOD officials typically serve no more than a single presidential term in office. Occasionally, they move to higher positions, but it is far more common that they leave the government altogether. They therefore rarely have to live with the long-term consequences of the policy decisions they make. The uniformed officers in a service, however, do remain. The mid-level officers frequently are given the task of implementing decisions made at higher levels and then may rise to general officer rank themselves years later, when they are faced with the consequences of the decisions made earlier. In the case of the shuttle decision, many Air Force officers who were colonels and lieutenant colonels at the time later rose to general officer rank when the true effects of the shuttle decision—particularly the higher costs and the schedule delays—were being felt. At that point, they were inclined to heavily resist any further cooperative efforts with NASA.

This was the legacy that NASA and DOD faced as the 1990s began. The situation was akin to what Mark Twain once said about a cat that sits on a hot stove top: it will never sit on a hot stove top again, but neither will it sit on a cold one. Thus, despite the change of the civilian political leadership at both DOD and NASA from both the change of administrations and simple personnel turnover, the institutional memory of the Air Force—its uniformed officers—remained highly distrustful of any cooperative agreement foisted on them by civilians.

Conclusion

The civilian-military relationship in space has been one that has evolved over time and continues to evolve to this day. Determining whether it has been a success or not is largely impossible, because the question depends on at what level one wants to look.

At the operational level, there has been much successful cooperation on all aspects of the space program. DOD provided facilities, material, and personnel in support of the civilian space agency. Navy ships conducted retrieval operations for NASA missions. Air Force personnel served in important positions in the Apollo program. DOD and NASA shared tracking and communications facilities for each other's programs. Even the highly secretive "black" intelligence programs have been used in the civilian space program. Optics developed for reconnaissance satellites found their way into Apollo and other space science missions. In fact, a reconnaissance satellite was even used to photograph the

Skylab space station soon after launch to assess the damage it incurred during liftoff. The photographs were used to train the NASA astronauts who flew the repair mission.[98]

At the policy level, it has been a different story. From the Air Force's perspective, the service has largely come up short—being relegated to less glamorous, but more vital roles in space, while also being forced to serve in a support capacity for NASA, which managed to take much of the credit. For the first decade of its existence, NASA reaped the fruits of much military spending and research on space and was frequently predominant in policy disputes. Beginning with the shuttle, NASA's dependence on the military for more than just operational support became blatantly clear. In the end, however, the Air Force seems to have suffered more from this situation as well.

By the early 1990s, the situation had become much more complex. Both NASA and DOD needed each other to find a solution to the problem of excessive launch costs. Perhaps more importantly, NASA began the painful transformation to a post–Cold War world much earlier than the military space program. Whether the military can learn from NASA's example awaits to be seen.

---

98.   Dwayne A. Day, "The Air Force in Space: Past, Present and Future," *Space Times: The Magazine of the American Astronautical Society* 35 (March-April 1996): 17.

Document II-1

Document title: Major General L.C. Craigie, Director of Research and Development Office, Deputy Chief of Staff, Materiel, to Brig. Gen. Alden R. Crawford, Air Materiel Command, Wright Field, Dayton, Ohio, "Satellite Vehicles," January 16, 1948, with attached: Memorandum for the Vice Chief of Staff, "Earth Satellite Vehicles," January 12, 1948, and General Hoyt S. Vandenberg, Vice Chief of Staff, United States Air Force, "Statement of Policy for a Satellite Vehicle."

Source: NASA Historical Reference Collection, NASA History Office, NASA Headquarters, Washington, D.C.

Following RAND's study titled "Preliminary Design of an Experimental World Circling Spaceship," published as Document II-2 in Volume I of Exploring the Unknown, *RAND conducted several more studies. The staff of Headquarters United States Air Force ordered the Air Materiel Command to evaluate RAND's studies. The Materiel Command returned a cautious report stating that the practicality of satellites was questionable and advised further study. As a result, the Air Staff authorized further study of the subject by RAND, and also stated that the Air Force was the logical service for developing satellite systems. This was the first definitive statement by the Air Force that it should have primacy in space systems.*

[no pagination]   16 January 1948

SUBJECT: Satellite Vehicles

TO: Commanding General
            Air Materiel Command
            Wright Field, Dayton, Ohio
            Attn: Brig Gen Alden R. Crawford

   1. Reference is made to memorandum dated 8 December 1947, file TSKON-9/MSR/loa, subject as above.

   2. In line with the contents of referenced letter, the attempted statement of policy covering this matter has been formulated and approved.

   3. It is requested that this policy be implemented by action under the RAND contract. This matter has been co-ordinated [sic] with the local RAND office.

   4. The classification of this subject may be considered confidential with the exception of the attached policy statement.

BY COMMAND OF THE CHIEF OF STAFF:

                        L.C. CRAIGIE
                        Major General, U.S. Air Force
                        Director of Research and Development
                        Office, Deputy Chief of Staff, Materiel

****

[no pagination]   12 JAN 1948

Memorandum for the Vice Chief of Staff

SUBJECT: Earth Satellite Vehicles.

DISCUSSION.
   1. Progress in guided missile research and development by the Air Force, the Navy and other agencies is now at a point where the actual design, construction, and launching of an Earth Satellite Vehicle is technically, although not necessarily, possible. The passage of time, with accompanying technical progress, will gradually bring the cost of such a missile within feasible bounds.
   2. It seems therefore, imperative, in order that the USAF maintain its present position in aeronautics and prepare for a future role in astronautics, that a USAF policy regarding Earth Satellite Vehicles be promulgated. A suggested policy is attached hereto.

RECOMMENDATION.
   That the inclosed [sic] policy be approved.

****

[no pagination]

Statement of Policy for a Satellite Vehicle

   The USAF, as the Service dealing primarily with air weapons—especially strategic—has logical responsibility for the Satellite.
   Research and development will be pursued as rapidly as progress is guided missiles are justifies and requirements dictate. To this end the problem will be continually studied with a view to keeping an optimum design abreast of the art, to determine the military worth of the vehicle—considering its utility and probably cost—to insure development in critical components, if indicated, and to recommend initiation of the development phases of the project at the proper time.

                                        HOYT S. VANDENBERG
                                        General, United States Air Force
                                        Vice Chief of Staff

Document II-2

Document title: Robert R. Bowie, Policy Planning Staff, Department of State, "Memorandum for Mr. Phleger," March 28, 1955.

Source: State Department Central Decimal Files (711.5/3-2855), Record Group 59, National Archives and Records Administration, Washington, D.C.

Document II-3

Document title: Robert R. Bowie, Policy Planning Staff, Department of State, to Secretary of State, "Recommendations in the Report to the President by the Technological Capabilities Panel of the Science Advisory Committee, ODM (Killian Committee): Item 2—NSC Agenda 10/4/56," October 2, 1956.

Source: Record Group 59, General Records of the Department of State: Records Relating to State Department Participation in the Operations Coordinating Board and the National Security Council, 1947-1963, Box 87, "NSC 5522 Memoranda," National Archives and Records Administration, Washington, D.C.

In February 1955, the Technological Capabilities Panel, headed by MIT professor James R. Killian, produced a report on the threat of surprise attack on the United States. The report made a number of recommendations on how to reduce this threat, including the development of radar early warning systems and better intelligence collection methods. One recommendation was the establishment of the concept of "Freedom of Space" by first orbiting a scientific satellite before orbiting an intelligence satellite. This recommendation resulted in the signing of NSC 5520, "Draft Statement of Policy on U.S. Scientific Satellite Program," published as Document II-10 in Volume I of Exploring the Unknown. *Prior to the signing of this document, the Department of State was requested to study the issue and report to the National Security Council (NSC), as stated in the recently declassified top secret letter by Robert Bowie to Assistant Secretary of State Herman Phleger. The Policy Planning Staff at the Department of State continued to study the issue, along with several other recommendations in the Technological Capabilities Panel's report, and issued further reports on their status, also recently declassified from "Top Secret status," including the "Freedom of Space" recommendation. "Freedom of Space" continued to be an issue for several years after Sputnik.*

Document II-2

[no pagination]   March 28, 1955

Memorandum for Mr. Phleger

At a recent meeting, the NSC considered a report to the President by a panel of the Science Advisory Committee on threat of surprise attack.

Recommendations No. 9 and B. 12b of the report read as follows:

"9. A re-examination be made of the following principles or practices of international law from the standpoint of recent advances in weapons technology:

"a. Freedom of the Seas. Radical extension of the 'three-mile limit' to permit control of surface and subsurface traffic from the coastline to beyond the likely striking range of sea-launched nuclear missiles.

"b. Freedom of Space. The present possibility of launching a small artificial satellite into an orbit about the earth presents an early opportunity to establish a precedent for distinguishing between 'national air' and 'international space,' a distinction which could be to our advantage at some future date when we might employ larger satellites for intelligence purposes."

"B. 12b. Studies should be made of appropriate changes in the concept of the 'three-mile limit' to permit actions in keeping with the threat; for realistic implementations of any policy changes, the missions of the Coast Guard and Navy must be amended and forces increased to equal the tasks of inspection and control."

The Department of State has been requested to study these recommendations, in coordination with the Departments of Defense, Treasury, and Justice, and to submit a report and recommendations to the NSC on or about May 15, 1955.

It seems clear that L should undertake the two studies involved, working with other interested divisions and offices of the Department.

Robert R. Bowie

Document II-3

[no pagination]   October 2, 1956

TO: The Secretary

THROUGH: S/S

FROM: S/P - Robert R. Bowie

SUBJECT: Recommendations in the Report to the President by the Technological Capabilities Panel of the Science Advisory Committee, ODM (Killian Committee): Item 2—NSC Agenda 10/4/56

1. The Council is asked to note the status of implementations of the Technological Capabilities Panel (TCP) recommendations on "Meeting the Threat of Surprise Attack," as presented in the several agency reports contained in NSC 5611 ("Status of National Security Programs on June 30, 1956"). Oral reports may be given to the Council by Defense, AEC, ODM, FCDA [Federal Civil Defense Authority] and CIA.

2. The draft Record of Action, which the Council will be asked to approve:

a) noted a number of changes in programs to carry out that is assigned to Defense;

b) requests Defense to supplement its Council briefing, in December, on the ICBM, with a report on the anti-missile missile program; and

c) defers decision on a follow-up study to the Killian Report, which the TCP recommended "within two years."

Defense and ODM differ as to the need for this: The Planning Board agreed to defer a recommendation to the Council until the ODM consults its Science Advisory Committee, the TCP parent, on whether technological advance in the past two years justifies initiation of another study at this time.)

3. Five TCP Recommendations were assigned as our primary responsibility by NSC Action 1355. We do not make an annual Status Report and therefore have not submitted an accounting. In the event that questions arise concerning their status, I am attaching a brief memorandum of comments you may care to use.

****

[1]
Status of Implementation of TCP Recommendations Assigned to the Department of State

General Recommendation 7 a - b - c:
"The NSC initiate preparatory studies of the problems of international negotiations in the following areas growing out of recommendations of this Report."

a. "*Atomic weapons in air defense negotiations with Canada* to provide our air defense forces with authority to use atomic warheads over Canada."
Status: Under current negotiation with the Canadian Government.
Comment: Preliminary negotiations were opened last month between the Department and the Canadian Ambassador to discuss the integration of atomic weapons in joint US-Canadian air arrangements. The Ambassador was informed of new weapons developments and their implications for air defense. We pointed out in particular that US forces must have advance authority to overfly Canada with atomic weapons and to use such weapons over Canadian territory in air defense. The conversations covered other aspects of the problem including the compatibility of Canadian aircraft for US weapons, the training of Canadian personnel, the storage of weapons on Canadian soil, and the availability of the weapons to Canadian forces. The Canadian Ambassador stressed the political sensitivity of the problem and stated that he would report to his government and reply to the US how it thought the matter might best be studied.

b. "*Extension of the Planned Early Warning Line* - International negotiations for the seaward extension of the distant Early Warning Line from Greenland via Iceland and the Faroes to join future NATO warning systems."
Status: a) *Denmark:* Under current negotiation with the Danish Government; b) *Iceland:* in abeyance pending political developments with respect to the base problem; c) *UK:* awaiting a Defense report of current conversations between the US and UK Chiefs of Staff.
[2] *Comment:* With respect to the requirements in Greenland (6 radar sites and their associated communication facilities), the Danish Foreign Office has recently granted approval for the conduct of technical and engineering surveys by US military authorities but has made clear that the approval is without prejudice to final decision of the Danish Government regarding the establish-ment location and operation of the proposed radar stations. With respect to the programmed Northwest radar site in Iceland, the present situation is obscure in view of the uncertain future status of US and NATO defense installations in Iceland. With regard to requirements in the Faroes, the Depart-ment has recently requested information from the Department of Defense of the details of these requirements in order that they may be considered from the political viewpoint. With respect to the termination of the DEW Line in the United Kingdom, the US Joint Chiefs of Staff have informed the British Chiefs of Staff of the general nature of this proposal, and are currently awaiting a reply. The Department of Defense has been requested to inform the Department of State as soon as the reply is received. The Department of State has also asked for information from Defense on the relation-ship of the proposed DEW Line extension both to SHAPE's plans and to SACLANT's plans, both of which contain NATO requirements for early warning facilities.

c. "*Remote Sea Monitor Line* - International negotiations for the installation of a submerged, sea traffic monitor line extending from Greenland to Iceland and to the United Kingdom."

Status: The Department is awaiting definitions of defense requirements, which, it understands, are now being worked out in service to service discussions.

General Recommendation 9 - b:

"*Freedom of Space* - The present possibility of launching a small artificial satellite into an orbit about the earth presents an early opportunity to establish a precedent for distinguishing between 'national air' and 'international space,' a distinction which could be to our advantage at some future time when we might employ larger satellites for intelligence purposes."

[3] *Status:* The Department's Legal Adviser has this problem under current review. State has participated with Defense, the National Science Foundation, and the National Academy of Science in planning the program for launching an earth satellite as part of the US participation in the International Geophysical Year 1957-58. Our studies are continuing in cooperation with the interested agencies.

Comment: So for as law is concerned, space beyond the earth is an uncharted region concerning which no firm rules have been established. The law on the subject will necessarily differ with the passage of time and with practical efforts at space navigation. Various theories have been advanced concerning the upper limits of a state's jurisdiction, but no firm conclusions are now possible.

A few tentative observations may be made: (1) A state could scarcely claim territorial sovereignty at altitudes where orbital velocity of an object is practicable (perhaps somewhere in the neighborhood of 200 miles); (2) a state would, however, be on strong ground in claiming territorial sovereignty up through the "air space" (perhaps ultimately to be fixed somewhere in the neighborhood of 40 miles); (3) regions of space which are eventually established to be free for navigation without regard to territorial jurisdiction will be open not only to one country or a few, but to all; (4) if, contrary to planning and expectation, a satellite launched from the earth should not be consumed upon reentering the atmosphere, and should fall to the earth and do damage, the question of liability on the part of the launching authority would arise.

General Recommendation 2B - 12-a:

"We recommend that comprehensive programs be instituted to provide effective control of surface and, so far as possible, sub-surface traffic in both oceans from the coastlines to beyond the likely striking range of sea-launched attacks. For proper implementation:

"a. international arrangements should be made for the establishment of information reporting procedures and of control measures."

[4] *Status:* The Department is awaiting the results of other studies, assigned to Defense, which will bear on the scope and type of the "international arrangements" desired. It is our understanding that Defense has recently consulted with Treasury to ascertain whether international arrangements for search and rescue operations could be expanded to satisfy defense requirements.

Document II-4

Document title: Percival Brundage, Director, Bureau of the Budget, to the President, "Project Vanguard," April 30, 1957.

Source: Bureau of the Budget Files, Dwight D. Eisenhower Library, Abilene, Kansas.

Project Vanguard was the result of NSC 5520 and was intended to establish "Freedom of Space"—the right to overfly foreign territory for future intelligence satellites. The initial estimate of its cost was $15 to $20 million, but by mid-1956 the program was already over budget, and estimates of its total costs continued to grow. In April 1957, the Director of the Bureau of the Budget, Percival Brundage, wrote President Eisenhower explaining the costs of the program and where additional funding had been found. His memorandum provides a good insight into the close relationship between the National Academy of Sciences and the Department of Defense. It also indicates that $2.5 million for the Scientific Satellite Program came from the Central Intelligence Agency. Finally, Brundage notes that work on the Air Force reconnaissance satellite was funded for the next fiscal year and that if the Vanguard satellite was not completed, satellite research would still continue.

---

[1]   April 30, 1957

Memorandum for the President

Subject: Project VANGUARD

The Department of Defense advises that developmental difficulties requiring additional time and effort have resulted in further revision of the estimated total cost of Project VANGUARD and that it will not be possible to complete the presently authorized six vehicle project within the January estimate of $83.6 million for the total cost. Arrangements have been made to fund approximately $70 million to date. Of this amount, some $50 million is being provided by the Department of Defense for the launching vehicles and related activities, of which $25 million was advanced from the fiscal year 1957 Department of Defense emergency fund and has not been replaced. A fiscal year 1956 supplemental appropriation for the National Science Foundation has provided funds for the satellites themselves and the scientific instrumentation and ground observations.

We have been advised that it is currently estimated that if no further major developmental problems are encountered, the project may be completed within a total of $110 million. With respect to the probability of success of the project within this level of funding, the Department of Defense has reviewed and reconfirmed its statement to the National Security Council at the meeting of January 24, 1957, that in the technical judgment of Defense scientists and their consultants at least one successful satellite should result from six launchings of the presently planned Project VANGUARD launching vehicle. Since arrangements have been made to fund approximately $70 million, an additional amount of $40 million would be required to complete the project on present assumptions.

While no further major technical difficulties are now anticipated, it must be recognized that flight tests have not yet been completed. We have been advised that in the event unforeseeable developments should make it necessary to incorporate fundamental changes in the present approach or to employ an alternative approach, substantial additional funds beyond the $110 million estimate might be required.

When continuation of the policy established under NSC 5520 [was] considered at the NSC meeting of May 8, 1956, it was decided that this policy should be continued "with the understanding that the program developed thereunder will not be allowed to interfere with the ICBM and IRBM programs but will be given sufficient priority by the Department of Defense in relation to other weapon systems to achieve the objectives of NSC 5520."

The use of Department of Defense emergency funds in late fiscal year 1956 as well as during fiscal year 1957 was necessary because costs of [2] development and procurement of the launching vehicles increased much higher than the original estimate. The Central Intelligence Agency had made $2.5 million available to the Department of Defense, and the National Science Foundation was able to transfer $5.8 million when the decision was made to plan for no more than six launchings. It is the position of the Department that use of its funds was not based on any understanding by the Department that it had a con-tinuing responsibility for funding this project but rather that the Department has used its funds thus far because no other clear-cut assignment of responsibility for funding the launching vehicles has been made and because it was assured that funds advanced to this project would be replaced, at least insofar as advances were made from fiscal year 1957 funds.

The Secretary of Defense has now concluded that it is not advisable for the Department to provide further support of the project in fiscal year 1957 or future years from the emergency fund. In addition to the fact that the Department does not consider that it has a continuing responsibility for the project, the Secretary's position is under-stood to result from the fact that the Department has not been reimbursed for fiscal year 1957 emergency funds already provided as well as from congressional criticism of the use of emergency funds for this purpose. In this connection it is noted that in view of estab-lished fiscal policies limiting supplemental appropriations to the most urgent cases, the Bureau of the Budget recently disapproved a request of the Department of Defense to reimburse the emergency fund.

The Bureau of the Budget has reviewed this problem with staff of the Department of Defense and the National Science Foundation. From the evidence at hand, the Bureau of the Budget believes that the project cannot go forward without additional funding. Taking into consideration the fact that this project has all the elements of a guided mis-sile development program together with additional problems of a novel and difficult char-acter, it is not surprising that substantial cost increases have occurred. However, inasmuch as the Department is now well into the project and states that it has already resolved a number of the technical problems, the present estimate of $110 million may be more reli-able than previous estimates.

On the other hand, in the light of past experience with this project and in the absence of flight test results confirming the soundness of the present approach, I believe that it should be recognized that the cost of the project may be as high as $150 to $200 million. In weighing the benefits deemed to be derived from the project and its priority in com-parison with all the other current projects, it was initially approved in the expectation that the cost would be between $15 and $20 million. I question very much whether it would have been authorized, at least on a crash basis, if the actual cost had been known at that time.

[3] It is hoped that in the future more careful estimates will be made as to the total cost or range in possible costs before such projects are initially approved. Furthermore, this seems to offer an opportunity to give up a desirable project for something else which is considered to be of higher priority in relation to cost and benefits to be derived. We are presently developing nine intercontinental and intermediate missiles with a range of over 1,000 miles, some of which involve comparable techniques and which will require difficult

priority decisions as to programming and funding. Some eliminations will have to be made.

The Department of Defense has indicated interest in this program to about the same degree it has shown on some other basic research projects, but has stated that its interest is not sufficient to justify the project's continuance with Department of Defense financing. Therefore, the Department believes that the program must be justified on the basis of the several national objectives stated in NSC 5520 rather than on the Department's interest.

The Department of Defense believes that to prosecute the balance of the program successfully, adequate financing should be arranged by supplemental requests submitted for appropriation to the National Science Foundation, which the Department considers to be the sponsor of the program. The Department would assist in justifying the supplemental requests of the National Science Foundation by assuming the burden of justification as to the technical difficulties encountered and the cost elements involved.

It should be noted that one of the important considerations has been and is the completion of the project during the period of the International Geophysical Year. If you desire the project to be continued in accordance with the existing policy under NSC 5520, it is suggested that the following actions could resolve the current financing problem:

1. The Department of Defense should be directed to provide immediately $5.8 million from the emergency fund to continue the project from May 1 through approximately August 1. The Department feels it must clear this use of the emergency fund with the Appropriations Committees who have questioned the propriety of its use for this purpose. It should be recognized that the Department would prefer that these funds be replaced.

2. A fiscal year 1958 budget amendment should be submitted requesting an additional $34.2 million for appropriation to the National Science Foundation to cover costs to completion of the project, assuming that current cost estimates are valid, that no further major difficulties are encountered in the course of completing the development, and that the [4] Department of Defense would continue to provide general support for which no special funding has been considered necessary. Upon availability to the National Science Foundation these funds would be transferred to the Department of the Navy to complete the program.

The National Science Foundation believes that in view of the national interests involved the program cannot be permitted to fail at this stage. If it were the only possible alternative to cancellation of the project, the National Science Foundation would consider it necessary in the total national interest to request a supplemental appropriation to cover the costs required to complete the responsibilities undertaken by the Department of Defense under NSC 5520. Moreover, the National Science Foundation recommends that the Department of Defense provide the necessary funds to complete the project for the following reasons: (1) the Department of Defense is responsible under the present terms of NSC 5520 for the portion of the program requiring additional funds; (2) the Department of Defense is best qualified to justify to the Congress the reasons for present cost increases.

Apparently, both the Department of Defense and the National Science Foundation are very reluctant to continue to finance this project to completion. But each is quite prepared to have the other do so.

General Cutler believes the following considerations are particularly relevant to a decision in this matter:

"1. The substantive scientific information concerning upper atmospheres which might be acquired by the launching of a successful satellite. Included in this information would be data as to the content of the upper atmosphere (such as invisible heavenly bodies) through which the very costly intercontinental ballistic missiles, if perfected, must pass.

"2. The world reaction to an abandonment by the U.S. in mid-stage of the satellite program. A conclusion that the richest nation in the world could not afford to complete this scientific undertaking would be unfortunate. Even more unfortunate would be an inevitable inference that American scientists were not up to bringing the project to a successful conclusion.

"3. The reaction of the scientific community to the abandonment by the U.S. in mid-stage of the satellite program. A time when the Free World is coming more and more to depend on advanced technology and scientific accomplishment is not a time to alienate the scientific community at home and lead it to believe that the Government has lost faith in scientific accomplishment. [5] From what I hear and read, the scientific community and those in highly technical industry who work with them are already sensitive in this regard.

"4. A final decision on the satellite program should be made by the President on an integrated presentation of the views of all concerned in this matter. The integrated process of presentation, such as is illustrated in the National Security Council, is a primary achievement of this Administration. Where so much, beyond financial considerations alone, is at stake, the President should have the benefit of an integrated presentation and discussion. This point of view is important, irrespective of what the President's decision might ultimately be."

It should be noted that the Air Force has already started its own project for a much larger reconnaissance satellite vehicle and is spending approximately $10 million in fiscal year 1957 and is currently planning additional funding of at least $10 million for fiscal year 1958. Therefore, whether or not the International Geophysical Year satellite project is completed, research in this area will not be dropped.

Percival Brundage
Director

Document II-5

Document title: Lieutenant General Donald L. Putt, Deputy Chief of Staff, Development, U.S. Air Force, to Dr. Hugh L. Dryden, Director, National Advisory Committee for Aeronautics, January 31, 1958.

Document II-6

Document title: Gen. Donald L. Putt, to Commander, Air Research and Development Command, "Advanced Hypersonic Research Aircraft," January 31, 1958.

Document II-7

Document title: General Thomas D. White, Chief of Staff, USAF, and Hugh L. Dryden, Director, NACA, "Memorandum of Understanding: Principles for Participation of NACA in Development and Testing of the 'Air Force System 464L Hypersonic Boost Glide Vehicle (Dyna-Soar I),' " May 20, 1958.

Source: All in NASA Historical Reference Collection, NASA History Office, NASA Headquarters, Washington, D.C.

Even before NASA was created, the Department of Defense (DOD) and National Advisory Committee for Aeronautics (NACA) were cooperating on space-related developments. The letter from Lt. General Putt, Air Force Deputy Chief of Staff, Development, opened the possibility for NACA participation in a potential X-series aircraft with the qualities of both a spacecraft and an airplane. Technically a hypersonic boost-glide vehicle, its flight characteristics were termed "dynamic soaring" for its ability to skim the thin air of the upper atmosphere. It was given the nickname of "Dyna-Soar." While the motivation from the DOD side was the development of technologies for an orbital bombing aircraft and related missions, NACA participation was intended to benefit civil applications. Dyna-Soar was not covered in the original agreements creating NASA that outlined transferring or sharing programs with DOD. Dyna-Soar's importance was its demonstration of the possibility of joint development of a major new system, despite widely differing reasons for cooperation. Although the program was canceled in 1963 for technical and cost reasons, it set a precedent for future cooperation.

Document II-5

[1]   31 January 1958

Dr. Hugh L. Dryden
Director
National Advisory Committee for Aeronautics
1512 H Street, N.W.
Washington 25, D.C.

Dear Dr. Dryden:

In the last few months the dimensions of the contest for superiority in aircraft and missile technology have suddenly and drastically expanded.

This letter is addressed to a particularly important event in this contest—the matter of a research vehicle program to explore and solve the problems of manned space flight. Specifically, the Air Force is convinced that we must undertake at once a research vehicle program having as its objective the earliest possible manned orbital flight which will contribute substantially and essentially to follow-on scientific and military space systems.

The Air Force has set up a design competition for a hypersonic boost glide vehicle nicknamed Dyna Soar I. The objectives of this program closely conform to the recommendations of the NACA report of last summer. It appears probable that this vehicle will be able to orbit as a satellite since the aerodynamic heating problems of re-entry appear less severe than those of the Dyna Soar I flight profile. However, it may be feasible to demonstrate an orbital flight appreciably earlier with a vehicle designed only for the satellite mission than would be possible with a vehicle capable of the boost-glide mission as well. It is necessary, therefore, to determine whether a research aircraft designed only as a satellite will give us an orbital flight of technical significance enough sooner than a vehicle designed for the glide mission to warrant a separate development.

Both the NACA and the Air Force are well along in investigations seeking the best approach to the design of a manned earth orbiting research vehicle. We earnestly believe that these efforts should be joined at once and brought promptly to a conclusion. Accordingly the NACA is invited to collaborate with the Air Research and Development Command [ARDC] in this important task. Because of the advanced stages to which the individual NACA and ARDC investigations have already [2] progressed and because of the urgency of getting on with the job, we believe that the evaluation should be confined to existing and planned projects, appropriate available proposals, and competitive approaches already under study. We visualize that any program growing out of this joint evaluation will best be presented, managed and funded along the lines of the X-15 effort, with the Navy being brought into the picture as soon as possible without delaying the evaluation.

To provide further insight into Air Force thinking on this matter, the concluding paragraphs of the letter directing ARDC to make this evaluation are quoted:

"4. . . . it is desired that the evaluation consider separately the following approaches:

"a. What is the best design concept, the minimum time to first orbital flight and the dollar cost of demonstrating a manned one-orbit flight in a vehicle capable only of a satellite orbit? Time is a primary consideration, but to qualify, an approach must offer prospects of tangible contributions to the over-all astronautics program.

"b. What is the minimum time to first orbital flight and dollar cost of demonstrating a manned one-orbit flight with a vehicle designed to utilize the boost-glide concept? In this approach it is not necessary that the first orbit flight be made within the atmosphere if an "outside" orbit offered the possibility of an earlier successful flight.

"5. The following additional guidance is provided:

"a. The program to meet the stated objective should be the minimum consistent with a high degree of confidence that the objective will be met. Maximum practical use must be made of existing components and technology and of the momentum of existing programs.

"b. The hazard at launch and during flight will not be greater than that dictated by good engineering and flight safety practice. If feasible, in order to save time and money, pilot safety may be provided by emergency escape systems rather than insisting on standards of component reliability normally required for routine repetitive flights of weapon systems. This statement is particularly pointed at the problem of qualifying boosters for initial orbital flights.

"6. It is requested that this Headquarters be furnished the results of your evaluation of each of the approaches specified in paragraph 4. Finally, your over-all conclusions and recommendations for accomplishing the objective stated in paragraph 1 are desired.

[3] "7. The requested information should be forwarded at the earliest practicable date, but in no event later than 15 March 1958."

It is hoped that the Air Force-NACA team relationship which has proven so effective in earlier programs of the X-airplane series can be continued in the conception and conduct of this and other research vehicle programs directed to the extension of our knowledge and capability in upper atmosphere and space operations.

We look forward to receiving your comments and suggestions to this proposed course of action.

Sincerely,

D. L. Putt
Lieutenant General, USAF
Deputy Chief of Staff, Development

Document II-6

[1]   31 January 1958

SUBJECT: Advanced Hypersonic Research Aircraft

TO: Commander
      Air Research and Development Command
      Andrews Air Force Base
      Washington 25, D.C.

1.   It is desired that ARDC in collaboration with the NACA expedite the evaluation of existing or planned projects, appropriate available proposals and other competitive approaches with a view to providing an experimental system capable of an early flight of a manned vehicle making an orbit of the earth. The Air Force-NACA team relationship which has proven so productive in earlier programs of the X-airplane series will be continued in the conception and conduct of this new program. A letter, copy attached, has been sent to invite NACA collaboration. It is contemplated that as soon as possible without delaying the evaluation, the Research Aircraft Committee will be convened to invite Navy participation.

2.   A manned orbital flight, whether by a glide vehicle or by a minimum altitude satellite essentially outside the earth's atmosphere is a significant technical milestone in the USAF space program. It is also vital to the prestige of the nation that such a feat be accomplished at the earliest technically practicable date—if at all possible before the Russians. However, it should be clearly understood that only those approaches to an early demonstration of manned orbital flight will be considered which can be expected to contribute information of a substantial value to follow-on systems.

3.   It is understood that the boost-glide test vehicle which will be developed under the Dyna-Soar I program will be able to orbit as a satellite. It is also understood, however, that the problems associated with a manned orbital flight as a satellite, [are] outside the stringent design requirements than the lower altitude, hypersonic Dyna-Soar I flight profile. Consequently, it may be feasible to demonstrate an orbital flight appreciably earlier with a vehicle designed only for the satellite mission than would be possible with a vehicle capable of executing the boost-glide mission as well. An important objective of the evaluation, then, will be to determine whether a test vehicle designed only as a satellite will give us an orbital flight of technical significance enough sooner than a vehicle designed for the glide mission to warrant a separate development. Consequently, it is desired that the evaluation consider separately the following approaches:

a.   What is the best design concept, the minimum time to first orbital flight and the dollar cost of demonstrating a manned one-orbit flight in a vehicle capable only of a satellite orbit? Time [2] is a primary consideration, but to qualify, an approach must offer prospects of tangible contributions to the over-all astronautics program.

b.   What is the minimum time to first orbital flight and dollar cost of demonstrating a manned one-orbit flight with a vehicle designed to utilize the boost-glide concept? In this approach it is not necessary that the first orbit flight be made within the atmosphere under typical boost glide conditions—it could be made outside the atmosphere if an "outside" orbit offered the possibility of an earlier successful flight. . . .

5.   The following additional guidance is provided:

a.   The program to meet the stated objective should be the minimum consistent with a high degree of confidence that the objective will be met. Maximum practical use must be made of existing components and technology and of the momentum of existing programs.

      b.   The hazard at launch and during flight will not be greater than that desired by good engineering and flight safety practice. If feasible, in order to save time and money, pilot safety may be provided by emergency escape systems rather than insisting on standards of component reliability normally required for routine repetitive flights of weapon systems. This statement is particularly pointed at the problem of qualifying boosters for inisial [sic] orbital flights.

      6.   It is requested that this Headquarters be furnished the results of your evaluation of each of the approaches specified in paragraph 4. Finally, your over-all conclusions and recommendations for accomplishing the objective stated in paragraph 1 are desired.

      7.   The requested information should be forwarded at the earliest practicable date, but in no event later than 15 March 1958.

FOR THE CHIEF OF STAFF:

Gen. Donald L. Putt

Document II-7

[1]

Memorandum of Understanding

Subject: Principles for Participation of NACA in Development and Testing of the "Air Force System 464L Hypersonic Boost Glide Vehicle (Dyna Soar I)."

1.   System 464L is being developed to:
    a.   Determine the military potential of hypersonic boost glide type weapon systems and provide a basis for such developments.
    b.   Research characteristics and problems of flight in the boost glide flight regime up to and including orbital flight outside of the earth's atmosphere.
2.   The following principles will be applied in conduct of the project:
    a.   The project will be conducted as a joint Air Force-NACA project.
    b.   Overall technical control of the project will rest with the Air Force, acting with the advice and assistance of the NACA. The two partners will jointly participate in the technical development to maximize the vehicle's capabilities from both the military weapon system development and aeronautical-astronautical research viewpoints.
    c.   Financing of the design, construction, and Air Force test operation of the vehicles will be borne by the Air Force.
    d.   Management of the project will be conducted by an Air Force project office within the Directorate of Systems Management, Hq ARDC. The NACA will provide liaison representation in the project office and provide the chairman of the technical team responsible for data transmission and research instrumentation.
    e.   Design and construction of the system will be conducted through a negotiated contract with a prime contractor selected by the USAF on the basis of the recommendations of the ARDC-AMC-SAC Source Selection Board, acting with the consultation of the NACA.

[2] f. Flight test of the vehicle and related equipments will be accomplished by the NACA, the USAF, and the prime contractor in a combined test program under the overall control of a joint NACA-USAF Committee, chaired by the Air Force.

General Thomas D. White
Chief of Staff, USAF
13 May 1958

Hugh L. Dryden
Director, NACA
20 May 1958

Document II-8

Document title: T. Keith Glennan, NASA Administrator, and Roy W. Johnson, Director, Advanced Research Projects Agency, "Memorandum of Understanding: Program for a Manned Orbital Vehicle," November 20, 1958.

Source: NASA Historical Reference Collection, NASA History Office, NASA Headquarters, Washington, D.C.

NASA Administrator Glennan and Advanced Research Projects Agency (ARPA) Director Roy Johnson agreed in mid-September 1958 that their two agencies would cooperate on a "man-in-space" program based on the development of space capsules; this program would complement the Air Force Dyna-Soar program. They established a joint NASA-ARPA Manned Satellite Panel, which included six representatives from NASA and two from ARPA, reflecting the Eisenhower Administration's desire to have NASA primarily responsible for manned spaceflight. This memorandum of understanding established guidelines for this early cooperation.

[no pagination]   November 20, 1958

Memorandum of Understanding

SUBJECT: Program for a Manned Orbital Vehicle

1.   The Administrator of NASA is responsible for management and technical direction of a program for a manned orbital vehicle to be conducted in cooperation with the Department of Defense. The objectives of the program are to achieve, at the earliest practicable date, orbital flight and successful recovery of a manned satellite and to investigate the capabilities of man in this environment. The accomplishment of the program is a matter of national urgency.

2.   In carrying out the program, the Administrator of NASA intends to make full use of the background and capabilities existing in the Department of Defense.

3.   The Department of Defense will support the program until it is terminated by the achievement of a sufficient number of manned orbital flights to accomplish the above objectives.

4. $8,000,000 of FY 1959 funds will be contributed by ARPA in support of the program and will be made available by appropriation transfer to NASA. NASA will budget for and fund all subsequent years' costs.

5.   A working committee consisting of members of the staff of NASA and ARPA will be established to advise the Administrator of NASA on technical and management aspects of the program. The chairman of the committee will be a member of the NASA staff.

T. Keith Glennan
Administrator
National Aeronautics and
Space Administration

Roy W. Johnson
Director
Advanced Research Projects
Agency

Document II-9

Document title: T. Keith Glennan, Administrator, NASA, and Wilber M. Brucker, Secretary of the Army, "Cooperative Agreement on Jet Propulsion Laboratory Between the National Aeronautics and Space Administration and the Department of the Army," December 3, 1958.

Document II-10

Document title: T. Keith Glennan, Administrator, and Wilber M. Brucker, Secretary of the Army, "Cooperative Agreement on Army Ordnance Missile Command Between the National Aeronautics and Space Administration and the Department of the Army," December 3, 1958.

Source: Both from NASA Historical Reference Collection, NASA History Office, NASA Headquarters, Washington, D.C.

In 1958, the Army was called on to transfer two major development agencies to the newly created NASA. The Jet Propulsion Laboratory (JPL) was part of the California Institute of Technology, with expertise in guidance, communications, telemetry, and rocket propellants. All agreed that JPL was a center of technical expertise important to the future of NASA and the space program, and the transfer was complete and immediate. In contrast, the transfer of the Army Ballistic Missile Agency (ABMA) generated substantial controversy, because it was the major development arm of the Army Ordnance Missile Command, and the Army leadership considered it too important to relinquish. ABMA's development work included weather satellite programs such as TIROS, rocket engine work such as the F-1 engine (which later powered the Saturn V), and the booster development group headed by Wernher von Braun. The Army was reluctant to lose von Braun and his team of talented engineers. The Department of Defense had been willing to transfer to NASA such research work as that performed by JPL, recognizing that it could still benefit from the research performed. However, the Army resisted losing a major development group such as ABMA, despite its unsure budgetary footing. The Army's initial intransigence eventually required presidential intervention to resolve the situation.

Document II-9

[1]
Cooperative Agreement on Jet Propulsion Laboratory Between the National Aeronautics and Space Administration and the Department of the Army

A. *AUTHORITIES*

This agreement is authorized by Public Law 85-568 as implemented by Executive Order 10793, dated 3 Dec 1958.

B. *PURPOSE*

The purpose of this agreement is to establish the relationships between the National Aeronautics and Space Administration (NASA) and the Department of the Army (Army) that will govern the following:

1. Implementation of Executive Order No. 10793 dated 3 Dec 58, which is incorporated herein by reference.

2. Planning for the orderly transition from current Army military operations and weapons systems development program to programs predominately in the field of exploration and exploitation of space science and technology for peaceful purposes under NASA direction.

3. Provision for certain Army administrative and logistical support desired by NASA in the operation of JPL.

C. *POLICY*

The Army states and NASA recognizes that an abrupt transfer or cessation of Army activities relating to military operations and weapons systems development programs performed at the JPL would be deleterious to both national defense and the accomplishment of NASA objectives. Both NASA and the Army recognize that NASA is not fully staffed to perform certain administrative functions and to provide the administrative and logistical support essential to the uninterrupted operation of JPL and that NASA may request that certain services and support be provided by the Army.

D. *OPERATING CONCEPTS*

1. NASA will provide for the general management and technical direction of the JPL, except as to projects relating to military operations and weapons systems development programs.

[2] 2. For Calendar Year 1959 the Army will continue its contractual relations with the California Institute of Technology for continued effort by the JPL on the following programs which are specifically related to military operations and weapons systems development programs:

a. The SERGEANT guided missile program.

b. Special intelligence investigations.

c. Secure communications research.

d. Aerodynamic testing and research.

It is expected that these specific Army activities will be largely phased out during CY 59; however, if it is necessary to continue certain activities for a longer period of time, this may be done by direct Army contract or through NASA as may be mutually agreed by NASA and the Army.

3. The Army budgets on a program basis and Army installations receive funds on the basis of assigned program activities. Traditionally, the Army has funded the activities performed at JPL on a Calendar rather than Fiscal Year basis. For these reasons, a firm 1959 program had been agreed to by the Army and JPL prior to the publishing of the Executive Order effecting transfer of JPL. NASA, through assumption of technical direction of the general supporting research portion of the program on 1 January 1958, can reorient the effort toward NASA objectives by the end of the first half of the Calendar Year 1959. Therefore, the Army and NASA reached prior agreement and the Executive Order provided for transfer of Army funds in the amount of $4,078,250 to NASA for this general supporting research program for the first half of Calendar Year 1959. The additional funds for general supporting research during CY 1958 will be provided by NASA.

4. NASA may request from time to time, and the Army agrees, that certain administrative and logistical support can and will be furnished to NASA on a non-reimburseable [sic] basis for servicing contract activities at JPL for Calendar Year 1959. Provision of this support may require in certain instances delegations of authority from NASA to the Army where appropriate to the service or support action requested. After Calendar Year 1959 such services and support may be provided in such scope and under such conditions as may be mutually agreed upon.

The following types of services and support are contemplated:

[3] a. Contract administration;

b. Property transfer; and

c. Such other matters as fall within the purview of this instrument.

The Administrator, NASA, and the Secretary of the Army hereby designate respectively the Director of Business Administration, NASA, and the Chief of Ordnance, Army, to jointly formulate the necessary teams to effectuate this Agreement.

5. It is understood and agreed that the Administrator will delegate to the Secretary of the Army, or his designee, such authority as may be required to authorize the Army to fulfill the intent and purposes of this Agreement.

Date: 3 December 1958
Washington, D.C.

T. KEITH GLENNAN
Administrator, NASA

WILBER M. BRUCKER
Secretary of the Army

Document II-10

[1]

Cooperative Agreement on Army Ordnance Missile Command Between the National Aeronautics and Space Administration and the Department of the Army

A. *AUTHORITY*

This agreement is authorized by public Law 85-583.

B. *PURPOSE*

This agreement is for the purpose of establishing relationships between the National Aeronautics and Space Administration and the Department of the Army for the efficient utilization of United States Army resources in the accomplishment of the purposes of the National Aeronautics and Space Act of 1958. This agreement is intended to provide for relationships in the national interest that will prevent undue delay of progress in the national space program, and prevent undesirable disruption of military programs. This agreement is also intended to contribute to effective utilization of the scientific and engineering resources of the country by fostering close cooperation among the interested agencies in order to avoid unnecessary duplication of facilities.

C. *POLICY*

The National Aeronautics and Space Administration (NASA) and the Department of the Army recognize the often inseparable nature of the efforts of this Nation in meeting military and scientific objectives in the missile and space field. Continuation of the organizational strength of the Army Ballistic Missile Agency (ABMA) of the U.S. Army Ordnance Missile Command (AOMC) and its established contractor structure and support from other elements of the Army has been stated by the Defense Department to he essential to the Defense mission. The proper provisions for asking the capabilities of this organization available for meeting objectives of NASA permit the application of these resources to the needs of both civilian space activities and essential military requirements. Accordingly, this agreement establishes relationships between NASA and the Department of the Army which make the AOMC and its subordinate organizations immediately, directly, and continuously responsive to NASA requirements.

D. *PROCEDURES*

1.   The CG, AOMC, will have full authority, as the principal agent of the Army, to utilize the resources of his Command, those organizations [2] directly under his control through contractual structure, and other elements of the Department of the Army with which he deals directly, for the accomplishment of assigned NASA projects.

2.   Key personnel of AOMC and appropriate subordinate elements, as may be requested by NASA, will serve on technical committees under the chairmanship of NASA, or on advisory groups, or will serve as individual consultants to:

a.   Assist in the development of broad requirements and objectives in space programs.

b.   Assist in the determination of specific projects and specific methods (including hardware development) by which NASA may accomplish its overall objective.

3.   Specific orders for projects to be accomplished for NASA will be placed direct by NASA upon AOMC with provision of funds for their accomplishment. AOMC will accept full responsibility for the fulfillment of the assigned projects as accepted from NASA.

4.   NASA will have direct and continuing access, through visits or resident personnel, for technical contact and direction of effort on assigned NASA projects. In this connection, NASA is invited to place a small staff in residence at AOMC. This staff will provide for a continuing exchange of information on all projects assigned by NASA, as well as exchange of information on supporting research in the entire missile and space field.

5.   On request by NASA, in connection with projects funded by NASA, the prime and sub-contractor facilities of the Army in weapons systems and other programs, including scientific and educational institutions and private industry, will be made available through identical procurement channels and with use of the special authorities delegated to the CG, AOMC, by the Secretary of the Army. In addition, resources of other elements of the Army, available to AOMC on a direct basis for space and missile system development, will be used as deemed necessary in the fulfillment of assigned NASA projects.

6.   The CG, AOMC, is responsible for scheduling the space and missile activities under his control to meet the priority requirements of NASA in a manner consistent with overall National priorities. He is further responsible for anticipating in advance any possible conflict in the commitment of effort to NASA and Defense programs, and for providing a timely report to NASA, as well as to the Department of the Army, for the purpose of resolving such conflicts.

[3] 7.   Public information and historical and technical documentation of assigned NASA projects will be under the direction and control of NASA.

8.   The CG, AOMC, is authorized to enter into specific agreements with the duly designated representative of the Administrator, NASA, in implementation of this agreement.

Date: 3 December 1958
Washington, D.C.

T. KEITH GLENNAN   WILBER M. BRUCKER
Administrator   Secretary of the Army

Document II-11

Document title: T. Keith Glennan, Administrator, NASA, and Thomas S. Gates, Acting Secretary of Defense, Memorandum for the President, "Responsibility and Organization for Certain Activities," October 21, 1959.

Source: Presidential Papers, Dwight D. Eisenhower Library, Abilene, Kansas.

The Army had been reluctant to transfer the Development Operations Division of the Army Ballistic Missile Agency (ABMA) to NASA; it required presidential intervention to settle the matter. This joint agreement finally settled the issue of the transfer of the Development Operations Division headed by Wernher von Braun and the assignment to NASA as the lead in developing a U.S. heavy-lift booster. President Eisenhower approved the proposals outlined in this memorandum on November 2, 1952.

---

[1]   October 21, 1959

Memorandum for the President

SUBJECT: Responsibility and Organization for Certain Space Activities

The Secretary of Defense and the Administrator of NASA have agreed upon, and recommend to the President, certain actions designed to clarify responsibilities, improve coordination, and enhance the national space effort. The actions recommended below are consistent with the steps taken by the Secretary of Defense to clarify responsibilities and assignments in the field of military space applications within the Department of Defense.

The Secretary of Defense and the Administrator have agreed upon and recommend to the President the following actions:

A.   The assignment to NASA of sole responsibility for the development of new space booster vehicle systems of very high thrust. Both the DOD and NASA will continue with a coordinated program for the development of space vehicles based on the current ICBM and IRBM missiles and growth versions of those missiles.

B.   The transfer from the Department of the Army to NASA of the Development Operations Division of the Army Ballistic Missile Agency, including its personnel and such facilities and equipment which are presently assigned and required for the future use of NASA at the transferred activity, and such other personnel, facilities and equipment for administrative and [2] technical support of the transferred activity as may be agreed upon.

C.   The provision by the Army to NASA of such administrative services as may be agreed upon to effect a smooth transition of management and funding responsibility of the transferred activity.

The Secretary of Defense and the Administrator of NASA are in agreement on the following:

1.   The nation requires and must build at least one super booster and responsibility for this activity should be vested in one agency. There is, at present, no clear military requirement for super boosters, although there is a real possibility that the future will bring military weapons systems requirements. However, there is a definite need for super boosters for civilian space exploration purposes both manned and unmanned. Accordingly, it is agreed that the responsibility for the super booster program should be vested in NASA. It is agreed that the recommendations to center this function in NASA and to transfer the Development Operations Division of ABMA to NASA are independent of any decisions on whether either or both of the super booster systems currently under development are continued in their presently conceived form.

2.   The transfer of the Development Operations Division of ABMA shall include transfer of responsibility for Saturn, together with 1960 funds allocated for the project, and transfer to the NASA 1961 budget of such amounts as may be approved for this project in the 1961 Department of Defense budget.

3.   In carrying out its responsibilities, NASA will keep the Department of Defense thoroughly and completely informed on its booster program and will [3] be fully responsive to specific requirements of the Department of Defense for the development of super boosters for future military missions as requested by the Secretary of Defense.

4.   It is NASA's intent to center at the transferred activity the bulk of its space booster vehicle systems work, including an appropriate research and development effort, and ultimately, substantial responsibility for NASA launch operations.

5.   It is agreed that NASA wall provide support to the Department of Defense and military services at the transferred activity in the same manner as it now does at all other field centers.

6.   The management and employment of the transferred activity will be the responsibility of NASA, and no commitment is possible with respect to levels of staffing or funding for the operation. NASA, however, will make every possible effort within its responsibilities and resources to utilize the capabilities of the Development Operations Division of ABMA.

7. The transfer of personnel, facilities, end equipment will he on a nonreimbursable basis.

8. The Department of the Army will provide and maintain on a reimbursable basis station-wide services as required by NASA within the Redstone Arsenal complex.

9. NASA will provide for continuation, transfer, or phasing out of military projects under way at the transferred activity as may be requested and to the extent funded by the Department of Defense, and will undertake at the transferred activity such additional military projects as may be agreed upon by NASA and the [4] Department of Defense.

10. The Department of Defense, the Department of the Army, and NASA, recognizing the value to the nation's space program of maintaining at a high level the present competence of ABMA, will cooperate to preserve the continuity of the technical and administrative leadership of the group.

11. The detailed implementation of the actions proposed will be accomplished through the subsequent negotiation of cooperative agreements between the Department of Defense and NASA.

The Secretary of Defense and the Administrator of NASA have reached agreement and recommend approval of the above actions in the firm belief that the national space effort requires a strong civilian agency and progress and a strong military space effort by the Department of Defense, and clear lines of responsibility and authority if the U.S. is to employ its best efforts in the exploration of outer space and to assure the defense of the nation.

If the President approves the recommended actions set forth in A, B, and C above, the Secretary of Defense and the Administrator of NASA will proceed immediately to form the necessary staff teams to develop the required implementing documents.

Administrator, NASA   Acting Secretary of Defense

                                                        OCT 30 1959

[Handwritten presidential note: "Approved Dwight Eisenhower 2 Nov 59"]

Document II-12

Document title: T. Keith Glennan, NASA Administrator, "DOD-NASA Agreement— Reimbursement of Costs," NASA Management Instruction 1052.14, November 17, 1959, with attached: Thomas S. Gates, Jr., Deputy Secretary of Defense, and T. Keith Glennan, NASA Administrator, "Agreement Between the Department of Defense and the National Aeronautics and Space Administration Concerning Principles Governing Reimbursement of Costs," November 12, 1959.

Source: NASA Historical Reference Collection, NASA History Office, NASA Headquarters, Washington, D.C.

As resources from other government agencies were being allocated to NASA, it became imperative to draw up policies outlining reimbursement procedures. These agreements represented the first comprehensive policy on reimbursement between the Department of Defense and NASA that did not apply to a specific program. They also demonstrated the dominant role of the Defense Department as a provider of various services to NASA in the early years.

---

[1]   November 17, 1959

Management Instruction

SUBJECT: *DOD-NASA AGREEMENT—REIMBURSEMENT OF COSTS*

1. *PURPOSE*
This Instruction incorporates into the NASA Issuance System an agreement entered into between the Department of Defense (DOD) end NASA for the reimbursement of certain costs incurred by either agency in providing services, equipment, supplies, personnel, and facilities for use by the other agency. Provisions of the agreement are effective as of November 12, 1959.

2. *AUTHORITY*
Section 203(b)(6) of the National Aeronautics and Space Act of 1958 (42 U.S.C. 2473(b)(6)).

3. *SCOPE*
   a. Principles set forth in the DOD-NASA agreement, which is included as Attachment A, shall govern the reimbursement of costs incurred by NASA or DOD in providing services, equipment, supplies, personnel, and facilities of the types and for the purposes described therein for use by the other agency.
   b. The agreement shall not apply to existing agreements or arrangements already [agreed] upon between NASA and the military departments or the Advanced Research Projects Agency (ARPA) which may not yet be formalized. However, all future arrangements, agreements, and amendments of existing agreements between NASA and the military departments or ARPA shall conform to the provisions of Attachment A.

[2] 4.   *IMPLEMENTATION* . . .

5.   *CANCELLATION*

NASA Management Manual Instruction 2-3-5 (TS 43), November 17, 1959.

Administrator

****

[1]

Attachment A to NMI 1052.14

Agreement Between the Department of Defense and the National Aeronautics and Space Administration Concerning Principles Governing Reimbursement of Costs

1. *Purpose.*

    Section 203(b)(6) of the National Aeronautics and Space Act of 1958, authorizes the National Aeronautics and Space Administration (NASA) "to use, with their consent, the services, equipment, personnel, and facilities of other Federal agencies with or without reimbursement, and on a similar basis to cooperate with other public and private agencies and instrumentalities in the use of services, equipment,and facilities." Federal agencies are also required to cooperate fully with NASA in making their services, equipment, personnel, and facilities available, and are authorized by this statute "to transfer to or to receive from NASA, without reimbursement, aeronautical and space vehicles, and supplies and equipment other than administrative supplies or equipment." It is the purpose of this Agreement to set forth the general principles governing the reimbursement of costs incurred by DOD or NASA in providing for use by the other of its services, equipment, personnel and facilities and in transferring equipment and supplies.

2. *Principles Governing Reimbursement.*

    Subject to the provisions of paragraph 3 hereof, DOD and NASA agree upon the following general principles governing the reimbursement of costs:

[2] A.   *Orders Contracted Out.* Where DOD or NASA places an order with the other which is contracted out (in whole or in part) to industry, reimbursement will be limited to the direct costs to the contracting agency of the contract, or the standard price established for the item being procured where procurement is accomplished through consolidated contracts covering the same or similar items (or components thereof) for the contracting agency. Except as otherwise provided in subparagraph E below, the agency placing the contract shall bear without reimbursement therefor the administrative costs incidental to its procurement of material or services for the ordering agency. As used in the foregoing sentence the term, "administrative costs" includes the normal administrative services performed in connection with placing, administering or terminating contracts, and such related administrative services as security, contract auditing, inspection, etc. (not all inclusive). Administrative costs are to be distinguished from the procurement costs of end items or services, the latter being appropriate for reimbursement under the provisions of this subparagraph.

B.   *Orders Performed "In-House."* Where DOD performs an "in-house order" for NASA and the order is performed (in whole or in part) in facilities using an industrial-type cost accounting system, the basis of billing will be the same as that used for all customers of the Federal Government. Where the order is performed in facilities not using an industrial-type cost accounting system, reimbursement [3] will be limited to the direct costs (including an allowance for annual and sick leave, holidays, contributions for group life insurance and civil service retirement, etc.) attributable to the performance of the order. In no case, however, will charges be made for depreciation or rent for use of facilities and equipment in connection with the performance of orders.

C. *Administration of Other Agency's Contract.* Where DOD or NASA assigns one of its contracts to the other for purposes of administration, the administering agency may be reimbursed for the cost of contract administration services performed in connection with the contract to the extent of the special direct costs incurred in providing these services to the other and mutually agreed upon as clearly identified added costs.

D. *Material.* Where DOD or NASA provides the other with materials, supplies or equipment from stock, reimbursement will be made in accordance with established agency pricing practice. DOD materials, supplies or equipment which are in excess of DOD requirements (called "transferable-nonreimbursable" property in the DOD), will be furnished without charge, except that the furnishing agency may require reimbursement for transportation and handling costs. DOD may loan equipment to NASA without charge, subject to return in the same condition as when loaned, normal wear and tear excepted. The return of such equipment may be waived by DOD under the circumstances set forth in paragraph 3 of this Agreement. Where the loaned equipment is not returned, DOD will [be] reimbursed for the value thereof, unless the return of the equipment has been specifically waived by DOD under the circumstances set forth in paragraph 3 of this Agreement. [4] Where the loaned equipment is returned in a damaged condition, DOD will be reimbursed for the cost of restoring it to the same condition as when loaned, unless such reimbursement has been waived under the provisions of paragraph 3 of this Agreement, or waived on the basis that the equipment, at the time of return, is excess to the requirements of DOD.

E. *Travel.* In connection with the services covered by subparagraphs A, B, and C above, special travel costs attributable to the performance of these services will be reimbursed.

F. *Construction or Public Works.* Construction or public works projects undertaken by the DOD for NASA will be charged directly to NASA funds (or where appropriate will be reimbursed) on the basis of "project costs," the customary basis used by the DOD for charging DOD sponsored projects.

G. *Tenancy on Installation.* Except where other arrangements are in existence or are agreed upon, where either DOD or NASA is a tenant on an installation of the other, all direct costs or increases in direct costs attributable to such tenancy will be reimbursed.

H. *Use of Government-Owned Facilities.* No charge will be made for rent or depreciation in connection with the use by either DOD or NASA of Government owned facilities under their cognizance whether operated by the government or by a contractor.

3. *Exceptions.*

The foregoing principles do not apply to work or services, materials, supplies or equipment furnished to NASA or DOD for use in connection with specific projects of either agency, which are mutual interest and benefit to each. In such cases, work or services, materials, supplies or equipment furnished by one agency to the other will be on a non-reimbursable basis to the extent of the furnishing agency's interest in the particular project.

4.   *Effective Date.*

This Agreement is effective immediately, but it does not apply to existing agreements or arrangements already agreed upon which may not yet be formalized between NASA and the military department or ARPA. However, all future arrangements, agreements and amendments of existing agreements between NASA and the military departments or ARPA shall conform to the provisions of the Agreement.

5.   *Duration of Agreement.*

The provisions of the Agreement may be revised at any time, based upon further experience of the two agencies.

Deputy Secretary of Defense

Administrator
National Aeronautics and
Space Administration

NOV 12 1959
Date

Document II-13

Document title: T. Keith Glennan, NASA Administrator, and James H. Douglas, Deputy Secretary of Defense, "Agreement Between the Department of Defense and the National Aeronautics and Space Administration Concerning the Aeronautics and Astronautics Coordinating Board," reprinted in: U.S. Congress, House, Committee on Science and Astronautics, Subcommittee on NASA Oversight, "The NASA-DOD Relationship," 88th Cong., 2d sess. (Washington, DC: U.S. Government Printing Office, 1964), pp. 10-11.

The drafters of the 1958 Space Act considered it necessary to have close coordination of activities between NASA and the Department of Defense; therefore, a liaison board was provided for in the Act. By 1960, this liaison board was no longer effective and was replaced by the Aeronautics and Astronautics Coordinating Board. Over the years since then, the board has varied in its importance in coordinating cooperation between NASA and the Defense Department.

---

[10]

Agreement Between the Department of Defense and the National Aeronautics and Space Administration Concerning the Aeronautics and Astronautics Coordinating Board

I. POLICIES AND PURPOSE

(a) It is essential that the aeronautical and space activities of the National Aeronautics and Space Administration and the Department of Defense be coordinated at all management and technical levels. Where policy issues and management decisions are not involved, it is important that liaison be achieved in the most direct manner possible, and that it continue to be accomplished as in the past between project level personnel on a day-to-day basis.

(b) It is essential that [a] close working relationship between decision-making officials within the National Aeronautics and Space Administration and the Department of Defense be developed at all management levels. Where policy issues and management decisions are involved, it is important that the planning and coordination of activities, the identification of problems, and the exchange of information be facilitated between officials having the authority and responsibility for decisions within their respective offices.

(c) To implement the forgoing [sic] policies it is the purpose of this agreement to establish the Aeronautics and Astronautics Coordinating Board.

II. ESTABLISHMENT OF THE BOARD

There is hereby established the Aeronautics and Astronautics Coordinating Board, which shall be responsible for facilitating

(1) the planning of activities by the National Aeronautics and Space Administration and the Department of Defense to avoid undesirable duplication and to achieve efficient utilization of available resources;

(2) the coordination of activities in areas of common interest to the National Aeronautics and Space Administration and the Department of Defense;

(3) the identification of problems requiring solution by either the National Aeronautics and Space Administration or the Department of Defense; and

(4) the exchange of information between the National Aeronautics and Space Administration and the Department of Defense.

III. COMPOSITION OF THE BOARD

(a) The Board shall be headed by the Deputy Administrator of the National Aeronautics and Space Administration and the Director of Defense Research and Engineering as Cochairmen.

(b) The other Board members shall consist of chairmen of panels as hereinafter established, and a minimum number of additional members as may be equipped to insure that each military department is represented and that the National Aeronautics and Space [11] Administration and Department of Defense have an equal number of members.

(c) The members of the Board, other than the Cochairmen, shall be appointed by the Administrator and the Secretary of Defense, jointly.

IV. PRINCIPLES OF OPERATION

(a) Panels of the Board shall be established by the Administrator and the Secretary of Defense and, initially, shall include the following:

(1) Manned Space Flight.

(2) Spacecraft.[1]

(3) Launch Vehicles.

(5) Supporting Space Research and Technology.

(6) Aeronautics.

(b) Terms of reference shall be prescribed for each panel by the cochairmen of the Board. The members of each panel shall be designated by the cochairmen of the Board.

---

1.   For purposes of clarity, the name of the Panel was changed to Unmanned Spacecraft by the Aeronautics and Astronautics Coordinating Board at the second meeting on July 26, 1960.

(c) The board shall meet at the call of the Cochairmen, at least bimonthly, and the cochairmen shall alternately preside over the meetings. Only Board members, and such others as the cochairmen specifically approve, may attend meetings.

(d) The cochairmen shall establish a small secretariat to maintain records of the meetings of the Board and of its panels and to perform such other duties as the cochairmen may direct.

(e) The board, its panels, and the secretariat shall make full use of available facilities within the National Aeronautics and Space Administration and the Department of Defense, and all elements of the Administration and the Department of Defense shall cooperate fully with the board, its panels, and the secretariat.

(f) Actions based on consideration of matters by the board may be taken by individual members utilizing the authority vested in them by their respective agencies.

For the National Aeronautics and Space Administration:

T. KEITH GLENNAN,
Administrator.

For the Department of Defense:

JAMES H. DOUGLAS,
Deputy Secretary of Defense.

Promulgated this 13th day of September 1960.

Document II-14

Document title: "General Proposal for Organization for Command and Control of Military Operations in Space," with attached: "Schematic Diagrams of Proposed Organization for Command and Control of Military Operations in Space," no date.

Source: White House Office, Office of the Special Assistant for Science and Technology, Records (James R. Killian and George B. Kistiakowsky, 1957-61), Box 15, "Space [July-December 1959] (7)," Dwight D. Eisenhower Library, Abilene, Kansas.

ARPA was created in February 1958 to manage all the military space programs. Once NASA was created, several programs were taken from ARPA and given to the civilian space agency. ARPA did maintain managerial control of the military space program, but this was not popular with the military services. The Army and the Navy were concerned, however, that if ARPA was eliminated, the Air Force would be given control of all space programs. In April 1959, Chief of Naval Operations Admiral Arleigh Burke urged the Joint Chiefs of Staff to create a single military space agency. The Army leadership agreed, but the Air Force chief of staff objected that this would remove the weapons systems from the unified commands. By July 1959, White House and Department of Defense officials began evaluating this separate military space agency. It would report directly to the Joint Chiefs, and command would rotate among the military services. It was to be known as the Defense Astronautical Agency. The authorship of this document is unknown, but it was probably presented to the President's Science Advisory Committee in the summer of 1959. The idea was ultimately rejected, and the space programs were returned to the services. The Air Force was given control of most of the military space program, with the Army and Navy responsible for developing payloads for their own use.

[1]
General Proposal for Organization for Command and Control of Military Operations in Space

Encl: (1) Schematic Diagrams of Proposed Organization for Command and Control of Military Operations in Space

The rapid advances achieved by our research and development agencies need to be exploited by the uniformed services. A whole family of militarily useful satellite vehicles is now coming into being. Facilities for launching, tracking, data acquisition and recovery of satellite and space vehicles are now in operation. In the very near future these new capabilities will become accepted operational techniques of the Army, Navy and Air Force units deployed over the oceans and land masses of the Free World. The military implications of these developments to the National Security dictate the command attention of the Joint Chiefs of Staff.

The basic facilities required for conducting satellite and space vehicle operations are: launching equipment with associated safety and control instrumentation, tracking, data acquisition and communication networks and coordinated vehicle recovery equipment located on land, sea and in the air. The compression of time in relation to the new space era, wherein satellites encircle the globe in 90 minutes, dictates the need for integrating all satellite and space vehicle facilities under one military commander. Each of the 3 national missile range commanders, presently has the facilities for conducting, in at least a limited capability, satellite and space vehicle operations. The global nature of military satellite and space vehicle operations, particularly satellite vehicle recovery operations, requires that the 3 national missile range commanders be incorporated into one over-all military command.

It is recommended that a joint command for the coordination of military operations in space be established incorporating the following features:
[2] 1.   That the commander report directly to the Joint Chiefs of Staff.

2.   That the command position be rotated among the services.

3.   That a Scientific Director be designated as a staff assistant to the Space Commander whose prime function would be scientific direction and the assurance of rapid military exploitation of technological breakthroughs in astronautics. The incumbent of this position would be designated by NASA or ARPA and would be satisfactory to those two agencies. The incumbent could fill joint positions on the Space Command and NASA similar to joint military and AEC billets.

4.   That the joint headquarters be located at the primary space surveillance control center to minimize time involved in receipt and processing of intelligence and the transmission of command decisions.

5.   That consideration be given to locating this control center within reasonable distance from Washington D.C. to simplify liaison with all the Services and with NASA.

6.   That the space surveillance control center be manned by a group consisting of personnel from the 3 services.

7.   That all the facilities of each particular service related to satellite and space tracking, data acquisition and communications continue to function within that service, but under the respective range commander for operational control by the joint command.

8.   That all research and development and training activity continue as heretofore on a not-to-interfere basis with the national security responsibilities of the joint space command.

9. That each national range commander report directly to the commander of the joint space command for operational control, and to his normal commander for other control.

[3] The commander of the space command force would perform the following 5 functions:

1. Under the direction of the Joint Chiefs of Staff, command the 3 national ranges in-so-far as they contribute operationally to our national security.

2. Review and approve the planned operation of the 3 national ranges to assure consonance with the operational requirements of over-all national security.

3. Review the annual budgetary requirements of the 3 national ranges for national priority, scope and adequacy in support of national security objectives and make recommendations accordingly to the Joint Chiefs of Staff.

The Joint Chiefs of Staff would submit the annual budgetary requirements to the Secretary of Defense who in turn would submit the requirements to the National Security Council for review concerning national priority, scope and adequacy for support of national security objectives and for financial coordination with the Atomic Energy Commission, National Aeronautics and Space Administration and the Bureau of the Budget.

4. Integrate satellite and space vehicle tracking, data acquisition and communications control into one centralized global system.

5. Provide for the participation by all services, as appropriate, for indoctrination and training in the field of satellite and space vehicle operations.

The following advantages would accrue for national security by the establishment of a joint task force:

1. A central command, responsive directly to the Joint Chiefs of Staff, would insure the earliest possible military effectiveness of satellite and space vehicles.

2. Parallel developments and duplicative installations for R&D with [an] expensive network of communications, launching facilities and logistics systems would be eliminated.

[4] 3. Indoctrination and training of the uniformed services in all aspects of space operations would be insured.

4. The evolution of sound military requirements would be improved.

5. The relative importance of military space operations in national security would be responsibly defined.

It is to be noted that since ARPA does not actually operate any facilities it is not involved in this type of operational chain of command.

Regarding the tie-in with NASA's facilities, it is proposed that consideration be given that NASA facilities be controlled in a manner similar to the relationship between the Coast Guard and the Navy. That is, in time of emergency operationally useful equipment and facilities would be at the disposal of the joint space commander.

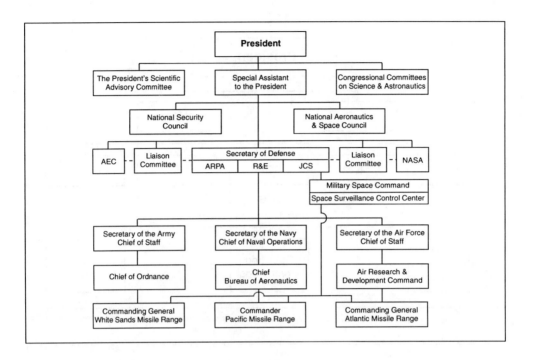

Orbital Space Operations							
Combat Systems	MET	GEO	Comm.	NAV	RECCO	EW	ECM
STRAC	●		●		●	●	●
Deployed Armies	●		●		●	●	●
Amphib	●		●	●	●	●	●
Carrier Strike	●	●	●	●	●	●	●
FBM	●	●	●	●	●	●	
ASW	●		●	●	●	●	●
Air Def	●		●		●	●	
Mis Def			●		●	●	
TAC	●		●		●	●	●
Strat Air	●	●	●	●	●	●	●
ICBM		●	●		●	●	
IRBM		●	●		●	●	

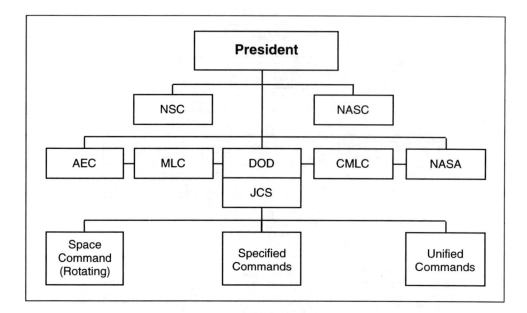

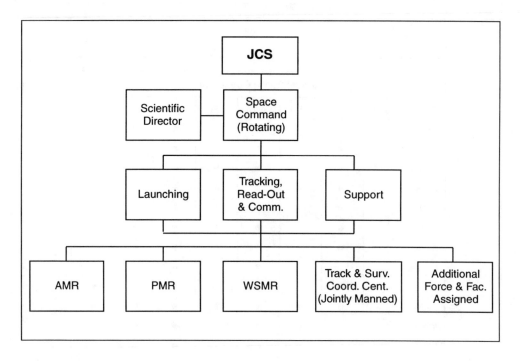

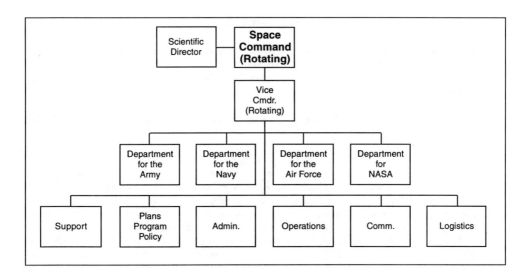

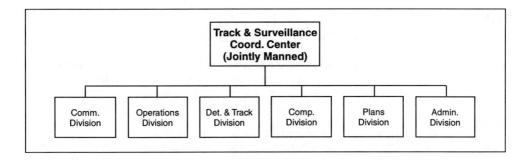

Advantages - JCS Desig. Command

- Insures Earliest Space Exploitation for U.S.A.

- Eliminates Duplicative Installations

- Insures Indoctrination and Training of Uniformed Personnel

- Would Improve Evolution of Sound Operational Requirements

- Would Define Relative Importance of Military Space Operations

Document II-15

Document title: "Military Lunar Base Program or S.R. 183 Lunar Observatory Study," Study Summary and Program Plan, Air Research and Development Command, Project No. 7987, Task No. 19769, Directorate of Space Planning and Analysis, Air Force Ballistic Missile Division, April 1960, pp. 1-9.

Source: NASA Historical Reference Collection, NASA History Office, NASA Headquarters, Washington, D.C.

The Space Act did not settle the issue of which organization—NASA or the Air Force—would conduct human spaceflight. The Space Act clearly indicated, however, that NASA would be responsible for most basic science in space. This created a much higher standard of justification of humans in space for the Air Force, which searched for practical missions requiring piloted spacecraft. In April 1960, the Air Research and Development Command completed a report on the feasibility of establishing a lunar base and argued that it should be recognized as an Air Force requirement. The base could serve as the site of a lunar-based Earth bombardment system capable of launching nuclear missiles with an accuracy of two to five nautical miles. Echoing the arguments made for many civilian manned space programs, the report noted that the cost of such a base ($8.14 billion over ten years) was less than the annual cost of the Farm Subsidy Program. It was more ambiguous about the need for such a base.

---

[1]

Study Summary

The purpose of this study was to "determine an economical and sound approach for establishing a manned intelligence observatory on the moon." Normally the end product of this type of study is an Evaluation Report. However, due to the importance of the study conclusions and the significance of time, it was decided to prepare a preliminary Program Plan, as part of the final Report.

The final report has been prepared in two volumes. Volume I includes this Study Summary and the Program Plan. Volume II consists of the Technical Requirements to support the Program Plan. The Technical Requirements are presented in "technical packages" that cover each of the major technical areas. Each package includes the characteristics and required development schedules for all known items within the specific technical area, as well as the development philosophy to be followed.

The "technical packages" have been prepared to assist the appropriate development agencies to initiate the required applied research and technical development programs. The complete Military Lunar Base Program Report is suitable for use by personnel in a Program Office to establish a Lunar Base Program, or to coordinate Air Force lunar requirements with the NASA.

Based on present knowledge, the study has concluded that it is technically feasible to establish a manned base on the moon. "Technically feasible" is not meant to imply that the equipments are available, or the techniques are completely known. Actually it means that the problems have been analyzed, and logical and reasonable extensions to the "state-of-the-art" should provide the desired techniques and equipments and this is comparable to the establishment of the original "design objectives" for the Ballistic Missile Programs in the year, 1954.

As the study progressed it became obvious that this is *not* a program "far off in the future." Actually the long lead development items should be started *immediately* if maximum military advantage is to be derived from a lunar program. If this is done the United States could send a man to the moon and return him to the earth during the last quarter of 1967.

The final decision concerning the types of strategic systems to be placed on the moon (such as a Lunar Based Earth Bombardment System) can be safely deferred for three to four years. However, the program to establish a lunar base must not be delayed and the initial base design must meet *military* requirements. For example, the base should be designed as a permanent installation, it should be underground, it should strive to be completely self-supporting, and it should provide suitable accommodations to support extended tours of duty. A companion study of Strategic Lunar Systems (SR-192) has shown that the lunar base is the most time-critical part of the system, so it is obvious that any delay in initiating the base development program will proportionally delay the final operational capability.

The subject of establishing a military lunar base is extremely complex and includes almost every known technical discipline. For the technical portion of this report the technical problems have been categorized as Propulsion, Secondary Power, Guidance, Life Support, Communications and Data Handling, Sensors, Materials and Resources, Lunar Base Design, and Environment. However, the general subject can be simply described as searching for the answers to the following four questions.

1. HOW can a manned base be established on the moon?
2. WHEN can a manned base be established on the moon?
3. HOW MUCH will it cost to establish a manned lunar base?
4. WHY should a manned base be established on the moon?

A majority of the study effort was expended on the question of "How can a manned base be established on the moon?" The first step was to perform a Transportation Analysis and determine [2] the most advantageous method of transporting men and materials to the moon and returning the men to earth. All conceivable chemical, nuclear and ion propulsion systems, using earth and lunar satellites, as well as "direct shot" trajectories, were considered. In addition, every reasonable technical perturbation was considered. *As a result of the analysis it was conclusively shown that the "direct shot" to the moon, using a five stage chemically propelled vehicle, is the most desirable.* This was not the expected conclusion since the establishment and use of a manned earth satellite-refueling station has been proposed for many years as the best way for man to travel to the moon. However, these original proposals did not have the benefit of a detailed analysis like the one performed in this study.

The analysis indicated the nuclear propulsion system could not be operational before 1970, so it was not advisable to rely on this system to establish the lunar base. However, if a nuclear system is available as expected in 1970, it could be used as indicated on the Master Program Schedule to logistically support the base.

With the "direct shot" determined to be the most desirable approach, it was possible to develop a vehicle concept. Based on technical and payload considerations, as well as the psychologists['] philosophy on "ideal crew size," it was concluded that a three-man aerodynamic re-entry vehicle would be the best method for transporting men to the moon and for returning them to the earth. This vehicle would weigh approximately 30,000 pounds as it enters the earth's atmosphere, and it would be capable of completely automatic-unmanned-10 day flights. The initial unmanned earth re-entry flights will require a landing area of 10 x 20 miles. When mail has been included in the system a more conventional landing strip will be usable, but to meet both of these requirements a facility like Edwards Air Force Base will be necessary.

The vehicle would be launched as the payload of a fire stage system that has six million pounds of thrust in the first stage. All stages of the system would use liquid hydrogen and oxygen for propulsion, since this combination has about a 3 to 1 payload advantage over the more conventional liquid oxygen and RP-1 combination. It was determined that the proposed NOVA vehicles using liquid oxygen and RP-1 in the first stage would not be adequate for supporting manned lunar base operations. Therefore, it is desirable to go completely to the use of liquid hydrogen and oxygen as soon as possible.

The first four stages of this same system will provide the capability of soft landing a payload of 50,000 to 80,000 pounds at a preselected lunar site. This provides a configuration suitable for transporting large cargo payloads to the moon for use in constructing the permanent lunar base. Approximately one million pounds of cargo will need to be delivered to the lunar surface in order to construct and support the permanent base. Part of this cargo will consist of telescopic and sensing equipment for performing "surveillance and control" of cislunar space.

An analysis of the functions that are necessary to operate a lunar base has shown that a base complement of 21 personnel will be required. The tour of duty for space personnel is extremely critical, since "personnel transport" is one of the most important cost factors in a space program. Present studies show the maximum tour of duty on an orbiting space satellite is in the neighborhood of 30 days. However, it seems reasonable to expect tours of 7 to 9 months on a lunar base due to the possibility of better living conditions, availability of a natural gravity environment, and greater protection from natural hazards while in the underground base.

Once the decision was made to use a "direct shot" chemical system and a vehicle configuration was determined, it became possible to outline a program for development equipments and a plan for establishing the lunar base. The program broke down into six logical phases with each phase designed to meet a specific secondary objective. These objectives all lead directly to the prime objective of establishing a manned military lunar base.

Basic to each phase of the program is our present knowledge of the environment in space and on the moon. Therefore, as part of this study all existing space and lunar environment knowledge [3] was surveyed, analyzed, summarized and applied to the program plan. The environmental data obtained from each phase of the program will add to this knowledge and assist in the design of equipments for the following phases.

Reliability and safety are of basic importance to each program phase. Reliability is equally essential to the unmanned as well as the manned flights. However, when man is placed in the vehicles safety becomes of prime importance. It was determined that the multi-engine vehicles should be capable of performing the mission even following the loss of one engine. Normally the loss in payload and efficiency to achieve an "engine out" capability is undesirable, but in this program where large quantities of hydrogen and oxygen are part of the regular payload to support the base, the corresponding loss in payload to provide extra fuel and oxidizer is not a disadvantage. Actually a "real" payload loss will only take place when a catastrophic engine failure occurs. In the cases of non-catastrophic failure, the mission will still be accomplished at reduced efficiency.

The following table presents the objectives and systems to be used in each of the six program phases.

PHASE	OBJECTIVE	BOOSTER	PAYLOAD (Pounds)	NO. OF SHOTS	METHOD
1. Lunar Probes	Obtain Lunar and Cislunar Environmental Data	ATLAS-ABLE	370	6	High Resolution Video System and Sensors.
2. Lunar Orbits	Map Complete Lunar Surface (10-15' Resolution)	ATLAS-CENTAUR	1,200	6	Solar energy and strip mapping.
3. Lunar Soft Landing	Soft Land on Moon and obtain environmental data	SATURN (4 stages)	2,000	9	Deceleration stage, terminal guidance alighting gear, core sampling devices.
		SATURN (5 stages)	4,500		
		NOVA-4 (5 stages)	25,000		
4. Lunar Landing and Return	Return First Payload from Moon (A core sample of the lunar surface)	SATURN (5 stages)	1,400	6	Core drilling and analysis package, lunar launching-atmospheric drag and retro-rocket re-entry, earth terminal guidance.
		NOVA-4 (5 stages)	10,000		
5. Manned Vehicle Development	Develop a Three Man Space Vehicle for Aerodynamic Earth Re-entry	NOVA-4 (5 stages) *ARAGO (5 stages) (Lunar Landing & Return with Man)	30,000 (Hi alt & Lunar Pass) 30,000	13	Extend Dyna Soar Techniques to Re-entry velocities of 37,000 ft/sec, fully automatic flight of manned space vehicle to moon and return to earth.
6. Lunar Base Development	Construct an Operational Permanent Base on the moon and support a 21 man crew.	*ARAGO (5 stages)	30,000 Man Space Vehicle) 57,000 80,000 (One Way Cargo Vehicle)	1/mo 1/mo	Construct temporary base, build underground permanent base, install operational surveillance equipment. Support of the completed base will require a total of 1 flight/month.

*ARAGO is the term used to describe the 6 million pound thrust, liquid hydrogen and oxygen, propulsion stage.

[4] Many items of equipment will be required for the lunar base program and wherever existing or programmed equipments would meet the requirements of the lunar base program they were scheduled for use. Where the item did not presently exist and none is programmed, a development schedule was provided. In addition, all necessary items are scheduled for use in the program as early as possible. This will improve reliability by use and growth, and allow the equipments to be "man-rated" by the desired time.

The major-pacing hardware items that require development to start immediately are as follows:

1. A liquid hydrogen and oxygen rocket stage which develops six million pounds of thrust.
2. A 30,000 pound, three man, earth return vehicle.
3. A 100 KW nuclear power unit capable of operating on the lunar surface for two years.
4. A suit/capsule capable of protecting personnel in the lunar environment.
5. A closed ecological system for use in the permanent lunar base.

6. A high definition video strip mapping system to map the lunar surface.
7. Suitable biopacks for use in the first three phases of the program.
8. A fully throttable, 6,000 pound thrust, liquid hydrogen and oxygen propulsion system.
9. A hydrogen-oxygen fuel cell.
10. A horizon scanner and altitude control system for lunar terminal guidance.
11. A command link midcourse guidance system.
12. A communications and terminal guidance package to be dropped on the lunar surface from orbiting vehicles.

The second major question concerning the establishment of a manned base on the moon is, "When can this be accomplished?" The Master Program Schedule for establishing a manned lunar base was obtained by scheduling the development of every known technical item and then integrating these individual schedules to determine when the base could become a reality. . . .

Five major milestones worthy of special mention are:

1. First lunar sample return to earth   November 1964
2. First manned lunar landing and return   August 1967
3. Temporary lunar base initiated   November 1967
    (This temporary base will be on the lunar surface and it will provide facilities while the permanent underground base is under construction.)
4. Permanent lunar base completed   December 1968
    (The permanent base will support a complement of 21 men.)
5. Operational Lunar Base   June 1969
    (Equipment will be installed to perform surveillance of earth-lunar space.)

The third major question is, "How much will it cost to establish a manned lunar base?" . . . These cost figures were prepared by the Air Force. After the technical program plan was completed, the Cost Analysis Panel "coated" the program using the best Air Force information available from present ballistic missile and aircraft programs.

[5] The important cost figures are summarized below:

Total Cost-Permanent Lunar Base $7,726 million
Total Cost-10 Year Program   8,146 million
(Includes installation of the permanent base and 6 months of operations.)
Annual Operating Cost   631 million

These costs are based on the following assumptions:

1. The major development engineering costs on the Saturn B and the NOVA 4 boosters has [sic] been assumed to be provided under independently funded programs. However, the actual cost of the boosters has been included and it was assumed that the first vehicle would be made available to the lunar program. If this is not the case, due to the "learning curve" it is expected that the vehicle costs would be decreased.
2. The costs include all shots in the program except the nuclear shots shown in the last half of 1970. The development costs for the nuclear system were not included because the lunar base program is not dependent upon the nuclear system. However, if the nuclear system is available and more economical it would be used to support the operational base.
3. Costs of all items normally considered as part of a weapon system (such as, launch pads and ground facilities) have been included.
4. It was assumed that adequate earth based tracking facilities will be available as the result of other programs. If they are not available the costs could increase by 300-600 million dollars in the later phases of the program.

When the average annual cost ($814 million) of the proposed program is compared to the Air Force efforts, it becomes apparent that this program is approximately equal to the output of just one of the major airframe companies normally supported by the Air Force. As a matter of information, the annual cost of the U.S. Farm Subsidy Program is approximately the same as the 9 1/2 year program required to install the permanent lunar base.

One point worthy of particular mention when considering costs, is the development of lunar resources. Analysis has shown that the development of lunar resources could decrease the cost of Strategic Lunar Operations by as much as 25 per cent [sic]. This is based on the fact that the moon's surface probably consists of many types of silicates. Since hydrogen and oxygen are used as propellants in the transport vehicles, as essential elements in the secondary power systems, as an element for personnel breathing, and when combined as water for life support, the value of obtaining these two elements on the moon is obvious. Should oxygen and hydrogen be obtained on the lunar surface they would be literally worth more than their weight in gold. This study has shown that it may be very possible to process lunar silicates to obtain water and then, by dissociation, the elements oxygen and hydrogen. It seemed very worthwhile to pursue this objective so a program schedule has been presented in the Environment section of Volume II. A glance at the lunar resource program schedule shows that the sample "core" of the lunar surface to be obtained in Phase IV, is critical to this effort. Although the process will require large quantities of power, solar energy is available in unlimited supply and nuclear power has been programmed for use on the lunar base.

The fourth major question, "Why should a manned base be established on the moon?," was not answered as a part of this SR-183 study. SR-192, the Strategic Lunar System Study was initiated on 29 August 1958 for the specific purpose of looking at this question. However, to provide a complete picture on the lunar base it seems necessary to consider the question in this report. Since the [6] final results of SR[-]192 are not yet available, the mid-term conclusions have been utilized. The Space Mission Analysis portion of this final SR[-]183 report briefly discusses these conclusions. The essential factors can be stated as follows:

1. The lunar base possesses strategic value for the U.S. by providing a site where future military deterrent forces could be located.
2. The decision on the types of military forces to be installed at the lunar base can be safely deferred for 3 to 4 years provided a military lunar base program is initiated immediately.
3. A lunar based earth bombardment system could have a CEP of two to five nautical miles.
4. The development of lunar resources could enhance the potential for strategic space operations in the cislunar volume.

[7] CONCLUSIONS

The most important conclusions of this study can be summarized by the following statements:

1. It is technically feasible to establish a lunar base by logical extension of present techniques.
2. Earliest lunar operations may be attained through the use of a direct shot chemically powered booster.
3. A 6 million pound thrust LOX/LH propulsion capability must be developed for the three-manned vehicle for lunar landing and return missions.

4. Investigation indicates that the payload penalty for using earth re-entry retro rockets is so great that the only logical re-entry approach is by means of aerodynamic braking. Therefore, the present Dyna Soar program is essential to provide re-entry vehicle design data.

5. A multi-phased program is essential to establish an operational lunar base. The Program Plan presented in this report included the following six phases:

Phase I	Lunar Probes
Phase II	Lunar Orbits
Phase III	Soft Lunar Landing
Phase IV	Lunar Landing and Return
Phase V	Manned Vehicle Development
Phase VI	Lunar Base Development

6. Based on the above program the following milestones have been established as reasonable objectives.

a. First Lunar Sample Returned to Earth	November 1964
b. Manned Lunar Landing and Return	August 1967
c. Temporary Lunar Base Initiated	November 1967
d. Permanent Lunar Base Completed	December 1968
e. Operational Lunar Base	June 1969

7. The initial pleases of the program can be undertaken for an investment which averages approximately 800 million dollars per year during the initial building phase. After the establishment of the base the annual costs will decrease to about 600 million dollars per year. This may be still further reduced when nuclear propulsion becomes available and as lunar resources are developed to provide oxygen and hydrogen to support space operations.

8. A lunar base is the initial and essential step in the attainment of a military capability in the lunar volume.

9. A military lunar system has potential to increase our deterrent capability by insuring positive retaliation.

10. The decisions regarding the type of military operations to be conducted in lunar and cislunar space can be safely deferred for several years provided a military lunar base is established which can be readily expanded to support lunar operations.

11. From a national viewpoint it is desirable that a lunar base be established as soon as possible. This conclusion is based on the strategic potential as well as the psychological, political and scientific implications.

[8] This page intentionally left blank.

[9] The following actions are recommended as a result of this study.

1. The program for establishing a military lunar base be recognized as an Air Force requirement.

2. Immediate action be taken to implement the early phases of the program.

3. Immediate action be taken to start the development of the critical long lead items listed below:
   a. Six million pound thrust LOX/LH propulsion system.
   b. Three-man space vehicle which can re-enter earth's atmosphere.
   c. There are smaller items that should be started before the end of 1960. These are listed in the separate technical areas.

4. A program office be established within ARDC to coordinate with NASA, all activities directed toward the establishment of the lunar base.

5. The military requirements and NASA's requirements be integrated into one national lunar program.
6. Responsibilities be assigned for the various phases of the integrated lunar program.
7. The establishment of the base be considered a military expedition.
8. The Air Force develop space operational know-how by being intimately involved in all phases of the lunar program. This is in keeping with the philosophy of concurrency and is necessary to shorten the development cycle.
9. Further study be initiated as explained in each section of the technical report. The follow-on SR-183 study will tie all of these together into a comprehensive systems study.

Document II-16

Document title: General Thomas D. White, Chief of Staff, United States Air Force, to General Landon, Air Force Personnel Deputy Commander, and General Wilson, Air Force Development Deputy Commander, April 14, 1960, reprinted in: *Defense Space Interests*, Hearings Before the Committee on Science and Astronautics, U.S. House of Representatives, 87th Cong., 1st sess. (Washington, DC: U.S. Government Printing Office, 1961).

Document II-17

Document title: Robert S. McNamara, Secretary of Defense, to the Secretaries of the Military Departments, *et al.*, "Development of Space Systems," March 6, 1961, with attached: Department of Defense Directive 5160.32, "Development of Space Systems," March 6, 1961, reprinted in: *Defense Space Interests*, Hearings Before the Committee on Science and Astronautics, U.S. House of Representatives, 87th Cong., 1st sess. (Washington, DC: U.S. Government Printing Office, 1961).

Document II-18

Document title: Overton Brooks, Chairman, Committee on Science and Astronautics, U.S. House of Representatives, to the President, March 9, 1961.

Source: NASA Historical Reference Collection, NASA History Office, NASA Headquarters, Washington, D.C.

Document II-19

Document title: President John F. Kennedy, to Overton Brooks, Chairman, Committee on Science and Astronautics, U.S. House of Representatives, March 23, 1961.

Source: NASA Historical Reference Collection, NASA History Office, NASA Headquarters, Washington, D.C.

After the Advanced Research Projects Agency (ARPA) returned control of the military space program to the individual services—primarily the Air Force—there was gradually increasing concern in Congress and the press that the Air Force was interested in expanding its power over other aspects of the civilian space program as well. In April 1960, Air Force Chief of Staff Thomas White wrote a memorandum to his staff stating that he wanted them to cooperate more fully with NASA and that it might be possible that NASA would eventually be combined with the military. Almost a year later, newly appointed Secretary of Defense Robert McNamara gave the Air Force control of the development of all military space systems. The other services would still conduct basic research, but after some predetermined point, the program would be turned over to the Air Force. White's memo "leaked" and, combined with the McNamara policy statement, led to hearings before the House Committee on Science and Astronautics, chaired by Overton Brooks. Before the hearings started, Brooks sent a letter to President Kennedy asking for clarification of the Air Force's role in conducting aspects of the national space program. By the last day of the hearings, Kennedy responded, declaring that manned and unmanned exploration of space and the application of space technology to peaceful activities were NASA missions, but that there also were exclusively military missions in space as well.

Document II-16

[no pagination]   14 April 1960

AFPDC (Gen Landon)
AFDDC (Gen Wilson)

1.   I am convinced that one of the major long range elements of the Air Force future lies in space. It is also obvious that NASA will play a large part in the national effort in this direction and, moreover, inevitably will be closely associated, if not eventually combined with the military. It is perfectly clear to me that particularly in these formative years the Air Force must, for its own good as well as for national interest, cooperate to the maximum extent with NASA, to include the furnishing of key personnel even at the expense of some Air Force dilution of technical talent.

2.   It has come to my attention that key personnel in NASA feel that there has been a shift in Air Force policy in respect to the type of cooperation stated above. I want to make it crystal clear that the policy has not changed and that to the very limit of our ability, and even beyond it to the extent of some risk to our own programs, the Air Force will cooperate and will supply all reasonable key personnel requests made to it by NASA.

3.   To meet the above requirements I have no doubt that some shifting of Air Force personnel within the Air Force will be necessary in order to feed new talent into [the Air Research and Development Command]. This should be done. In addition, while late, we must increase the number of slots in civil technical institutions for Air Force officers. I want this type of technical education to be given the highest priority in our civil educational program and the percentage of slots in this respect to be radically increased, effective as early as possible.

THOMAS D. WHITE
Chief of Staff

cc: Under Secretary of the Air Force Dr. Perkins
       General LeMay
       General Schriever

Document II-17

[no pagination]

THE SECRETARY OF DEFENSE
Washington, D.C., March 6, 1961

Memorandum for the Secretaries of the Military Departments
The Director of Defense Research and Engineering
The Chairman, Joint Chiefs of Staff
The Assistant Secretaries of Defense
The General Counsel
The Assistants to the Secretary of Defense

SUBJECT: *Development of Space Systems*

Having carefully reviewed the military portion of the national space program, the Deputy Secretary and I have become convinced that it could be much improved by better organization and clearer assignment of responsibility. To this end, I directed the General Counsel of the Department of Defense to obtain your comments on a new draft DOD Directive, "Development of Space Systems."

After careful consideration of the comments and alternate plans that were submitted, the Deputy Secretary and I have decided to assign space development programs and projects to the Department of the Air Force, except under unusual circumstances.

This assignment of space development programs and projects does not predetermine the assignment of operational responsibilities for space systems which will be made on a project by project basis as a particular project approaches the operational stage, and which will take into account the competence and experience of each of the Services and the unified and specified commands.

We recognize that all the military departments, as well as other Defense agencies, may have requirements for the use of space equipment. The directive expressly provides that they will continue to conduct preliminary research to develop specific statements of these requirements, and provides a mechanism through which these requirements may be fulfilled.

Attached is a directive incorporating this decision. We expect all elements of the Department of Defense to support it fully and to help develop the military portion of the national space program in the most effective manner.

Robert S. McNamara

Encl. DOD Dir. 51GO.32

****

[1]   *March 6, 1961*
                                                  Number 5160.32

Department of Defense Directive

SUBJECT: *Development of Space Systems*

References:
 (a) Memorandum (Conf) from Secretary of Defense to Chairman, Joint Chiefs of Staff, subject: Satellite and Space Vehicles Operations, September 18, 1959
 (b) Memorandum from Director, Advanced Research Projects Agency to Secretary of the Army, Secretary of the Navy, and Secretary of the Air Force, subject: Study Contracts for Projects Assigned to the Advanced Research Projects Agency, September 14, 1959
 (c) Memorandum from Director, Advanced Research Projects Agency to Secretary of the Army, Secretary of the Navy, and Secretary of [2] the Air Force, and Director, Advanced Research Projects Agency, subject: ARPA Programs, June 11, 1959

I. *Purpose*
 This establishes policies and assigns responsibilities for research, development, test, and engineering of satellites, anti-satellites, space probes and supporting systems therefor, for all components of the Department of Defense.

II. *Policy and assignment of responsibilities*
 A. Each military department and Department of Defense agency is authorized to conduct preliminary research to develop new ways of using space technology to perform its assigned function. The scope of such research shall be defined by the Director of Defense Research and Engineering in terms of expenditure limitations and other appropriate conditions.
 B. Proposals for research and development of space programs and projects beyond the defined preliminary research stage shall be submitted to the Director of Defense Research and Engineering for review and determination as to whether such proposals, when transmitted to the Secretary of Defense, will be recommended for approval. Any such proposal will become a Department of Defense space development program or project only upon specific approval by the Secretary of Defense or the Deputy Secretary of Defense.
 C. Research, development, test, and engineering of Department of Defense space development programs or projects, which are approved hereafter, will be the responsibility of the Department of the Air Force.
 D. Exceptions to paragraph C, will be made by the Secretary of Defense or the Deputy Secretary of Defense only in unusual circumstances.
 E. The Director of Defense Research and Engineering will maintain a current summary of approved Department of Defense space development programs and projects.

III. *Cancellation*
 Reference (a), except as to the assignments of specific projects made therein, and references (b) and (c) are hereby cancelled.

IV. Effective date

This directive is effective upon publication. Instructions implementing this directive will be issued within thirty (30) days.

ROBERT S. MCNAMARA, *Secretary of Defense*

Document II-18

[1]   March 9, 1961

The President
The White House
Washington, D.C.

My dear Mr. President:

I am seriously disturbed by the persistency and strength of implications reaching me to the effect that a radical change in our national space policy is contemplated with some areas of the executive branch. In essence, it is implied that United States policy should be revised to accentuate the military uses of space at the expense of civilian and peaceful uses.

Of course, I am aware that no official statement to this effect has been forthcoming; but the voluminous rash of such reports appearing in the press, and particularly in the military and trade journals, is, it seems to me, indicative that more than mere rumor is involved.

Moreover, I cannot fail to take cognizance of the fact that emphasis on the military uses of space is being promoted in a quasi-public fashion within the defense establishment. Nor can I ignore the suggestion, implicit in the unabridged version of the Wiesner report, that the National Aeronautics and Space Administration role in the development of space systems will be predominant. Such an assertion not only seems to disregard the spirit of the law but minimizes the values of peaceful space exploration and exploitation.

I have hesitated to call this to your personal attention. However, since the National Aeronautics and Space Council, whose duty it is to advise on the formulation of United States space policy, remains unformed, I feel constrained to broach the matter directly.

May I point out that the National Aeronautics and Space Act of 1958 passed the Senate in which both you and the current Vice President so ably served without a record-ed dissenting vote. It was unanimously approved by the House. In that Act Congress took great pains to declare that space activities "shall be the responsibility of, and shall be directed by, a civilian agency exercising control over aeronautics and [2] space activities sponsored by the United States. . ." Space activities "peculiar to or primarily associated with the development of weapons systems, military operations, or the defense of the United States" were quite properly made the responsibility of the Defense Department, but this was a literal "exception" to the proclaimed procedure.

As you know, I served twenty-two years on the House Armed Services Committee. I would be the last person to attempt to weaken our defense posture. But neither do I intend to sit by and, contrary to the express intent of Congress, watch the military tail undertake to wag the space dog.

The law makes it crystal clear that the prime American mission in space is toward peaceful purposes. It specifically enjoins NASA to promote space science and technology with a view to the "application thereof to the conduct of *peaceful activities* within and out-side the atmosphere." The law not only does not limit NASA's functions to scientific research; it affirmatively directs NASA to make peaceful use of space and to develop oper-ational space systems, manned and unmanned. This is a legislative requirement imposed. To place the prime operational responsibility for space exploration and use with the mil-itary, particularly when no military requirement for men in space yet exists, would be to disorient completely the space program as contemplated by Congress and as set forth in the law.

As Chief Executive of the United States charged with the conduct of foreign affairs and as a former member of the Senate Foreign Relations Committee, you are, I know, aware of the great importance of preserving the peaceful image of the United States with-in the international community. At the same time, few areas of national endeavor today serve better to reflect the American attitude in world politics than what we intend and how we behave in this new dimension of human activity. I do not see how we can square an exclusive, or even a predominant, military exploitation of space with our announced aspirations for peace and disarmament. To sublimate military operations in space would thus seem to be inconsistent with our foreign policy and, in my judgment, would serve to impede our future negotiations for world-wide disarmament.

There is another extremely important aspect of this picture, namely participation by private enterprise in the space venture. If we are to reap genuine economic pay-off through space exploration, we must find ways of eliciting and using the resources of pri-vate capital. While I recognize that the armed services have a legitimate interest in such space enterprises as communications, weather prediction, navigation and the like, I sub-mit that these concepts have a predominant use for peaceful activities. In my judgment, we will lag seriously in any efforts to bring private enterprise into space if we turn control of research, development and operation of such endeavors over to the military whose [3] needs are highly specialized and whose research methods tend to be restricted in scope and concept.

To amplify: if we envision the military in control of world-wide space communications, it is difficult to understand in advance what basis would be provided for world-wide media of communications such as television, radio and telephone systems. If we concede military control of weather satellites, how shall such control be reconciled to the needs of farmers, merchants, and business generally? If we permit military domination of space navigation devices, are we fulfilling our obligations to the merchant marine and the commercial air fleets operating on and above the high seas? I think not. In fact, many of the benefits which humanity could expect to reap from the exploration of space may easily be lost unless they are made available on a non-military basis. If the fruits of our efforts to con-quer space are to enrich people's lives and raise standards of living throughout the world, they must be handled through a civilian peace-time agency, not by the military which nec-essarily is governed by its particular objectives.

In conclusion, I feel obliged to point out that in view of the recent Defense Department decision to concentrate all military space research in a single service, this question of civilian preeminence in space exploration becomes paramount. Space explo-ration involves much research, basic and applied, and it is axiomatic that the rate of research pay-off is accelerated many times when a variety of approaches, ideas and con-cepts are explored simultaneously. Testimony before our Committee permits no doubt whatever that the United States space effort, civilian and military, has achieved what it has during the past three years only because of an imaginative and *diversified* approach.

If NASA's role is in any way diminished in favor of a space research program con-ducted by a single military service, it seems unlikely to me that we shall ever overtake our

Soviet competition—a competition which, by the way, has been peculiarly effective because of its public emphasis on scientific and peaceful uses of space.

It is my hope that you will find it feasible to clear up this matter, and, coincidentally, to reassure me in the very near future.

Very sincerely yours,

OVERTON BROOKS
Chairman

Document II-19

[1]   March 23, 1961

Dear Overton:

Recently you wrote to me concerning my attitude toward the conduct of our national space effort. I appreciate your comments and have given considerable thought to the problems of this program which you have raised. I hope that this letter will serve to reassure you that there is no basic disagreement between us.

It is now, nor has it ever been, my intention to subordinate the activities in space of the National Aeronautics and Space Administration to those of the Department of Defense. I believe, as you do, that there are legitimate missions in space for which the military services should assume responsibility, but that there are major missions, such as the scientific unmanned and manned exploration of space and the application of space technology to the conduct of peaceful activities, which should be carried forward by our civilian space agency. Furthermore, I have been assured by Dr. Wiesner that it was not the intention of his space task force to recommend the restriction of the NASA to the area of scientific research in space. One of their strongest recommendations was, in fact, that vigorous leadership be provided by NASA in the area of non-military exploitation of space technology.

As you have pointed out, there are programs which have strong implications in both the military and civilian fields. In making policy decisions on such programs, I intend to rely heavily on the advice of the Vice President, based on his invaluable experience with the Senate Committee on Aeronautical and Space Sciences. We are also moving ahead with plans to reactivate the Space Council, and to make it an active and effective organization. As [2] you know, I have nominated Dr. Edward C. Welsh to be executive secretary of the Space Council. I believe that he, working under the Vice President, can assemble a top-flight staff that will make the Space Council more than just a box on an organization chart as it has been heretofore.

I agree wholeheartedly with you that there are highly important benefits to be realized from the civil applications of space technology and that private enterprise must play an important role. I am confident that with the help of the Vice President, the Space Council, the Senate Committee under Senator Kerr, and the House Committee under your able leadership, we can assure that the proper policy decisions will be reached.

Again, may I thank you for your comments and also express my appreciation for the outstanding job you are performing as Chairman of the Committee on Science and Astronautics.

Sincerely yours

John F. Kennedy

Document II-20

**Document title: "Summary Report: NASA-DOD Large Launch Vehicle Planning Group,"
September 24, 1962, pp. ii-iv, I-1–III-13.**

**Source: NASA Historical Reference Collection, NASA History Office, NASA
Headquarters, Washington, D.C.**

*After President Kennedy's May 25, 1961, speech to Congress, which committed the United States to a
lunar mission, there was an attempt to establish military and civilian requirements for large launch
vehicles, with the hope of establishing a single national launch vehicle fleet. On July 7, 1961, NASA
Associate Administrator Robert Seamans proposed a joint study to determine mission models and
requirements affecting the selection of large launch vehicles; the study was headed by NASA's Nicholas
Golovin. As the study progressed, the different requirements and institutional interests of NASA and
the Department of Defense (DOD) became clear and both agencies quickly distanced themselves from
the contents of the report. By the time this report was released on September 24, 1962, almost a year
after the group had completed its work, it had been obvious for some time that there would be very little cooperation between NASA and the DOD on large launch vehicles.*

[ii]

Foreword

Early in 1961 numerous studies relative to our space programs were undertaken
under a variety of auspices. In one of the initial efforts the Space Exploration Program
Council of NASA revived in detail the various aspects and approaches to manned lunar
landing including both rendezvous and direct ascent. This review culminated in the decision that NASA would proceed toward the manned lunar landing on a broad base.
Accordingly, studies were initiated to aid in formulating an approach to the task. At about
the same time, the Secretary of Defense requested a comprehensive study by his staff of
our total national space program and a comparison of these with what was known of Soviet
undertakings in this field.

Early in May 1961, NASA presented its plans for accomplishing a manned lunar
landing and estimates of the cost to the Department of Defense for the purpose of coordinating the resources and efforts of these two agencies to accomplish this mission. These
discussions culminated in a NASA-DOD report submitted to the Vice President, in his
capacity as Chairman of the National Aeronautics and Space Council, entitled
"Recommendations for Our National Space Program: Changes, Policies, Goals." This
report, dated 8 May 1961, was submitted jointly by the Administrator of NASA and the
Secretary of Defense.

The most important of these recommendations was that the achievement of manned
lunar landing before the end of the decade be established as a national goal. In addition,
it was recommended that scientific exploration of space be intensified; that operational
communications and meteorological satellite systems be developed at the earliest reasonable time; that large scale boosters be developed for potential military use as well as to
support the civilian space program; and that an increased effort be placed on advanced
technology, particularly with regard to the development of chemical and nuclear rocket
propulsion. It was recognized, of course, that further analysis would be required to develop more detailed program plans in each of the recommended areas. It is important to
note, however, that the basis for such planning was clearly [iii] specified in the 8 May 1961

report; it was recognized that long-range planning, especially for launch vehicles (which account for more than half the cost of space programs), must first be successfully accomplished, and that such planning was essential to insure that national resources would be properly harnessed to national tasks on a national scale.

The 8 May 1961 report gave renewed support to the "building block" concept. It was stated in this connection: "It is absolutely vital that national planning be sufficiently detailed to define the building blocks in an orderly and integrated way. It is absolutely vital that national management be equal to the task of focusing resources, particularly scientific and engineering resources, on the essential building blocks."

The budgetary and policy recommendations of the 8 May report were adopted by the President and presented to the Congress in his message of 25 May. Subsequently, virtually all the recommended authorizations and appropriations were passed by the Congress.

Simultaneously with these acts, NASA and DOD continued an intensified effort to define mere explicitly and explore more thoroughly the actions that would have to be taken to implement these recommendations. New studies were begun in both agencies and the talents of specialists were harnessed in organizations in many parts of the country.

Within NASA the first of a series of major study efforts was begun on 2 May 1961, initiated to define in greater detail the feasibility, schedule, and costs of accomplishing manned lunar landing, giving attention to the various possible approaches for accomplishing the mission. An ad hoc task group was then established and assigned the responsibility for defining in detail a feasible approach for accomplishment of an early manned lunar landing. A second ad hoc group was assigned the task of conducting a broad survey of the feasible ways for accomplishing manned lunar landing.

One of the results of these studies was establishment of the need for further information on the rendezvous approach to manned lunar landing and the associated launch site planning and resources required. Accordingly, two [iv] additional ad hoc task groups were established. One group established on 20 June 1961 was assigned responsibility for studying in detail the plans and supporting resources needed to accomplish manned lunar landing by the rendezvous technique. The other group established 23 June 1961 conducted a joint NASA-DOD study of national launch site planning and of the resources required to accomplish the manned lunar landing mission—as it was defined by the earlier studies.

Early in July it became apparent that a very major effort was necessary to aid in defining the large launch vehicles which would be needed for the Manned Lunar Landing Program. Numerous mechanisms were considered for this purpose. The idea of establishing a committee of scientists and technologists somewhat analogous to the Von Neuman Committee, which in 1954 recommended the initiation of our ICBM program, was considered. The possibility of establishing a contractor or a group of contractors charged with responsibility for this analytical and planning effort was also considered. From such considerations emerged the concept of the Large Launch Vehicle Planning Group (LLVPG). This group was to be comprised of representatives from both NASA and DOD reflecting equally the experience, viewpoints and special knowledge of both agencies. The group was to be responsible jointly to a senior official in each agency and empowered to draw upon scientific and technological resources wherever they might be found and needed. . . .

[I-1]

Chapter I
Summary and Recommendations

I. Introduction

   A. *Formation of the Large Launch Vehicle Planning Group*

The Secretary of Defense and the Administrator of NASA, in an exchange of correspondence on 7 July 1961, established the DOD-NASA Large Launch Vehicle Planning Group (LLVPG), to provide the necessary joint planning leading to future specification and development of large launch vehicles required as a result of the expansion of the national space effort outlined by the President on 25 May 1961.

The LLVPG was headed by Dr. Nicholas E. Golovin of NASA, and by Dr. Lawrence L. Kavanau of OSD [the Office of the Secretary of Defense], who served as Deputy Director of the group. They reported jointly to Dr. Robert C. Seamans, Jr., Associate Administrator of NASA, and Mr. John H. Rubel, Assistant Secretary of Defense (Deputy Director, Defense Research and Engineering) of DOD. A total of nine DOD representatives and ten NASA representatives made up the membership of the LLVPG. Their names appear in Appendix A, which also describes the assistance the group secured from other agencies, such as the Marshall Flight Test Center, Aerospace Corporation, Jet Propulsion Laboratory, and industrial contractors. As noted in Appendix A, the LLVPG commenced its operations in July and continued through the month of November 1961, and utilized the equivalent full-time services of approximately 150 people during this period.

Representatives from many NASA and DOD components served as members of the LLVPG. Although it was desired and necessary to insure that organizations charged with on-going responsibilities for the execution of space programs would have the opportunity to participate in this planning effort, the principal criterion used in selecting members of the LLVPG was their personal technical ability and experience. The objective was not to attain a compromise between the preconceived notions of DOD representatives on the one hand and NASA representatives on the other, but to harness, through cooperative study, the best capabilities available for the task of laying out a long-range plan for a National Launch Vehicle Program.

[I-2] The initial instructions to the LLVPG were comprised [sic] in a memorandum dated 7 July 1961 to the Administrator of NASA from Dr. Seamans. This document was approved by the Administrator of NASA and the Secretary of Defense. While these documents served as the important starting point and the principal framework for LLVPG deliberations, the LLVPG was responsive to considerable detailed guidance furnished by Dr. Seamans and Mr. Rubel, immediately following its establishment, and from time to time during the course of its deliberations. Since the objective of the LLVPG was to formulate plans, it was natural to expect that ideas and concepts would be changed as their studies and analyses evolved. This was indeed the case, and some of the notions with which this undertaking began were significantly modified before the completion of the group's effort.

Based on the direction received by the LLVPG, the following frame of reference for the study was adopted:

   a. The launch vehicle configurations and the operational procedures to be developed and recommended by the LLVPG were to take into account the current and anticipated needs of DOD and of NASA and be guided by the following national objectives for large launch vehicles:

      (1) Early successful landing of manned spacecraft on the moon to return to earth.

    (2) Manned scientific missions in earth orbit and circumlunar flight as well as on the lunar surface.

    (3) Launch vehicle developments for advanced military missions.

    (4) Increased reliability and economy of effort achieved by multiple use of vehicle components, vehicle stages, and complete launch vehicles.

    b. The principal specific allegation of the LLVPG was the explicit development, in useful detail, of a technically well-established planning basis in which coordinated action could be taken leading to the development [I-3] and use of the recommended launch vehicles and the necessary facilities for their test and launching. The group was charged with the identification and preparation of preliminary specifications for long lead time items for which development should be initiated immediately, and was directed to review and recommend a suitable balance between early achievement of major goals, over-all costs, and growth potential of large launch vehicles.

    c. Guidelines provided the LLVPG included the following:

    (1) Both direct ascent and rendezvous operations with respect to the lunar landing were to be considered.

    (2) Plans were to be based on components within the present state-of-the-art but not restricted to on-the-shelf items. When the scheduled development of a new component appeared questionable, a duplicate approach was to be included.

    (3) Although only liquid and solid motors were to be employed, proposal designs should facilitate exploitation of nuclear and electric propulsion for follow-on systems if feasible.

    (4) The group was to concern itself only with large launch vehicle systems. The word "large" was to mean those vehicles whose capability to accelerate payloads on spacecraft to escape velocity would be greater than the capability of the Atlas-Agena B system. (This guidance was subsequently modified as a result of the booster requirements arising in connection with the NASA Gemini program, and the group was reconvened by a memorandum dated 18 November 1961, from the Associate Administrator of NASA and the Assistant Secretary of Defense (Deputy Director of Defense Research and Engineering) to extend its study of vehicle systems with the range of payloads down to 5000 pounds.) The term "vehicle system" was to include not only propulsion elements, but guidance, control and those instrumentation, telemetry, and command/communication subsystems which are normally physically part of the vehicle system and are employed for maneuvering the payload or spacecraft into a desired sequence of position and velocity coordinates.

[I-4] B. *Approach*

    The general approach followed by the group was that of defining stage and vehicle combinations which could reasonably be expected to become available within the next 5 to 8 years: executing a systematic quantitative analysis of their relative performance, schedule, cost and reliability characteristics; and comparing resultant launch vehicle capabilities, with the projected national missions requirements. In developing national launch vehicle requirements for the period 1962 - 1970, the LLVPG utilized forecasts of launch needs prepared by DOD and NASA reflecting programmed and anticipated mission needs.

    It was not considered an assignment of the LLVPG to establish preferred mission modes where alternative operational concepts were involved as, for example, in the case of various approaches to accomplishment of manned lunar landing. However, it was the aim of the group to define the launch vehicle configurations (and their availability, cost, reliability and performance characteristics) associated with such alternative mission modes and thereby provide inputs which could be used for decisions by DOD and NASA.

In defining building block combinations of boosters and upper stages, consideration was extended to major subsystems including guidance systems, control systems, power supplies, telemetry and the like. Quantitative preliminary design analyses were made by Aerospace Corporation and/or the Marshall Space Flight Center (MSFC), and were carried through to a sufficient depth of technical detail to substantiate the operational feasibility of each prospective launch vehicle. These studies included: propulsion system performance characteristics; controllability; structural behavior in typical trajectories; detailed development schedules for the engines, stages, and vehicles, structured in the form of PERT diagrams incorporating all significant milestones throughout the development cycle up to first vehicle availability for flight test, and detailed cost estimates for each phase of the development process for each stage of every vehicle, including manufacturing facilities, static and dynamic testing facilities, as well as the necessary launch complexes.

[I-5] In this effort one of the early decisions of the LLVPG was that no recommendation, consistent with the guidelines given to the group was likely to be of practical utility as a basis for management decisions in NASA or DOD unless the prospective reliability of the vehicle systems involved was estimated. Accordingly, arrangements were made for detailed reliability analysis of each stage, each major subsystem, and each over-all vehicle system.

In view of the fact that durations of testing programs, both static and flight, are dependent on the engineering and testing philosophies employed in the development process, substantial attention was also given by the group to these matters. The experience of qualified staffs at MSFC and Aerospace Corporation, as well as of members of the LLVPG, were melded in sharpening the concepts involved and in applying them, to establish vehicle development and flight schedules for various mission-vehicle combinations later considered by the group.

Further details of the participation of the LLVPG, the manner in which the group proceeded in its activities, and the contents of the final report are included in Appendix A.

As stated previously, it was the objective of the LLVPG to develop recommendations for a National Launch Vehicle Program that would satisfy NASA and DOD flight mission requirements for the remainder of the decade. Therefore, one of the initial steps taken by the group was to obtain the mission requirements of the two agencies and analyze them with a view to developing a systematic mission requirement base to serve as a foundation for the vehicle studies to follow. Spacecraft development and mission attempt schedules were assumed to be paced by the availability of vehicles, the derivation of vehicle types and their development schedules.

For convenience in analyzing the characteristics of the various vehicles considered, the mission requirements were divided into four classes. These mission classes are:

>   Class I - Unmanned NASA and DOD missions plus early manned flight
>   Class A - Low earth orbit missions for large manned spacecraft systems (Apollo, Dyna-Soar, Orbiting Laboratory)
> [I-6]   Class B - Manned lunar missions involving lunar circumnavigation, lunar orbit and lunar landing by earth orbit or lunar orbit rendezvous
>   Class C - Manned flight to the moon by direct ascent

The launch vehicles studied were correspondingly divided into four classes. The performance characteristics, reliability and development schedules for these vehicles are summarized in Appendix B.

A total of over 800 missions was projected for NASA and DOD to the end of the decade. These missions were distributed among the four mission classes as follows:

	NASA	DOD	Total
Class I			
Class A	277	523	800
Class B	69	Undefined	69
Class C	10		10
Total	356	523	879

[II-1] II. Principal Recommendations
    A. *Recommendations*
    The following are the principal recommendations of the LLVPG with a brief discussion of each.
    1. *Class A Launch Vehicle Development*
Recommendation: *Development of the Saturn C-1 should continue.*
    The Saturn C-1 is the only vehicle (A-1) available in time to meet the present development schedule of the Apollo program. Therefore, the Saturn C-1 vehicle (A-1) should be developed, flown, and man-rated as soon as possible as a matter of high priority. Such development will not only allow initiation of Apollo spacecraft tests at the earliest possible date, but will also generate experience in the operation of large hydrogen-oxygen stages and will provide definition of the problems and the potential of multiple engine clusters for such stages.
    2. *Titan Launch Vehicle System*
    At the conclusion of the LLVPG studies in October 1961, the following recommendation was made by the group with regard to the Titan III:
    Recommendation: *The 120-inch diameter solid motor and the Titan III launch vehicle should be developed by the Department of Defense to meet DOD and NASA needs, as appropriate in the payload range of 5000 to 30,000 pounds, low earth orbit equivalent.*
    Of the various considerations taken into account in evaluating the advisability of proceeding with development of the Titan III system, the principal arguments leading to the conclusion of the group were: (1) the anticipated large number of DOD missions during this decade justify the development of the Titan III family because of its substantially lower cost per launch than for Saturn based vehicles; (2) the importance to DOD of having a launching system not dependent on the use of cryogenic propellants; (3) the Titan III, by virtue of the way its building blocks can be combined, permits greater flexibility; [II-2] (4) the Titan III uses DOD experience with Titan II, making logistics and training easy for DOD; and (5) development of large solid motor technology would be part of the development effort and cost of this vehicle system. Such development would be in accord with prior governmental policy decisions that advancement of large solid rocket technology would be vigorously pursued.
    Following the adjournment of the LLVPG in October 1961, unresolved questions still remained relative to the role of the Titan III. The LLVPG had given little or no attention to the Titan II-1/2, the Department of Defense had initiated a Phase I development on the Titan III, and NASA was soon to make a decision on Gemini and was considering the Titan II-I/2. To assist the pending decision by DOD and NASA relative to these vehicles, the LLVPG was reconvened for analysis of the National Launch Vehicle Program in the 5000 to 30,000 pound low earth equivalent range.

At the termination of this reconvened session of the group. the recommendation on the Titan III was as follows:

Recommendation: *The Titan III space launching system should be developed by the Department of Defense providing that the Phase I study now under way confirms the technical feasibility and desirability of the system.*

The further review of the Titan III by the group did not result in the introduction of any additional factual evidence either for or against the prior recommendation. Thus, the arguments outlined above in favor of Titan III suggest again the recommendation to proceed with development of the vehicle should the Phase I study confirm the technical feasibility.

3.   Saturn Upper Stage

Recommendation: *Develop the S-IVB stage as promptly as possible using it as an alternate stage for the 5-IV in the Saturn C-1. This stage is necessary for the Class B vehicle recommended and its combination with the S-I stage (A-2 vehicle) will constitute another potential Class A vehicle.*

[II-3] An examination of the various Class B vehicles considered shows that all of the interesting versions have as their third stage the S-IVB which is powered by one J-2 engine. In view of its almost certain use in Class B vehicles it is considered extremely desirable that plans be made for early flight tests of the S-IVB on the S-I stage in order to build up its reliability as rapidly as possible.

4.   Class 13 Launch Vehicle

Recommendation: *Develop as promptly as possible a Class B vehicle (B-8, B-10, or B-15) consisting of a four or five F-1 engine first stage, a four or five J-2 engine second stage and a one J-2 engine third stage (S-IVB). This vehicle should be designed for use as a two-stage vehicle for low earth orbit missions and a three-stage vehicle for escape missions with a minimum performance capability of 180,000 pounds in a low earth orbit and 70,000 pounds to escape.*

It is felt that a Class B vehicle can be developed with relatively little delay and that this development should be pursued with the highest priority. This conclusion results from recognition that both earth orbit rendezvous and lunar orbit rendezvous are attractive mission concepts and that they can be achieved with Class B vehicles. Furthermore, lunar orbit rendezvous offers the chance of the earliest accomplishment of manned lunar landing. It is quite likely that the pacing item for any rendezvous approach is development of the Class B vehicles, hence the high degree of urgency recommended.

5.   Use of Solid Motors in Class B Launch Vehicles

Recommendation: *The design of the second and third stages of the Class B vehicle recommended (B-8, B-10, or B-15) should, if practicable, provide potential for economical and early substitution of a solid motor first stage for the four or five F-1 engine first stage. Substitution of such a solid motor stage may permit the construction of a vehicle (B-5 or B-14) of comparable but somewhat lower capability than the recommended all-liquid Class B vehicle.*

[II-4] The group examined the question as to whether a solid first stage should be developed for the Class B vehicle in parallel with the recommended liquid first stage. It was considered that while LOX/RP is a familiar propellant combination and the F-1 engine appears to be progressing satisfactorily thus far, there is considerable merit in a backup development that exploits large solid rocket motors. This is particularly the case if the manned lunar landing program is to be considered a high priority program aimed at accomplishing the mission at the earliest possible date. Therefore, it was recommended that the upper stages of the all-liquid Class B vehicle should be designed for possible substitution of a solid first stage. Such a solid first stage vehicle appears to be attractive in terms of a low cost, high reliability and operational simplicity if there are sufficient continuing needs for Class B vehicles in the late 1960's and early 1970's.

6. *Rendezvous Operations Techniques*

Recommendation: *A major engineering effort should be made to develop rendezvous operations techniques in both earth and lunar orbits as possible approaches for accomplishing the manned lunar landing mission at the earliest possible date.*

The Class B vehicle required for manned lunar landing by rendezvous operations will be available earlier than the Class C vehicle necessary to carry out the mission by direct ascent. Thus, if the development of rendezvous operations are not the pacing item, use of the Class B vehicle offers the earliest possibility of a manned lunar landing.

It is therefore important to determine the feasibility of rendezvous at the earliest possible date. Accordingly, efforts should be initiated as soon as possible to develop techniques for both earth and lunar orbital rendezvous. A detailed discussion of these rendezvous techniques is included in Chapter VI, Volume III.

[II-5] 7. *Class C Launch Vehicles*

Recommendation: *Since it is by no means certain that the development of rendezvous operations will advance rapidly enough to provide earliest accomplishment of manned lunar landing, it is recommended that the direct ascent capability be developed on a concurrent basis.*

For that purpose the following specific steps are recommended:

a. On a concurrent and urgent basis a thorough engineering analysis should be made of attractive Class C vehicles (C11, C-16, C-20, C-24, C-25), their constituent building blocks and other related possible configurations to enable selection of the most desirable NOVA vehicle for manned lunar landing.

b. The large solid rocket motor and the large hydrogen/oxygen engine development also recommended should be pursued in a manner that will permit their potential use in a NOVA configuration for planetary missions.

The group felt that a Class C vehicle program must be carried forward on an urgent basis and concurrent with development of orbital rendezvous. Nevertheless, the group also felt that initiation of Class C stage and vehicle development at this time was inappropriate because of the lack of sufficient information to select a specific Class C vehicle.

The initial step that the group felt should be taken is to analyze in detail the potential Class G vehicles. This analysis should take into consideration the large solid motors, the M-1 engine, and the stages of the recommended Class B vehicle (B-8, B-10, B-15) that would potentially be available as building blocks. It was also considered important to study in greater depth the technical problems and schedule implications involved in producing very large solid motors.

8. *Large Liquid Hydrogen Engine for Class C Launch Vehicles*

Recommendation: *Initiate promptly the development of a hydrogen-oxygen engine having a nominal thrust of 1.5 million pounds.*

[II-6] Studies made by the group to date do not support a specific thrust level recommendation at this time but do suggest that a level above 1.2 million pounds is necessary to provide for follow-up programs after a manned lunar landing.

Although it had been concluded that insufficient information was available to initiate development of a specific Class G vehicle it was recommended by the group that development be initiated or continued on certain components of attractive Class G vehicles that might prove useful in the development of a Class G vehicle. One such component on which the group felt development should be initiated was a large hydrogen-oxygen engine, the M-1.

9. *Large Solid Motors for Class C Vehicles*

Recommendation: *Initiate promptly a program aimed at the development and production of solid propellant motors up to 300 inches in diameter and 3,000,000 pounds in weight. The program should be associated initially with a thorough study of the advantages and disadvantages of the segmented type assembly, with particular attention given to clustered motor configurations.*

Emphasis in the initial phase of the program should be to produce an early test firing of a unitized motor of at least 240-inch diameter and to utilize to a maximum the existing solid motor facilities for the development of 156-inch diameter segmented motors for test firing as promptly as possible. This solid motor program should be conducted concurrently with development of the Class C liquid propellant vehicle.

The effort should be incrementally funded so as to reduce the total funds that must be committed before definitive engineering information is available on the suitability of large solid motors of various dimensions and before the requirements are established for the number of motors needed in each size class. These recommendations are made in full awareness of the fact that a new facility on water for case fabrication, propellant mixing, casting and curing and for static firing purposes must be committed at the beginning of this development effort.

[II-7] 10. *Launch Facilities*

Recommendation: *The following launch complex plans should be implemented:*

    a.   *The complex for the Saturn C-1 should be built so that it is compatible for use with the S-1 and S-IVB stage versions of Saturn (A-2).*

    b.   *Develop an Integrate-Transfer-Launch (ITL) complex for solid-boosted Class A vehicles.*

    c.   *Construct an ITL complex to handle the all-liquid Class B vehicle (B-8, B-10, or B-15) and initiate the A and E work necessary to permit use of the complex for launch of a solid-boosted Class B vehicle (B-5 or B-14).*

    d.   *Initiate A and E studies on a Class C vehicle launch complex designed to accommodate either liquid or solid first stage boosters and using all or part of the ITL concept.*

Consideration has been given to the launch facilities required for all three vehicle classes. In general, where new facilities are to be constructed, the group favors an ITL type complex. This type of complex provides for an integration building near, but not on, the launch pad in which the launch vehicle and spacecraft integration and checkout are performed. After completing the checkout, the vehicle is moved to the pad, where it is fueled and launched. By utilizing this technique, the on-pad time can be cut drastically and overall cost reduced while high launch rates are simultaneously achieved.

For Class A vehicles it is clear that the Saturn configuration should be launched from the existing pad and others of similar design. An ITL is not worthwhile for these vehicles because of the urgent program schedule. On the other hand, for a workhorse Class A vehicle, which would have a solid first stage, the ITL concept should be used.

[II-8] B. *Discussion of Recommendations*

The basis for the principal recommendation, briefly discussed above are [sic] amplified in the following paragraphs.

    1.   *Class A Vehicles*

Considerations of the group relating to the Class A vehicles led to a study in some detail of the reasons for supporting development of a Titan III-C vehicle in addition to the Saturn C-1. It was projected that during this decade there would be over 200 missions, largely for DOD, in which 12,000 to 30,000-pound payloads will be required in low earth orbit. In addition to the Titan III-C there are two versions of the Saturn that have this payload capability. These vehicles are the Saturn C-1 and a possible variation of Saturn (A-2) using the S-1 and the S-IVB stages. The Saturn C-1 is already in the National Launch Vehicle Program and the A-2 version of Saturn was recommended to provide early flight development of the S-IVB stage.

Although the two versions of Saturn have performance capabilities that are comparable or superior to the Titan III-C, the Saturns are likely to have a somewhat higher cost; they do not have the militarily desirable feature of employing solid and storable propellants, which permit fast reaction times, or, stated differently, permit long waiting periods on-pad; and they do not contribute to the development of large solid motor technology. On the other hand, the two Saturn vehicles appear to offer more growth capability than members of the Titan III family for a general program of space exploration. In this regard the Saturn permits larger diameter payloads than the Titan III-C. The Saturn can accommodate payloads of up to 20 feet in diameter, which also makes them suitable for launching a nuclear stage. This will be an important advantage if it is eventually desirable to use them for the development of such stages or to provide increased payload capabilities for future missions by the use of a nuclear upper stage. It was after weighing these factors that the LLVPG recommended that the Saturn C-1, which is scheduled for use in Apollo manned orbital flights prior to completion of Titan III development, should be improved (Recommendation 1) for the purpose of its continued use in support [II-9] of Apollo. Furthermore, the Titan III should be developed primarily for support of other scientific and military missions (Recommendation 2).

In considering the Titan III vehicle family it was noted that the Titan III-A and the Titan III-B, which are closely associated with the Titan III-C development, have payload capabilities that can he provided by the second generation Centaur on an Atlas. There was, therefore, some question as to whether the Titan Ill-A and III-B should be developed. However, development of the Titan BI family will enable the economical introduction of the Titan III-A and III-IS; also those versions of Titan III will serve as backup to cryogenic based boosters or as a substitution for them in cases where fast reaction time is needed.

2. *Class B Vehicles*

In connection with the development of the Class B vehicle, it was recommended that development of the upper stage, the S-IVB stage, be initiated immediately with a view toward flight testing it on an S-1 stage (Recommendation 3). This procedure would insure most rapid development of the Class B vehicle.

The Class B vehicle recommended (Recommendation 4) will provide a minimum payload capability of 180,000 pounds in low earth orbit and 70,000 pounds to lunar escape velocity. This capability is sufficient to enable manned circumlunar flight using the Apollo spacecraft and with a single rendezvous operation in earth orbit, to perform the manned lunar landing mission. It was strongly recommended that the rendezvous approach be pursued vigorously.

An item that received particular attention with relation to the Class B vehicles was whether a large solid rocket motor should be developed in parallel or as backup to the liquid first stage booster. Two configurations were examined, both of which utilized 156-inch diameter solids on the first stage and J-2 engines on the second and third stages.

[II-10] It was concluded that a vehicle such as these might be attractive in terms of low cost high reliability and operational simplicity if there are sufficient continuing needs for Class B vehicles in the late 1960's and early 1970's (Recommendation 5). By reducing the size of the solid motor first stage and thus significantly reducing vehicle cost, a useful vehicle can be provided that will cover the payload range between 30,000 and 180,000 pounds in a low earth orbit. While the total cost of the solid motor rocket development program cannot be justified on the basis of this application alone, there are other applications for solid motors in the development of Class C vehicles.

3. *Class C Vehicles*

The principal reason for recommending development of direct ascent capability concurrently with rendezvous, and thus the development of a Class C vehicle was so that success of the manned lunar landing mission is not solely dependent on the timely success of rendezvous techniques (Recommendation 7). It is also important to recognize that the

national space program should be projected beyond the initial manned lunar explorations to the problems of a more thorough exploration of the moon, possible establishment of a moon base and the initiation of a manned planetary exploration program.
Aside from the obvious direct importance of early attainment of U.S. capability for planetary exploration there is also the consideration that failure to develop a Class C vehicle
at an early date could, if our rendezvous capability is delayed, leave this country in a particularly difficult posture if the USSR should be first to achieve a successful manned lunar
landing.

After reviewing the various configurations for the Class C or NOVA vehicle it was concluded that insufficient information exists to permit selection of a specific NOVA configuration to be developed or to support a recommendation that development of a specific
stage be initiated. It was felt that more must be known about the performance and design
feasibility of the various vehicles considered and about the development risk of such
important elements as the very large solid rocket motors. Another consideration was
recognition of the cost and management difficulties of undertaking development of a
Class IS and C vehicle simultaneously and with equal urgency.

[II-11] The best approach appeared to he initiation and continuation of component
development which may be applicable to the NOVA vehicle. It would also be desirable to
intensify detailed engineering studies of the most promising Class C configurations. One
component that should be developed is the large hydrogen/oxygen engine, the M-1,
which is visualized as having a thrust between 1 and 2 million pounds (Recommendation
8). Such an engine could replace the five J-2 engines of the S-II stage with a single engine.
It would also permit the design of two or four engine stages considerably larger than the
present S-II stage, thus providing greater payload capability.

The other major component possibly useful for NOVA class vehicles is a very large
solid motor, and thus the recommendation that development be initiated on solid motors
up to 300 inches in diameter and weighing up to 3,000,000 pounds (Recommendation 9).
From the study of the various vehicle configurations it appears possible to make a Class C
vehicle by clustering 4 to 10 solid motors in the first stage on top of which would be placed
a complete Class B all-liquid vehicle (B-8, B-10, or B-15). The Class B vehicle would
require suitable modifications to the first stage to provide for altitude starting of the F1
engine and increased strength to withstand the structural loads that it would experience
as a second stage. If such a vehicle is not feasible, two other approaches are offered. One
is to be a cluster of solids for the first stage and new upper stages based on the M-1 and
J-2 engines (C-16 and C-21). The other approach is to make an all-liquid vehicle with all
of the stages different from those of the Class B vehicle. The C-11 is an example of such a
vehicle.

The development of large solid rocket motors was examined quite thoroughly by the
group. Of course, the generally claimed advantages of solids are high reliability, low cost,
and short development time. The group, however, found it very difficult to establish any
clear superiority in reliability or development time for the solid over the liquid rocket
booster. From the standpoint of cost, the solid motors appear relatively most attractive in
the Class A vehicles, less attractive in the Class B, and least attractive in the Class C. Since
the Class A vehicles require smaller diameter solids [II-12] (100 to 120 inches), which present the least development risk and earliest availability, the group favors the development
of a solid first stage Class A vehicle as a workhorse. A solid first stage Class B vehicle
appears attractive from the viewpoint of operating convenience, cost and perhaps reliability (based on the use of 156-inch diameter clustered solids). However, this vehicle is not
sufficiently attractive in itself to justify development of solid motors larger than 120 inches in diameter. For Class C vehicles, the 240 to 280-inch diameter solids are considered the

most attractive size.

If solid motors are selected for use in large vehicles, it therefore appears that the two most attractive sizes are 156 and 240-inch diameter. The 156-inch diameter motors are favored because, if segmented, they can be fabricated, tested and shipped with presently existing facilities and transportation methods. The basic factor limiting the size of a segmented motor is the limit for railroad transportation. Present manufacturing facilities permit research and development motors to be made in the 156-inch diameter size but they are inadequate to supply production quantities. Development of the capacity to supply production demands for this size motor would require a new propellant mixing, casting and curing plant. One unattractive feature of the 156-inch diameter motor is the fact that as many as 7 to 10 motors must be clustered together to provide the first stage of a Class C vehicle. This means that the reliability of each motor must be very high and of each segment even higher if the over-all stage reliability is to be satisfactory.

The advantage of going to larger diameter solid motors, those in the range of 240 to 280-inch diameter, is that only a few motors need be clustered in the vehicle first stage. For example, four motors of this size appear to be adequate for the first stage of a Class C vehicle. Fewer motors favor higher stage reliability and also simplify the intrastage structural design and vehicle bending load analysis.

There are three principal disadvantages of the larger motors. The first is that a greater chance exists for the occurrence of developmental problems, although at this time no such problems can be identified by scaling [II-13] analysis. The second disadvantage is that production of even the early test motors must await construction of new plant facilities. In order to facilitate transportation to the launch site such facilities should be located on navigable waterways. Thus, it would require from 6 to 18 months longer to develop these motors than those of 156-inch diameter. Finally, such large motors, particularly if unitized, are extremely heavy, weighing about 2,500,000 pounds. Thus, new problems in handling, transportation and assembly must be faced.

Whether large solid motors will actually provide the advantages of early availability, flexibility of configuration, simplicity of operation and high reliability in Class C vehicles cannot yet be predicted with any assurance. However, the importance of developing a Class C vehicle at the earliest possible date is so great that initiation of a large solid motor program, including development of integrated motors up to 300-inch diameter, is called for. Furthermore, the design studies of various Class C vehicles with solid propellant first stages should be intensified. It is felt that such an effort will insure availability of a Class C vehicle at the earliest possible date with a relatively modest additional development effort.

In connection with the possibility of using large solid motors, a NOVA vehicle comprised of all solid stages was considered. The most carefully investigated vehicle in this class was conceived by the Jet Propulsion laboratory and proposed for the manned lunar landing program in JPL-TM33-52, "A Solid Propellant Nova Injection Vehicle System," 3 August 1961. The report proposed a four-stage vehicle consisting entirely of solid propellant motors with a liftoff gross weight of 25,000,000 pounds, and an estimated capability of placing 130,000 pounds in a lunar escape trajectory. This design was considered sufficiently interesting to warrant careful review by qualified and disinterested organizations. Accordingly, it was requested that Space Technology Laboratories and the Boeing Company review the JPL report. After completion of these studies, the group arrived at the conclusion that the all-solid NOVA development constituted a very high risk program and thus should not receive further consideration.

[II-14] 4. *Future Decisions*

There are three major future program decisions that are implied by the conclusions and recommendations of the group. These are:

a.   Whether or not the S-I stage - S-IVB stage version of Saturn, which is recommended to provide early flight testing of the S-IVB stage, should be fully developed as another Class A vehicle.

b.   Selection of the Class C or NOVA vehicle design.

c.   Establishment of the diameter and other pertinent specifications of the large solid rocket motors to be developed, and definition of the stages in which the large solids are to be used.

Replacing the S-IV stage of the Saturn C-l with the S-IVB stage will provide a substantial increase in performance capability. In addition, the single engine of the S-IVB stage offers the potential for greater ultimate reliability than does the six engine S-IV stage.

Even though the S-IVB stage is successfully flown on an S-I stage for test purposes, considerable additional design and development effort would be required to fully develop such a vehicle for operational use. Therefore, the decision as to whether the development of such a vehicle should proceed must be based on the degree of success achieved in developing the Saturn C-l and the Titan III and on their ability to fulfill the mission requirements for Class A vehicles.

The decisions on the Class C vehicle design and the selection of the large solids to be developed can interact strongly. The first opportunity for a decision on Class C vehicle configurations will occur when the recommended design studies are completed in about mid-1962. Probably a better decision can be made if it is postponed until late 1962, by which time significantly more should be known about the performance of the F-1, the J-2, the cluster of eight H-1 engines in the S-1 stage, and about 156-inch diameter solid motors. More may also be known about the feasibility of orbital operations. If the solid motor development program and stage engineering studies [II-14] proceed as recommended, probably no appreciable time will be lost in the Class C vehicle operational date by delaying the configuration decision for a year. This viewpoint is based on the premise that the final configuration selected would use a large solid motor first stage and the modified upper stages of the Class B vehicle based on the M-1 engine. If the configuration chosen is the all-liquid Class C vehicle (C-11), some time will probably have been lost.

If the solid motor diameter decision is not made as part of the vehicle configuration choice, but is kept open among 156, 240 and 280-inch or greater, it will probably be an additional six months to a year before enough is known from actual tests of the large solid motors to enable selection of a diameter with confidence.

[III-1] III.   Supplemental Recommendations

In addition to the primary recommendations there were several supplementary recommendations made by the group. One subject of particular importance, which NASA requested the group to consider at the end of its study efforts, was a possible launch vehicle for the Gemini spacecraft. Other supplementary considerations and recommendations concern largely technical problems which stand out as requiring further detailed study to maximize vehicle system usefulness and to minimize time and costs. These supplemental recommendations of the group are summarized in the following paragraphs.

A.   *Supplemental Launch Vehicle for NASA's Gemini Program*

Recommendation: *A minimum modification version of the Titan II ballistic missile should be used for the Gemini program.*

In the studies of launch vehicle requirements for Gemini it was found that there were four alternative vehicles that might be used. These four vehicles are the Titan II, Titan II-1/2, Titan IIIAJ, and the Saturn C-1. The development schedule indicates that the Titan IIIAJ will not be available until a year later than the two versions of the Titan II. In addition the need for all of the Saturn C-1 vehicles scheduled for production to support the Apollo program, as well as launch facility scheduling problems associated with an

increased Saturn C-1 launch rate, indicate that consideration of this vehicle is purely academic, since it would not be available for use in the Gemini program. Thus the only useful alternatives are the Titan II and Titan II-1/2. The principal difference between these
two vehicles is that the Titan II-1/2 provides subsystem redundancy leading ultimately to
higher reliability but with a penalty in payload weight.

Since the performance differences between these two vehicle configurations are not
striking. vehicle reliability and development schedules were the areas of consideration in
making a choice between them. Safety of the crew will be insured by malfunction detection and abort systems in either vehicle. Thus the greater reliability offered by additional
redundancy of the Titan II-1/2 is a factor that can be supported principally as a need on
the basis [III-2] of launch vehicle economy. However, the initial planning of the Gemini
program calls for only about 18 flights and the Titan II will have attained a reasonable reliability by the time this program begins. Therefore, little weight can he given to possible
economic gains that might be realized with the Titan II-1/2. In addition, the inherent
uncertainties in reliability estimates as well as uncertainties in projected reliability growth
during the brief life span of the Gemini program suggest that even the economic arguments based on greater reliability of Titan II-1/2 may not be well founded.

Other major factors that were weighed in determining the relative suitability of these
two vehicles are: (1) availability in 15 to 18 months after program go-ahead; (2) the degree
to which either may interfere with DOD programs; and (3) relative cost.

Considering the many factors pertinent to a choice between these two vehicles, it was
the judgment of the group that use of the Titan II ICBM with minimum modifications in
the Gemini program would provide greatest assurance of timely availability of a vehicle
that has adequate reliability and performance, best utilization of DOD engineering and
management resources associated with the Titan II weapon system, and minimum vehicle
cost for the program.

B.   *Reliability and Reliability Growth*

Recommendation: *A vigorous theoretical study and experimental program must be implemented to determine the degree to which redundancy, engine-out and manned monitoring and control should be used in each vehicle and subsystem. The LLVPG believes that, in the size booster vehicles considered for the Apollo missions, it is practical and desirable to use such techniques to a far
greater extent than was possible in previous booster systems.*

The reliability to be expected in early flights of vehicles used for the manned lunar
landing program has an extremely important bearing on the time required to accomplish
the mission and on the cost of the over-all program. In addition, reliability will have an
effect on crew safety and on the [III-3] possibility of program stretch-out or cancellation.
Indeed, it might be said that the chances of being first to the moon are very small indeed,
unless a significant step forward can be made in obtaining high reliability earlier in the
life of the vehicles than has been experienced to date.

From an examination of the results of the calculation of mission success data analyzed
by the LLVPG, it was found that it would take two to three years of flight test and about
25 to 60 launchings to man-rate a Class B or C vehicle using the reliability growth estimates
of this study. As previously indicated, it is important to note that "man-rate," as used in this
entire study, refers to a vehicle having an absolute reliability of 50 percent or more. This
level of reliability should not be confused with "man-safety" which is sought to be maintained at a relatively much higher level by providing abort subsystems for crew escape in
case of catastrophic malfunction.

If a significant improvement in early reliability were achieved, the date for mission
accomplishment could be advanced about a year. In addition, 20 to 30 flight vehicles
could be eliminated from the program at a cost savings of the order of one billion dollars.

The reliability growth curves used in the analysis were based on past experience primarily with ballistic missiles and space adaptations thereof. The data from previous flight test programs were smoothed and interpreted in terms of the number of such systems, stages, restarts, etc., involved, and in terms of the number of redundant and non-redundant elements such as engines, thrust vector control systems and the like. The fundamental assumption underlying the argument is: It will be possible to obtain about the same early reliability on an absolute basis with the new, large launch vehicles as we have done in the past on smaller, primarily ballistic, missiles (Atlas, Titan, Thor, Jupiter, Polaris).

It could be argued that the reliability growth should be much better because so much has been learned from past failures and mistakes, and because weight and performance are not quite as critical as they were for such vehicles as the Atlas. Conversely, it could be argued that the reliability growth will be [III-4] worse because of the greatly increased size of these new vehicles, the use of a new propellant (hydrogen), the clustering of 4, 5, and even 8 liquid engines per stage, and the simultaneous development of so many large stages and multi-stage vehicles by the same organization. In the group's deliberations it was agreed that these two sides of the argument just about offset one another and that reliability growth about equal to that of past vehicle development programs might reasonably be expected. Nevertheless, it was recognized that there is a very wide range of uncertainty in reliability growth projections.

It is important to examine very carefully the question as to whether, and how it might be possible, to improve significantly the reliability growth rate of the new vehicles to be developed. In order to be somewhat more specific about the major problems, the LLVPG had specific studies made in the technical areas of redundancy, the role of man in complex systems, and engine-out capability.

C.   *Reliability Budgeting*

Recommendation: *The iterative use of the "reliability budget" during the design phase is probably the most practical means of achieving an optimum approach in reliability engineering of complex systems.*

Because of the large number of stages involved in the total lunar mission, the requirement for a much higher level of redundancy should be anticipated than has been normal in the past. This redundancy will vary from conservative design margins and state-of-the-art engineering to the use of completely redundant subsystems in some cases. The iterative use of reliability budgeting provides a basis for establishing the amount of redundancy to be employed in a given system or subsystem.

Reliability budgeting is a general approach toward reliability which has been used on some programs and which can be extended and improved for application to the manned lunar landing program. It is an iterative approach which must be run repeatedly until the design converges or is frozen for other reasons. Underlying the whole process is a recognition and an acceptance of the fact that there are gains to be made by the judicious employment of [III-5] redundancy but that such employment in no way diminishes the need for a sound analytical approach to design.

The process of reliability budgeting begins with the system engineer. The first step is for the system engineer to block out the total system design and translate it into a reliability budget. Each subsystem is assigned a level of reliability which in combination with those of the other subsystems will produce the desired system reliability. Where the assumed reliability is not feasible with the simplest system configuration, redundancy is added judiciously until it is attained. Costs and schedules must be evaluated in parallel to assist in weighing the merits of the particular design choice.

The reliability budgeting task is then given to the subsystem designer who carries through the same processes for his subsystem against the assigned reliability target given to him by the system engineer. Should the subsystem designer find the reliability target impossible to meet, even with optimum redundancy, he must obtain a new target from the system engineer, thus requiring that the system engineer rebudget the reliability requirements among the various subsystems. Conversely, if the subsystem designer finds ways of obtaining reliability higher than the target value, the system engineer can likewise take this information into account together with the cost and schedule implications to rebudget his reliability among the various subsystems.

By carrying this process on through to the lowest level of component design and by maintaining the over-all design relatively fluid in its early stages and freezing it as late as possible, the maximum number of iterations can be made and thus the optimum use of redundancy can best be approached.

D.   *The Role of Man in System Operation*

Recommendation: *It is recommended that prompt steps be taken to initiate further detailed studies concerning the role of man in system operation employing well-definitized systems and subsystems of the launch vehicles intended to be used for future manned missions. Furthermore, launch vehicle systems designs compatible with crew participation in vehicle control, but not solely dependent on it, should be investigated in detail. It would be desirable* [III-6] *for these studies to be conducted by organizations having experience and capability in the manned aircraft and missile design field.*

The problem which is of concern here is the establishment of the role of the flight crew during the launch phase of manned flight operations. That is, whether the crew should be given an active role in the control and management of the launch vehicle systems or whether they should maintain a completely passive role with all functions being programmed automatically. Because of man's inherent ability to perceive, reason, and judge in even unrehearsed situations, it is believed that the idea of a completely passive role for the crew is unreasonable.

There are several modes of manned participation which could be considered, namely:

a.   Direct control
b.   Monitor, switching, and override
c.   Monitor, adjustment, and maintenance

The direct control mode would provide the crew with the primary path for control inputs to the given system in much the same manner as our present day aircraft are designed. In this mode, the automatic controls would be provided for crew convenience for use during reasonably uneventful periods. The second mode—monitor, switching, and override—would provide the crew with a generally subordinate control approach with the option for primary control. In this mode, the crew would normally monitor a system and, in the event of some malfunction, they could exercise direct control by manually switching to a redundant system or by manually overriding the automatic system. The third mode provides for the lowest degree of crew participation. In this mode, the crew would monitor certain function displays and would make only minor adjustments, such as gain settings, gyro realignment, etc. In addition, the crew could perform certain maintenance functions, such as changing fuses and small components.

[III-7] Of the three modes of participation cited, it is believed that the direct control mode is probably too drastic in view of our present, very limited experience in this area. On the other hand, some real gains in the over-all reliability, or mission success achievement, are likely to be made by the judicious adoption of the second and third modes of crew participation for certain launch vehicle systems.

After reviewing this problem and the various possible approaches, the following general conclusions were reached:

    a.   From an environmental standpoint, no evidence exists indicating that the vehicle control task cannot be handled by man as an integral control element.

    b.   Considerable evidence exists to show that man, having been given adequate instrumentation and training, has the capability of successfully completing booster trajectory control during launch.

    c.   It is believed that appreciable gains in mission success can be achieved through crew participation, particularly during the early development stages, where the demonstrated reliability of launch vehicles is generally quite low.

In the light of the foregoing, it is believed that the role of the spacecraft crew should be one of active participation during the launch phase of flight. The exact degree of crew participation cannot, of course, be definitely specified at this time. However, available evidence suggests that the crew should be provided with more than merely monitor capability.

E.   *Engine-Out Capability*

Conclusion: *While engine-out capability appears attractive on the basis of the engine and control system redundancy considerations, a detailed engineering study of the implications on the remaining portions of the vehicle system is required.*

1.   *Performance Degradation Versus Reliability Increase*

The LLVPG has made some estimates of the losses in payload that would result from stage designs with engine-out capability. The major points revealed by this study are as follows:

    a.   For a given number of engines, the performance loss with one engine out is about one half as great in a second stage as in a first stage.

    b.   Engine-out performance loss is serious in the first stage, particularly if the number of engines is small (four or five compared to eight) and engine shutdown occurs early in the stage burning time.

    c.   Engine-out penalties in first and second stages are a non-linear function of time; one engine shutdown at the halfway point results in about one-fourth to one-fifth as much performance loss as when the engine shutdown occurs just after ignition.

    d.   Operation with an engine out does not result in a significant performance loss in a third (escape) stage.

These performance degradation results are based upon reasonably well-designed vehicles and therefore should not be assumed as applying to off-optimum or unique vehicle designs.

Another approach to engine-out redundancy would be to add extra or spare engines. The performance loss for the Saturn C-4 class vehicles using such an approach has been examined and found to be acceptably low. It would be possible to design a stage carrying a true spare engine which would not be started unless required; however, this "delay-until-needed" design philosophy would appear undesirable in the lower stages.

Preliminary analyses of the over-all problem by the LLVPG has led to the following stage-by-stage design philosophy:

    a.   First Stage—the design should probably be based upon hold-down and engine-out. One engine out in this stage could extend to two if a large number of unreliable engines are used. Similarly, if a stage [III-9] contains a small number of very reliable engines, the engine-out design approach should not be used.

All engines should be started before liftoff and should be able to operate through the thrust/weight instability and high q regions. The use of hold-down aids the reliability since over one-half (68 per cent) of the engine failures occur in the first few seconds of engine operation.

    b.   Second Stage—all engines in this stage should be started. The q problem is not important in this stage since staging occurs at a very high altitude. Engine-out capability should be provided for in all multiple engine stages. The performance reliability "map" is attractive in this stage since engine-out performance losses are about one-half to one-third as severe as in the first stage and the improvement in reliability with one engine out capability is attractive.

    c.   Third Stage—a two-engine stage seems attractive here from a reliability point of view. The reliability of a two-engine stage is typically raised from 0.90 to 0.95 by the use of one engine redundancy.

The third stage problem is somewhat unique. First examinations seem to indicate that a two-engine system should be used by starting just one engine, with the second engine started only if the first should fail. The guidance control problems in the third or higher stage, if both engines are initially started, seem to be quite severe. Therefore, the delay-until-needed approach is suggested.

    2.   *Effect on Other Systems*

The previous suggestions are based upon considerations of the reliability of engine and control systems and their associated failures and performance. For a stage to have engine-out capability a number of modifications of other subsystems may be required. These modifications will affect system reliability and performance. The autopilot, for instance, may be required to have provisions for automatic reprogramming when an engine is shut down. Similarly, the control system may be required to have faster [III-10] response and to operate with larger gimbal angles and increased actuation forces. Vehicle structure will be subjected to new load distribution which may necessitate a different design. The implications of engine-out operation will vary between stage designs. It is anticipated that, in some cases, significant modification (by present standards) of autopilot and/or structure and control systems will be required to accommodate the engine-out feature. Other stages may conceivably require little or no change in these systems.

    F.   *Automatic Vehicle Checkout and Countdown Considerations*

Conclusion: *The significance of the considerations concerning automatic checkout and count-down of vehicles is twofold:*

    a.   *A most intimate relationship is needed among design criteria of the vehicle and its subsystems, of ground support equipment and the launch complex of the spacecraft, its propulsion and other subsystems, and of the payload. The extent of this relationship, and the amount of preplanning needed cannot be fully envisioned at this time.*

    b.   *The necessity to standardize specifications of interrelated components will require a level of systems engineering, both in comprehensiveness and in detail, far surpassing in complexity previous technological undertakings of any kind.*

Among the factors strongly influencing the probability of mission success is the efficacy of checkout procedures used just prior to launch. The checkout procedures may require the testing of all essential components, subsystems and systems and thus involve measurement of up to 1500 functions in research and development vehicles. The concept of automatic checkout has been advanced primarily for two reasons: (1) to reduce the amount of time required in using launch facilities; and (2) to enhance the reliability of the entire checkout operation.

No figures can be quoted on time savings or operational vehicle reliability improvements achievable with automatic checkout procedures, since numbers and types of measurements, amount of data processing, manner of [III-11] presentation and use of processed data are not currently defined. It has been estimated, however, that manual operations for a large vehicle might require two to three weeks as contrasted with three days for an automatic system.

The employment of automatic checkout equipment will require a high level of advance planning effectiveness and good over-all system engineering, in the following areas:

    a.   Design criteria of the vehicle, the ground support equipment and the launch complex must contain automatic checkout requirements so that the automatic checkout concept is extended back to, and properly accounted for, at the stage and subsystem manufacturing level.

    b.   Management arrangements between contractors involved in development of the vehicle, the ground support equipment and the launch complex.

    c.   The planning and stocking of spares, up to and including individual stages.

    d.   Equipment modification and change control. The potential conflict between research and development or operational changes and automatic checkout compatibility requires that the changes be carefully scheduled.

Even with effective preplanning, 30 to 40 flights may be needed to perfect the launch vehicle automatic checkout system. It is possible that the spacecraft checkout system can be perfected in fewer flights, since it is in some ways less complex than the vehicle system, but this implies extensive systems engineering coordination at the earliest stages between spacecraft and vehicle contractors. In this view, conceptual separation between spacecraft and launch vehicles is largely artificial and has significance or convenience principally for administrative rather than substantive engineering purposes.

The difference between automatic checkout of solid and liquid motors is not entirely clear due to unknowns affecting solid motor design and assembly. Checkout procedure on solid motors may be shorter and less complex but the [III-12] loading process may be longer, since by some estimates the motors must be perhaps assembled at the launch area instead of the assembly area. The estimated installation and checkout time required for solids may be as long as several weeks. There is little doubt, however, that the advantages of automatic checkout will be required for solids as well as liquids.

The high level of design unification which will be required for the launch vehicle, ground support equipment and launch complex must also be extended to include the spacecraft and all of its essential subsystems. Since the demands on the crew in flight should be minimized, the spacecraft system must incorporate design provisions permitting not only automatic checkout on demand but also containing continuous reliability and damage assessment checks. These checkout provisions must naturally be compatible with the ground-based launching checkout system. In addition, limited but effective and compatible provisions must be included for in-flight maintenance, based on modular design, at least for those components with the lowest reliability and for those most subject to in-flight damage.

G.   *Technical Manpower Requirements*

Recommendation: *Because the preliminary study of technical manpower requirements for DOD and NASA programs during the remainder of the decade suggests that a potential shortage of technical manpower may be in store, becoming critical in CY 64, it is recommended that a more thorough and complete inquiry in this area be initiated by DOD and NASA as expeditiously as possible. It may also be desirable to begin developing plans promptly for appropriate action by DOD and NASA in case the difficulties predicted by the LLVPG are confirmed.*

In view of the large scale and long duration of the research and development efforts needed to accomplish the manned lunar landing mission, and the need to superimpose them on the already large and growing requirements of the Department of Defense for scientific and technical manpower, a study was undertaken by the group to provide information on whether such [III-13] manpower resources might be a limiting factor to early accomplishment of national apace exploration objectives.

This study compared an estimate of the supply of scientists and engineers for each year through 1967 with three estimates of the need for such manpower. The supply for any year was computed by beginning with an inventory for 1960 (as reported by the National Science Foundation), increasing it by the number of college graduates and non-degree personnel entering the field each year, and decreasing it by losses due to retirement because of age or death and transfers to other fields.

Three estimates for manpower need were developed in an effort to insure realism in the final comparisons of supply with demand. One estimate was based on the projections by industry of the ratio of scientists and engineers to total employment, the latter itself being estimated from gross national projections. The other two estimates were based on building up the total national need for scientists and engineers from estimates of total research and development and other dollar expenditures using "experience" ratios for numbers of scientists and engineers per million dollars for various types of such expenditures.

The conclusion of the study is quite clear. No matter what projection of the national needs for scientists and engineers is chosen as the probably correct one, the supply does not appear adequate; the lowest reasonable estimate of requirements approximates the projected supply. This lowest reasonable estimate includes, however, a substantial number (many tens of thousands) of scientists and engineers engaged in writing proposals and brochures, and in advancing state-of-the-art through engineering overhead, and may, therefore, be subject to adjustment if appropriate national policies and implementation procedures are developed.

It is also of interest that the most stringent problem in adjusting demand and supply for scientific and technical manpower will probably occur during 1964 if LLVPG estimates for program growth turn out to be valid.

Document II-21

Document title: James E. Webb, Administrator, to The Honorable Robert S. McNamara, Secretary of Defense, January 16, 1963.

Document II-22

Document title: James E. Webb, Administrator, Memorandum for Dr. Robert Seamans, Associate Administrator, January 18, 1963.

Document II-23

Document title: James E. Webb, Administrator, NASA, and Robert S. McNamara, Secretary of Defense, "Agreement Between the National Aeronautics and Space Administration and the Department of Defense Concerning the Gemini Program," January 21, 1963.

Source: All in NASA Historical Reference Collection, NASA History Office, NASA Headquarters, Washington, D.C.

As an Earth-orbiting program that would develop capabilities for in-orbit rendezvous and human observation of the Earth from space, the Gemini program was of high interest to the Department of Defense as well as the program's sponsor, NASA. In late 1962, Secretary of Defense Robert McNamara, with the support of Presidential Science Advisor Jerome Wiesner, attempted to seize control of the program from NASA, or at least share in its management. This initiative set off an intense conflict between NASA and Department of Defense (DOD) top management. Several documents give a sense of the issues at stake. The January 16, 1963, letter from James Webb indicates the depth of NASA concern, while the January 18 Webb memorandum to Associate Administrator Robert Seamans suggests Webb's desire to find a way to settle the dispute. The January 21 NASA-DOD agreement resolved the conflict. NASA would retain management control over the Gemini program, but a joint NASA-DOD Program Planning Board would ensure that the program's activities were responsive to DOD's interests and requirements. Mentioned are Deputy Secretary of Defense Roswell L. Gilpatric, Director of Defense Research and Engineering Harold Brown, Deputy Director of Defense Research and Engineering John H. Rubel, and NASA Deputy Administrator Hugh Dryden.

Document II-21

[1]   January 16, 1963

The Honorable Robert S. McNamara
Secretary of Defense
Department of Defense
Washington 25, D.C.

Dear Bob:

I cannot agree that your proposed version of an agreement would set up management arrangements suitable to a national Gemini program. Nor do I consider its basic pattern one which can be made acceptable through a series of negotiated changes.

In the recent discussion in which you, Mr. Gilpatric, Dr. Brown, and Mr. Rubel participated, with Dr. Dryden, Dr. Seamans and me, I presented in detail the reasons why we here in NASA consider it a serious mistake to proceed with any plan to transfer the Gemini program to the jurisdiction of the Department of Defense as raised by Dr. Wiesner. Following the subsequent receipt of your suggested agreement, Dryden, Seamans, and I have consulted with our senior associates involved in the manned space flight program. We are unanimous in the view that for us to proceed with the arrangements you suggest would jeopardize our ability to meet our manned lunar landing target dates, would disrupt or certainly impair the effectiveness of an organization that is functioning in a magnificent way on a very tight schedule, and would raise a public and Congressional storm of protest that the language and intent of the National Aeronautics and Space Act of 1958 was being violated.

The scientific knowledge and technologies we, as a nation, need are being rapidly accumulated. An effective capability to continue this activity has been created. It is operating in close co-operation with the military services, and we have recently, through the establishment of a Deputy Associate Administrator for Defense Affairs, strengthened our effort to make available all that is of use to them. We should not risk this hard-won progress.

The policies we have been following in this agency have been directed toward the establishment of a broad national participation by and stimulation of the utilization of increased resources in the universities to meet present and future national requirements. Similarly, increasingly important programs of international cooperation involving both governmental and scientific agencies have been successfully established and are a valuable asset to the Nation's space program, both operationally and scientifically. To mix military and civilian activities to the extent proposed would appear to us to have the most serious implications for the future success of these important national and international activities.

Further, the clear and repeated pronouncements which have been made by the President, the Vice President, the Secretary of State, and [2] other leaders concerned with space, would be compared here and abroad with the action taken, with the inevitable conclusion drawn that there had been a major change in policy with regard to the objectives and purposes of the United States in space activity. Such a conclusion could have a far-reaching influence upon this country's relationships with both the neutral and hostile blocs and upon their policies.

As an alternative to your suggestion, I enclose a brief agreement with an attached suggested plan for increased Air Force participation in Project Gemini. It is about as far as we in NASA feel we can go at this time.

Permit me to close with the suggestion that the agreement I enclose will retain for the President a flexible military program including manned space flight, with the ultimate growth of that program dependent on the knowledge both NASA and the Department of Defense gain as we go along. It facilitates the closest co-operation in obtaining and utilizing this knowledge. The President can as a matter of policy increase this military program or decide not to go forward with it. Likewise, the proposed agreement, taken with the program which he is recommending to Congress in his 1964 budget for NASA, gives him a civilian program to develop the scientific and technological base for pre-eminence in space with a vigorous program to make the manned lunar landing and the incident gaining of experiences in extended manned space flight on a fast schedule. Here again the President retains the flexibility, dependent on the needs of the Nation, for speeding up or slowing down the NASA program. To join the DOD and NASA programs in a monolithic effort would inevitably cause the total program to be characterized as military with substantial loss of flexibility in our international posture.

Sincerely yours,

James E. Webb
Administrator

Document II-22

[1]   January 18, 1963

Memorandum for Dr. Seamans—AA

After thinking overnight about the suggestions made by Secretary McNamara, it seems to me that in reality he is coming back with the same pattern of joint management. I do not see how this is possible under the law. However, I think it is essential that we explore every possibility of working with him and retaining his support. Further, we certainly must go into the question very carefully of why he feels he needs a voice in our management to be sure we accomplish the things that are required in the interest of the Department of Defense.

Further, it seems to me that when he says he does not know what is going on in these programs, we could suggest some way that he could find out and keep abreast without having to actually participate in the decisions. Somehow, we must convince him that we can operate this program better as it is now being operated, producing more value for the total national interest, including the military, than under any other system, but are perfectly prepared to have any system that helps identify the things that are in the national interest and facilitates their accomplishment.

I got the impression last night that somehow the clause about extending the arrangements we now have about launch vehicles—that neither of us will start another one without a sign-off by the other—to the manned space flight field is of great importance. It may be that he feels his situation would be seriously impaired if we should start a manned orbiting station, and that he would then be expected to support it as having value for the military services.

On the other hand, I do not see how we can discharge our responsibilities and give him a veto over this. We could do it with respect to the launch vehicles because each of us was [2] developing some, each of us had authority to develop others, and we needed some device to insist on a national program. It may be that there are some elements of this situation in the manned orbital problem, and if so we should explore them with great care.

It may be that he will tell the Bureau of the Budget what he has not yet told us—his real reasons for wanting a jointly-managed effort.

While I believe the instructions to the Bureau of the Budget should be as I mentioned them to Harold Brown—that the last paper drawn by McNamara represents something on which he and I would like to try to find agreement, provide there is a basis without destroying fundamental values for either of us or impairing the President's position, requirements and responsibilities. I think there are many elements in the draft that do not correspond with this. However, it seems to me that some agency experienced in handling Presidential problems must put these forward perhaps more forcefully than I have been able to do so.

It seems to me that you, Hugh, and I should bear in mind that we have signed, as you said last night, and sent over a paper that truly represents our views. While we want to go just as far as we can to meet Mr. McNamara, we must not recede from this position except as we reach a settlement that all of us can live with.

I wonder if Harold Brown would be willing to list what it is they want from Gemini?

I have no doubt whatever that McNamara is underrating the problems that will be created with Congress if he insists on the participation in our management or that we participate in the management of the development of military equipment such as weapons systems. We can contribute a great deal, but when it come [sic] to the actual development, this is not our function under the law.

There is another element which we must consider. Under the proposed arrangement, we would lose control of the research which we will do. The basic policy from NACA days is that we would determine the research which was necessary, would fund it, and would do it. This made us independent of those who wanted us to undertake contract research, but of course, we were always [3] sensitive to their needs. I believe this principle is one that has made for advance, has given the nation strength, and that even though Mr. McNamara does not seem to be able to understand it today, we must not lightly put it aside. After all, we do not know how long he or I or any of the principal actors will be on the stage, and we must keep a system that others can operate under.

These are just early morning thoughts as I leave for the airport.

James E. Webb
Administrator

Document II-23

[1]

Agreement
Between the National Aeronautics and Space Administration and the Department of Defense Concerning the Gemini Program

This document defines the policy agreement for arrangements to insure the most effective utilization of the GEMINI Program in the national interest.

1. Objectives of the GEMINI Program
   The GEMINI Program constitutes a major portion of the current near-earth manned space program in the United States. It is the intent of this agreement to assure that the scientific and operational experiments undertaken as a part of the GEMINI Program are directed at the objectives and requirements both of the DoD and the NASA manned space flight program.

2. Establishment of the GEMINI Program Planning Board
   A GEMINI Program Planning Board is hereby established reporting jointly to the Administrator of the NASA and the Secretary of Defense. The Associate Administrator of the NASA and the Assistant Secretary of the Air Force for Research and Development will serve as Co-Chairmen of the Planning Board. The Board will include two additional representatives of each of the two agencies. Members will be named by the Co-Chairmen and approved by the Administrator of the NASA and the Secretary of Defense.

3. Functions of the GEMINI Program Planning Board
   The Board hereby created is intended to assure that the GEMINI Program is planned, executed, and utilized in the over-all national [2] interest, in accordance with policy direction from the Secretary of Defense and the Administrator of the NASA, so as to avoid duplication of effort in the field of manned space flight and to insure maximum attainment of objectives of value to both the NASA and the DoD. The functions of the Board in carrying out this responsibility shall include the delineation of NASA and DoD requirements and program monitoring to insure that they are met in:
   1. The planning of experiments.
   2. The actual conduct of flight and in-flight tests.
   3. The analysis and dissemination of results.
   Should actual project plans fail to meet the requirements specified by the Board, or should competing requirements produce resource or schedule conflicts, the Co-Chairmen shall so inform the Administrator of the NASA and the Secretary of Defense.

4.   GEMINI Project Management

NASA will continue to manage the GEMINI project. It is, however, agreed that the DoD will participate in the development, pilot training, pre-flight check-out, launch operations and flight operations of the GEMINI Program to assist NASA and to meet the DoD objectives.

5.   Funding

In recognition of its interest in the program, the DoD will contribute funds to assist in the attainment of GEMINI Program [3] objectives. The amount of such support will be determined on the basis of recommendations submitted by the Board.

6.   Additional Programs

It is further agreed that the DoD and the NASA will initiate major new programs or projects in the field of manned space flight aimed chiefly at the attainment of experimental or other capabilities in near-earth orbit only by mutual agreement.

James E. Webb   Robert S. McNamara
Administrator, NASA   Secretary of Defense
Date: January 21, 1963   Date: January 21, 1963

Document II-24

Document title: Robert S. McNamara, Secretary of Defense, Memorandum for the Vice President, "National Space Program," May 3, 1963.

Source: NASA Historical Reference Collection, NASA History Office, NASA Headquarters, Washington, D.C.

In April 1963, President Kennedy asked Vice President Johnson to conduct, as chairman of the National Aeronautics and Space Council, an overall review of the national space program, published as Document III-17 in Volume I of Exploring the Unknown. *Secretary of Defense McNamara's reply suggests the many ways in which the programs of NASA and Department of Defense had become intertwined.*

[1]   3 May 1963

Memorandum for the Vice President

SUBJECT: National Space Program

This memorandum will respond to Dr. [Edward C.] Welsh's memorandum to me of April 10, requesting information on which to base replies to the questions in the President's memorandum to you of April 9. I should point out first that most of the points raised by the President deal with matters for which NASA has primary or exclusive responsibility. My comments will, therefore, be confined to the military aspects to questions 2 and 3, and to question 5.

Question 2: What specifically are the principal benefits to the national economy we can expect to accrue from the present, greatly augmented program in the following areas . . . military technology?

I have attempted to measure these benefits by estimating the extent to which [the] DoD budget would be increased, in each of the DoD budget categories corresponding to the major NASA budget categories, if the present greatly augmented program had not been undertaken by NASA. It should be borne in mind that a part of the augmented program, including, for example, the TITAN III development, has been undertaken directly by the Department of Defense. The military justification for this portion of the program is such that it would have been undertaken without regard to the other objectives of the National Space Program. The great bulk of the augmented program, $4,388 million out of $4,696 million, is in the NASA budget; the Department of Defense space program for FY 1964 is about 7% higher than the 1 January 1961 projection for FY 1964, after adjustment for comparability. A comparative tabulation of DoD and NASA budgets for the National Space Program appears at Table I.

Research. Although it is difficult to assess the direct military value of space research, it appears likely about $20 million of NASA's $100 million research budget proposed for FY 1964 would be undertaken by DoD in the absence of a NASA program.

[2] *Exploratory and Advanced Development* (corresponds to NASA's Supporting Research and Technology). While the military value of this category of expenditures is almost equally difficult to determine in advance, I estimate that some $100 million of NASA's augmented program might be supported by DoD if NASA were not supporting it. In addition, the Department of Defense would probably support the entire NASA "base" program in this area under like circumstances.

Engineering Development

Launch Vehicles. The major NASA development activity in this field is focused on the use of liquid hydrogen to lift the extremely large payloads required for the lunar mission. This technology is probably not of much military value because of severe operational restraints on its handling and storage. Some of this development work will undoubtedly have incidental military benefits, but they cannot be estimated in advance, and would not merit DoD expenditures in the absence of the NASA program. Primary DoD reliance is on the TITAN III as the standardized workhorse building block for military applications in space. It is important to point out, however, that the concept of a single National Launch Vehicle Program dates back to the first agreement between NASA and the DoD signed in the new Administration, by Mr. Webb and Mr. Gilpatric, in February 1961, and that the Department of Defense includes in its consideration of launch vehicles for new military missions any vehicles under development by NASA for non-military space missions.

Manned Space Craft. The APOLLO space craft, designed for the lunar mission, has no predictable military applications. The GEMINI space craft, however, is in a different category, and if it were not under development, the Department of Defense would probably undertake a GEMINI-type program. The NASA GEMINI program has a critical early flight date as a part of the over-all lunar project. This condensed scheduling cannot be supported as a military requirement, and, therefore, an additional Defense program of $150-$200 million in FY 1964 might be justified in lieu of the $300 million level of effort proposed by NASA for FY 1964.

Unmanned Space Craft. In part because DoD was active in this area before the organization of NASA, there are no vehicles under development by NASA which would have been undertaken or would be taken over by DoD in the absence of the program.

[3] *Mission Applications.* A number of the special mission applications of NASA space vehicles, such as meteorological satellites and communication satellites are of military interest. If they were not undertaken by NASA, the Defense budget might be increased by $25-$50 million in these particular mission application areas. Most of these applications

stem from the pre-1961 NASA program, and their present level of effort cannot easily be apportioned between the "base" program and the augmented program. These essentially experimental mission applications, however, do not include the necessity for extensive military development activity, since the technology for military operations is increasingly distinct from the technology for experimentation.

Other. Most of the increase in the augmented NASA effort classified in Table I as management and support reflects the lunar program directly and has no demonstrable military value. We have found, for example, that military use of GEMINI could very likely be fitted into our existing DoD tracking facilities for current classified programs, without major increases in funds. Of course, if space becomes very much more important from a military standpoint—if many more laboratories, tracking sites, launch facilities and the like were needed over and above what we already have in the Defense Department[—]then NASA's extensive facilities could be combined into indirect military assets. On the other hand, based upon what we presently foresee, the Defense Department would not pay for the large augmented management and support effort, or any appreciable fraction of it, if NASA did not.

Summary

The NASA budget estimate for FY 1964 totals approximately $5.7 billion. It is about $4.5 billion larger than the NASA budget for FY 1962 as of January 1961. The NASA budget for FY 1964, as projected at that time was somewhat less than the present amount.

In the foregoing paragraphs, I have identified approximately $600-$675 million of NASA effort which appears to have direct or indirect value for military technology. Of that amount, about $275-$350 million stems from the augmentation of NASA programs since January 1961.

Question 3: What are some of the major military problems likely to result from continuation of the National Space Program as now projected in the fields of . . . government. . . ?

[4] While the detailed answer to this question will come more appropriately from NASA, some comments from the special vantage point of the DoD may be appropriate.

The concerns suggested in this question were foremost in our minds two years ago when Mr. Webb and I submitted our report to you of 6 May 1961. On page 10 of that document, urging the importance of planning at the national level, we noted that the decade of 1950-1960

"has witnessed a great expansion in U.S. government sponsored research and development especially for large scale defense programs. Enormous strides have been made, particularly in our space efforts and In the development of related ballistic missile technology on a 'crash' basis. We have, however, incurred certain liabilities in the process. We have over-encouraged [sic] the development of entrepreneurs and the proliferation of new enterprises. As a result, key personnel have been thinly spread. The turnover rate in U.S. defense and space industry has had the effect of removing many key scientific engineering personnel from their jobs before the completion of the projects for which they were employed. Strong concentrations of technical talent needed for the best work on difficult tasks have been seriously weakened. Engineering costs have doubled in the past ten years.

"These and other trends have a strong adverse effect on our capacity to do a good job in space. The inflation of costs has an obvious impact, and they are still rising at the rate of about seven per cent [sic] per year. This fact alone affects forward planning. It has often led to project stretch-outs, and may again in future years. The spreading out of technological personnel among a great many organizations has greatly slowed down the evolution of design and development skills at the working level throughout the country."

Earlier in the same report we also stated again in connection with planning, that

"it is absolutely vital that national management be equal to the task of focusing resources, particularly scientific and engineering resources, on the essential building blocks. It is particularly vital that we do not continue to make the error of spreading ourselves too thin and expect to solve our problems through the mere appropriation and expenditure of additional funds."

[5] The concerns expressed in the report of 8 May 1961 were related to the impediments to and opportunities for success in undertaking an expanded space program. The concerns implicit in Question 3 of the President's memorandum relate to impact of this program on the non-government sector. These two concerns are opposite sides of the same coin. Moreover, the same trends that were of concern two years ago are, in many cases, of equal or greater concern today.

For example, it turns out that federal expenditures for research and development, although they exhibit fluctuations from year to year, seem to have been following a long-range trend for the last fifteen years, at least. This trend rises much more steeply than the total federal budget or total [gross national product]. In fact, if extrapolated, federal expenditures for research and development can be predicted to equal the entire gross national product by about the year 2000. It is obvious, therefore, that the slope of the curve must flatten out over the next few years.

The Department of Defense, along with every other agency of government and the private sector of the economy, is in increasingly sharp competition for the research and development dollar. The elimination of waste and inefficiency in the National Space Program, whether it occurs in NASA, in DoD, or in overlaps between the two agencies, is essential to our national security.

Question 5: Are we taking sufficient measures to insure the maximum degree of coordination and cooperation between NASA and DoD in the areas of space vehicles development and facility utilization?

The adequacy of coordination and cooperation between NASA and DoD must be measured by the extent to which such efforts support the policy of creating and maintaining a single National Space Program. That policy has governed our actions since the beginning of this Administration. In our report of 8 May 1961, Mr. Webb and I stated, in summary:

"Clearly, then, the future of our efforts in space is going to depend on much more than this year's appropriations or tomorrow's new idea. It is going to depend in large measure upon the extent to which this country is able to establish and to direct 'an Integrated National Space Program'."

[6] We pointed out then (page 12) that:

"It will be necessary, therefore, to find a way to formulate and apply plans and policies aimed at insuring the success of an Integrated National Space Program. Top level scientific and policy direction must be forthcoming from the top management echelons. The mere statement of broad objectives will not be enough. Periodic budget reviews and their intensification in the spring of each year will not suffice. It will be necessary to impose policy and management actions which will alter many of the trends of the past ten years, particularly in the management of research and engineering resources on a national scale."

In my view, it is essential that all major space programs be integrated with military requirements in the early stages of their development. This integration has been fostered through the organization and operation of the Aeronautics and Astronautics Coordinating Board and its six panels. A series of written agreements between NASA and DoD spells out this general policy in such fields as development of launch vehicles and space craft, administration of range facilities, and planning for communications satellites.

I am not satisfied, and I am sure that Mr. Webb is not satisfied, that we have gone far enough to eliminate all problems of duplication and waste in administration. We are engaged in a continuing joint effort in this area. But I am more concerned with the potential dangers in the divergence of our efforts in the study and planning of potential new large projects.

Take, for example, the proposed space station being considered by NASA and DoD, and still in the planning phase. While it is not yet clear that the project is justified, either on a military or nonmilitary basis, it is clear that it should be undertaken only as a national program, which meets the requirements of both NASA and DoD, and that it must be jointly planned from its inception.

Coordination and joint planning of our efforts must extend to all so-called "advanced studies." Experience has demonstrated that if many or sizeable [sic] studies are supported throughout industry, the expectation of a new project grows rapidly until such expectations are translated into public debate and controversy. Mr. Webb and I agreed on this matter in recent discussions.

In the National Launch Vehicle Program, to take another example, we must be constantly alert to consider new vehicles for inclusion as standard "building block" vehicles meeting the requirements of both agencies. We must refrain from undertaking unnecessary new developments, and we must limit the scope of adaptations of standard devices to unique projects. Both NASA and DoD continue to be exposed to proposals for additional launch vehicles or modifications of those that are already a [7] part of the National Launch Vehicle Program. It is even conceivable that within a year or two pressures will arise to develop vehicles using new materials and techniques on the sole ground that "no new launch vehicle projects have been undertaken" in a long time. This is not to say that we should abandon the continuing examination of new technological achievements in these areas. But development projects must be jointly planned and development decisions jointly taken.

Coordination and joint planning of our efforts must extend to all so-called "advanced studies." Experience has demonstrated that if many or sizeable [sic] studies are supported throughout industry, the expectation of [a] new project grows rapidly until such expectations are translated into public debate and controversy. Mr. Webb and I agreed on this matter in recent discussions.

I am also concerned with the potential dangers in the divergence or unnecessary duplication of our efforts in fields where technology and other factors are rapidly changing. Communications and meteorological satellites are two examples. I have already canceled some major programs in the communications area, and I do not propose to launch any additional projects until the roles of NASA, DoD and the Communications Satellite Corporation have been clearly defined.

The heads of the two agencies must constantly be sensitive to the dangers of duplication and waste. The problem is of sufficient importance to require continuous monitoring at a level above that of the agencies themselves. I suggest that responsibility for this monitoring be assigned to the Bureau of the Budget and to the Director of the Office of Science and Technology. Only by assigning specific responsibility in this fashion can the integrity of the National Space Program be protected.

Robert S. McNamara

[8]

TABLE I

SPACE AND SPACE RELATED PROGRAMS

BREAKDOWN BY DOD RESEARCH AND DEVELOPMENT PROGRAM CATEGORY OF AMOUNTS APPEARING ON PAGE 404 OF BUREAU OF THE BUDGET SPECIAL ANALYSIS G, "Research and Development and Selected Scientific and Technical Activities of the Federal Government," January 1963.[1]

NEW OBLIGATIONAL AUTHORITY (MILLIONS OF DOLLARS)

DOD R&D Program Category	DOD			NASA		
	FY 1962 (act.)	1963 (est.)	1964 (est.)	1962 (act.)	1963 (est.)	1964 (est.)
Research	4	4	6	23	65	99
Exploratory Dev.	140	159	166	-	-	-
Advanced Dev.	535	509	405	-	-	-
Supporting R'sch. & Technology	675	668	571	236	439	647
Engineering Dev.	112	382	437	845	1,858	3,297
Operational Sys. Dev.	26	39	40	-	-	-
Mgt. & Support	467	525	614	692	1,261	1,621
TOTAL	1,285	1,618	1,668	1,796	3,623	5,664

[1] Special Analysis G states: "The amounts show for the National Aeronautics and Space Administration cover all the activities of that agency except those specifically identified with aircraft technology. The estimates for the Department of Defense include all the principal amounts identifiable with the Department's space programs, but exclude certain amounts which cannot be feasibly separated from other military expenses, such as the development of missiles which are also used in the space programs, military personnel costs, and various other operating costs."

Document II-25

Document title: W.F. Boone, to Mr. Webb, Dr. Seamans, Dr. Dryden, "DOD-NASA Relations," July 12, 1963.

Source: NASA Historical Reference Collection, NASA History Office, NASA Headquarters, Washington, D.C.

NASA's Office of DOD Liaison, headed by retired Admiral W. Fred Boone, performed the difficult task of attempting to keep communications open between NASA and the military when Secretary of Defense McNamara and NASA Administrator Webb were at odds. Boone's July 12, 1963, memorandum to top NASA officials attempted to place in perspective NASA's views of the military intentions and the military's view of NASA intentions. It highlights the problems of developing collaborative programs with such widely differing needs.

[no pagination]   July 12, 1963

A/Mr. Webb
AA/Dr. Seamans
AD/Dr. Dryden
AAD-3

DOD-NASA Relations

In response to your desire expressed at a recent staff meeting, the attached paper is submitted.

The paper has been prepared with the thought that it would be used as a "talking paper" rather than one to be given to Mr. McNamara.

The whole paper has been coordinated with [D. Brainerd] Holmes and has his concurrence.

The section on GROUND SUPPORT OPERATIONS has been coordinated with [Edmond C.] Buckley [special assistant to the administrator] and has his concurrence.

If the DOD agrees that NASA and the DOD should work together primarily on the basis of coordination rather than joint action, I suggest that we might want to ask the AACB to agree on the meaning of "coordinate" in this context.

It is suggested that this be held on an "eyes only" basis among Dr. Dryden, Dr. Seamans, and yourself, until all or part of the paper is released by you.

W.F. Boone

2 Enclosures
DOD-NASA Relations
Definition

****

[1] *PRIVATE—Eyes Only for Mr. Webb, Dr. Dryden, and Dr. Seamans.*

DOD-NASA Relations

1.   The purpose of this paper is to bring into focus the divergent philosophies, attitudes, and interpretations of the Department of Defense and the National Aeronautics and Space Administration with respect to the implementation of the National Aeronautics and Space Act of 1958. Delineation of certain differing points of view may suggest guidelines for their resolution, and closer agreement as to principles involved will permit the two agencies to work more harmoniously, economically, and effectively together in the national interest.

2.   This discussion will be presented under the headings of *NATIONAL POLICY, PLAN-NING,* and *GROUND SUPPORT OPERATIONS* as these pertain to space activities, and *AERONAUTICAL RESEARCH.* These are the areas in which the principal problems appear to lie.

NATIONAL POLICY

3.   A difference of opinion exists as to the proper function and status of the Space Administration under the Space Act of 1958.

NASA Position

4.   The National Aeronautics and Space Act was responsive to national requirements in two categories: (1) general welfare, and (2) security. The objectives set forth in the Act were formulated after thorough deliberation by the Executive and Legislative Branches, and extensive correlation with the scientific community. The Act provided that the scientific exploration and exploitation of space shall be the responsibility of and directed by an independent *civilian* agency, while stating the major exception that "activities peculiar to or primarily associated with the development of weapons systems, military operations or the defense of the United States (including the research and development necessary to make effective provision for the defense of the United States) shall be the responsibility of, and shall be directed by, the Department of Defense. . . " Thus, the Congress clearly recognized the need for two mutually supporting but separately directed space programs. The Act established a liaison mechanism (the Civilian-Military Liaison Committee, later superseded by the Aeronautics and Astronautics Coordinating Board) through which the DOD and NASA were required to "advise and consult with each other on all matters within [2] their respective jurisdictions relating to aeronautical and space activities" and to "keep each other fully and currently informed with respect to such activities." The Act provided that the President shall determine which agency shall have responsibility for the direction of a space activity.
5.   In drafting the Act, the Congress stressed the peaceful purposes of our space activities. It was apparently recognized that the exploration of space was more than an area of future significance to the defense of the United States, and that the scientific, political, and economic benefits to be derived from a space program might be subordinated if space exploration were conducted solely under military auspices.
6.   NASA sees the national space effort as a spectrum encompassing three areas: (1) acquisition of basic scientific knowledge and the development of basic technologies and operating techniques; (2) the application of space knowledge, technologies, and techniques to the development of prototype space systems; (3) the production and operation of commercial and military space systems to meet national requirements. A necessary adjunct to this total effort is the establishment of a government in[-]house capability supported by a broad industrial base competent in the space field.
7.   NASA's assigned functions lie primarily in category (1) above. The DOD has research and development responsibilities in this category to the extent that such research and development pertains to the defense of the United States. NASA's responsibilities do not extend to the area of category (3). Category (2) is a gray area in which the responsibilities of DOD and NASA overlap to a considerable extent. NASA of necessity becomes an operating agency in those cases where basic subsystems and operating techniques can best be developed by means of an experimental operational flight system, and where NASA is called upon to furnish operational services to another agency. NASA recognizes that some

programs to meet the requirements of DOD and NASA in category (2) are of such magnitude as to require that a single program serve the needs of both agencies; i.e., a manned orbiting laboratory. Where predominant interest is at issue in such cases, a Presidential decision as to management responsibility would be needed. Presumably, the decision, in addition to the matter of relative interest in terms of experiments to be accommodated, would take into account additional factors such as management competence, operational experience, and political impact.

8.   Consideration of national policy and national interests dictates that the civilian space program under NASA should be an "open" program with maximum dissemination of derived information "for the benefit of all mankind," whereas these same considerations require that a military space program be conducted essentially under security restrictions.

[3] DOD Position—as it appears to NASA

9.   The attitude of the DOD with respect to the roles of the two agencies in the national space effort differs from that of NASA in that the DOD sees the civilian and military space programs as one program which should be jointly conducted to attain both civilian and military objectives. They believe that the military requirements in space were not as well foreseen when the Space Act was passed in 1958 as they are now. In the intervening years, it has become apparent that the Soviet space program is directed primarily toward the gaining of a military advantage through space operations, forcing the United States to build a military defense in space. Because of this increasing role, the military should have a stronger voice in shaping and direction of the total national space program than was recognized and provided for in the Space Act.

10. This attitude has led to efforts on the part of the DOD to have segments of the NASA program transferred to the DOD (i.e., Gemini, bio-astronautics, training of astronauts, MILA). The desire to control is especially strong within the Air Force, as the Service of primary interest in the field of space, is disproportionately small and has not received the proper public recognition. The Air Force considers that space operations are simply an extension of flight operations in the atmosphere, and therefore that they should be under Air Force control. Lacking greater support for this position at the DOD level, the Air Force has made an "end runs" [sic] to members of Congress and the White House staff, and has launched an intensive and well organized public relations campaign to convert the public to the Air Force point of view. The Air Force is inclined to look upon NASA as a competitor rather than a partner in the field of space.

Proposed Basis of Agreement

11. The Secretary of Defense and the Administrator should agree in principle along the following lines, and should join in a vigorous effort to indoctrinate subordinate staffs and agencies in acceptance of these principles:

(a) It was the intent of Congress, and remains in the national interest so far as possible without jeopardizing national security, that the United Sates maintain in the eyes of the world the peaceful image of our space program.

(b) As a corollary to (a), NASA should remain a fully independent, civilian agency.

[4] (c) There are certain advantages to the national space effort, and in the long run specifically to the Department of Defense, which accrue the virtue of civilian agency management of a major portion of the total space effort; i.e., international cooperation; and relations with the research and development organizations of industry, with the civilian scientific organizations, and with the university community.

(d) At the same time, the unfolding military requirement in space demands an expanding role for the Department of Defense in the total space effort.

(e) For the present, this increasing role will be accommodated by earlier and stronger concerted DOD-NASA action on the basis of *coordination* rather than *joint control,* and in a manner which will not compromise the civilian character of NASA's activities.

(NOTE: "Coordination" as used in this paper, will have the following meaning: An agency having responsibility to "coordinate" with another agency on a specified project (1) will recognize the interest of the other agency in the project, (2) will initiate a full exchange of information and consultations early in the conceptual phase, (3) will encourage the active participation of the other agency in the planning from the very outset, and (4) will make an earnest effort to meet the requirements and objectives of the other agency. Concurrence of the other agency will be sought in the planning and execution of the project. Concurrence is not required as a pre-condition to further action. However, matters on which agreement is not reached may be referred for resolution to the next higher authority in which both participants have a voice.)

(f) It is expected that the decision as to management responsibility for a major new program will be made by the President primarily on the basis of predominance of interest, but also taking into account other factors such as capability to conduct the program, relation to other major programs, international aspects, security considerations, etc.

[5] (g) There will be maximum cross-servicing in the use of support resources and technical know-how.

(h) Except in unusual cases, joint management responsibility is not favored on the basis that the requirement for concurrence at every step [is] inefficient, uneconomical, and tends to impede rapid progress.

PLANNING

12. A difference of opinion exists as to the desirability of joint versus coordinated planning.

NASA Position

13. NASA's assigned mission is to maintain a national position in the vanguard of space exploration. In its quest for scientific knowledge and its efforts to develop the basic techniques necessary for space operations, NASA must constantly seek to advance man's space frontier further into the unknown. In pursuing this mission, NASA should not be restricted by a limitation that its advanced exploratory studies must be related to established operational requirements of either a military or commercial nature. At the same time, NASA should ever be alert to discern those areas of research which appear to offer the most promising potential for the solution of military problems and for otherwise contributing to the national welfare, and be prepared to orient its efforts responsively to these objectives.

14. There should be a thorough, inter-agency exchange of ideas and information as to requirements and problems early in the process of formulating advance studies in an area of mutual interest, but to impose the restriction that the formal concurrence of another agency is required before NASA may proceed with such a study would seriously obstruct NASA's ability to discharge its statutorily assigned functions.

15. Major future progress in space are [sic] likely to be so costly that the nation will be able to afford only one program in each category. Consequently, each such program should be designed to meet, in so far as possible, the requirements of all government agencies for space research and development.

16. Once the decision is made to embark upon a multi-interest project, the agency responsible for its direction should be designated. Thereafter, the planning and execution should be coordinated between interested agencies to assure that in so far as practicable the requirements of all agencies are fulfilled in the national interest. The primary responsibility for that coordination should reside with the agency directing the project.

[6] *DOD Position—as it appears to NASA*

17. The DOD view with respect to planning differs from that of NASA in that the DOD feels all planning relating to NASA programs or projects which are of interest to DOD should be jointly conducted from inception. This view has led DOD to seek inflexible agreements concerning the manner in which NASA's advance exploratory studies may be initiated, including sign-off authority for DOD.

Proposed Basis of Agreement

18. (a) Requirements and objectives in any particular area of space research and development will, as a general rule, be developed unilaterally by DOD and NASA. Subject to security restrictions, general knowledge of each other's requirements and objectives must be assumed.

   (b) Prior to the approval by either agency of a study project in an area of mutual interest, inter-agency coordination will be accomplished. This will take the form of a free exchange of information concerning requirements, objectives and plans for the study, and an earnest attempt to cast the study in such manner as to be responsive to the requirements and objectives of both agencies in so far as practicable. Provisions will be made so that in the event an agency feels that its needs are not being adequately met in formulating the study, recourse may be had to higher authority for resolution of differences, initially to the Co-chairmen of the [Aeronautics and Astronautics Coordinating Board].

   (c) Results of studies in an area of mutual interest will be made available to both agencies.

   (d) Upon approval of a new major project of mutual interest, the agency responsible for its direction will also be charged with insuring that adequate arrangements for coordinated planning and coordinated monitoring of execution are made. Again, provision will be made for recourse to higher authority to resolve differences.

GROUND SUPPORT OPERATIONS

19. There are some conflicting views in the matter of control of ground support operations.

[7] *NASA Position*

20. NASA fully subscribes to the concept of national launch ranges operated by the DOD for the benefit of all government user agencies. NASA has levied known requirements on the ranges for over 140 future launches, over 40 of which involve tracking ships. However, the requirements which NASA must place on the ranges have become so large, complex,

and exacting that NASA feels it must actively participate in planning the manner in which the ranges are to be equipped and operated to provide the project-peculiar services required by NASA.

21. The magnitude and nature of the Manned Lunar Landing Program are such as to require that the assembly, check-out, and launch area (Merritt Island) be under NASA control.

22. The world network of land-based orbital tracking and data acquisition stations should, to the maximum practical extent, be under NASA control for NASA missions. This applies to planning of facilities, specification and installation of instrumentation, training and maintenance of proficiency activities, communication links, and operational control during a flight. As high a degree as possible of standardization among stations is necessary in order to permit the most effective operational flexibility and casualty control during an operation. Exceptions to this doctrine can be accepted in the case of a few DOD stations that are already in existence, strategically located, and responsive to NASA requirements. This doctrine is made necessary by the indivisible relationship between program management and the operations control organization.

23. If and when stations of the NASA world network are utilized to track DOD missions, NASA would be willing to place these stations temporarily under DOD operational control if DOD considers this necessary to the mission and to the extent permitted by international agreements. (Nearly half of the spacecraft being tracked by the NASA satellite network are DOD spacecraft.)

24. Arrangements for the procurement, preparation, and operation of the project peculiar tracking ships required to occupy the critical stations for insertion into orbit and injection into the moon transfer in the Apollo operation must be such as to give NASA a high degree of control through relatively direct administrative channels.

25. To this end, NASA's present intention is to employ MSTS [Military Sea Transportation Service (NAVY),] a DOD agency experienced in the operation of special purpose ships, to prepare the hulls and machinery and to operate the ships themselves as differentiated from the instrumentation installed therein. In the interest of standardization, NASA plans to use the same contractor for installation and operation of the instrumentation as is used in the case of other NASA stations in the net. [8] While these ships will be required nearly full time for the Apollo mission, NASA has no objection to adding general purpose instrumentation to the extent this will not compromise the project peculiar instrumentation, and to make the ships available for general purpose use when not required in connection with Apollo. Generally speaking, these ships should basically be special purpose ships, with a general purpose secondary mission, rather than vice versa.

26. The priority assigned to the Apollo program and considerations of safety are such that where other agencies are depended upon to furnish facilities or perform essential services in the loop, NASA must have the prerogative of monitoring the provisions for rendering such services to the extent necessary to assure itself that all recognizable potential limitations which might delay the schedule or increase the risk of the mission are eliminated.

DOD Position—as it appears to NASA

27. The DOD takes the position that the launch ranges are a national asset which would be used to capacity by other agencies of the government, and on which requirements should be levied without voice as to the manner in which these requirements are to be met. The range facilities, including tracking ships, should be primarily "general purpose" in nature, with "project peculiar" provisions added. The DOD fears that NASA, by establishing the Merritt Island Launch Area and seeking to acquire its own project-peculiar

tracking ships, wishes to depend less and less upon the DOD ranges for services, becoming a range operator instead of a range user.

Proposed Basis of Agreement

28. The differences in this area stem more from a lack of mutual trust than from differing concepts. Each agency sees the other as seeking control of segments of its operations. This has at times inhibited a free exchange of information. In order to dispel any such fears, it is proposed that DOD and NASA agree in principle to the following, and that all subordinate organizations be informed accordingly:
  (a) The concept of national launch ranges operated by the DOD is to be fully accepted and implemented. NASA will depend upon the facilities and services of these ranges to the extent that they can meet NASA requirements.
[9] (b) The principle of "primary assignment" will be applied in accordance with priorities established by mutual agreement or by higher authority.
  (c) Where NASA specialized requirements exceed the capacity of the national range, the range will be given an opportunity to augment its capacity if desired, before NASA proceeds to make its own provisions to meet the excess requirements.
  (d) NASA will continue to be responsible for operating the world networks required for tracking NASA spacecraft in orbit and in lunar and planetary transfers. In the interest of avoiding unwarranted duplication, the DOD will utilize these NASA networks for DOD orbital missions where feasible.
  (e) Generally speaking, the point of demarcation between the ranges and the world tracking nets will be the point of insertion into orbit.
  (f) Each agency will participate actively on a coordination basis in the other's plans for equipment development and facilities with the objectives of achieving the maximum practicable degree of standardization and permitting such facilities and equipments to meet the needs of both agencies to the maximum practicable extent.
  (g) All tracking [of] the data acquisition ships, once ready for service, will be assigned to the national ranges who will utilize MSTS to operate and maintain the ships generally under the same arrangements that currently govern the MSTS to operation and maintenance of special purpose ships for various agencies of the government. Under this arrangement, there will be a mutually agreed upon scheduling authority who will assign the ships to the operational control of the user agency on a prime assignment basis as necessary to meet the requirements of the user agency as to training, calibration check-out, minor modifications to instrumentation, and tracking and data acquisition operations.
  (h) Operation of instrumentation aboard each ship will be contracted for directly by the user having primary interest.

[10]   *AERONAUTICAL RESEARCH*

29. There is a difference of opinion as to the relative importance time-wise of aeronautical research programs utilizing new prototypes and the flight test programs of these prototypes.

RS-70 Program

30. In a letter dated May 3, 1962, the Administrator proposed to the Secretary of Defense that one of the three XB-70 prototypes be made available to NASA for use in conducting an advanced aeronautical research program in the area of supersonic cruise flight. NASA considers that this program is essential to our country's progress in the field of aeronautics, and that the information desired cannot be obtained by any other means. No official response to this request has been received. Recently, after an elapse of over a year, the proposal has been revived by NASA in the [Aeronautics and Astronautics Coordinating Board].

31. NASA considers it to be of the utmost importance that the opportunity presented by the XB-70 for flight research in the supersonic range be fully exploited as early as practicable. The data to be gained thereby will have special application in the development of the supersonic transport now advocated by the Administration.

32. A proper flight research program cannot be conducted simultaneously with and as a part of the flight tests of this aircraft. Since the instrumentation for the research program should be installed during the fabrication of the designated aircraft, an early decision by the Secretary of Defense to make one of the XB-70 prototypes available on loan to NASA is required if the valuable data to be derived from such a research program is to be available in time to be used in designing the supersonic transport.

TFX Program

33. By letter dated January 15, 1963, the Administrator requested that one of the early TFX prototypes be made available on loan to NASA to be used in conducting a flight research program to obtain basic data concerning the variable swept wing concept incorporated in the aircraft. This concept originated at the Langley Research Center, and much of the supporting ground research data were gathered there. On 1 March 1963, the Secretary of Defense responded by disapproving the request, making the alternate suggestions that:

    (a) NASA participate jointly with the Air Force by combining the research program with the flight test program,

[11] (b) NASA acquire one of the prototypes upon completion of the flight test program, or

    (c) NASA purchase an additional prototype at a cost of about $10 million.

Alternative (c) appeared to involve unwarranted duplication, and neither alternative (a) nor (b) would permit the accomplishment of an adequate flight research program in a timely manner.

34. Following personal negotiations with the Secretary of the Air Force by the Deputy Associate Administrator for Defense Affairs, discussions were commenced between NASA and the TFX Project Office at Wright-Patterson Air Force Base to find ways and means of meshing an adequate flight research program under NASA control with the flight test program on the TFX prototypes. It is too early to say whether satisfactory arrangements for meeting the requirements of both agencies will evolve from these negotiations. The Secretary of Defense and the Administrator should agree to review this matter again about six months hence.

Proposed Basis of Agreement

35. In layout [of] a program for the acquisition of a new military aircraft type incorporating a new concept or a substantial projection of a current design concept, provision will be made to make an early prototype available to NASA for the purpose of accomplishing an "in-flight" research program designed to obtain advanced technical data in the field of aeronautics.

****

[no pagination]

Definition of
"Coordination With" and "In Coordination With"

This expression means that agencies coordinated with shall participate actively; and concurrence shall be sought; and that if concurrence is not obtained the disputed matter shall be referred to the next highest authority in which all participants have a voice.

(The above information from JCS Publication "Dictionary of U.S. Military Terms for Joint Usage" and Army Regulation 320-5)

Document II-26

Document title: James E. Webb, Administrator, NASA, and Robert S. McNamara, Secretary of Defense, "Agreement Between the Department of Defense and the National Aeronautics and Space Administration Covering a Possible New Manned Earth Orbital Research and Development Project," August 17, 1963, with attached: "Procedure for Coordination of Advanced Exploratory Studies by the DOD and the NASA in the Area of Manned Earth Orbital Flight Under the Aegis of the Aeronautics and Astronautics Coordinating Board."

Document II-27

Document title: Robert S. McNamara, to Honorable James E. Webb, NASA Administrator, September 16, 1963.

Source: Both in Administrators Files, NASA Historical Reference Collection, NASA History Office, NASA Headquarters, Washington, D.C.

Among the more important areas on which NASA and the Department of Defense (DOD) agreed to cooperate was the development of future orbital space stations. This agreement, signed on August 17, 1963, was to cover the development of a joint national space station. Although Secretary of Defense Robert McNamara signed the agreement, in a September 16, 1963, letter to Administrator Webb, he expressed his reservations, focusing particularly on the need for both agencies to concur on, not just coordinate, their future activities related to future station design and development.

Document II-26

[1]

Agreement Between the Department of Defense and the National Aeronautics and Space Administration Covering a Possible New Manned Earth Orbital Research and Development Project

Objective

It is the purpose of this agreement to ensure that in the national interest complete coordination is achieved between the National Aeronautics and Space Administration and the Department of Defense in approaching a possible new project in the area of manned earth orbital research and development vehicles.

Basic Considerations

The National Space Program has now advanced to the point that further significant progress in the areas of scientific research, space exploration, basic space technology, and defense applications may well require the operation of a manned orbital research and development system involving spacecraft larger and more sophisticated than Gemini and Apollo. Such a system would be a major technical and financial undertaking. For this reason, and while recognizing that the National Aeronautics and Space Act of 1958 assigns to their respective Agencies separate and distinct responsibilities in the planning, directing, and conduct of aeronautical and space activities, the Secretary of Defense and the Administrator of the NASA agree that advanced exploratory studies and any follow-on actions in this area should be most carefully coordinated through the Aeronautics and Astronautics Coordinating Board (AACB), successor to the Civilian-Military Liaison Committee established by the Space Act. They further agree that in so far as practicable all foreseeable future requirements of both agencies in this area should be encompassed in a single project.

A system involving a manned earth orbital research and development vehicle capable of prolonged space flight would provide basic scientific and technological knowledge and basic design and operational criteria which would have across-the-board application to both military and civilian operational programs. Such a developmental system would be a mandatory forerunner of any long duration manned space operational system. Based upon present knowledge, it appears that the requirements of the DOD and the NASA, as well as of all other interested governmental agencies, can be met in a single national program. It is necessary that the NASA and the DOD take steps to ensure that their total effort is directed to this end.

Agreement

Pursuant to the foregoing, the Secretary of Defense and the Administrator of NASA agree to a common approach to this project through the steps set forth below. In the event that agreement is not reached on [2] any issue considered by either party adversely to involve the responsibilities of his Agency, the issue of disagreement will be jointly referred to the President for resolution.

a. The DOD and the NASA will continue advanced and exploratory studies in this area as considered necessary by the Secretary of Defense and the Administrator, NASA, respectively, to develop data as to Agency requirements, possible design concepts, feasibility, and costs; these studies will be coordinated under the AACB in accordance with the procedures set forth in the attachment hereto.

     b.   The AACB will include the evaluation of various concepts from the standpoint of productiveness, feasibility, and estimated costs.

     c.   The Secretary of Defense and the Administrator, NASA, will then attempt to arrive at a joint recommendation as to whether to proceed with a new project in this area, evaluating the national need by comparing potential returns to returns which could be realized by an extension of current on-going projects.

     d.   If the recommendation under c., above, is affirmative, the DOD and the NASA will jointly formulate an agreed project description for submission to the President together with

     e.   A recommendation as to responsibility for the direction of the project based on predominant interest and consideration of other pertinent factors, such as management competence, relation to other programs in progress, and international political implications.

     f.   If and when a decision is made by the Administration to proceed with such a project, the appropriate timing determined, and responsibility for direction assigned, a joint DOD/NASA board will be established to formulate the specific objectives to be obtained by means of the project and to approve the experiments to be conducted.

     g.   Acting in accordance with the results of f., above, the Agency assigned responsibility for direction will prepare a definitive project plan for approval by the Administration and submission to Congress for funding.

     h.   On provision of the necessary funding, the project will be implemented under single management but with joint DOD/NASA participation and monitorship.

James E. Webb   Robert S, McNamara
Administrator, NASA   Secretary of Defense
Aug. 17, 1963

Attachment
    Procedure for Coordination of
    Advanced Exploratory Studies

****

[1]

Procedure for Coordination of Advanced Exploratory Studies by the DOD and the NASA in the Area of Manned Earth Orbital Flight Under the Aegis of the Aeronautics and Astronautics Coordinating Board

(Attached to McNamara/Webb Agreement dated August 17, 1963)

1.   As a general procedure, there will be the maximum practicable interchange of ideas and information at all levels within the two Agencies beginning early in the conceptual or planning stage of the advanced exploratory studies in this area.

2.   Within fifteen (15) days after the signing of the Basic Agreement, each Agency will present to the Manned Space Flight Panel a list of studies which have been completed during the past three (3) years. Detailed information relating to these studies will be furnished to the non-sponsoring Agency on request.

3.   Within fifteen (15) days after the signing of the Basic Agreement, each Agency will present to the Manned Space Flight Panel a status report concerning:

(a) All studies which are in progress under contract and in-house;

(b) All studies which already have been formalized in a Statement of Work but not yet approved; and

(c) All additional new studies under active consideration or development.

4.   Within thirty (30) days after the signing of the Basic Agreement, the Panel will:

(a) Institute a review of the studies under category (a) above, and will effect such coordinating action as is deemed appropriate and practicable in the light of their on-going status;

(b) Designate to the AACB those studies in categories (b) and (c) above which either Agency considers should be formally coordinated to incorporate requirements of both Agencies and to avoid unwarranted duplication.

5.   Thereafter, the Panel will be kept informed of all new studies taken under active consideration or development by either Agency, and will promptly designate to the AACB any new study which either Agency considers should be formally coordinated as above.

[2] 6.   In the case of each study designated to the AACB for coordination, the non-sponsoring Agency will, within fifteen (15) days of such designation, indicate in writing its concurrence in the study without change, its reasons for not concurring, or submit in writing a list of the requirements of the non-sponsoring Agency which are desired to be considered for incorporation in the study. If no comments are received within the fifteen (15) day limit, satisfactory coordination may be assumed.

7.   Within thirty (30) days of the receipt of notification from the Panel of the designation of a study for coordination, the Co-Chairmen of the AACB will either:

(a) Certify in writing that satisfactory coordination has been accomplished, or

(b) Jointly submit to the Secretary of Defense and the Administrator, NASA, an explanation of any areas of disagreement arising out of the coordinating action. At that point, the sponsoring Agency may, if desired, proceed with the study.

8.   In all of the foregoing steps, the responsibility for taking the initiative in the coordinating process will rest with the Agency sponsoring the study in question.

Document II-27

[1]   16 Sept. 1963
Honorable James E. Webb
Administrator
National Aeronautics and Space Administration
Washington 25, D.C.

Dear Jim:

Thank you for your correspondence of August 17, 1963, and your proposed agreement covering a possible new manned earth orbital research and development project. I appreciate your constructive and earnest efforts to develop a method which will insure a sound, coordinated approach to this potentially important national effort. I am fully aware that, since we began our discussions on this matter, there have been many actions implemented which have already gone a long way toward improving the exchange of information between our agencies and the coordination of our study efforts. I concur in your proposed agreement in many respects and I feel that it is an excellent contribution to improved understanding and mutually useful effort of our agencies. I do, however, have certain reservations.

As I have expressed several times, my greatest current concern is to insure that advanced engineering studies are properly integrated and phased so that the requirements and design constraints of each agency can be really incorporated from the beginning. For this reason, and because of the potential scope and national importance of this program, I have continued to insist on the principle of concurrence of one agency in the proposed actions of the other vice simple coordination and possible subsequent unilateral action in the face of disagreement. As an example of the type of problem we are confronting, I refer to your proposed $3.5 million for contractor effort for the design of a Manned Orbital Research Laboratory (MORL). I believe that an effort of this magnitude is premature by eight months to a year since it will not be possible prior to that time for us to provide properly for the incorporation of Defense Department judgements [sic] and thoughts on military requirements into the design. You must realize that if ongoing DOD studies provide justifiable military objectives for a space station development, there may be the necessity for a significantly different design approach which will be responsive to both agency's needs.

[2] I further note that the proposed agreement does not define specifically the level of study effort required to qualify for interagency coordination [in] an "advanced exploratory study," although provision is made for the *coordination* of all such studies. I believe that an annual level of effort of $100,000 defines a reasonable threshold for initiating such action.

I concur in your view that the AACB is the proper medium for interagency coordination. I would observe, however, that while coordination has always been a prima facie AACB function, this has been accomplished in the past largely by other means, through other channels. I believe the AACB can serve as an effective coordinating body as long as proper attention continues to be accorded to the membership of the Board and its panels and the formulation and execution of meeting agendas, and as long as we both emphasize the resolution of issues at the Board level.

There remains, of course, the subject of recourse in the event that you and I cannot reach agreement on any issue referred to us. In the unlikely event that this should occur, I feel that, as a matter of practice, we should inform the Director of the Bureau of the Budget concerning the nature and extent of disagreement before initiating unilaterally any program actions which might later be subject to criticism in Congress or elsewhere.

Finally, I believe that at the present time it is not essential that we define the procedure for implementing the possible development program. It is inevitable that this procedure will be influenced by the nature and extent of each agency interest in such a program. Our final determinations of these procedures, therefore, may be somewhat different from what we now envisage.

I believe we have discussed this matter as much as is useful and that it is most important to insure continued harmonious accord between our agencies. Therefore, hoping you can accept my reservations as expressed in this letter, I have signed the agreement as you have prepared it. I believe that we can proceed constructively on the basis of this agreement and our mutual desire to formulate a recommended course of action in the best national interest.

Sincerely

Robert S. McNamara

Document II-28

Document title: John S. Foster, Jr., Director of Defense Research and Engineering, to Dr. Robert C. Seamans, Jr., NASA Associate Administrator, March 19, 1966, with attached: Robert Seamans, Jr., NASA Deputy Administrator, and John Foster, Jr., Director of Defense Research and Engineering, DOD, Memorandum of Agreement, "Establishment of a Manned Space Flight Experiments Board (MSFEB)," no date.

Source: Deputy Administrators Files, NASA Historical Reference Collection, NASA History Office, NASA Headquarters, Washington, D.C.

This agreement established the principle of reciprocity and sharing of flight opportunities between NASA and the Department of Defense, and it applied to the Apollo and the Manned Orbiting Laboratory programs. When the Space Shuttle agreement was formulated, the agreement in this memorandum was not renewed. When the subject of human spaceflight experiments arose again in the mid-1980s, the approach taken in the earlier agreement was modified to fit with the shuttle management process and was handled by the Air Force.

[no pagination]

DIRECTOR OF DEFENSE RESEARCH AND ENGINEERING
Washington 25, DC 20301

March 16, 1966

Dr. Robert C. Seamans, Jr.
Associate Administrator
National Aeronautics & Space Administration
Washington, DC 20546

Dear Bob:

In response to your letter of 1 March 1966, I have concurred in the NASA-DoD Memorandum of Agreement establishing the "Manned Space Flight Experiments Board (MSFEB)."

Based on discussions of our staff, and with the understanding that it would be acceptable to you, I have added to paragraph 6 of the Memorandum the following sentence:

"Similar technical advice will be made available from appropriate DoD agencies."

A copy of the revised Memorandum is attached.

Sincerely,
"Johnny"
John S. Foster, Jr.

Enclosure

****

[1]

Memorandum of Agreement

Subject: Establishment of Manned Space Flight Experiments Board (MSFEB)

General Guidelines

This Memorandum of Agreement is implemented in order to provide a means of coordination of the DoD and NASA manned space flight experiments program. These experiments, of a scientific, technological, or non-military operational nature, will be carried as a secondary objective on a space-available basis on selected DoD flight missions and as primary or secondary objectives on NASA flight missions.

It is anticipated that experiments will be submitted from a variety of sources to both DoD and NASA where they will be reviewed and, if approved, submitted to a joint experiments review board whose functions are defined in this agreement. In general, those experiments which are related primarily to basic space science, technology, and applications will be assigned to NASA programs. Similarly, those experiments which are peculiar to or primarily associated with the development of weapons systems, military operations, or the defense of the United States would normally be assigned to DoD programs, whenever possible. This is not to preclude, however, the assignment of any experiment to a program of either Agency when this appears desirable on the basis of economy, timeliness, or other considerations of national interest.

[2] 1. *PURPOSE*

This agreement established a Manned Space Flight Experiments Board (MSFEB) to coordinate experiment programs which will be conducted on DoD and NASA manned space flights.

2. *AUTHORITY*

The MSFEB is advisory to the Associate Administrator for Manned Space Flight, NASA, and the Deputy Director for Strategic & Space Systems, DoD.

3. *FUNCTIONS*

The MSFEB will have the following functions:

a. Recommend the approval or disapproval of experiments to be conducted under DoD and NASA Manned Space Flight Programs.

b. Recommend assignment of experiments to specific flight programs.

c. Recommend relative priorities of experiments to be implemented, and periodically review the numbers of experiments scheduled for specific missions.

d. Review the status of approved experiments.

As used herein, "experiment" means an investigation which is not essential to the primary mission, launching, navigation, or recovery of the space vehicle or the spacecraft. Experiments normally will be under three general classifications: scientific, applications, and technological or non-military operational. MSFEB recommendations will be based on analyses which show that it will be operationally and technically feasible to conduct the experiment, and that the basic experimental objectives of the investigation can be satisfied within the framework of the primary mission objectives of the program to which the experiment is assigned.

[3] 4. *MEMBERSHIP*

The following personnel will serve as members and alternate members of the MSFEB:

Members	*Alternates*
Dr. Homer E. Newell Associate Administrator for Space Science & Applications, NASA	Mr. Edgar M. Cortright Deputy Associate Administrator for Space Science & Applications, NASA
Dr. Mac C. Adams Associate Administrator for Advanced Research & Technology, NASA	Dr. Alfred J. Eggers, Jr. Deputy Associate Administrator for Advanced Research & Technology, NASA
Dr. George E. Mueller Associate Administrator for Manned Space Flight, NASA	Mr. James C. Elms Deputy Associate Administrator for Manned Space Flight, NASA
Mr. Daniel J. Fink Deputy Director for Strategic & Space Systems, DOD	Mr. John E. Kirk Assistant Director for Space Technology, DOD
Gen. Bernard A. Schriever Commander of the Air Force Systems Command, USAF	Brig. Gen. Harry L. Evans Vice Director, MOL Program Office of the Secretary of the Air Force

5. *CHAIRMANSHIP AND VOTING PROCEDURES*

The Associate Administrator for Manned Space Flight, NASA, will act as Chairman. In his absence the DOD member will act as Chairman.

MSFEB recommendations will not be based on majority and minority voting. Where recommendations are not unanimous, the views of all members will be recorded.

6. *STAFF SUPPORT*

A technical advisor to the Board will be appointed from the staff of the Associate Administrator for Manned Space Flight to provide an independent source of advice to the Board on the feasibility and technical merit of proposed experiments submitted for Board approval, and on such other matters as the Board may deem desirable. Similar technical advice will be made available from appropriate DoD agencies.

[4] A member of the staff of the Associate Administrator for Manned Space Flight will serve as Executive Secretary to the Board and will be responsible for the management of the Board operations and maintenance of records. Additional support will be provided to the Board, as required, by the Director, Advanced Manned Missions Program.

7. *SUBMISSION OF EXPERIMENTS*

Experiments will be reviewed within the sponsoring NASA or DOD Program Offices for scientific and technical merit prior to their submission to the MSFEB Secretariat for consideration by the Board. This review should include a recommendation of the priority of an experiment relative to others submitted by the sponsoring office.

8. *COORDINATION*

It is the responsibility of the sponsoring office to accomplish appropriate coordination of experiment proposals within its program. The Executive Secretary, MSFEB, in conjunction with his coordination duties for the NASA Advanced Manned Missions Program, will effect overall coordination of experiments among the NASA Program Offices and a designated point of contact in DOD prior to placing them on the agenda for MSFEB consideration.

9.   *GENERAL*
The Executive Secretary, MSFEB, will document the recommendations of the MSFEB for presentation to the Associate Administrator for Manned Space Flight, NASA, and to the Deputy Director for Strategic & Space Systems, DOD.

Robert C. Seamans, Jr.   John Foster, Jr.
Deputy Administrator, NASA   Director of Defense Research &
                                           Engineering, DOD

Document II-29

Document title: Thomas O. Paine, NASA Administrator, to Honorable Robert C. Seamans, Secretary of the Air Force, April 4, 1969, with attached: "Terms of Reference for Joint DOD/NASA Study of Space Transportation Systems."

Source: Administrators Files, NASA Historical Reference Collection, NASA History Office, NASA Headquarters, Washington, D.C.

President-elect Richard Nixon appointed a Space Task Group, chaired by Vice President-elect Spiro Agnew, to oversee American space policy. At a March 22, 1969, Space Task Group meeting, the membership discussed joint development of a Space Transportation System (STS). Less than two weeks later, on April 4, NASA Administrator Paine formally invited Secretary of the Air Force Seamans to study jointly the possibility of building a national STS.

[no pagination]   April 4, 1969

Honorable Robert C. Seamans
Secretary of the Air Force
Department of Defense
Washington, DC 20301

Dear Dr. Seamans:

Enclosed is a draft of Terms of Reference for a joint DoD/NASA study of space transportation systems. I understand this draft has been coordinated between our staffs, and I have signed it. Upon notification of your approval and signature, we are prepared to proceed immediately to implement the terms of the study.

Sincerely yours,

T.O. Paine
Administrator

Enclosure

****

[1]

Terms of Reference for Joint DoD/NASA Study of Space Transportation Systems

OBJECTIVE:

The objective of the joint DoD/NASA study of Space Transportation Systems is to assess the practicality of a common system to meet the needs of both the DoD and the NASA. Emphasis will be placed on the economic sensibility and technical feasibility of such a system.

BACKGROUND:

The need for a joint DoD/NASA group to study space transportation was discussed at the Space Task Group meeting of March 22, 1969. The Space Task Group was established by the President to recommend by September 1, 1969, a National Space Program for the post-Apollo period. It is expected that submissions by each participating agency will occur in June or July 1969. The joint DoD/NASA Study Group should provide timely results for these submissions.

FUNCTIONS:

The study shall be accomplished in two parts, the first part to be done separately by the two agencies, DoD and NASA, and the second part to be done jointly.

   1.   The first part of the study shall proceed as follows:

       (a) Each agency, DoD and NASA, shall study its own [2] needs, present and future, for a new space transportation system.

       (b) On the basis of its own needs, each agency shall make a preliminary determination of the characteristics of the transportation system that would best meet its needs.

   2.   The second part of the study shall be done jointly and shall proceed as follows:

       (a) The Joint Study Group shall assemble and correlate the needs of both agencies for a space transportation system.

       (b) The Joint Study Group shall assess the technical feasibility of various systems to meet the needs of both agencies.

       (c) The Joint Study Group shall compare the relative costs and assess the economic sensibility of systems meeting the needs of both agencies.

       (d) The Joint Study Group shall recommend a preferred concept and, if appropriate, alternative concepts of a space transportation system and provide the supporting rationale for each concept.

RESULTS:

A report shall be provided to the President's Space Task Group on June 15, 1969.

APPROACH:

The Staff Directors of DoD and NASA serving the Space Task [3] Group shall each designate a co-chairman for the Joint Study Group. Theses [sic] co-chairmen shall appoint members from each agency to form the group. The Staff Directors shall be responsible for providing a report to the Space Task Group on June 15, 1969.

APPROVAL:

Dr. Robert C. Seamans   Dr. Thomas O. Paine
Secretary of the Air Force   Administrator, NASA

Document II-30

Document title: George M. Low, NASA Deputy Administrator, Memorandum for the Record, "Space Shuttle Discussions with Secretary Seamans," January 28, 1970.

Source: Deputy Administrators Files, NASA Historical Reference Collection, NASA History Office, NASA Headquarters, Washington, D.C.

By early 1970, NASA had recognized that Department of Defense (DOD) support would likely be essential if White House approval for the Space Shuttle program were ever to be obtained. In this memorandum, NASA Deputy Administrator George Low records an early policy-level discussion with Secretary of the Air Force Robert Seamans and Assistant Secretary for Research and Development Grant Hansen on NASA-DOD cooperation in shuttle planning.

---

[1]   Jan. 28, 1970

Memorandum for the Record

SUBJECT: Space Shuttle Discussions with Secretary Seamans

On January 27, 1970, I met with Bob Seamans to discuss the Phase B shuttle effort, shuttle classification, and the proposed DoD/NASA agreement on the Space Shuttle.

I informed Bob of our plans to move out with a Phase B effort in the near future and told him of our general Phase B plans. I mentioned that, to my knowledge, the Air Force was in basic agreement with these plans except possibly on the questions of gross weight versus payload weight and cross-range requirements. I explained the reasons for going with a 3 1/2 million-pound gross weight and pointed out that the studies could be redirected at mid-point if this was the wrong weight. I also pointed out that the cross-range question would be handled by having two point designs, one with low cross-range and the second with high cross-range. Seamans agreed that the basic objective of the shuttle program should be to develop a low-cost transportation system and that requirements, such as cross-range, go-around capability, etc., must be tested in the light of this objective. Although he made no specific commitment, I believe that he has no significant objections to the points that I made. Grant Hansen was also present and raised a question concerning the use of gross weight instead of payload weight. However, he voiced no strong objections to our approach. A letter from Paine to Seamans on this subject was given to Seamans.

On the subject of classification, there was agreement that the Space Shuttle program should be conducted on a generally unclassified basis. The justification for specific DoD performance requirements can, of course, be presented internal to the government on a classified basis, but the resulting Space Shuttle system should be unclassified in the same sense that the Apollo Program was unclassified. Seamans agreed to these points and fully recognized the international flavor of the program.

[2] I left copies of the proposed NASA/DoD agreement on the Space Shuttle. . . . Bob Seamans pointed out that the Air Force had no money to spend on shuttle development this year, but nevertheless was very much interested in developing the shuttle as a national capability. He strongly urged the establishment of a co-chaired board for DoD requirements. Although he did not have time to read the agreement while I was there, I read

pertinent excerpts to him and received a favorable response. I would expect that the Air Force will sign the agreement in short order.

In response to a direct question, Bob Seamans pointed out that the Air Force was indeed an agent for the DoD on the Space Shuttle program and that he had discussed this with both Secretary Laird and John Foster.

Following the discussions on the Space Shuttle, we talked about aeronautics, with Seamans emphasizing the need to move forward on an aeronautics program. Al Eggers [Dr. Alfred J. Eggers, Jr., special assistant to the Administrator] had, earlier in the day, discussed with him our dealings with [the Department of Transportation (DOT)] on the VTOL/STOL [vertical takeoff and landing/short takeoff and landing] aircraft program. Seamans indicated that the Air Force would like to participate in this effort as a third party, with the principal effort coming from NASA and DOT.

We also discussed the DoD/NASA funding picture, and Seamans pointed out that Secretary Laird is most interested in getting NASA to "pay its own way." He felt that Tom Paine should have lunch with Secretary Laird in the near future to discuss this in more detail. We agreed that the immediate problem is that of ETR [Eastern Test Range] and KSC and that some joint study in this area may be called for. At the present time the Air Force is conducting its own study on whether or not it should maintain a capability at ETR.

Bob Seamans' last point concerned the direction of NASA programs. He mentioned that, in the 1960's, NASA was fully supported because of the competition with Soviet Russia. This type of support should not be expected in the 1970's. NASA should therefore help solve the problems of the natural environment and thereby help pay for itself.

George M. Low
Deputy Administrator

Document II-31

Document title: Thomas O. Paine, NASA Administrator, and Robert C. Seamans, Jr., Secretary of the Air Force, "Agreement Between the National Aeronautics and Space Administration and the Department of the Air Force Concerning the Space Transportation System," NMI 1052.130, Attachment A, February 17, 1970.

Source: NASA Historical Reference Collection, NASA History Office, NASA Headquarters, Washington, D.C.

During 1969, it became clear that there was great interest within the Department of Defense (DOD) as well as NASA with respect to a reusable space launch system. Reflecting this, a joint NASA-Air Force (USAF) Space Transportation System Committee was formally created on February 17, 1970, and was given primacy among all joint activities pertaining to the Space Transportation System. Important concepts established in the agreement included the unclassified nature of the program, the possibility of international cooperation, and equal participation of NASA and DOD in shuttle development, in terms of both investment and operations. This equality of investment was later used as the basis for subsequent shuttle pricing agreements.

[1]

Agreement Between
the National Aeronautics and Space Administration
and the Department of the Air Force
Concerning the Space Transportation System

This document establishes an agreement between NASA and the Department of the Air Force, acting as the agent of DoD, to insure that the proposed National Space Transportation System will be of maximum utility to both NASA and the DoD.

I.   Objective of the Space Transportation System
The objective of the Space Transportation System (STS) is to provide the United States with an economical capability for delivering payloads of men, equipment, supplies, and other spacecraft to and from space by reducing operating costs an order of magnitude below those of present systems.

The program may involve international participation and use. The development of the STS will be managed by NASA. The project will be generally unclassified. For purposes of this agreement, the STS will consist of the earth-to-orbit Space Shuttle.

II.   NASA/USAF STS Committee
A.   Organization

In order that the STS be designed and developed to fulfill the objectives of both the NASA and the DoD in a manner [2] that best serves the national interest, a NASA/USAF STS Committee is hereby established that will report jointly to the Administrator of the NASA and the Secretary of the Air Force. The Committee will consist of eight members, four to be appointed by the Administrator of the NASA and four to be appointed by the Secretary of the Air Force. The Co-Chairmen of the Committee will be the Associate Administrator for Manned Space Flight (NASA) and the Assistant Secretary for Research and Development (Air Force). Any proposal for changing the composition or functions of the Committee will be referred to the NASA Administrator and the Air Force Secretary for their joint consideration.

B.   Function

The Committee will conduct a continuing review of the STS Program and will recommend steps to achieve the objectives of a system that meets DoD and NASA requirements. Specifically, the Committee will review and make recommendations to the Administrator of NASA and to the Secretary of the Air Force on the establishment and assessment of program objectives, operational applications, and development plans. This will [3] include, but not be limited to: Development and operational aspects, technology status and needs, resource considerations, and interagency relationships.

THOMAS O. PAINE   ROBERT C. SEAMANS, JR.
Administrator, NASA   Secretary of the Air Force
Date: Feb. 17, 1970   Date: Feb. 17, 1970

Document II-32

Document title: William F. Moore, NASA STS Secretary, and Lt. Col. Donald L. Steelman, USAF STS Secretary, "Space Transportation System Committee: Summary of Activities for 1970," June 1971.

Source: NASA Historical Reference Collection, NASA History Office, NASA Headquarters, Washington, D.C.

The Space Transportation System (STS) Committee was established as the policy-level coordination forum between NASA and the Department of Defense (DOD) for developing the Space Shuttle. It drew its authority from the February 17, 1970, NASA-DOD agreement on the STS. It was through the forum of the STS Committee that DOD's requirements for the Space Shuttle were first transmitted to NASA; DOD indicated the conditions under which it would place exclusive reliance on the shuttle. This report summarizing the committee's first year activities, endorsed by the NASA and Air Force secretaries to the STS Committee, demonstrates the considerable groundwork that was laid during that time for the joint program. The acronyms MSC, MSFC, and KSC are NASA centers and stand for the Manned Space Center, the Marshall Space Flight Center, and the Kennedy Space Center, respectively. The acronym OSSA refers to NASA's Office of Space Science and Applications, and AFSC stands for the Air Force Space Center.

[1]

Introduction

The NASA/USAF Space Transportation System Committee was formed for the purpose of providing a policy level interface between NASA and the USAF on the problems of developing the Space Shuttle. An agreement was formally signed on February 17, 1970 by Dr. Robert C. Seamans, Jr., Secretary of the Air Force, and Dr. Thomas O. Paine, Administrator, NASA. The agreement specified the objective of the Space Transportation System (STS), defined its limits, and established a committee to perform a continuing review of the program and recommend steps to achieve the objectives of the system that would meet the needs of both NASA and the DOD. The committee consists of eight members, four from each agency, and is co-chaired by the Associate Administrator for Manned Space Flight (NASA) and the Assistant Secretary for Research and Development (USAF). A copy of the agreement is attached to this summary.

The original members were:

USAF

Mr. Grant L. Hansen	Co-Chairman
General Walter Hedrick	Member (HQ USAF)
General Raymond Gilbert	Member (AFSC)
General F. M. Rodgers	Member (AFSC)

NASA

Mr. Dale D. Myers	Co-Chairman
Mr. Vincent Johnson	Member (OSSA)
Mr. Lee James	Member (MSFC)
Dr. Chris Kraft, Jr.	Member (MSC)

By separate correspondence the Secretary of the Air Force and the Acting Administrator of NASA invited the Executive Secretary of the National Aeronautics and Space Council [NASC] to participate on the committee as an official observer. During the year the membership has changed in the Air Force representation and provision was made for specific alternates to attend when the principal was unable to make a called meeting.

[2] The current membership and alternates are:

USAF
    Mr. Grant IL. Hansen—Co-Chairman
        Alternate—Mr. Frank Ross
    MGen Paul Cooper—Member (AFSC)
        Alternate—Col Paul Atkinson
    BGen Kenneth Chapman—Member (AFSC)
        Alternate—Col Ralph Ford
    Col John Albert—Member (AF/RDS)
        Alternate—Col Frank Knolle

Official Observer
    Mr. William Anders—NASC
        Alternate—M. Raymond Gilbert

NASA
    Mr. Dale D. Myers—Co-Chairman
        Alternate—Mr. Charles Mathews
    Mr. Vincent Johnson—Member (OSSA)
        Alternate—Dr. Robert Wilson
    Mr. Lee James—Member (MSFC)
        Alternate—Dr. William Lucas
    Dr. Chris Kraft, Jr.—Member (MSC)
        Alternate—Lt Gen (Ret) Frank Bogart

The following summary of the Space Transportation System Committee's activities covers the period from the initial meeting on May 28, 1970, through the sixth meeting on December 15, 1970.

[3]

USAF Personnel Participation in the Space Shuttle Program

One of the first questions at the initial meeting of the STS Committee was the extent to which the Air Force would participate in the Space Shuttle activities. The discussion focused on USAF personnel participation in the NASA program offices particularly at MSC and MSFC. SAMSO [Space and Missile Systems Organization] on an ad-hoc basis was already covering early integration meetings by travel assignments (TDY). NASA stressed that the activities were beginning to accelerate and that a more permanent arrangement would be welcome if the Air Force wanted to participate actively. It was emphasized that very close coordination between NASA and the Air Force at the center level was critical to the Phase B definition effort and that this was the most effective way to facilitate the exchange of technical data and program activity status. NASA preferred direct involvement (or detailing) of Air Force personnel in the program activity, but that as a minimum, immediate liaison was recommended.

At the second meeting of the STS Committee in June 1970, NASA presented its plan based on non-reimbursable assignment of USAF officers to NASA Centers and Headquarters. The plan requested five officers each for Headquarters and MSFC, ten officers for MSC and two officers for KSC during the Phase B activity. When Phases C and D were begun the Air Force could augment these assignments with additional officers as the need arose. The USAF accepted the plan for further study and stated that ten qualified officers would be assigned to SAMSO with five placed at MSFC and five on duty at MSC to participate in the Phase D activity. They expected to have the officers on site by fall. In the meantime SAMSO would continue covering the two centers by TDY until the assignments were executed. No assignments were made to KSC but [the] Air Force agreed to reappraise its manpower situation and report to the Committee in 90 days. They would also investigate the possibility of establishing a point of contact in the 6555th Aerospace Test Wing at Patrick AFB to coordinate activities with KSC.

At the sixth meeting in December the Air Force reported that the two officers requested for KSC would be assigned to SAMSO with duty at KSC and that they should be on board by July 1971.

[4] As a result of these actions, the following officers are currently participating in the Space Shuttle activities at the two Centers.

MSFC	MSC
LCol Thomas Moore	Maj Patrick Crotty
Maj James A. Feibleman	Maj Gary H. Minar
Capt Byron Thurer	Maj Charles T. Essmeier

Implementation of Phase B
Space Shuttle Management Plan

NASA reported to the STS Committee at the first meeting its management plan for implementing the Phase a definition studies. The organization chart attached shows the relationship of the three Manned Space Flight Centers (MSFC, MSC and KSC) to each other and to the Headquarters Space Shuttle Office. Also shown were the Phase B contractor management assignments to the centers and the Vehicle System Integration Activity (VSIA) function between MSC and MSFC with Headquarters participation.

Main points relative to the management of Phase efforts were the assignment of the North American Rockwell vehicle contract to MSC and the McDonnell Douglas vehicle contract to MSFC. Houston would have the overall orbiter technical responsibility for both contractors and Huntsville would have the overall booster technical responsibility for both contractors. The three Phase B engine contracts with Pratt and Whitney, Aerojet and Rocketdyne are being managed by MSFC. KSC has representatives in both center program offices and participates in the integration activity. Program integration activity takes place on a regular basis and includes representation from the Air Force (SAMSO).

Space Shuttle Facilities Planning

A briefing on the Master Facilities Planning Study was presented to the STS Committee at its first meeting. Basically the NASA Facilities Office is managing a $380K study by the Ralph M. Parsons Co. The study is to survey the candidate facilities as to their adequacy to support the Space Shuttle [5] Program and the costs of modifications or new construction required to meet criteria established as necessary for the launch, recovery and refurbishment of the Space Shuttle. The twelve month study is to culminate in a report to NASA setting forth the plan having the most favorable overall features as measured against the "ideal facilities matrix."

The Committee was concerned as to how this study was tied into the facilities activity of AACB [Aeronautics and Astronautics Coordinating Board], but was assured that close personnel liaison and information exchange would prevent any duplication of effort. It was stated that the AACB effort is an across-the-board national facility activity whereas the Parsons study is specifically oriented toward Space Shuttle requirements. It was also pointed out that the Air Force had personnel participating in the Space Shuttle Facilities Planning Group and therefore would be kept fully aware of the progress of the study.

As a part of the discussion the question of industrial funding was raised by NASA, in particular, as it relates to the use of AEDC [Arnold Engineering Development Center] test facilities at Tullahoma and the Rocket Propulsion Laboratory Test Stand I-56 at Haystack Butte. The present policy of DOD requires user funding for such facilities and the Air Force did not have FY 71 funds available to support shuttle testing at AEDC. An alternative would be to reprogram funds within the DOD to support Space Shuttle testing. However, military priorities for project funding precluded this. Therefore, any Phase B Space Shuttle testing at AEDC facilities would have to be on a cost reimbursable basis in accordance with the DOD policy.

Space Tug or Orbit-to-Orbit Shuttle

A discussion of the expected similarities and differences between the DOD and NASA requirements for the space tug or orbit-to-orbit shuttle (OOS) was presented to the STS Committee by NASA at the first meeting. The main point emphasized was that a single design may be possible, but that further conceptual study and definition of mission requirements were needed.

NASA informed the Committee that it was proceeding with a pre-Phase A study of the space tug which it hoped would define its requirements. The Air Force reports that it also was planning to conduct a concept and requirements analysis for the OOS. The Committee felt that the two studies would be [6] complementary.

The Air Force Co-Chairman indicated that it might be appropriate for the development of the OOS to be undertaken by the Air Force. The NASA Co-Chairman stated that they would like the Air Force to consider that approach. Also the NASA Co-Chairman reported that the European Launcher Development Organization (ELDO) had contracted with two groups of foreign contractors for a pre-Phase A study to determine the feasibility and derive a simple definition for a space tug design. The costs of the contracts are approximately $500K. The STS Committee agreed that ELDO should be encouraged to continue in their space efforts.

At the sixth meeting in December 1970, NASA briefed the Committee on the ELDO tug studies and the NASA pre-Phase A Space Tug studies. The various configurations being studied by ELDO were discussed and the observation was made that nothing different from U.S. findings on the space tug had emerged. NASA concluded their presentation on the pre-Phase A studies briefing with the following list of findings:

  a.   Reusable tug synchronous mission performance is extremely sensitive to mass fraction.

  b.   Ground based tugs will not be recovered for most synchronous missions.

  c.   Synchronous payload recovery will require tug staging or orbital propellant loading.

d.   Moderate increase in shuttle payload capability (above the 25K reference payload) will not affect general conclusions or tug utilization for synchronous missions.

e.   Current upper stages may serve as effective interim expendable tugs for synchronous missions.

f.   Shuttle economic model should assume no synchronous tug or payload recovery—at least for [the] early operational years.

The NASA Co-Chairman stressed that in the tug studies we want to make sure that the payload and Space Shuttle interface is minimized in order to keep the system complexity and cost down. NASA also covered the expendable stages for use with the Space Shuttle in lieu of an OOS or space tug. This included the current state of "kick" stages such as Agena, [7] Centaur, Burner II, and the Titan Transtage as well as the potential modified Agena and modified Centaur stages all of which could serve as interim tugs.

The Air Force gave a status report on the DOD orbit-to-orbit shuttle and expendable stage study efforts. In FY 71 the DOD effort has involved both contractor and in-house activity to define an OOS that would meet unique DOD requirements. Contracts for conceptual designs of a reusable OOS were let with two contractors in February 1971. This effort is directed toward meeting DOD needs with an assessment being made to see if [the] vehicle couldn't meet the needs of both agencies with a minimum of modification. The Air Force also was specifying that deployment/retrieval considerations for the earth orbital shuttle/orbit-to-orbit shuttle (EOS/OOS) and payload interfaces be examined. Engine design studies to define a light weight, high performance propulsion system for potential use in a high energy upper stage/OOS were being conducted at the same time.

General Security Guide

The development of a general security guide for the Space Shuttle program was assigned to both NASA and the USAF at the first meeting in May 1970 of the Space Transportation System Committee. A draft of the security guidelines was presented to the Committee for review and comment at the second meeting. It was requested that the guidelines for their comments and a report be made to the Committee at a later date. The Committee also suggested that the draft be as short as possible. A condensed version was submitted at the third meeting for consideration and coordination.

Comments were incorporated and the general security guidelines were accepted by the Committee at the fourth meeting in October 1970. The Co-Chairmen instructed the Secretariat to prepare the document for their signature. The guidelines were signed on November 19, 1970 and distributed through channels to all elements participating in the Space Shuttle Program.

[8]
Space Shuttle Payload Size

The Air Force briefed the STS Committee on DOD payload size and weight requirements at the second meeting in June 1970.

Payload physical size has a definite influence on development and operational costs; however, in order to make the decision, mission utility to both NASA and the DOD must be considered in the analysis as well. From the baseline the size and weight of future payloads was projected for missions to be flown eight to ten years hence when the shuttle would be operational. Also the growth history of launch vehicle payload capabilities and the length of payload fairings were shown as indicators of the need to plan for the accommodation large payload mission requirements that would utilize [a] 60 foot by 15 foot cargo bay and carry an equivalent payload weight of 40,000 pounds to low earth polar orbit.

The diameters of current launch vehicles restrict their payload diameters which in turn causes design complications and the attendant high costs for packaging and reliability. Furthermore, analysis of available data shows that the pressing need for improved capability and mission use demands larger diameters and greater payload weight capabilities. Increased lifetime, power and minimum design cost are additional parameters for consideration.

Based on required improvements to the present systems, mission needs and payload growths predicted for the 1980's an equivalent payload weight capability of 40,000 to 53,000 pounds is required to low earth polar orbit; 40,000 to 50,000 pounds is required to low earth polar orbit. A 60 foot cargo bay length is necessary for current and projected missions and a 15 foot diameter is needed for high energy missions if the 60 foot length is not to be exceeded.

It was pointed out that studies [have] shown a Space Shuttle with a 40,000 to 50,000 pound capability coupled with sufficient payload volume (the baseline requirement) is the most economical size for DOD and national mission projections. NASA studies were in agreement and also indicated that the larger vehicle was more economical from a total dollar standpoint but there was the problem of securing the annual funding levels required for this type [of] development.

[9] The Air Force emphasized that if a shuttle of reduced payload capability was developed then NASA could expect the Air Force to retain an inventory of expendable launch vehicles to satisfy their mission needs and this would cause the shuttle to lose some of its economic attractiveness and probably degrade the utility of the shuttle. It was also noted that DOD has not been considering any upgrading of its current stable of expendables because it is intended that the shuttle, if properly sized and with the proper capability, would replace them.

NASA suggested that cost tradeoff studies for retaining a limited expendable launch vehicle capability and developing a smaller Space Shuttle versus the development of a large Space Shuttle should be considered. This suggestion was accepted and a report was requested for the next meeting.

At the fourth meeting the Committee was informed by NASA that the 60 foot by 15 foot cargo bay should be retained and that the 25,000 pound payload to reference orbit (55° x 270 nm) with air-breathing engines in [it] could be increased to 40,000 pounds to low earth polar orbit by removing the air-breathing engines. The USAF emphasized that operational and safety considerations must be analyzed before such a proposal would be accepted. NASA indicates that the airbreathers would be retained for all development/test flights and also for the early operational flights.

International Participation

At the second meeting of the STS Committee, the Office of International Affairs discussed the possibility of foreign industry and governments participating in the Space Shuttle Program. This would require a technology exchange between the parties involved. The STS Committee received a request from the Chairman, Interagency Ad Hoc Group on NSDM 72 for assistance in establishing procedures for the exchange of technical data with those nations desiring to participate in the development program. The Air Force indicated they had been studying this and therefore was assigned the task of drafting a technology sensitivity guidelines document for review by the Committee.

While the sensitivity guidelines document was being coordinated in both the DOD and NASA, the Phase A and B contractors were advised by NASA to control foreign representatives [10] within the contractor's system on the same basis as any foreign visitor. At the fifth meeting the STS Committee learned that the Grumman agreement with Dornier of West Germany and the North American Rockwell (NAR) agreement with Messerschmidt, Boelkow and Blohm (MBB) and British Aircraft Company (BAC) had been approved in two phases. The first phase provides for transfer of general data and the second phase provides for the transfer of more specific data after the U.S. contractors and their foreign participants have defined the areas of interest and government-to-government agreements have been approved.

The STS Committee requested copies of the coordinated sensitivity document be supplied to each member at the sixth meeting with comments to be forwarded to the Secretariat by December 28, 1970. The STS Committee also decided that the sensitivity document, when approved, would be subject to semi-annual reviews. (The document was subsequently approved and forwarded to the NASA Office of International Affairs—Code I).

Other Government Agency/Other Military Service Space Shuttle Mission Requirements

NASA was requested by the STS Committee to check with other civilian agencies and the Air Force was requested to check with other military services for all possible mission requirements that might be factored into the Space Shuttle mission model being formulated for the Phase B study contractors. NASA reported at the third meeting that mission requirements from other government agencies are coordinated by the Meteorological Satellite Program Review Board and provided to NASA planners when these requirements are firm. The Air Force reported that Army and Navy mission requirements have been validated and are reflected in the extended DOD mission and traffic models provided to NASA on 4 June 1970. These models cover projected missions and traffic through 1990. The Air Force will keep the model data current by updating or revising when necessary.

[11]
Early Flight Payload Identification

NASA informed the Air Force at the third STS Committee meeting that they were attempting to identify meaningful specific payloads that could be candidates for the early orbital shuttle flights. Primary emphasis was being placed on identifying payloads for low altitude missions, particularly those which would not require high energy stages. Payloads for high energy missions [that] would require additional propulsive stages would also be identified but in a separate category. It was suggested that the Air Force also identify a number of specific payloads that could be candidates for early flights.

At the fourth meeting the STS Committee was briefed on the results of a joint NASA/USAF-SAMSO study leading to the selection of specific payloads that could be carried on early shuttle flights. The STS Committee requested that USAF and NASA field installations be provided copies of the study for review and comment. Guidance for the review was given by USAF (Hdqtrs) and [the NASA Office of Manned Space Flight].

A briefing on the in-depth review of the first ten Space Shuttle missions was presented in December at the sixth meeting of the STS Committee. The NASA portion of the briefing provided data on the constraints that must be placed upon the early payloads and the capabilities that the crew and orbiter will have on the first few flights. With these limitations in mind, several prospective payloads were discussed but no hard schedule was proposed nor desired. The USAF portion was classified and provided alternate payloads to those first proposed in the original package. They stressed that the Air Force data was for planning purposes and as such could change as mission requirements changed during the next eight years of shuttle development.

The STS Committee decided that NASA and the USAF should continue the study since it had proven a good mechanism for learning about some of the expected operational and interface problems.

[12]

Phase B Cost/Design Performance Management Plan

NASA presented its Cost/Design Performance Management Plan which was implemented during Phase B at the third meeting of the STS Committee in August 1970. The plan resulted from the need to assure NASA that they could afford to build the Space Shuttle and that the contractors were aware of the limitations of the NASA projected budget. By establishing objectives early in the program, NASA hoped to give the contractors "bogeys" which they could use in their definition studies and that the studies would produce a realistic program that NASA and the nation could afford.

These cost objectives or "bogeys" are in fact specific cost estimates established as a target or baseline reference to accomplish the specific goal. The bogeys which the Phase C vehicle contractors are using now related primarily to that portion of the Space Shuttle program for which they are responsible. It is important to realize that other cost elements such as main engines, facilities, special test handling equipment, etc., will have to be taken into account in addition to the vehicle contractor cost in order to arrive at a total Space Shuttle program cost. Cost objectives for these other elements of the program have been set and will be used at the appropriate time in the phased program plan.

The fundamental principal of the cost objective plan is to provide working cost targets as a cost reference in the design selection process during the Phase B definite effort. Cost thus becomes a major design criteria in the same sense as performance. The high cost elements and influence will be identified and consideration can be given to alternate design approaches or a modification of the requirements if necessary, e.g., the decision to make GLOW [gross liftoff weight] a tradeoff variable and baseline the payload weight as a means of lowering costs and simplifying design.

The necessity to stay within the cost objective can then be an incentive to find and adopt new ways of doing business including subsystems tradeoff. This method thus becomes the shared responsibility of both the government and the contracts to keep costs as low as possible while at the same time maintaining the high quality and reliability that have been a hallmark of the space program to date.

[13]

Crossrange Requirements

Operational requirements of the DOD and refinement of NASA studies have resulted in the crossrange of the Space Shuttle being baselined at 1100 nm.

In a classified briefing at the fourth meeting of the STS Committee the Air Force pointed out that the military need for a high crossrange is based on DOD dedicated missions requiring a fast response in the event of a national crisis, a quick return from orbit, [or] abort to orbit[,] and return to a high crossrange, the order of 1100 nm, is necessary to provide the operational flexibility required by these types of mission.

One way of achieving this requirement is to trade payload weight for the added Thermal Protection System (TPS) weight which will protect the vehicle in the hypersonic maneuvers that produce the desired crossrange. A study to determine the merit of such a trade was initiated by the Air Force. At the sixth STS meeting the Air Force gave a classified briefing covering the preliminary findings of the study.

Of the DOD applications, the near polar missions were shown as the ones requiring the 1100 nm crossrange if the orbiter is to return to the launch site after once around. This high crossrange requirement could be reduced if alternate landing sites were used. However, the orbiter would have to be ferried from the alternate recovery site to the launch site for refurbishment prior to its next launch. Use of alternate sites then would require additional handling and servicing equipment. Since the orbiter ferry range is limited to about 700 nm, either in-flight refueling or several flight legs night be required depending on the location of the alternate site.

About 30% of the DOD missions require the orbiter to carry an equivalent payload weight of about 40,000 pounds to low entry orbit and still have a high crossrange capability. This equivalent payload weight does not include the propellant weight of 11,000 pounds required for abort to orbit using the then currently baselined engine size. (Engine size has subsequently been increased to 550,000 pounds of thrust at sea level.)

The briefing concluded that, for some DOD missions, high crossrange requirements are coincident with heavy payloads. Therefore, unless alternate recovery sites and ferrying [14] capabilities are shown to be operationally attractive, the shuttle orbiter must have both the 1100 nm crossrange capability and the ability to deliver 40,000 pounds to low earth polar orbit. This capability will enable the Space Shuttle to capture the type of mission discussed above.

Air Force Phase B Study Tasks

The Air Force briefed the Committee on their FY 71 STS study tasks at the third meeting. Their primary emphasis was a study effort to identify the functions and operating modes peculiar to the support of DOD missions. Contract tasks were proposed as add-on effort to the two NASA Phase B vehicle contracts. This would provide an assessment of NASA Phase B candidate Space Shuttle system capabilities to support missions unique to DOD.

The contractors would perform tradeoff studies and cost analysis to determine the impact of specific DOD needs on baseline system design and operations and to determine the modifications necessary to the baseline configuration in order to capture the DOD missions. The Air Force assured NASA that this study effort would identify those DOD missions that the current NASA baseline configuration would satisfy. It was emphasized that contractor teams supporting the DOD study effort would be identifiable and separate from the teams performing work under the NASA Phase B contract. The contracting alternatives were discussed and the STS Committee recommended that the NASA Phase B Space Shuttle contracts be amended to accomplish the specified Air Force tasks. Also recommended was a management approach which assured the close integration of the SAMSO and NASA study efforts. NASA agreed with this approach and felt that the addition of the two $300K tasks would contribute significantly to the Phase B effort.

Document II-33

Document title: John S. Foster, Jr., Director of Defense Research and Engineering, to Dr. James C. Fletcher, Administrator, NASA, April 13, 1972.

Source: Administrators Files, NASA Historical Reference Collection, NASA History Office, NASA Headquarters, Washington, D.C.

Because a large number of military and national security payloads are placed into polar orbits and the launch sites at Cape Canaveral are unsuitable for this purpose, the military has launched satellites into high-inclination orbits from Vandenberg Air Force Base in California since February 1959. The use of Vandenberg as a shuttle launch and landing site was one of the primary drivers of shuttle design, determining cross-range requirements and abort modes. In April 1972, the Department of Defense officially concurred with the selection of both Kennedy Space Center and Vandenberg as launch and landing sites for the Space Shuttle.

[no pagination]   13 April 1972

Dr. James C. Fletcher
Administrator, National Aeronautics and Space Administration
Washington, D.C. 20546

Dear Dr. Fletcher:

This is to advise you that the Department of Defense concurs in the selection of the Kennedy Space Center (KSC), Florida, and Vandenberg Air Force Base, California, as launch and landing sites for the Space Shuttle, as follows:

1.   The initial launch and landing site will be at KSC and be used for research and development launches and for all easterly operational launches feasible from KSC. General purpose shuttle facilities for all users will be provided by NASA at KSC on a time schedule compatible with the shuttle development program.

2.   A second operational site for missions requiring high inclination launches not feasible from KSC is planned at Vandenberg Air Force Base toward the end of the 1970's. General purpose shuttle facilities for all users will be provided by the Department of Defense at Vandenberg AFB on a time schedule compatible with progress in the shuttle development program and timely utilization of the shuttle for operational missions requiring high inclination launches.

Sincerely,

John S. Foster, Jr.

Document II-34

Document title: George M. Low, NASA Deputy Administrator, to NASA Associate Administrator for Manned Space Flight, "Space Tug Decision," October 3, 1973.

Source: Deputy Administrators Files, NASA Historical Reference Collection, NASA History Office, NASA Headquarters, Washington, D.C.

This memorandum from NASA Deputy Administrator George M. Low reflected NASA thinking regarding management of the space tug. Low's reasoning included Department of Defense (DOD) funding of part of the development costs of the overall Space Transportation System, so NASA could reduce its costs and peak funding requirements. Furthermore, Low considered it important that the Air Force get more involved in the shuttle's development. DOD had committed to use the shuttle conditionally, requesting further study of its performance and technology and demonstration of both its cost savings and operational status. Deeper involvement by the Air Force, it was assumed, would lead to its stronger commitment to the shuttle. Don Fuqua, mentioned in the memorandum, was a Florida congressman active on the House Committee on Science and Astronautics. Jim Wilson was a committee staff member.

---

[1]   October 3, 1973

Memorandum

TO: M/Associate Administrator for Manned Space Flight

FR: AD/Deputy Administrator

SUBJECT: Space Tug Decision

Don Fuqua asked to see me privately after the ASTP [Apollo-Soyuz Test Project] briefing. During the private meeting he asked, "Does NASA intend to develop the Tug or do you intend to let the Air Force take it away from you?"

I told Don that this decision had not yet been made but that NASA management was quite interested in having the Air Force develop the Tug for two reasons:

1.   to minimize NASA's peak funding requirements, and

2.   to get the Air Force (DOD) more deeply involved in the Space Shuttle development.

Don voiced a number of concerns, most of which are expressed in the attached document, which, I believe, was prepared by Jim Wilson. I promised two things:

1.   Phil Culbertson would get together with Jim Wilson soon to discuss some of the points raised in the document. Specifically, the question of the applicability of the Space Act would be discussed.

2.   [NASA Associate Administrator for Manned Space Flight Dale] Myers and Low would get together with Fuqua toward the end of October to discuss the entire issue.

[2] I am not sure whether the end of October date needs to be firm, but certainly we ought to talk to Fuqua about it before a final decision is made.

By copy of this memo, I am asking Gerry Griffin to keep track of setting up this meeting.

George M. Low

Document II-35

Document title: James C. Fletcher, Administrator, NASA, to Honorable James R. Schlesinger, Secretary of Defense, June 21, 1974.

Document II-36

Document title: W.P. Clements, Deputy Secretary of Defense, to Honorable James C. Fletcher, Administrator, NASA, August 7, 1974.

Source: Both in Administrators Files, NASA Historical Reference Collection, NASA History Office, NASA Headquarters, Washington, D.C.

NASA and the Office of Management and Budget had agreed on January 3, 1972, that the Space Shuttle would have a large payload bay, capable of handling the largest U.S. military satellites being planned. This did not mark a policy decision of exclusive use of the shuttle, however, as is evident in this letter from NASA Administrator James Fletcher to Secretary of Defense James Schlesinger and the reply from Deputy Secretary of Defense W.P. Clements. By 1974, the Department of Defense (DOD) was examining the wisdom of a complete phaseout of expendable launch vehicles, which raised serious concern within NASA and Congress. Later budgetary decisions would make abandoning expendable launch vehicles a de facto *policy because of the cost of maintaining both options. This* de facto *policy, however, was never explicitly stated; DOD continued to favor a prudent expendable launch vehicle backup policy. The handwritten note on the Clements letter is from NASA Deputy Administrator George Low to Fletcher. In the Fletcher letter, Mal Currie was the Director of Defense Research and Engineering; his name was misspelled by Clements.*

Document II-35

[1]   June 21, 1974

Honorable James R. Schlesinger
Secretary of Defense
Washington, DC 20301

Dear Mr. Secretary:

I had hoped to see you before having to leave town for two weeks, but since this has not been possible I am taking this way to alert you to the matter I wanted to talk to you about.

It concerns the Space Shuttle. Through our regular contacts with DOD, we understand that in the present review of the DOD five-year plan questions are being raised on the DOD participation in the shuttle program which had been agreed to for planning purposes at the time the program was approved by the President. Questions are being raised on the DOD's provision of launch and landing facilities on the West Coast, on future DOD procurement of orbiters for DOD use, and on the planned phase-out of DOD's use of expendable launch vehicles.

We have discussed these problems with the Air Force and Mal Currie and they are working on ways to reduce the cost of the facilities planned at Vandenberg Air Force Base and to minimize the budgetary impact on DOD procurement of orbiters. Neither the VAFB facilities nor the procurement of orbiters are matters requiring actual decisions now or in the FY 1976 budget.

My concern is that a decision in the DOD planning process to back away from previously planned DOD participation in the shuttle program, or a decision which implies that the DOD will not rely on the shuttle for its space activities in the 1980's, could be used by Congressional opponents of the program to attack and perhaps even cut back the shuttle development program.

As you know, the Space Shuttle is an Administration program that is national in scope, and decisions to proceed with the shuttle were based, in part, on previous DOD studies which indicated [2] very substantial benefits to DOD through use of the shuttle. I'm sure you would plan to consult with me in advance if you believed that any decisions making significant changes in DOD's previously planned role and use of the Space Shuttle are necessary at this time. However, I was afraid that due to the press of other DOD business such consultation might have been overlooked and therefore was most anxious to see you before I left.

In my absence George Low will be available to meet with you whenever convenient.

With best wishes,

Sincerely,

James C. Fletcher
Administrator

Document II-36

[no pagination]   Aug 7 1974

Honorable James C. Fletcher
Administrator
National Aeronautics & Space Administration
Washington, D. C. 20546

Dear Dr. Fletcher:

The Secretary and I were pleased to have the opportunity to discuss with you and Dr. Low the Space Shuttle program and the concerns which you raised earlier in your June 21, 1974, letter.

The Department of Defense is planning to use the Space Shuttle, which NASA is developing, to achieve more effective and flexible military space operations in the future. Once the Shuttle's capabilities and low operating cost are demonstrated we expect to launch essentially all of our military space payloads on this new vehicle and phase out of inventory our current expendable launch vehicles.

Recent budget actions assure that adequate outyear funding will be available to develop a low cost modified upper stage for use with the Shuttle. This stage will be ready for operational use at Kennedy Space Center concurrently with the Shuttle in 1980. Funding is also included now in out budget for establishing a minimum cost Shuttle launch capability at Vandenberg Air Force Base consistent with realistic DOD and NASA needs. This addition should be available around December 1982; however, funding constraints could cause some delays. As we made clear in our conversation, overall budget constraints force us to defer any consideration of orbiter buys at this time.

Dr. Curry [sic] has been very much involved in our budgetary deliberations on the use of the Shuttle and will be available to discuss these points further with you at any time.

Sincerely,

W. P. Clements

Document II-37

Document title: John F. Yardley, NASA Associate Administrator for Space Flight; John J. Martin, Assistant Secretary of the Air Force (Research and Development); James C. Fletcher, NASA Administrator; William P. Clements, Jr., Deputy Secretary of Defense, "NASA/DOD Memorandum of Understanding on Management and Operation of the Space Transportation System," January 14, 1977.

Source: NASA Historical Reference Collection, NASA History Office, NASA Headquarters, Washington, D.C.

In the mid-1970s, NASA and the Department of Defense (DOD) began to discuss the management and operations of the Space Shuttle system. These discussions resulted in a memorandum of understanding, which expanded earlier principles of cooperation between NASA and DOD. The document avoided asserting that the Space Transportation System would be the exclusive launch vehicle for DOD, referring to it instead as the primary launch vehicle.

---

[1]

NASA/DOD
Memorandum of Understanding on
Management and Operation of
the Space Transportation System

1.0 PURPOSE: This Memorandum of Understanding establishes the broad policies and principles that will govern the relationships between the DOD and NASA relevant to the development, acquisition and operation of the national Space Transportation System. The Memorandum of Understanding shall be used as the basis for more detailed documentation between the NASA and the DOD further delineating Space Transportation System management and operations concepts and the specific roles and responsibilities of each agency.

For purposes of this Memorandum of Understanding, the national Space Transportation System consists of an earth-to-orbit Space Shuttle, the upper stage(s) required for orbital velocities exceeding the Shuttle capability, and the ground support equipment and facilities necessary for operation of the system. A DOD-developed expendable Interim Upper Stage (IUS) will be available concurrently [2] with the operational Space Shuttle for use by both agencies. There is planning for development of Spinning Solid Upper Stage (SSUS) to supplement the IUS which would be available concurrently with the operational Space Shuttle for use by both agencies.

2.0 BACKGROUND: On February 13, 1969, the President appointed a multi-agency Space Task Group to develop recommendations on the direction which the U.S. Space Program should take in the Post Apollo period. The Space Task Group recommended that a reusable Space Transportation System be developed to allow more economical and effective use of space.

On February 17, 1970, NASA and the Air Force, acting as the designated agent for DOD, established by joint agreement the NASA/USAF Space Transportation System Committee to provide an instrumentality for joint review and recommendations concerning development and evolution of a Space Transportation System which fulfill the objectives of both NASA and DOD in a manner that best serves the national interest.

[3] On January 5, 1972, the President decided that the United States should proceed at once with the development of a space transportation system capable of providing routine access to space and taking the place of all present launch vehicles except the very smallest and the very largest.

On April 13, 1972, the selection of J.F. Kennedy Space Center, Florida, and Vandenberg Air Force Base, California, as launch/landing sites for the Space Shuttle was agreed upon.

3.0 GENERAL POLICIES AND PRINCIPLES: The Space Transportation System (STS) is a national program designed to serve all users—both civil and defense. The evolution of a viable, cost effective system requires the efficient use of extensive national resources, primarily those of NASA, DOD and the aerospace industry. The overall planning and coordination to insure the most effective utilization of these resources in the development, acquisition and operation of the STS are the responsibility of NASA. The DOD will use the STS and participate as a partner in development, acquisition, and operation activities as specifically defined herein.

[4] Effective and efficient use of the national STS requires an environment of understanding and cooperation between the agencies. To this end, there shall be maintained a free and effective interchange of essential technical, financial, and managerial information between the two agencies. This interchange shall be accomplished primarily throughout the NASA/USAF Space Transportation System Committee. Coordination will be maintained with the Aeronautics and Astronautics Coordinating Board and other joint groups established by mutual agreement.

It is anticipated that interest in the National Space Transportation System will continue to grow as more and more agencies recognize the merits and benefits associated with a non-expendable means for placing and retrieving payloads in space. The STS should provide benefits for many varied space requirements. Fulfillment of requirements from actual and potential users of this system must be given careful consideration. Insofar as their fulfillment does not compromise other priority requirements to an unreasonable degree, they will be accommodated.

[5] The cooperation and coordination required will be implemented so as to assure consistency with applicable policy with respect to the relationship between civil and military space activities.

4.0 MANAGEMENT AND OPERATIONS CONCEPTS: The overall objective is to ensure that the national Space Transportation System will be of maximum utility to both agencies. The accomplishment of this objective will be under the purview of the joint NASA/USAF STS Committee.

The following concepts, policies and principles, and the associated roles and responsibilities are agreed to:

4.1 NASA RESPONSIBILITIES: The NASA is responsible for developing the overall STS operations concepts and plans for serving as overall financial manager for the STS. In addition:

4.1.1   The NASA is responsible for the development of the Space Shuttle, to include the orbiter and its propulsion systems, the solid rocket boosters, the external tank and general purpose ground support equipment and facilities.

[6] 4.1.2   The NASA will make every effort to incorporate the DOD requirements into the Space Shuttle, with due consideration for schedule and cost impacts, in order that the STS be designed and developed to fulfill the objectives of future uses of the STS.

4.1.3   The NASA is responsible for providing the general purpose Shuttle equipment and facilities to perform the ground, launch and landing activities for all Space Shuttle operations at the Kennedy Space Center (KSC). NASA will plan for an initial operational capability at KSC in 1980.

4.1.4   The NASA will plan to use the Interim Upper Stage (IUS) for appropriate missions and is responsible for providing to DOD those requirements affecting the IUS design which are considered important to meet NASA objectives. NASA will provide the USAF with funds for their peculiar IUS requirements.

[7] 4.1.5   The NASA will plan to use the IUS for all of its planetary missions for those earth orbital missions that are not more economically achieved by the SSUS. The SSUS will be used primarily for geo-synchronous missions of the type currently flown by the expendable Delta and Atlas-Centaur vehicles.

4.1.6   The NASA is the responsible agency for Space Shuttle flight planning and interacting all flights and users. NASA will provide for management, integration, flight operations, and control for all Shuttle flights regardless of launch or landing site used. For DOD dedicated missions DOD will provide the mission director. STS users will provide to NASA their requirements in the format and to the detail required by NASA to allow the hardware and software integration of the payload or combined upper stage payload combination. Payload mission planning and operations are the responsibility of the payload agency. Funding for these activities will be in accordance with the reimbursement sub[-]agreement referred to in 4.1.8.

[8] 4.1.7   NASA with USAF assistance will develop integrated STS logistics and training plans encompassing, JSC [Johnson Space Center], KSC, and VAFB.

4.1.8   NASA, as financial manager of the STS, is responsible for establishing an STS pricing and reimbursement policy for all non-DOD users for the STS operational era. Because of DOD's heavy investment, large usage, and the operation of VAFB, the DOD pricing and reimbursement arrangements will be jointly negotiated between NASA and DOD and will be set forth in a more detailed NASA/DOD sub[-]agreement.

4.2 DOD RESPONSIBILITIES: The DOD will plan to use the STS as the primary vehicle for placing payloads in orbit. In addition:

4.2.1   The DOD is responsible for providing to NASA those requirements affecting the Space Transportation System which are the responsibility of NASA and considered essential to meet the DOD objectives.

[9] 4.2.2   The DOD will develop the IUS including the general purpose ground support equipment. The DOD will insure that both DOD and NASA requirements are considered in the current IUS validation phase.

4.2.3   The USAF is the responsible agency for planning the mission integration of users involving DOD programs and international military activities covered by government-to-government agreements. The USAF is the focal point for providing the necessary data to NASA for the STS integration of the integrated DOD payload upper stage combination.

4.2.4   The USAF is responsible for providing the general purpose Shuttle equipment and facilities to perform the ground, launching and landing activities for all Space Shuttle operations at VAFB. The USAF will operate VAFB and plan for an initial operational capability at VAFB of 1982.

[10] *4.3 OTHER RESPONSIBILITIES*

4.3.1   The resources of both agencies which can contribute to the development, testing, production, training and operations for the STS will be used to the maximum extent possible. The plans and agreements on agency roles and responsibilities for use of these resources will be developed as required.

4.3.2   To the maximum extent possible, ground support equipment and ground operating procedures developed for use at KSC by NASA will be used by DOD at VAFB. NASA will consider the DOD operational needs at VAFB in the development of KSC equipment and procedures.

4.3.3   Each agency is responsible for providing its own payload facilities external to the launch pad area. Launch pad payload facilities will be provided by the developing agency to satisfy the normal mode of payload operations at that launch site. Other payload peculiar facilities and [ground support equipment] will be provided [11] by the agency responsible for the peculiar payload. Mutual usage of facilities will be considered where feasible and appropriate.

4.3.4   Orbiter flight control for all missions will be the responsibility of the NASA JSC Mission Control Center (MCC) unless mission traffic changes or security needs require that a DOD MCC be developed. DOD and NASA will agree on DOD peculiar security provisions required at NASA facilities. Such provisions will be subjected to negotiated reimbursement.

4.3.5   STS flight elements procured will be interchangeable for use on either agency's missions, and capable of being operated at all designated sites.

4.3.6   A procurement strategy for acquisition of STS production items will be jointly developed by NASA and the USAF for both initial investment and continuing procurement.

4.3.7   The STS will be compatible with the communications, command, and control systems of both agencies.

[12] 4.3.8   An operating/using agency(ies) mission model, to include expendable booster transition and phase-out plans, will be maintained to provide the basis for program and operational analyses and planning.

4.3.9   This Memorandum of Understanding represents the current status of agreements between NASA and the DOD on development, acquisition and operation of the Space Transportation System. Revisions and/or amendments will be made as required to maintain the currency of this document.

5.0 EFFECTIVE DATE: This Memorandum of Understanding is effective on the last day of the signatures below:

John F. Yardley
Associate Administrator
for Space Flight

John J. Martin
Assistant Secretary of the
Air Force (Research and
Development)

Date: 13 October, 1976

Date: 13 October 1976

APPROVED:

James C. Fletcher   William P. Clements, Jr.
National Aeronautics and   Deputy Secretary of Defense
Space Administration

Date: December 6, 1976   Date: 1-14-77

Document II-38

Document title: John J. Martin, Assistant Secretary of the Air Force (Research and Development), Department of Defense; John F. Yardley, NASA Associate Administrator for Space Flight; Robert N. Parker, Acting Director, Defense Research and Engineering, Department of Defense; A.M. Lovelace, NASA Deputy Administrator, "Memorandum of Agreement Between NASA and DOD: Basic Principles for NASA/DOD Space Transportation System Launch Reimbursement," March 7, 1977.

Source: NASA Historical Reference Collection, NASA History Office, NASA Headquarters, Washington, D.C.

John J. Martin, Assistant Secretary of the Air Force (Research and Development), and John F. Yardley, NASA Associate Administrator for Space Flight, signed an agreement in March 1977 that determined what the Department of Defense (DOD) would pay for shuttle launch services. For the first six years of operation, DOD would pay NASA what amounted to the incremental costs of materials and services. This agreement later caused much public discussion about the favorable price allowed DOD payloads, but it is important to note that this decision had been based on the recognition of equal involvement established in the original Space Transportation System agreement of February 1970 (Document II-21 in this volume). VAFB is the acronym for Vandenberg Air Force Base, and KSC stands for Kennedy Space Center.

---

[1]

Memorandum of Agreement Between NASA and DOD

SUBJECT: Basic Principles for NASA/DOD Space Transportation System Launch
            Reimbursement

1. The intent of this reimbursement agreement is to encourage efficient operation, early transition from expendable launch vehicles to the Space Shuttle, provide pricing stability and to establish a mutually acceptable price for STS launch and flight services. This agreement applies to DOD sponsored US payloads and DOD cooperative agreement payloads.

2. It is agreed that:
    (a) The DOD should pay a fair share price to have payloads placed in orbit by the Space Transportation System.
    (b) The price to the DOD should recognize that both the DOD and NASA will incur STS investment, operating and support costs.
    (c) NASA, as financial manager of the STS, is responsible for establishing an STS pricing and reimbursement policy for all non-DOD users which should recover

appropriate support and depreciation of investment costs. NASA will reimburse the DOD for appropriate use charges paid to NASA under NASA's reimbursement policy (reference Federal Register, dated January 21, 1977) in addition to any other changes as may be specifically required by law at the time of contract.

(d) The DOD reimbursement to NASA will be based on the costs of materials and services, to be mutually agreed upon. The DOD will provide the VAFB Space Shuttle launch support for all non-DOD users in return for provision by NASA of all Shuttle launch operations support from KSC and Shuttle flight operations support for all DOD flights. These services are projected to be of approximately equal value to each agency.

[2] 3. In line with the above, we agree that:
(a) The DOD should be charged a fixed price for the first six years of operations.
(b) The initial six year price per launch should be a realistic projected materials and services cost per launch averaged over the first six years. The materials and services costs definitions are set forth in Appendix A.
(c) There should he no recoupment of prior years costs ever or under the mutually agreed upon projected costs of part 3b.
(d) For launches after the first six years of STS operations, the price to DOD will be adjusted annually based on actual costs projected each year for materials and services. The adjustment is intended to insure meeting the goals established in parts 2a and 3c of this Agreement.
(e) The DOD and NASA agree to establish the price of STS launches for the DOD. The specific price for materials and services will take into consideration the programmatic, operational and technical services uncertainties in providing STS launch services during the six year fixed price period. The mutually agreed to price is $12.2M in FY 1975 dollars escalated according to a mutually agreed to economic index.

4. This agreement is contingent on the DOD meeting the VAFB STS launch site activation schedule agreed to in the MOU dated January 14, 1977, that NASA meet the [initial operational capability] dates for the KSC launch site and the Shuttle, and that NASA provide an adequate orbiter fleet.

5. DOD agrees to reimburse NASA for STS launches in the fiscal year prior to the fiscal year of launch and at least twelve months prior to the planned launch date. The reimbursement will be made in dollars escalated to the fiscal year of payment (reference paragraph 3e above). If after payment [3] for a DOD launch, the launch is slipped or cancelled, the DOD will receive credit on a future launch. The DOD and NASA will develop a launch schedule three years prior to launch based on the most probable launch requirements. The schedule will be updated annually.

6. This agreement becomes an integral part of the NASA/DOD Memorandum of Understanding on Management and Operations of the Space Transportation System dated January 14, 1977.

John J. Martin
Assistant Secretary of the Air
Force (Research and Development)
Department of Defense

John F. Yardley
Associate Administrator
for Space Flight
National Aeronautics
and Space Administration

Date: 7 MAR 1977

MAR 7 1977

Robert N. Parker
Acting Director, Defense Research
and Engineering
Department of Defense

A.M. Lovelace
Deputy Administrator
National Aeronautics
and Space Administration

Date: 7 MAR 1977   MAR 7 1977

****

[no pagination]

Appendix A

The total of all costs incurred by the government for the procurement of all expended hardware; refurbishment hardware and all flight spares and provisions excluding external tank propellants, the maintenance and support costs included in the $12.2M are:

Space Shuttle Main Engine (SSME)
    Refurbishment
    Spares
    Engine Overhaul and Test
    Transportation

Solid Rocket Boosters (SRB's)
    Solid Propellants
    Refurbishment of SRB's
    Spares
    Procurement of Replacement Units
    Transportation

External Tank (ET)
    Production
    Spares
    Transportation (excludes West Coast Port to Launch Site Transportation)

System Support
    ET, SRB and SSME Sustaining Engineering Support Services

Orbiter Spares
    Replenishment and Transportation of LRU's and Shop Replaceable Units to
    Support Orbiter [Hardware] Maintenance and Replacement

Crew GPE
    Replacement and Replenishment Hardware and Field and Maintenance Support
    for all Crew Related GPE

Contract Administration
    Costs Associated with Contract Administration of all Shuttle Direct Support
    Contractors

Document II-39

Document Title: George M. Low, NASA Deputy Administrator, Co-Chairman, Aeronautics and Astronautics Coordinating Board (AACB), and Malcolm R. Currie, Director of Defense Research and Engineering, Department of Defense, Co-Chairman, AACB, "Joint NASA/DOD Position Statement on Space Shuttle Orbiter Procurement," January 23, 1976.

Source: Documentary History Collection, Space Policy Institute, George Washington University, Washington, D.C.

The initial launch rate for the shuttle was set at 60 flights per year, with 40 from Kennedy Space Center and 20 from Vandenberg Air Force Base. NASA soon determined, however, that this flight rate was unachievable without a five-orbiter fleet, and in 1976 the space agency began to ask for a fifth orbiter. NASA expected the Air Force to pay for this vehicle. Department of Defense (DOD) leadership refused to acknowledge that its mission dictated the need for the fifth orbiter and feared it would have to procure the vehicle on its own. NASA and DOD agreed that a fifth orbiter was needed, but both agencies deferred the decision to budget funds for the fourth and fifth orbiters, as well as the decision on who would pay for them. Ultimately, only four orbiters were built initially. A fifth orbiter was not built until after the loss of the Challenger.

[no pagination]

Joint NASA/DOD Position Statement on Space Shuttle Orbiter Procurement

The National Aeronautics and Space Administration and the Department of Defense agree that five Space Shuttle Orbiters are needed to meet our national traffic model requirements. Orbiters are funded by NASA within the [design, development, testing, and evaluation] and [the] production programs. Neither agency has budgeted funds for the remaining two Orbiters. While this is a current interagency Space Shuttle issue, NASA has evolved a production plan which does not require an FY 1977 funding increment. Therefore, NASA and DOD agree to work together to resolve this issue as part of the FY 1978 budget cycle activities.

George M. Low	Malcolm R. Currie
Deputy Administrator	Director of Defense Research
National Aeronautics and	and Engineering
Space Administration	Department of Defense
Co-Chairman, AACB	Co-Chairman, AACB
23 Jan 1976	January 23, 1976
Date	Date

****

Plan for NASA-DOD Orbiter
Procurement Decision

1.	Fletcher-Clements Exchange of Letters	Dec 75/Jan 76
2.	Currie/Low sign position paper	Jan 76
3.	Currie/Low prepare detailed request for NASA/DOD issues paper to be prepared by STS Committee	Jan 76
4.	STS Committee address the following issues:	By Aug 76
	a. Verify need for 5 orbiters	
	b. Develop detailed budget plans, using various delivery assumptions, and assuming either NASA or DOD funding	
	c. Prepare draft issues paper for Fletcher-Rumsfeld meeting	
5.	STS Committee prepare monthly progress reports addressed to Currie and Low. Currie and Low meet as necessary	Feb, Mar, Apr, May, Jun, Jul
6.	Fletcher-Rumsfeld meeting	Aug 76
7.	If Fletcher-Rumsfeld cannot agree on which agency funds orbiter, prepare joint Presidential issues paper	Aug 76
8.	Fletcher-Rumsfeld-Lynn discuss joint issues paper	Aug 76
9.	Fletcher-Rumsfeld-Lynn meet with President	Sep 76

Document II-40

Document title: General Robert T. Marsh, Commander, Air Force Systems Command, to General Charles A. Gabriel, Chief of Staff, USAF, August 5, 1982, with attachment.

Source: NASA Historical Reference Collection, NASA History Office, NASA Headquarters, Washington, D.C.

By the time the Space Shuttle became operational, it had changed considerably from what the Air Force had originally anticipated. The Air Force faced launch costs totaling nearly $300 million per flight. Air Force Systems Command Commander General Robert T. Marsh, who was in charge of Air Force participation in the Space Transportation System, felt it was necessary to inform Air Force Chief of Staff General Charles A. Gabriel of rising shuttle costs. His information package provides a detailed comparison of launch costs for a variety of Titan and Space Shuttle vehicle mixes.

[no pagination]

General Charles A. Gabriel   5 AUG 1982
Chief of Staff
United States Air Force
Washington, DC 20330

Dear Chief

Although many of us are familiar with projected costs of conventional weapons systems, the understanding of space systems and support costs, as well as future predictions, is not as clear. To enhance this understanding, I've provided a macro-perspective of where launch costs in the Shuttle era are headed.

I am emphasizing launch costs in this package because I want to alert you to the significant Air Force requirements we will see when the STS at Vandenberg AFB and [the Consolidated Space Operations Center] become operational. The effective cost to ride [the] Shuttle will be about $300M per launch in the late 1980s. These costs are based on an optimistic launch plan, and due to the high fixed costs involved, reducing the number of flights will increase the cost per flight.

The amounts in this package do not reflect our approved program. They are merely intended to convey the message that costs for access to space are increasing. Although most of our near-year requirements are founded, I think you'll agree that we face a significant budgeting challenge in the out-years when these systems become operational.

I think the attachments help generate a clearer understanding of the space arena. We will provide additional information should you desire.

Sincerely

ROBERT T. MARSH, General USAF   2 Atch
Commander   1. Titan and Shuttle Costs Per Flight
                                            2. Launch Costs w/ Investments
                                               Amortized

TOTAL AMORTIZED LAUNCH PROFILE WITH AMORTIZED INVESTMENTS COSTS
(THEN YEAR $M)

SEGMENT	76	77	78	79	80	81	82	83	84	85	86	87	88	89	90	91
HARDWARE	55.6	58.2	66.2	65.2	84.2	100.4	145.7	220.9	409.3	380.3	266.0	290.2	408.2	725.5	646.1	743.3
LAUNCH SERVICES	19.9	24.1	28.3	29.2	34.8	38.9	42.0	46.3	107.8	123.9	132.1	133.0	137.6	149.8	163.4	178.1
RANGE SUPPORT	63.1	68.9	74.0	80.7	88.9	98.1	107.2	118.4	244.8	271.8	256.0	305.4	309.3	336.9	367.4	400.4
SCF SUPPORT	65.7	57.7	64.9	69.5	96.1	90.3	99.2	114.1	144.5	225.2	167.1	255.2	243.7	265.4	289.5	315.4
ORBITER FLIGHT CHARGE[1]	-	-	-	-	-	-	-	-	106.5	193.0	383.6	597.0	520.8	1063.5	1082.2	1263.0
ORBITER FLIGHT CHARGE[2]	-	-	-	-	-	-	-	-	-	-	165.2	258.0	224.8	460.5	467.6	547.5
STS OPERATIONS	-	-	-	-	-	-	-	-	120.0	314.2	375.9	424.1	445.8	444.2	484.3	527.9
CSOC OPERATIONS	-	-	-	-	-	-	-	-	15.9	57.1	245.9	288.1	316.9	376.3	410.4	447.2
STS SECURITY	-	-	-	-	-	-	-	-	36.9	40.5	45.2	49.2	51.3	55.9	61.0	66.5
AMORTIZED COSTS[3]	-	-	-	-	-	-	-	89.9	89.8	109.4	263.3	245.7	245.7	245.7	245.7	245.7
TOTAL	204.3	208.9	233.4	244.6	304.0	327.7	394.1	589.6	1275.5	1715.4	2300.3	2845.9	2904.1	4123.7	4035.6	4735.0
DOD LAUNCHES[4]	2	2	2	2	2	2	2	3	6	7	7	10	8	15	14	15
COST P/LAUNCH WITH AMORTIZED COSTS	102.2	104.5	116.7	122.3	152.0	163.9	197.1	196.5	*	*	328.6	284.6	363.0	274.9	288.3	315.7

1 BASED ON $16.0M/FLT FY 84-85 AND $20.8M/FLT FY 86-91 (CONSTANT FY 75 $).
2 DELTA COST IF DOD REQUIRED TO PAY $29.8/FLT FY 86-91 (CONSTANT FY 75 $).
3 INCLUDED IUS DEVELOPMENT, AND CSOC, DELTA SYSTEMS MODERNIZATION, AND STS-VAFB INVESTMENT.
4 LAUNCH FORECASTS ARE HISTORICALLY OPTIMISTIC. REDUCTION IN LAUNCHES WILL INCREASE COST PER FLIGHT.

*NOT APPLICABLE: OVERLAP YEARS FOR TITAN AND SHUTTLE.

TOTAL AMORTIZED LAUNCH PROFILE WITH AMORTIZED INVESTMENTS COSTS
(THEN YEAR $M)

SEGMENT FY	84	85	86	87	88	89	90	91
HARDWARE	130.0	155.1	266.0	290.3	408.2	725.5	464.1	743.3
LAUNCH SERVICES	55.4	67.0	132.1	133.0	137.6	149.8	163.4	178.1
RANGE SUPPORT	244.8	271.8	256.0	305.4	309.3	336.9	367.4	400.4
SCF SUPPORT	144.5	225.2	167.1	255.2	243.7	265.4	289.5	315.4
ORBITER FLIGHT CHARGE[1]	106.5	193.0	383.6	597.0	520.8	1063.5	1082.2	1263.0
ORBITER FLIGHT CHARGE[2]	0	0	165.2	258.0	224.8	460.5	467.6	547.5
STS OPERATIONS	120.0	314.2	375.9	424.1	445.8	444.2	484.3	527.9
CSOC OPERATIONS	15.9	57.1	245.9	288.1	316.9	376.3	410.4	447.2
SECURITY	36.9	40.5	45.2	49.2	51.3	55.9	61.0	66.5
AMORTIZED COSTS[3]	89.8	109.4	263.3	245.7	245.7	245.7	245.7	245.7
TOTAL (THEN YR $)	943.8	1433.3	2300.3	2845.9	2904.1	4123.7	4035.6	4735.0
(FY 84 $)	943.8	1315.8	1936.9	2197.0	2056.1	2680.4	2405.2	2590.0
DOD LAUNCHES[4]	3	5	7	10	8	15	14	15
COST/FLIGHT (THEN YR $)	314.6	286.7	328.6	284.6	363.0	274.9	288.3	315.7
(FY 84)	314.6	263.2	276.7	219.7	257.0	178.7	171.8	172.7

AVERAGE COST PER FLIGHT: $307.1M (THEN YR $), 232.0M (FY 84)

1   BASED ON $16.0M/FLT FY 84-85 AND $20.8M/FLT FY 86-91 (CONSTANT FY 75 $).
2   DELTA COST IF DOD REQUIRED TO PAY $29.8/FLT FY 86-91 (CONSTANT FY75 $).
3   INCLUDED IUS DEVELOPMENT, AND CSOC, DELTA SYSTEMS MODERNIZATION, AND STS-VAFB INVESTMENT.
4   LAUNCH FORECASTS ARE HISTORICALLY OPTIMISTIC. REDUCTION IN LAUNCHES WILL INCREASE COST PER FLIGHT.

AMORTIZED COSTS BACK-UP[1]
(THEN YEAR $M)

SEGMENT FY	83	84	85	86	87	88	89	90	91
IUS AMORTIZED[2]	72.2	72.2	72.2	72.2	72.2	72.2	72.2	72.2	72.2
STS SECURITY AMORTIZED[3]	17.6	17.6	17.6	17.6	17.6				
DATA SYSTEMS MOD AMORTIZED[4]		19.6	19.6	19.6	19.6	19.6	19.6	19.6	19.6
STS VAFB AMORTIZED[5]			—	93.2	93.2	93.2	93.2	93.2	93.2
			—	60.7	60.7	60.7	60.7	60.7	60.7
AMORTIZED COSTS[6]	89.8	943.8	109.4	263.3	263.3	245.7	245.7	245.7	245.7

1   AMORTIZING BEGINS UPON IOC (STRAIGHT LINE)
2   TOTAL DEVELOPMENT ($722.2M)/PROJECTED LIFE (10 YEARS)=$72.2M/YR
3   TOTAL DEVELOPMENT ($87.9M)/PROJECTED LIFE (5 YEARS)=$17.6M/YR
4   TOTAL DEVELOPMENT ($391.5M)/PROJECTED LIFE (20 YEARS)=$19.6M/YR
5   TOTAL DEVELOPMENT ($2797.9M)/PROJECTED LIFE (30 YEARS)=$93.2M/YR
6   TOTAL DEVELOPMENT ($1213.5M)/PROJECTED LIFE (20 YEARS)=$60.7M/YR

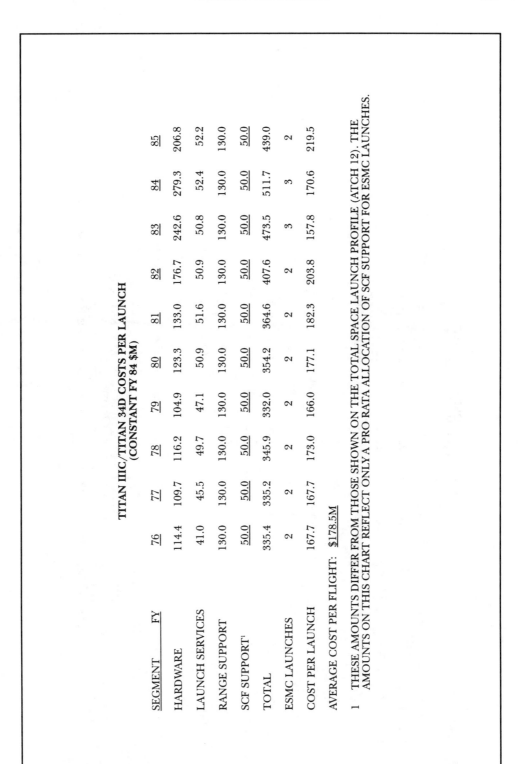

TITAN IIIC/TITAN 34D COSTS PER LAUNCH
(CONSTANT FY 84 $M)

SEGMENT FY	76	77	78	79	80	81	82	83	84	85
HARDWARE	114.4	109.7	116.2	104.9	123.3	133.0	176.7	242.6	279.3	206.8
LAUNCH SERVICES	41.0	45.5	49.7	47.1	50.9	51.6	50.9	50.8	52.4	52.2
RANGE SUPPORT	130.0	130.0	130.0	130.0	130.0	130.0	130.0	130.0	130.0	130.0
SCF SUPPORT[1]	50.0	50.0	50.0	50.0	50.0	50.0	50.0	50.0	50.0	50.0
TOTAL	335.4	335.2	345.9	332.0	354.2	364.6	407.6	473.5	511.7	439.0
ESMC LAUNCHES	2	2	2	2	2	2	2	3	3	2
COST PER LAUNCH	167.7	167.7	173.0	166.0	177.1	182.3	203.8	157.8	170.6	219.5

AVERAGE COST PER FLIGHT: $178.5M

1   THESE AMOUNTS DIFFER FROM THOSE SHOWN ON THE TOTAL SPACE LAUNCH PROFILE (ATCH 12). THE AMOUNTS ON THIS CHART REFLECT ONLY A PRO RATA ALLOCATION OF SCF SUPPORT FOR ESMC LAUNCHES.

STS LAUNCH COST PER FLIGHT*
(CONSTANT FY 84 $M)

SEGMENT FY	84	85	86	87	88	89	90	91
HARDWARE	130.0	142.4	224.0	224.0	289.0	471.6	276.6	406.6
LAUNCH SERVICES	55.4	61.5	111.2	102.7	97.4	97.4	97.4	97.4
RANGE SUPPORT	244.8	249.5	215.6	235.8	219.0	219.0	219.0	219.0
SCF SUPPORT	144.5	206.7	140.7	197.0	172.5	172.5	172.5	172.5
ORBITER FLIGHT CHARGE[1]	106.5	177.5	322.7	461.0	368.8	691.5	645.4	691.5
ORBITER FLIGHT CHARGE[2]	0	0	139.3	199.0	159.2	298.5	278.6	298.5
STS OPERATIONS	120.0	288.4	316.5	327.4	315.6	288.7	288.7	288.7
CSOC OPERATIONS	15.9	52.4	207.0	222.4	244.6	244.6	244.6	244.6
SECURITY	36.9	37.2	38.1	38.0	36.3	36.3	36.3	36.3
TOTAL	854.0	1215.6	1715.1	2007.3	1902.4	2520.1	2259.1	2455.1
DOD LAUNCHES[3]	3	5	7	10	8	15	14	15
COST PER FLIGHT	284.7	243.1	245.0	200.7	237.8	168.0	161.4	163.7

AVERAGE COST PER FLIGHT: $218.8M

* RECURRING COSTS ONLY. TREATS DEVELOPMENTS/ACQUISITIONS AS SUNK COSTS.
1 BASED ON $16.0M/FLT FY 84-85 AND $20.8M/FLT FY 86-91 (CONSTANT FY 75 $).
2 DELTA COST IF DOD REQUIRED TO PAY $29.8/FLT FY 86-91 (CONSTANT FY 75 $).
3 LAUNCH FORECASTS ARE HISTORICALLY OPTIMISTIC. REDUCTION IN LAUNCHES WILL INCREASE COST PER FLIGHT.

TOTAL SPACE PROGRAMS
(FY 84 $B)

SECTOR	84*	85	86	87	88	89**	90	91	92	93	94	95
Current												
Space***	5.4	6.3	6.3	5.9	6.3	6.4	6.6	6.7	6.8	7.0	7.1	7.3
Launch	0.9	0.9	1.0	1.0	1.2	1.2	1.2	1.3	1.3	1.3	1.3	1.4
Support	0.9	1.0	0.9	0.9	0.9	0.9	0.9	0.9	0.9	0.9	1.0	1.0
Subtotal	7.2	8.2	8.4	7.8	8.4	8.5	8.7	8.9	9.0	9.2	9.4	9.6
Future												
Space	-	-	-	-	0.1	0.2	0.6	0.8	0.8	1.0	1.1	1.2
Launch	0.1	0.3	0.2	0.2	0.3	0.7	1.0	0.8	0.6	0.5	0.2	0.1
Support	–	–	–	–	–	–	0.1	0.1	0.1	0.3	0.3	0.3
Subtotal	0.1	0.3	0.2	0.2	0.4	0.9	1.7	1.7	1.5	1.8	1.6	1.6
TOTAL	7.3	8.5	8.6	8.0	8.8	9.4	10.4	10.6	10.5	11.0	11.0	11.2

*   Current Systems—FY 84 AF POM through FY 88.
**   Current Systems—FY 89 and out assumes 2% per year real growth.
***   Includes PE 34111F.

Document II-41

Document title: Lt. General Richard C. Henry, Commander, Air Force Space Division, to General Robert T. Marsh, Commander, Air Force Systems Command, March 4, 1983.

Source: NASA Historical Reference Collection, NASA History Office, NASA Headquarters, Washington, D.C.

By the early 1980s, the Space Shuttle program was considerably behind schedule and was not meeting its promised flight rates or cost targets. Various leaders in the U.S. Air Force were increasingly uneasy with relying on the shuttle. In March 1983, Lt. General Richard C. Henry, in this letter to General Robert T. Marsh, expressed growing concern that carrying humans aboard a vehicle designed to merely deliver payloads to orbit created an unnecessary expense. This indicated the changed status of human spaceflight initiatives in the military, which was later reflected in the Department of Defense's (DOD) position on the proposed NASA space station. Henry's letter also gave a broad overview of a proposed military launch strategy, which eventually evolved into what was called a "mixed fleet" after the Challenger *accident.*

---

[1]   4 March 1983

General Robert T. Marsh
AFSC/CC
Andrews AFB, DC 20334

Dear General Marsh

Last year the AF committed to using [the] Space Transportation System exclusively and according to current planning, we will close down the Titan production line this spring and expand all Titans and Atlas' in the 1987, 1988 time frame. I believe this plan is seriously deficient from the DOD standpoint both operationally and economically.

Current estimates of STS mission model requirements have been reduced to where they can be satisfied with a launch capability of about 20 per year, 16 at KSC and 4 at VAFB. Thus, there is a debate underway as to whether a fifth orbiter should be procured. This situation coupled with a phase of Expandable Launch Vehicles, might lead to (an economically irreversible) loss of *all* U.S. capabilities to produce space launch vehicles in the 1985 time period.

A four orbiter only fleet, experiencing problems similar to those of Challenger, would develop a backlog of launches that would take months to years to work off. This presents a considerable threat to the continued vitality of the national space program and in particular, could impact national security through inadequate launch support of priority DOD spacecraft.

In the past, it has been argued that the shuttle would achieve economy by launch rate. A high launch rate is not materializing, and is unlikely to come forth; therefore, we should seek alternative ways to achieve best return on investment. An example is the acceptance of orbiter refurbishment and checkout at KSC prior to Vandenberg launch as a permanent procedure to restrict work force build up on the west coast. Another example is to re-look at the economics of using the shuttle on missions where its unique capabilities are not needed.

The current cost estimate ($FY 83) for shuttle launch to place a payload in geosynchronous equatorial orbit (GEO) is 165 million dollars. Similar estimates ($FY 83) for a commercial version of the Titan/Centaur and a modified Atlas/Centaur are 125/115 million dollars and 120/90 million dollars respectively, where the first number includes the amortization of development costs over nine launches and the second is the cost per launch thereafter. Launch of a stretch version of the Titan/Centaur is estimated at 145/120 million dollars. The major driver in the higher STS costs is the cost of carrying man on a mission which does not need man. The costs shown here for expendable vehicles, launched, [2] are slightly less conservative than we would have used in the past (possibly by 10 or 15 million per launch). However, the important point is that the GEO mission can be accomplished at less cost with an expandable booster. I have not included the not insignificant costs to our spacecraft to enable their carriage on a manned vehicle (the orbiter).

Assuming that commercialization of expandable launch vehicles does occur, I believe their most important use by DOD would be the transport of spacecraft to GEO, namely [the Defense Support Program], MILSTAR, [Defense Satellite Communication System] III and other special missions.

From the DOD standpoint, either the Titan/Centaur or the modified Atlas/Centaur launch vehicles would meet most of the performance requirements through the 1980's. Thus, a DOD commitment to commercial launches of either vehicle could provide an expendable launch vehicle capability for critical DOD programs through the late 1980's (in the longer term, the growth Titan/Centaur presents the option for launching larger payloads than does Atlas). DOD launch rate requirements for this time period, are expected to be about four or five per year.

Another opportunity for DOD participation in commercialization of expandable launch vehicles exists for the Delta class launch vehicles. The GPS [Global Positioning System] and the DMSP [Defense Meteorological Satellite Program] programs are currently being launched on Atlas. Both payloads are relatively small and lightweight and, therefore, both require manifesting with other payloads for effective Shuttle launching. To date, no other appropriate DOD payload has been found for manifesting with either GPS or DMSP. Although manifesting with non-DOD payloads may prove feasible, single payload launches, when needed, are necessary for effective systems operation. Thus, it would be highly desirable to have a dual capability for launching these payloads; the Satellite replacement rate for the GPS and DMSP programs is expected to be about three or four per year.

We estimate that the 20 flight per year STS requirement would include 6 flights per commercial GEO satellites and 7 for government. If commercial launch vehicles captured these flights, the yearly STS flight rate would be reduced to about 7. Most, if not all, of these would require the unique capabilities of the Shuttle.

100 flights have been postulated for the useful life of a Shuttle. Thus, a four orbiter fleet flying 20 flights per year could be expected to wear out in about 2 decades. Reducing Shuttle flights to those for which it has unique capabilities could significantly expand the life of the fleet.

The orbiter is necessarily an essential element of a space station program which NASA proposes to initiate. Therefore, if the nation embarks on a space station program in the near future, it will be argued that more orbiters should be procured for the construction and sustaining of the station. This would be an investment of about $2 billion per orbiter above and beyond the non[-]recurring and recurring space station costs.

The question of requirements for a space station is now under debate. I suggest that this debate is premature. The more fundamental question is the utility of [3] man in space and whether we first, need him in a hostile environment; then, if we do, how can we sustain him in a more affordable way than we do today. I believe strongly that these questions can be addressed and answered with the existing four orbiter fleet on spacelab type missions.

In summary, I believe that the orbiter is a marvelous machine, but it is better used for those missions where the utility of man is clear or needs further exploration. It is clear that man is not needed on the transport mission to GEO and is, in fact, the more expensive alternative. I recommend an investment strategy in a mixed fleet, preferably with commercialization. The primary DoD mission is on orbit, not in getting there. I recommend the government endorse commercialization, and commit to commercial launches to GEO. This will assure the success of commercialization.

I recognize that these are issues that transcend the Air Force and DoD, and need NASA, OMB and National Security Council involvement, but I suggest that Air Force leadership is not inappropriate.

I urge your serious consideration of my recommendations before we burn our bridges behind us and stand ready to give any additional support that you may need.

Warm regards

Richard C. Henry
Lt. General, USAF
Commander

Document II-42

Document title: Caspar Weinberger, Secretary of Defense, Memorandum for Secretaries of the Military Departments; Chairman of the Joint Chiefs of Staff; Under Secretaries of Defense; Assistant Secretaries of Defense; General Counsel, "Defense Space Launch Strategy," February 7, 1984, with attached: "Defense Space Launch Strategy," January 23, 1984.

Source: NASA Historical Reference Collection, NASA History Office, NASA Headquarters, Washington, D.C.

The Department of Defense continued to support the Space Shuttle despite reservations about its performance and reliability. The Air Force, however, wanted a back-up expendable launch vehicle until the shuttle's problems had been solved. In early 1984, Secretary of Defense Caspar Weinberger issued a directive that established a need for a "complementary expendable launch vehicle" to supplement the Space Shuttle. The vehicle developed to meet this requirement became known as the Titan IV.

Memorandum for Secretaries of the Military Departments
Chairman of the Joint Chiefs of Staff
Under Secretaries of Defense
Assistant Secretaries of Defense
General Counsel

SUBJECT: Defense Space Launch Strategy

On 23 January 1984, I approved the attached Defense Space Launch Strategy. The approach described in this document will be used to guide future defense space launch planning. Please ensure. maximum distribution to all those affected within your departments and agencies.

Caspar Weinberger

Attachment

****

[1]

Defense Space Launch Strategy

POLICY

Defense space launch strategy has been developed in response to validated DoD assured space launch requirements and implements the launch policies contained in the National Space Policy and the Defense Space Policy. The National Space Policy identifies the Space Transportation System (STS) as the primary U.S. government space launch vehicle, but recognizes that unique national security requirements may dictate the development of special purpose launch capabilities. The Defense Space Policy states that:

"While affirming its commitment to the STS, DoD will ensure the availability of an adequate launch capability to provide flexible and operationally responsive access to space, as needed for all levels of conflict, to meet the requirements of national security missions."

REQUIREMENTS

The DoD has a validated requirement for an assured launch capability under peace, crisis and conflict conditions. Assured launch capability is a function of satisfying two specific requirements: the need for complementary launch systems to hedge against unforeseen technical and operational problems, and the need for a launch system suited for operations in crisis and conflict situations. While DoD policy requires assured access to space across the spectrum of conflict, the ability to satisfy this requirement is currently unachievable if the U.S. mainland is subjected to direct attack. Therefore, this launch strategy addresses an assured launch capability only through levels of conflict in which it is postulated that the U.S. homeland is not under direct attack. Additional survivability options beyond an assured launch capability are being pursued to ensure sustained operations of critical space assets after homeland attack.

STRATEGY

Near Term: Existing Defense space launch planning specifies that DoD will rely on four unique, manned orbiters for sole access to space for all national security space systems. DoD studies and other independent evaluations have concluded that this does not represent an assured, flexible and responsive access to space. While the DoD is fully committed to the STS, total reliance upon the STS for sole access to space in view of the technical and operational uncertainties, represents an unacceptable national security risk. A complementary system is necessary to provide high confidence of access to space particularly since the Shuttle will be the only launch vehicle for all U.S. space users. In addition, the limited number of unique, manned Shuttle vehicles renders them ill-suited and inappropriate for use in a high risk environment.

The solution to this problem must be affordable and effective and yet offer a high degree of requirements satisfaction, low technical risk, and reasonable schedule availability. Unmanned, expendable launch vehicles meet these criteria [2] and satisfy DoD operational needs for a launch system which complements the STS and extends our ability to conduct launch operations further into the spectrum of conflict. These systems can provide unique and assured launch capabilities in peace, crisis and conflict levels short of general nuclear war. These vehicles are designed to be expendable and the loss of a single vehicle affects only that one mission and would not degrade future common, national launch capabilities by the loss of a reusable launch system.

The President's policy on the Commercialization of Expendable Launch Vehicles [ELVs] states that the goals of the U.S. space launch policy are to ensure a flexible and robust U.S. launch posture, to maintain space transportation leadership, and to encourage the U.S. private sector development of commercial launch operations. Consistent with this policy, the DoD will pursue the use of commercially procured ELVs to meet its requirements for improving its assured launch capabilities. For requirements that cannot be satisfied by commercially available ELVs, unique DoD developments may be undertaken for special purpose launch capabilities.

The STS will remain the primary launch system for routine DoD launch services. Unmanned, expendable launch vehicles represent a complementary capability to the STS and will be maintained and routinely launched to ensure their operational viability. To accomplish this, selected national security payloads will be identified for dedicated launch on ELVs, but will remain compatible with the STS.

Long Term: While commercial expendable launch vehicles represent an available solution to the unique DoD space launch requirements into the early-1990s, the need for other DoD launch capabilities to meet requirements beyond that must be evaluated and validated. This effort must be initiated immediately in order to ensure that future national security space missions are not constrained by inadequate launch capability. The evaluation should examine potential DoD launch requirements, such as the need for a heavy lift vehicle, and should attempt to take maximum advantage of prior investments in the U.S. launch vehicle technology base.

IMPLEMENTATION

As Executive Agent for launch vehicles, the Air Force will take immediate action to acquire a commercial, unmanned, expandable launch vehicle capability to complement the STS with a first launch availability no later than FY 1990. These vehicles must provide a launch capability essentially equal to the original STS weight and volume specifications.

In addition, the Air Force, in conjunction and coordination with other Services, affected agencies and departments, will:

a)   identify specific national security systems that will be used on the commercially procured expendable launch vehicles and the proposed peacetime launch rate required to maintain an operationally responsive posture.

[3] b)   develop a comprehensive space launch plan to meet projected national security requirements through the year 2000. This strategy will be submitted to the Secretary of Defense for approval and validation.

The Defense Space Launch Strategy will be reflected in the FY-86 Defense Guidance Plan.

Document II-43

Document title: Charles W. Cook, Executive Secretary, Defense Space Operations Committee, Memorandum for Defense Operations Committee (DSOC) Principals, "DoD Position on Shuttle Issues," November 19, 1984.

Source: NASA Historical Reference Collection, NASA History Office, NASA Headquarters, Washington, D.C.

The Defense Space Operations Committee was a Department of Defense (DOD)-wide internal policy-making and coordination group composed of the leading space individuals in each military service, the Office of the Secretary of Defense, and the organization of the Joint Chiefs of Staff. President Reagan had directed NASA and DoD to determine what steps were necessary to make the shuttle fully operational. The Defense Space Operations Committee was the mechanism to coordinate the DOD definition of the steps necessary to attain operational status. On October 19, 1984, the committee principals were briefed on the issues identified by the Air Force. Their comments were included in the operational plan. The committee met again on October 29, 1984, and the recommendations were finalized on November 19, in a memorandum representing the first coherent statement by DOD of what it meant by an "operational Space Transportation System." DOD felt that a number of require-ments for the Space Transportation System had not been adequately addressed by NASA, and the out-standing issues were stated as changes needed in the Space Transportation Master Plan.

[no pagination]   19 November 1984

Memorandum for Defense Operations Committee (DSOC) Principals

SUBJECT: DoD Position on Shuttle Issues

Attached is a revised copy of the DoD Position resulting from the DSOC meeting of 29 October 1984. Changes have been incorporated to reflect the comments received. I would like to touch base with each of you personally early next week to go over the final position.

CHARLES W. COOK
Executive Secretary
DSOC

2 Attachments
1. Revised DoD Position
2. Summary—Issues Not Discussed

****

[no pagination]

Memorandum for the Defense Space Operations Committee (DSOC)

SUBJECT: DoD Position on Shuttle Issues

During the 19 October 1984 meeting of the DSOC, DoD positions were established on several key Shuttle issues.

Attachment 1 summarizes the DoD positions on issues discussed during the DSOC meeting. Attachment 2 summarizes less controversial issues which were coordinated with you.

I am requesting that the Executive Secretary coordinate with NASA in revising the Space Transportation System Master Plan to reflect DoD positions prior to the Master Plan being approved.

2 Attachments
1. Summary—Issues Discussed
2. Summary—Issues Not Discussed

****

[1]

Attachment 1

Defense Space Operations Committee, 29 October 1984 DoD Position on Key STS Master Plan Issues

Continued Orbiter Production

The Space Transportation System (STS) Master Plan must include a viable, long-term plan for the Space Shuttle System. Since the STS is the primary means of transportation to space for all U.S. programs, including national security programs, it is essential that the STS Master Plan contain a NASA program for providing continued orbiter capability.

The current NASA budget and financial program does not include plans for a fifth orbiter, follow-on orbiter, continuing spares production, requalifying and restarting production lines, or qualifying the orbiter fleet beyond 100 flights. In view of the national policy for the use of the Shuttle system, the plan would not be complete without a specific program for viability of the orbiter fleet through continued orbiter production. Therefore, the DoD takes the position:

"In accordance with National Policy, the STS is the primary means of access to space for all U.S. programs, including National Security programs. The STS Master Plan should include provisions for continued orbiter fleet capability. Specifically, NASA should develop definitive plans with adequate budgetary funding for continuing spares production and qualification of the orbiter fleet beyond the current 100 flights. Since the loss of an orbiter would have a significant impact on the STS overall mission capability, NASA should develop a plan to address that contingency."

Interoperability of Orbiters

Space launch operational flexibility is restricted by the fact that each of the orbiters in the current fleet has different characteristics and capabilities. [2] Therefore, the DoD takes the position:

"The STS Master Plan should include Provisions to increase interoperability of the orbiter fleet. Specifically, additional orbiters should be fully capable of meeting all existing and documented DOD mission requirements. NASA should modify existing orbiters as follows:

(1) Orbiter 103 modified to be Centaur-capable.
(2) Orbiter 099 upgraded to allow operating from Vandenberg.

All launch facilities should be interoperable with all orbiters. Therefore, the Air Force and NASA should modify the shuttle launch facilities to accommodate the configurations of all orbiters."

Payload Performance (Shuttle Lift Capability)

The STS Master Plan should include the Level I requirement of 32,000 pounds of payload lift capability for a Vandenberg Reference Mission 4 or equivalent. Of concern to the DoD is the fact that even with filament-wound-case solid rocket boosters and main engines operating at 109% thrust, maximum performance is approximately 28,000 pounds of payload to low earth orbit. Additionally, there is not a specific program (aside from hopeful flight experience, demanifesting, etc.) to attain the 4,000 pounds needed to reach the NASA "goal" of 32,000 pounds. Therefore, the DoD takes the position that:

"The STS Master Plan should include a definitive technical plan with appropriate budgetary funding which, with a high degree of confidence, will meet the commitment of a lift capability of 32,000 pounds for Reference Mission 4 or equivalent."

Orbiter Crossrange Capability

The Shuttle orbiter crossrange requirement of 1100 nautical miles cannot be met with the current design. This shortfall will prevent a Vandenberg Shuttle launch from aborting once-around back to Vandenberg. Current orbiter capability is approximately 800 nautical miles. This impacts DoD payloads by involving increased exposure to landing at abort and contingency landing sites outside the Continental United States.

Complying with the 1100-mile crossrange requirement would appear to entail a costly orbiter redesign. [3] Therefore, the DoD takes the position:

"The Level I crossrange requirement of 1100 miles remains unchanged. The STS Master Plan should include extension of the current Shuttle crossrange beyond 800 miles. This extension should be accomplished through flight test and analysis. Until the crossrange requirement of 1100 miles can be met, NASA should develop definitive plans with adequate budgetary funding for a capability to provide air transportation of payload and orbiter from contingency landing sites to the launch site. NASA should also assure that the design of any future orbiters or Thermal Protection System (TPS) meet the needed 1100 mile crossrange capability."

Orbiter/Cargo Transportation Capabilities

The STS Master Plan should include specific steps to be taken to provide payload and orbiter transportation capabilities.
The DoD takes the position:

"NASA should provide a second Shuttle Carrier Aircraft (SCA) and should install refueling capabilities on both SCA. The Air Force should plan to procure outside airborne cargo transportation capability. Both NASA and the Air Force should develop definitive plans with adequate budgetary funding to accomplish these tasks."

It is noted that the Air Force is examining a way that they may provide a Civil Reserve Air Fleet (CRAF) 147 that could be modified by NASA for use as a backup SCA.

Orbiter Bay Contamination

Since orbiter bay contamination could have a significant effect on the design of future payloads, the orbiter bay contamination environment must be accurately characterized. The DoD takes the position:

"The STS Master Plan should reflect the NASA and the Air Force Contamination Working Group plan to provide pre-flight cleanliness specifications and procedures, and inflight measurements to define the orbiter bay environment. NASA should provide quantitative contamination data to the payload community for design consideration."

[4] *Future Shuttle Management*

On the issue of future management of the Space Transportation System (STS) the DoD position is:

"The status quo with the current NASA-led, joint NASA/DoD management arrangement is the preferred management option for the foreseeable future. NASA should identify and separately account for the Shuttle budget (e.g. budgetary fencing) to distinguish that funding from other NASA Programs. Transfer of the STS to another government agency in the foreseeable future is not recommended."

Additional DoD comment[s]:

"The DoD should not be the sole operator of the STS."
"An STS operational organization within NASA might be acceptable to DoD if the following conditions are met:

- DoD participation in organizational implementation
- DoD participation in operational management
- Specific NASA commitments are made to complete the necessary Shuttle system enhancements as specified in the STS Master Plan's Baseline Operation Plan."

****

Attachment 2

Issues
STS Baseline Operations Plan

INFORMATION ITEMS

ISSUE	COMMENT
DOD SECURITY COSTS	– NON-SECURITY CHANGES TO SECURITY SYSTEMS
	– IN WORK BY NASA AND SYSTEMS COMMAND
OIL LEASE OFF VANDENBERG COAST	– COULD LIMIT LAUNCH AZIMUTH
	– SENSITIVE "POLITICAL" ISSUE
FUTURE FLIGHT CHANGES	– IAW REIMBURSEMENT MOA NEW PRICE DETAILED IN 1985
	– EXPECT $63-100M PRICE (FY 84 $)

RECOMMENDATION

CONTINUE WORKING THESE ITEMS SEPARATELY

NON-CONTROVERSIAL CAPABILITIES SHORTFALLS

ISSUE	SPECIFICATION SHUTTLE SYSTEM	CAPABILITY	COMMENT
MISSION DURATION	30 DAYS	10-12 DAYS	DOD REQ'T IS 7 DAYS + 2 DAYS CONTINGENCY
RESCUE CAPABILITY	SUITS & PERSONAL RESCUE SYSTEM	NONE	NO DOD REQUIREMENT
DOCKING MODULE	INTERNATIONAL REQUIREMENT FOR RENDEZVOUS & DOCKING CAPABILITY	NONE	NO DOD REQUIREMENT

OPERATING LIFE	10 YEARS, 500 USES	CERTIFIED TO 100 USES	SATISFIES PROJECTED 20-YEAR MISSION MODEL
ADDITIONAL	ORBITAL	NONE	NO DOD REQUIREMENT
PROPELLANT	MANEUVERING SYSTEM (OMS) KITS		

RECOMMENDATION

CONCUR WITH NASA POSITION TO CHANGE REQUIREMENTS SPECIFICATION
TO BE CONSISTENT WITH CAPABILITY

NATIONAL SECURITY/CRISIS CONSTRAINTS

ISSUE	SHUTTLE SYSTEM SPECIFICATION	CAPABILITY	COMMENT
LANDING WEATHER CONSTRAINTS AND AUTOLAND	NONE	NO PRECIPITATION 15,000 FT CEILING 7 MILE VISIBILITY 8 KNOT CROSSWIND	– RTLS & EOM* ALTERNATE LANDING SITES PLANNED – AUTOLAND DEMO ON STS 51-E (FEB 85)
ORBITER AUTONOMY	NONE	TACAN FOR NAV AND DEORBIT TARGETING UNTIL 1992	– GPS PLANNED – ORBITER COMPUTER UPGRADE APPROVED
LAUNCH FROM	WITHIN 2 HRS	6.5 HRS (KSC) 4.5 HRS (VAFB)	– ACCEPTABLE CONSTRAINTS
ORBITER TURN-AROUND TIME	14 DAYS BETWEEN FLIGHTS	28 DAYS IS GOAL	– ACCEPTABLE CONSTRAINT (DOD HAS PRIORITY)

RECOMMENDATION

ACCEPT FACT THAT STS WILL NOT MEET TRADITIONAL MILITARY SYSTEMS
REQUIREMENTS (ALL WEATHER, RAPID DEPLOYMENT, SURVIVABILITY, ETC.)

* RETURN TO LAUNCH SITE AND END OF MISSION

PAYLOAD MISSION FLEXIBILITY CAPABILITIES

ISSUE	SHUTTLE SYSTEM CAPABILITY	DOD REQUIREMENT	COMMENT
NAVIGATION ACCURACY	1000'–ALL AXIS	45'–ALL AXIS WITH GPS	– GPS WILL MEET REQUIREMENT – NASA/AF PLAN FOR JOINT IMPLEMENTATION OF GPS CAPABILITY (FY 87 BUDGET $30-40M)
REDUNDANT PAYLOAD SERVICES – Ku BAND ANTENNA CONTROL – PAYLOAD DATA SYSTEM (PDI) – MANIPULATOR ARM (RMS)	NO REDUNDANCY	REDUNDANCY IN MISSION CRITICAL SYSTEMS	– AF PAYLOADS RELUCTANT TO USE SERVICES – REDUNDANT ANTENNA CONTROLLER OR MECHANICAL STOPS NECESSARY – REDUNDANT PAYLOAD DATA SYSTEM, MORE RELIABILITY IN ARM NEEDED – COSTS HIGH: PDI (40 POUNDS, $2M); RMS (900 POUNDS, $20M)
EXTRA VEHICULAR ACTIVITY (EVA) PROVISIONS			
– IMMEDIATE EVA	MINIMUM SEVERAL HOURS	NO CURRENT REQUIREMENT	– ACCEPTABLE CONSTRAINT
– CARGO BAY ENVELOPE	56 FT TO 60 FT	60 FT	– REQUIRES CONTINUED MISSION-BY-MISSION COORDINATION

RECOMMENDATION

– NASA/DOD AGREE ON EFFECTIVITY OF GPS
– NASA/DOD EVALUATE ON MISSION-BY-MISSION BASIS, COST AND WEIGHT TRADES OF REDUNDANT SYSTEMS
– ACCEPT EVA CONSTRAINT

Document II-44

Document title: National Security Decision Directive 164, "National Security Launch Strategy," February 25, 1985.

Source: NASA Historical Reference Collection, NASA History Office, NASA Headquarters, Washington, D.C.

Under Secretary of the Air Force Edward C. Aldridge wanted to keep expendable launch vehicle production lines open, because he was concerned that valuable manufacturing expertise would be lost. Having completed a commercial competition to select the complementary expendable launch vehicle, Aldridge needed NASA to concur with the Air Force's selection of a Titan derivative. Negotiations at the staff level had little success. Aldridge called NASA Administrator James Beggs to discuss the matter. They reached an agreement, which was transcribed and taken to the National Security Council to be processed for the President's signature. The result was the National Security Launch Strategy, which, after the Challenger *disaster, resulted in the Department of Defense transferring most of its payloads off the shuttle.*

---

[1]   February 25, 1985

National Security Launch Strategy

NSDD 144, National Space Strategy, states that the Space Transportation System (STS) will continue as the primary space launch system for both national security and civil government missions. It also directs DoD to pursue an improved assured launch capability that will be complementary to the STS. This NSDD provides a launch strategy to implement these two provisions, as well as initiate a study to look toward the future development of a second-generation space transportation system.

The National Aeronautics and Space Administration (NASA) and the Department of Defense (DoD) will work together to insure that the STS is fully operational and cost-effective at a flight rate sufficient to meet justified needs. (The target rate is 24 flights per year.)

The Air Force will buy ten expendable launch vehicles (ELVs) and will launch them at a rate of approximately two per year during the period 1988-92. A competitive decision will be made between the Titan derivative vehicle and the SBR-X before March 1, 1985.

DoD will rely on the STS as its primary launch vehicle and will commit to at least one-third of the STS flights available during the next ten years. NASA and DoD will jointly develop a pricing policy for DoD flights that provides a positive incentive for flying on the Shuttle. The pricing policy will be based upon the principle that an agreed reimbursement rate per flight will be comprised of a fixed and variable component. This will result in an annual fixed fee and a charge per flight at marginal or incremental cost. NASA will propose a pricing policy based upon this principle by April 15, 1985.

DoD and NASA will jointly study the development of a second-generation space transportation system—making use of manned and unmanned systems to meet the requirements of all users. A full range of operations will be studied, including Shuttle-derived technologies and others. It would be anticipated that NASA would be responsible for systems management of civil manned systems and DoD would be responsible for [2] systems management of unmanned systems. DoD and NASA will jointly define the terms of reference of this effort for issuance as a National Security Study Directive (NSSD).

Any disagreements regarding implementation of this Strategy should be referred first to the Assistant to the President for National Security Affairs and subsequently, if necessary, to the President for resolution.

Chapter Three

The NASA-Industry-University Nexus: A Critical Alliance in the Development of Space Exploration

by W. Henry Lambright

The National Aeronautics and Space Administration (NASA) is and always has been more than a simple, conventional government organization. When NASA Administrator James E. Webb spoke in 1966 about the organization he headed, he referred proudly to an enterprise involving some 420,000 men and women involved in the single-minded purpose of leading the United States into space. At the time, however, less than 10 percent (34,000) of those employees were civil servants. NASA extended its reach through contracts and grants to numerous external organizations, chiefly industry and universities. The government-industry-university team constituted a powerful institutional partnership throughout NASA's history.[1]

The 1960s—the Apollo years—were the time when this partnership reached its peak in terms of scale.[2] It was also the period during which NASA established or refined most of its innovative management practices. Since then, NASA has consolidated and built further on the foundation it created for itself; few fundamental changes were made in the character of the relations between NASA and its nongovernment partners during the 1970s and 1980s, even as the partners attempted to adjust to diminished budgets and a lower national priority for space. As an agency, NASA still represents one of the more effective government-industry-university systems in existence. This essay focuses on how this system came into being after Sputnik I, was expanded, was pushed to its limit during the 1960s, and was altered in the post-Apollo era of spaceflight since the decade of the 1960s. Most of NASA's interactions with industry and academia since Apollo have been an extension of the approaches put in place during that earlier time.

Origins: The Glennan Era, 1958-1961

Because NASA was formed from existing components based elsewhere within the U.S. government, especially from among the various components of the defense organization, it inherited a strong "in-house" tradition of technical expertise (referring to the idea that

---

1.   For biographical information on James E. Webb, especially as it relates to his management philosophy for large-scale technological systems, see W. Henry Lambright, *Powering Apollo: James E. Webb of NASA* (Baltimore, MD: Johns Hopkins University Press, 1995). For a condensed analysis of Webb's leadership in this effort, see W. Henry Lambright, "James E. Webb: A Dominant Force in 20th Century Public Administration," *Public Administration Review* 53 (March/April 1993): 95-99; W. Henry Lambright, "Past and Present in Powering Big Technology," *Space Times: Magazine of the American Astronautical Society* 34 (November-December 1995): 11-13.

2.   For a critique of this administrative approach from one who sees in it too great an aggregation of power, see the Pulitzer Prize-winning book by Walter A. McDougall, . . . *The Heavens and the Earth: A Political History of the Space Age* (New York: Basic Books, 1985).

most engineering and even some production work would be performed by a government entity rather than an industry or university contractor). An emphasis on "in-house" technological skill had been bequeathed from various weapons laboratories, becoming known collectively as the "arsenal system." Clearly, Wernher von Braun's Army Ballistic Missile Agency team at the Redstone Arsenal in Huntsville, Alabama, was an organization firmly rooted in this culture of in-house capability. In addition, such government organizations as the nonmilitary National Advisory Committee for Aeronautics (NACA), first established in 1915 as a means of improving the quality of airplanes in the United States to help offset foreign competition in the commercial market, developed strong "in-house" technical expertise in aeronautical research and development.[3]

The first NASA Administrator, T. Keith Glennan, on leave from his position as president of the Case Institute of Technology, appreciated the legacy of "in-house" engineering capability that the organizational components incorporated into NASA had developed, but it did not mesh well with the mission of the new agency as he understood it. Accordingly, he determined that most of NASA's work would be performed externally by industry. This was required in part by the need to "scale up" rapidly for Project Mercury, NASA's first human spaceflight program, but it was also a matter of ideology. As he wrote in his diary:

> . . . having the conviction that our government operations were growing too large, I determined to avoid excessive additions to the federal payroll. Since our organizational structure was to be erected on the NACA staff, and their operation had been conducted almost wholly "in-house," I knew I would face demands on the part of our technical staff to add to in-house capacity. . . . But I was convinced that the major portion of our funds must be spent with industry, education, and other institutions.[4]

Glennan, as an Eisenhower Republican, believed that government's role should be kept small and that the federal government should rely on private enterprise for getting the public's work done whenever possible.[5]

To a very real extent, Glennan was both an Eisenhower Republican with a fiscally conservative inclination and an aggressive businessman with a keen sense of public duty. He also possessed a strong opposition to government intrusion into the lives of Americans. But he was also an administrator and an educator with a rich appreciation for the role of science and technology in an international setting.[6] As historian Roger D. Launius has written of Glennan:

3.   This legacy of "in-house" engineering capability has been explored in detail in Howard E. McCurdy, *Inside NASA: High Technological and Organizational Change in the U.S. Space Program* (Baltimore, MD: Johns Hopkins University Press, 1993), pp. 34-50.

4.   T. Keith Glennan, *The Birth of NASA: The Diary of T. Keith Glennan*, J.D. Hunley, ed. (Washington, DC: NASA SP-4105, 1993), p. 5.

5.   Robert L. Rosholt, *An Administrative History of NASA, 1958-1963* (Washington, DC: NASA SP-4101, 1966), pp. iii-vii.

6.   These themes are well developed in Glennan's diary, *The Birth of NASA*. See also "Glennan Announces First Details of the New Space Agency Organization," October 5, 1958, NASA Historical Reference Collection, NASA History Office, NASA Headquarters, Washington, DC; James R. Killian, Jr., *Sputnik, Scientists, and Eisenhower: A Memoir of the First Special Assistant to the President for Science and Technology* (Cambridge, MA: MIT Press, 1977), pp. 141-144; James R. Killian, Jr., Oral History, July 23, 1974, NASA Historical Reference Collection. Eisenhower's concerns about this aspect of modern America are revealed in "Farewell Radio and Television Address to the American People," January 17, 1961, *Papers of the President, Dwight D. Eisenhower 1960-61* (Washington, DC: U.S. Government Printing Office, 1961), pp. 1035-40.

While he was an ardent cold warrior and understood very well the importance of the space pro-gram as an instrument of international prestige, Glennan emphasized long-range goals that would yield genuine scientific and technological results. Second, he believed that the new space agency should remain relatively small, and that much of its work would of necessity be done under contract to pri-vate industry and educational institutions. This was in line with his concerns about the growing size and power of the federal government. Third, when it grew, as he knew it would, Glennan tried to direct it in an orderly manner. Along those lines, he tenaciously worked for the incorporation of the non-military space efforts being carried out in several other federal agencies—especially in the Department of Defense—into NASA so that the space program could be brought together into a mean-ingful whole.[7]

Glennan fostered the replication of his values and perspectives in NASA as he began to direct its affairs in the fall of 1958, and by the time of his departure from Washington in January 1961, they had been placed on the road to adoption.

Little attention was given to universities, *per se*, during the Glennan era. There was interest in nurturing space science and research projects sponsored at universities, but Glennan did not develop a master plan for the incorporation of a partnership with uni-versities. What he did establish in 1958 was a University Research Program Office at NASA Headquarters under the direction of the Office of Aeronautical and Space Research. This organization, at the behest of the technical program offices, oversaw a small "research by contract" program. [III-1]

In May 1960, Glennan reorganized this structure and created the Office of Research Grants and Contracts as an administrative unit of NASA to coordinate research conduct-ed by nonprofit institutions. This effectively made the new organization the liaison between NASA and most universities, acting on behalf of program offices for work per-formed outside the agency. All such research activities, therefore, were approved by NASA Headquarters, even though the agency's field centers might still manage the actual work once it was put into place by the Office of Research Grants and Contracts.[8]

The Department of Defense Framework

Also during the Glennan period, the basic structure of NASA-industry relations was established. The National Aeronautics and Space Act of 1958, which laid out the legal groundwork for NASA's creation, anticipated that the agency would contract with indus-try for much of its activities.[9] In a significant policy action, it extended to NASA the pro-curement authority contained in the Armed Services Procurement Act of 1947 (ASPA). The importance of this legislation was that it recognized that NASA would be establishing a partnership with many of the same companies with which the Department of Defense (DOD) already had long-standing relations.

The ASPA provisions, which had been amended frequently over the preceding decade, provided the government with the flexibility to address work based on research and development (R&D). ASPA allowed the federal government to divert from the tradi-tional practices of advertising for competitive bids and awarding contracts to the lowest responsible bidder. Instead, the government could use negotiation, a technique developed largely in World War II to meet the war crisis and institutionalized subsequently by DOD.

---

7.   Roger D. Launius, "Introduction," in Glennan, *The Birth of NASA*, p. xxii.
8.   Rosholt, *Administrative History of NASA*, pp. 128-29.
9.   This act is available as Document II-17 in John M. Logsdon, gen. ed., with Linda J. Lear, Jannelle Warren-Findley, Ray A. Williamson, and Dwayne A. Day, *Exploring the Unknown: Selected Documents in the History of the U.S. Civil Space Program, Volume I: Organizing for Exploration* (Washington, DC: NASA SP-4407, 1995), 1: 334-45.

On October 30, 1958, Glennan underlined NASA's intent to extend ASPA and DOD practices into its jurisdiction. He announced that NASA's procurement and contracting regulations "would conform in every practicable way" to ASPA. Defense contractors would not have to learn how to work under new rules.[10] This was important for many reasons, including the fact that much of what NASA was doing in its initial years was work DOD had pioneered and then transferred to NASA. With the work came the defense contractors. As a principle of his policy, Glennan maintained continuity between the defense realm and NASA's civilian realm as much as possible, and this continuity increased NASA's contracting options.

Glennan also helped smooth relations with DOD and industry by hiring many former DOD officials to work specifically on NASA-industry relations. In January 1959, he appointed Ernest Brackett, a DOD procurement specialist, to head NASA's Procurement Division. John Johnson, Glennan's General Counsel appointee, also came from DOD. These men brought others to NASA, many of whom had learned government-industry relations in the DOD setting.[11]

Once its staffing was well under way, NASA began holding conferences with industry to discuss NASA hardware needs and the legal/administrative relations governing procurement. [III-2] NASA stressed that (1) it expected most of its work to be performed by industry and (2) it intended to make it easy for industry to work with the agency by maintaining a principle of continuity between DOD and NASA contracting procedures.

The Patent Problem

One problem in NASA-industry relations in the Glennan era loomed very large: the question of how to ascertain and assign the rights to patented inventions. In the Space Act, there is a lengthy provision (section 305) requiring that inventions (and their patents) made in performance of contracts for NASA become the property of the U.S. government, unless waived (in which case the government retained a royalty-free license for the use of the invention). The responsibility of waiving U.S. rights to an invention was retained by the NASA Administrator, assisted by the Invention and Contributions Board (a body established by the Administrator). Waivers were to be made only in the public interest.

This statutory policy was similar to the statutory policy guiding the Atomic Energy Commission (AEC), but it was very different from the policy that DOD had promulgated administratively. DOD, the nation's largest buyer of R&D and the agency with an industrial clientele similar to NASA's, followed a more liberal policy from the contractor's point of view. The invention remained the property of the contractor, with the provision that the government would have a royalty-free license for the invention's use. In other words, NASA would have to invoke the waiver procedure to grant the same privileges to a contractor that DOD could grant outright in the contract itself.[12]

Glennan was stymied in regard to his "continuity" policy, at least in this area. The patent issue mattered in two ways. First, there might well be *tangible* stakes involved. Significant inventions might derive from working for NASA, and industry could therefore make additional money from marketing them in other contexts. For the federal government, these financial stakes did not exist, because the government itself did not commercialize inventions. However, at a second, *symbolic* level, there were two political issues:

10.   Rosholt, *Administrative History of NASA*, p. 62.
11.   *Ibid.*, pp. 62-63; Glennan, *Birth of NASA*, pp. 108, 120.
12.   Rosholt, *Administrative History of NASA*, p. 92.

whether the federal government might end up "paying twice" for its work and whether the government was "giving away" the public's property rights to an invention.

These symbolic concerns were intensely expressed by certain legislators with a populist bent. As government expanded its R&D work beyond DOD, they made it a point to push their "title policy" (the AEC model) into other fields. In atomic energy, which began as a government-created monopoly, industry did not have the opportunity to fight for the license policy (the DOD approach). In the case of NASA, industry did fight—but only after the National Aeronautics and Space Act had been passed.

What the patent issue also illuminated was the degree to which terms such as "procurement" and even the broader concept of "acquisitions policy" masked a dynamic institutional relationship between "buyers" and "sellers." The federal government is a buyer, but the buyer does not always hold all the bargaining advantages. In the pre-Apollo days, NASA was distinctly at a disadvantage—or so it perceived itself—vis-à-vis DOD in acquisition of the best contractors. Many firms quietly spread the word that "NASA would take your patents." They also pointed out that NASA was an R&D agency and would not provide the lucrative production runs industry could get through DOD. Moreover, the leaders of NASA, such as Glennan and the DOD transplants, did not want an AEC type of policy. They wanted continuity with DOD and an equal chance to get the best and most enthusiastic contractors—the contractor companies that would allow their most creative (and inventive) people to work on NASA contracts.

Although it is all but forgotten today, this matter of patent policy was a significant political issue in the Glennan era. NASA was trying to initiate a close relationship with industry; it was a rival of DOD for many space missions, and by no means an advantaged rival. The patent issue grated on these relations. Glennan felt that the section 305 legislation tied his hands, and the result was his decision to get the legislation amended. Legislators favoring the title policy took a stand, and a legislative struggle ensued. [III-3] This created a controversial backdrop to the more staid NASA-industry relations in other areas during 1959 and 1960. NASA fought for an amendment throughout the Glennan period and was finally able to get legislation through the House of Representatives that would shift the law from title-oriented policy to license-oriented policy. This legislation failed in the Senate, however, where such powerful and populist Democratic senators as Clinton Anderson of New Mexico, Estes Kefauver of Tennessee, and Russell Long of Louisiana favored the title approach.

Other Early NASA-Industry Developments

Having decided that most of its work would be done with industry, and using DOD practices as much as possible, the next question for NASA was how to award major contracts. Glennan concluded that he should be the decision maker for large contracts. In October 1959, NASA promulgated a formal procedure for selecting the recipients of very large NASA contracts. The procedure provided that the NASA Administrator would select all contractors when the intended contract exceeded $1 million. Glennan indicated that the Administrator would be advised on these decisions by ad hoc source selection boards, primarily composed of technical specialists.[13]

For instance, Glennan's personal diary discussed the dilemmas involved in the bidding and selection process. In one particularly poignant section, he described the process leading to his selection of Rocketdyne to build the J-2 engine, which powered both the Saturn 1B first stage and the Saturn V second stage:

---

13. *Ibid.*

At 9:00 o'clock the Source Evaluation Board on the 200 K engine reported. This is a stinker, in the vernacular—five companies bid and three of them are very close together at the top. In fact, they are so close in the technical evaluation that it is almost impossible to choose between them. The same is essentially true in the business evaluation except that one of them bid $138 million, a second bid $69 million, the third bid $44 million. These bids are really estimates of the total cost of the project since this research and development work is always handled on a cost plus a fixed fee basis. The costs do give an indication of the extent of experience of a company in undertaking a difficult task of this sort. For instance, one of the companies which was not in the running bid only $24 million dollars. While the highest one is undoubtedly high, the lowest indicates a complete lack of understanding of the difficulty of the job.

I took the reports and will now have to sit down with myself in an attempt to find a proper answer to this question.

Later he met with his chief advisers, and they agreed that Rocketdyne would receive the contract. This process was ticklish at best. As Glennan concluded: "It is a fact that if 10 people bid, 9 of them are going to be unhappy because only 1 can win. With the 9 having representatives in Congress, it is almost inevitable that some charges of favoritism, lack of objectivity, etc., will be tossed our way."[14]

Throughout 1959 and 1960, Glennan had various management consultants take a look at the agency's administrative issues, including NASA-industry relations. The general thrust of these reports confirmed Glennan's view that as much of NASA's work as possible should be contracted out. [III-4, III-5] However, the reports also pointed out the need for balancing the external work with internal competence. NASA's centers pressed on Glennan their need to grow and build competence. By the end of the Glennan years, 85 percent of NASA's $1 billion budget was going to industry. But the agency was also expanding its in-house work and capabilities.

Under Glennan in the 1958-1960 period, NASA established a strong relationship with industry based on the principle of continuity with DOD contracting practices. As a Republican, Glennan's conservative values helped create a sense of partnership critical to jumpstarting NASA-industry relations. The "closeness" with industry bothered some critics, including NASA civil servants who wanted more work to be performed in-house. This was particularly true of those who had come to NASA from NACA and the Army Ballistic Missile Agency. The patent issue was also left unsettled. All in all, however, Glennan left a solid foundation on which the next Administrator of NASA could build.

The Webb Era, 1961-1968

Under James E. Webb, NASA Administrator from 1961 to 1968, the NASA-industry-university relationship expanded tremendously. Webb continued the basic philosophy of Glennan—to contract out most of the agency's work to industry—but he surpassed Glennan by consciously seeking innovation in these relations. Both continuity *and* change were objects of NASA policy. Whereas only modest efforts had been made on the academic front under Glennan, Webb established a "university program" that went beyond, in its goals, anything seen in government before—or since. Glennan was rather cautious in his approach to external institutions. Webb used government with an eye toward reform. Glennan was a *technical* engineer. Webb—a lawyer-administrator with exceptional political skills—had the instincts of a *social* engineer. When he took the oath of office, Webb stated that "my purpose would be to work toward creating an environment within which

---

14.   Glennan, *Birth of NASA*, p. 137.

NASA could be as innovative in the management of its programs as it was in aeronautics and space science."[15] [III-6]

When first asked by the White House whether he would accept a nomination as NASA Administrator, Webb made it clear that he wanted to inherit the two principal NASA officials from the Eisenhower administration: Deputy Administrator Hugh L. Dryden, who was a physicist and former NACA chief, and Associate Administrator Robert C. Seamans, Jr., who was also known as the "General Manager." Webb wanted to make the major decisions in conjunction with these two men, and together they would form a "triad" for the administrative leadership of the agency.

The May 1961 decision to go to the Moon—just three months after Webb took office—had tremendous impacts on NASA in many ways. The question facing Webb, Dryden, and Seamans after the Apollo decision was how well Glennan's contracting system would serve to organize the lunar landing program. In general, they accepted that system, even strengthening the procedures instituted by Glennan. The principle of contracting out for R&D was reaffirmed, and the role of in-house staff in technical direction was stressed. Headquarters officials took it upon themselves to make procurement policy more uniform, yet flexible enough for NASA to obtain space hardware whose main features could not be specified in advance.[16]

Given Webb's orientation, there would have been changes in the way NASA dealt with universities and industry even without Kennedy's decision to go to the Moon. However, what that decision did was enlarge the scope of NASA's effort and give it a new urgency and many more resources.[17] Also, problems that were important for Glennan were less significant under the impetus of Apollo. For example, the patent issue was one problem Glennan felt had to be resolved with new legislation. Webb decided that he could handle the problem administratively, using the waiver clause. In effect, Webb used an administrative strategy to bring NASA patent policy in line with DOD policy. This infuriated title policy advocates, but Webb pushed ahead and absorbed intense (sometimes very personal) criticism thereafter from particular legislators. This made for easier NASA-industry relations, though. It also made it possible for NASA to move on other issues in Webb's agenda.

Among the issues Webb wanted to address was the role of universities and industry in economic and social development. Just two days prior to the announcement of the Apollo goal, on May 23, 1961, Webb sent Vice President Lyndon B. Johnson a memorandum in which he revealed that NASA-industry-university relations would have a new flavor.[18] [III-7] For Webb, Apollo was both an end and a means. As an end, it served as an arena for a technological race with the Soviet Union for pride and prestige. As a means, it would provide an impetus that would allow NASA to spend a large amount of R&D money in

---

15. James E. Webb, "Foreword" in Rosholt, *Administrative History of NASA*, p. iv.

16. Arnold S. Levine, *Managing NASA in the Apollo Era* (Washington, DC: NASA SP-4102, 1982), pp. 65-105; Sylvia D. Fries, *NASA Engineers and the Age of Apollo* (Washington, DC: NASA SP-4104, 1991), pp. 174-83; McCurdy, *Inside NASA*, pp. 134-41.

17. On the Kennedy lunar decision, as well as the ramifications of it, see John M. Logsdon, *The Decision to Go to the Moon: Project Apollo and the National Interest* (Cambridge, MA: MIT Press, 1970); Charles Murray and Catherine Bly Cox, *Apollo: The Race to the Moon* (New York: Simon and Schuster, 1989); Roger D. Launius, *Apollo: A Retrospective Analysis*, Monographs in Aerospace History, No. 3 (Washington, DC: NASA, 1994); Andrew Chaiken, *A Man on the Moon: The Voyages of the Apollo Astronauts* (New York: Viking, 1994).

18. James E. Webb, Administrator, Memorandum for the Vice President, May 23, 1961, Administrator's Files, NASA Historical Reference Collection.

ways that would help the country, including having government work with universities and industry in terms of regional economic development.[19]

Webb also told Johnson in this wide-ranging memorandum how he thought about Apollo. He mentioned the prospect, for example, of a new NASA facility to manage the Apollo program; this eventually was the Manned Spacecraft Center in Houston (later named Johnson Space Center). He suggested its possible linkage with Rice University. Webb noted that Lloyd Berkner, Chair of the Space Science Board of the National Academy of Sciences, was establishing a Graduate Research Center in Dallas, Texas, with industrial backing and that this new organization might also be brought into the alliance. Senator Robert Kerr of Oklahoma (Chair of the Senate Space Committee) also had interests in this area, and he and others saw a development potential for Arkansas and Mississippi. Even without NASA's involvement, Webb anticipated a scientific-industrial complex in California, running from San Francisco south through the new University of California at San Diego. Webb saw another center emerging around Chicago, as a pivot, and a strong northeastern arrangement with Harvard, MIT, and similar institutions. He envisioned work in the Southeast, perhaps revolving around the Research Triangle in North Carolina, in which Charlie Jonas as the ranking minority member on Albert Thomas's House Appropriations Subcommittee (which controlled the NASA budget) would have an interest. To fill out the picture, he thought NASA could help make the possibility of a southwestern complex into a reality.

It was clear that Webb thought about NASA-industry-university relations both as a procedure to secure the Apollo goal and also as a way toward advancing regional socioeconomic development. The latter end would also be a means for Apollo in terms of winning congressional support. For Webb, it was one mosaic, with each part contributing to the whole design. Thus, on May 25, 1961, when Kennedy announced the Moon decision, Webb had an institutional strategy in mind, and he was ready to go at full speed.

Early Decisions Involving Contractors

Like Glennan, Webb believed the big decisions on procurement should be made at the top. However, with NASA's budget soaring, the $1 million level established by Glennan as the threshold for the Administrator's personal involvement was raised to $5 million. As Webb wrote:

> Dr. Dryden, Dr. Seamans, and I determined that we would personally examine, in detail, the results of the work of all source evaluation boards on competitively negotiated contracts that amounted to 5 million dollars or more. We expected these boards to appear before us personally in a formal setting and make a full and complete presentation of (1) the method chosen to break down for evaluation the contractor proposals, (2) the results achieved in the application of this method, and (3) the judgment of the board on each of the categories of the breakdown.
>
> The fact that the three senior officers of the agency would take the time to conduct what amounted to a thorough hearing and question-and-answer period on each contractor selection action enabled

---

19.   This approach toward handling Apollo has been explicitly laid out in Loyd S. Swenson, Jr., "The Fertile Crescent: The South's Role in the National Space Program," *Southwestern Historical Quarterly* 71 (January 1968): 377-92; Robert A. Divine, "Lyndon B. Johnson and the Politics of Space," in Robert A. Divine, ed., *The Johnson Years, vol. II: Vietnam, the Environment, and Science* (Lawrence: University Press of Kansas, 1987), pp. 217-53; Robert Dallek, "Johnson, Project Apollo, and the Politics of Space Program Planning," unpublished paper delivered at a symposium titled "Presidential Leadership, Congress, and the U.S. Space Program," sponsored by NASA and American University, March 25, 1993.

all levels of management, in Headquarters and in our Centers, to get their questions out on the table before all three of us for debate and clarification. Another important result was that when the presentation to the three of us was over, everyone involved had a clear understanding of the elements basic to a proper decision and everyone in NASA concerned with the matter was aware of this. The burden then passed to Dryden, Seamans, and me to make the final decision, and the personnel of the boards were in position to form their own judgments as to whether the three of us did in fact arrive at the best decision as indicated by the facts and analysis. Further, an important element of a NASA-wide and pervasive self-policing system was thereby established. This has had an important effect on maintaining high standards throughout the agency.[20]

In the months following the Apollo announcement, NASA made one decision after another involving contracts to companies for the Moon program. The most controversial decision, made late in 1961, was the award to North American Aviation for the construction of the Apollo spacecraft. This was controversial because the Source Evaluation Board recommended in favor of a company other than North American. A number of astronauts and Manned Spacecraft Center Director Robert Gilruth believed that North American Aviation, which the Source Evaluation Board also ranked highly, was more qualified and said so to Webb, Dryden, and Seamans.

Webb had an informal policy to spread contracts around so NASA would not overly depend on any one organization. North American had already been awarded the contract for the second stage of the Saturn rocket (S-II). Hence, given the Source Evaluation Board recommendation, there was a second reason not to give this critical contract to North American. However, pressured by Gilruth and others, Webb and his senior colleagues decided in favor of North American. It was the largest single contract of the entire Apollo program.[21]

In early 1962, with most of the big hardware contracts for Apollo signed, NASA made two other industrial decisions of policy significance. These involved contracts for supporting NASA in Apollo management. One was the Bellcomm contract with AT&T. Bellcomm was a profit-making subsidiary established by AT&T in March 1962 at NASA's request to conduct analytical studies in support of Apollo. The second, a General Electric contract signed in February 1962, was to assist with the integration, reliability, and checkout of hardware at the three large spaceflight centers (Houston, Marshall, and Cape Canaveral). These two contracts, negotiated on a sole-source basis, helped NASA with the total Apollo system, whereas other contractors worked only on Apollo components.[22]

One other development involving industry during the start of the Webb era is worth noting. In a November 1961 reorganization, a small Industrial Applications Office was established as part of the Office of Space Applications. The larger office was concerned with communications satellites, weather satellites, and large hardware programs, while the Industrial Applications unit concentrated on NASA technology "spinoffs" to industries outside the space arena. [III-8] This highlighted Webb's interest in the socioeconomic mission for NASA, as mentioned in his memorandum to Vice President Johnson. In the university field, Webb was similarly seeking to achieve multiple goals in parallel.

---

20. James E. Webb, "Foreword," in Rosholt, *Administrative History of NASA*, p. v.

21. Courtney G. Brooks, James M. Grimwood, and Loyd S. Swenson, Jr., *Chariots for Apollo: A History of Manned Lunar Spacecraft* (Washington, DC: NASA SP-4205, 1979), pp. 41-44.

22. Levine, *Managing NASA in the Apollo Era*, pp. 88-93.

Launching a New University Relationship

With the overall expansion of the space program, NASA's interactions with universities grew enormously. Most of these were in the field of space science. The November 1961 reorganization established an Office of Space Science to organize and sponsor most of this work.[23]

Academic participation in Project Apollo was relatively modest compared to that of industry. However, one of the most critical contracts for Apollo did go to a university—the Apollo guidance and navigation contract awarded to MIT's Instrumentation Laboratory on August 9, 1961. This was a sole-source award, much to the annoyance of industry. NASA justified the selection because the laboratory's director, C. Stark Draper, was viewed as the country's leading expert on guidance systems.

However, the most striking aspect of NASA's university relationship came with the advent of the Sustaining University Program.[24] In November 1961, the Office of Research Grants and Contracts was moved under the Office of Space Science, and a new program was launched under this organization's aegis. This program aimed to use universities for socioeconomic goals.

Until 1961, NASA, like most other federal agencies supporting research in universities, concentrated on specific projects. The agency's interest in allocating resources for the best research was paramount. The consequence was that a relative handful of universities in the nation received most federal research grants and contracts. Webb inherited the "project system" and did not interfere with this basic pattern of NASA relations to academic science. Most of NASA's science money was spent on projects directed by leading academic investigators.

But Webb did not believe this was enough. [III-9] In late 1961, following considerable discussions within and outside NASA, the Administrator directed the agency to establish the Sustaining University Program.[25] This was intended to complement the project system model with an approach that would relate NASA to universities as institutions, rather than to specific individuals and projects. The program had three basic components: fellowships, research grants, and facilities.

The program also embodied a number of policy thrusts. One thrust was human resources, with the goal of enlarging the number of Ph.D.s in selected technical fields through fellowships. A second thrust was geographical spread, to nurture new centers of strength (as well as new talent) in university science throughout the country. NASA provided funds to universities, not to individual students. The fellowships were then awarded by those universities; hence, students had incentives to enroll there, rather than going to a few elite schools. The third thrust was the interdisciplinary principle. NASA provided research funds to support broad areas of research and involve a cross section of disciplines, including social scientists, who would study the impacts of science and technology. A fourth thrust focused on regional socioeconomic development. NASA would provide laboratory facilities—buildings—if the presidents and faculty of a university receiving a NASA facility pledged to work actively with private enterprise and community leaders in their local area, using the scientific, technological, and managerial advances being generated by the space program to benefit their regions and communities. Finally, there was

---

23.   John E. Naugle, *First Among Equals: The Selection of NASA Space Science Experiments* (Washington, DC: NASA SP-4215, 1991), pp. 107-11.

24.   For more on this subject, see W. Henry Lambright, *Launching NASA's Sustaining University Program* (Syracuse, NY: InterUniversity Case Program, 1969).

25.   W. Henry Lambright and Laurin L. Henry, "Using Universities: The NASA Experience," *Public Policy* 20 (Winter 1972): 61-82.

a fifth principle that was implicit in all the rest—the enhancement of the university as an institution. NASA wanted a coherent response from the university; this meant that the university administration, especially the president, had to be a proactive force—a leader—in implementing the objectives of the Sustaining University Program.

This program went well beyond anything any other department or agency was doing (or even considering) at the time. NASA had no specific legislative authority to do what it did, but, in the environment of Apollo, it was possible for an individual with Webb's goals and skills to innovate in ways that would be impossible later. [III-10] The president and key legislators gave Webb enough leeway to start the Sustaining University Program in 1961. Once under way, the program's geographical spread attracted a considerable constituency. There is no doubt that the ability of NASA to reach most states through the Sustaining University Program helped build support for the agency. However, the broader notions of using universities for a NASA-based socio-industrial policy mattered to Webb. The historian Walter McDougall contends that Webb aimed at building a "Space Age America."[26] If so, a major part of the leverage was to be supplied by the Sustaining University Program and the "space age university." Thus, by early 1962, the NASA-industry-university partnership had been forged anew. Although an extension of the Glennan period, it bore the distinct stamp of James Webb, especially with respect to NASA's university relations. [III-11]

"Incentivizing" Contracts

One of the problems of R&D contracting was that technical uncertainties made it difficult to judge how much it would cost to create a particular item of hardware. Hence, most of the industrial contracts NASA awarded in the late 1950s and early 1960s were cost plus fixed fee. In 1962, Administrator Webb participated in an interagency task force headed by David Bell, Director of the Bureau of the Budget. The report of this group (the "Bell Report") examined various aspects of the public-private relationship between government and industry.[27] Webb was a major participant on the task force, and the report emphasized areas in which NASA was already moving.

One of these was the notion of "incentivizing contracts" so that industry would have some motivation to perform well and save money. Following some internal studies and the advice of Robert Charles, who served as Webb's special assistant for procurement in 1963, NASA established more and more contracts with incentive provisions. The basic notion of NASA contracting would claim that "significant improvement in product quality . . . timeliness and cost can be achieved if the procurement process is saturated with competition before contract execution, and with performance and cost reduction incentives thereafter." In late November 1963, NASA directed that the number of cost-plus-fixed-fee contracts be reduced substantially and that incentives be considered for all contracts. Many existing contracts were subsequently converted to incentive arrangements, including the North American contract for the Apollo spacecraft. Doing so was difficult. In some cases (such as the North American contract), the process was achieved over a period of years.[28]

26.   McDougall, *Heavens and the Earth*, p. 361.
27.   This has been published as Document IV-9, *Report to the President on Government Contracting for Research and Development*, Bureau of the Budget, U.S. Senate, Committee on Government Operations, 87th Cong., 2d sess. (Washington, DC: U.S. Government Printing Office, 1962), pp. vii-xiii, 1-24, in Logsdon, gen. ed., *Exploring the Unknown*, 1: 651-72.
28.   Levine, *Managing NASA in the Apollo Era*, p. 77.

Bringing Projects Under Greater Control: Phased Project Planning

During 1964-1965, NASA leaders made an effort to bring industrially managed projects under greater control. Webb's view was that "when you let the contract, all you've done is started a process that with the greatest of care, and ability, and drive will produce a bird. All you've done is put in motion forces that have the capability but which could fail at any point along the line."[29]

In the mid-1960s, NASA received increasing criticism from members of Congress and others who believed too much was being spent on space versus the Great Society or Vietnam. Webb believed that NASA had to be especially careful to avoid even the appearance of mismanagement of its industrial contracts (cost overruns, schedule slippages, and so on), because this would give critics a wedge to attack the entire program.[30]

In 1964, the deputy of Robert Seamans, Earl Hilburn, studied NASA's methods of scheduling and project cost estimation. In 1965, the results of Hilburn's analysis were implemented in the form of a new agency policy, "phased project planning," to define programs more explicitly. This policy was aimed at conducting R&D contracts in a number of sequential phases with maximum competition characterizing the "phase-by-phase increments of project execution," with each phase allowing for "the fundamental concept of agency top management participation at all major decision points."[31] Presumably, government could terminate the contract at each phase and go elsewhere if dissatisfied. It also permitted better opportunities for an agency to keep track of costs and schedules.

Phased project planning was "predicated on the assumption that NASA employees would be responsible primarily for defining programs and providing technical direction to agency contractors."[32] The concept that government would direct industry in large-scale development programs was also a critical principle of the Bell Report. For the most part, NASA felt exceedingly capable of exercising technical management. But there was one area where the agency did not, and this caused NASA to create a new in-house center.

The Electronics Research Center

As the space program grew, it became evident that electronics was a crucial disciplinary area, cutting across virtually every NASA field. As one scholar, Thomas Murphy, concluded:

NASA specialists estimated that forty percent of the cost of the space boosters would be accounted for by electronics components. The figure was even higher with respect to spacecraft, where it was estimated that fifty percent of the cost involved electronics. In the tracking and data acquisition elements of the program, as much as ninety percent of the resources were electronics-oriented.[33]

---

29. *Ibid.*, p. 80.
30. See Document III-17, *Summary Report: Future Programs Task Group,* January 1965, in Logsdon, gen. ed., *Exploring the Unknown,* 1: 73-90.
31. Levine, *Managing NASA in the Apollo Era,* p. 84.
32. *Ibid.*, pp. 84-85.
33. Thomas Murphy, *Science, Geopolitics, and Federal Spending* (Lexington, MA: D.C. Heath, 1971), p. 226.

NASA leaders were increasingly feeling the need to have more in-house competence to direct the vast electronics work being performed by industry and universities. "NASA management was very sensitive to avoiding some of the problems the Air Force had experienced," Murphy added, "in relying too heavily on contractors whose work it lacked the ability to evaluate."[34]

Webb, Dryden, and Seamans decided in 1962 that NASA needed an Electronics Research Center and that the best place in the country to put it was the Boston area. They wanted it located where frontier research was going on in universities and where there was a concentration of the electronics industry. In their view, the Harvard-MIT-Route 128 complex made the Boston area a natural. There was also relatively less NASA work in this region, compared to California, another possible site. Finally, Webb no doubt viewed the Boston area as an ideal site to test his vision of government, industry, and university cooperation.

The problem was that the Boston area was Kennedy territory. Not only was the president from Massachusetts, but his younger brother Edward (Ted) was running for senator in the fall of 1962 with the slogan: "I can do more for Massachusetts." Webb kept quiet about the Electronics Research Center decision, informing the president, but not making it known even in preliminary discussions with the Bureau of the Budget. He feared a leak that would mix NASA interests with the Massachusetts election. After Ted Kennedy's election, the decision was made known to the Bureau of the Budget and became official when NASA submitted its budget to Congress in early 1963.[35]

The protests were large and immediate, with most of the criticism coming from Midwest legislators. The "taint" of political favoritism was charged, and Webb denied it. However, those against the siting choice prevailed in Congress to the extent that the Electronics Research Center's approval was made contingent on NASA conducting a nationwide search for sites. NASA conducted the required search, and this did not change the final outcome, but it did delay the start in building the center (in Cambridge, Massachusetts) by a year. By that time, Lyndon Johnson had become the president.[36]

NASA's Controversy With the California Institute of Technology's Jet Propulsion Laboratory

In January 1964, Ranger 6—one of the space vehicles designed to study landing sites on the Moon prior to NASA's sending astronauts—failed. This was the sixth Ranger flight in a row to fail, and so much effort had been invested in this particular flight to make it succeed that its failure brought many festering issues to light. The Ranger failure raised questions about the relationship between NASA and the California Institute of Technology's (Caltech) Jet Propulsion Laboratory (JPL).[37]

The primary issues were the responsiveness to NASA of JPL, which was in charge of Ranger, and JPL's capacity to manage large technology projects. JPL was different from all other NASA centers in that it was not a civil service organization. The laboratory grounds, buildings, and equipment belonged to the government, but the laboratory itself—as an

---

34.   *Ibid.*
35.   Lambright, *Powering Apollo*, p. 121.
36.   Jan Van Nimmen and Leonard C. Bruno, with Robert L. Rosholt, *NASA Historical Data Book, Vol. I: NASA Resources, 1958-1968* (Washington, DC: NASA SP-4012, 1988 ed.), pp. 285-94.
37.   See Homer Newell, *Beyond the Atmosphere: Early Years of Space Science* (Washington, DC: NASA SP-4211, 1980), pp. 258-73.

organization—was part of Caltech, and its staff were Caltech employees. JPL identified with the academic values of Caltech, and Caltech charged NASA for managing JPL.

JPL had been accustomed to near-total autonomy under its previous sponsor—the Army—and had expected the same under NASA. There were special provisions in JPL contracts—a mutuality clause—indicating that JPL could refuse to perform certain kinds of work that did not suit its interests. However, two factors created seeds for change. The first was the Apollo decision, which gave a special urgency to Ranger and changed it from a research-oriented lunar science project to an enabling mission for Apollo. In response, NASA wanted to install a general manager under the JPL director who would instill project management values and skills. JPL needed to give more attention to deadlines, costs, and tight engineering procedures. The second factor was Webb's desire for more response from Caltech to Washington's management directives.

As part of its management responsibility, according to Webb, Caltech should be more involved with JPL, getting the laboratory to interact with other universities and industry in California. JPL should set an example for the universities under the Sustaining University Program to follow. Caltech president Lee DuBridge, however, was not interested in doing what Webb wanted, and he told Webb that.[38]

This institutional struggle continued into the early 1960s. JPL had on its side both prestige and a history of independence. NASA, however, supplied the money, and the Caltech-JPL contract was up for renewal. What tipped the scales in favor of NASA was Ranger 6. The Ranger disaster first produced a NASA investigation and then a congressional inquiry. Because of these inquiries, Caltech's Board of Trustees became involved.

Webb protected Caltech and JPL from congressional actions that might have gone too far in punishing these institutions. At the same time, he bargained with Caltech's Board of Trustees to get more control over JPL. The chairman of Caltech's Board of Trustees, Arnold Beckman, became a Webb ally, and the pressure on Caltech and JPL to change became too strong to resist. [III-12, III-13] The mutuality clause was removed, there was agreement by Caltech and JPL that a general manager would be appointed, and the Caltech fee was made subject to performance evaluation. [III-14, III-15] Webb was unable to get DuBridge to go along with his vision of a "space age university," but Webb never stopped trying. Most importantly, from the standpoint of buffering the NASA-Caltech-JPL partnership from a congressionally mandated restructuring, Ranger 7 was launched on July 28, 1964, and was successful.[39]

Problems With the Sustaining University Program

Starting in 1966, Webb initiated several studies on how the Sustaining University Program was doing.[40] What he found was that by most "standard" measures of a successful government-university program, the Sustaining University Program was doing very well indeed. The fellowship program was highly regarded in the academic community. The facility grants provided badly needed buildings. The research money was put to work in ways that could be described as interdisciplinary, in comparison to traditional research groupings, although in most cases this involved relations among physical and life scientists rather than between such "hard" scientists and social scientists.

---

38.   Clayton R. Koppes, *JPL and the American Space Program* (New Haven, CT: Yale University Press, 1982), Ch. 8.

39.   R. Cargill Hall, *Lunar Impact: A History of Project Ranger* (Washington, DC: NASA SP-4210, 1977), pp. 256-70.

40.   See Lambright and Henry, "Using Universities"; Lambright, *Launching NASA's Sustaining University Program.*

Webb was disappointed, however, with the university response to his desire for innovative approaches to complex problems. He had signed a memorandum of understanding (MOU) with each university president receiving a facility grant. These MOUs included commitments by the presidents to work on the Sustaining University Program's broader goals. For example, they were to seek new and more effective ways to make research results available to external clientele. There were reports on some campuses of industrial advisory committees, conferences on applications of new findings, outside consulting relationships of individual faculty members, and so on. But most of these initiatives seemed trivial to Webb. He was seeking a more profound response, basic attitude changes, a major restructuring of campuses, and new external relationships for academic professionals.[41]

Webb badgered his staff and eventually reorganized and changed the leadership of his Office of Research Grants and Contracts. However, the more fundamental problems were on the campuses of the United States. A task force that he appointed to study NASA-university relationships told him in 1968: [III-16]

The failure of the universities to respond to the explicit agreements of the memorandums—technology transfer and multidisciplinary research—suggests that the [Sustaining University Program] goals, which they contained implicitly, were not achieved. Thus, the [Sustaining University Program] facilities program cannot claim to have developed concern for societal problems, capability for institutional response, awareness of a service role, or strengthened ties with industry and the local and regional community.

The major criticism that must be made of the universities' response to the Memorandum of Understanding is that they did not try. They clearly committed themselves to make an "energetic and organized" effort to implement the memorandums, and then did not make it.[42]

The year 1967 was the turning point for the Sustaining University Program, as well as a turning point for NASA in general. The reasons behind this shift reflected Webb's policy dissatisfaction, but they were more closely related to budget constraints. President Lyndon Johnson, in putting together the federal budget that went to Congress that year, looked everywhere for budget savings to finance the conflict in Southeast Asia, which was now becoming his dominant preoccupation. The Sustaining University Program was nice to have, but not really essential, in the president's view, and he ordered Webb to terminate the program.

Given his own frustration with the program's results, Webb was not in a good position to defend the universities. Indeed, as Vietnam protests on campuses heated up, Johnson was not anxious to listen to any defense of academia. The best thing Webb could accomplish was to get permission to curtail, rather than terminate, the program and to do so over time. Webb had just a few more initiatives he wished to try before closing the program—initiatives that included research in administration and management, engineering systems design, and aid to historically black colleges and universities.

The $31 million budget for the Sustaining University Program was slashed to $10.9 in fiscal year 1968 (calendar year 1967). As Webb left NASA in November 1968, the program was scaled down even further, and it was eventually terminated completely by President Richard M. Nixon. The program's funding ended officially in 1970.

---

41.   Laurin L. Henry, *The NASA-University Memoranda of Understanding* (Syracuse, NY: InterUniversity Case Program, 1969).
42.   Homer Morgan, *et al.*, *A Study of NASA University Programs* (Washington, DC: NASA SP-185, 1968), p. 58.

The Sustaining University Program's lifetime ran from 1961 to 1970 (fiscal years 1962 to 1971). It obligated more than $200 million to research, training, and facilities that complemented and facilitated NASA's larger research project effort. Some of the program's accomplishments are as follows:

- More than 4,000 graduate students at more than 100 universities were financed in space-related disciplines.
- About 1,400 faculty members participated in research and design projects at NASA centers during the summers.
- Thirty-seven research laboratories were built on university campuses.
- More than 3,000 space-related endeavors were carried out under the research portion of the program.

Successful by almost every customary standard, the Sustaining University Program enlarged the personnel base from which to draw aerospace scientists and engineers, brought new universities into aeronautics and space research, facilitated regular participation by scientists in NASA project research, consolidated disparate research endeavors into space "centers" on campus, and served as a model for other agencies with regard to institutional grants, geographical spread, and other features. It even stimulated many social scientists to focus on science policy and technology as a dominant concern. What the program did not do was meet the broader criteria set by Webb:

He hoped to see more innovation and change in universities—broader capabilities for multidisciplinary research, university concern with the technology transfer process, increased involvement with industry and community and regional problems, developing capability for institutional response to societal need. These hopes were largely disappointed. By the late 1960s, there was evidence on some campuses of movement in the directions Webb sought, but just as these were appearing [the Sustaining University Program] ended.[43]

Problems With Industry: NASA's Relationship With North American Aviation

Without question, the NASA-industry-university partnership had produced the successes of the Mercury program. This partnership was so effective in the Gemini program that it won an award for achievement in 1966 as an example of excellence. In January 1967, however, the Apollo fire occurred, taking the lives of three astronauts while they conducted tests in a space capsule on the launch pad.[44] This served to focus attention on problems in the relationship of NASA with North American Aviation, the builder of the Apollo spacecraft. No doubt, some of the issues involved were present where other government-industry interactions were concerned. However, the NASA-North American problems were especially significant, given the central role North American played in

---

43.   Lambright and Henry, "Using Universities," p. 73.
44.   On the Apollo 204 capsule fire, see "The Ten Desperate Minutes," *Life*, April 21, 1967, pp. 113-14; Erik Bergaust, *Murder on Pad 34* (New York: G.P. Putnam's Sons, 1968); Mike Gray, *Angle of Attack: Harrison Storms and the Race to the Moon* (New York: W.W. Norton and Co., 1992); Erlend A. Kennan, and Edmund H. Harvey, Jr., *Mission to the Moon: A Critical Examination of NASA and the Space Program* (New York: William Morrow and Co., 1969).

Apollo, and the resulting managerial solution was an example of what Webb called "innovating our way" out of a problem.[45]

The North American controversy went back to the original award of the Apollo spacecraft contract. As noted, this was one of those rare occasions when Webb, Dryden, and Seamans overruled the Source Evaluation Board. Charges of "politics" were hurled at the time, and not forgotten in subsequent years by NASA critics. What made the North American Aviation award stand out was its size and the fact that it made the corporation the single most important contractor for NASA in terms of sheer work.

The nature of the Apollo program was such that it entailed a relatively small number of huge awards. North American received two of these. The six largest NASA contract awards made to industry all involved Project Apollo. The expenditures on these contracts through fiscal year 1969 are shown in the following table.

Major NASA Contracts
(cumulative awards through 1969)

Contract	Contractor	Cost (in billions)
Apollo Spacecraft	North American Aviation	3.345
Lunar Excursion Module	Grumman Aerospace	1.914
S-IC Stages of Saturn V Rocket	Boeing Company	1.377
S-II Stage of Saturn V	North American Aviation	1.269
S-IVB Stage of Saturn V	McDonnell Douglas	1.097
Apollo Integration and Systems Support	General Electric	0.754

Source: NASA, *Annual Procurement Report,* FY 1969, p. 30. Cited in "R&D—The Government-Industry Relationship," Thomas P. Murphy, *Science, Geopolitics, and Federal Spending* (Lexington, MA: D.C. Heath, 1971), p. 173.

Hence, from 1961 on, NASA knew it had an unusually dependent relationship with North American Aviation. Marshall Space Flight Center managed the S-II contract on behalf of NASA, and the Manned Spacecraft Center managed the Apollo spacecraft contract. NASA worried that North American was not always giving the agency's work the attention required.

---

45.   Numerous inquiries took place concerning the Apollo 204 capsule fire. See U.S. House, Committee on Science and Aeronautics, *Apollo and Apollo Applications: Staff Study for the Subcommittee on NASA Oversight of the Committee on Science and Astronautics, U.S. House of Representatives, Ninetieth Congress, Second Session* (Washington, DC: U.S. Government Printing Office, 1968); U.S. House, Committee on Science and Astronautics, *Apollo Program Pace and Progress: Staff Study for the Subcommittee on NASA Oversight, Ninetieth Congress, First Session* (Washington, DC: U.S. Government Printing Office, 1967); U.S. House, Committee on Science and Astronautics, Subcommittee on NASA Oversight, *Apollo and Apollo Applications: Staff Study, Ninetieth Congress, Second Session* (Washington, DC: U.S. Government Printing Office, 1968); U.S. House, Committee on Science and Astronautics, Subcommittee on NASA Oversight, *Investigation into Apollo 204 Accident, Hearings, Ninetieth Congress, First Session,* three volumes (Washington, DC: U.S. Government Printing Office, 1967); U.S. Senate, Committee on Aeronautical and Space Sciences, *Apollo Accident Hearings, Ninetieth Congress, First Session,* seven volumes (Washington, DC: U.S. Government Printing Office, 1967); U.S. Senate, Committee on Aeronautical and Space Sciences, *Apollo 204 Accident: Report of the Committee on Aeronautical and Space Sciences, United States Senate, with Additional Views* (Washington, DC: U.S. Government Printing Office, 1968).

During the early years of the relationship, North American Aviation developed a negative reputation within NASA. The company, for its part, thought NASA's criticism unfair. By 1965, the delays on both the S-II and the spacecraft were long enough for NASA Headquarters to become truly concerned. Late in 1965, the director of the Apollo program within the Office of Manned Space Flight, U.S. Air Force General Samuel C. Phillips, organized a "tiger team" of NASA specialists who went to North American to investigate what was going on. Phillips prepared a highly critical report that would later become notorious as the "Phillips Report." In the report, a series of extreme criticisms were pointed directly at North American. [III-17, III-18]

During 1966, North American worked to respond to the NASA criticisms; however, problems continued. The most visible ones were recounted by *Aviation Week and Space Technology*, a trade journal, on November 21, 1966. It reported on a "crisis" threatening the U.S. Moon landing venture.[46] The specific problems reported in the article included the structural failures of both a North American command module fuel tank and the S-II stage. They were indeed serious problems, so much so that Webb felt obliged to alert President Johnson to them.

NASA and North American did in fact quickly address these known issues. By the end of 1966, the situation was looking so good that optimism prevailed among NASA's technical people. However, one technical issue that was not addressed was the possibility of a fire in the pure oxygen atmosphere of the space capsule. The fire problem did not become an issue until it actually occurred in January 1967. Indicative of the tangled state of NASA-North American Aviation relations at the time was the circumstance that NASA and its contractor were haggling over a renegotiation in their basic agreement at the turn of the year. This most significant of all the NASA-industry partnerships was actually held together only by a letter contract as 1967 began.

The fire took place January 27, 1967, and threw NASA-North American relations into turmoil. NASA established an internal accident review board, which was followed by a series of congressional investigations. With the benefit of hindsight, it can be seen that the first six months after the fire was a period of crisis management, with the succeeding months a time of recovery. During the crisis management period, media attention was searching and accusatory. The NASA-North American partnership was a target, as was the performance of the NASA Administrator in particular.

There were charges that the original award to North American Aviation was a result of political pressure led by North American lobbyist Fred Black and a former Lyndon Johnson associate and Washington insider named Bobby Baker. Meanwhile, the NASA investigation showed that NASA and North American were both at fault, with many errors of both omission and commission. Webb concluded that the basic relationship was sound; however, "surgical" changes would have to be made. This meant key personnel changes; the head of NASA's Apollo spacecraft project office in Houston was replaced. At Webb's adamant insistence, his counterpart at North American was also replaced. The NASA-North American contract was renegotiated so that the contractor was penalized financially for the accident. And most importantly, a new contract was negotiated with Boeing to certify that "the whole unit, vehicle and payload, does function together, is compatible, and is ready for flight." The Boeing contract was announced by Webb in congressional testimony on May 9, 1967.[47] All these actions were taken rapidly, largely at the command of Webb, and sometimes after bitter discussions between Webb and North American

---

46.   "Problems Force Drastic Apollo Rescheduling," *Aviation Week and Space Technology*, November 21, 1966, p. 36
47.   Levine, *Managing NASA in the Apollo Era*, p. 90.

President, J. Leland Atwood, with Webb threatening to take the Apollo contract away from North American unless the company went along—which it ultimately did.

All this happened while Webb defended publicly and before Congress the basic strength of the NASA-North American system. Congress and Webb engaged in a major struggle over the right of Congress to see the aforementioned "Phillips Report" that had been so critical of North American. Webb regarded NASA's ability to deal frankly and privately with contractors as critical to its ability to root out problems at an early stage and then address them. In the end, Webb let Congress see the Phillips Report only in executive sessions of the Senate and House space committees.

After six months, the crisis decision making gave way to recovery. The wounds between NASA and North American Aviation began to heal. For everyone, the Apollo lunar landing in 1969 marked the final evidence of successful recovery. The Apollo fire, while not forgotten, became much less significant in the wake of this triumph. The issues in the NASA-North American relationship became matters for historians rather than policy makers. The successful lunar landing quite properly refocused attention on the positive aspects of NASA's industrial and university partnerships.

Other Organizational Innovations: Research Institutes

The basic relationship NASA had with industry and universities was a direct one. NASA addressed a university or corporation one-on-one. However, the agency experimented in its early years with other approaches to getting its work done. One approach worth documenting was the creation of a research institute. Its earliest manifestation originated in the Glennan years and grew under Webb. A different version came into being at the end of the 1960s, and a third variation was born in the 1970s.

One of these was the Institute for Space Studies. Robert Jastrow, a NASA physicist and scientific administrator, was concerned that NASA needed to have a close relationship with the best scientific minds in the country for its theoretical space science work. He proposed to Glennan that a special institution be established. In December 1960, Glennan approved setting up the Institute for Space Studies in New York City. It was established as an arm of the Goddard Space Flight Center, but with considerable autonomy over the choice of its research activities. The institute would have a small in-house staff and be a place where notable scientists could come and work for relatively brief stays. It would also work closely with Columbia University and other institutions in the New York City area. The institute flourished in the 1960s and evolved various programs of interaction with universities, succeeding in its prime objective of linking NASA more closely to the very best space science theorists. Such individuals came to NASA via fellowship and other arrangements with the institute.[48]

Another organization NASA created was the Lunar Science Institute, which was founded on a different kind of model—the university consortium. The origins of the Lunar Science Institute lay in the realization in the late 1960s that as Apollo flights brought lunar samples and other data back to the Manned Spacecraft Center, there was a need to maximize the use of these samples and other data by non-NASA space scientists.

The Institute for Space Studies was obviously a model, but NASA's Manned Spacecraft Center in Houston, in contrast to Goddard (which ran the Institute for Space Studies), was not oriented toward science. Instead of an institute managed by a NASA center, Webb turned to the possibility that an institute might be managed by a university or a group of

---

48.   Van Nimmen and Bruno, with Rosholt, *NASA Historical Data Book, Vol. I,* pp. 314-25.

universities. With the help of the National Academy of Sciences, NASA established the Lunar Science Institute, based near the Manned Spacecraft Center. [V-19] Then, on March 12, 1969, NASA formed a university-based consortium, called the University Space Research Association, to manage the institute, which remained in Houston.

However, the Lunar Science Institute was launched at a time of budget shrinkage, whereas the Institute for Space Studies had been born during a time of growth. The new entity was not greeted with enthusiasm by civil servant-led NASA entities that were hard-pressed to defend existing resources. Personality issues exacerbated the situation. The Lunar Science Institute survived, but it left a legacy that was controversial.[49]

A somewhat later and entirely different approach to these institutions was the Space Telescope Science Institute. By 1970, NASA had a number of ambitious space science projects on its agenda, but because of budget cutbacks and government-academic rivalries, relations between NASA and the scientific community had deteriorated. The agency consciously searched for better ways to deal with the community. The space telescope, a high-priority program for scientists as well as for NASA, became a vehicle for finding a solution to what Homer Newell has described as a "love-hate relationship." Astronomers, those scientists most concerned with the telescope, had Kitt Peak National Observatory and other national facilities in mind. They called for an institute that would be managed by a university consortium and located at a university to maximize their control over the telescope's observation agenda. NASA, which had its own in-house scientists, did not wish to relinquish such control. NASA insisted that it was a mission agency, not the National Science Foundation.[50]

University astronomers and NASA scientists (chiefly at Goddard Space Flight Center) fought for the next few years. By 1975, an important inside ally of academia emerged. [III-20] This was Noel Hinners, Associate Administrator for Space Science. For Hinners, "an institute could solve two problems: one, pacify, if you will, the ground based astronomy community, so that they'd be all the more supportive of the Space Telescope, and two, really provide an external advocate for a good operations program." In short, Hinners concluded that unless NASA had a united constituency outside NASA to help promote the telescope, the agency could not get the necessary resources to have a telescope at all. This meant giving the astronomers what they wanted: the Space Telescope Science Institute.[51]

Hinners arranged for the National Academy of Sciences to study the plan and eventually added its blessing to the institute in 1976. [III-21] In 1978, NASA Administrator Robert Frosch followed suit. NASA Headquarters backed the academic astronomers over the NASA scientists, and Hinners announced the NASA decision to Congress, pointing out that the agency would retain operational control of the telescope in orbit. [III-22] On January 16, 1981, following a vigorous competition, Frosch announced that a university consortium based at Johns Hopkins University in Baltimore, Maryland, would receive the contract to operate the Space Telescope Science Institute; it has been in operation since that date.[52]

---

49.   Newell, *Beyond the Atmosphere*, pp. 240-42.
50.   The story of this institute's creation has been told in Robert W. Smith, with contributions by Paul A. Hanle, Robert H. Kargon, and Joseph N. Tatarewicz, *The Space Telescope: A Study of NASA, Science, Technology, and Politics* (Cambridge, England: Cambridge University Press, 1989), pp. 207-8, 226, 337-39, 342-52.
51.   *Ibid.*, p. 187.
52.   *Ibid.*, pp. 187-220.

Changing NASA-University Relations

During the latter 1970s, there were several efforts to improve the efficiency and effectiveness of NASA's administration of research grants and contracts to colleges and universities. In 1977 and 1978, a review of the entire program led to several reforms to improve accountability, ensure quality, and establish mission criticality for university research supported by NASA.[53] [III-23, III-24]

Even if these reforms were successful, other difficulties had emerged by the early 1980s as the launch rate of scientific satellites by NASA had dropped from its peak of four to five missions a year to only one or two annual flights. Moreover, the Sustaining University Program had disappeared a decade earlier, and NASA's graduate fellowship program had been terminated. Indeed, the purchasing power of the space science budget had been cut almost in half over a two-decade period. Contrary to expectations, frequent opportunities to carry out scientific investigations on the Space Shuttle were not emerging.[54]

In this context, NASA in 1983 undertook a comprehensive re-examination of its relationship with American universities. This review validated the perception that there were serious problems in the relationship and proposed a series of steps that NASA might take to address those problems. [III-25] However, most of those steps fell victim to continuing pressures on the Office of Space Science and Applications budget; only the recommendation to reinstitute a Graduate Fellowship Program was fully implemented. By the mid-1980s, the NASA space and Earth sciences program, including its university-based component, perceived itself in a crisis situation; the intimate and mutually productive relationship that had developed over the past quarter century required revitalization.[55]

A new wrinkle to NASA-university relations took place in 1988, when Congress passed the National Space Grant Act, which established a national program of space grant colleges and universities eligible for a major fellowship program. [III-26] With the first competitive awards for fellowships in 1989, 21 independent space grant consortia began operation. Three years later, the number of consortia stabilized at 52. The intent of this program was to:

- Continue to strengthen the national network of colleges and universities with interests and capabilities in aeronautics, astronautics, Earth systems, space science and technology, and related fields
- Encourage cooperative programs and collaborations among colleges, universities, business and industry, and federal, state, and local governments
- Promote programs related to aeronautics, astronautics, Earth systems, and space science and related technology in the areas of research, education, and public service

---

53.   Robert A. Frosch, NASA Administrator, to Frank Press, Director, Office of Science and Technology Policy, Executive Office of the White House, December 12, 1977; Walter C. Shupe, NASA Director of GAO Liaison Activities, to Distribution, "GAO Survey of NASA's Administration of Research Grants and Contracts to Colleges and Universities," March 2, 1978; W.A. Greene, NASA Chief of Policy Coordination, to Director of Contract Pricing and Finance Office, "Policy on Advance Payment of Contract Financing," June 2, 1978; Walter C. Shupe, NASA Director of GAO Liaison Activities, to Distribution, "Comments on GAO 'discussion paper' on NASA's administration of research grants ad contracts with universities," June 28, 1978, all in University Affairs Files, NASA Historical Reference Collection.

54.   NASA/University Relations Study Group, Office of Space Science and Applications, "The Universities and NASA Space Sciences," Initial Report, July 1983, p. 1, copy in NASA Historical Reference Collection.

55.   Space and Earth Science Advisory Committee, NASA Advisory Council, *The Crisis in Space and Earth Science: A Time for a New Commitment*, November 1986, copy available in NASA Historical Reference Collection.

- Recruit and train U.S. citizens for careers in aeronautics, astronautics, and space science and related technology, placing special emphasis on diversity by recruiting women, underrepresented minorities, and persons with disabilities
- Support the national agenda to develop a strong science, mathematics, and technology education base from elementary through university levels[56]

This partnership infused various educational institutions in the United States with funding from NASA to further aerospace science and technology in the same way that the National Land Grant College Act of 1862 made federal resources available for higher education in the nineteenth century.

A New Role for NASA—Supporting U.S. Industry

One of the themes that President Ronald Reagan's administration brought to Washington in 1981 was increased reliance on the U.S. private sector, rather than the government, to take the lead in developing new areas of economic and societal activity. With respect to the space program, there was a flurry of interest in "privatizing" various elements of the government's activities, including the Landsat program, the operation of expendable launch vehicles, and even the construction and operation of additional Space Shuttle orbiters. Another area of emphasis was the potential for substantial economic returns from space; one influential projection was that by the year 2000, the annual revenue from commercial activities in space could reach $65 billion.[57] The White House issued a National Commercial Space Policy in 1984; in response, NASA developed a "NASA Commercial Use of Space Policy" during the same year. [III-27] This policy was intended to implement a new goal for the space agency—partnerships with U.S. industry to "expand opportunities for U.S. private sector investment and involvement in civil space and space-related activities."[58]

In response to this emphasis on space industry, NASA established in September 1984 an Office of Commercial Programs, to be overseen by an associate administrator at the NASA Headquarters level. This new entity was intended to provide "a focus for and facilitate efforts within NASA to expand U.S. private sector investment and involvement in civil space related activities." Specifically, NASA Administrator James M. Beggs intended the office to foster:

- New commercial high-technology ventures
- New commercial applications of existing space technology
- Unsubsidized initiatives aimed at transferring existing space programs to the private sector[59]

---

56. *National Space Grant College and Fellowship Program: The First Five Years, 1989-1994* (Washington, DC: NASA Education Division, n.d.), p. 17.

57. For one version of this forecast, see David Lippy, as quoted in *The New York Times*, June 24, 1984, Sec. 3, p. 1. For a sample of the discussion about the commercial potentials of space, see Aerospace Industries Association of America, *A Current Perspective on Space Commercialization*, (Washington, DC: Aerospace Research Center, 1985); Business-Higher Education Forum, *Space: America's New Competitive Frontier*, April 1986; John M. Logsdon, "Space Commercialization: How Soon the Payoffs?," *Futures* 16 (February 1984); John M. Logsdon, "Status of Space Commercialization in the U.S.A.," *Space Policy* 2 (February 1986).

58. James M. Beggs, NASA Administrator, Memorandum to Officials-in-Charge of Headquarters Offices and Field Installations, "NASA Commercial Use of Space Policy," October 29, 1984, Administrators Files, NASA Historical Reference Collection.

59. James M. Beggs, NASA Administrator, Special Announcement, "Establishment of the Office of Commercial Programs," September 11, 1984, Administrators Files, NASA Historical Reference Collection.

Since it was first established in 1984, the Office of Commercial Programs has enjoyed mixed success in meeting the objectives laid out in the original charter.[60]

NASA, as an R&D organization created to carry out national science and exploration objectives in space, found its new relationship with an emerging but uncertain commercial space sector difficult to incorporate into its long-established patterns of institutional behavior. While a number of the initiatives contained in the 1984 Commercial Use of Space Policy were formally put into practice (perhaps most notably a network of university-based Centers for the Commercial Development of Space that brought industry and university researchers together with funding from both NASA and industry), a combination of mixed returns from early commercially oriented experiments, the *Challenger* accident and the resultant dramatic decrease in Space Shuttle flight opportunities, and institutional resistance at NASA meant that space commercialization never got very high on the agency's list of priorities for its future.

The emphasis on government-industry cooperation in commercializing space had another implication for NASA; other government agencies began to take a more active role in space-related issues that NASA had previously thought were its exclusive purview. During the 1980s, the Department of Commerce created an Office of Space Commerce, while the Department of Transportation formed its own Office of Commercial Space Transportation. Operating through the Executive Branch interagency process, these organizations were often critical of how NASA was carrying out its new partnership with industry. At other times, they pushed for new roles for NASA in the commercialization process. By the last year of the Reagan administration, commercialization advocates within the government were able to delay the release of a new statement of national space policy until it was accompanied by a set of commercially oriented initiatives. [III-28] The proliferation of space organizations within the government was not a comfortable development for NASA.

The efforts toward greater commercialization of space activities did not abate with the change of administrations in 1989. In January of that year, George Bush succeeded Ronald Reagan as president, with whom he had served as vice president. Bush continued to emphasize the development of space industry. During the Bush administration, the shaping and articulation of space policy were the work of the National Space Council, a descendant of the National Space Council first established in 1958 under the National Aeronautics and Space Act (Public Law 85–568). Chaired by Vice President Dan Quayle, the council consisted of the heads of all federal departments or other high-level offices having either a programmatic role or legitimate concern in federal government space activities, including NASA, the Department of Commerce (which contains the National Oceanic and Atmospheric Administration), the office of the director of the Central Intelligence Agency, and the office of the chairman of the Joint Chiefs of Staff, among others.

Several of the National Space Council's policy declarations, designated "National Space Policy Directives" (NSPD), related directly to commercial space policy. NSPD-2 ("Commercial Space Launch Policy") [III-29] reflected the administration's commitment in 1990 to developing a thriving commercial space sector by establishing "the long-term goal of a free and fair [space launch] market in which the U.S. industry can compete" internationally. NSPD-3 [III-30] elaborated the administration's commercial space policy

---

60. NASA Management Instruction 1103.38, "Role and Responsibilities—Assistant Administrator for Commercial Programs," November 6, 1984; James M. Beggs, NASA Administrator, to Edward P. Boland, Chair, Subcommittee on HUD-Independent Agencies, House of Representatives, December 27, 1984; NASA Press Release, "New NASA Initiatives Encourage Commercial Space Activity," November 3,1987, all in NASA Historical Reference Collection.

with specific guidelines "aimed at expanding private sector investment in space by the market-driven Commercial Space Sector." Each of these documents emphasized a strong presidential commitment to commercial space activity. Each also helped redefine the relationship of NASA to the space industry.[61]

Privatizing the Space Shuttle

One of the most potentially significant developments in NASA's history of private sector relations has been the privatization effort for the Space Shuttle. The Space Shuttle has been seen as a momentous technological innovation that has gone from R&D to operations. As it has made that transition, many observers have suggested that NASA, whose mission is R&D, should "spin off" the shuttle to the private sector. In the early 1980s, NASA administrators spoke of this eventuality taking place by 1990.

The transition did not take place in light of the 1986 *Challenger* accident, which showed the shuttle to be far less routine than NASA officials believed. The Space Shuttle is a piloted vehicle, and it is utterly indispensable for many of the most important NASA activities, including the space station. The notion of "operational" has, therefore, had to be redefined. In a real sense, the shuttle is not routine, and safety must be foremost in everyone's mind. After 15 years of flights, learning has taken place, however, and a new structural relationship has been proposed in the mid-1990s as desirable and possible. Also, budget pressures have forced NASA to take a hard look at shuttle management. The key document in privatization decision-making thus far is the *Report of the Space Shuttle Management Independent Review Team* (February 1995). Chaired by Christopher Kraft, former director of the Johnson Space Center, the review team called for replacing much of NASA's shuttle bureaucracy and many contractors with a single contractor possessing broad decision-making authority.

The report led to a decision by NASA in 1996 to negotiate a contract with a new company called United Space Alliance (USA), formed by a partnership of Rockwell International and Lockheed-Martin. It is believed that such a move would save $1 billion annually in present shuttle costs and require far fewer employees to service shuttle operations. The actual details of what would remain governmental and what would be private are to be worked out over time. Scheduled to begin by September 1996, the transition of the shuttle from public to private would take years. Privatization of the Space Shuttle would break new ground in NASA-business relations—indeed governmental-private sector relations in general. There has been talk and some action at NASA in terms of privatization in the past, but never has an activity so central to NASA been privatized, or one so overladen with risk to human life.

Privatization of the shuttle makes NASA a showcase for the Clinton administration's call for "Reinventing Government." However, the move is a controversial change in public-private relations. It entails marrying private profit, cost reduction, and public purpose in shuttle utilization. At the same time, privatization is expected to maintain a virtually perfect record in preventing loss of human life. The combination of requirements is unprecedented.

Conclusion

This essay has discussed NASA-industry-university relationships—a research partnership. The basic infrastructure for this partnership was established in the period from 1958 to 1969. Changes subsequent to this era have been variations on the models of this time

---

61.   This information was obtained from Sylvia K. Kraemer, Office of Policy and Plans (Code Z), "Explanation of Executive Branch Policy Directives," September 1995, copy available in NASA Historical Reference Collection.

frame, modified by the need to address funding constraints. Attempts to expand NASA's role to support private sector commercial space initiatives have had difficulties. Nonetheless, NASA overall has been an important pioneering agency in terms of industry and university relations. NASA's innovations in contracting and emphasis on spinoff technologies have been adopted by other government functionaries. NASA-university relations in the Sustaining University Program, while disappointing to Webb, were precursors of the current emphasis on government-university-industry relations today. NASA's geographical spread and institutional development policies certainly have been emulated elsewhere.

The problems in these relationships are more than balanced by their positive features. NASA's basic problem in these relationships since 1969 has been how to maintain some of their more successful features that were seen earlier in NASA's history—such as the balance between in-house capability and contractor expertise. It is easier to innovate when funds are growing rather than declining. Also, as the space program matured, it has become increasingly necessary to determine what activities must remain governmental and what can be privatized. The division of labor based on concepts of what is R&D and what is operational in space can be controversial, as the shuttle case indicates. Still, the basic infrastructure has proved itself robust and resilient. During the 1960s, NASA built a base that could last and a set of partnerships that could be renewed. The NASA-industry-university relationship today remains one of the more adaptive and important policy concepts when applied to national purposes.

Document III-1

Document title: T. Keith Glennan, Memorandum from the Administrator, "Functions and Authority—Office of Research Grants and Contracts," April 6, 1959.

Source: NASA Historical Reference Collection, NASA History Office, NASA Headquarters, Washington, DC.

When T. Keith Glennan became the NASA Administrator in the fall of 1958, just as the space agency began operations, he had the opportunity to frame relations with other federal and private organizations as he wished. Recognizing that universities held much scientific and technical expertise, he naturally sought a formal alliance that would allow a mutually beneficial relationship. Much of what he put in place was carried out by the Office of Research Grants and Contracts, the formal entity at NASA Headquarters charged with caring for this relationship. The following memorandum provides a statement of functions and authority for this office, as well as a rationale for action.

[1]   April 6, 1959

Memorandum from the Administrator

Subject: Functions and Authority—Office of Research Grants and Contracts

1. *Purpose of this Memorandum.*

   a. To redesignate the University Research Program Office as the Office of Research Grants and Contracts.

   b. To provide a statement of functions and authority for the office.

2. *Functions.* The Office of Research Grants and Contracts is assigned the following functions.

   a. Developing the NASA basic research program to be conducted in educational, scientific and industrial organizations, except for research directly related to or accomplished under the Space Flight Development Program.

   b. Assisting other offices and divisions in identifying basic research projects which justify NASA support.

   c. Serving as NASA contact point for research scientists and administrators of other organizations concerning research grants and contracts.

   d. Advising educational, scientific and industrial organizations of NASA basic research needs.

   e. Providing procedures for handling all unsolicited research proposals received by NASA.

   f. Obtaining and coordinating the review and evaluation of all research grant and contract proposals, with other interested and responsible offices and divisions.

[2] g. Providing the Procurement and Supply Division with recommendations and necessary justifications for all research grant and contract actions.

h. Ensuring, or providing when necessary, proper technical monitoring of sponsored research.

i. Coordinating the sponsored basic research program with related programs of the National Science Foundation and other Government agencies.

j. Ensuring, and assisting in, the publication of research information arising from the sponsored research program.

k. Providing administrative services for all approved research grants and contracts, including recommending type of contracts or grant instrument forms, maintenance of official agency files and records, handling of all correspondence, receipt and processing of vouchers for payment, etc., but not including such services for industrial research sponsored with Space Flight Development funds.

3. *Reporting Responsibility.* The Chief, Office of Research Grants and Contracts reports directly to the Director, Aeronautical and Space Research.

4. *Scope of Authority.* The Chief, Office of Research Grants and Contracts is authorized and directed to take such action as is necessary to carry out the responsibilities assigned to him within the limitations of this and other official NASA issuances and communications.

5. *Limitations on Authority.* The authority of the Chief, Office of Research Grants and Contracts, does not include technical cognizance of research activities funded in the Space Flight Development Program or research conducted in NASA facilities, but does include administration of university and non-profit institution grants and contracts to ensure conformance to administrative policies and procedures.

6. *Relationships* With Other NASA Officials. In performing the functions assigned to him, the Chief, Office of Research Grants and Contracts is responsible for recognizing the delegations of authority and responsibility of other NASA officials and for seeing that instructions he may issue are properly coordinated with the offices and divisions having joint interests.

T. Keith Glennan
Administrator

Document III-2

Document title: Walter D. Sohier, NASA Assistant General Counsel, "Legal Framework of NASA's Procurement Program," *NASA-Industry Program Plans Conference, July 28, 1960* **(Washington, DC: NASA, 1960), pp. 105-108.**

At a first-of-its-kind NASA-industry conference in mid-1960, NASA presented its thinking regarding future spaceflight plans to the industries that would play a key role in implementing those plans. In anticipation of increased contracting with industry, NASA's Assistant General Counsel, Walter D. Sohier, provided an overview of the space agency's procurement policy. In it he emphasized the legal aspects of the procurement policy being implemented by NASA. This policy served the space agency during the earliest period of its contracting for spacecraft, ancillary components, and support infrastructure in the lunar landing program of the 1960s.

[105]

Legal Framework of NASA's Procurement Program

It is my purpose to discuss with you the legal framework of NASA's procurement program. Since many of you are familiar with the basic statutory and regulatory authority under which the procurement operations of the Military Departments function, particular emphasis will be given in this discussion to similarities and differences between the rules which we in NASA must follow and those which govern the military. The subject of NASA's statutory patent policy is presented in the paper by Mr. Gerald D. O'Brien, our Assistant General Counsel for Patent Matters, and therefore will be omitted entirely from this discussion.

The question of what statutory procurement authority to give such a new agency in order for it to be able to carry out its rather unique program within the tight schedules necessarily involved was given considerable thought during preparation and enactment of what is now the National Aeronautics and Space Act of 1958. The original bill which was submitted to Congress by the Executive Branch contained a broad grant of substantive authority for NASA to enter into such contracts or other transactions as might be necessary in the conduct of its work and on such terms as the new agency might deem appropriate. This bill also proposed making applicable to the new agency the provisions of chapter 137 of title 10 of the U.S. Code, formerly known as the Armed Services Procurement Act of 1947.

There were both history and practical reasoning behind choosing this legislative approach to NASA's procurement authority. Historically, NASA's predecessor organization, the National Advisory Committee for Aeronautics (NACA), had been included along with the military departments and the Coast Guard as an agency to which the provisions of the Armed Services Procurement Act applied. Hence, this set of rules was already familiar to NASA people. From a practical standpoint, it was felt that the research and development procurement activities of the new agency were likely to involve the same general industry as that which was engaged in military research and development programs. To require this agency to follow about the same set of procurement rules as the military followed might avoid needless confusion on the part of industry and might cut to a minimum delays created by unfamiliarity with the practices of the new agency.

Essentially, this formulation of procurement authority was in the end enacted into law [in] the Space Act by the Congress, requiring NASA to follow the same statutory rules governing procurement procedures as the military. In spite of this fact, there are a number of differences between the statutory authority available to NASA and to the military departments that have an impact on the procurement process and which will be apparent to industry in its dealings with NASA.

The first, and most serious, difference relates to NASA's lack of authority to indemnify research and development contractors against unusually hazardous risks. The military departments have had such indemnification authority since 1952, but unfortunately this authority was not extended to NASA. We have been able largely to surmount this problem where nuclear material is involved, since the Atomic Energy Commission can extend indemnification coverage under the Atomic Energy Act to NASA contractors covered by operating licenses of the AEC. There remain, however, instances of other unusually hazardous risks that are involved in the performance for work for NASA. These risks by very definition are not normally insurable unless exorbitant premiums are paid. NASA has sought to rectify this lack of authority by proposing in [106] our legislative program to Congress that NASA be given the same authority to indemnify research and development contractors as is available to the military. We have hopes that when Congress comes back next month this, along with other items of the legislative program, will get favorable action.

A second difference between the legal authority available to the military departments and to NASA that has procurement implications relates to the authority of the military to exempt foreign purchases from the payment of duty under 10 U.S.C. 2383. This statutory provision provides that the Secretary of a military department may make "emergency purchases of war material aboard." It is clearly inapplicable to NASA. The immediate practical effect of this difference in authority is obvious.

A less obvious effect arises in connection with the Buy American Act and the handling of purchases in Canada. Defense currently provides that the purchases in Canada of supplies appearing on certain departmental lists will, in effect, be exempt from the Buy American Act. Supplies purchased in Canada that do not appear on such lists are likewise exempt, except that duty will be added to the price offered by the Canadian supplier, whether or not a duty free entry certificate is provided pursuant to 10 U.S.C. 2383.

For NASA to treat certain listed supplies in the same way as the Department of Defense would mean that, in spite of the actual payment of duty, the Canadian firm would be treated as if no duty were to be paid. Thus, under such a procedure, a Canadian firm would be awarded a contract even though the ultimate cost to NASA, when duty is considered, might be considerably more than the next lower bid or proposal. The taxpayer would come out the same in the end, but NASA appropriations would suffer.

We have recently determined that because of this difference in the application of duty to purchases in Canada by NASA we cannot adopt precisely the same policy and procedures as the military in dealing with Canadian companies. Moreover, American industry will be involved in this problem since the duty situation, so far as it affects subcontractors in Canada as well as other countries, must be taken into consideration. As you can see, this is a pretty complicated subject. Suffice it to say that NASA has tried to minimize the procedural differences in this area between dealing with the military and dealing with NASA. But certain differences must remain, since our authority to exempt from duty purchases from foreign sources is not the same as that of the military.

A third difference in legal authority available to DOD and NASA has been resolved by Executive Order. At the outset, NASA was not an agency authorized by the President to include the so-called "no set-off" provision in its contracts pursuant to the Assignment of Claims Act. By Executive Order No. 10824, dated May 29, 1959, the President remedied the situation, thus placing NASA in the same position in this respect as the Department of Defense, the Atomic Energy Commission, and the General Services Administration. Prior to the issuance of the executive order, this lack of authority of NASA had posed difficulties for contractors that were in need of financing.

An effort to eliminate a fourth difference between DOD and NASA is being made in the NASA legislative program for this year. This difference relates to the bonding requirements of the Miller Act. Under this act, the Secretaries of the military departments and of the Treasury are authorized to waive the requirements of the Miller Act in the case of cost-type contracts. NASA is not an agency to which this authority to waive has been extended.

When Title II of the First War Powers Act was in existence, this lack of authority posed no difficulty for NACA, NASA's predecessor, because Title II afforded a similar authority to waive this bonding requirement. However, with the repeal of Title II by Public Law 85-804, this failure to be specifically authorized by the Miller Act to waive bonds under cost-type contracts became significant. This was so because Public Law 85-804 clearly states that it is not to be construed as authorizing the waiver of "any bid, payment, performance, or other bond required by law."

We have had a recent contract situation arise where this has posed an awkward and, seemingly, unnecessary situation for us. It would [107] appear that the end result was never intended by the Congress, and we hope the situation will be remedied by the Congress in August by adding NASA to the agencies authorized by the Miller Act to waive bonds in cost-type contracts.

A final difference in authority, a remedy for which is also in our legislative submittal presently before the Congress, relates to our authority to outlease property. It has procurement implications to the extent we lease out industrial facilities to companies in connection with the performance of NASA contracts. The heart of this problem lies in the fact that the military departments have express statutory authority in 10 U.S.C. 2667(b)(5) to outlease property even though the consideration for such leasing is no more than maintenance of the property by the lessee. The absence of such express authority requires the charging of additional consideration. NASA has, in the Space Act, the authority to lease out property but does not have express authority to accept maintenance of the property as sole consideration. We are seeking such authority in our legislative program. Without it, NASA must treat industry differently in this respect than does the military.

I have discussed in some detail certain differences in authority which must be borne in mind in doing business with NASA as distinguished from the military departments. These are exceptions to the general rule that the rules involved in doing business with NASA are not appreciably different from doing business with the military. But I do not wish to overemphasize the differences in procurement authority between NASA and the military. Essentially, the same set of rules applies. This may be illustrated by turning briefly to a discussion of the regulations governing NASA procurement.

There are two main bodies of government Procurement Regulations at the present time—the Armed Services Procurement Regulation [ASPR], and the Federal Procurement Regulations. Generally speaking, the Federal Procurement Regulations are followed by the civilian agencies of the government, the Armed Services Procurement Regulation by the military agencies. This would seem to be a logical division, since most of the civilian agencies are governed by Title III of the Federal Property and Administrative Services Act of 1949, under which the Federal Procurement Regulations are issued, whereas the military is governed by a different statute establishing procurement procedures. The difficulty arises with respect to NASA, however, since—civilian as it is—it is governed by the same procurement law as the military departments and not the civilian agencies of the government.

Where does this lead to in terms of which procurement regulations to apply to NASA? A compromise was worked out on this question which, we feel, achieves the laudable objectives of the Federal Procurement Regulations System to eliminate the multiplicity of government procurement regulations while also achieving the objective of not requiring contractors of NASA to learn a different set of rules from those which they must follow in contracting with the military departments.

In essence, this compromise permits NASA to adopt the policies and procedures set forth in ASPR rather than in the Federal Procurement Regulations [FPR]. In practice, it seems unlikely that there will be many differences of substance between the FPR and the ASPR. If a policy or practice is not covered by the FPR, NASA will follow any existing ASPR policy or practice unless the Administrator of NASA determines that the objective of uniformity between NASA and DOD is outweighed by other considerations. Of course, in the area of patents, NASA must adopt special policies and procedures because of the unique patent provisions contained in the Space Act. It is also contemplated that, in the area of grants and contracts falling generally within the purview of Public Law 85–934, NASA will not be required to conform to any future FPR coverage on these matters.

NASA has already published a considerable portion of its procurement regulations. In the near future, the balance will be published. These will appear, as they are published, in the Federal Register as part of the FPR System. However, they will read and look very much like ASPR.

One practical effect of these arrangements concerning ASPR and the FPR and our manner of proceeding in the adoption of procurement regulations is that, when a military department is faced with administering a NASA [108] contract, it will not be unfamiliar to it. We can, in effect, tell the military: "Just follow your normal procedures for contract administration; our contracts are pretty much the same as yours." And, of course, for industry—ordinarily the same industry with which the military departments deal—a new set of rules need not be learned.

This simplifies the problems of negotiating contracts, too. If a company wants to change a NASA clause or form, the first question asked is whether a similar deviation has been granted by the military departments. If not, why is NASA any different? If so, NASA will certainly give the request careful consideration and will ordinarily grant the request.

NASA cannot afford to hash over old arguments with respect to some of the policies now set forth in ASPR and to open up these matters for extensive negotiation. If we have ideas as to changes that should be made in standard clauses or in major policies, we would prefer to work these changes out with the other government agencies as a normal course of proceeding. Of course, our special mission may give rise to the adoption of some different procurement policies and procedures in fields other than patents. In addition, it must be recognized that at the present time NASA's contracting is largely of a research and development nature: hence, it must orient its procurement methods to this fact. We do not wish to abandon flexibility where this is needed to get our special job done. But we feel that the present arrangements under applicable statutes and regulations are, in general, well suited to meet our needs. We are hopeful that the few deficiencies in authority which were noted earlier will be remedied by the Congress when it returns to finish up its work in August.

Document III-3

Document title: U.S. Congress, House, Committee on Science and Astronautics, Subcommittee on Patents and Scientific Inventions, "Property Rights in Inventions Made Under Federal Space Research Contracts," Hearings on Public Law 85–568, August 19-20, November 30, December 1-5, 1959, Report No. 47, 86th Cong., 1st sess. (Washington, DC: U.S. Government Printing Office, 1959), pp. 1-36.

Responding to a drive spearheaded by NASA Administrator T. Keith Glennan, the House of Representatives voted to amend NASA's title-oriented patent policy to reflect the Department of Defense's license policy. This legislation died, however, when the Senate failed to pass a similar version. These hearing excerpts capture the issues underpinning the patent policy question.

---

[1]

Property Rights in Inventions Made Under Federal Space Research Contracts

Wednesday, August 19, 1959

House of Representatives,
Subcommittee on Patents and Scientific Inventions,
Committee on Science and Astronautics,
Washington, D.C.

The subcommittee met, pursuant to call, at 10:15 a.m., in room 214-B, New House Office Building, Hon. Erwin Mitchell (Chairman of the subcommittee) presiding.

MR. MITCHELL. The subcommittee will be in order.

As the witnesses know, this is the first general session of the special Subcommittee on Patents and Scientific Inventions. I feel—and I know that the members of the subcommittee feel—that we are certainly considering a most important problem—not only one that is important currently, but which will have a great significance in the future. I think each of us feels that we can, by very slow and thorough study, possibly set a course of action in the patent field insofar as the Government is concerned.

We are certainly most privileged to have two distinguished specialists in this field to testify this morning—Mr. John A. Johnson, the General Counsel of NASA, and Mr. Gerald D. O'Brien, the Assistant General Counsel for Patent Matters, NASA.

Mr. Johnson, do you have a prepared statement?

Statement of John A. Johnson, General Counsel, National Aeronautics and Space Administration; Accompanied by Gerald D. O'Brien, Assistant General Counsel for Patent Matters, National Aeronautics and Space Administration

MR. JOHNSON. Mr. Chairmen, I do not have a prepared statement.

MR. MITCHELL. We would like you to just give us a general outline of NASA's activities insofar as patents are concerned.

MR. JOHNSON. I will be glad to, Mr. Chairman.

I should at the outset say, despite the chairman's very generous introduction, that I am not a specialist in patent matters, but Mr. O'Brien, our Assistant General Counsel for

Patent Matters, is and I would hope on more technical aspects of the patent problems that he will be our witness this morning. However, I am acquainted with and responsible to the Administrator of NASA for the patent policies of the National Aeronautics and Space Administration.

[2] As the committee knows, the National Aeronautics and Space Act of 1958 contains a section, section 305, which deals in quite elaborate detail with the subject of inventions which are made in the performance of contracts for the National Aeronautics and Space Administration.

The overall effect of section 305 is to require that such inventions become the property of the U.S. Government if they are made under the conditions specified in section 305(a) unless the Administrator of NASA determines that the public interest is better served by a waiver of rights to those inventions. In that case, however, the Government would still remain a royalty-free license to the use of the invention.

This policy which is expressed in section 305, the statutory policy, is at fundamental variance with the policy followed by the Department of Defense. It is rather similar to that followed by the Atomic Energy Commission—not identical with that, but it is quite apparent that the statute does, in its overall substance, follow the Atomic Energy Act rather than the practice of the Department of Defense.

As you know, the Department of Defense policy is one of ordinarily acquiring only a royalty-free license to inventions that are made in the course of research and development work sponsored by the Department of Defense agencies.

This policy of the Department of Defense is not the result of legislation. It is the result of policy determinations made in the executive branch of the Government, which have been well known to the legislative branch for many years and evidently acquiesced in by the Congress.

MR. MITCHELL. Mr. Johnson, just to pinpoint that, what is the underlying philosophy insofar as the NASA point of view is concerned?

Why the difference?

What is your thinking in NASA?

Why should there be the difference in the patent policy in DOD and NASA?

MR. JOHNSON. Mr. Chairman, the reason there is a difference is because the Congress so decided a year ago.

This was not the result of any determination within the agency. As a matter of fact, the agency didn't exist when Congress passed this law. Therefore, it has not been an open question for NASA as it has been for the Department of Defense.

The Department of Defense, being unhampered by legislation on this subject, has determined its policy on its own, but with congressional knowledge and acquiescence.

MR. MITCHELL. Is there any existing policy in NASA now insofar as this matter is concerned?

MR. JOHNSON. The existing policy in NASA is to do our best to implement the provisions of law passed by the Congress a year ago.

This has really been the only thing we could do, and it has been our task.

Now, if you are asking, Mr. Chairman, whether the agency has yet evolved a position on whether this legislation should be continued, this hasn't been formally developed yet. I am not really in a position to express either the agency's or the administration's point of view yet on that. It will be developed in time for the Congress to consider [3] at the next session because we are now in the process of preparing our legislative program for the next session of Congress. As you know, this must be submitted to the Bureau of the Budget. It may be transmitted to Congress only after we have the approval of the executive branch on it. I can, however, express some personal points of view on the matter, if you wish.

MR. MITCHELL. I would like for you to do so.

MR. JOHNSON. It is my own personal point of view—and I have expressed this publicly several times over the past few months—that it is undesirable for an agency such as NASA to be compelled by legislation to follow a patent policy that is fundamentally divergent from that of the Department of Defense.

Now, I say this without entering upon the question of whether it is good Government policy to take title to inventions that arise from Government-sponsored research or not. This is a question which, as you know, has been much discussed in the Congress and the legislative branch for many, many years. The Congress has never chosen to enact uniform legislation on this subject for the entire executive branch of the Government. We have some piecemeal legislation; we have legislation for the National Science Foundation; we have legislation for the TVA; we have legislation for NASA; we have legislation for the Atomic Energy Commission and probably others too. All of these are different. We have no legislation for the Department of Defense, which is the biggest agency of all spending money on research and development contracts.

What I would say is this: That, leaving aside for the moment the ultimate question of what is good Government policy as a whole, until a uniform legislative plan is devised by Congress for the entire executive branch of the Government—it is desirable that in the field of patents, as in all other legal aspects of our procurement program, we should be free to follow the Department of Defense policies.

I say that for this reason: All of our principal contracts are with the very same companies and will be with the very same companies that are principal contractors for the Department of Defense.

We are not really like the Atomic Energy Commission, which had to embark on an entirely new field of technology and where the major work was done within the Government—at least at the beginning. Here we are right in midstream as far as the whole aeronautics and space industry is concerned.

The space industry, as you know, is the aeronautics industry in transition.

MR. FULTON. I can't agree to that.

MR. MITCHELL. Mr. Fulton.

MR. FULTON. I can't have that go by unchallenged.

MR. MITCHELL. Go ahead

The record will show Mr. Fulton's objection.

MR. FULTON. Yes. I just can't have that as a general comment. I don't think you mean it.

MR. JOHNSON. Well, may I elaborate a bit on it?

MR. FULTON. Go ahead, sir.

MR. JOHNSON. At the present time the companies that have expressed the greatest interest—this applies to all parts of the country—in our leading contracts, of course, that are producing the boosters [4] for the space program are the same companies that have been in the aeronautics and missile business down through the years.

The years are of sort of recent origin because this is a fast moving industry.

There are probably some companies that may be confined solely to space business, but this is, I would say, not very much in evidence yet.

In any event, our contracting, by and large, is with the same companies that have substantial business with the Air Force and the Navy in particular, and the Army to some extent.

I might just cite.

Well, I won't mention names. That is beyond your investigation, I think, this morning.

In those cases where research and development work is involved we have had to request that our patent clauses be inserted in the contracts placed by the military departments.

This might not technically be called for by the terms of the statute because the statute speaks of contracts of the administration and I suppose it is an arguable question whether a contract placed by the Air Force with X Company at the request of NASA is a contract with the administration, but we felt, as a matter of policy, it would open the door wide to a type of evasion which the Congress certainly could not have contemplated if it were possible for NASA to place contracts through the military departments and evade section 305.

So, we have required, as a matter of policy, that our patent clauses be inserted in all of those contracts.

This means that a contract is placed by the Air Force at NASA's request for work that is substantially similar to the very work they would be placing themselves with that same company, the patent results of the first contract are essentially different from the patent results of the second contract, and yet this is the U.S. Government dealing with this company with the right hand and with the left hand. It is our feeling that this is not a good position for the Government to be in.

Now, I would like to say something more in that connection. Congress has been quite careful in every other respect in recognizing that we must do business essentially as the Department of Defense does it. NASA is the only nonmilitary agency that is under the terms of the Armed Services Procurement Act of 1947, now codified as chapter 137, title 10, of the United States Code.

NACA was under that act when it was first enacted 11 years ago. It was actually passed in 1948, I believe. Last year when the National Aeronautics and Space Act of 1958 was passed, Congress in section 301(b), specifically amended chapter 137 of title 10, which appears in that portion of the code that applies to the Defense Department, to make it applicable to NASA. This doesn't apply to the Atomic Energy Commission; it doesn't apply to the General Services Administration or any other Federal agencies of the Government. Thus, unlike all the other civilian agencies, NASA alone is under the terms of the Armed Services Procurement Act.

[5] One of the first official acts of the Administrator after NASA came into existence last fall was to announce that NASA would, insofar as practicable, follow the policies and procedures of the Armed Service Procurement Regulations, which is an elaborately developed set of regulations implementing the Armed Services Procurement Act of 1917.

We thought it would be undesirable, since Congress has determined that we should be under the same corpus of legislation, to be developing an essentially different group of regulations.

As you know, the General Services Administration has a responsibility for achieving maximum uniformity in procurement regulations in the executive branch of the Government and they recently published the Federal Procurement Regulations, or the first portions of it at least.

NASA has secured from the General Services Administration authority to deviate from the Federal Procurement Regulations. Insofar as the Armed Services Procurement Regulations are not consistent with them, we have the authority to follow the Armed Services Procurement Regulations rather than the FPR's when the Administrator determines that to be in the public interest.

There have been a number of other instances during the past year in which we have striven for legal uniformity with the Department of Defense to carry out what clearly seemed to be the intention of Congress in amending the Armed Services Procurement Act to include NASA.

We obtained by executive order the so-called V-loan authority to guarantee loans to contractors under the Defense Production Act. We obtained the authority to use what is called the no set-off clause under the Assignment of Claims Act. Both of these could be accomplished by Executive order. They were available to the Department of Defense and the President extended them to us.

We also joined with the Department of Defense in seeking identical pieces of legislation which would grant both NASA and the Department of Defense authority to indemnify contractors to a very large amount against certain extra hazardous risks involved in the kind of business they are doing for us.

One of those bills came to this committee; the other one to the Armed Services Committee. There has been no action on them at this session of Congress.

All of these actions, which we deemed to be in accordance with the intent of Congress, expressed in the portion of the act I referred to, have been designed to put us in a posture of legal equality or parity with the Department of Defense.

The one outstanding exception to that is in the field of patents and this, of course, is a field of great importance to industry.

Now, we are sort of the tail on the dog in this. Our program is not as big as the Department of Defense program; yet in the development of much of this hardware, it is quite indistinguishable so far as the technology is concerned from the kinds of things that the Department of Defense is doing.

MR. MITCHELL. If I may interrupt you at this point, I think I should state to the members of the committee the gentlemen were not requested to give any official position as far as NASA is concerned, but [6] merely to brief us on the existing law. However, I think it is most important—and I appreciate deeply, Mr. Johnson, your willingness—to give us your own personal views because that is exactly what the subcommittee wants. We want to hear opinions concerning the existing law and the operation of the law that you and Mr. O'Brien are so familiar with.

I can see, as a matter of convenience, why NASA would want to operate similarly to DOD, but, in your personal opinion, if you care to give it, is there any uniqueness about the R. & D. field so far as NASA is concerned that would cause the Government to have more interest in the result of these inventions?

Is there some difference between DOD and NASA in the R. & D. field?

MR. JOHNSON. Well, Mr. Chairman, there is a difference in the results so far as some of the ultimate product is concerned.

I would think so far as the technology is concerned and so far as the public interest is concerned that they are substantially identical. There is no significant difference.

As you know, the Department of Defense is way out in forward-looking research in space technology. It has to be because, while NASA is given a very extensive statutory responsibility by the first sections in the National Aeronautics and Space Act of 1958, section 102(b) excerpts from that and gives to the Department of Defense those activities peculiar to or primarily associated with the development of weapons systems, military operations or the defense of the United States, including the research and development necessary to make effective provision for the defense of the United States.

So far as inventions are concerned, the same kind of inventions can very well be made in the course of developing these advanced weapons systems that are utilized in space, as might be the case on the civilian side.

Now there are some uniquely civilian applications this might not be true of.

MR. MITCHELL. Any questions by any members of the committee?

MR. KING. Yes.

MR. MITCHELL. Mr. King.

MR. KING. Mr. Johnson, you expressed your opinion, unofficially, and we understand the spirit in which it was given, but I am interested in pursuing it just a bit.

Your reasoning has followed pretty much in the line that you think uniformity is a good thing inasmuch as NASA's practice, as indicated by the Congress, is not uniform and not consistent with the practice developed by the DOD: therefore, that creates an anomaly. You feel it might be well to bring the two together.

That, as I understand it, is the burden of your reasoning.

That, of course, avoids the question, the big question, which is: What is a desirable policy here?

If the NASA policy, as expressed by Congress, is inherently correct and sound and if the DOD policy, which has not received congressional approval, but has just grown up, is inherently unsound, then it seems to me the movement should be in the other direction.

Even though NASA may be the tail and the DOD is the dog, if the tail happens to be right and the dog happens to be wrong, then the movement would be in the other direction.

[7] So, I get to the more ultimate question as to what is a sound policy.

Now, that, I realize, is a tremendous question. You may not want to comment on it, but if you would like to I would be interested in hearing your comments.

MR. JOHNSON. Mr. King, I agree with your analysis entirely. That is the ultimate question.

I would rather not express an opinion on that because I am sure that we in NASA have a lot more to learn about this.

We have been in the process during the past several months of administering this just as objectively and fairly as we can, and I would like to assure the committee that I feel confident that the views I have just expressed have not impaired our objectivity in the administration of this provision of the law.

This is the ultimate question that Congress has been discussing off and on, and so has the executive branch for I don't know how many years. It would be, I think, a great public service if it could be decided wisely and finally.

I think I would rather at this stage of things simply say that until that question is settled and the Congress itself is able, through the processes you have of bringing together so many different points of view and the practices of the different agencies, to settle this thing, it is undesirable for an agency like NASA, given the kind of business we have to do, to be compelled to be essentially different from the Department of Defense.

MR. KING. Mr. Johnson, don't you feel, though, that this ultimate question is inevitably before us?

We can't evade it, and I, personally, would be most reluctant to predicate any decision of mine simply on the grounds of uniformity without coming to grips with this more ultimate question, and I haven't made up my mind on it and I don't want anything that I am saying to you to intimate that I have. I just recognize that as the ultimate problem, and I would be loath to take an action simply for the sake of uniformity if that action actually represented a step away from what I would otherwise consider to be the more desirable objective.

So, my comment is this: Don't you feel that this subcommittee still must face this ultimate question and predicate its action on the basis of the ultimate question rather than on the basis of uniformity alone?

MR. JOHNSON. I'm sure that the committee can't avoid facing the ultimate question.

I do think, though—and I suppose maybe I must differ with you fundamentally on this—that if a problem like this can't be settled with some reasonable degree of uniformity, here is an area where equal treatment by Government agencies is a principal that is perhaps even paramount to the question that you are concerned with. I don't mean by this you have to, say bring TVA into this picture, but you have here two agencies, NASA and the Department of Defense, that are doing a very similar kind of business with a similar segment of U.S. industry.

I don't think that you will find that this is a question that lends itself to a very clear black and white solution.

The very fact that Congress, itself, has dealt with it in such a variety of ways before, and the fact that it is argued by people who [8] have spent their lives in the patent field without clear-cut answers, I think would indicate that there is probably a lot to be said on both sides of the question, so that I don't think you would be committing a really gross error by at least achieving uniformity before you have solved the ultimate problem.

I think there is something essentially wrong with the U.S. Government, which, after all, is one legal person, dealing with a company through two different agencies on essentially the same kind of contracts and taking an invention with one hand and leaving it there with the other, or say, two different companies—one that happens to be only contracting with NASA at that time and the other one with the Department of Defense, but on essentially similar kinds of business and involving inventions that are in the same field of technology.

I think equality is still the basic principle of equity; and it is more desirable here to have equality of treatment than it is to perpetuate inequality for fear you might depart temporarily from what would appear ultimately to be the best principle.

MR. DADDARIO. Will the gentleman yield?

MR. KING. Yes.

MR. DADDARIO. If that is so and you feel there should be this equality, why is it that you put a limitation on some of these departments, which are not under this restriction, when they make contracts in behalf of NASA, that [th]is patent infringement type of restriction should apply?

MR. JOHNSON. Mr. Daddario, as I said before, we are administrating this law as objectively as we can without regard to the personal opinions that I have been asked to express here this morning. As we see the law, it could not be intended that NASA, simply by placing an order with the Defense Department rather than entering into a contract with X Company directly, would cause an arbitrary difference in patent results.

We know that when Congress writes a law, even as complicated as this, they can't say everything, and we have to try to determine what the intent of Congress was.

We read in section 305(b):

"Each contract entered into by the Administrator with any party for the performance of any work shall contain effective provisions——"

and so forth.

This is the basis for our patent clauses.

It was our conclusion that Congress must have intended that when any work is placed as a result of a NASA requirement by the Government it is within the intention of Congress that the patent provisions of section 305 apply.

You wouldn't get uniformity, anyway, because you would still have the NASA contracts as distinguished from the DOD contracts. So, you are already faced with the lack of uniformity. You have the contracts placed directly by NASA. You have the contracts placed directly by the Defense Department for its own business. Those are already nonuniform by virtue of the legislation.

Now, you have this intermediate category of contracts placed by the Defense Department at the request of NASA with our funds and for our proposes, and this is the question: Should we throw these into the pot with the Defense Department contracts or should [we] throw them into the pot with our contracts?

[9] We felt that if we didn't throw them into the pot with our contracts this would be just an open-sesame to an evasion of the patent requirements of section 305.

MR. DADDARIO. In some cases you have departments which enter into contracts for the benefit of NASA without necessarily utilizing funds obligated to NASA. They are essentially using their own funds.

Isn't that so?

Mr. JOHNSON. There is—yes—a small amount of that.

I don't know of any new contracts of that kind being placed.

MR. O'BRIEN. No.

MR. JOHNSON. I think all of our new business—you see, we have a certain number of contracts that were commenced originally by the Defense Department. We had an executive order last October transferring a number of projects from the Air Force and from ARPA, and also the Vanguard project from the Navy Department, as an example.

Now, in the case of those projects, contracts were already in existence and we have taken them over. That is a case where we clearly didn't feel it would be legally proper for us to amend the contract to change the patent situation, because if a contract means anything at all it means what is says when the contractor signs it.

At the present time I don't know of any cases where other Government agencies are continuing to place contracts with their own funds for our benefit.

MR. DADDARIO. That is all.

Thank you.

MR. KING. Mr. Chairman, I don't want to belabor this, but I would like to say for the record I think we would be derelict in our duty as a subcommittee if we did not consider the problem of uniformity in the context of the larger problem; that is, whether or not the Government's retaining patents is inherently a good policy or a bad policy.

I feel that that problem is before us, and I just wanted to state that for the record.

MR. MITCHELL. Thank you, Mr. King.

I think you have stated the purpose of the creation of this subcommittee, and that is the problem that we are going to look to.

Mr. Bass, I am going to recognize you, but I have just one question first.

What difficulty, if any, have you encountered as a result of the wording in the NASA Act and the difference in the policy of DOD?

Has it concerned you or made it more difficult to obtain contracts?

MR. JOHNSON. This is a difficult question, Mr. Congressman, to answer categorically.

We have had a number of contractors express reluctance at first to enter into contracts with us and have even requested additional compensation because of the loss of what would otherwise be their patent rights.

I think that in every case, even though it has taken time, we have negotiated this problem successfully and have not, I believe, to date been faced with a known situation of unwillingness to do business with NASA.

We have also taken a firm stand against any additional compensation for the loss of what they regard as their patent rights, but which we [10] regard under the terms of the law as the patent rights of the United States.

On the other hand, we have had a number of reports—these, I should say, are unauthenticated and it is not the kind of thing we can trace down easily—of companies that have put out the word to their own personnel that they will not accept any work for NASA; they will not do any work either as a prime contractor or a subcontractor which involves the loss of patent rights which they would otherwise retain if they were doing business for the Department of Defense.

Now, it is very difficult to know the extent of that because when we put out requests for proposals we don't know whether a company that doesn't respond is not responding for that reason or for some other reason.

Also, when you get down to the subcontractor level, in the lower tiers, this is something which some of our prime contractors might know more about than we do; but we have had information to that effect.

Perhaps Mr. O'Brien could be more specific on this.

MR. O'BRIEN. No; I don't think I could add very much to that, Mr. Johnson. I think that is the extent of my information.

MR. JOHNSON. We have to recognize, too, that we have been only in the beginning phase of this thing. It takes quite a long while for the impact of these things really to be felt.

We have been in the beginning stage of our contracting program, and the whole NASA program is still pretty young.

So far I don't believe we have had yet the first report, have we, of an invention made in the performance of one of our contracts?

MR. O'BRIEN. No.

MR. JOHNSON. The ultimate—

MR. MITCHELL. In negotiating these contracts if I may interrupt, have you had indication, without going into specifics, that if the contractor had the patent right this contract could be let less cheaply to the Government?

Have they indicated that, knowing the existing law and knowing it could be done?

Has there been some such indication when you negotiate on these contracts?

MR. JOHNSON. There have been some indications of contractors that wanted extra compensation for this thing, but it has been refused and they have taken the contract.

The answer is that, from my personal knowledge—and, of course, there are many of these that I have not had personal knowledge of—I don't know of that kind of case.

MR. MITCHELL. Mr. O'Brien.

MR. O'BRIEN. I know of none where they have placed a premium or said they would do it for a lesser amount if the patent provisions of their contract were similar to the Department of Defense patent provisions.

I only have the instances where they had tried to make additional charges for taking the patent rights provisions of the NASA patent clauses, and this was not permitted and they didn't take the contract with the original pricing.

MR. MITCHELL. Mr. Bass.

[11] MR. BASS. Then, Mr. Johnson, I assume you base your feeling in regard to this question primarily on the grounds of equity and what is fair rather than on any matter of impeding or hindering the defense effort because of this unequal treatment?

MR. JOHNSON. I do base it on that, Mr. Bass, primarily.

I think, too, with reference to Mr. King's comments, I am looking at this mainly as a lawyer rather than as a person concerned, as the committee has to be, with the ultimate question of Government policy. I think that, as a lawyer, in the negotiation of contracts with industry, it is basically unfair for two Government agencies, both representing, after all, the same U.S. Government, to be dealing in essentially different ways on a matter of this importance with the same contractors or with two contractors similarly situated.

MR. KING. Would the gentleman yield at that point of one question?

MR. BASS. Yes.

MR. KING. Right in connection with that, Mr. Johnson, do you not feel that the waiver provisions in the law allowing the Administrator to waive them under certain circumstances—that if he exercised that rather liberally, that that might not bring about the uniformity that you desire?

MR. JOHNSON. From the strictly legal point, Mr. King, the waiver provisions could be exercised to achieve absolute uniformity, but that would only be, I think, by disregarding the main intent of Congress in enacting section 305.

I mean by that it would be necessary for the Administrator to adopt almost a policy of automatic waiver in every case, because typically the Department of Defense does not acquire title to inventions.

Of course, this is what industry would like to have us do. It has been proposed to us. This is only natural. They would say, "Why don't you just utilize this very extensive authority granted here and, if you think uniformity is desirable, announce a policy of automatic waiver in almost every case?"

It certainly doesn't seem to us that Congress could have taken the trouble to enact a provision as elaborate and detailed as this is and expect that to be the result.

We haven't gone into our waiver regulations at this hearing today. We do have interim waiver regulations out and, while we think they are reasonably liberal, they don't begin to go that far.

MR. BASS. Mr. Johnson, you pointed out a little earlier in your testimony that in the Atomic Energy Commission they are governed by the same policy as NASA.

Is that not correct?

MR. JOHNSON. Not precisely, but more like ours than like the Department of Defense.

MR. BASS. And you justified that on the ground that in the atomic energy field this was a brand new field and, therefore, perhaps there was no inequity involved; is that right?

MR. JOHNSON. Mr. Bass, I was simply explaining the difference between AEC and NASA. I don't want to be in the position of justifying that legislation either. I don't know enough about the atomic energy business. I do know, I think, enough about it to know that it is quite different from our business.

[12] MR. BASS. I always thought of your business as pretty much pioneering, too.

That is the point I am coming to.

MR. JOHNSON. Mr. Bass, here, I think, is the reason we have section 305 in the act if I can speculate a little bit, because, as you know, this is a rather unique piece of legislation and has no significant legislative history behind it that we can read in the reports and the debates of Congress. In the establishment of a new civilian agency to carry on a very forward-looking program of research and development and a new and expanding technology, it must have seemed that the Atomic Energy Commission was the best precedent, the most analogous field of Government activity. But I think when you look at the kind of technology we are involved in, the kinds of contracts we are making, the very fact that most of the business we initially have had was transferred to us from the Defense Department, we must conclude that while we are out in a very forward field of technology, it is a field that has been in process of development a long, long time. You can't just drive a sharp line between space technology and missile technology and between missile technology and aeronautical technology.

It is a field in which the Department of Defense has already had a long and well-understood patent practice, which the Congress has at least acquiesced in, because it has been well known and is one of the big features of our economy.

I think the atomic energy field is quite different. It was developed originally as what you might call a Government-housed effort through the Manhattan project. This was done in large Government laboratories and installations segregated from private industry. We have a rule of Government monopoly in that field that pervades the whole thing which we don't have in the space and aeronautical field. We must not forget either that this agency is the National Aeronautics and Space Administration and the act is the National

Aeronautics and Space Act. It isn't just space technology we are talking about. This is the National Aeronautics and Space Act of 1958, and this section 305 isn't confined to space technology either. It applies to the whole field of activity of NASA.

MR. BASS. Now, for instance if I may interrupt, we are in the process of developing the nuclear-powered engine. Is that done by the Atomic Energy Commission or us?

MR. JOHNSON. We are participating in this.

I don't know how much—

MR. O'BRIEN. Yes.

MR. BASS. I was wondering—

MR. JOHNSON. I am not technically equipped to describe the division of effort between NASA and AEC on that sort of thing.

MR. FULTON. We have the Rover program.

MR. O'BRIEN. Under the Rover program funds are transferred to AEC.

MR. FULTON. If I may comment on that, under Rickover, of course, the AEC has the atomic nuclear engine and we have it under NASA under the Rover and other allied projects. There is a lapover.

MR. JOHNSON. In that area I know that Mr. O'Brien has worked out some patent procedures with the Atomic Energy Commission's patent counsel. Perhaps you would like to have him explain those.

MR. BASS. Yes; I would like to have him explain that.

[13] MR. O'BRIEN. In connection with Project Rover, the funds were transferred from NASA to the Atomic Energy Commission, which placed the contract with North American, and in this contract we had both the Atomic Energy Act of 1954 and the National Aeronautics and Space Act of 1958 to consider. The contract terms provide that the inventions which emanate from the research work undertaken pursuant to that contract will be subject to both acts and, without going into any details of the patent article which was included in that contract, it does attempt—and we hope it achieves that purpose—to make the inventions which were made in carrying out the research under that contract subject to both acts.

That is about the gist of the situation, I would think.

MR. BASS. One final question: If this committee and the Congress should decide it would be better to change the patent policy with regard to NASA, would we not be forced into applying the same rules with respect to the patents of the Atomic Energy Commission?

MR. JOHNSON. Mr. Bass, I don't think so at all because they don't have the same situation of relative uniformity in all these other respects with the Department of Defense that we have.

Congress, as I have mentioned before, has already decided that in the field of general procurement regulations NASA is to follow the Department of Defense.

This decision was made last year.

No similar decision has ever been made with the Atomic Energy Commission.

It has been a unique operation from the beginning.

So, whereas NASA is a separate agency, it doesn't have the same kind of uniqueness in its manner of doing business. Congress has recognized that in the legal field it is desirable for us to be as uniform with the Department of Defense as possible.

MR. BASS. Thank you.

I have no further questions.

MR. MITCHELL. Mr. Fulton.

MR. FULTON. We are glad to have you here, and I would like to go over this field rather widely so that we can check into and see what the problems are, and I would say to you, rather than have some of the answers directly today, I would rather have you give it some more thought, because I have been a member of the previous select committee and was on the committee at the time of the conference report, and I was also one of the conferees when the patent provision was put in

The question first comes up in this field, as it does in any field: What are the limits that we are talking about?

For example, are we going to talk simply about patents in space?

Are we going to talk about them in the field of aeronautics?

Are we going to talk about them in both fields?

For my part, I could see there would be a distinction between the patents fields in aeronautics and in space. One, the aeronautics field, has been developed under the National Advisory Committee for Aeronautics over a period of time under established principles. The other is an entirely new field.

Now, would you agree with that?

Would you agree that you could have a distinction between aeronautics' patents and space patents?

[14] Then I have some other distinctions I would like to make.

The question is: In your mind, must the aeronautics field always apply to space in the patent field?

I don't think they should.

MR. JOHNSON. I agree, Mr. Fulton, there can be a distinction between patents in the field of aeronautics and space.

I would like to define "space" rather restrictively in that connection if I could, and recite the fact that we have already made this distinction in our waiver regulations.

As you know, the law doesn't make any distinction between aeronautics and space.

MR. FULTON. I am going to point out the defects in the law, as I see it.

MR. JOHNSON. Yes.

MR. FULTON. Likewise, I am going to point out the defect, possibly, in not distinguishing between research and development contracts as regards patents and ordinary supply contracts either in space or aeronautics.

MR. JOHNSON. We made the distinction also.

MR. FULTON. You see, our section we made in the previous select committee just applies across the field in aeronautics and space as well as on every type of contract.

Isn't that right?

MR. JOHNSON. That is correct.

I feel I must say a few words in self defense at this point because—

MR. FULTON. No. I am just inquiring. I am not criticizing you.

MR. JOHNSON. May I say something in explanation?

We did make that distinction. We have made it administratively—and we were without any published legislative history on this to help us—because we simply could not believe, in the context of this section, that every time we entered into a contract for the supply of some office supplies or something of that kind it was intended that this kind of patent clause should go into it. We have confined the use of the patent clause to—we have a rather elaborate formula in our regulations; but, to oversimplify it, it is basically a research and development type contract. We felt, after all, that this is the only reasonable intention we could read into this section of the law; but the language is so broad that some of the initial commentators on this section made it appear more horrible than it actually is in practice.

MR. FULTON. The point I am making is: The law is too broad, and in that connection I disagree with it and believe it should be more carefully written, so that, as a matter of fact, I would compliment the NASA, the Administrator, and the people who have been advising him on making the distinction as to the type of contract that the patent provisions apply to.

Of course, when you come to a situation where there is a Defense Department type of contract, the Defense Department for years has had the provision that the particular person, the inventor, or the company with the contract has the exclusive right to the patent, subject to a free license or, rather, a free use by the Government, unless the inventor or the particular person who made the discovery is an employee of the Government.

[15] Now, that brings me to the next question: Should we not have a distinction under the patent provisions of the NASA law as between the contractors and the employees?

I would say to you I see no particular reason why there should be a difference as to employees in this connection, Government employees in this connection, especially when we have the Executive Order 10096 of 1950 covering all Government employees.

When there has been such an Executive order and we have the Government Patent Board, why do we make a distinction in this particular act?

I think the act might be deficient in that regard.

What do you think?

MR. JOHNSON. Mr. Fulton, we have taken the position that section 305(a) does not apply to our employees, but that they are still under Executive Order 10096—

MR. FULTON. I think that is fine.

MR. JOHNSON (continuing). Because it says:

"Whenever any invention is made in the performance of any work under any contract of the administration. . ."

The term "contract" is a broad one, and I admit it would be arguable to construe it so broadly as to include our employment contract with our own employees. But in view of the fact the Congress has, for example, in its TVA legislation dealt specifically with employees, we couldn't believe it was intended to work a distinction between the NASA employees and, say, the employees of the Department of Defense in view of Executive Order 10096.

MR. FULTON. But you specifically limit yourself to the determination of what the Chairman of the Government Patent Board has decided and the decisions of that Board, and under no circumstances do you go outside that and try to apply direct court decisions?

You are restrained administratively, are you not?

MR. JOHNSON. Right.

MR. FULTON. I will ask the other gentleman that question.

MR. O'Brien. Yes; this is true, Mr. Fulton. We are bound by the decisions of the Government Patent Board.

MR. FULTON. So, the particular agency of the Government—and you are representing NASA here—makes its own determination and then forwards that determination to the Chairman of the Government Patent Board for his decision to see if it is right, doesn't it?

MR. O'BRIEN. This is correct, sir.

MR. FULTON. But even there the Chairman doesn't decide whether the inventor is entitled to the invention unless the inventor, himself, appeals; isn't that correct?

MR. O'BRIEN. The Chairman of the Government—

MR. FULTON. The particular person aggrieved must appeal?

MR. O'BRIEN. The Chairman of the Government Patent Board has the inherent right to either agree or disagree with the initial determination of the agency, but—

MR. FULTON. Yes; but he doesn't review the particular ownership of the patent unless the inventor, himself, appeals; isn't that right?

MR. O'BRIEN. He may review the initial determination; yes.

[16] I can't agree with you, Mr. Fulton.

I think he may review the initial determination.

MR. FULTON. That is the practice.

I am trying to get the practice.

MR. O'BRIEN. Oh, yes; I think this is generally true.

MR. FULTON. As a matter of fact, when it comes to the Chairman of the Government Patent Board, he then is the one who construes this Executive Order 10096 of 1950 in accordance with the court decisions and not particularly in reference to its strict legal language.

MR. O'BRIEN. That is correct.

MR. FULTON. Isn't that correct?

MR. O'BRIEN. That is correct.

MR. FULTON. So that you have to go through this system to get a determination?

Is that not the case?

MR. O'BRIEN. That is true.

MR. FULTON. Let's go a little bit further. Let's look particularly to section 203(b)(3), where it says "to acquire (by purchase, lease, condemnation, or otherwise), construct, improve," and so on, and then, in the same sentence, includes "such other real and personal property (including patents)," and then it gives the right "to sell and otherwise dispose of real and personal property (including patents and rights thereunder)...."

Actually, to me, that portion of the section referring to condemnation is completely unnecessary in this provision because we have other provisions that will take care of it.

Is that not right?

MR. O'BRIEN. With respect to patents, I believe this is true.

MR. FULTON. With respect to patents.

MR. O'BRIEN. I don't know about other properties.

MR. JOHNSON. We wouldn't want to delete that wording because it applies to other things.

MR. O'BRIEN. A lot of other property.

MR. FULTON. Yes; but I am referring only to patents—

MR. O'BRIEN. I agree.

MR. FULTON. And I think we should exclude the wording in that section applying to patents because under title 48 of the United States Code there is also the provision that takes care of that administrative authority for patents.

MR. JOHNSON. This will simply not be used.

MR. FULTON. My point is: it is overlapping and redundant in respect to patents. So, the act is poorly written in that regard in that particular section.

Is that not right?

MR. O'BRIEN. I agree.

MR. FULTON. I would say when no condemnation is necessary, because the Administrator can acquire the use of any patents there existing upon payment of reasonable compensation to the patentee, it would then further cloud the title of anybody and make it harder for the individual patentee.

Is that not right?

MR. O'BRIEN. It would be if the authority were so exercised.

[17] MR. JOHNSON. I think it would be just inconceivable this authority would be exercised.

MR. FULTON. Why shouldn't we have a provision that gives to the inventor or the company that hires him the exclusive right to the ownership of the patent in commercial situations that have no direct relation either to military or security uses?

MR. JOHNSON. I think several of your questions have come pretty close to the waiver regulations we have developed under the present law.

MR. FULTON. Yes; that is correct, but I am trying to set what the law should be changed to because actually your regulations are based upon what the legislative intent of Congress must have been without any hearings on the patent provision and no legislative history.

Is that not correct?

MR. JOHNSON. Well, of course, the law itself gives us considerable discretion. So we haven't really had to justify everything in terms of what Congress might have foreseen.

We felt that congress certainly expected the Administrator to use his best judgment, but at the same time you are quite right in saying that we have had to sort of look in the dark here in trying to stay consistent with what Congress must have intended.

We have tried to do that.

MR. FULTON. Then where supply contracts are concerned and there is either background information, trade secrets, or previous patent rights—in that case, it would seem to me this particular NASA Act of 1958 is burdensome and restrictive.

You see, it doesn't give credit to the company which has a patent and experience built up in a particular field; does it?

MR. O'BRIEN. I don't know that I exactly follow you, Mr. Fulton.

MR. FULTON. Here is the point—

MR. O'BRIEN. I don't think we acquire rights under background patents.

MR. FULTON. Suppose some person, some inventor, or some company has the background information, the trade secrets and previous patent rights in a particular field; the question is: Should these all be made available to the Government without reasonable compensation?

MR. O'BRIEN. They should not and they are not under the act.

MR. FULTON. Secondly, when there is a new patent or patent in that field or a substantial discovery that would require the company to disclose these or make them available to the Government, does the mere fact of an additional discovery in the field require them to come up with all this other background, patent and trade information?

MR. O'BRIEN. Certainly not with respect to background patents. There is some question about the acquisition of technical data in order to practice the invention which is made under a contract with NASA.

MR. FULTON. That is the question I am raising, and I wish you would submit some sort of statement on it to get the line of demarcation as to where that might be.

(The information requested is as follows:)

The first question concerns the issue of whether or not the operation of section 305 of the National Aeronautics and Space Act of 1958 is burdensome or restrictive upon contractors with respect to the Government's acquisition of background patents or trade secrets.

[18] With regard to background patents, the NASA Patent Regulations, subpart A (24 F.R. 3575), specifically states that it is the policy of the National Aeronautics and Space Administration to pay reasonable compensation for the acquisition of "rights in background patents" and that the same will be acquired only by "specific negotiation," not by the automatic operation of the contract clauses used to implement section 305 of the act.

To the same effect, the special NASA "property rights in inventions" clause, which appears as appendix IX-A in these regulations, also provides in paragraph (g)(i) that any license granted to the Government does not imply the granting of any license under any dominating "background" patent.

Accordingly (excluding those inventions made by Government employees), NASA does not acquire, except by direct purchase, any rights in an invention that has been reduced to practice prior to and independently of a NASA contract.

With regard to trade secrets as they may be involved in normal patent acquisition, the special NASA "property rights in inventions" clause, referred to above, requires that the contractor shall furnish to the contracting officer a written report containing full and complete technical information concerning any invention made in the performance of any work under the contract. Compliance with this clause may require the contractor to reveal background technical information of a proprietary nature. Ordinarily, however, the type of information required for the preparation of a patent application is not that type of "background information" which would be susceptible to protection as a trade secret. Moreover, the NASA Patent Regulations, subpart A, paragraph 1201, 101 3(b), states that the contractor may initially furnish to the contracting officer only such technical information as may be required for the purpose of identifying an invention made by the contractor and in determining its utility in the conduct of aeronautical and space activities. When requested by the contracting officer, the contractor shall, however, prepare and furnish such additional technical descriptions of the invention as will be adequate for ready transposition to patent specification form and for effective prosecution of the patent application.

With regard to the matter of acquiring trade secrets directly, NASA's practice is like that of the Department of Defense concerning the acquisition of technical data and of rights in technical data. In those NASA contracts which have as one of their purposes the performance of technical or scientific work directed toward the development of models of equipment or practical processes, NASA requires that there be delivered such technical information as may be necessary for the manufacture of the equipment or the performance of the process. To this end NASA has adopted the data clauses as set forth in sections 9-203.1 and 9-203.4 of the Armed Services Procurement Regulation.

MR. FULTON. I think that is a defect of the act at present. With respect to research and development, I think, we could make a distinction there on that type of contract because generally industry is willing to give the background information, especially when it is for a military or security purpose.

Is that not right?

MR. O'BRIEN. Generally, I think so.

MR. FULTON. All right. Then let me disagree with the former gentleman here a little bit.

When you were speaking, I was making some notes.

You had spoken of this being the creation of a civilian agency and remarked that this was a new field of patent law that is being developed for a civilian agency when, as a matter of fact, under the Department of Defense the provisions for patents were otherwise.

I would like to point out to you in the TVA Act of 1933, under 16 United States Code, as well as in the National Science Foundation Act of 1950, there were two civilian agencies created, each of which had patent provisions different from the Department of Defense.

At the time this act of 1958 that we are speaking of for NASA was passed, we were within the emergency conditions, which may now be forgotten, of the first orbit of the sputnik. Secondly, no one then [19] knew as much about space as we do now and we thought that it was a new field, that it was much over and beyond anything that was then covered by the National Advisory Committee for Aeronautics.

I say in that connection, as a member of the former Space Committee, that is why the National Advisory Committee for Aeronautics was not just continued and the space put as a subdivision under that particular body, but the whole new concept was set up that it was to be the NASA rather than the NACA.

One of the great differences between Dr. Dryden and myself—I will speak for myself, although I know that Mr. McCormack felt a little bit along the same lines that I did—was that, as it was discussed so many times in the bearing before the select committee, space was just a buildup of aeronautics. Now, our feeling was that it was a new field and should be treated as such; secondly, that it had a good bit of the security requirements of the Atomic Energy Act because at that time we thought that either the sputnik or a space platform could cause us to lose everything we had. Under those circumstances, we wanted no one company to find and get the key to space and then everybody else in the country or the Government have to go through that one particular source in the approach to space.

So, I think you should take that philosophic background into account when we are now, at a later time, looking at the past history. For example, I had written down here my recollection that the inventions or discoveries of any employee of the U.S. Government or by any employee of the TVA corporation, together with any patents on those discoveries, are the sole and exclusive property of the corporation and the corporation is authorized to give licenses to various people.

MR. JOHNSON. May I comment on that?

MR. FULTON. That is the provision of the TVA Act, as I recall it, in 1933, so that we do have a precedent for NASA.

MR. JOHNSON. I would like to comment on that, Mr. Fulton.

That provision you refer to applies to TVA research by its own employees.

As I recall the report rendered by the Senate Judiciary Committee earlier this year on the TVA patent practices, they had acquired no patents as a result of Government-sponsored research with private industry.

As we said before, section 305 does not apply to NASA employees. It applies solely, on the other hand, to Government-sponsored research in private industries, and TVA is not a precedent for this situation at all. The TVA situation is taken care of under Executive Order 10096, which imposes a certain regime on it. TVA is different from other Government departments, but it is not a precedent for this kind of treatment of contractors.

The National Science Foundation, on the other hand, is not a precedent either because there the legislative provision merely is that the Foundation shall take such interest in patents as the public interest requires and, as you know, the National Science Foundation has followed the same practice as the Department of Defense in requiring only a royalty-free license.

MR. FULTON. As I recall it, the National Science Foundation provision requires that the contracts shall contain provisions regarding [20] the disposition of inventions produced under those contracts in a manner calculated to protect the public interest.

MR. JOHNSON. That is correct.

MR. FULTON. And the discoveries and patents must be directly related to the subject matter of the contract, and in the case of either the contractor or the inventor being an employee it must be directly in connection with the assigned duties or the purpose of the contract.

Is that not right?

MR. O'BRIEN. Yes.

MR. JOHNSON. But the legislative provision does not say anything about the taking of title to those inventions being the rule in the case of the National Science Foundation. In carrying out that particular provision of law the National Science Foundation ordinarily does the same thing as the Department of Defense does and only acquires a royalty-free license.

MR. FULTON. Yes, but don't you think when there is a specific legislative provision under the National Science Foundation Act of 1950 that the contracts that are let shall contain provisions governing the disposition of inventions produced under the contracts in a manner calculated to protect the public interest that that certainly is a provision relating to the title and use and licensing of the patents?

MR. JOHNSON. It relates to that subject matter, but it doesn't require the Government to take title to the patents, by any means.

MR. FULTON. No, but it certainly governs—

MR. JOHNSON. If it does, the National Science Foundation has been in gross disregard of the law for a number of years.

MR. FULTON. No, but it certainly limits the use of the patents, doesn't it?

Doesn't it limit the use, because every contract that is made with the National Science Foundation has to have these provisions in it that they are to be handled in a manner calculated to protect the public interest?

MR. O'BRIEN. It seems to me it would certainly lead to some interest of the Government or some governmental interest being acquired, but—

MR. FULTON. So, it is an extension in the act of NASA, but it is not contrary to those other two agencies and some of their actions.

I think it is certainly a like comparison to compare these two previous civilian agencies—one, the TVA in 1933 and the other the National Science Foundation Act of 1950, as well as the Atomic Energy Act.

Now, let us look at that for a minute. The Atomic Energy Act has been changed by the act of 1954. Would you please comment on what you think of the present state of the art in the Atomic Energy Act with the amendment of 1954 put in?

MR. JOHNSON. I am not competent to do that at all.

MR. FULTON. Would you please state that—

MR. JOHNSON. I know the Atomic Energy Commission has had testimony recently before the Joint Committee, but I don't feel competent to talk on that.

MR. FULTON. I believe they appeared before subcommittee of the Judiciary Committee as well.

If you will give us a short statement on that, I would like to have that.

[21] (The information requested is as follows:)

The question raised by Congressman Fulton concerns the statutory concept of aeronautical and space activities as it is used in section 305(c) of the National Aeronautics and Space Act.

Section 305(c) imposes a responsibility upon the Commissioner of Patents to determine which applications for patents disclose inventions having significant utility in the conduct of aeronautical and space activities. It was suggested by Mr. Fulton that the concept is too broad and that it does not permit a distinction between the field of governmental interests and the field of private interests regarding the area in which patents may not be issued without first having the applicant submit written statements of the circumstances under which the invention was made. It was suggested that the responsibility of the Commissioner should be delimited and proposals for doing so were requested.

It appears that the foregoing objective could be effected by statutory language basing the selection criterion to be used by the Commissioner of Patents on the concept expressed in the NASA Patent Waiver Regulations, subpart 1 (24 F.R. 8788), of inventions—

(1) primarily adapted for and especially useful in the development and operation of vehicles, manned or unmanned, capable of sustained flight without support from or dependence upon the atmosphere, or

(2) of basic importance in continued research toward the solution of problems of sustained flight without support from or dependence upon the atmosphere.

MR. FULTON. Could we make a distinction, then, between patents that are not being used for what we would call the welfare of the Government?

Suppose you had a patent discovery where its prime importance or effect was relating to the welfare of our Government or some important governmental functions; would you make some distinction there in trying to eliminate and put into the private field such patents?

Would it not have that effect?

MR. JOHNSON. I didn't hear the last.

MR. FULTON. For example—I will simplify it—to protect private industry in the private field, where there are nongovernment usages chiefly.

MR. O'BRIEN. Yes. I think that we have tried to make such distinctions in patentable inventions in our waiver regulations.

We have tried to reserve an entire area of patentable inventions, with respect to which no waiver would generally be granted, as those inventions which become perhaps associated with the public interest, so that it wouldn't be to the public benefit to grant rights in these inventions, inventions used almost exclusively in outer space, solar sails, or something of that character, because to grant rights in these inventions or patents on these inventions might carry the inference that private industry or private parties were authorized to go into outer space under no governmental regulation.

We have also in our waiver regulations identified a class of inventions as those inventions which have predominant commercial utility and only incidental utility in space and aeronautics.

MR. FULTON. Yes. Now, there is a comment there—

MR. O'BRIEN. As to this type of invention, we are granting or proposing to the Administrator to grant waiver of rights so that the contractor who made these inventions can exploit the invention to the public benefit, to bring these inventions into the hands of the public and to use the patent for that purpose.

MR. FULTON. So, my comment is: Section 103, when it makes the definitions that are very broad covering both equipment that is usable [22] and possible exclusively in outer space, as well as commercial-type equipment, is, therefore, too broad in its coverage and should be changed.

So, I would make a change in the definitions in section 103 to make the field of private enterprise larger and to protect what we in Government are deeply interested in, that is, the things that are related to Government uses, exclusive outer space uses or weapon purposes.

What do you think of that?

To summarize, that is to change the definitions and restrict them in section 103.

MR. O'BRIEN. Well, I think the definition of aeronautical and space activities, as set forth in section 103, is broad and probably could be more carefully defined.

I haven't given much thought to that, Mr. Fulton.

MR. FULTON. Would you look into that and submit us some sort of recommendation along the lines I have been trying to point up here?

I would rather not do it here because the time is running out.

MR. JOHNSON. That definition, Mr. Fulton, is only of significance in connection with section 305(c) insofar as patent matters are concerned.

MR. FULTON. That is correct.

MR. JOHNSON. That is where the term appears.

MR. FULTON. It has to be taken in connection with section 305(c).

Just one more point and I am through.

I was just trying to think back.

The question comes up of the development of the space field in relation to time. I can see that when we were passing and preparing for the passage of the act of 1958 we were under emergency conditions. The question now occurs: Are we in the same emergency conditions in space and are we in the same relative place where we have such a lack of knowledge that we have to keep the field open and, therefore, have a larger Government interest in these patents or has the time come where we now see more about the field and we should, therefore, say, as I would recommend, that the field of private enterprise and individual initiative and private rights should be more stressed?

Would you comment on that?

Where are we in point of time in relation to a transition period that is different from the Atomic Energy Commission in its development?

MR. JOHNSON. Well, I think we are in a substantially different position than we were a year ago.

I know our Administrator has made several statements to that effect—that we are able to shake down, in a sense, into a more orderly program and know where we are going and the worthwhileness of the things we are doing in a much better way than we were a year ago when it was necessary to try to do everything at once.

I don't think that I can compare this very profitably, Mr. Fulton, with the Atomic Energy Commission.

MR. FULTON. As you remember, the patent section of the American Bar Association at its 1956 meeting had recommended the repeal of the provisions of the Atomic Energy Commission patent sections.

They wanted them repealed.

They haven't taken any action since.

Then at their 1958 meeting they recommended the outright repeal of the patent sections of the NASA Act.

[23] To me, that probably goes too far and my disposition would be to try to go over it, as we are today, and pick out the places where the language is too broad and the provisions cover more than we intended because at the time we passed it, at that stage of the act, we couldn't make definite provisions that would account for all these variations.

Now, which approach would you use?

Would you use the ABA approach or would you use the approach that some of us on this committee recommend of revision, and move toward the private ownership and the private field?

MR. JOHNSON. Mr. Fulton, this, of course, is a question that we are all sweating over a good deal in NASA right now in preparing our legislative recommendations for the next session of Congress.

I would not expect Congress to repeal outright section 305, and I wouldn't think, speaking personally now, that NASA would make any such recommendation.

It seems to me that—

MR. FULTON. You, therefore, disagree with the patent section of the American Bar Association at its 1958 meeting?

MR. JOHNSON. I read that. I don't recall the detail now, but if it is true that they recommended simply an outright repeal I would disagree with that.

On the other hand, there are two ways of approaching it, and I think—

MR. FULTON. Actually, while you are on that point, while we are commenting on what they did do, they had a resolution opposing Government ownership of the patents and inventions arising from Government-financed research and development as well as repeal of the patent sections.

I must say that to you.

MR. JOHNSON. Yes.

MR. FULTON. 1958.

MR. JOHNSON. You have mentioned as a precedent the National Science Foundation provision. I would think that would be probably the minimum that the

Congress ought to do, if you were to undertake a radical treatment of section 305. Substitute something of that kind, which would express the concern of Congress in the protection of the public interest in patents in this field, but would leave to the Administrator great discretions as to how to do it, without imposing the kind of rules from which we now have to depart by means of waiver.

This is quite a different thing from section 305.

MR. MITCHELL. Would the gentleman yield?

MR. FULTON. I would like to have him continue. I am very interested in this point.

MR. JOHNSON. I say this without regard to whether the National Science Foundation has or has not carried out its legal responsibilities.

I don't have any opinion on that either because I don't know enough about their business.

In the alternative—and it is my guess, if I must do some forecasting now—this is probably the way we will present our legislative proposals.

MR. FULTON. I will be glad to hear it.

[24] MR. JOHNSON. In the alternative, we would propose a cleanup of this legislation along the lines you have mentioned this morning.

There are some things that obviously were done in haste, it seems to us, in this section and, on the basis of the past year's experience, even in line with what one might call the overriding philosophy of this section, you can make a lot of changes in it and make it more understandable and easier to administer.

Certainly I think the ultimate choice, as far as patent philosophy is concerned, is going to be one that the Congress will have to make and ought to make, I think, with this question of uniformity in mind, as well as Mr. King's ultimate question.

These two things have to be balanced, and whether you give one the greater consideration or the other I think is a very serious legislative problem.

My own personal preference would be to substitute for section 305 something very much like the general principles in the National Science Foundation Act and then hold us responsible for the way we protect the public interest.

MR. FULTON. How would that then correlate with your previous statement on the Department of Defense?

Why do you now say you would correlate this with the civilian agency, the National Science Foundation Act of 1950, when previously I thought you were going to say correlate it with the military and Department of Defense practices?

MR. JOHNSON. I am not suggesting the National Science Foundation just for the sake of making NASA uniform with a civilian agency. The National Science Foundation practice is actually the same as the Department of Defense practice at the present time. Now, that practice could be changed. If it seemed to be desirable in the public interest to change the practice under the broad terms of the National Science Foundation Act, they could do it. Under that kind of authority from Congress we could, as a matter of administrative policy, make our policies as uniform with those of the Department of Defense as we felt the situation demanded, and we could examine the results of that on the case-by-case basis to see whether the public interest was adequately protected.

In order to achieve uniform practice with the Department of Defense, you don't have to have uniformity in statutory language. The Department of Defense has no statutory language. The broad grant of authority to the Administrator to take such action as is in the public interest, which is really what the National Science Foundation Act says, could result in uniformity of practice, although not in uniformity of statute.

MR. FULTON. But you would still have that assertion of title under the section 305(d) and (e) remain subject to the Board of Review of the Patent Interferences, wouldn't you, and you would also have the final decision on the authority of the Administrator of NASA, wouldn't you, that is, the final decision on waiver?

MR. JOHNSON. If we did the thing I was just suggesting, you would eliminate all of this portion of section 305 that relates to title.

The National Science Foundation Act has nothing about title in it, Mr. Fulton.

[25] MR. FULTON. I know, but I am still saying: Wouldn't you still retain a waiver provision of some type or a title provision and keep it under the Review Board of Patent Interferences and leave some final authority on that particular type of thing in the Administrator of NASA?

MR. JOHNSON. I don't see how they are compatible.

It seems to me what you are suggesting now is that you really retain a rule that says title will ordinarily vest in the Government with the power of waiver vested in the Administrator.

This is radically different from what the National Science Foundation Act says. The National Science Foundation Act doesn't impose a rule of title acquisition.

MR. FULTON. That is right.

MR. JOHNSON. It leaves all the discretion to the head of the agency.

MR. FULTON. So, you would then have the complete title provision cut out in the NASA Act?

MR. JOHNSON. This is what I would say my personal preference would be at the present time in view of the fact I feel very strongly about the inequity that now exists between the DOD practice and ours.

MR. FULTON. That is all.

Thank you.

I appreciate very much both of your comments, which have been excellent and very interesting.

MR. MITCHELL. What you are saying, Mr. Johnson, in substance, is that you are suggesting legislation which would give to the Administrator the right to determine the specific phraseology that would go into the contract insofar as whether the Government would retain title or not; is that it?

MR. JOHNSON. This is correct, which is the way I read the National Science Foundation Act.

MR. MITCHELL. Mr. Daddario.

MR. DADDARIO. Mr. Johnson, taking the present posture of the space program into consideration and also last year's experience, do you find any need that NASA have greater protection in inventions than the Department of Defense?

MR. JOHNSON. I don't think so.

By this, I am not meaning to say I agree entirely with the Department of Defense policy as a matter of policy either; but on this question I would say—and I might hark back to Mr. Fulton's remarks about the great interest in such things as space platforms and security interests, and so forth; naturally, all of this applies to intercontinental ballistics missiles, too—you have got the most urgently needed things with the greatest security considerations right over in the Department of Defense.

Our work by and large, is unclassified. Not all of it, but the greater portion of it is in the nonmilitary side of the program. I think I would have to say, honestly, that I cannot see any reason why there is a need for acquisition of title to inventions under our contracts if such a need does not exist under Department of Defense contracts.

MR. DADDARIO. Following that further, if such a need does not exist and, therefore, we can assume from that there is an imposition of a greater need than is necessary on these companies which might wish to enter into contracts with the Government, is this added prohibition, if we can put it that way, affecting the space movement?

[26] Are companies not contracting with you as a result?

MR. JOHNSON. I testified earlier that it is very difficult to get definite information on that.

MR. DADDARIO. What is your thought?

MR. JOHNSON. We have not encountered so far in any of our negotiations with contractors a turndown because of this.

We have encountered a lot of resistance, but they have all been negotiated successfully.

We cannot be sure, of course, that some of the things we hear about the complete unwillingness of some companies to do business with NASA may not be true.

We have had rumors and reports particularly at the subcontractor level that some companies have put out the word they don't want the business; they will not do any business that involves the vesting of title to any of their inventions in the Government and, hence, their people are not to bid on NASA contract proposals.

This kind of thing is hard to get definite information on because you just don't know about the people who don't respond to your invitations or requests for bids and proposals.

Some may be doing it because they don't want the business; they are completely booked up or they aren't interested; or they may be staying away for this reason.

You cannot be sure of this.

MR. DADDARIO. Mr. Johnson, if you have a company which is in the aeronautical field and, because of the great interest there is in space, it has a strong research and development section, couldn't you assume they would look very carefully into putting the endeavors that they have already put into this field to the use of the Government, when that whole program could then be taken by the Government and then passed off into commercial enterprises or to other countries or to other companies, and this could be research and development which they have built up to this point with their own means and without any Government assistance whatsoever?

MR. JOHNSON. I could speculate along those lines. That sounds quite reasonable and, of course, we are told by industry this is exactly their reaction to it.

MR. DADDARIO. Wouldn't you say this must be the reaction because this is traditional way which many companies, those with great tradition, have operated?

MR. JOHNSON. Yes.

MR. DADDARIO. They have looked ahead; they have research and development programs to keep themselves apace with progress?

MR. JOHNSON. That is true.

MR. DADDARIO. And it must necessarily, as a result, be something that they would look into very carefully, and if they are doing so, this need that you have tagged on here and which you, yourself, say is not necessary, is probably slowing down the whole space program because companies are staying away from it?

MR. JOHNSON. Mr. Daddario, I simply cannot say I know the program is being slowed down by this. I couldn't honestly say that.

Everything you say sounds reasonable, and we are told that there are companies that are reluctant—in fact, even unwilling—to do business, particularly with our prime contractors on the sublevel.

[27] I couldn't document it by saying I know X company or Y company or Z company has refused to do business with us or has slowed down their participation because of this.

MR. DADDARIO. Let me ask you this: Let's assume there is a situation where you have a company that does enter into a contract with NASA and, in the performance of this contract, it uses other inventions which it has produced to increase its technical superiority or potential. What would be the situation involving the utilization of these other inventions?

MR. JOHNSON. Do you want to comment on that?

MR. O'BRIEN. Yes.

I will comment on this, Mr. Daddario.

We do not, by acquiring a right to use or acquiring title to an invention made under contract with NASA, acquire also rights under patents on inventions developed independently of a Government contract. These are called background patent rights on inventions. The owner or right to practice the invention under contract, where we acquired rights, does not automatically give the Government rights under these background patents.

MR. DADDARIO. Who is to decide whether it follows within one patent or the other?

You have no way of waiving, do you, under the present provisions, these rights to inventions before a contract is signed?

So, if you sign the contract, then it is up to your Administrator to determine whether or not they are background inventions or whether or not they fall within the area under which they can then be separated from Government control?

MR. O'BRIEN. I think I misunderstood you perhaps as to what you regard as a background invention.

We regard as a background invention an invention which has been made by a contractor prior to the entering into a contract with NASA, and by "made" we mean actually reduced to practice.

As to those inventions, NASA would acquire no rights merely because an improvement on that invention was made in the pursuance of research work under a NASA contract.

MR. DADDARIO. Let me ask you this: Is there any provision at the present time under which a waiver can be granted before a contract is entered into?

MR. O'BRIEN. The law so provides.

Our regulations do not provide for granting of any waivers prior to entering into a contract.

MR. DADDARIO. Then, under the act, the situation is this: Under all circumstances, even though the Administrator would have the authority, as the chairman has pointed out previously, you would first have to give him the complete control and he would then have to decide whether or not it fell within the categories set forth?

MR. O'BRIEN. That is right, sir.

MR. JOHNSON. I think we might mention the prima facie case for waiver, though, in this connection.

MR. O'BRIEN. Yes.

Mr. JOHNSON. I think this is related to the questions you asked.

MR. O'BRIEN. Yes.

[28] In this regard, we have established certain categories of invention. If an invention which is made by a contractor falls within these categories, and he can show to the Administrator or to the Inventions and Contributions Board that this is so, then the contractor has established a prima facie case for waiver of title or the waiver of the right of the United States to acquire title.

Now, these classes of inventions are, one, those inventions which a contractor may have conceived prior to entering into a contract with NASA and upon which he has filed a patent application, but which was first actually reduced to practice in the performance of the contract.

That is the first class.

MR. DADDARIO. Before you go further, because there isn't much time and there may be others who have questions, there is one thing which bothers me here and I am sure you can give me the answer.

When a waiver is granted under any circumstances, are there minimum requirements?

MR. O'BRIEN. There are.

MR. DADDARIO. Therefore, there is no such thing as a complete waiver?

No matter what the situation might be, once an invention comes under the jurisdiction of NASA, whatever waiver is granted, there are minimum requirements and, therefore, a sort of a cloud on the title of whatever the invention might be?

MR. O'BRIEN. The first class of invention which I gave you—there are very minor
requirements.

MR. DADDARIO. But some?

MR. O'BRIEN. With respect to this first class of invention, the requirements would
not place any cloud on title.

We have certain requirements in our waiver instrument, but as to those requirements
they would not place a cloud on title.

As to other categories of invention, the requirements are provisional; that is, title is
provisional, the retention of title is provisional, upon the satisfying of certain requirements, those requirements being that the invention should be developed to the point of
practical application, which means that it must be developed so that it is put into the
hands of the public. We believe that the granting of rights to inventions to a contractor by
waiver must carry some assurance that the contractor will not shelve the patent on this
invention or not let the public have the benefit of it. If this were to be permitted the waiver would not be in the public interest. For that reason, we have placed compulsory working provisions upon the grant of these waivers. So, if the invention has, in our view or in
the view of the contractor, to which we agree, predominant commercial interest and only
incidental interest and utility in space and aeronautics and we give him the right to
acquire title in the invention and the right to acquire a patent on it, then we say, "You shall
practice this invention; you shall put it into the hands of the public within a period of years
or you shall make it available for license to anyone who desires to do so."

MR. DADDARIO. Does that include foreign governments and foreign countries, anyone who would do so?

MR. O'BRIEN. No; I think not.

[29] Mr. DADDARIO. You think that would be restricted to the continental limits of the
United States?

MR. O'BRIEN. Yes.

MR. DADDARIO. That is all.

MR. MITCHELL. Mr. Quigley.

MR. QUIGLEY. I have no questions.

I do regret my inability to be here on time. I occupied the witness stand in another
committee and on a matter which was controversial. I couldn't quit under fire. So I had
to stay, and I deeply regret it, because I wanted to get here and get the benefit of this background presentation. So I will have to study the record.

MR. MITCHELL. Mr. Yeager.

MR. YEAGER. Mr. Johnson, did I understand you to say in the recommendations for
legislation next year there will be some recommendations for a change in section 305?

MR. JOHNSON. No, Mr. Yeager. I didn't predict that positively—

MR. YEAGER. There might be?

MR. JOHNSON (continuing). As to what NASA's position would be, I said we are hard
at work in developing this as a part of our entire legislative program, and I said that, so far
as a personal prediction was concerned. I would predict that we might submit even alternative provisions as means of treating this problem.

I suppose, if we did that, we would have a clear-cut recommendation as to which one we preferred.

I certainly think it is fair to say we will have some legislative recommendation to amend section 305.

I don't see how we could help but have that. This is one of our major legal problems.

MR. MITCHELL. You are going to have to live with this law, and certainly you should give us the benefit of your experience and your recommendation.

MR. JOHNSON. Yes, sir.

MR. YEAGER. Might that include section 306, too, on the—

MR. JOHNSON. That is an entirely separate question. At the present time, I don't personally have any—

MR. YEAGER. This doesn't give you concern at the moment, then?

MR. JOHNSON. No.

MR. YEAGER. As 305 does?

MR. JOHNSON. No, this is an entirely separate question.

MR. YEAGER. I would like to develop just one brief line here.

You have interpreted in section 305(a) the phrase "any work" to exclude procurement contracts; is that correct?

MR. O'BRIEN. Yes, sir.

MR. YEAGER. And according to the memorandum, I think, of May 6, which you submitted to this committee, you indicated that you are requiring your patent clause in contracts where the work is of a technical or scientific or engineering type. Does this extend to subcontracting?

MR. O'BRIEN. Yes

MR. YEAGER. It does?

MR. O'BRIEN. Yes.

MR. JOHNSON. The description is a little more elaborate than that.

MR. YEAGER. Yes.

[30] MR. JOHNSON. I think you are giving it sort of a shorthand characterization.

MR. YEAGER. Yes; but what I wanted to get at is not in direct reference to that provision. What I am getting at is how you arrived really at the intent of Congress on this, and again in section 305(c), where apparently you have interpreted this to mean that this section applies only in the case of work, done under a contract with NASA. You say NASA has concluded that this was not intended; this section was not intended to give the Government rights under inventions outside the contractual situation with NASA.

MR. O'BRIEN. We regard this provision of the act as a policing provision.

MR. YEAGER. How did you reach that conclusion?

MR. JOHNSON. Mr. Yeager, you have asked several questions, I am not sure just which one I am answering first.

MR. YEAGER. How did you reach the conclusion that Congress did not intend for section 305(c) to apply to situations other than those where a work contract was under NASA? That is what I was getting at.

MR. JOHNSON. Section 305(c).

MR. YEAGER. The record, as I recall the previous testimony, is pretty skimpy on this.

MR. JOHNSON. Yes.

MR. YEAGER (continuing). And I was just wondering whether you perhaps didn't have to just play it by ear.

MR. O'BRIEN. I think a resolution reading of 305(c) and a reading of 305(d) answer that the information on the material, which, under these provisions of the act, the Commissioner of Patents is required to secure from the applicant for a patent, is that information and material which bears directly upon the circumstances surrounding the making of the invention and whether or not it was made under any contract with NASA.

So, if all of the information and material which is submitted in these statements which the Commissioner secures, bears on the question of whether or not the invention was made under any contract of NASA, what other purpose could this provision of the law have other than to make inquiry as to whether or not it was made under a contract?

Therefore, we believed this provision generally to have two purposes, the first providing a policing provision for our contracts and the second providing an opportunity to have the Administrator's determination, subject to a review by another independent agency; namely, the Board of Patent Interferences of the Patent Office or, ultimately, the Court of Customers and Patent Appeals.

We tried to derive from this subsection of the act some incidental advantage to NASA, from a technological point of view, that in bringing this information to the attention of the NASA technical staff, where the inventions are of significant utility to space and aeronautics NASA might derive some technology benefit from its disclosure.

It hasn't proven to be of much value in this respect, but—

MR. YEAGER. What I was trying to get at was: You say you believe this to have been the case, and this seemed reasonable to you. [31] But as far as the record shows, there isn't much to go on, since there were no bearings and very little debate on it in Congress, and the conference report was very meager.

MR. O'BRIEN. The conference report has—

MR. YEAGER. It says something about it, but my question is: Wouldn't you agree this is susceptible of a different interpretation?

MR. JOHNSON. Mr. Yeager, I would like to answer that.

You mentioned before, I think, three or four important interpretations we have given to section 305. In the absence of any legislative history, all of these have been rather arbitrary. I have to admit that.

This is the problem you are faced with in giving an initial interpretation to any important piece of legislation.

I don't think in any of these cases that we have done violence to the statutory language, and we have always tried, as well as we could, to discover from reading the sections as a whole what we felt the legislative intent was.

Mr. O'Brien has just explained how we think the interpretation we have given to section 305(c) does derive from a study of the section as a whole.

MR. YEAGER. Sure.

MR. JOHNSON. This is true of all the rest of it, but we would admit these are arguable propositions.

We have tried in each case also, while not doing violence to the language, to try to reach an interpretation which we thought was a most workable one and one that we could administer.

MR. YEAGER. Yes.

I wasn't suggesting there was any violence done to it. The only point I was driving at was: Unless these sections are clarified, perhaps at some point in the future a future administrative body might very well construct them differently than you have.

MR. JOHNSON. That is quite possible.

I would like to say, too, that we have tried, each step along the way, to keep the committee fully informed of the administrative interpretations we have given this act.

I think you have been constantly supplied with our regulations and contract clauses and have been informed of all our significant steps just as soon as we have taken them.

MR. YEAGER. Thank you.

That is very helpful.

MR. MITCHELL. Mr. Bass.

MR. BASS. Mr. Johnson, I just want to say I have been very much impressed with your presentation, and particularly the grasp that you and Mr. O'Brien have shown of this very complicated technical field.

Would you give us, very briefly, a biographical sketch of yourself?

It might be interesting.

MR. JOHNSON. I am a graduate of DePauw University, and University of Chicago Law School, and have a graduate degree from Harvard Law School, LL.M. I am a member of the Illinois bar, practiced law in the general counsel's office of the Chicago, Burlington, & Quincy Railroad and with the law firm of Wilson & Mellvaine in Chicago before World War II.

MR. BASS. I know the firm very well.

MR. JOHNSON. I have 3 years of active duty in the Navy.

[32] My Government service—I have been with the Department of State in the Office of United Nations Affairs and with the Department of the Air Force where I was General Counsel for the last 6 years before assuming the position of General Counsel of NASA last October.

MR. BASS. How old are you?

MR. JOHNSON. Forty-three.

MR. MITCHELL. Any further questions?

MR. DADDARIO. No further questions, Mr. Chairman.

MR. MITCHELL. Let me express appreciation on behalf of the committee for the appearance of both you, Mr. Johnson, and you, Mr. O'Brien. Certainly the information you have given us will be of help. As I stated previously, we are in no hurry on this matter and we will be looking forward to seeing you back with an official recommendation.

Thank you very much.

MR. JOHNSON. Thank you very much, Mr. Chairman.

MR. MITCHELL. The committee will be in recess until 10 in the morning.

(Whereupon, at 12:04 p.m., the meeting was recessed, to reconvene at 10 a.m., Thursday, August 20, 1959.)

[33]

Property Rights in Inventions Made Under
Federal Space Research Contracts

Thursday, August 20, 1959

House of Representatives,
Subcommittee on Patents and Scientific Inventions,
Committee on Science and Astronautics,
Washington, D.C.

The subcommittee met, pursuant to adjournment, at 10:10 a.m., Hon. Erwin Mitchell (chairman of the subcommittee) presiding.

MR. MITCHELL. The subcommittee will be in order.

This morning we are privileged to have Mr. Ray M. Harris, Assistant Patent Counsel, National Aeronautics and Space Administration, who formerly was chairman of the Armed Services Procurement Regulations Committee, and procurement and patent specialist, Department of Defense.

Mr. Harris is presently with the Space Administration, as I pointed out. The purpose of his appearance today is to brief the members of the committee on patent policies followed by the Department of Defense and other Government agencies.

We are happy to have Mr. O'Brien back again this morning.

Do you have a prepared statement, Mr. Harris?

Statement of Ray M. Harris, Assistant Patent Counsel, NASA; Accompanied by Gerald D. O'Brien, Assistant General Counsel for Patent Matters, NASA

MR. HARRIS. Mr. Chairman, as announced, my subject was supposed to be the patent policy of the Department of Defense and other Government agencies, but I felt they were discussed pretty extensively yesterday and probably if the members have any more interest in those policies than was brought out yesterday, it could be handled by questions.

On the other hand, in view of some of the questions raised yesterday, I thought the members might be interested in a discussion of some of the more fundamental aspects of the patent problem and system as an aid to arriving at a determination of what the Government's patent policy should be. Mr. King particularly raised that question.

I thought if the committee would care to, I would discuss that aspect.

MR. MITCHELL. I think it would be most benefiting.

MR. HARRIS. My prepared statement here is a couple of pages of introduction. The first paragraph is what I have already said and then the second paragraph:

[34] I would like to say at the beginning that these are my personal views and have not been coordinated with my superiors, Mr. O'Brien and Mr. Johnson. I agree with Mr. Johnson, who spoke yesterday, that this problem is so complex that it is difficult to give a categorical answer.

As Mr. Johnson said, this problem has been with us for many, many years. One might be justified in arriving at different answers to the question with respect to Government employees' inventions versus Government contractors' inventions, with respect to different Government agencies, and with respect to different fields of technology.

The problem is currently being studied by the staff of the Patents, Trademarks, and Copyrights Subcommittee of the Senate Committee on the Judiciary and by an interdepartmental working group under the chairmanship of the Commissioner of Patents, Study No. 14 of the Interagency Task Force for Review of Government Procurement Policies and Procedures.

Mr. O'Brien is a member of this study group 14 and I was while I was with the Department of Defense.

The problem has been the subject of numerous studies in the past, most notable being the Attorney General's report of 1947 to which there was a sequel report of November 9, 1956. In the sequel, the Attorney General pointed out that the Department of Defense patent license policy was permitting the concentration of patents in the hands of big business.

I would like also to mention, in the interests of what has been done on this subject, that Dr. Howard L. Forman, who is a personal friend of mine, got his Ph.D. degree on the subject as a result of his investigations into what should be the patent policy of the Government with respect to its employees' inventions and has written a book on the subject: "Patent—Their Ownership and Administration by the United States Government" published by Central Book Co., Inc.

I think the above introduction indicates the extent of the problem. Nevertheless, I have a conviction that the people concerned with this problem have spent too much time attacking it from the standpoint of who should have the rights to patents as a matter of law or equity, and not nearly enough time as to what is the purpose of ownership of a patent, and from the Government's viewpoint, what should it do with the patents it owns. If the Government doesn't have a good program of administration of its patent property, why should it be so concerned with getting title to the patents, and getting more patent property?

I might say also, Dr. Forman takes that position, that we have the cart before the horse. We have been concerned with deciding who should get the rights to the patents and we haven't decided first what we are going to do with the patents we've got.

MR. KING. As a matter of fact, what does happen to Government patents? Do they go into the public domain or are they locked up for 17 years?

MR. HARRIS. The practice largely with Government-owned patents is, in effect that they come under the public domain because the Government does not have a policy of enforcing its patents. In order for a patent to be used as the patent law intends it to be, it must be exercised—the exclusivity provided by the patent must be exercised which [35] means that you must use it for yourself or your licensees and not permit others to use it. The Government's policy is exactly the opposite. When it gets a patent, most of the Government agencies will grant a revocable, not an irrevocable, royalty-free license to anyone who asks for it. If you don't ask for it, it is all the same because they won't sue you for infringement.

Mr. O'Brien, would you like to add to that?

MR. O'BRIEN. I would only mention that one of the reasons for the Government's patent policy, as Mr. Harris has stated, is that the major executive branches of our Government have no authority to grant rights in patents which that agency of the Government may own. The Congress has never provided the executive branch of the Government with that authority except in a few instances such as the Tennessee Valley Authority and the AEC.

MR. HARRIS. And our own organization.

MR. O'BRIEN. And the NASA.

MR. MITCHELL. Mr. King, will you yield?

MR. KING. Yes, I am through, Mr. Chairman.

MR. MITCHELL. I understand you to say that in most agencies you do not have the authority to grant licenses?

MR. O'BRIEN. The authority to grant any irrevocable or exclusive license.

MR. MITCHELL. The policy has been to grant these licenses but they are revocable?

MR. O'BRIEN. That is right.

MR. BASS. Does that also mean the Government cannot collect royalties and enter into that kind of agreement?

MR. HARRIS. I think it would mean that except in the case of these agencies which have the authority such as NASA, TVA, and AEC, I believe.

MR. BASS. They have the authority?

MR. HARRIS. Yes.

MR. BASS. Do they exercise it?

MR. HARRIS. No, sir.

We haven't developed our policy on the subject. We are in the process of trying to formulate a policy but one of the difficulties that one is going to have in trying to grant royalty-bearing licenses is that it is obligatory on the licensor in such cases to defend that patent against infringers because it is unfair to the person who takes a license and agrees to pay royalties if somebody else would start to manufacture the thing and not pay royalties and have it royalty-free.

So, therefore, in private practice, it is incumbent upon the patent owner who grants a license to undertake to sue infringers. In the Government's case, if it were to adopt a policy of granting royalty-bearing licenses it would mean the Department of Justice would have to sue infringers of patents.

MR. KING. May we pursue this, Mr. Chairman.

MR. MITCHELL. Yes.

MR. QUIGLEY. May I ask a question here just for clarification?

Do I gather, sir, that the Tennessee Valley Authority and the AEC, those two, have in the past granted exclusive licenses, not with the royalties attached?

MR. HARRIS. The Tennessee Valley Authority has granted at least one royalty-bearing license.

[36] MR. QUIGLEY. At least one royalty-bearing?

MR. HARRIS. Yes. The AEC has never granted more than a revocable license, which the Department of Defense also grants. They have never exercised the authority of their act.

MR. QUIGLEY. In other words, while the AEC has the authority to grant exclusive irrevocable licenses, they have not in fact exercised it?

MR. HARRIS. That is right.

MR. QUIGLEY. What you are saying in effect, then, is that the only Government agency that has done that would be the TVA?

MR. HARRIS. That is right, and also that license was to a British concern and it may be that they didn't know the situation over here, as well as ourselves, because had I been representing an American client or them, I would have advised then not the enter into a royalty-bearing license.

MR. QUIGLEY. Even though this authority has existed on the books for a number of years, in fact it has not been exercised?

MR. HARRIS. That is right.

MR. QUIGLEY. With this one exception?

MR. HARRIS. That is right, sir.

MR. O'BRIEN. I would like to add one comment.

The Tennessee Valley Authority does grant licenses which are irrevocable, but not royalty-bearing. It has granted exclusive licenses.

MR. QUIGLEY. That would be the only agency of the Federal Government that has done that. AEC has the authority to, but hasn't.

MR. O'BRIEN. Yes, sir, except for a few instances of vested property of the Alien Property Custodian where licenses have been granted under those vested patents.

MR. MITCHELL. Mr. Harris, this example you gave of the British concern obtaining a license was later canceled. . . .

Document III-4

Document title: T. Keith Glennan, Administrator, Memorandum for Distribution, "Appraisal of NASA's Contracting Policy and Industrial Relations," February 29, 1960, with attached: "Preliminary Outline of Plan for Appraising NASA's Contracting Policies and Industry Relationships," February 26, 1960.

Source: NASA Historical Reference Collection, NASA History Office, NASA Headquarters, Washington, D.C.

Document III-5

Document title: McKinsey and Company, Inc., "An Evaluation of NASA's Contracting Policies, Organization, and Performance," October 1960 (a report prepared under contract for NASA).

Source: NASA Historical Reference Collection, NASA History Office, NASA Headquarters, Washington, D.C.

NASA Administrator T. Keith Glennan began 1960 with an eye to the future, concerned with establishing policies to guide NASA's external relationships with other government agencies and private industry. Knowing NASA would be contracting out the majority of its work through various field centers with differing characteristics, and recognizing that his actions would set a precedent for the agency in years to come, Glennan felt it important to acquire outside advice on these issues. Consequently, he hired the management consulting firm of McKinsey and Company to undertake an extensive study of how NASA might best establish these external relationships. Reporting back eight months later, McKinsey laid out a number of recommendations. The first chapter of the firm's report summarized them.

Document III-4

[1]   February 29, 1960

Memorandum for Distribution

Subject: Appraisal of NASA's Contracting Policy and Industrial Relations

A contract has been entered into with *McKinsey and Company*, Management Consultants, for a comprehensive study of (1) how NASA should utilize industry and private institutions, (2) method of utilizing in-house research capabilities, and (3) the extent and manner of sharing responsibility and authority between government and industry. Now that our field organizations are shaping up, it seems particularly important to study very carefully how NASA can best conduct its business with industry in carrying out the program planned for the next 10 years and in a context decentralizing the major elements of industry relationships to the development centers.

The study will follow three basic approaches: (1) an examination of our experience to date in handling several major contracting actions; (2) an appraisal of experience of other government agencies; and (3) an analysis of approaches and techniques used or advocated by our own centers. I urge all elements of NASA to be fully and completely cooperative in working with the McKinsey staff.

In the past NASA has found that it obtains the greatest results from such studies if the outside consultant group has a close liaison with responsible program areas most involved. In this instance, our plan is to assign one NASA staff member to work virtually full time with the McKinsey staff. This person in turn will be assisted by and will head up a task group of people from various parts of NASA Headquarters. The task group will be composed of the following people:

Leader—William P. Kelly, Jr.—Office of Business Administration
Member—Newell Sanders—Office of Space Flight Programs
Member—Col. D. H. Heaton—Office of Launch Vehicle Programs
Member—Emerson V. Conlon—Office of Advanced Research Programs
Member—Walter D. Sohier—Office of General Counsel
Member—John R. Scull—Office of Program Planning and Evaluation

In addition, it is requested that each NASA research and development center, including the Jet Propulsion Laboratory, designate a top level technical person with management responsibility as a point of contact for the study group with that center. The name of the individual so designated should be supplied in writing to the leader of the NASA task group, Mr. William P. Kelly, Jr., Chief, Procurement Assistance Branch, Procurement and Supply Division, Office of Business Administration, NASA Headquarters, as soon as possible.

[2] Attached for your information is a brief summary of the study purposes and objectives. I believe this is a timely study of one of our major problem areas and can result in a major contribution toward improved program management if it is properly and enthusiastically pursued.

> T. Keith Glennan
> Administrator

Attachment as stated

****

[Attachment p. 1]

February 29, 1960

Preliminary Outline of Plan for Appraising NASA's Contracting Policies and Industry Relationships

NASA is now a principal source of government contracts and may be expected in the future to contract for the requirements of an even larger space program. It is dependent upon its ability to contract effectively for the industrial and scientific resources of the nation to carry out the national space program. NASA has now (and probably only within the next year) the opportunity to appraise objectively and to revise imaginatively its contracting policies and relationships with private industry and institutions.

Scope and Objectives of the Study

This study is to be primarily concerned with an analysis of the basic concepts of (1) how NASA should utilize industry and private institutions, (2) the method of utilizing in-house research and development capabilities, and (3) the extent and manner of sharing responsibility and authority between government and industry.

The answers that this study seeks must be reconcilable with (1) the ten-year planned program, (2) the present order of magnitude of in-house development resources (at least through Fiscal Year 1961), and (3) NASA's basic policy of decentralizing major elements of the contracting job (and related industry relationships) to the development centers. These factors establish a basic frame of reference against which the feasibility of recommendations must be tested.

A study of NASA contracting at this time should be designed to provide factual and reasoned answers to the following (and related) questions:

1. What role should the space development centers—Goddard, Huntsville, and JPL—play in contracting? How does this role relate to the need for in-house development and engineering capabilities? Which of several approaches should be followed by the development centers in contracting, e.g., contracting with a single company for a major system as [2] contrasted with contracting sub-systems and components with several companies? To what extent should the approach be varied in terms of the type of project involved? What are the implications of various approaches to contracting in terms of laboratory requirements for personnel and facilities, and in government-industry relationships?

2. Under what circumstances, and for what reasons, should NASA employ each of the following in systems management?
   a. NASA space development center.
   b. Industrial contractor as solely a systems manager.
   c. Industrial contractor as systems manager and prime contractor.
   d. University or other type of nonprofit contractors as systems manager with an industrial prime contractor as in the *Vega* case.

3. What approaches and techniques should NASA use in supervising contractor operations and in evaluating contractor performance—from both a technical and administrative point of view? How should these techniques be varied in terms of (a) contractor capabilities, (b) amount of advanced research and development involved, (c) priorities, and (d) similar factors? What decisions should be made by the development centers and various elements of the headquarters staff in contractor supervision? What information is required to make these decisions effectively and how should it be provided?

4. How and to what extent should NASA encourage elements of United States industry not now interested in or involved in space technology to enter the field?

5. What new approaches can be developed to provide effective incentives to industry to control costs and increase performance? On what types of contracts, and under what circumstances, can these innovations to contracting be employed?*

[3] 6. What problems does NASA's present approach to contracting cause in terms of the agency's internal processes, particularly program planning, integration, and control? What changes are indicated in terms of either contracting policies or internal processes to increase the agency's over-all effectiveness?

7. To what extent is NASA limited by the government frame-work in making desirable changes in its approaches to contracting and in its relationships with contractors? What steps should be taken to modify or remove these limitations?

An Approach to the Study

To answer these questions a three-pronged approach to fact finding and analyses will be undertaken:

1. Appraise NASA's contracting experience by examining a sample of representative contracts NASA has executed.** The analysis of the actions taken on each contract should provide effective insights as to actual experience. To this end the Study Team will, with the aid of NASA's staff, select contracts that provide

---

* Recognition must be given to the difficulties involved in providing effective incentives in rapidly evolving areas of research and development.
** This technique has been tested in an extensive study of "Weapons Acquisition" now under way at the Harvard Graduate School of Business Administration in collaboration with the Rand Corporation.

contrasting approaches to contracting, e.g., the McDonnell Corporation contract for the Mercury Capsule and the role of the Langley Research Center; *Vega* and the role of JPL; and the North American contracts for the "big engine" and the role of headquarters. The Study Team would not expect to derive answers to the questions listed above from the analysis of any sample of contracts alone.

2. Appraise the experience of other government departments and agencies in contracting for research and development projects. Evidence would be sought as to the advantages and disadvantages of the differing approaches employed, e.g., AEC in reactor development; the Army in a program such as *Jupiter*; the Air Force on *Atlas*; and the Navy on *Polaris*. In addition, the contracting practices of one [4] large laboratory outside the NASA and AEC orbit will be reviewed, e.g., Lincoln Laboratory of the Massachusetts Institute of Technology.

3. Analyze the contracting approaches and techniques now being employed by the Development Operations Division, the Space Project Group, and JPL. This approach will include assembling specific illustrations of the advantages and disadvantages of the various approaches to contracting represented by these three groups. In addition, review and appraise the procedures followed by one or more of NASA's Research Centers to make certain that the contracting requirements and procedures at these Centers will not be incompatible with the policies to be recommended.

Specific Steps Involved in the Study

More specifically, the Study Team proposes to proceed as follows:

Approximate Timing   *Steps*

Feb. 29–Mar. 18   1. **Finalize Detailed Study Plan:** To make more precise the types of information and analyses required, the ideas of key personnel in NASA headquarters, Langley, Goddard, and at JPL as to materials and experience relevant to the questions listed above will be assembled. This step will also involve establishing criteria for the selection of contracts to be studied. At the completion of this step, the Study Team will:

a. Formulate, in terms of outlines and questionnaires, the specific detailed inquires to be made at NASA headquarters, NASA development and research centers, successful industrial contractors, unsuccessful contractors, and other government departments and agencies (Army, Navy, Air Force, and AEC).

[5] *Approximate Timing*   *Steps*

Feb. 29–Mar. 18
(continued)   b. Make a detailed presentation to the top staff of NASA—both headquarters and field—picturing the study objectives and plans. This will be done to ensure understanding of the kinds of issues and problems the study seeks to resolve, and the kinds of evidence, experience, and opinion that will be required to resolve these problems. It will be important that this step result in a consensus among key personnel as to the desirability of the study objectives and the feasibility of the approach. The Study Team will evaluate with the Administrator, at this point, the adequacy of the study plans, and the reactions of NASA's staff to these plans.

Approximate Timing	*Steps*

Mar. 21–May 13   2. **Assemble Contracting Experience:** This step will involve three simultaneous efforts:

    a. In assembling and analyzing NASA's contracting experience, the Study Team will be seeking information on such questions as:

       (1) Where did the idea for the project come from? What program decisions gave rise to it? Was its feasibility adequately considered?

       (2) Were in-house capabilities available for all or part of the project? What factors such as cost, were considered in making the decision to place the contract with an industrial firm or private institution?

       (3) What criteria or guidelines were used to select organizations to submit proposals?

       (4) What factors were considered in evaluating proposals and what was the relative significance of each factor in negotiating and awarding the contract?

[6] *Approximate Timing*	*Steps*

Mar. 21–May 13

       (5) What major technical, timing, and cost modifications were required in the contract and for what reason? Who made these decisions and on what basis? What has been the impact of these changes in NASA (e.g., reprogramming of available funds) and on the contractor?

       (6) How are the contractor's operations supervised and his performance evaluated?

    b. In assembling and analyzing the experience of other government departments and agencies, the Study Team will want to determine why certain approaches have been selected for the contracting of specific research and development programs rather than others, e.g., the Special Projects Office in the case of Polaris; the separation of technical and management supervision in the case of certain Air Force contracts; the management services contract for systems management on the Atlas; and the Army approach of in-house systems management.

    c. In assembling and analyzing the contracting approaches employed within the NASA centers at Huntsville, Langley, and JPL, the Study Team will want to determine what circumstances created or accounted for the different approaches to contracting and the specific advantages and disadvantages of the varying approaches, in terms of concrete illustrations.

Approximate Timing		*Steps*

May 16–June 24   3.   ***Develop Preliminary Findings, Conclusions, and Recommendations:*** This step will involve (a) preparing a series of discussion papers on each of the study's major objectives, and (b) subjecting these discussion papers to the review and criticism of key headquarters and field personnel. [7] This step has a dual purpose—(a) to refine the conclusions and recommendations, and (b) provide a basis for achieving a consensus among key NASA personnel as to the approaches NASA should take to contracting and government-industry relationships in the future.

June 27–June 29   4.   ***Prepare Final Report:*** The Study Team's objective will be to present a final report that sets forth recommendations and implementing action steps that have, for all practical purposes, been agreed to by key headquarters and field personnel. The previous study steps are designed with this objective in mind.

Document III-5

[1]

An Evaluation of NASA's Contracting Policies, Organization, and Performance
National Aeronautics and Space Administration

1—How Better to Perform NASA's Contracting Job—
A Summary of Recommendations

Importance of Contracting to NASA's Total Job

No single element of NASA's management is as essential to the accomplishment of NASA's job as the ability to contract effectively for the research, development, production, and services required. The volume of work to be done and the fast range of scientific and engineering skills involved require that NASA utilize effectively through contracts those enterprises—universities or business firms—that possess the skills required.

Approximately 85 percent of NASA's annual appropriations, hence, are spent on contracts. This fact is illustrated by the following table:

	Estimated Obligations FY 1960 (millions)		*Budget Estimate FY 1961 (millions)*	
	Dollars	*Percent*	*Dollars*	*Percent*
Contracts	468	85.2	770	84.2
Personnel	81	14.8	145*	15.8
Total	549	100.0	915	100.0

---

*   Increase due largely to added personnel costs resulting from transfer of Development Operations Division (Marshall Space Flight Center) from Army to NASA effective beginning with Fiscal Year 1961.

[2] Factors That Condition NASA's Job

The manner in which the contracting job is carried out is conditioned by four factors—(1) the unique characteristics of NASA's job, (2) the legislative framework within which NASA operates, (3) the political sensitivity of contracting, and (4) the manner in which NASA came into being.

(a) Characteristics of NASA's job

NASA's ultimate objective is the acquisition, evaluation and dissemination of scientific information. Space vehicles and associated hardware provide the tools to achieve this objective. This means that most of NASA's contract dollars go for never-before-produced experimental equipment and systems, requiring diverse engineering and scientific skills.

The bulk of NASA's contracting, hence, is carried out on a cost-plus-fixed-fee basis. This method of contracting demands a closer day-to-day working relationship between NASA's technical and procurement specialists than other methods of procurement in such areas as the preparation of work statements, analyses of costs, in selecting suppliers, and in progress reporting and evaluation.

Contracting for such efforts is complicated further by the fact that many projects utilize industrial resources on what is essentially a "one time basis." The enterprise that contracts to carry out a NASA project may have to assemble scientists, engineers, technicians, and facilities especially adapted to an unprecedented undertaking. Upon completion of the project the "team" and facilities may no longer be required. There is little need for the repetitive production of a succession of items (e.g., as in aircraft or even military missile systems) but for the production of a single or very limited number of launch vehicles and space craft. Procurement of a small number of unique items places major stress on the reliability of each item.

The high reliability requirements, plus the small number of similar units that are used, are central characteristics that distinguish and complicate NASA's procurement job. These characteristics mean that the normal cost and performance incentives are often not available to NASA and contractors. Therefore, NASA must substitute for the self-discipline of such incentives continual and effective technical supervision of contractor's efforts.

[3] Over and above its own immediate needs for the services of industrial enterprises, NASA has a longer-run obligation in a free enterprise society to provide industry opportunities to take advantage of the commercial aspects of research and development.*

The goods and services that NASA contracts for and the distribution of contracts among suppliers inevitably condition the capacity of American industry and of individual enterprise to participate in those areas where (a) commercial applications are foreseeable, e.g., communications, and (b) where space research and development has an indirect impact on industrial technology and commercial products, e.g., electronics.

These factors also determine the extent of economic concentration or dispersion that will characterize the supplying industry in the decades ahead. At present, relatively few industrial concerns possess the engineering and scientific skills requisite to the successful completion of a total space vehicle subsystem such as the launch or space vehicle. However, unless industrial contractors are encouraged to round out their capabilities, NASA will find it necessary to expand its in-house capabilities—facilities and personnel wise.

---

*   Some of the problems involved were set forth in an address by Ralph J. Cordiner, Chairman of the Board, General Electric Company, entitled "Competitive Private Enterprise in Space" at the University of California, Los Angeles, May 14, 1960.

(b) The Legal Framework of Contracting

The National Aeronautics and Space Act of 1958 provided NASA broad authority "to enter into . . . and perform such contracts . . . or other transactions as may be necessary to the conduct of its work and on such terms as it may deem appropriate." The Act also made applicable to NASA the provisions of the Armed Services Procurement Act of 1947.

These legislative grants of procurement authority were designed (1) to grant NASA the same flexibility to procurement as is available to the military and (2) to avoid the imposition of an additional set of procurement regulations with which industry would have to cope. This latter point is of particular [4] significance since a substantial proportion of NASA's requirements are similar to those of the military departments and are produced by the same companies.

The contracting authority granted by the Congress has made it possible for NASA to depend on the military departments during its first two years of existence for substantial assistance in contracting. Without this assistance it would have been impossible for NASA to have achieved as much in the time that has elapsed. However, this dependence has influenced the speed and effectiveness with which NASA has developed its own organization and contracting processes. It has also limited the extent to which NASA has been able to initiate new approaches and techniques for contracting for research and development.

(c) Political Sensitivity of Contracting

No aspect of NASA's job is more politically sensitive than the contracting process. In substantial part this political sensitivity arises out of the large value of the contracts being let and their significance to individual contractors and to the communities in which their plants are located. A second cause of this sensitivity is the fact that the contracting activities of large government agencies have become instruments for achieving indirect objectives. These include (1) assisting small business, (2) channeling public funds into depressed and labor surplus areas, (3) maintaining a broad national industrial based for mobilization, and (4) supporting academic and institutional programs.

NASA's public and Congressional relations will depend, in considerable part, upon the manner in which the contracting process is carried out.

(d) NASA's Organizational Inheritance

NASA's organization was built on the foundations of the NACA laboratories. The traditional job of these laboratories had been in-house supporting research for the military departments and the aircraft industry. Their staffs had little experience in contracting for complex development projects.

The Jet Propulsion Laboratory, prior to its transfer to NASA, had been primarily concerned with the in-house development of Army missile systems. Although this laboratory had spent approximately half of its annual budget via contractors and vendors, the items contracted for consisted primarily of raw materials, parts, components, and similar items. Laboratory [5] personnel possessed little or no experience in contracting with industry for major subsystems of the nature involved in NASA's program.

The individuals making up these groups had been primarily concerned with in-house development and had had little experience in utilizing non-governmental contractors for development of subsystems as distinguished from components. The staff of the Development Operations Division of the Army Ballistic Missile Agency had had a markedly different experience but this staff was similarly oriented toward in-house development.

A further factor conditioning NASA's contracting processes was the inheritance by the Agency of a number of projects that had already been initiated by other agencies. These include the Vapor Magnetometer Project, initiated by the Naval Research Laboratory; the Saturn Launch vehicle by the Advanced Research Projects Agency of the Department of

Defense and the Development Operations Division of the Army Ballistic Missile Agency; the Centaur launch vehicle initiated by the Air Force; Tiros I, a project conceived and initiated by the Army Signal Corps; and Echo, a project developed by the Langley Research Center of NACA.

Each of these projects involved differing approaches to (a) the division of effort between government and private resources, (b) project management, (c) technical supervision of contractor efforts, (d) contract administration, and (e) progress reporting, including financial and procurement control processes.

Method of Analysis

In studying NASA's approach to its contracting job, we took the pragmatic approach of analyzing step-by-step twelve significant space flight and launch vehicle projects. The projects studied are identified in Table 1—"Framework for Analyzing NASA's Contracting Policies."* For each project, we studied the

[6] 1. Division of effort between NASA and private contractors in terms of the major elements (e.g., detailed design) that comprise each project.

2. Varying approaches employed in contracting, i.e., relying for the project on a single contractor, procuring subsystems from various contractors, and procuring components to be assembled with NASA.

3. Varying approaches employed in project management.

4. Techniques employed in technical supervision and administration of contracts.

In addition to these analyses of NASA's experience, we:

1. Studied the working relationships between technical and Procurement staffs in the headquarters and in the field centers.

2. Acquainted ourselves with the comparable contracting experience of our agencies, i.e., the Departments of the Air Force, Navy, Army, and the Atomic Energy Commission.

Summary of Recommendations

The results of these analyses are set forth in the following chapters of this report. Here we summarize those recommendations on which action has already been initiated or on which we urge that action be taken.

1.   NASA has made significant progress in reorienting staffs that had been oriented toward in-house research and development and in increasing the utilization of industrial enterprises and other non-governmental contractors. To stimulate further contracting out, we recommend that NASA approve and generally promulgate the following criteria to govern what work shall be done in-house, and what shall be contracted out:

(a) NASA should retain in-house the conceptual and preliminary design elements of a major project, or its equivalent, in each major program.**

---

*   In addition to the project listed, we examined various aspects of contracts of the F-1 engine; Minitrack; research Grants and Contracts at Johns Hopkins and Stanford Universities and at the Massachusetts Institute of Technology; Atlas—Able Space Probe; Snap 8; GE Plug Nozzle engine; nuclear rocket plump; and Deep Space Net.
**   Major programs include—(1) Applications, (2) Manned Space Flight, (3) Lunar and Planetary, (4) Scientific Satellite, (5) Sounding Rocket, and (6) Launch Vehicle. . . .

[7]
Table 1
Framework for Analyzing NASA's Contracting Policies

Space Flight Projects	Estimated Obligations FY1960* (Millions of Dollars)		Distribution of Responsibilities			
	In-House	Out-of-House	Program Mgmt	Project Mgmt	Contract Admin.	Principal Contractors
Mercury	3.8	87.2	OSFP	STG	Navy/Air Force	McDonnell, Convair, Western Electric**
Ranger	5.5	10.9	OSFP	JPL	Air Force	Convair, Lockheed
OAO	0.5	0.3	OSFP	GSFC	Air Force	Convair, Lockheed
S-16	0.05	2.1	OSFP	GSFC	Air Force	Douglas, Ball Brothers
P-14	0.7	0.2	OSFP	GSFC	Air Force	Douglas, MIT, Varlan
Echo	0.05	3.2	OSFP	GSFC	Air Force	Douglas, Bell Telephone, General Mills, MMM
Launch Vehicle Projects						
Saturn	43.0	135.3	OLVP	MSFC	Air Force	Douglas, Convair, Rocketdyne
Centaur	0.2	36.5	OLVP	MSFC	Air Force	Convair, Rocketdyne
Agena-B	0.1	7.3	OLVP	JPL	Air Force	Convair, Lockheed
Delta	0.7	11.8	OLVP	OLVP	Air Force	Douglas
Scout	0.05	2.5	OLVP	Langley RC	Navy	Chance Vought
Vega	0.1	3.5	OLVP	OLVP/JPL	Air Force	Convair
Total $	56.1	300.8				
Total %	18.8	81.2				

*   The in-house estimates include obligations from the Salaries and Expenses Appropriation; out-of-house obligations from the Research and Development Appropriation. The estimates were obtained from the various project managers and reflect the general magnitudes only.

**   The Western Electric contract for the Mercury tracking system is supervised by the Langley Research Center.

[8]   (b) NASA's in-house efforts in the conceptual and preliminary design elements of space flight and launch vehicle projects should be supplemented extensively through the use of study contracts.

(c) NASA should retain in-house the detailed design, fabrication, assembly, test and check out elements of a single advanced launch vehicle* and spacecraft unique to each major program.

(d) Each center should contract out the detailed design, fabrication, assembly, test, and check out elements of all launch vehicles and spacecraft except the relatively few required to meet the criteria set forth in item (c) above.

(e) NASA's centers should contract all production manufacturing efforts including the standard or relatively standard parts and components used for in-house launch vehicles and spacecraft of an advanced developmental nature.

(f) NASA should contract out total space vehicles including the physical integration of subsystems, i.e., the launch vehicle and spacecraft.

---

*   Or stage in the case of a project such as the Saturn Launch Vehicle, i.e., the S-I Stage.

       (g) NASA should contract with the external scientific community for a preponderant proportion (70 to 85 percent) of all space flight experiments.

Adoption of these criteria will ensure the retention in-house of the capability required to enable NASA effectively to contract for the bulk of the research and development services needed. Adoption of the criteria will curb the tendency to do all that can be done in-house and contract out what remains.

    2.   To utilize its in-house facilities to the fullest, we recommend that NASA:

[9]   (a) Place responsibility for a limited number of development projects in the research centers where they have the capabilities required, and these capabilities are needed by NASA for the particular project.

       (b) Establish project management teams in the Research Centers where this means a center's capabilities can best be utilized to provide needed development assistance.

    3.   The complex character of space vehicle subsystems makes inevitable the distribution of responsibility among several NASA centers and among industrial contractors. To resolve more effectively the technical (in matching up one space vehicle subsystem with another) and jurisdictional problems (headquarters staffs vs. center staffs) that arise, we recommend that NASA:

       (a) Assign as full responsibility as practicable for the execution of each project to a specific center.

       (b) Clarify the relative responsibilities of the headquarters staff and the space flight centers by concentrating the efforts of the headquarters staffs on reviewing and approving:

           (1) Development plans for each space flight project, including conceptual and preliminary designs and allocation of responsibilities in- and out-of-house.

           (2) Schedules in terms of major procurement actions and technical milestones.

           (3) Budget justifications and financial operating plans.

In addition, the headquarters technical staffs would evaluate projects and approve changes in the project plans which significantly alter objectives, schedules, and/or costs.

    4.   Strengthen the capabilities of the space flight centers to manage projects, particularly those in which major systems or total space flight vehicles are developed by contractors. To this end, we recommend that NASA:

[10]   (a) Improve the competence of its project managers. Steps must be taken to ensure that project managers develop the full complement of technical and managerial skills essential for this task. The "custom-tailored" training program for project management personnel that has been initiated is a promising step toward this end.

       (b) Improve the project organizational arrangements that now exist. Each project management team responsible for a major space flight project should be headed by a full-time project manager reporting directly to the director or deputy director of the responsible center.* Each project management team should include sufficient technical and administrative (e.g., financial procurement) personnel to make the project manager effective in mobilizing the resources of the whole center, of other centers, and of the contractors.

---

*   Because of the inability to attract senior project managers at the salary level NASA is able to offer, achievement of this objective will require, in a number of cases, a considerable period of time.

5. NASA is faced with a major and complex task of developing, under cost-plus-fixed-fee contracts, working relationships with contractors which neither stifle the contractor's capabilities, nor relieve them of their obligations to use public funds wisely and economically. To this end, we recommend that NASA:

(a) Develop a guide for preparing and evaluating statements of work to be done and service to be rendered under research and development contracts.

(b) Institute a continuing program to assemble and study cost data as a basis for improving funding estimates.

(c) Provide a single point of ultimate technical authority for each contractor on a given project—the project manager.

(d) Establish guidelines as to the approaches and techniques to be used in technical supervision of contractors.

[11]   (e) Establish guidelines as to staff action on the analysis and control of costs in terms of pre-award analyses of price, costs, and profits, and post-award costs control techniques.

(f) Continue to make its own source selections, handle its own contract negotiations, and provide its own technical supervision.

(g) Supplement use of the military services for "field service functions" by periodic evaluation of services rendered, direct handling when required in special situations, and approval of subcontracts within clearly prescribed criteria.

6. To overcome apparent deficiencies in the functioning of the headquarters Procurement and Supply Division, we recommend:

(a) Approval of the organizational plan prepared by the Director of the Procurement and Supply Division with one major exception; that is, focus all activities related to facilities planning and utilization in a separate division in the Office of Business Administration rather than in a branch of Procurement and Supply Division.

(b) Development of a system of field center procurement reviews with will involve key personnel from each of the branches of the headquarters Procurement and Supply Division. This step plus the one recommended in item (a) above will make it possible to abolish what is presently termed the Field Installations Branch in the Procurement and Supply Division.

(c) Establishment of a position of Assistant Director in the Procurement and Supply Division.* The person appointed to fill this position should be given primary responsibility for the day-to-day internal management of the Division.

(d) Additional staff be made available, particularly in the Policies and Procedures Branch, for the Procurement Committee, and in the Procurement Assistance Branch.

[12] 7. NASA's technical staff have reflected lack of understanding of the processes that must be carried out if their needs for research and development services are to be translated into contracts with qualified suppliers and NASA's resources are to be conserved. To overcome this lack, we recommend that steps be taken to aid the technical staffs—in headquarters and in the centers—in expanding their understanding of the:

(a) Succession of actions that the procurement staff must take to negotiate and administer a contract.

---

*   Action has been taken to establish such a position.

(b) Importance of keeping procurement staffs advised of needs that will affect procurement actions.

(c) Importance of recognizing what constitutes contractual commitments and refraining from making them without advice from NASA procurement staffs.

(d) Importance of cost analysis and negotiation and tolerance of the time that is required.

There is no simple nor established method of creating understanding and acceptance of these points by technical personnel. The primary obligation falls on NASA's management. It is to establish in day-to-day practice—at headquarters and in the field centers—the concept of team action on procurement matters.

To implement this concept requires the availability of procurement personnel who are strongly program oriented, while at the same time possessing outstanding experience in, and a clear understanding of, the contracting processes associated with complex research and development projects—including their financial and program implications.

8.   Most of the development contracts that are still being awarded and supervised by NASA headquarters can be associated either with a specific project or with the technical skills available in one of the field centers. Wherever this is the case these contracts should be technically supervised and administered from a given field center rather than from headquarters. In a very limited number of cases it may be appropriate for NASA headquarters to award and supervise contracts related to the development and feasibility of future programs. This should knowingly be the exception to the general rule.

[13] 9. All contracts now supervised from headquarters that can be associated either with a specific project or with the specific skills of one of the field centers should be technically supervised and administered from the field centers; for example, those advanced technology studies for the development of solid rocket motors which are technically supervised from headquarters and administrated by the Goddard procurement office.

Document III-6

Document title: James E. Webb, Address at Graduation Exercises, Advanced Management Program, Graduate School of Business Administration, Harvard University, December 6, 1966.

Document source: Administrators' Files, NASA Historical Reference Collection, NASA History Office, NASA Headquarters, Washington, D.C.

Experienced in public management, NASA Administrator James E. Webb considered the development of new approaches to management an important goal of the Apollo project. His emphasis called for the assimilation of concepts and processes from government, industry, and academia into a usable form. In this 1966 graduation address at the Harvard Business School, Webb took the opportunity to explain his view of the interaction of various communities on space flight management, as well of NASA's broader contribution to public administration.

---

[1] During the time spent here, you have been studying the present state of the management art as it has developed in recent years. You have brought yourselves up to date, and I am certain that you hope that what you have learned will last you for at least a few years to come.

On the other hand, you came here because you are not complacent. You recognize that the world is changing and the requirements you have to meet on the job and off the

job are changing. I am sure you want to continue to keep abreast of the times.
[2] That being the case, let me take this opportunity to talk about some of the changes I see going on that challenge any new complacency you might be tempted to develop.

Let us start with some new kinds of management problems that all of us are going to be dealing with in the days and years immediately ahead. Secondly, let us move on to talk about some new approaches, new techniques and new solutions that are being tested and that have proved productive in dealing with these new kinds of problems. Some of these are too new to be written into the literature or even into the case studies generally available.

As I see it, there are new ways of thinking about management problems, new ways of doing things or getting them done in an organization, new styles of management.

I.   The Changing Dimensions of the Challenge

During the years since World War II, we have all been mindful of the magnitude of the changes going on around us. The numbers needed to describe the growth in our [3] gross national product or our national income, or the magnitude of our private investment or public debt are all enormous numbers. We have heard a great deal, too, about the pace of change and about its acceleration. Much of our attention, therefore, has been given to size and speed, and to how these affect the requirements for good effective management.

I want to talk about some other dimensions of the challenge we face. As I see it, the problems that we are going to be dealing with in the days ahead of us are not just bigger than the problems our parents or grandparents were faced with. They are different in a number of important ways.

First, they are going to be more complex, in many bewildering ways.

How complex our environment is was brought forcibly to my mind in a recent article in *Business Week* on the wood product industry. Some years ago, companies in the industry who owned timberland became aware of the fact that they really had to farm their land if they wanted [4] to stay in business. They had to grow new crops of trees to replace those they cut down. Then the timber companies began to diversify, as they realized that the closer they got to the end product, the more control they had over their markets and their customers. And so timber companies began to go into all kinds of businesses. Some went one way and some another. Some went into building products and others into paper products and one into retail stationery stores. This article in *Business Week* talked about the furniture business and it told how one furniture manufacturer was building diningroom chairs of wood, except that the legs were made of plastic, because that had proved to be much stronger than wood for that purpose. In some of these companies, production of both wood and plastic parts is now controlled by punch tape and by optical scanners that trace cutting patterns electronically. As good wood gets scarcer, some companies are using thin veneers backed with aluminum foil coated with vinyl. This article then went on to describe some of the production techniques the furniture industries have borrowed from the aerospace companies, resulting in highly automated production lines [5] that produce new kinds of raw materials, and then shape them and mold them under electronic control. One company has adapted the technology of textile and paper mills to bleach natural wood to a neutral color and then stain it to produce a more uniform finish than can be found in natural timber. One company is working with epoxy impregnation of wood that has been treated with nuclear radiation to change its molecular structure. The purpose of this is to make hard wood out of pine, according to this manufacturer.

I cite this example only to illustrate one aspect of the complexity of what might appear to be a relatively simple business. It serves to illustrate kinds of decisions that the managements of even relatively small companies are faced with today, and will be faced with increasingly in the days ahead.

An interesting reflection for me as I read this article was the viewpoint of the TVA I had gained back in 1947 when, as Director of the Budget, I had made an inspection of each major river system which was being developed with Federal funds. In addition to its demonstration farms which were experimenting with various new [6] phosphates and other fertilizers developed through TVA research, experiments were being carried out to determine how the small farmer could "tree-farm" his wood lot with highest yield. Another reflection is that recently I read a report on the research which led to the radiation hardening of treated wood which had been partly financed by the Atomic Energy Commission and sponsored by the Southern Interstate Nuclear and Space Board. A wise utilization of an accumulation of technology based on research does pay off—in the health of a regional economy or in the profitability of a business. It pays off in the field of management too.

Certainly you are mindful of the fact that very few of the companies that make up *Fortune's* list of our 500 largest industrial corporations can be said to be in any one industry, or even in two or three industries. The logic of events and of circumstances have led them to diversify all across the industrial spectrum. And most of them are just as far flung geographically as they are industrially. The search for raw materials and markets and labor supply have caused them to set up shop in one [7] country after another all around the world. Each of them has at its command many different kinds of raw materials, natural and synthetic, and many different production technologies. Products are proliferating and markets are fragmented and all of this requires different entrepreneurial skills which require new kinds of management approaches.

What is going on in the private industrial sector of our economy is also going on throughout our society. Our universities are no longer the simple "halls of ivy" they used to be. Every major university is a large complex of different and diverse highly specialized schools, and centers, and institutes, and research laboratories.

Our cultural institutions have become similarly complex. Instead of a Metropolitan Opera House or a Carnegie Hall, New York now has a Lincoln Center and a similar cultural complex is emerging in each of our metropolitan areas, or will soon emerge there. Or think of our approach to the problem of poverty. Not so many years ago, we thought of poverty in terms of incompetence or charity, in terms of drives to support charitable institutions. Now we recognize that poverty is a much [8] more complex fact, requiring a much more fundamental approach involving many different disciplines. Management of efforts to apply new approaches can only be elaborately intricate.

Not only are the challenges facing us much more complex than they used to be, but they are also involved increasingly with new sciences and new technology. Whether you think of the wood product business or the Lincoln Center complex, those who occupy the positions where important decisions are made are more and more dealing with a rapid pace of scientific and technological progress. The furniture executive has to make decisions involving optical scanners and radiation. The management of Lincoln Center finds itself dealing with scientists who are experts in acoustics one day and on the next day with engineers who are masters of the technology involved in the giant rotating mechanism that operates the center stage of the new Opera House, and with the problems posed when that breaks down the night before the new Opera House was to be the scene of its first public performance. We in NASA face the same problem when a diesel engine refuses to start and a gantry [9] cannot be lowered to accommodate a major rocket launching.

Similarly, those who work in the field of poverty are involved in the latest findings of behavioral scientists and economists. The same is true of those who are dealing in the problems of mass transportation or air pollution or management of vast health and welfare programs to serve our major communities. We in NASA are similarly involved when we have to translate a supersonic transport design into pilot performance or into a predicted return on invested capital for an airline.

And our affluent society is becoming day by day a more impatient society. Those who hold positions of responsibility are expected to be able to cope with the most complex of new scientific findings and their potential at the very frontier of technology. It was only a few years ago that Henry Ford made his contribution by putting to productive use the proven engineering practices involved in assembly line mass production. Production of things is no longer the major challenge of our society. We are dealing with problems and with solutions that involve high [10] elements of creativity and, associated with them, high degrees of uncertainty and risk. Management must be able to assess these in its decision-making. And to solve these problems we find ourselves involved with creating and learning to use different kinds of skills and talents and training.

I am reminded of the fact that not so many years ago one of our major corporations was faced with the challenge of shifting from the assembly of electrical components to the manufacture of products involving the latest developments in solid-state physics. The electrical assembly operation required long lines of women with nimble fingers. The new production line was peopled entirely by physicists with advanced degrees. This involved a different kind of recruiting, a different kind of motivation, and a much different kind of supervision. And, of course, it meant a different kind of management at the higher levels of the company. These are some of the new dimensions that we are facing in our private sector and in the public sector of our society. They define a new challenge and they require a new kind of management.

[11] **II. New Perspectives on Available Resources**

I believe we can accept the fact that today's furniture manufacture has to think of the new world of plastics as well as new kinds of treated wood. We have at our command, in other words, a much wider range of natural and synthetic materials to take into account in our critical decisions as managers.

But more important, I suspect, are the human resources we have to work with.

Our generation of managers grew up in a world in which there were some rather nice distinctions between the world of commerce, the world of the university, and the world of government. We came to think of these as quite separate, peopled with quite different kinds of human beings, with different value systems and different sets of capabilities. To some extent, at least, we thought of these as worlds in conflict with each other. One was the world of the practical man of action, the other the world of the intellectual. One was a profit motivated world and the other a world motivated by a desire to teach and to learn.

But as we look at the kind of problems facing us and accept the challenge of dealing forthrightly with these, [12] it becomes increasingly apparent that we need to learn how to work with or draw on each of these resources and learn how to meld them together and balance them in proper proportion.

Certainly we have seen this at NASA where our successes can be traced to our learning how to relate our needs and resources to the needs and resources of these great segments of our society. We have labored hard to set up a partnership in which each contributes its capabilities to and receives its rewards from the effort to master and use the air and space environments.

The first industrial revolution put to practical use the principle of standard or interchangeable parts. I suspect that the world we are making will be characterized by mobility, but also by interchangeability of people, by people who can transfer their work and talents from the university into industry or from industry into government, a mobility in any direction. The first name that comes to my mind is Robert Seamans, who was an associate professor in the Department of Aeronautical Engineering at MIT, and [13] moved from there into industry where he had a distinguished career from which he was drafted into government and is now the Deputy Administrator of NASA. There are many other examples, and the number of people who can move easily and comfortably from one of these spheres to another is increasing day by day. In dealing with the problems that you will be working on in the years ahead, you will be drawing more and more on people with this kind of talent.

III. New Kinds of Organizations

One thing that is becoming increasingly clear to students and practitioners of management is that the classic approaches to organization are inappropriate for dealing with the kind of problems we are talking about.

The earliest attempts to increase the effectiveness of organizations followed the prevailing concepts of the division of labor. The work to be done was broken down into identifiable tasks or functions, and a specialist was put at the head of each major element. This had some obvious advantages, but it also had the disadvantage of dividing responsibility into pieces that really did not [14] correspond to the reality of everything required to get the total job done. Everyone had only partial responsibility so no one had the total responsibility.

This led to the idea of decentralization, which divided the organization into units, each of which had an identifiable task, for which the head of the unit could be held responsible. This proved to have some advantages, but it had the disadvantage of weakening the leadership contribution of those responsible for giving the entire organization its direction and its momentum.

I believe we have learned that neither of these broad-brush concepts, nor any other rules of thumb, work for all organizations. They fail particularly to meet the needs and challenges we face. What we see going on today is the tailoring of new types of organizational structures and new kinds of assignments of authority and responsibility. We are hearing more and more about free-form management, which connotes the development of specific organizational approaches designed to serve a particular unit of a large complex organization. Return to earth is so important to each astronaut and to NASA that we tailor [15] to each his re-entry support or couch to give him maximum support at the time he needs it most.

In modern management, we are seeing increasing use of organizational concepts like product management and project management in which the responsibility for the development and marketing of a product, or the completion of an important project are [sic] put in the hands of one individual who has all required elements of command over all of the resources he needs to get the job done. What characterizes these new kinds of organizational structures is that they cut across the traditional proverbs used to express concepts of authority and responsibility. They utilize, rather than accept as limits, the differences of function or discipline or the division of work into bits and pieces. At NASA, the concept of project management has been applied successfully to large and complex efforts in which one individual is responsible for integrating all of the capabilities and resources necessary to get the job done. Whenever possible, even while exercising very broad authority associated with his responsibility for performance, cost, and schedule, we leave him attached to [16] the laboratory or technical group within which his technical competence was demonstrated and where the forward thrust of current research keeps him up-to-date. This also gives him easy access to colleagues who know how to wring out the facts needed for the difficult trade-off decisions.

The kind of challenges that we in management are facing today do, therefore, call for new and experimental approaches to organization. One that I think worth commenting on in detail is the question of the chief executive function. In traditional organizational thinking, the structure of an organization peaked in the chief executive, who was positioned at the top of the organizational hierarchy. This concept goes back to some of the origins of modern organizational theory and practices, to the Catholic church, and to the Prussian military, which are the prototypes of much of modern organizational thinking. However, as organizations have become more complex and their challenges more interdisciplinary, it is becoming increasingly apparent that there is nothing sacred about the notion of a single chief executive. Accordingly, there has been an increasing tendency to experiment with the idea of the multiple executive, [17] usually in the form of the "office of the president" concept. I understand that a number of important companies, including Union Carbide, General Mills, Metropolitan Life Insurance, Boise Cascade, and others, have experimented with this pragmatic approach to the requirements of managing the kind of far-flung and diverse activities over which some form of executive authority is necessary. We saw this kind of need at the very beginning of NASA's history. We evolved, therefore, a partnership arrangement which included Dr. Hugh Dryden, Dr. Robert Seamans, and myself. We all had many common ideas, and yet each brought to our work on the critical decisions affecting the nation's space effort certain specialized experience. To do it any other way would have deprived the organization of critical inputs needed for important decisions. To do it any other way would have deprived us of the kind of mutual support and broadly-based leadership that I think we achieved.

The point I want to make is that there is need for innovation and risk-taking as well as seasoned judgment in the structuring of organizations to face the challenges [18] of today. This is true in the business world. It is equally true in managing many of the other undertakings in our increasingly complex society.

IV. New Approaches and New Techniques

There are, then, no pat or ready-made organizational devices for structuring these efforts which will substitute for analysis and judgment. Neither are there approaches or techniques that can be taken off the shelf. We are in the midst of a period of innovation and experimentation in both, and there is the same need for creativity that there is in science and technology.

I find this going on in many efforts at the kind of complex problem-solving and decision-making I am talking about. Some specific examples from NASA may be helpful.

To begin with, every aspect of the aeronautical and space effort draws on many different disciplines and many different contractors and suppliers of services. Some of our sources are within NASA itself. Others are in other agencies, and still others in universities. Altogether we have over 20,000 prime, first, and second tier contractors [19] in industry, each of whom is making its contribution to the total effort.

From the beginning of the Space Act, we realized that this effort could achieve its objectives only if each of the contributions to it fit into a carefully designed, fully integrated, totally engineered system. Each of the 200 or more major projects could achieve its objectives only if its elements similarly fit together into a desired whole. In this sense, the space effort represents what is probably the greatest experiment to date in the design, development, test, and use of large complex systems and sub-systems. In this effort, we were concerned, of course, with the performance and cost of each element. We were also concerned that all could be delivered and used on a very short time-phased schedule. Ranger had to precede Surveyor, and Orbiter had to follow. Apollo needed the knowledge to be gained from each. We knew that the perfection of the parts would not guarantee the success of the effort. The interfaces among the elements were at least as important as the elements themselves, and to manage this kind of achievement we found little in the textbooks or in the case [20] histories. We did find men in our military services and in industry who had experience in the management of large projects such as Minuteman and Polaris. From the beginning we worked at developing new approaches and new techniques appropriate for the design and management of this kind of systems effort, in the open, without the protection of military security classification. One of the techniques we had to develop involved the gathering, processing, and dissemination of large amounts of information. We had to collect information on the state of each scientific and technical field in which we or our suppliers were working, and we had to make sure this information was used where appropriate. We had to establish techniques for collecting and distributing information on the state of each of our programs, so that everyone with responsibility or need-to-know could be kept informed.

Sometimes the collection and processing of data had to meet some rather strenuous deadlines. For example, a few seconds after the launch of a manned vehicle, a decision had to be made to abort or to continue the flight. [21] Thus we became involved in developing techniques for real time information processing.

Similarly, some of our projects involved many thousands of discrete activities, all of which had to be coordinated and controlled at a central point. We had to develop display techniques so that the progress of each of these elements could be displayed to teams of people working on different aspects of the same project, in a manner that made it possible for everyone to know where everyone stood at a particular moment in time. PERT in its original form was only a starting point to the development of the control technique we use at Houston and at Cape Kennedy. Again, we had to experiment and to innovate. It is gratifying that the techniques we developed have already found application outside of the space effort.

One of the principles underlying a number of our management techniques is the principle of visibility. We decided it was important that as far as possible problems be identified in a manner visible to everyone involved and that the people responsible for solving these problems be [22] visibly identified to their colleagues. A number of management techniques we have developed serve the purpose of achieving this kind of visibility of information and responsibility.

Similarly, we wanted to achieve an approach to management in which everyone with responsibility was aware that on any decision he could consult both colleagues and superiors without delay and without an involved system to assure a common basis for almost instantaneous identification of the important elements requiring attention. We had to build individual competence and confidence that work could go on with full knowledge of the individual that his superiors were literally "looking over his shoulder" at all times. We had to do this without discouraging initiative and innovation. In this kind of an effort, there was no room for protectiveness or self-consciousness. Accordingly, we developed a number of techniques to achieve this kind of real time "over the shoulder" supervision.

[23] These are only a few of the management techniques we have developed. As a result of this period of experimentation and testing, there are now available a number of techniques of proven usefulness that may well have applicability to problems in other areas of our economy and our society, in our country and around the world.

V. New Breeds of People

What kind of people do we need to manage and to carry out this kind of effort? What qualities identify the individual with this kind of temperament and capability, and how do we go about developing such people to their full potential? Very little is known about this. It is all too new. The only thing we can be sure of is that they are different kinds of people than those that have succeeded in management in the past. One characteristic we have always depended on is that of a strong urge to compete and the urge to excel. In the kind of complex challenges we are talking about, it is rarely possible to attribute a solution or an achievement to one individual. In this kind of effort the boundaries between disciplines is all [24] but erased and the skills of individuals fuse with each other. It is all but impossible to identify who has contributed some key element to the final outcome. I suspect that it is in this area of identifying the new manager and developing him to his full potential that we have the most to learn and in which the greatest progress is yet to be made. This may well be the greatest challenge to those of you who are dedicated to the art of management.

Document III-7

Document title: James E. Webb, Administrator, Memorandum for the Vice President, May 23, 1961.

Source: NASA Historical Reference Collection, NASA History Office, NASA Headquarters, Washington, D.C.

After delivering to President Kennedy a recommendation supporting an American-piloted lunar landing program on May 8, 1961, Vice President Lyndon B. Johnson departed on a tour to review the military and political situation in Southeast Asia. Given Johnson's interest in the space program, NASA Administrator Webb prepared this memorandum for him upon his return. This memorandum is an excellent example of the broad context in which Webb was contemplating the mobilization that would be required to accomplish the Apollo program. The memorandum refers to Edward Welsh, the executive secretary of the National Air and Space Council, Secretary of Defense Robert McNamara, Deputy Secretary of Defense Roswell Gilpatric, and Glen Seaborg, the head of the Atomic Energy Commission. Webb also mentions Albert Thomas, a Democratic congressman from the Houston area and chair of NASA's Appropriations Subcommittee; George Brown, one of the principals in the Houston construction firm of Brown and Root; Jon Erik Jonsson, chairman of the board of Texas Instruments; Cecil Green, a Dallas business leader; Senator Robert Kerr of Oklahoma, chair of the Senate Committee on Aeronautical and Space Sciences; and a Charlie Jonas, Republican House member from North Carolina.

---

[1]   May 23, 1961

Memorandum for the Vice President

By way of a brief report, as you return to Washington, let me set down the following:

1.   The President has approved the program you submitted, with very few changes, and the message will go up on Wednesday.

2.   In working out this program and all of the details involved, there has been an absolutely splendid spirit of teamwork not only with Ed Welsh but with the Defense Department, the Atomic Energy Commission, and the Bureau of the Budget.

3.   Considerable interest has been expressed in this program by members of the Congress, following your consultations with them, and as I have followed up, I have

impressed on them the need you have felt for action and the importance we have placed on the operating responsibilities to be carried by McNamara, Seaborg, and myself. Without exception, all have responded well to this, and many have pledged fullest cooperation and assistance.

4. In preparing for the hearings on the original Kennedy submission before the House Appropriations Committee, and in other discussions with Congressman Thomas, Thomas made it very clear that he and George Brown were extremely interested in having Rice University make a real contribution to the effort, particularly in view of the fact that some research funds were now being spent at Rice, that the resources of Rice had increased substantially, and that some 3 00 [sic] acres of land had been set aside for Rice for an important research installation. On investigation, I find that we are going to have to establish some place where we can do the technology related to the Apollo program, and this should be on the water where the vehicle can ultimately be barged to the launching site. Therefore we have looked carefully at the situation at Rice, and at the possible locations near the Houston Ship Canal or other accessible waterways in that general area. George Brown has been extremely helpful in doing this. No commitments whatever have been made, but I believe it is going to be [of] great importance to develop the intellectual and other resources of the Southwest in connection with the new programs which the Government is undertaking. Texas offers an unusual opportunity at this time due to the fact that Dr. Lloyd Berkner, Chairman of the Space Science Board of the National Academy of Sciences, is establishing a Graduate Research Center in Dallas with the backing of Erik Jonsson, Cecil Green, and others in that area (estimated at about one hundred million dollars), and in view of the fact that Senator Kerr and those interested with him in the Arkansas, White, and Red River System have now pushed it to the point that it is opening up the whole area related to Arkansas, Oklahoma, and in many ways helping to provide a development potential for Mississippi. If it were possible to get a combination where the out-in-front theoretical research were done by Berkner and his group around Dallas in such [2] a way as to strengthen all the universities in the area, and if at the same time a strong engineering and technological center could be established near the water near Houston and perhaps in conjunction with Rice University, these two strong centers would provide a great impetus to the intellectual and industrial base of this whole region and would permit us to think of the country as having a complex in California running from San Francisco down through the new University of California installation at San Diego, another center around Chicago with the University of Chicago as a pivot, a strong Northeastern arrangement with Harvard, M.I.T., and like institutions participating, some work in the Southeast perhaps revolving around the research triangle in North Carolina (in which Charlie Jonas and the ranking minority member on Thomas's Appropriations Subcommittee would have an interest), and with the Southwestern complex rounding out the situation. I am sure you know that the decisions relating to this must await the completion of the work on our program by the Congress, but I am convinced, and believe you should consider very carefully, that will attract the kind of strong support that will permit the President and you to move the program on through the Congress with minimum political in-fighting. I think this is important in the present situation and particularly to avoid the kind of end-runs that some of our friends related to the Pentagon, directly or industrially, have pursued in the past.

5. To get clearly before the country the idea that this is a national effort, the appearance which will introduce the new program to the Senate Committee on Aeronautical and Space Sciences will be made by Gilpatric, Seaborg, and myself, all three sitting together at

the witness table, and each of us presenting a brief statement to start the discussion. I believe this is the kind of image of unity and drive in the Executive Branch that you would like to see.

6.   In all of the work that has gone on while you have been doing such a great service in Southeast Asia, we have emphasized the important place you and the Space Council have occupied in pressing forward for the necessary decisions. In view of this you may wish to consider some form of statement or public expression in connection with the presentation of the program to the country and to the Congress.

7.   In order to discharge our obligation to give both the general public and the scientific community a report on the Shepard flight, we are having a session sponsored by NASA, the National Institutes of Health, and the National Academy of Sciences, in the State Department Auditorium on June 6th. All the people concerned with the program, and particularly those in the scientific and technological side, will be present, as will Commander Shepard. Secratary [sic] Connaly of the Navy is giving a lunch that day for Commander Shepard and Robert Gilruth, Director of the Space Task Group. Would you like to give a lunch or join with me in giving a lunch to the scientists and others on the program? Generally we have tried to avoid getting up any large lunch but could have a small one right in the [3] State Department for those actually on the program and perhaps one or two of the other leaders here that day.

James E. Webb
Administrator

Document III-8

Document title: James E. Webb, Administrator, NASA, Memorandum to NASA Program Offices, Headquarters; Directors, NASA Centers and Installations, July 5, 1961.

Source: Presidential Papers, Agency Records, John F. Kennedy Library, Boston, Massachusetts.

One justification for spending money on space is the benefit derived from "spinoffs"—knowledge or technology developed for a specific space purpose that yields benefits in different fields altogether. In this letter, Webb made an early effort to encourage NASA personnel to facilitate this process, not only to justify space spending but on the grounds that it would help the United States in its Cold War endeavor to outstrip the Soviet economy.

---

[1]   July 5, 1961

Memorandum

To: Program Offices, Headquarters
       Directors, NASA Centers and Installations

One of the most important aspects of the space program is the possibility of the feedback of valuable, new technological ideas and know-how for use in the American economy.

Our economy is expected to grow to something over 700 billion dollars per year by 1970. In the next ten years Dodge Reports estimates that something over 700 billion dollars will be spent for building all kinds of things—highways, bridges, houses, airplanes, trains, and so forth. It also estimates that some 360 billion dollars will be spent for maintenance and repairs in this period. This means that something over a trillion dollars will be spent in America to build or repair or maintain capital items.

Under the above circumstances, any technological gains from our program, if rapidly inserted into the stream of the above activity, can yield great benefits. We must obtain this yield at the most rapid rate to stay ahead of the USSR economy, which is constantly seeking to gain from the technological ideas and know-how which are emerging from its military and space effort. Our problem is to get the feed-back into our normal stream of activity in a better manner than they are able to do.

I will appreciate your sending me any ideas you or your staff have as to specific areas connected with our program where the feed-back can be accelerated or the method of obtaining the feed-back improved.

James E. Webb
Administrator

Document III-9

Document title: James E. Webb, Administrator, Memorandum for Dr. Dryden, Deputy Administrator, "University Relationships," August 4, 1961.

Source: Presidential Papers, Agency Records, John F. Kennedy Library, Boston, Massachusetts.

In assuming the leadership of NASA, a key goal for James E. Webb was to foster space-oriented academic institutions in each of the nation's major geographic areas, with the ultimate goal of stimulating the general academic environment of each region. This plan, which would eventually be encompassed within the Sustaining University Program, broke new ground for the relationship between the federal government and universities. In this memorandum, Webb targets Rice University in Houston, Texas, as such a facility in the Southwest. A little over a month later, he recommended to President Kennedy that Houston be chosen as the site for the Manned Spacecraft Center, which became the Johnson Space Center in 1973, and thereby a focal point for the entire Apollo program. As identified in Document III-7 above, Lloyd Berkner was the chair of the National Academy of Sciences's Space Science Board.

---

[1]   August 4, 1961

Memorandum for Dr. Dryden—AD

Subject: University Relationships

As I believe we agreed before you started on your vacation, the whole area of developing university relationships is of very vital importance to our future, particularly the development of some centers capable of greater efforts in the space science field. Of course we must supplement this with some work with universities who can generally raise the level of education in the basic sciences, and the great reservoir remaining in the country seems to be the Middle West and the Southwest.

There are signs of stirrings in the Upper Middle West, around Minnesota, and some in the Central Middle West, around Kansas City and the general South Illinois-Missouri-Kansas area, and then quite a bit of stirring in the Southwest.

Also, the Research Institute, based on the North Carolina University complex, is making some presentations as to the things they can do in the space program. And Lloyd Berkner has suggested some activities for the Graduate Research Center.

In line with the above, I got a call yesterday from Hugh Odishaw, who says that the Provost of Rice University will be here on Tuesday of next week, and I am to meet the two of them for lunch at the National Academy of Sciences to talk over what Rice can contribute to the program. I believe we already have an active program there and have been told that the new president, Dr. Pitzer, is quite an outstanding man around which a real effort could be built.

By copy of this memorandum, in the absence of Dr. Dryden, I would like to have such information about Rice as will be helpful in conducting the above conference and endeavoring to develop the most constructive lines of interest for the agency with Rice.

James E. Webb
Administrator

Document III-10

Document title: James E. Webb, Administrator, to Dr. Lee A. DuBridge, President, California Institute of Technology, June 29, 1961.

Source: President's Science Advisory Committee Files, John F. Kennedy Library, Boston, Massachusetts.

As a master politician, NASA Administrator James E. Webb realized the need for a broad national consensus in support of the Apollo program. Recognizing that the university science community was likely to be critical, Webb reached out to explain the program as he envisioned it. This letter is one example of his approach. William Pickering, whom Webb mentions, was the director of the Jet Propulsion Laboratory at the California Institute of Technology.

---

[1]   June 29, 1961

Dr. Lee A. DuBridge
President
California Institute of Technology
Pasadena, California

Dear Lee:

Last night the Senate passed the full requested authorization of $1,784,000,000 for our 1962 budget, which is the first formal endorsement of the program suggested by President Kennedy. I believe this means that we will get an approval of our program somewhat earlier than I had expected and with a broader base of acceptance throughout the country than seemed indicated even two or three weeks ago.

Even so, I know the ultimate commitment to the program will depend on the way we go at the job and the results we achieve. Therefore I have been wondering if it might not be helpful if some of the leaders of American science, such as yourself, might not like to have a rather complete briefing on exactly where we stand with respect to our planning. We did have a task force drawn from our ablest people all over the country who have put together a program that appears to be capable of accomplishment, and we are now considering alternatives to see whether we can better this plan. There are several areas where competition exists, such as between the liquid and solid approach.

Would you feel it helpful to take the time, when you are next in Washington, for a quite complete briefing as to how we expect to carry out our entire ten-year program, including the lunar landing? I am taking the position that this program must be so complete and so useful that even if we never make the lunar landing, or do it after the Russians have done so, we still will have obtained outstanding value for the time and money invested. Your own judgement [sic] as to whether the program we have fits this requirement would be helpful.

[2] Another possibility, which I have discussed with several, including Bill Pickering, is that of asking a group of outstanding scientists who have expressed concerns about the program to come in for a group briefing. In this way no one would be singled out, and we would not have present anyone except those who were explaining the program. We would not have those who are in favor of the program and who might want to argue on its behalf. The purpose of this would be to facilitate the understanding which we hope everyone concerned with the program will endeavor to achieve before they take their firm and final positions on it.

As I told you by telephone when we first discussed this program, I certainly have no desire whatever to suggest that anyone who wishes to oppose the program soften his criticism. However, I do feel it quite important, under the conditions that exist in the world today, that the program be quite thoroughly understood before strong adverse positions are taken by our national leaders in any field.

Sincerely yours,

James E. Webb
Administrator

Document III-11

Document title: Hugh L. Dryden, "The Role of the University in Meeting National Goals in Space Exploration," *NASA and the Universities: Principal Addresses at the General Sessions of the NASA-University Conference on the Science and Technology of Space Exploration in Chicago, Illinois, November 1, 1962* **(Washington, DC: NASA, 1962), pp. 87-91.**

NASA Deputy Administrator Dryden gave this presentation at a NASA-university conference in 1962. This meeting, which was patterned after the NASA-industry conference of 1960, was the first meeting in which NASA attempted to convey to the academic world the role envisioned for universities in the Apollo program. This represented the principal address at the general sessions of the conference and pronounced formal NASA policy on the issue. As such, it was especially important as a statement of government position on the interactions of various scientific and technical organizations in conducting space exploration.

The Role of the University in Meeting
National Goals in Space Exploration

[87] The last half century has brought forth a succession of new technologies, sparked by advances in scientific knowledge but brought to maturity by the interaction of scientists and engineers in an environment of national needs for national defense or social and economic development. I need only mention the technologies of aeronautics, communications, radar, nuclear energy, and, now, space. These scientific and technological developments have affected our individual lives as citizens and as professional men and women, and our social institutions, including universities, industry, and other segments of the Nation, as well as government itself. Our international relations, our social and economic development, our military strength—all have been profoundly modified by the powerful forces of science and technology.

It is my purpose to discuss the role of the university in our present-day environment, specifically its responsibilities in space exploration, the responsibilities of NASA, and our joint responsibility for promoting the national welfare.

What is the role of the university today?

There is, I think, general agreement that the university's primary objectives are the education and guidance of students and the promotion of scholarly and scientific inquiry. The ideal university is a community of scholars engaged in research and teaching. In particular, graduate education at its best rests on research, the students learning as apprentices to teachers engaged in advancing knowledge in their professional field.

Yet to state these principles is not to provide a sufficient basis for determining the role of a university. Better than I, college officials and faculty members know that this statement of principles merely indicates where the university's ultimate identity and integrity lie; it does not indicate how this state of affairs is to be achieved in the modern world.

So many at least superficially contradictory demands must be met: the requirements of teaching our swollen enrollments·as opposed to those of research; the desire of the individual scholar to wend his solitary way as opposed to the rising tide of programmatic and team work; the necessity, from an institutional point of view, for drawing a balance between scholarly withdrawal—from which perspective may be gained—and an involvement with on-going life that provides both intellectual stimulation and humane feelings.

The truth is, of course, that in the modern world the university must—for its own survival, and I think for the survival of all that we hold dear—face both inward and outward; it must somehow contain the contradictory forces that threaten to tear our world apart. Because of this, university administration and faculty members bear one of the most difficult burdens of our time. We in NASA—sharing many of the same problems—are aware of this fact; and our aim is to remain aware of it in all of our activities.

In a Commemoration Day Address at the Johns Hopkins University on February 22, 1936, Isaiah Bowman presents this picture of a university which is, I think, equally applicable today:

A university is like a state in the variety of the forces that determine its life: clash of divergent opinion, power to inspire men with exalted purpose, association of distinctive personalities, ordered procedure in a self-governing system, financial perils, and even treasury crises. A citizen in a university-state is not a recluse [88] trending daily a well-worn path of routine. True, he may deal one day with quite petty details of courses and classes; but the next day finds him standing, as it were, on the rim of the universe, analyzing the spectrum of a beam of starlight that left its remote source two hundred million years before the tree-dwelling precursors of man passed their first anxious nights on the ground. The range of the university's interest extends from microscope to telescope, from a student's minute personal problem to the nature and impact of social forces that are rocking the world.

I suggest that the exploration of space is a social force which is rocking the world. I feel no hesitation whatever in saying that the university cannot ignore this force, that it has an inherent responsibility entirely apart from any thought of governmental support to contribute to this major task. Like the small nations of the world which many never launch a satellite, but which must find ways of participation in space exploration, the smallest university must contribute some of its intellectual resources and active interest. Again quoting from Bowman:

To keep research in pure science in the University actively related to social needs and national strength is a duty which cannot be evaded. Pasteur's dream of a private research institute was interrupted again and again by waking realities. There was a national need for knowledge about the silkworm disease and for an understanding of the fermentation problem. His flaming sense of social responsibility was the source of energy and inspiration in his attack upon national problems. As men of privileged education we are not being trained and equipped for isolated and protected living, playboys in the land of dreams.

Our educational institutions bear a major responsibility for the success of our national effort to explore space. Our universities and colleges are called upon to produce a body of scientists and engineers of unexcelled competence. Some of these graduates will enter governmental service with NASA and other agencies participating in the space program: some will join private research organizations and industrial corporations; but some must remain at the universities where they continue to advance knowledge and produce new talent. This last function, as previously mentioned, should receive high priority. The government laboratory, industry, the research foundation, all are users of creative and talented men without reproducing this vital national resource. The university alone is the producer of new engineers and scientists.

The university is not only a center for the development of men with eager, trained, self-starting minds but also a center of creative activity in research. The Summer-Study Committee on NASA/University Relationships of the Space Science Board of the National Academy of Sciences points out that:

. . . the opportunities for developing new fundamental knowledge and technical applications may very well equal or exceed those which have existed in the atomic and nuclear physics fields during the past thirty years. . . . A vigorous academic program in all appropriate aspects of the space endeavor must be developed. Such a program must enjoy a visible relationship to that of the federal establishment itself: but it is of utmost importance that it preserve the essential virtues of universities—a devotion to scholarly and scientific inquiry, a primary concern for the guidance and eduction of students, full freedom of discussion and publication, and essential autonomy in the formulation of research objectives and of programs of work directed toward such objectives.

Other aspects of the independent role of the university in the environment of a national program of space exploration will be discussed subsequently. Consider now NASA's specific needs for assistance from the university community. The NASA program comprises four main areas—space sciences, manned space flight, applications of earth satellites to communications and meteorology, and advanced research and technology. What help do we expect to get from the university in each of these areas?

The term "space sciences" is a shorthand expression to describe investigations in any field of science carried out by apparatus carried into space by sounding rockets, earth satellites, and lunar, planetary, or interplanetary probes. Sometimes the term is extended in meaning to include laboratory or earth-based observations related to the flight experiments. The fields of science included are, in the main, astronomy and solar physics; geophysics, including aeronomy, ionosphere physics and energetic particles and fields; interplanetary investigations; lunar and planetary investigations; and biosciences.

The NASA program in space sciences is being built on the participation by the competent scientific community. It is freely recognized that the U.S. would have no space science program worth talking about if at least some of the [89] most competent scientists of the Nation were not deeply involved in it. The importance of the creative activity of the individual working scientist in the program is paramount. It is necessary to make use of scientific competence wherever that competence may be found. Although there is significant participation by scientists within NASA, scientists in other government agencies, in the industrial community, and the international scientific community, the major element in the participating scientific community is the university community of the U.S.

The university scientist who participates in satellite and space-probe experiments finds an environment different from that to which he has become accustomed. Traditionally, a scientist conceives an experiment, builds the apparatus himself or has it built under his supervision in the university shop or by contract, carries out his experiment, analyzes the data, and publishes his results. This relatively simple procedure is not possible in satellite and space-probe experiments, although a fair approximation to it is feasible for experiments with small sounding rockets. Satellite launching requires large rockets, special launch sites, a worldwide tracking and data-acquisition network, sharing by many experimenters in a single flight, and a large team of cooperating specialists. The scientist becomes involved in scheduling his work to meet a flight date, once that date is set. His apparatus must be engineered to meet severe environmental requirements of vibration, temperature, exposure to radiation and charged particles, and so forth. Some universities are able to provide this service; others must depend on industrial help. Thus, the role of the university scientist often reduces to concept of the experiment, development of laboratory prototypes of the equipment, analysis of the data and publication, plus participation in a large team to design the actual satellite, launch it, and receive the data. NASA policy is to support the tradition of responsibility and freedom of the experimenter to the maximum extent consistent with the nature of the operation. Selection of experiments to be flown is made by a Space Sciences Steering Committee composed of scientists and engineers in NASA Headquarters who are not contenders for payload space and who have the advice and guidance of outside consultants.

In the space sciences area, NASA supports by grants the development of scientific and technical information in areas broadly related to space science as well as specific project tasks. Examples of current specific tasks are: develop, construct, and test four magnetometer instruments suitable for use on a satellite to determine the magnitude and direction of the earth's magnetic field and analyze telemetered data from the instrument; design, construct, and test a Cerenkov counter and associated circuitry to measure the energy spectrum of high energy gamma rays; test and calibrate the equipment by synchrotron or balloon techniques; and assemble instrument packages suitable for use in satellites. Examples of broader tasks in areas related to space science are: research in solar and cosmic-ray physics; theoretical research on low-energy electronic, ionic, and atomic impact phenomena; and the magnetohydrostatics of the magnetosphere of the earth and problems in theory of orbits of space vehicles.

In the field of advanced research and technology not directly connected with the flight program of sounding rockets, satellites, and space probes, NASA is interested in and supports a wide range of research activities from basic research to technological applications, from theoretical investigations to laboratory experiments. Some are related to problems of immediate operational concern; others endeavor to extend the present limits of knowledge and broaden the research capabilities available for such extension. Our quarterly program report for July 1, 1962, shows about 450 active grants and research contracts. A few of these are related to the manned space flight and the applications program of NASA but the majority are in the fields of advanced research and technology and space sciences.

Although NASA does place demands for direct assistance on the universities, we consider that we have an obligation to conduct the space program in such a way as to help strengthen the university. We wish to work within the existing university structure rather than to set [90] up independent contract-operated activities that tend to draw the university research scientist or engineer away from the teaching of students in the course of the research he performs and directs. We seek to share in a joint responsibility to add to our national strength. It is clear that NASA cannot meet all the desires or even needs of the universities or mount a program of general support to education. We have neither the responsibility nor the resources to do this. But like the logger who has a responsibility of replacing for the future the trees which he harvests, NASA, as a user of university trained talent, has an obligation to carry a fair share of the load of replacing the resources consumed. The universities must bear their share of responsibility for the success of the space program, as previously discussed, and must allocate an appropriate fraction of their own material and human resources to the effort. But NASA stands ready to invest substantial resources in partnership with the university.

Thus, in addition to direct project support, NASA initiated in fiscal year 1962, a program of enlarged scope for utilizing more fully the abilities of our universities. The program is frankly NASA-oriented but planned in such a manner to recognize the acute needs of the university as well. In brief, to meet the space program needs, we are proceeding to strengthen university participation in four ways: (1) to utilize university resources for specific research projects under grant or contract as appropriate; (2) to encourage the establishment of interdisciplinary groups for research in broad areas to be supported by grants; (3) to support the training of people in the field of space science and technology through grants; and (4) in certain cases to provide research facilities.

The first method is the traditional support of projects; the other three are new so far as NASA is concerned. The broad grants are intended to encourage the establishment of creative multidisciplinary investigations, the development of new capabilities, and the consolidation of closely related activities. As will be discussed subsequently, multidisciplinary is here intended to include not only cooperative effort among branches of the physical sciences but also between physical and biological sciences and with some participation from the social sciences, all as appropriate to the selected broad areas in which a given university possesses high competence.

The third method comprises research training grants to increase the supply of scientists and engineers in space-related science and technology. It has been estimated that by 1970 as many as one-fourth of the Nation's trained scientific and engineering manpower will be engaged in space activities, although I cannot confirm the accuracy of this estimate. For planning purposes only, we have suggested as a goal the support of about 4,000 graduate students per year in 150 qualified universities, to yield an annual output of about 1,000 new Ph.D.'s in space-related fields. In selecting universities, we consider such factors as accreditation ratings, resources, previous and current efforts in developing research activity in the space sciences, location and extent to which the region already is provided with advanced training opportunities, and so forth.

The fourth method is the provision of grants for facilities in certain cases. Consideration is given to the urgency of the need, the nature and extent of the university's involvement in space-related research, the relative importance of the research to the national space program, the demonstrated competence, past achievements, and potential future accomplishments of the research groups, and similar factors. In general, we attempt to consider a total university situation and use an appropriate mix of the several methods for the specific circumstances, subject of course to the total resources available for the program.

In FY 1962 the commitments for the support of project research at universities were of the order of $28 million and the estimate for FY 1963 indicates an increase to about $55 million. A few interdisciplinary grants date back to FY 1961. In FY 1962 eleven such grants were made, amounting to a total of about $3 1/2 million. Training grants were made to ten institutions amounting to a total of about $2 million, and facilities grants to five institutions amounting to $6 1/2 million, all of which have existing interdisciplinary activities. This total [91] of $12 million for the last three categories will be increased to about $30 million in FY 1963. The many proposals on hand are under evaluation at the present time. We recognize that a larger effort needs to be made and hope to move toward the desirable goals in succeeding years.

In recognizing the separate responsibilities and specialized interests of the universities and NASA and their interrelationships, we cannot forget other parties at interest in the space-exploration program. The major fraction of the effort, as measured by dollar value or manpower, is conducted under contract by private industry. There are many aspects of university-industry and NASA-industry relations which lie outside the province of the present discussion. Here we note only that NASA, the universities, and the aerospace industries have a collective responsibility for the conduct of the space program.

The collective responsibility goes far beyond that for the success of the technical aspects of the program, if the greatest benefit to the nation is to be realized. We have previously discussed at some length the conduct of the program in such a manner as to strengthen the universities as an element of national strength. Similar conditions apply to the aerospace industry, but our obligations extended further to every aspect of our social, economic, and political life.

Space research and development, like the predecessor fields of rapid scientific and technical advance at the frontiers of knowledge—aeronautics, electronics, and nuclear research and development—produce corollary benefits in the form of new knowledge, new products, new methods, and new materials which can be employed in the development and manufacture of countless articles for human use. In the past the transfer process proceeded in a laissez-faire manner at a relatively slow pace. We believe that it is incumbent on all of us to try to accelerate this process. We have suggested that universities participate in promoting wider use of the information obtained by associating members of the faculties in economics, business administration, and political science in the activities of the interdisciplinary groups.

It is our feeling that the universities should go still further to assert leadership in attacking the totality of problems affecting the welfare of man within their sphere of influence, whether this be a community, a region, or the entire nation. Abraham Horwitz, in discussing "The Changing Scene in Latin American Medical Education" in the *Journal of Medical Education* for April 1962, made some observations which, in the following paraphrased form, are applicable to the current situation in the United States: There is a new spirit abroad in the U.S. today, a spirit imbued with the determination to create more wealth, to distribute it more equitably, and to promote the well-being of man. The focal point of this signal endeavor should, we believe, be the universities for the primary need is for experts to put to work the capital that will be invested in systematic programs. Equally pressing is the need for a deep and searching examination of the problems that beset us and the establishment of the procedures for their solution. A debate of this kind can best be carried on in the university, which is wedded to the free examination of all problems affecting the life of man in society, and where culture, in the sense of perfection of man, has its wellspring. . . .

In summary, all of us who participate in the conduct of the space-exploration problem should endeavor to discharge our task in the light of these broader considerations of human welfare. The university has a unique opportunity, not only to perform basic research and train new talent in new areas of science and technology and to carry a large share in the scientific aspects of the space flight programs, but also to provide leadership in the wide discussion and practical solution of the broader aspects of extracting from our space effort the greatest possible contributions to human welfare within its sphere of influence. For its part NASA is attempting to give due consideration to its responsibility in these major questions of the social impact of the space program.

Document III-12

Document title: Edgar M. Cortright, Memorandum for Mr. Webb, "NASA-CIT/JPL Relations as they pertain to the present contractual arrangements of operating conditions and the future role of JPL in the NASA Program," June 1964.

Document III-13

Document title: Arnold O. Beckman, Chairman, Board of Trustees, California Institute of Technology, to James E. Webb, Administrator, NASA, June 26, 1964.

Source: Both in NASA Historical Reference Collection, NASA History Office, NASA Headquarters, Washington, D.C.

One of the persistent challenges faced by NASA managers in the agency's earliest years was the relationship with the California Institute of Technology's (Caltech, or CIT, as stated in Document III-12) Jet Propulsion Laboratory (JPL) in Pasadena, California. JPL had been established during World War II as a contractor facility developing rockets and other technologies for the U.S. Army. Since the war, it had expanded its capabilities, and by the time of NASA's establishment in 1958, JPL was a major location not only for the development of rocket technology but also space science. Because of this, NASA leaders secured the transfer of JPL from the Army and re-emphasized in the late 1950s a JPL effort already under way—Project Ranger, an effort to send satellites to the Moon. Following the failure of the Ranger 6 spacecraft in January 1964, NASA Administrator James E. Webb pressed Arnold O. Beckman, chair of the Caltech Board of Trustees, to alter the methodologies of management at JPL. These two documents describe this situation and propose changes. They successfully set in motion a number of activities that affected the relationship for more than a decade thereafter. Edgar Cortright was NASA's Deputy Associate Administrator for Space Science and Applications.

Document III-12

[1]   # Memorandum for Mr. Webb

Subject: NASA-CIT/JPL Relations as they pertain to the present contractual arrangements of operating conditions and the future role of JPL in the NASA Program.

Although this memorandum is designed as a position paper it is necessary to review
certain aspects of the history in working with Cal Tech and JPL.

I.   History
      A.   Contract Provisions
The initial NASA contract placed in late 1959 with CIT for the operation and management of JPL was quite broad and free from constraint and provided for minimum
control over the activities of the Lab. The current contract, executed in December 1961,
reflects the experience gained in the two preceding years, of dealing with JPL, but still permits JPL considerable latitude for independent operation. This operating latitude results
primarily from the necessity of mutual agreement between NASA and CIT/JPL on substantive changes in program or administration. During negotiation of the current contract, NASA officials suggested a change in the requirement for mutuality in certain
aspects of JPL operation. However, this change was not successfully negotiated.
[2] B.   JPL Assignments
Since the beginning of our working relationship with JPL, the Laboratory has been
assigned functions in the areas of flight projects, deep space instrumentation, and supporting research and technology. Among the flight projects, the assignments have included Ranger (Blocks 1, 2, 3, 4, and 5), Mariners (A, R, C and B), Surveyor, Surveyor Orbiter
(study phase), Voyager (study phase), and Prospector (study phase—cancelled). In addition, the launch vehicle, Vega, was assigned to JPL and subsequently cancelled. JPL has
carried out the buildup of the deep space instrumentation facility on a worldwide basis. It
has carried out research in fluid mechanics, structures, propulsion, electronics, telemetry
guidance and control, and other areas, many of which were not covered at other NASA
Centers. JPL has, through a master planning board initiated by NASA, undertaken to
expand and upgrade the existing laboratory facilities for the Government.
      C.   JPL Organization
The JPL organization was originally structured as a research laboratory in propulsion
fuels, materials, etc., and subsequently assigned one large project, e.g., Corporal, then
Sergeant. This meant that research people were intermixed with project people; the laboratory was strictly a matrix organization and a loose one at that. With the assignment of
multiple projects, JPL began a series of reorganizations. [3] Basically, they created a
Systems Division to do systems engineering for all of the projects, and two program
offices—the Lunar Program Office, and the Planetary Program Office. These program
offices, the Systems Division, and all of the other laboratory divisions reported to the
Director's office. The Program Offices contained the project managers with small staffs.
To assist in the management of this matrix, Dr. Pickering [JPL Director] hired a Deputy
Director (Brian Sparks). This early configuration has recently been modified to combine
the two program offices into a single program office; to strengthen the coordination
among projects, deep space instrumentation facility, research and development, and busi-
ness administration; and to strengthen the reliability and quality assurance effort.
Although the laboratory was not projectized, all employees working for the projects have

been identified and fixed to a project. In brief then, the laboratory has moved in the direction of strengthening its project management and correcting its faults after they have become apparent to the laboratory. They still retain a matrix organization with many important individuals reporting to Pickering and Sparks directly and with project offices which are marginal in strength and quite dependent on strong front office leadership to insure a smoothly functioning total laboratory team.

[4] **II. NASA Direction**

Initially, NASA direction to JPL was almost exclusively from the Office of Space Flight Programs. With the advent of the NASA matrix organization JPL receives direction from many offices in Headquarters, e.g., OSS, OART, T&DA, Office of Programs, Procurement and Contracting, and the Office of Administration. The quality and depth of direction have varied from situation to situation and many have been inadequate to the situation existing within JPL on several occasions. The rapid growth of the laboratory of 2400 to 4000 has certainly contributed to developing problem areas. The rescheduling of projects necessitated by the Vega cancellation and the Centaur slippage have been serious perturbations. The overloading of the laboratory by NASA Headquarters and its own management had caused problems which might have been avoided if we had used better judgement [sic]. Lastly, the changing interface between JPL and NASA has caused communications problems and misunderstanding with regard to direction functions and authority.

III. Strength of CIT/JPL Performance

From the positive point of view, JPL represents a collection of highly imaginative and skilled engineers and technicians. This scientific and engineering team has been attracted to JPL, at least partially, because of the outstanding technical reputation of CIT. [5] They have shown considerable flexibility and have been able to roll with the number of reprogramming punches which have been forced upon them by circumstances. They have shown a keen interest in the space program and, despite frequent internal wranglings, they have never carried their arguments with NASA to the public. The working relationships have grown steadily better and excellent communications links exist among individuals in certain areas. The Project performance has generally been spotty, having varied from outstanding on Mariner to poor on Ranger. Similarly, the quality of business performance has varied ranging from excellent on source evaluation procedures used on Surveyor to inadequate administration of the resulting contract.

IV. Weaknesses of CIT/JPL Performance

In general, the performance of Cal Tech and JPL can be summarized as follows: Cal Tech has provided almost no visible leadership to JPL and has generally proven to be a poor communication link between NASA policy makers and JPL policy makers, e.g., at the DuBridge-Pickering level. Also lacking is action by the CIT Board of Trustees to clearly define the Institute's responsibility in the management of JPL, and to assign specific responsibility to designated positions or individuals. The CIT/JPL top leadership has been weak in terms of attention to substantive program issues in the [6] laboratory and in terms of responsiveness to official NASA guidance and direction. At times, the leadership has almost obstructionist. This has primarily been the case when NASA suggestions have been made with a view to improving laboratory management. The top management has consistently taken the attitude that the management of their laboratory is their business, and that unless the the contract terms specifically cover items discussed they have no interest in our compulsion to perform functions or take actions demanded by NASA managements [sic]. The most serious concern on the part of those of us doing business with the laboratory, however, has been the lack of involvement of the top management in direction of the day-to-day operations. The organization is structured so that it requires such involvement yet little or no evidence of such management direction is apparent. The members of the team operate much of the time with no apparent leadership. Many of the problems which JPL is now struggling to solve might have been avoided or at least recognized earlier had JPL management been more involved in the day-to-day execution of the major laboratory assignments or had they worked with NASA to correct those weaknesses detected and pointed up. One might say that it took the Ranger situation to make JPL face up to its many problems. I might add that NASA is having to face up to a few of its own by the same token.

[7] **V. Actions That Can Be Taken Prior to Contract Renewal**

Some of the things that can be done under the present arrangement for operation of JPL are:

1. The CIT Board of Trustees should, by formal action, define the responsibility of CIT for direction of JPL.

2. CIT should designate a top University official to whom NASA can direct its requests for corrective actions. This official should have clearly assigned authority to effect changes in all areas (management, technical, and business administration). In this regard, it may be desirable for NASA to offer to present its views to the CIT Board of Trustees.

3. An understanding should be reached whereby CIT/JPL will be responsive to NASA suggested changes in management and organization. For example, there is still a need for a strong General Manager at the Laboratory.

4. The "Task Order" problem should be resolved. The contract provides for separate task orders covering major NASA projects and these have not yet been negotiated.

5. The business management practices at JPL should be made compatible with NASA policies and practices. Examples of areas where business management practices can be improved are:

    a. Procurement policies and procedures
    b. Budget programming, financial management, and reporting systems
[8]   c. Management of facilities property and supply
    d. Travel and other fringe benefit policies

VI. Alternatives for Consideration Before Present Contract Expires

Since the present contract expires December 31, 1964, it is not too early to think about the relationship of NASA-CIT/JPL after that date.

Several alternative arrangements are possible.

A. The contract could be allowed to expire and the Government owned Laboratory could be operated by civil servants.

Advantages

1. True center of NASA would operate under same NASA policies and regulations as other NASA Centers.

2. Problem of salary differential for similar work would disappear.

3. One echelon of management would be eliminated, i.e., CIT.

Disadvantages

1. Loss of effort and drive for some period while change takes place (6-12 months) Projects disrupted.

2. Loss of hardcore of key personnel—would probably move to industry.

3. NASA recruitment problem to be faced.

[9]   4. NASA public image problem.
         a. Cal-Tech
         b. Scientific community-industry
         c. Congressional
       5. Loss of flexibility laboratory enjoys as contractor operated, e.g., not bound by all Government rules and regulations.
    B. A non-profit corporation could be substituted for CIT/JPL management.
       Advantages
       1. Single purpose of Board of Directors. Minimizes possibilities of conflict of interest situations. Only serve one customer, NASA.
       2. Provide NASA ability to have direct influence on management selected or replaced.
       3. Provide flexibility of wage and fringe benefit allowances—not tied to campus scale or limitations.
       Disadvantages
       1. Project disruption while changes take place.
       2. NASA public image problem.
       3. Higher cost operation.
       4. Magnification of differences between the Lab and other NASA Centers.
    C. An industrial contractor could be selected to operate the Laboratory for the Government.
[10]   *Advantages*
       1. Initial selectivity from range of industrial capabilities.
       2. Flexibility of industry management policies and practices.
       3. Responsiveness to changes in direction or level of effort.
       Disadvantages
       1.. Project disruption while changes take place.
       2. NASA public image problem.
       3. High cost operation.
       4. Loss of relationship of Lab to other NASA Centers.
       5. Possible conflict of interest situations.
       6. Loss of active and direct control.
    D. A form of the present contract with CIT/JPL could be continued if the following improvements can be worked out.
       1. Clearly defined management responsibilities and accountability for CIT and JPL.
       2. Clearly defined communication links between CIT-JPL-NASA Managements.
       3. Acceptance of NASA contractor relationship by CIT/JPL.
       4. JPL responsiveness to NASA direction and control.
          Advantages
          a. No major disruption to programs and projects.
          b. No major loss of hardcore key personnel.
[11]   c. No public image problem.
          d. No loss of flexibility of operating outside Government rules and regulations.
          Disadvantages
          a. Continued status of "almost NASA Center" concept.
          b. Continued problem of campus-off campus status.
          c. Management layer between lab and NASA-CIT.
          d. Conflight [sic] of interest situations.

VII. Summary

These observations on JPL and Cal Tech do not begin to tell the whole story, either good or bad. However, I think they can provide background for a position which is rather firm toward CIT in terms of demanding stronger management of the laboratory. In being fair, however, I think we can be responsible in terms of time required to implement some of the more radical changes we have suggested, such as hiring a general manager to supplement Pickering and Sparks or breaking up the Systems Division to strengthen the project offices. In reviewing our own judgements [sic], it might help to point out that these opinions of the laboratory are held rather widely throughout industry and among many of the JPL staff. The staff itself, I believe, hopes for continued NASA pressure which will result in stronger management by evolution rather then revolution. I consider it desirable that JPL continue in its past role of performing much the same function as a NASA center. The laboratory will be of most use to NASA if we can truly develop [12] the working relationships to make this possible.

Edgar M. Cortright

Document III-13

[1]   June 26, 1964

Mr. James E. Webb
Administrator
National Aeronautics and Space Administration
Washington, D.C.

Dear Mr. Webb:

About three months ago, at a meeting in your office, we discussed the NASA-JPL-Caltech relations. This meeting was the first opportunity since I assumed the chairmanship of the Caltech Board of Trustees to hear directly from you and members of your staff about a number of problems related to JPL. I promised then to do everything possible to assist in eliminating the causes of past complaints, improve management operations at JPL in the light of suggestions made by your staff and others and attempt to find new ways in which NASA and Caltech could be mutually helpful in expanding fundamental research in space. Substantial progress has been made, I believe, and I thought you would be interested in hearing about it. In the following pages and attached appendices I have outlined briefly some of the highlights in the areas of management, technical coordination, and Caltech-JPL research activities.

Management

Prior to January 1, 1964 Price-Waterhouse management advisory services department had been retained to study the organizational structure of JPL. At my request, the McMurry Company was called in to evaluate the top dozen or more administrators at JPL, and to make an independent study of the organization. This work, performed personally by Dr. McMurry, has been completed.

One of Dr. McMurry's principal recommendations was the procurement of a new Deputy Director at JPL. A detailed job description was prepared and two leading executive recruiting firms were retained to find suitable candidates. Many candidates were screened, including persons recommended by Mr. Hilburn and others in NASA. We were very fortunate, we believe, in being able to secure General A. R. Luedecke, currently General Manager of the Atomic Energy Commission. I believe that General Luedecke is an extraordinarily fortunate choice. His experience in handling large operations in the Air Force and the AEC has given him an excellent background in governmental procedures and requirements. In addition to his demonstrated high level of competence, the fact that the AEC carries much, if not all, of its research and development through university-type contracts has given the General very valuable experience which especially qualifies him for the NASA-JPL-Caltech operation.

[2] Dr. McMurry's report recommends that certain organizational changes be made at JPL. He recommends, however, that these changes be made after a new Deputy Director has assumed his duties. In the meantime, several changes have already been made which should improve management.

In December of 1963, the Lunar and Planetary Projects at JPL were consolidated under Mr. R. J. Parks, who [was] appointed Assistant Laboratory Director for Lunar and Planetary Projects.

In February 1963 the Director of the Jet Propulsion Laboratory formed an Executive Council consisting of the Deputy Director, Assistant Laboratory Directors, and the Special Assistant for Advanced Technical Studies. This group will advise the Director on all major policy matters, develop long-range plans, and to recommend preferable courses of action relative to major Laboratory questions and problems.

JPL management has consolidated all quality assurance and reliability activities into one office, reporting directly to the Director/Deputy Director of the Laboratory. The chief of this office, Mr. Brooks Morris, has been delegated the responsibility for all quality assurance and reliability activities related to JPL projects and to evaluate the probable reliability of the designs and plans for Laboratory missions.

A Management Information Office has [been] established in March 1963 to provide accurate and timely information to JPL top management, and to the appropriate elements within NASA Headquarters.

As suggested by certain people in NASA Headquarters, the Laboratory had taken a very close, hard look at the advisability of modifying the matrix organization in favor of a strict project structure. The results of this review has been a high degree of projectizing within the technical divisions. The majority of the professional staff, working on the flight projects, have been assigned full time and their efforts restricted to specific projects. The management of JPL is continuing to move in this direction in the establishment of new projects, as well as in the strengthening of existing projects.

The Financial Management Division has been transferred. The manager of that division now reports to the Deputy Director, giving that office increased stature and authority in keeping with the Laboratory's growth, and the increased emphasis on fiscal and contractual activities.

The Procurement Division has been transferred. It now reports directly to the Deputy Director in order to provide more complete integration of the technical and managerial problems associated with the increasingly large procurement actions entered into by the Laboratory in carrying out NASA's projects.

The Technical Studies Office, headed by Dr. Homer J. Stewart, has been established to direct, coordinate and to originate all JPL advanced mission studies for the unmanned lunar and planetary exploration.

[3] To accommodate the increasing number of outside projects utilizing the DSIF and the JPL SFOF, an Assistant Laboratory Director has been appointed to head the Deep Space Network activities at JPL. Dr. Rechtin, who is in charge of this office, is responsible

for coordinating all Laboratory actions relative to the DSIF, SFOF and the JPL technical divisions in order to assure that project requirements are understood and met.

In January of this year, the Facilities Office was reorganized and given responsibility for developing the implementing [of] a technical and supporting facilities program that will provide those facilities required for the accomplishment of its assigned tasks and for coordinating and integrating the inputs from the JPL technical divisions and other sources into a single approved long-range master facilities plan.

The Laboratory has engaged the services of the Harbridge House organization to make additional detailed studies of the procurement process, and to recommend procedures and policies to be adopted by JPL in this area.

Internal audit groups reporting to CIT and to the top management of the Laboratory are being established to review and ascertain the degree to which Laboratory policies and procedures are being complied with in order to adequately inform management of need for corrective actions.

In addition to the organizational changes delineated above, which are aimed at strengthening the decision-making processes by which JPL conducts its affairs, the management of the Laboratory has requested a review of the Procurement Division operations by a panel of NASA procurement specialists and has responded to all suggestions offered by this group; the majority of the substantive recommendations have been carried out.

To insure that JPL will receive that best possible guidance and assistance from Caltech, two new and influential working committees, reporting to the Chairman of the Board of Trustees of the Institute, have been formed. This will bring the knowledge and experience of many business executives and scientists to bear on the problems concerning the tasks to be performed by the Jet Propulsion Laboratory.

A Trustees Committee composed of the Chairman of the Board, the President of the Institute and four other trustees has been established. They bring to the Committee a vast background of experience in the management of industrial organizations operating in the aerospace field.

[4] The members of the Trustees Committee are:

Dr. Arnold O. Beckman (Chairman)
    President—Beckman Instruments, Inc.
Dr. Lee A. DuBridge
    President—California Institute of Technology
Mr. John G. Braun
    President—C. F. Braun & Co.
Mr. Thomas V. Jones
    President—Northrop Corporation
Dr. Augustus B. Kinzel
    Vice President, Research—Union Carbide Corp.
Mr. Herbert L. Hahn
    Partner—Hahn & Hahn
Mr. William E. Zisch
    President—Aerojet General Corporation

Mr. Robert B. Gilmore and Dr. William H. Pickering are ex officio and nonvoting members. This group has already met several times. Its principle role is that of advisor to the Laboratory top management on major policy matters, and to keep the Executive Committee and Board of Trustees of the Institute informed on important matters at the Laboratory.

A committee of appropriate facility members has also been formed to deal with the very important interrelationships between the academic and scientific staff of the Institute and the technical staff of the Laboratory. The membership of the Facility Committee is as follows:

Dr. Clark B. Millikan (Chairman)
    Director—Graduate Aeronautic Laboratories
Dr. Robert F. Bacher
    Provost
Dr. Norman Horowitz
    Professor, Biology
Dr. Robert B. Leighton
    Professor, Physics
Dr. Frederick C. Lindvall
    Professor, Electrical & Mechanical Engineering
Dr. Robert P. Sharp
    Professor, Geology
    Chairman, Division Geological Sciences
Dr. William H. Pickering
    Director, Jet Propulsion Laboratory

This committee is principally concerned with the technical problems in which the experience of the scientific and technical staff of the Institute can be of support to the Laboratory. It will meet frequently to review [5] activities at the Laboratory and the Campus, to provide the Director of JPL with advice and support on important technical decisions, and to arrange for the exchange of technical information and to advise on the selection of highly qualified scientific personnel at the Laboratory.

Mr. Robert B. Gilmore, Vice President for Business Affairs at Caltech, has submitted several reports to Mr. Hilburn, stating in some detail the corrective measures that have been taken upon the recommendations of the Army Audit Report number LA 64-581, date of issue February 26, 1964 entitled "Report on Financial Management and Related Operations for the Period Ended June 30, 1963." A brief summary of some of the principal items is attached as Appendix A to this letter.

Technical Problems

As you know, there has been some criticism of JPL concerning technical matters such as design features, quality control, and testing. Some have stated their opinion that JPL scientists have not been adequately responsive to suggestions made by others. Not all suggestions are necessarily good, of course. To assist JPL in evaluating suggestions and to make sure that JPL's technical problems will receive the attention of the best research people at Caltech, the Caltech-JPL Facility Committee referred to above meets from time to time. This group has given Dr. Pickering and his associates probably the best advice available today, in the respective fields of the committee members, on the suggestions and the recommendations in the Kelly and Hilburn reports. To the best of my knowledge, every technical suggestion that has been received by JPL has either been adopted or, if not adopted, sound reasons for the rejection have been given.

With respect to Ranger 7, I have been informed and believe there has been the utmost cooperation between JPL and NASA officials. So far as I know, JPL has performed every task and made every test that has been requested by Dr. Seamans, Dr. Newell and Mr. Cortright. I have been unable to find any indication of unresponsiveness or lack of cooperation on the part of JPL. If something less than complete agreement on technical matters existed in the past, that situation does not exist today.

Caltech-JPL Research Activities

I have been aware, and I heartily applaud, your great personal interest in expansion of basic research. The Institute certainly shares your interest and desires to do all it can to assist in the development of a vigorous research program, not only at Caltech but in other universities capable of carrying on fundamental research. To give me an idea of what Caltech is doing in research related to space, Dr. Bacher and the various [6] division chairmen at Caltech have provided the information that is attached as Appendices B through I. The following items are included:

A proposal dated April 18, 1962 for NASA support of research in certain fields of physics and astronomy. This report gives a broad outline of important fields of research in which Caltech and the Mount Wilson and Palomar Observatories are engaged, together with a specific recommendation for a 3-year research program. This has been brought up to date by the attached list of staff members, post-doctoral research fellows and graduate students now working in the various fields of research.

A proposal dated July 20, 1962 for pre-doctoral training grants for Caltech. This proposal covers work that would be carried on in the divisions of biology, chemistry and chemical engineering, engineering and applied sciences, geology, physics, mathematics and astronomy.

A proposal dated July 22, 1963 for pre-doctoral training grants for a 3-year period. This proposal is essentially a duplicate of the previous one, with certain added programs.

Extracts of memoranda given by the various division chairmen at Caltech, which contain information of direct interest. Your attention is called particularly to the special summer program which started June 22, 1964 with 36 students selected from institutions all over the country for an intensified course in problems of space technology.

I hope I haven't burdened you unduly with this rather lengthy letter. I feel that Caltech and JPL both have done excellent jobs in getting on top of their problems and in taking steps to insure that NASA will receive the type of managerial and technical competence and performance that it desires. I believe that most of the sources of annoyance in the past have been eliminated and that developments of the past three of four months, while not entirely to the liking of any of us, have actually resulted in a substantial improvement in understanding and in over-all operations.

If you would care to make any comments or suggestions, I should be pleased to receive them.

Cordially yours,

Arnold O. Beckman, Chairman
Board of Trustees

Document III-14

Document title: Raymond Einhorn and Robert B. Lewis, Memorandum to Mr. Hilburn, "Review of Purposes and Application of CIT Fee and Overhead for the JPL Contract," with summary of report, October 20, 1964.

Source: NASA Historical Reference Collection, NASA History Office, NASA Headquarters, Washington, D.C.

One of the issues of contention in the NASA-Caltech relationship during the early 1960s was the fee Caltech (referred to below as "CIT") charged NASA for managing the Jet Propulsion Laboratory

(JPL). When the failure of the Ranger 6 spacecraft in January 1964 brought the terms of contract renewal into question, the management fee amount became a significant issue. Consequently, NASA Assistant Deputy Administrator Earl D. Hilburn instructed Audit Division Director Raymond Einhorn and Financial Management Division Director Robert B. Lewis to review the fee. The following memorandum discusses this investigation and transmits a lengthy, five-part report. Also included here is the summary from the larger, 41-page report that was used to provide information for the space agency's effort to reorient relations between the NASA Headquarters and JPL.

[1]

Memorandum

October 20, 1964

TO: Mr. Earl D. Hilburn
FROM: Raymond Einhorn [initialed]
        Robert B. Lewis [initialed]

SUBJECT: Review of Purposes and Application of CIT Fee and Overhead for the JPL Contract

In accordance with your request to us and your discussion with Mr. Robert Gilmore, we visited the California Institute of Technology to review a current statement of CIT's reasons for a management fee for the Jet Propulsion Laboratory contract, and how CIT applied the fee and other income received by the Institute. We also were to determine the kinds of indirect expense which CIT charged to the Jet Propulsion Laboratory contract, the amounts charged, and how reimbursements for these charges were applied by CIT. . . .

[2]   **I. SUMMARY**
1. CIT Refused to Discuss Fee
    CIT was unwilling to discuss the reasons for requesting or paying a fee for the JPL operation. It thought it was inappropriate for NASA to ask about the application of the fee, inappropriate for CIT to give the information, and dangerous for NASA to have the information even if it should get it. Despite the recent discussions that have been held with NASA officials concerning the fee, we were informed that Dr. DuBridge could not present a list of reasons immediately since the statement had to be carefully drawn up and reviewed by CIT officials and the Board of Trustees. The statement probably will not be submitted until after modification 10 is signed, and apparently will not contain dollar or other measurement factors.

2. CIT Reasons for a Fee
    We summarized in Section II of this report the reasons previously given by CIT for the fee, and have made comments based on our analyses. Of the reasons given, we believe only four are suitable for consideration: (1) the benefits which NASA derives from a competent technical team at JPL, attracted and retained by CIT's reputation and academic environment, and from the availability of eminent faculty scientists to advise and consult with the JPL technical team; (2) to compensate CIT for the risk of possible injury to its reputation and damage to the future of the Institute, due [3] to technical failures in JPL projects, which are beyond the control of CIT; (3) to provide a "buffer" or a reserve of funds to help absorb the economic shock of the loss of fee and campus overhead pay-

ments in the event the JPL contract expires; and (4) to assist in the current financing of higher campus costs, such as higher salaries and operation of expanded facilities caused by the operation of JPL.

These reasons should be re-evaluated by NASA and CIT. Presumably Dr. DuBridge's additional statement of CIT's fee reasons will be of assistance in accomplishing this evaluation. The assignment of dollar or other measurements to fee factors is difficult, except for the "buffer" and "higher campus cost" items. These two factors can be measured if CIT would cooperate in the effort.

Other factors given by CIT in support of a fee do not have merit, such as unallowable costs and intangible and other costs no longer recoverable under current cost principles prescribed by the Federal Government (see Section IV). In addition, it should be noted that several factors given to justify a fee relate to operations for which NASA is already paying a very large proportion of the costs.

3.   Source and Application of CIT Funds

We are fairly certain that the fee is used to finance current CIT expenditures and to provide the "buffer" that CIT stated it would need in case of the expiration of the JPL contract. We also believe [4] it may be used to supplement the plant fund and other special purpose funds. We are also fairly certain that the "buffer" is included in CIT's income stabilization reserve, a general reserve established for the purpose of smoothing out the "peaks and valleys" in the Institute's income. Our analysis of CIT's financial statements confirms the fact that the reserve is broader than just for the "buffer."

The only information CIT was willing to provide on the overall source and application of its operating income is summarized in Section II. However, the statement gives little guidance on the application of the fee.

4.   CIT Indirect Expenses

NASA pays CIT for about one-half of its general and administrative expenses and about 65% to 78% of all major categories of general and administrative expenses that are applicable to on and off campus activities. These expenses are summarized in Exhibit A and Paragraph 2 of Section III. Our analyses of these payments showed not only that the allocation basis of salaries and wages is not suitable in all instances, but that the benefits to the JPL contract do not in many instances flow in this direction. A review of data available in selected areas, such as the Office of the Comptroller, showed that the vast majority of the effort is for on campus activities rather than for the JPL contract. Studies by the Army Audit [5] Agency indicate that there are many general and administrative areas that will be questioned by the contracting officer and the auditors when the preliminary audit report is discussed with CIT officials in late December.

NASA also pays around 78% of the operation and maintenance expenses of the CIT administration buildings and a corresponding proportion of the use charge and depreciation on these buildings. To the extent that the allocation of general and administrative expenses to JPL is high, operation and maintenance expenses are correspondingly high.

For other categories of overhead, NASA pays small amounts related to JPL's usage of students and other educational facilities.

5.   Practices of Other Agencies

We made a limited examination of the practices followed in similar contracts with respect to the payment of a fee, as described in Section V. The only university situations which appear to us to be truly comparable were the AEC contracts with the Universities of California and Chicago. In these cases, the AEC pays a management allowance which

AEC policy states may exceed a conservative estimate of indirect cost, provided the allowance is not greater than the lower of the university's overhead requests or the fee that would be payable to a commercial contractor operating a government-owned plant.

[6] It should be noted that as adjustments are contemplated or made in the amount of overhead paid to CIT there will be even greater pressure for a fee. CIT has stated that the sum of the overhead and fee is the payment it requests for the operation of the JPL contract, and that this payment can be measured only partially by assignments of cost incurred. The balance, however determined, is the price tag CIT places on the contributions it makes, including the privilege given to the Government of using the University to conduct research. . . .

Document III-15

Document title: Contract Briefing Memorandum: Contract NAS7-100 With California Institute of Technology, January 12, 1965.

Source: Jet Propulsion Laboratory Archives, California Institute of Technology, Pasadena, California.

In response to both NASA and congressional investigations following the failure of the first six Ranger spacecraft, NASA's relations with the Jet Propulsion Laboratory and Caltech underwent a difficult period in 1964 and 1965. As demonstrated in earlier documents, NASA leaders demanded a number of changes in the nature of the agency's relationship with both Caltech and JPL. Because the 1962 NASA-Caltech contract was due to expire in December 1964, contract renewal was contingent upon these changes. The result was a two-year extension to the 1962 contract (NAS7-100), but with a number of significant changes, which are documented in this briefing memorandum.

---

[1]   January 12, 1965

CONTRACT BRIEFING MEMORANDUM
Contract NAS7-100 With
California Institute of Technology

A. General

1. Contract NAS7-100 was originally entered into effective January 1, 1962, between NASA and the California Institute of Technology for the performance of Research & Development activities at Jet Propulsion Laboratory. NAS7-100 continued the effort performed under NASW-6 which expired on December 31, 1962 and which was originally entered into on May 1, 1959 when NASA took over the facilities at JPL from the Department of the Army (Los Angeles Procurement District, Pasadena). The facility was then administered under Army Contract No. DA-04-495-ORD-18.

Total costs under NASW-6 approximated	$ 166,516,043.31
Total Obligations Under NAS7-100 to Date, Approximates	$ 776,183,640.01

2. NAS7-100 was scheduled to expire on December 31, 1964 and negotiations commenced in late 1963 and concluded in early 1964 for both a contract extension as well as desirable management changes to be effected both contractually and organizationally. During the latter part of 1964, CIT instituted many organizational changes, principal ones

included (1) the hiring of General A. R. Luedecke, (Ret) formerly with the Atomic Energy Commission and assigned him as Deputy Director of the Jet Propulsion Laboratory (actually as General Manager) and (2) projectizing its major programs from what was originally a matrix type organization. During this same period, NASA developed Task Orders, setting out for the first time in the almost five years since NASA started work at JPL the specific areas the programs to be covered in separately identifiable and funded Task Orders.

3. NASA Headquarters, satisfied that CIT had instituted mutually desirable changes, approved on December 16, 1964, a two-year extension to Contract NAS7-100 to expire on December 31, 1966. The extended contract, issued as modification No. 10 to NAS7-100, actually is a completely revised contract superseding in its entirety the terms and conditions of the original contract, as amended.

4. All contract management and monitoring activities are administered by the NASA Resident Office at JPL under the direction of the NASA Institutional Director, the Associate Administrator for Space Science and Applications.

[2] **B. NAS7-100 (mod. 10) Principal Provisions**
1. Scope of Work
a. Both NASA and the CIT have agreed that the Contractor shall perform only those specific tasks as may be designated in *unilaterally* issued Task Orders which fall within the following broad areas of activity:

(1) Exploring the moon and its environment and the planets and interplanetary space, including earth-based investigations and operations related thereto.

(2) Conducting (i) a program of supporting research and (ii) a program of advanced technical development, designed to make contributions to space science, technology, and exploration.

(3) Developing and operating the Deep Space Instrumentation Facility and Space Flight Operations Facility in support of NASA programs.

(4) Carrying out investigations and providing services in the field of aeronautics.

(5) Assisting NASA in the formulation and execution of its programs by providing NASA with technical advice, studies and reports of investigations.

(6) Providing technical direction or project management in connection with contracts for work falling within the broad areas defined above which are awarded by NASA to other contractors.

The principal change between the old contract and that part revised as indicated above is that NASA now may issue unilateral direction for CIT to perform within the areas noted whereas CIT had the right previously to reject NASA's directions or insist on changes before it would accept any specific task. The old provisions (commonly referred to as "mutuality") served to restrict the Government on the work or services it could demand of CIT and was the cause of much friction between technical counterparts of both NASA and JPL. It is believed that the present arrangement will prove more satisfactory and follows more closely the normal task order type contract which allows for unilateral issuance of task orders. The contract does include, however, a safeguard against the Government issuing technically unfeasible or otherwise unworkable tasks. The contractor has an obligation to advise the Contracting Officer, within 10 days, of any Task Order it (the Contractor) does not consider feasible. Such an occurrence, will, of course, be investigated by NASA.

b. If NASA desires any work performed by JPL which is not included in the broad areas agreed to, it will be issued in a Task Order which requires acceptance by CIT. This type of work is expected to be insignificant.

[3]   c.   Preliminary to the award of Modification No. 10 (the revised contract) both NASA and CIT have agreed upon the specific task orders to be issued to cover the work then in progress. Other than for a relatively few and minor tasks, all definitive task orders required to cover the work in progress prior to effective date of Modification 10, have now been issued and will be maintained on a current basis.

        d.   In addition to specific scopes of work included in separate Task Orders, there are also included provisions concerned with Technical Direction and Guidance, Operation of a Technical Plan, Reliability and Quality Assurance, Specific Reporting Requirements, Manpower Utilization Plans and Project Management Responsibilities—all of which are spelled out and included in contractual directions for the first time.

        e.   Concurrently with the issuance of Definitive Task Orders, NASA has negotiated with CIT and the contract has now been amended to provide for an "authorized manpower" clause. This clause permits the Contracting Officer, for the first time, to establish a manpower ceiling on the total number of persons which the Contractor may employ at JPL and provides a penalty in the form of disallowing costs of persons employed in excess of the ceiling. Under the old contract, although informed ceilings were established, they were usually exceeded without any penalty placed on the Contractor. The initial ceiling under the revised contract was established on October 31, 1964 as 4,275 persons. JPL reduced to 4,245 persons as of November 30, 1964 and to 4,225 persons as of December 31, 1964. The ceiling has been reduced to 4,100 by June 30, 1965, and to 4,000 by December 31, 1965. JPL is expected to be sustained at about 4,000 persons. Adequate controls have been established at our Resident Office at JPL to preclude JPL from exceeding its established ceiling. New work is being monitored through the Resident Office relative to adequacy of JPL manpower resources without disturbing the manpower ceiling.

    2.   Contract Resources

        a.   CIT must provide all of the management, personnel, labor and services necessary for performance of all work under the contract except that work which it is authorized to subcontract for. NASA furnishes or CIT acquires for the Government's account, all property, including facilities, necessary for performance of work under the contract. This includes all real and leased property at JPL and buildings authorized for construction by JPL and/or Army Corps of Engineers. Now, for the first time, all property of a facilities nature, including real estate, comes under the cognizance and control of a separate Facilities Contract and removes it from the Research and Development area under which it was formerly controlled.

        b.   The Facilities contract (No. NAS7-270(F)) provides for periodic reporting, control, protection and maintenance of the Government property as well as a vehicle for authorized new construction.

[4] 3.   Reporting

        a.   Under the revised contract, CIT is obligated to furnish management, financial, technical, progress and other reports as the Contracting Officer may direct. Under the old contract each report had to be mutually agreed to be furnished before it could be placed into effect. Here again, "mutuality" has been removed to provide for prompt response from the Contractor. Under the revised Contract, however, CIT may initiate additional unclassified reports to disseminate scientific and technical knowledge to the scientific community. Distribution and costs of publication of such additional reports are furnished annually to the Contracting Officer for his review.

    4.   Fiscal and Other Management Requirements

        a.   CIT is required to segregate and separately maintain the costs of each Task Order and each program so that costs for each program are readily identifiable.

        b.   JPL's financial management system must be compatible with NASA's system including integration of the NASA Agency-wide coding structure. NASA-PERT and the NASA Financial Management Reporting System for cost type contracts have been imposed on JPL and its major subcontractors in implementation of an integrated time-cost management control and reporting system.

    c.   The Contractor is required to make maximum use of Department of Defense Audit and Administrative Services to preclude duplication of effort. Audit services are being utilized to the fullest extent. Property and Inspection Services performed by DOD Agencies are constantly being expanded to meet requirements. It is expected that proposed Defense Contract Administrative Services District when established in Los Angeles will be used to the maximum extent practicable.

    d.   The Contractor is required to submit annual budget estimates for the work it anticipates will be performed for each succeeding fiscal year of a particular program. Revised estimates will be also be furnished as program requirements change, are reduced or are increased. Periodic guidelines are furnished to CIT for use in projecting its estimates.

    e.   The usual "Limitations of the Government's Obligation" and "Estimated Cost" clauses limiting the Contractor's expenditures to funds allotted and estimated costs set forth in Task Orders are included in the contract to control unauthorized expenditures by the Contractor.

    5.   Allowable Costs

    a.   The allowability of all costs for purposes of determining amounts payable to the Contractor is determined by the cost principles set forth in [5] Part 3 of Section XV of the Armed Services Procurement Regulation applicable to Educational Institutions (negotiations with CIT will be accomplished to convert the ASPR reference to the appropriate part of NASA Procurement Regulation). The contract also lists specific items of direct costs for purposes of agreement on an "advance understanding" as to the allowability of certain costs by the Government. The types of costs listed are compatible with the ASPR Cost Principles.

    b.   The revised contract calls for negotiation of overhead rates to cover institutional indirect cost. These provisions follow the standard procedure for agreement on final overhead rates as contained in most Government cost-type contracts. The old contract provided for a fixed allowance for "indirect costs" which generally could not be changed. This was fixed after the beginning of the fiscal year regardless of the actual overhead expenditures which might be incurred during the year. The present procedure fixes rates only after the completion of the fiscal year and is based upon actual audited overhead expenditures. The present arrangement is more equitable to both parties.

    6.   Fixed Fee

    a.   A fixed fee is negotiated for each full fiscal year (or part of year included in term of contract) and the amount agreed upon is included in an amendment to the contract. The old contract did not provide any contractual incentive for raising or lowering the fee whereas the revised contract contains a schedule of fee ranges from a stated minimum to a maximum range according to the NASA approved Financial Operating Plan. The fee ranges are listed below:

Schedule of Fee Ranges

NASA Approved Financial Operating Plan		Fee Ranges	
		Minimum Fixed Fee	Maximum Fixed Fee
$	$	$	$
150,000,000	175,000,000	948,700.00	1,423,050.00
175,000,000	200,000,000	1,045,000.00	1,567,500.00
200,000,000	225,000,000	1,127,500.00	1,691,250.00
225,000,000	250,000,000	1,210,000.00	1,815,000.00
250,000,000	275,000,000	1,288,250.00	1,931,875.00
275,000,000	300,000,000	1,361,250.00	2,041,875.00
300,000,000	325,000,000	1,430,000.00	2,145,000.00
325,000,000	350,000,000	1,498,750.00	2,248,125.00

The basis for determining fee is the total JPL financial operating plan first approved by NASA following passage and approval of the NASA Appropriation Act for a particular fiscal year. Although the Plan may be [6] subsequently amended or revised, the fee remains unchanged and avoids any aspect of a cost-plus-percentage-of-cost situation. Within the range of a particular Financial Operating Plan, negotiations may then take place within the stated minimum and maximum ranges. Consideration is given by the Contracting Officer, among other factors, in negotiations, to:

    (1) Extent of subcontracting

    (2) Complexity of the work

    (3) Past performance evaluations conducted under a new clause of the contract entitled "Evaluation of Contractor's Performance."

    b. Using the current fiscal year (1965) for an example, the first approved Plan issued by the Contracting Officer, NASA Resident Office, totaled $216,195,000. This then falls within an operating plan range of $200,000,000 to $225,000,000 and a fee range of $1,127,500 to $1,691,250, which is subject to negotiations. The fixed fee negotiated for FY 1964 under the old contract amounted to $1,250,000.

    7. Patent, New Technology and Related Clauses

The contract contains the appropriate Patents, New Technology, Data Rights and Licenses clauses prescribed by NASA Procurement Regulations. The NASA Resident Staff includes a qualified Patents Attorney who monitors all of the patent type activities of JPL.

    8. Subcontracts

    a. The contract contains provisions for review of selected subcontracts by the Contracting Officer to ensure compliance with good business practices, NASA Procurement Regulations and special requirements placed upon the Contractor.

    b. All of the Contractor's procurement policies and procedures are subject to approval by the Contracting Officer. Included are Source Evaluation Board procedures which the Contractor has agreed to use for procurements in excess of $1,000,000.

    c. The Contractor, by contract terms, has established and maintains a "Small Business Subcontracting Program" in accordance with current statutes and regulations. It is also obligated to include "Small Business Program" requirements in all of its subcontracts which offer substantial small business subcontracting opportunities.

[7] 9. Advance Payments

CIT is permitted to receive, on an interest-free basis, advance payments usually permitted in the case of Educational Institutions. The advance payments are sufficient to pay current payroll and operating costs. Under negotiation, however, is a letter-of-credit procedure designed to replace the advance payments provisions. This procedure has been promulgated by the U.S. Treasury Department and simplifies the advance payment process. Its primary advantage is to reduce the time that cash is in the hands of the Contractor and save Treasury the interest cost of idle money in the hands of a Contractor. This new procedure should be in effect shortly.

    10. Safety and Plant Protection

The Contractor is obligated to maintain maximum safety conditions at all times and comply with applicable Federal, State and local laws and ordinances including Government regulations applicable to handling and storage of potentially dangerous fuels and propellants. CIT must also maintain plant protection devices, a security force and enforce applicable rules and regulations regarding Security and Classified matters. It must coordinate all Security matters with the cognizant Department of Defense Agency.

    11. Equal Opportunities for Employment

The Contractor has agreed to comply with all nondiscrimination policies of the Government and administratively enforce compliance by its subcontractors.

12. Key Personnel, Wages and Salaries

 a. Key personnel assigned to a particular program may not be reassigned without the consent of the NASA Headquarters Program Director.

 b. CIT is obligated to keep the Contracting Officer fully informed as to JPL's wage and salary policies including notice of any action to an employee involving a rate of compensation in excess of $15,000 per annum.

13. General Services Administration Supply Services

Under the terms of the Contract, the Contracting Officer has required JPL to utilize GSA sources for any property which can be furnished from either warehouse stock or from GSA contractors. The use of this authorization has resulted in savings in procurement costs.

[8] 14. Non-Renewal of Contract

Appropriate provisions are made for the settlement of closing costs which might reasonably be expected to occur in the event NASA should decide not to further extend the contract.

15. Evaluation of Contractor's Performance

The revised contract includes provisions, for the first time, for the Government to evaluate the Contractor's (JPL's) performance both semi-annually and at the close of each fiscal year. An Evaluation Board will be composed of representatives appointed by the NASA Administrator. Conclusions will be reached after consideration of all the facts and after giving CIT the opportunity to submit such information and material as it desires. The conclusions reached by the Board will influence, in part, subsequent fee negotiations.

16. Government Property

The Contractor receives, issues, maintains and protects all Government property under its control, in accordance with NASA Procurement Regulations and the NASA Industrial Property Control Manual. The Contractor's activities in this area are continuously monitored by a NASA Property Administrator assigned to the NASA Resident Office Staff. Property in the control of subcontractors is monitored by DOD Agencies assigned secondary property administration.

17. Other Requirements

Other contract clauses required by statute or regulation are included in the contract.

Document III-16

Document title: Office of Technology Utilization, Task Force to Assess NASA University Programs, *A Study of NASA University Programs* **(Washington, DC: NASA Special Publication-185, 1968), pp. 1-8.**

Source: NASA Historical Reference Collection, NASA History Office, NASA Headquarters, Washington, D.C.

Between 1962 and 1968, the NASA-university relationship expanded considerably. This document is the report of a task force assigned to review the totality of that relationship, which had resulted from NASA's attempt to use Apollo funding to effect a change in academic America. This report lent support to the decision to curtail drastically and eventually even to cancel the Sustaining University Program.

[1] **Precis**

This study examines the results of the total NASA university program. It is an assessment of the program based on goals publicly expressed by NASA managers as recorded in the literature and correspondence with universities. Foremost among the goals has been the intent of NASA to accomplish its aeronautics and space mission while at the same time strengthening the universities involved; NASA-sponsored research was to be conducted in the traditional atmosphere of instruction and learning in order to maximize the indirect returns from the mission-oriented programs. The study was approached through selected sampling of NASA-university interactions by interviews, university visits, and in-depth case studies. The significant limitations of the study are those imposed by the lack of sufficient time to collect and analyze data on such a huge and diverse program. However, the Task Force believes this report to be indicative of the total NASA university program.

Impact on NASA, Universities, and the Nation

The returns from all NASA university programs fall into the categories of new knowledge, trained people, or new capability for research, education, and service. The major impact of these returns is upon the participants. However, since NASA and universities are both parts of the Nation, anything that affects them also affects the Nation. The results of programs that affect the Nation outside the immediate areas of the participants generally are too obscure to be identifiable. Therefore, the emphasis of this study is on the new knowledge, trained people, and new capability that have impacted NASA and universities and, through them, the Nation.

General.—NASA's university programs have made major contributions to the aeronautics and space program. Research sponsored by university programs has generated new concepts, has developed new technology, and has created unique facilities for further education and research. Over 50 percent of all experiments flown on NASA satellites have been generated by university programs. Universities have awarded at least 500 graduate degrees and provided continuing education opportunities to thousands through NASA employee graduate training programs. Even management of the aerospace program has been influenced, since university consultants have given policy, scientific, and engineering advice to NASA at all levels. These contributions demonstrate that NASA university programs have been successful in their first and most important objective—obtaining the expertise of the university community to help meet the aeronautics and space goals of NASA and the Nation.

NASA university programs have had a significant impact on the university community. About 250 universities have been responsive to opportunities to become involved in the aeronautics and space program made available by NASA. [2] They have welcomed NASA support and have used it to strengthen and build research and education capability. Centers of excellence exist that were created with NASA support. Entire departments and graduate degree programs have grown out of NASA involvement, many new courses have been developed, and countless science and engineering courses have had their content altered by NASA programs. The national capability for education and research has been both broadened and strengthened.

In general, universities have not taken advantage of the opportunities offered by NASA to innovate in research management, multi-disciplinary research, and government-industry-university relations. There is little evidence that the long-range goals of NASA university programs, such as the development of a university capability to respond as an institution, capability for multi-disciplinary research, concern with societal problems, and acceleration of technology transfer, are being achieved. The examples that were identified—an Urban Laboratory at UCLA, the Industrial Development Division at the University of Michigan, Cornell's new Department of Environmental Systems

Engineering, etc.—are only loosely tied to NASA programs. Sometimes they were unknown to, or unrecognized by, the scientists administering the NASA grants. It should be pointed out, however, that the dollars NASA has used to encourage change have come mostly from the Sustaining University research and facilities programs and have amounted to less than 1 percent of the total Federal support to universities. From this perspective, the changes that NASA university programs have stimulated in universities appear more significant.

NASA's university programs have built up a reservoir of good will within the university community toward the agency. University administrators generally perceive that NASA is sensitive to their needs and has undertaken a program to assist them with facilities, graduate student support, and institutional support grants. Generally, faculty members appreciate the opportunities for research and education that have been made available to them.

Industry has benefited from NASA university programs through the increased availability of trained people, new knowledge, and new capability. For the most part, however, industry-university relations do not appear to have been altered by NASA programs. Little evidence was found that universities were working harder at transferring technology to industry or have been successful in increasing industry support for university research.

Although NASA's stated policy is to conduct its programs in such a manner as not to draw faculty away from teaching, some of the research institutes, centers and laboratories in universities have very few graduate students involved in the ongoing research. Some have full-time staffs of research professionals who neither teach nor supervise graduate students. Most universities that have such special research groups are aware of the problem and are attempting to find mechanisms to bring research closer to the educational process. Some are successful; some are not. Significant numbers of groups with little educational involvement still exist. NASA violates its own policies when it supports groups that continue to divorce themselves from the educational function of the university.

[3] *Project research*—About 70 percent of NASA funds obligated to universities has been by the project research method. This system of supporting the research of principal investigators within universities is serving both NASA and the universities well. Abuse of the system sometimes occurs (e.g., overcommitment by an aggressive university researcher, demands for industrial-type response by a NASA contract monitor, or too little educational involvement). However, on balance, these are excellent programs that have contributed directly to the aerospace objectives of NASA. Project research also involves large numbers of faculty and graduate students and generates about three out of four of the space-science publications from all NASA programs. A large amount of education at all levels—undergraduate, graduate, and postdoctoral—is supported by these NASA programs. More than 10 percent of all funds supporting project research have been invested in equipment, which is available in university laboratories for further education and research.

Small project grants, which involve only one or two faculty members and their graduate students, have often led to productive interactions with NASA center personnel. Research on optimal control of nuclear rockets at the University of Arizona and ablation-material research at Louisiana State University are examples of projects through which NASA has received new concepts and techniques, the university has improved curricula and research and increased the number of publications, and technology is being transferred from universities to other segments of society. Larger project research grants, while producing valuable research, do not seem to foster development of as close a tie to the ongoing NASA program.

Space-science flight experimentation represents an area of significant accomplishment in NASA university programs. University scientists have been eager to take advantage of the opportunities made available by NASA to conduct experiments in space. More than 98 percent of balloon-borne experiments, more than 40 percent of sounding rocket experiments, and more than 50 percent of satellite experiments flown on NASA vehicles had principal investigators or coinvestigators in universities. For the satellite experiments, this is five times the level of participation of industry and about the same as the participation of all government laboratories. For the Orbiting Geophysical Observatory program alone, 50 percent of the flight experiments and almost two-thirds of the early scientific publications came from universities. A large share of the significant discoveries in space science were made in university-originated experiments.

Although the university community appears to have an effective voice in flight programs and selection of experiments through advisory committees, some university people complain about favoritism in the selection of flight experiments. Another continuing problem with university participation in flight experiments is involvement of graduate students. Long lead times and project uncertainties limit the suitability of flight programs for thesis projects. Universities have adopted various approaches to circumvent the difficulties, but NASA must continue to be aware of them and continue to seek administrative mechanisms that encourage participation of graduate students.

A university research program in R. & D. management and socioeconomics in aerospace-related areas has been NASA's only significant support of the social [4] science disciplines. This program has been quite productive as measured by publications and involvement of faculty and students. Capability for research on management of large technological programs has been created in several universities and is now available to the Nation. However, few if any management or policy decisions or processes within NASA appear to have been influenced by the research. While some of the research may have had potential usefulness, NASA has no mechanism for utilizing its results. The program has had no centralized direction or policy and almost no involvement of the centers where many management problems occur. It may be significant that NASA has sponsored a university research program in these disciplines without a corresponding in-house research capability—a position it has carefully avoided in engineering and physical-science disciplines.

Sustaining University Program.—The Sustaining University Program, which provided about 30 percent of NASA funds obligated to universities and provides support to institutions rather than to principal investigators within universities, has generally been successful. Its short-range objectives—increasing the supply of trained manpower, increasing university involvement in aeronautics and space, broadening the base of competence, and consolidating closely related activities—have been achieved. However, the long-range goals that require innovation and change by universities—capability for multidisciplinary research, university concern with the technology-transfer process, increased university involvement with community and societal problems, developing capability for institutional response—have not been successfully attained. There are a few indications of change in the direction of long-range goals that may lead to future developments.

The aims and operation of the Sustaining University Program are poorly understood within NASA outside the Office of University Affairs. Only in the Office of Space Science and Applications, which formerly directed the program, are they reasonably well understood and felt to have value to NASA as a supplement to project research. In other Headquarters offices and in the Centers, no benefit to NASA is seen in the program. The Sustaining University Program grants are viewed as giveaways to help universities. The quality of research sponsored by the program is regarded as not good enough to obtain support in open competition. The impact on both NASA and universities would have been greater if the in-house managers had been involved and committed to the programs.

The Sustaining University Program has made grants for multidisciplinary space-related research to 50 universities. These grants were about 10 percent of the total research funds provided to universities by NASA. The grants achieved the objective of broadening the base of involvement and capability in aerospace research. They have contributed to the establishment of new departments (e.g., aerospace engineering or space sciences) and strengthened old ones (e.g., astronomy). Capabilities were nourished that since successfully competed for research support from NASA project research and other Government agencies.

The multidisciplinary aspect of Sustaining University Program research grants has generally not been taken seriously by universities. The universities perceive the grants as institutional support in a conventional sense that does not require innovation in the administration of research. A contributing [5] factor to this attitude is the lack of "systems" administrators in universities with broad views of real-world problems and the capability for breaking problems into small subsystems for attack by individual researchers. A small amount of multidisciplinary research that involves physical and life scientists and engineers is supported, but little of it was initiated under the grants. Research involving individuals from multiple disciplines, including social sciences, jointly attacking a multidisciplinary problem is nonexistent.

NASA has encouraged universities to involve social scientists in their research with little response. The small amount of social-science involvement that does exist is usually on a subproject that does not interact with other research.

Many of the individual researchers supported by Sustaining University Program research grants have no direct contact with NASA. If they know their counterparts in NASA, it is only by chance. While some of the scientists and engineers relish independence, many would welcome closer relations with NASA peers. Examples of interactions in project research illustrate the benefits that close relations could have for both universities and NASA.

A Sustaining University Program research grant in a university gives a focus to its aeronautics and space program that is not present in universities without such a grant. The steering committee which administers the grant seems to give identity and visibility to the total NASA program. The existence of this committee appears to give credence to NASA's concern for doing its business in a way that strengthens the university and is a step toward interdepartmental cooperation for multidisciplinary research. Key members of these committees tend to dominate the direction of the program for the total university.

The Sustaining University Program predoctoral traineeship grants to 152 universities accounted for about 15 percent of total NASA obligations to universities and have supported more than a thousand students who have earned Ph.D. degrees in space-related areas. By 1970, over 4,000 doctorates will have been earned by trainees. More than half of these highly trained scientists and engineers are remaining in universities and will contribute to the Nation through education and research for years to come. About a third of the former trainees are seeking industrial careers. Many of their skills are transferable to areas other than aerospace and will continue to benefit society and science whether or not they engage in aerospace research. Some evidence exists that traineeship grants have accelerated (as well as increased) the production of doctorates, but it is not conclusive except in the obvious cases of students who otherwise would have held part-time jobs.

The trainees tend to be isolated from NASA and have little opportunity to identify with the Agency. Since the program is administered by the individual universities, not even the stipend checks come from NASA. The Agency has overlooked an opportunity to communicate with the students, which is reflected by the statistic that only 1 percent of the Ph.D. recipients have been hired by NASA. This indicates very little direct impact on

NASA by the traineeship program.

[6] The traineeship-grant program has had little impact on large established graduate schools. Ten or 12 additional traineeships tend to get lost in universities such as Cornell or Michigan. However, traineeships were awarded to 152 universities, most of whom [sic] do not have the size or reputation of the two universities just mentioned. The grants have enabled the smaller and less well established universities to recruit more and better graduate students and to strengthen their graduate education programs.

The Sustaining University Program has made 35 facilities grants to 32 universities that have already resulted in 27 completed laboratories. The grants account for over 6 percent of NASA obligations to universities. The facilities are enabling universities to participate in aerospace programs more effectively by providing working space and by consolidating aerospace-related activities. They are being used to house interdisciplinary activities, usually in the form of an aerospace-related institute, center, or laboratory. Little evidence was found that technology-transfer processes or university interaction with the local or regional community had been stimulated by the facilities visited.

Little evidence was found that the Memorandums of Understanding associated with Sustaining University Program facilities grants have led to anything but talk. Usually only a few administrators with a university even knew about the Memorandum. They had not attempted to use it as a tool to induce changes in procedures or attitudes; they did not regard it as requiring them to do anything new or different. The major criticism which must be made is that universities have not made "energetic and organized" efforts to implement the Memorandums, which they clearly agreed to do.

Personnel development programs.—The temporary in-residence faculty programs (NASA-ASEE [American Society of Electrical Engineers] summer faculty fellowships, NASA-NRC [National Research Council] resident research associates) are among the most rewarding of NASA university programs. NASA managers feel that the participants bring new talent and ideas into NASA projects and develop continuing relationships with NASA after they return to their schools. The participants like the programs for the exposure to real problems, for new ideas for research, and because they often provide a sponsor for their own research. Almost a thousand NASA-ASEE summer faculty fellows have spent 10 weeks during the summer working on real-world problems at a NASA center. More than 300 NASA-NRC postdoctoral research associates have had the opportunity to conduct research in a NASA center for at least 1 year. These programs have led to new research projects, curriculum modifications, and the creation of new centers of excellence. The acoustics program at North Carolina State University is just one outstanding example of impact on NASA, the university, and the Nation resulting from participation in these programs.

The employee training program has contributed in a major way to upgrading the capabilities of NASA personnel. Employees have earned about 400 master's degrees and 100 Ph.D. degrees by this method in recent years. Simultaneously, in meeting training needs, NASA centers have strengthened old departments and accelerated the creation of new departments in nearby universities. The graduate program in physics at the College of William and Mary is one example of stimulation of regional graduate-education capability to meet Langley Research Center's graduate training needs.

[7] **Alternatives for Future Consideration**

The results of the study suggested many changes in procedures, policies, or approaches that would lead to more effective university programs. Many of these involve operational details and have been called to the attention of appropriate NASA managers. Only those of broad scope and general interest will be discussed here.

A substantial portion of Government-supported R. & D. management research within the country has been sponsored by NASA. However, NASA is not reaping full benefit from it because there is no mechanism for translating research into applications. In physical-science and engineering disciplines, university researchers interface with research-oriented NASA personnel who know how to disseminate and use their results. In the R. & D. management area, university researchers interface with NASA management practitioners with whom the researchers have difficulty communicating. Research-oriented management-science groups within NASA would be one approach to improving utilization of the sponsored research.

The Memorandums of Understanding associated with facilities grants have been ineffective in accomplishing change. The facilities may be a permanent symbol and reminder of NASA support, but NASA loses all leverage once the grant is awarded. Memorandums of Understanding might be more effective in inducing change if used in conjunction with institutional or multidisciplinary grants that have a renewal feature. University administrators could then use the threat of failure of renewal to influence faculty. NASA has recently begun to experiment with Memorandums associated with research and training grants and their effectiveness should be carefully evaluated.

Many NASA-university interactions have demonstrated that synergism occurs when personnel are in close communication. The element of close working relations has been missing from research sponsored by the Sustaining University Program. Therefore, the benefits to both NASA and universities from this research would be increased by closer ties with ongoing NASA programs. Individual researchers in universities need to communicate with their NASA peers and university administrators need more data on real NASA problems for decision-making in allocating grant resources. Therefore, centers and program offices should be participants—not advisors—and share responsibility in administration of Sustaining University Program research grants.

The mechanisms that have been established for bringing university faculty into NASA on a temporary basis are valued highly by NASA managers and by the participating university people. It is noteworthy that equivalent mechanisms permit NASA employees to enter the university community on a short-term basis but are not widely known or used. Many highly qualified NASA scientists, engineers, and managers could make significant contributions to universities in research, education, and administration, as well as increase their own understanding of university problems, if mechanisms could be developed for them to spend 6 months or a year as active participants—not students—in university programs. Exchange programs between universities and NASA should be encouraged.

[8] Employee graduate-training programs should be considered as another method for meeting the Nation's need for highly educated scientists, engineers, and managers. Innovations in these programs could help offset the reduction in Ph.D. production that will come after 1970 as a result of decreases in Sustaining University Program traineeships. If the employee graduate-training programs could be expanded, NASA would benefit from the services of highly motivated and capable employees while at the same time giving them educational opportunities. In addition, if NASA's requirements for employee graduate training at nearby universities are large, financial support to the universities for facilities and faculty augmentation should be considered.

A requirement that annual reports on all grants and contracts summarize numbers of graduate students given full or partial support, theses supported, technical reports published, curriculum changes, facilities acquired, and degrees earned by students being supported would emphasize to universities NASA's desire to support research in an educational environment and would provide data to assess the program.

Continuous feedback on the effectiveness of university programs is needed by NASA management at all levels. A better management information system and reporting of educational impact of NASA programs would satisfy many requirements. However, periodic use of ad hoc groups, university consultants, and regularly scheduled conferences of the Office of University Affairs, Centers, and Program Offices will probably all be required.

Document III-17

Document title: Major General Samuel C. Phillips, USAF, Apollo Program Director, to J. Leland Atwood, President, North American Aviation, Inc., December 19, 1965, with attached: "NASA Review Team Report."

Document III-18

Document title: George E. Mueller, Associate Administrator for Manned Space Flight, to J. Leland Atwood, President, North American Aviation, Inc., December 19, 1965.

Source: Both in NASA Historical Reference Collection, NASA History Office, NASA Headquarters, Washington, D.C.

In late 1965, at the request of NASA Associate Administrator for Manned Space Flight George E. Mueller, Major General Samuel C. Phillips, Apollo Program Director at NASA Headquarters, initiated a review of NASA's contract with North American Aviation, Inc. (referred to as "NAA" below), to determine why work on both the Apollo spacecraft and Saturn V second stage was behind schedule and over budget. This highly critical study, known as the Phillips Report, took on added significance when in the aftermath of the Apollo 204 capsule fire (just over one year later), it was discovered that NASA Administrator James E. Webb was apparently unaware of the existence of the report. General Phillips provided a set of the notes which comprised the study to North American President J. Leland Atwood, and George Mueller added his views in a separate letter.

Document III-17

[1] IN REPLY REFER TO: MA   December 19, 1965

Mr. J. L. Atwood
President
North American Aviation, Inc.
1700 E. Imperial Highway
El Segundo, California

Dear Lee:

I believe that I and the team that worked with me were able to examine the Apollo Spacecraft and S-II stage programs at your Space and Information Systems Division in sufficient detail during our recent visits to formulate a reasonably accurate assessment of the current situation concerning these two programs.

I am definitely not satisfied with the progress and outlook of either program and am convinced that the right actions now can result in substantial improvement of position in both programs in the relatively near future.

Enclosed are ten copies of the notes which was compiled on the basis of our visits. They include details not discussed in our briefing and are provided for your consideration and use.

The conclusion expressed in our briefing and notes are critical. Even with due consideration of hopeful signs, I could not find a substantive basis for confidence in future performance. I believe that a task group drawn from NAA at large could rather quickly verify the substance of our conclusions, and might be useful to you in setting the course for improvements.

[2] The gravity of the situation compels me to ask that you let me know, by the end of January if possible, the actions you propose to take. If I can assist in any way, please let me know.

Sincerely,

SAMUEL C. PHILLIPS
Major General, USAF
Apollo Program Director

[Attachment p.1]

NASA Review Team Report

I. Introduction

This is the report of the NASA's Management Review of North American Aviation Corporation management of Saturn II Stage (S-II) and Command and Service Module (CSM) programs. The Review was conducted as a result of the continual failure of NAA to achieve the progress required to support the objective of the Apollo Program.

The scope of the review included an examination of the Corporate organization and its relationship to and influence on the activities of S&ID [Space and Information Systems Division of North American], the operating Division charged with the execution of the S-II and CSM programs. The review also included examination of NAA off-site program activities at KSC and MTF [Mississippi Test Facility].

The members of the review team were specifically chosen for their experience with S&ID and their intimate knowledge of the S-II and CSM programs. The Review findings, therefore, are a culmination of the judgements [sic] of responsible government personnel directly involved with these programs. The team report represents an assessment of the contractor's performance and existing conditions affecting current and future progress, and recommends actions believed necessary to achieve an early return to the position supporting Apollo program objectives.

The Review was conducted from November 22 through December 6 and was organized into a Basic Team, responsible for over-all [2] assessment of the contractor's activities and the relationships among his organizational elements and functions; and sub-teams who [sic] assessed the contractor's activities in the following areas:

Program Planning and Control (including Logistics)
Contracting, Pricing, Subcontracting, Purchasing
Engineering
Manufacturing
Reliability and Quality Assurance.

Review Team membership is shown in Appendix 7.

Team findings and recommendations were presented to NAA Corporate and S&ID management on December 19.

II. NAA's Performance to Date—Ability to Meet Commitments

At the start of the CSM and S-II Programs, key milestones were agreed upon, performance requirements established and cost plans developed. These were essentially commitments made by NAA to NASA. As the program progressed NASA has been forced to accept slippages in key milestone accomplishments, degradation in hardware performance, and increasing costs.

A. S-II
  1. Schedules
     As reflected in Appendix VI key performance milestones in testing, as well as end item hardware deliveries, have slipped continuously in spite of deletions of both hardware and test content. The fact that the delivery [3] of the common bulkhead test article was rescheduled 5 times, for a total slippage of more than a year, the All System firing rescheduled 5 times for a total slippage of more than a year, and S-II-1 and S-II-2 flight stage deliveries rescheduled several times for a total slippage of more than a year, are indicative of NAA's inability to stay within planned schedules. Although the total Apollo program was reoriented during this time, the S-II flight stages have remained behind schedules even after this reorientation.
  2. Costs
     The S-II cost picture, as indicated in Appendix VI has been essentially a series of costs escalations with a bow wave of peak costs advancing steadily throughout the program life. Each annual projection has shown either the current or succeeding year to be the peak. NAA's estimate of the total 10 stage program has more than tripled. These increases have occurred despite the fact that there have been reductions in hardware.
  3. Technical Performance
     The S-II stage is still plagued with technical difficulties as illustrated in Appendix VI. Welding difficulties, insulation bonding, continued redesign as a result of component failures during qualification are indicative of insufficiently aggressive pursuit of technical resolutions during the earlier phases of the program.
[4] B. CSM
  1. Schedules
     A history of slippages in meeting key CSM milestones is contained in Appendix VI. The propulsion spacecraft, the systems integration spacecraft, and the spacecraft for the first development flight have each slipped more than six months. In addition, the first manned and the key environmental ground spacecraft have each slipped more than a year. These slippages have

occurred in spite of the fact that schedule requirements have been revised a number of times, and seven articles, originally required for delivery by the end of 1965, have been eliminated. Activation of two major checkout stations was completed more than a year late in one case and more than six months late in the other. The start of major testing in the ground test program has slipped from three to nine months in less than two years.

2. Costs

Analysis of spacecraft forecasted costs as reflected in Appendix VI reveals NAA has not been able to forecast costs with any reasonable degree of accuracy. The peak of the program cost has slipped 18 months in two years. In addition, NAA is forecasting that the total cost of the reduced spacecraft program will be greater than the cost of the previous planned program.

[5]   3. Technical Performance

Inadequate procedures and controls in bonding and welding, as well as inadequate master tooling, have delayed fabrication of airframes. In addition, there are still major development problems to be resolved. SPS engine life, RCS performance, stress corrosion, and failure of oxidizer tanks has resulted in degradation of the Block I spacecraft as well as forced postponement of the resolution of the Block II spacecraft configuration.

III. NASA Assessment—Probability of NAA Meeting Future Commitments

A. S-II

Today, after 4 1/2 years and a little more than a year before first flight, there are still significant technical problems and unknowns affecting the stage. Manufacture is at least 5 months behind schedule. NAA's continued inability to meet internal objectives, as evidenced by 5 changes in the manufacturing plan in the last 3 months, clearly indicates that extraordinary effort will be required if the contractor is to hold the current position, let alone better it. The MTF activation program is being seriously affected by the insulation repairs and other work required on All Systems stage. The contractor's most recent schedule reveals further slippage in completion of insulation repair. Further, integration of manual GSE has recently slipped 3 weeks as a result of configuration discrepancies discovered during engineering checkout of the system. Failures in timely [6] and complete engineering support, poor workmanship, and other conditions have also contributed to the current S-II situation. Factors which have caused these problems still exist. The two recent funding requirements exercises, with their widely different results, coupled with NAA's demonstrated history of unreliable forecasting, as shown in Appendix VI, leave little basis for confidence in the contractor's ability to accomplish the required work within the funds estimated. The team did not find significant indications of actions underway to build confidence that future progress will be better than past performance.

B. CSM

With the first unmanned flight spacecraft finally delivered to KSC, there are still significant problems remaining for Block I and Block II CSM's. Technical problems with electrical power capacity, service propulsion, structural integrity, weight growth, etc. have yet to be resolved. Test stand activation and undersupport of GSE still retard schedule progress. Delayed and compromised ground and qualification test programs give us serious concern that fully qualified flight vehicles will not be available to support the lunar landing program. NAA's inability to meet spacecraft contract use deliveries has caused rescheduling of the total

Apollo program. Appendix VI indicates the contractor's schedule trends which cause NASA to have little confidence that the S&ID will meet its future spacecraft commitments. While our management review indicated that some progress is [7] being made to improve the CSM outlook, there is little confidence that NAA will meet its schedule and performance commitments within the funds available for this portion of the Apollo program.

[8] IV. Summary Findings

Presented below is a summary of the team's views on those program conditions and fundamental management deficiencies that are impeding program progress and that require resolution by NAA to ensure that the CSM and S-II Programs regain the required program position. The detail findings and recommendations of the individual sub-team reviews are Appendix to this report.

A. NAA performance on both programs is characterized by continued failure to meet committed schedule dates with required technical performance and within costs. There is no evidence of current improvement in NAA's management of these programs of the magnitude required to give confidence that NAA performance will improve at the rate required to meet established Apollo program objectives.

B. Corporate interest in, and attention to, S&ID performance against the customer's stated requirements on these programs is consider[ed] passive. With the exception of the recent General Office survey of selected functional areas of S&ID, the main area of Corporate level interest appears to be in S&ID's financial outlook and in their cost estimating and proposal efforts. While we consider it appropriate that the responsibility and authority for execution of NASA programs be vested in the operating Division, this does not relieve the Corporation of its responsibility, and accountability to NASA for results. [9] We do not suggest that another level of program management be established in the Corporate staff, but we do recommend that the Corporate Office sincerely concern itself with how well S&ID is performing to customer requirements and ensure that responsible and effective actions are taken to meet commitments.

C. Organization and Manning

We consider the program organization structure and assignment of competent people within the organization a prerogative of the manager and his team that have been given the program job to do. However, in view of what we consider to be an extremely critical situation at S&ID, one expected result of the NASA review might be the direction of certain reorganizations and reassignments considered appropriate, by NASA, to improve the situation. While we do have some suggestions for NAA consideration on this subject, they are to be accepted as such and not considered directive in nature. We emphasize that we clearly expect NAA/S&ID to take responsible and thoroughly considered actions on the organization and assignment of people required to accomplish the S-II and CSM Programs. We expect full consideration, in this judgement [sic] by NAA, of both near and long term benefits of changes that are made.

Frankly stated—we firmly believe that S&ID is overmaned and that the S-II and CSM Programs can be done, and done better, with fewer people. This is not to suggest that an arbitrary [10] percentage reduction should be applied to each element of S&ID, but we do suggest the need for adjustments, based on a reassessment and clear definition of organizational responsibilities and task assignments.

It is our view that the total Engineering, Manufacturing, Quality, and Program Control functions are too diversely spread and in too many layers throughout the S&ID organization to contribute, in an integrated and effective manner, to the hard core requirements of the programs. The present proliferation of the functions invites non-contributing, "make-work" use of manpower and dollars as well as impediments to program progress.

We question the true strength and authority of each Program Manager and his real ability to be fully accountable for results when he directly controls less that 50% of the manpower effort that goes into his program. This suggests the need for an objective reappraisal of the people and functions assigned to Central versus Program organizations. This should be done with full recognition that the Central organization's primary reason for existence is to support the requirements of the Program Managers. Concurrently, the Program Manager should undertake a thorough and objective "audit" of all current and planned tasks, as well as evaluate the people assigned to these tasks, in order to bring the total effort down to that which truly contributes to the program.

[11]   It is our opinion that the assignment of the Florida Facility to the Test and Quality Assurance organization creates an anomaly since the Florida activities clearly relate to direct program responsibilities. We recognize that the existence of both CSM and S-II activities at KSC may require the establishment of a single unit for administrative purposes. However, it is our view that the management of this unit is an executive function, rather than one connected with a functional responsibility. We suggest NAA consider a "mirror image" organizational relationship between S&ID and the Florida operation, with the top man at Florida reporting to the S&ID President and the two program organizations reporting to the S&ID Program Managers.

   D.   Program Planning and Control
        Effective planning and control from a program standpoint does not exist. Each organization defines its own job, its own schedules, and its own budget, all of which may not be compatible or developed in a manner required to achieve program objectives. The Program Managers do not define, monitor, or control the interfaces between the various organizations supporting their program.

        Organization—S&ID's planning and control functions are fragmented; responsibility and authority are not clearly defined.

[12]   Work Task Management—General Orders, task authorizations, product plans, etc., are broad and almost meaningless from a standpoint of defining end products. Detailed definitions of work tasks are available at the "doing level"; however, these "work plans" are not reviewed, approved, or controlled by the Program Managers.

        Schedules—Each organization supporting the programs develops its own detailed schedules; they are not effectively integrated within an organization, nor are they necessarily compatible with program master schedule requirements.

Budgeting System—Without control over work scope and schedules, the budget control system cannot be effective. In general, it is an allocation system assigning program resources by organizations.

Management Reports—There is no effective reporting system to management that evaluates performance against plans. Plans are changed to reflect performance. Trends and performance indices reporting is almost nonexistent.

E. Logistics
  The CSM and S-II Site Activations and Logistic organizations are adequately staffed to carry out the Logistics support. The problems in the Logistics area are in arriving at a mutual agreement, between NAA and NASA, clearly defining the tasks required to support the programs. The areas requiring actions are as follows:
  [13] 1. Logistics Plan
        2. Maintenance Manuals
        3. Maintenance Analysis
        4. NAA/KSC Relationship
        5. Common and Bulk Item Requisitioning at KSC
        6. Review of Spare Parts, Tooling, and Test Equipment Status
F. Engineering
  The most pronounced deficiencies observed in S&ID Engineering are:
        1. Fragmentation of the Engineering function throughout the S&ID organization, with the result that it is difficult to identify and place accountability for program-required Engineering outputs.
        2. Inadequate systems engineering job is being done from interpretation of NASA stated technical requirements through design release.
        3. Adequate visibility on intermediate progress on planned engineering releases is lacking. Late, incomplete, and incorrect engineering releases have caused significant hardware delivery schedule slippages as well as unnecessary program costs.
  [14] 4. The principles and procedures for configuration management, as agreed to between NAA and NASA, are not being adhered to by the engineering organizations.
G. Cost Estimating
  The "grass roots" estimating technique used at S&ID is a logical step in the process of arriving at program cost estimates and developing operating budgets. However, there are several aspects of the total process that are of concern to NASA:
        1. The first relates to the inadequate directing, planning, scheduling, and controlling of program work tasks throughout S&ID. While the grass roots estimates may, in fact, represent valid estimates (subject to scrubbing of "cushion") of individual tasks by working level people, we believe that the present deficiencies in Planning and Control permit, and may encourage, the inclusion in these estimates of work tasks and level of efforts that are truly not required for the program.
        2. The second concern is that the final consolidation of grass roots estimates, developed up through the S&ID organization in parallel through both Central functional and Program organizations, does not receive the required [15] management judgements [sic], at successive levels for (a) the real program need for the tasks included in the estimate, or (b) adequate scrubbing and validation of the man-hours and dollars estimates.

3.   The third concern, which results from 1 and 2 above, is that the final estimate does not represent, either in tasks to be done or in resources required, the legitimate program requirements as judged by the Program Manager, but represents total work and dollars required to support a level of effort within S&ID.

Several recommendations are made in the appended reports for correcting deficiencies in the estimating process. The basic issue, however, is that an S&ID Management position must be clearly stated and disciplines established to ensure that the end product of the estimating process be only those resources required to do necessary program tasks. In addition, the Program Management must be in an authoritative position that allows him to accept, reject, and negotiate these resource requirements.

H. Manufacturing Work Force Efficiency

There are several indications of less than effective utilization of the manufacturing labor force. Poor workmanship is evidenced by the continual high rates of rejection and MRB actions which result in rework that would not be necessary if the workmanship [16] had been good. This raises a question as to the effectiveness of the PRIDE program which was designed to motivate personnel toward excellence of performance as a result of personal responsibility for the end product. As brought out elsewhere in this report, the ability of Manufacturing to plan and execute its tasks has been severely limited due to continual changing engineering information and lack of visibility as to the expected availability of the engineering information. Recognizing that overtime shifts are necessary at this time, it is our view that strong and knowledgeable supervision of these overtime shifts is necessary, and that a practical system of measuring work accomplished versus work planned must be implemented and used to gauge and to improve the effectiveness of the labor force. The condition of hardware shipped from the factory, with thousands of hours of work to complete, is unsatisfactory to NASA. S&ID must complete all hardware at the factory and further implement, without delay, an accurate system to certify configuration of delivered hardware, properly related to the DD 250.

I. Quality

NAA quality is not up to NASA required standards. This is evidence[d] by the large number of "correction" E.O.'s and manufacturing discrepancies. This deficiency is further compounded [17] by the large number of discrepancies that escape NAA inspectors but are detected by NASA inspectors. NAA must take immediate and effective action to improve the quality of workmanship and to tighten their own inspection. Performance goals for demonstrating high quality must be established, and trend data must be maintained and given serious attention by Management to correct this unsatisfactory condition.

J. Following are additional observations and findings that have resulted from discussions during the Review. Most of them are covered in most detail in the appended sub-team reports. They are considered significant to the objective of improving NAA management of our programs and are therefore highlighted in this section of the report:

1.   S&ID must assume more responsibility and initiative for carrying out these programs, and not expect step-by-step direction from NASA.

2.   S&ID must establish work package management techniques that effectively define, integrate, and control program tasks, schedules, and resource requirements.

3. S&ID must give concurrent attention to both present and downstream tasks to halt the alarming trend of crisis operation and neglect of future tasks because of concentration on today's problems.

4. A quick response capability must be developed to work critical "program pacing" problems by a short-cut route, with follow-up to ensure meeting normal system requirements.

[18]   5. S&ID must maintain a current list of open issues and unresolved problems, with clear responsibility assigned for resolving these and insuring proper attention by Program and Division Management.

6. Effort needs to be applied to simplify management systems and end products. There must be greater emphasis on making today's procedures work to solve today's problems, and less on future, more sophisticated systems. The implementation and adherence to prescribed systems should be audited.

7. NAA must define standards of performance for maintaining contracts current then establish internal disciplines to meet these standards. Present undefinitized subcontracts and outstanding change orders on the S-II prime contract must be definitized without delay.

CONCLUSIONS AND RECOMMENDATIONS

The NASA Team views on existing deficiencies in the contractor's management of the S-II and CSM Programs are highlighted in this section of the report and are treated in more detail in the appended sub-team reports. The findings are expressed frankly and result from the team's work in attempting to relate the end results we see in program conditions to fundamental causes for these conditions.

[19] In most instances, recommendations for improvement accompany the findings. In some cases, problems are expressed for which the team has no specific recommendations, other than the need for attention and resolution by NAA.

It is not NASA's intent to dictate solutions to the deficiencies noted in this report. The solution to NAA's internal problems is both a prerogative and a responsibility of NAA Management, within the parameters of NASA's requirements as stated in the contracts. NASA does, however, fully expect objective, responsible, and timely action by NAA to correct the conditions described in this report.

It is recommended that the CSM incentive contract conversion proceed as now planned.

Incentivization of the S-II Program should be delayed until NASA is assured that the S-II Program is under control and a responsible proposal is received from the contractor.

Decision on a follow-on incentive contract for the CSM, beyond the present contract period, will be based on contractor performance.

It is recommended that NAA respond to NASA, by the end of January 1966, on the actions taken and planned to be taken to correct the conditions described in this report. At that time, NAA is also to certify the tasks, schedules, and resource requirements for the S-II and CSM Programs.

[20] It is further recommended that the same NASA Review Team re-visit NAA during March 1966 to review NAA performance in the critical areas described in this report.

Document III-18

[1]   December 19, 1965

Mr. J. L. Atwood, President
North American Aviation, Inc.
1700 E. Imperial Highway
El Segundo, California

Dear Lee:

In my letter of October 27, 1965, I conveyed to you the seriousness with which I viewed the state of affairs in both the Apollo and S-II Programs at your Space and Information Systems Division. Phillips' report has not only corroborated my concern, but has convinced me beyond doubt that the situation at S&ID requires positive and substantive actions immediately in order to meet the national objectives of the Apollo Program.

Since I am not sure that you see the performance of S&ID in the same light that I do, let me give you a perspective from my point of view.

When I joined NASA in the Fall of 1963, I restructured the Apollo Program to bring its several elements into balance and to establish a schedule that could be achieved based on the state of development at that time. Since that time, in the spacecraft project, we have found it necessary to:

a.   Omit several sub-systems from 009.

b.   Delay flight of 201 from November 65 until probably February or March 66 due to late delivery of 009 and its GSE together with the many difficulties of getting things [2] to work together at the Cape.

c.   Reschedule the first manned flight from 203 to 204 to relieve the spacecraft schedule. NAA ability to support the 204 flight scheduled in October 66 now looks doubtful.

d.   Reschedule 202 from April to June 1966 because 011 is several months behind schedule. NAA ability to support the June schedule now looks doubtful.

e.   Reschedule the first Block II spacecraft flight from 206 in April 67 to 207 in July 67. Late last year, when the Block II Program was defined, your people agreed that they could and would do a better job on Block II engineering and that they would meet their design review and drawing release schedules. I'm very disturbed to learn now that Block II engineering has been neglected and that it is some months behind schedule. To me, considering performance to date, it looks like the danger flags portend delay of the critical 207 flight.

f.   Delay the delivery of 008 by several months. This is a critical vehicle to perform thermal vacuum tests in the Houston Chamber as a prerequisite to manned flight. People will argue that the Chamber isn't ready, but we urgently need that spacecraft to get it working as a system vehicle and with its ground equipment and crews.

g.   Delete seven boilerplate and flight spacecraft from the Block I Program to reduce cost growth and relieve the schedule to minimize slippage.

I could go on; there are other things that we've had to accommodate such as cost growth, but I believe this list gives you some insight into my evaluation of performance in the spacecraft project. Now, regarding the S-II Project:

a.   I am facing the probability that the flight of 501 will be delayed between three and nine months. I [3] assure you that this is due entirely to the status of the S-II stage. It is clear to me that it didn't have to come out this way, and I regret now that I wasn't more insistent a year and a quarter ago when you and I discussed the danger flags then flying, and the possibility of such far reaching actions as transferring the project to your Los Angeles Division.

b. The cost proposal which S&ID presented to MSFC in October of this year was shocking in light of cost projections reported only one month earlier. Perhaps I should even go so far as to suggest that it was irresponsible; in any case, it surely was a gross demonstration of management shortcomings.

c. The Battleship Program is another significant case. You got behind it personally and an ignition test in November 64 resulted; but that achievement was one year behind the original schedule and the test fixture was so devoid of systems as to be little more than a facade. Further, the firing record indicates that only about one-third of the firings really achieved their objectives. The firing program was stopped last April to incorporate flight systems; it has not yet resumed firing.

d. S-II-T is a real problem. It was delivered late with what was stated to be approximately 21,000 manhours of work to incorporate EO's and perform work that was not completed in the factory due to parts shortages. Today, the work stands at over 50,000 manhours and the firing scheduled for January 66 will most likely occur in March or April. Based on what I have seen so far, I am very concerned that the engineering on which S-II-I is based will require many changes when S-II-T is fired, and further delays of 501 will result.

It is hard for me to understand how a company with the background and demonstrated competence of NAA could have spent 4 1/2 years and more than half a billion dollars on the S-II Project and not yet have fired a stage with flight systems in operation.

[4] Again, I could go on and enumerate additional problems, but the points I have discussed should show you how I see the performance of NAA on these two programs.

I have been in this business long enough to understand quite well the difficulties and setbacks that occur and manifest themselves in many forms in government-industry programs which have as their objective the development, building, and operation of sophisticated systems involving advanced technology and real forward projection of thought. My experience indicates that results are a function of management and technical competence. I submit that the record of these two programs makes it clear that a good job has not been done. Based on what I see going on currently, I have absolutely no confidence that future commitments will be met.

I can see no way of improving future performance, and meeting commitments which NAA must meet if we are to achieve the national objectives of Apollo, except to improve the management and technical competence of your Space and Information Systems Division.

Sam Phillips is convinced that S&ID can do a better job with less people. He and his team discussed the reasons why they believe this in their briefing.

I suggest that you can go even further to concentrate management and technical talent on the two programs that constitute 98 percent of the business of S&ID. For example:

a. Eliminate or transfer to another Division those activities at S&ID that are not contributing directly to the progress of the Spacecraft and S-II projects. Examples are the Federal Programs Group, parts of the Information Systems Division, and parts of the Advanced Systems Division. This should make possible a substantial consolidation of central engineering and insure that [5] available talent concentrates on the two important programs.

b. Take a hard look at the competence and effectiveness of individuals, especially in the upper echelons of the organization; and move out those who are not really contributing, due either to the organization or to their own competence.

I urge you to consider the potential payoff of extending the project management principle beyond the "designated subsystems project manager" as now practiced in Dale Meyers' organization. I am convinced that there is no substitute for clear assignment of responsibility and accountability to individuals for delivering results. Work packages can be defined quite clearly in both projects and I am sure it is possible to assign responsibility to individuals who are given control of the applicable budget and who are held accountable for delivering on schedule and within budget.

I had hoped that a letter such as this would not be necessary. However, I consider the present situation to be intolerable and can only conclude that drastic action is in the best national interest. I assure you that I have only one purpose, and that is to carry out the Apollo Program on schedule and within planned costs.

I have instructed Sam Phillips to keep his team together so that they can visit S&ID again in March to see if progress is consistent with that required to achieve program objectives.

Sincerely,

George E. Mueller
Associate Administrator
For Manned Space Flight

Document III-19

Document title: James E. Webb, Administrator, to Dr. Frederick Seitz, President, National Academy of Sciences, December 20, 1967.

Source: NASA Historical Reference Collection, NASA History Office, NASA Headquarters, Washington, D.C.

In this letter, NASA Administrator James Webb thanks the National Academy of Sciences for its advice regarding the establishment of a Lunar Science Institute to be a central location for the analysis of samples returned from the Moon. He also attempts to clarify NASA's reasoning behind its decision to establish such an institute. Essentially, Webb sought the creation of this institute under NASA funding but with academic management. This arrangement, he believed, was critical if the institute were to achieve the stature Webb wanted for it.

———————————

[1]   December 20, 1967

Dr. Frederick Seitz
President
National Academy of Sciences
Washington, D. C.

Dear Fred:

We have your letter of November 1, 1967 and the report of the Academy of NASA/University Relations Committee. Will you please give them my thanks for the work they have done so far to help resolve the problems of the "Lunar Science Institute" we are thinking of establishing near the Manned Spacecraft Center (MSC) in Houston. We also deeply appreciate the help you and the National Academy are giving us in this matter.

We asked the Lunar and Planetary Missions Board (LPMB) to review the needs and plans for this "Institute," which they did at their last meeting. The Board did not take any formal action pending further clarification and discussion of the nature of the Institute at their next meeting in January. Several of the members still have grave reservations about the usefulness of the proposed Institute; its method of operation; and its effect on academic scientists interested in lunar exploration. Specifically, they are concerned that the establishment of the Institute might weaken the position of the university scientist either by encouraging him to participate only if he is a member of the Institute or by forcing him to come to the Institute to do his research. They were also concerned about the name, "Lunar Science Institute," which they felt implied a more substantial institution with a larger staff than [2] that described to them. In the discussion at the Board and in later consultations with the concerned members of the Board, it seems that we can alleviate much of this concern if it is made clear that:

1. NASA plans to continue the policy of encouraging all competent scientists to compete for participation in the lunar exploration program and that membership or non-membership of his parent institution in the Institute will not be permitted to affect the standing of his proposal in NASA's evaluation of it and others in that competition.
2. The selection of the principal investigators in the Lunar Exploration Program or for lunar sample analysis will continue to be made as they have in the past by the highest level and most competent personnel in NASA.
3. The "Lunar Science Institute" is being established to help those scientists who consider it desirable to come to MSC from time to time either to plan or conduct their research and to provide an easy access to scientists who have an interest in considering participation or in the pattern of relationships which will grow from this pilot model experiment in the continuing NASA effort to find the most satisfactory basis for scientists to participate in its programs.
4. Selection as a principal investigator automatically makes the facilities of the Institute available to him when he needs to come to Houston.
5. NASA will continue to follow the policy of encouraging an academic scientist to conduct his research at his home institution to the fullest extent possible and with as little interference with his academic responsibilities as possible.

[3] It is also apparent that what we are thinking of is not so much an "Institute" as it is a "Facility for Continuation Study" in a location that provides some benefits over and beyond those heretofore available. Therefore, we should seek a name which more accurately describes such a facility and its functions.

An interim arrangement whereby the National Academy of Sciences has a prime contract from NASA for the operation of the facility, and where it, in turn, negotiates a sub-contract with Rice University to operate the facility seems a reasonable arrangement provided the following matters, in addition to those above, are worked out to our mutual satisfaction and to the satisfaction of the LPMB and specified in the appropriate contracts or memoranda of understanding:

1. The administrative arrangements and agreements necessary to bring the facility into being and operate it. Careful attention must be given to the role of the LPMB, which is the principal group we look to for advice on the content of the lunar program and to represent the interests of the scientists involved in that program. Careful attention must also be given to the role of the Science and Applications Directorate at the Manned Spacecraft Center.
2. The size and type of staff required (should be small).
3. The location, size, and nature of the buildings and equipment to be utilized. Presumably, this would be the West Mansion located on the Rice property adjacent to MSC.

4. If it is the West Mansion, the nature and cost of the modifications which will be required.
5. If the arrangement with the Academy and Rice is to be regarded as temporary, then plans [4] leading to a permanent arrangement should be outlined.

Even though the arrangement for Rice to operate the facility may be temporary, these arrangements should specify the role that Rice will play in the administration, the fee considered proper, and any plans or actions which Rice expects to take to help evolve new and better relationships between graduate education in the disciplines involved and the space program.

Dr. John E. Naugle, Associate Administrator for Space Science and Applications, will be my representative in working out these arrangements with you as President of the Academy and Dr. Kenneth S. Pitzer in his dual role of Chairman of the Academy Committee on NASA/University Relations, and as President of Rice. Dr. Newell and I will be following these matters very closely.

Sincerely yours,

James E. Webb
Administrator

Document III-20

Document title: John E. Naugle, Associate Administrator, NASA, Memorandum to Administrator, "Space Astronomy Institute," February 4, 1976.

Source: NASA Historical Reference Collection, NASA History Office, NASA Headquarters, Washington, D.C.

This memorandum to Administrator James C. Fletcher from NASA Associate Administrator John E. Naugle reflects the lengthy debate over form and control of the proposed Space Astronomy Institute, soon to be renamed the Space Telescope Science Institute. The astronomy community was concerned about playing a role in the telescope project, and envisioned an institute separate from NASA and managed by a university or a university consortium. Dr. Hinners is Noel W. Hinners, NASA's Associate Administrator for Space Science and the key individual in deciding the institute's final form.

———————————

[1]   February 4, 1975

Memorandum

TO: A/Administrator
FROM: AA/Associate Administrator
SUBJECT: Space Astronomy Institute

On February 2, Dr. Richard Goody called on behalf of the International Astronomy Group with which we met on January 29 in Williamsburg, Virginia. Goody said a matter had come up after we left which he had been asked to discuss with NASA on behalf of the group.

The group discussed the so-called "Space Astronomy Institute" (SAI) and concluded that SAI would very likely become the key or certainly one of the two or three key astronomy institutions of the western world in the 1980's. The astronomers assembled in Williamsburg wanted NASA to know of their interest in the SAI and also they were concerned that there would be the necessary interaction of astronomers with NASA in developing the plan so this would indeed become such an institution. Goody said he had been empowered by that group to approach NASA to offer to help in this matter and he felt a group could be organized to represent the National Academy of Sciences (NAS), the American Astronomical Society [AAS] and the European Science Foundation (ESF).

I told Goody that: we certainly felt that the SAI was exceedingly important; we hoped it became precisely the kind of institution he envisioned; and Dr. Hinners had laid out a very careful approach in planning for the SAI which allowed for considerable interaction and review of our plans with and by astronomers. I told him that Dr. Hinners and I would need to discuss this matter with you before any commitment could be made, but that it would be helpful to have a small group [2] of senior astronomers designated as the spokesman for NAS, AAS and ESF. I told him there were precedents for NAS helping to organize such a facility—noting that Mr. Webb had worked closely with the then President of NAS, Dr. Seitz, in establishing University Space Research Association (USRA) and the Lunar Science Institute, and that AEC had also worked closely with Dr. Seitz in creating the Universities Research Association, Inc. (URA) in getting the big accelerator under way.

At our meeting with you on February 9, Dr. Hinners will outline the present strategy and plan of action for bringing the ASI [sic] into being. I told Dr. Hinners of Dr. Goody's call and asked that he consider how a group such as the one proposed by Goody could be brought into that plan of action.

John E. Naugle

Document III-21

Document title: Memphis Norman, Budget Examiner, SET, to Mr. Loweth, "National Academy of Sciences Report Regarding Institutional Arrangements for the Space Telescope," April 6, 1977.

Source: NASA Historical Reference Collection, NASA History Office, NASA Headquarters, Washington, D.C.

In attempting to determine the best form of management organization for the Space Telescope Science Institute, NASA requested that the National Academy of Sciences (NAS) study the issue and provide a recommendation. The resulting report would play an important role in the decision to have a university-led consortium manage the institute. This document, an internal Office of Management and Budget (OMB) memorandum, contains a summary of the report, as well as additional comments that reflect OMB's favorable disposition toward a non-NASA arrangement. Hugh Loweth was the head of that portion of OMB that oversaw the NASA budget, and Memphis Norman was one of his staff members. The name of the OMB division was Science, Engineering, and Technology (SET). PSAC stands for the President's Science Advisory Committee.

[1]
National Academy of Sciences Report Regarding Institutional Arrangements for the Space Telescope

The National Academy of Sciences conducted a Woods Hole Conference between July 19-30, 1976 to examine the institutional arrangements for the operational phase of the Space Telescope—the report was released in January 1977. The report was prepared in response to a request by NASA's Office of Space Science for the Space Science Board of NAS to examine organizational and management features of a possible Institute and to make recommendations for NASA's consideration. This memo summarizes the report, and provides NASA['s] and our comments on the subject matter. There is no action for use to take at this time, although we should keep it on our "watch list."

Background

- The Space Telescope will have the most complex organizational arrangement ever experienced on a NASA mission. The project will involve over a ten-to-twenty year period, two NASA centers, three headquarters program offices, NSF [National Science Foundation], the European Space Agency, other national and international organizations, and the complex of ground based observatories outside of NASA (the Space Telescope and ground telescopes will complement each other).

- NASA talked to us last fall about an Institute for the Space Telescope, but the details were sketchy. To assess the need for an Institute and plan it, NASA conducted an internal study last year, and asked the National Academy to conduct an additional study involving spaced-based and ground-based astronomers. NASA is establishing a working group (chaired by a NASA individual) to examine inputs from various groups regarding the Institute, and to make recommendations to NASA management.

Report Summary

- The fundamental point addressed by the report is how to maximize scientific return from a large investment for R&D and operations. The report proposes a strong role for the Institute and concludes that ST operations should move from engineers to the scientists and that central responsibility (a focal point) should be placed in a highly visible independent institute (free from organizational restrictions) and staffed by full-time astronomers.

[2] • Key recommendations include:
  - The institute should organize and manage itself, and pick its own location—off a NASA installation.
  - The new organization should include space-based and ground based astronomers (including foreigners) and provide for extensive coordination.
  - The Institute should have direct involvement in the development and operation of the Telescope. The Institute should have its own laboratories, facilities, and computers, and plan and manage the science program (observations and instruments); participate in technical development by developing hardware and software systems for data handling and control capability on-board the ST, and being involved in contract negotiations, trade-off decisions and design modifications; perform data analysis at its own laboratories; and checkout the ST before and after launch.
  - Operational decision-making should be the responsibility of the Institute since the participation of all astronomers should make possible decisions in the context of a comprehensive astronomy program (overall strategy).

- The Institute should be similar to a university consortium with a Director, Board of Trustees, scientific staff, and advisory committees. Staffing would build-up to 90 positions by 1983 (ST launch year), and to 150-200 positions during operations.
- The Institute should be established as soon as possible and a Director appointed.
- Funding should be provided under contract with NASA with contributions from NSF, other public agencies, foreign governments, and private organizations. *Costs were not provided.*
- The Institute should maintain close liaison with NASA headquarters, NASA centers, engineering groups, contractors, and scientists.

NASA's Position
- NASA has not made a decision about the Institute—even whether to have one. A working group will be set up (chaired by Warren Keller—NASA Headquarters) shortly to review the NAS report as well as other inputs.
- [3] • The study group will include NASA engineers, scientists, and operations specialists, and advisers from NSF and NAS.
- The review will probably start with the NAS concept, since scientists should be in a nucleus position and many aspects of the concept are good. However, NASA views the concept as too large and expensive—particularly if NASA should fund. NASA will cost out the NAS proposal and alternatives.
- NASA views the NAS report as an expression (particularly by ground-based astronomers such as Kitt Peak) to curtail NASA's influence because of fear of NASA. Astronomers want an NSF-type operation (independence).
- NASA plans to complete the study by July/August and recommend to NASA management whether to have an Institute and its size, structure, management, operations, budget, and timing. A budget request for the Institute would likely be in the 1980 budget—not 1979.

Staff Comments
- It appears that NASA is correct in sensing that astronomers (particularly ground-based) are afraid of NASA. We have heard numerous accounts before from PSAC members and NSF—perhaps there are good reasons for fear, particularly about the Marshall Center which will manage ST development.
- However, we do believe that an Institute is a good idea, particularly the involvement of ground-based astronomers. We have often talked about the need for coordination and a comprehensive strategy for astronomy. The Institute may be the beginning.
- We also believe that once the Shuttle becomes operational its new capabilities should allow for greater participation by scientists. Institutional arrangements to bring in more people is a consideration for NASA in the future—these new programs will not be "normal" NASA programs.
- We will need, obviously, to watch the funding arrangements and level of costs.
- Leo Goldberg feels very strongly that the science community should have a strong hand in the organization and management of the Space Telescope—you may want to formally ask NASA by letter to report to us on the NAS report and the Institute when they are ready. We can prepare a letter for the Director or Mr. Cutter, if you wish.

Document III-22

Document title: U.S. Congress, House, Committee on Science and Technology, Subcommittee on Space Science and Applications, "Space Telescope Program Review," 95th Cong., 2d sess., Report No. 85 (Washington, DC: U.S. Government Printing Office, 1978), pp. 3-7, 11-14.

Dr. Noel W. Hinners, NASA's Associate Administrator for Space Science, was the key player in resolving the dispute over whether a Space Telescope Science Institute should be operated by NASA or by a university consortium. Opting for the latter, Hinners presented to Congress NASA's reasoning behind its plans for the institute. This explanation was delivered in a filled hearing room before the Subcommittee on Space Science and Applications of the Committee on Science and Technology in the House of Representatives. Called by Hinners the "Space Telescope Program Review," his report on July 13, 1978, presented well the planning for the NASA-university partnership that governed the Hubble Space Telescope.

---

[3]

Statement
of
Dr. Noel W. Hinners
Associate Administrator for Space Science
National Aeronautics and Space Administration
before the
Subcommittee on Space Science and Applications
Committee on Science and Technology
House of Representatives

Mr. Chairman and Members of the Subcommittee:

I welcome the opportunity to review with you the status of the Space Telescope Program. Following my brief overview including some discussion of our planned approach to science operations, Mr. William C. Keathley, the NASA Space Telescope Project Manager at Marshall Space Flight Center, will give a more detailed description of the development program and its cost performance, and schedule status.

The Space Telescope is being designed as a general-purpose, astronomical observatory in space with an anticipated lifetime of more than a decade. To be launched in late 1983 by the Space Shuttle, it will be the first long term national astronomical observatory in space. . . . The availability of the Space Shuttle will allow in-orbit repair of the observatory, exchange of experiments by Shuttle crew members, and, if necessary, return of the entire system to Earth for refurbishment and subsequent relaunch.

The Space Telescope, by being outside the Earth's atmosphere, will enable us to image objects that are ten times smaller than possible with ground-based optical telescopes. This will permit us to study nearby objects in much greater detail or to detect stellar counterparts at about ten times greater distance than is now possible from Earth.

If the universe has a beginning, we should be able to see some objects as they were near the beginning of time. The Space Telescope will allow us to observe light over the entire range from the far ultraviolet to the far infrared (from wavelengths of approximately 1100 angstroms to about 1 millimeter = 10,000,000 angstroms). Most of this range is inaccessible from ground observatories.

[4] The spacecraft facility is a cylinder of about 14 meters (46 feet) in length and 4.3 meters (14 feet) in diameter, weighing about 9,000 kilograms (10 tons). The mirror size will be 2.4 meters (94 inches), comparable to the larger Earth-based telescopes. The mirror is sufficiently large that experiments requiring large, light-gathering power can be carried out with this Telescope that have been impossible with smaller predecessors. The high resolution of the Space Telescope will permit the detection and measurement of stars as faint as the 27th or 28th magnitude, some fifty times fainter than those which can now be detected from Earth. Spectra will be obtainable from objects as faint as 25th magnitude, which is 9 1/2 magnitudes (factor of approximately 7,000) fainter than is possible with the International Ultraviolet Explorer and 13 magnitudes (factor of approximately 100,000) fainter than with the Orbiting Astronomical Observatory.

Five versatile scientific instruments (four American and one European) have been selected for flight at the focal plane of the Telescope to carry out a wide range of observations. The Space Shuttle in-orbit maintenance capability, mentioned earlier, will permit the replacement of failed or outdated equipment at a small fraction of the cost of a new scientific mission. Thus, the Space Telescope can be operated with the best scientific instruments as they become available.

Preliminary design efforts of the scientific focal plane instruments for the Space Telescope are being carried out by Investigation Definition Teams, composed of participating scientists who were tentatively selected by NASA on November 8, 1977. Final evaluation and confirmation of the payload selection will occur in early FY 1979, based on the results of the preliminary design reviews. We are confident of our ability to develop the instruments on a time scale consistent with the Space Telescope Project schedule, which assumes a late 1983 launch.

As has been indicated in previous testimony, negotiations with the European Space Agency (ESA) covering their participation in the Space Telescope Program have resulted in a Memorandum of Understanding, signed on October 7, 1977, by the NASA Administrator and the ESA Director General. ESA will supply, without cost to NASA, one of the scientific instruments, the Faint Object Camera; the solar array, which will provide power for the spacecraft facility; and, a number of personnel for science operations support. In return, observing time on the Telescope will be provided for European Scientists.

NASA's Marshall Space Flight Center is responsible for overall management of the Space Telescope Project. . . .

[5] As you are aware, the Space Telescope Program was approved as a new start in NASA's FY 1987 budget. The Program, because of its very complex and interactive nature, has been carefully planned and well defined. Currently, the major hardware contracts have been awarded, and all elements of the development work are on schedule and within the cost estimate.

As indicated in the past testimony, the development program for the Space Telescope does not include funding for the operation and maintenance of the Telescope beyond thirty days after launch nor for the establishment of the hardware and software capabilities required for science operations. As I indicated in February, during the Hearings on the FY 1979 Budget, we must begin to budget for science operations in FY 1980, if we are to have the required operational capability at the time of launch. In the remainder of my time, I would like to discuss this area of science operations which has been left open in previous testimony. I promised that we would get back and discuss our plans with you.

As indicated earlier, the Space Telescope is planned for operation for more than a decade with attendant in-orbit maintenance; recovery, refurbishment, and relaunch; and update of the focal plane scientific instruments. . . . During the operational period, the Space Telescope will be used the majority of the time by "general observers" who will be selected on the basis of proposals submitted in response to periodic solicitations. In developing observing schedules for the Telescope, the requirements of these observers will be integrated with those of investigators who are involved with development of specific focal plane instruments. The Space Telescope operations, including the investigation selection, scheduling, maintenance, refurbishment, etc., can be viewed as quite analogous to the operation of a large, ground-based telescope.

An important consideration with respect to the science operations for the Telescope has been the question of whether or not a Space Telescope Science Institute will be established, rendering the operation similar to that for a number of large, ground-based facilities. Over the years, a growing number of astronomical groups have studied the question of Space Telescope science operations. While these considerations have been carried out to widely varying depths, all such groups have made recommendations in favor of the Science Institute approach. . . . The 1976 National Academy of Sciences study group, chaired by Professor Donald. F. Hornig, studied the problem at our request. This group, which consisted of an ad hoc group of independent scientists, strongly recommended the establishment of a Space Telescope Science [6] Institute and outlined, in some detail, the functions, structure, and implementation mode for the recommended Institute. This study served as the point of departure for our in-house study group in considering the possible establishment of a Science Institute.

After studying this question at considerable length, using inputs from both the in-house and external study groups, we have come to a conclusion that the most efficient and scientifically satisfactory approach to science operations would involve the establishment of [a] Space Telescope Science Institute which would be operated under a long-term contract with NASA. Our approach, however, would not be identical to the National Science Foundation's approach to operation of large, ground-based facilities, since NASA must retain operational responsibility for the spacecraft/observatory. . . .

We feel that the science operations concept for the Telescope must reflect a long-term commitment as would be accomplished by a dedicated "independent" institute, giving astronomers and science operations personnel access to computer and other facilities, based on Space Telescope priorities. There is no doubt that the science impact of the Space Telescope will be comparable to that of major laboratories, which are being operated efficiently as national facilities in the "institute" mode. Such laboratories have proven to be responsive to the user community and, at the same time, able to work well with the funding Agency. The Space Telescope is the first planned, long-life, NASA science flight project, and we feel that operational procedures used on past flight projects do not necessarily constitute the most efficient way to handle this program. We are, in a sense, taking our cue from the people who have been successfully operating the analogous, ground-based observatories over a large number of years. Another obvious advantage of the "institute" mode is that it is an operational mode with which the world-wide astronomical community is familiar and confident. . . .

The Space Telescope Science Institute would have independent management and staff and its own computer hardware, which, in order to minimize cost, would begin operation using software developed by NASA.

[7] The Science Institute would conduct science operations activities in three major areas: planning and management, Space Telescope scheduling, and data activities. Within the planning and management function, the Institute would implement those policies

established by NASA which pertain to Space Telescope use. In this endeavor, the Institute would solicit, evaluate, and select observational proposals received from the scientific community and would formulate, for NASA review and approval, yearly activity goals which are in consonance with the overall policy established by NASA.

The Science Institute, in addition to a long-term planning function, would generate the generalized observing schedules. To accomplish this function, computers would be located at the Institute to develop the target selection sequence while, at the same time, observing such factors as target availability, sky constraints, and spacecraft design constraints. The Institute would generate observing instructions as required. In turn, the Space Telescope Operations and Control Center (STOCC) at Goddard Space Flight Center would convert the observing instructions into space commands that would properly point and control the Space Telescope.

In the data activities areas, the Institute would provide equipment enabling visiting scientists and staff to perform analyses of Space Telescope data, as well as to conduct basic research in the field of astronomy. Further, it would evaluate science productivity of the Space Telescope research program. It would help coordinate both correlative research with ground observing facilities and international participation in the overall activity. Finally, it would be responsible for informing the public of research results, as well as for archiving all Space Telescope data for dissemination as requested.

In view of the use, initially, of NASA-generated software, the computer complex would be designed to be compatible with the computers in the STOCC at Goddard Space Flight Center. Consequently, the complex at the Institute would be designed by NASA prior to the establishment of the Institute.

No compelling reasons have been identified for locating the Science Institute close to any existing NASA facility, so long as appropriate Institute personnel are collocated at Goddard Space Flight Center to interface with the STOCC. It is anticipated that the operational site of the Institute would be included as part of the proposals for its operation. Only general site criteria would be specified by NASA. These criteria might include such factors as proximity to an active astronomical center of excellence, a major airport, etc.

We currently would anticipate release of a Request for Proposals early in FY 1980 for the operation of the Institute. The Institute would be built up slowly to full strength prior to launch of the Telescope in the first quarter of FY 1984. . . .

[11] [Briefing Charts]

RECOMMENDATIONS ON THE METHOD OF
SPACE TELESCOPE SCIENCE OPERATIONS
(ALL IN FAVOR OF THE SCIENCE INSTITUTE APPROACH)

<u>Date</u>   <u>Organization</u>

1966   REPORT OF STUDY HEADED BY NORMAN RAMSEY

APRIL 1975   ST SCIENCE WORKING GROUP—BODY OF SENIOR SCIENTISTS
                       SELECTED BY ANNOUNCEMENT OF OPPORTUNITY TO PAR-
                       TICIPATE IN THE PRELIMINARY DESIGN PHASE OF ST

AUGUST 1975   COUNCIL OF THE AMERICAN ASTRONOMICAL SOCIETY—
                       ELECTED COUNCIL OF THE ONLY PROFESSIONAL ASTRO-
                       NOMICAL SOCIETY IN THE UNITED STATES

NOVEMBER 1975   LST STUDY GROUP—AD HOC BODY OF SCIENTISTS CON-
                       VENED BY THE ASSOCIATE ADMINISTRATOR TO PROVIDE AN
                       OVERVIEW OF THE ST PROGRAM

FEBRUARY 1976   SHUTTLE ASTRONOMY MANAGEMENT AND OPERATIONS
                       WORKING GROUP—SCIENTIFIC WORKING GROUP CHAIRED
                       BY THE CHIEF OF THE ASTRONOMY AND RELATIVITY OFFICE
                       OF THE OFFICE OF SPACE SCIENCE

DECEMBER 1976   NATIONAL ACADEMY OF SCIENCES—REPORT OF THE SPE-
                       CIAL STUDY HEADED BY PROFESSOR HORNIG

[12]

SPACE TELESCOPE SCIENCE INSTITUTE
CHARACTERISTICS

- LONG-TERM COMMITMENT TO SCIENCE OPERATION
- COMPUTERS AND OTHER FACILITIES ACCESSIBLE TO ASTRONOMERS AND
  SCIENCE OPERATIONS PERSONNEL
- EFFICIENT MODE OF OPERATION WHICH HAS PROVEN TO BE RESPONSIVE
  TO THE USER COMMUNITY AND AT THE SAME TIME TO WORK WELL WITH
  FUNDING AGENCY
- ANALOGOUS TO THE OPERATING MODE EMPLOYED AT LARGE GROUND-
  BASED OBSERVATORIES OVER A LARGE NUMBER OF YEARS
- OPERATIONAL MODE WITH WHICH THE WORLDWIDE ASTRONOMICAL
  COMMUNITY IS FAMILIAR AND CONFIDENT

[13]
SPACE TELESCOPE SCIENCE INSTITUTE
SOME KEY OPERATIONAL FUNCTIONS

- ·SCIENCE PLANNING AND MANAGEMENT
    - IMPLEMENT NASA ST SCIENCE POLICY
    - SOLICIT, EVALUATE, AND SELECT OBSERVATIONAL PROPOSALS
    - COORDINATE CORRELATIVE RESEARCH
    - COORDINATE INTERNATIONAL PARTICIPATION

- SCHEDULING
    - GENERALIZED OBSERVING SCHEDULES
        - TARGET SEQUENCE
        - TARGET AVAILABILITY
        - SKY CONSTRAINTS
        - SPACECRAFT CONSTRAINTS
    - GENERATE OBSERVING INSTRUCTIONS

- DATA ACTIVITIES
    - REDUCE AND ANALYZE DATA
    - CONDUCT BASIC RESEARCH
    - EVALUATE SCIENCE
    - INFORM THE PUBLIC
    - ARCHIVE AND DISSEMINATE ST DATA

[14]
SPACE TELESCOPE SCIENCE INSTITUTE
SUMMARY

CHARACTERISTICS—
- INDEPENDENT MANAGEMENT AND STAFF
- DEDICATED FACILITIES (INCLUDING COMPUTERS)
- INITIAL SOFTWARE DEVELOPED BY NASA
- LONG-TERM CONTRACT WITH NASA

LOCATION—
- NO COMPELLING DATA-HANDLING, MANAGERIAL, OR COST REASONS FOR LOCATION AT ANY EXISTING FACILITY
- SITE TO BE INCLUDED AS PART OF PROPOSALS FOR INSTITUTE OPERA-TION/GENERAL SITE CRITERIA

IMPLEMENTATION—
- FIRST BUDGET YEAR—FY 1980
- RFP RELEASED—EARLY FY 1980
- FULLY OPERATIONAL—AT LAUNCH—FIRST QUARTER OF FY 1984

Document III-23

Document title: R. W. Gutman, Director, General Accounting Office, to Robert A. Frosch, NASA Administrator, November 11, 1977.

Document III-24

Document title: Robert A. Frosch, NASA Administrator, to Associate Administrator for Space and Terrestrial Applications, *et al.*, "NASA/University Relations," May 18, 1978, with attached: "Policy for Academic Involvement in the NASA R&D Program."

Source: Both in University Affairs Files, NASA Historical Reference Collection, NASA History Office, NASA Headquarters, Washington, D.C.

These two memoranda discuss the review and reform of NASA's university relations efforts during the latter 1970s. In these documents, NASA Administrator Robert Frosch, in concert with others, sought to delineate the relationship between NASA and academia, as well as the activities that were appropriate for each to undertake. Essentially, Frosch directed that NASA rely on university expertise to provide basic research relative to the mission of the agency, and he interpreted NASA's role in this arena as being one of facilitator. He was also responding to concerns expressed by the General Accounting Office (GAO) that NASA was conducting its university affairs program as basically open-ended support for scientists and engineers without clear program definition. By tying the research sponsored by NASA much more closely to aerospace research and development activities under way at the agency, Frosch helped resolve many of these concerns.

Document III-23

[1] The Honorable Robert A. Frosch   November 11, 1977
Administrator, National Aeronautics and Space Administration

Dear Dr. Frosch:

The General Accounting Office just completed a survey under assignment code 952174 of NASA's administration of research grants and contracts to colleges and universities. During this survey, several aspects of NASA's university research program were identified which we believe could be improved. Before planning additional work in this area we believe it would be mutually beneficial to both NASA and GAO to meet with you and your representatives. The purpose of the meeting would be to present to you our survey results and observations and to obtain your views thereon.

The specific areas we would like to discuss are:
— increasing university competition for research projects;
— improving the negotiation process and detailed support for the number of hours included in a proposal;
— the possibility of requiring NASA technical monitors to visit research sites to see what progress is being made;

— corrective action that NASA could take when the cognizant Federal audit agency reports accounting system deficiencies at universities having NASA research grants and contracts; and

— acquiring and disseminating technical information.

Another area to be discussed relates to administrative differences between grants and contracts. It is not always clear as to whether a grant or a contract is the proper instrument to fund a project. In the case of grants, grantees are not required to report how funds were spent, grants are not audited prior to closing, and grantee-acquired equipment is not entered in NASA's Equipment Visibility System although NASA has the option to obtain title to this equipment upon completion of a grant.

In summary, NASA's grant and contract administration practices give the appearance that a university assistance program is being conducted rather than a mission-oriented research program to further the agency's [2] mission. Several NASA officials told us that a grant is a gift and that if a university fails to comply with grant provisions, action taken by NASA is limited to "friendly persuasion." It may be a valid position that universities should be treated differently than commercial entities dealing with the Government; however, this should be balanced against the responsibility Federal agencies have for stewardship of public funds entrusted to them.

We would like to schedule a meeting at your convenience soon after the first of December. Arrangements for the meeting can be made with Mr. Chester S. Daniels, Assistant Director of this Division. He can be reached by telephone on 275-3191.

Sincerely yours,

R. W. Gutman
Director

Document III-24

[1]   May 18, 1978

Memorandum

TO: E-1/Associate Administrator for Space and Terrestrial Applications
          R-1/Associate Administrator for Aeronautics and Space Technology
          S-1/Associate Administrator for Space Science
          L-1/Associate Administrator for External Relations

FROM: A-1/Administrator

SUBJECT: NASA/University Relations

We have completed our review of the role of academic institutions in the NASA R&D Program, and it is our intention to continue to have strong academic involvement in the NASA R&D Program.

NASA intends to enhance and strengthen the academic participation in its research program, particularly in those disciplines supporting our aeronautical and applications programs.

It is NASA policy to involve academic scientists primarily in basic research. NASA encourages a growing independent academic research program; in particular, academic scientists will be given the opportunity to help advance the frontiers of science and technology in all disciplines of interest to the Nation in aeronautics and space. Cooperative programs between academia and in-house NASA research groups are beneficial and will continue. NASA will encourage the use of facilities at NASA centers by university scientists.

NASA's policy shall be to encourage centers of excellence in universities and to cooperate with academic groups to strengthen them as required in research and education in aerospace science, engineering, and management.

[2] NASA's relations with the university community will be conducted in a manner that reflects concern and understanding for the role of universities in education and research; avoids undue imposition of burdensome requirements; and does not tax an institution's financial resources.

Enclosed is a draft of policy guidelines for university relationships which will be converted into an appropriate policy statement by the Office of External Relations.

Each Associate Administrator, working with the appropriate Center Directors, shall prepare and submit by July 1, 1978, an action plan for my approval for accomplishing the goals of this policy. The action plan should define the current program with academic institutions, the plans to increase emphasis on independent research, and the management approach designed to place these policies into action. After acceptance of these action plans, the Associate Administrators and the Center Directors will be held accountable for the conduct of all academic activities under their control and, in particular, for strengthening academic programs in basic creative, and independent research in the area of applications, aerospace science and engineering.

The Associate Administrator for External Relations will be responsible for necessary coordination activities among program offices and should be kept appropriately informed.

It is my firm belief that judicious application of these policies will result in a combined stronger in-house and academic research establishment, and a stronger and more creative NASA research program in the decades ahead.

Robert A. Frosch

[Attachment 1]

POLICY FOR ACADEMIC INVOLVEMENT
IN THE NASA R&D PROGRAM

- Academic scientists will conduct a substantial portion of the basic research in all disciplines in the NASA program.
- Academic scientists will participate directly, or through advisory groups, in all phases of the basic research activity: conception, planning, programming, execution, analysis and interpretation of the data, and publication of the results.
- Academic basic research groups will be encouraged to show independence and creativity in their work which will be subject to periodic peer evaluation.
- Basic research opportunities using specified NASA spacecraft and/or specified instruments aboard a NASA spacecraft will be available to academic scientists on the basis of open competition, evaluation of their proposal by their peers and selection by the appropriate Associate Administrator.
- NASA's research facilities will be available for basic research by academic scientists. The appropriate Associate Administrator and Center Director will assure access of suitable facilities, broad notification, and proper selection of academic research projects.

- Cooperation in basic research between academic research groups and NASA in-house groups will be encouraged.
- Continuing research programs will be subject to peer evaluation at least once every three years involving reviews by a group of academic and in-house scientists with recognized research competence in the discipline.
- NASA's relations with the university community will be conducted in a manner that reflects concern and understanding for the role of universities in education and research; avoids undue imposition of burdensome requirements; and does not tax a university's financial resources.

Document III-25

Document title: NASA/University Relations Study Group, "The Universities and NASA Space Sciences," Initial Report of the NASA/University Relations Study Group, July 1983.

Source: University Affairs Files, NASA Historical Reference Collection, NASA History Office, NASA Headquarters, Washington, D.C.

By the early 1980s, the entire NASA-university space science relationship was still experiencing difficulties. In March and April of 1983, a special group of NASA and university representatives met to discuss the problems in the relationship and to discuss possible short- and long-term policy solutions. The study group was co-chaired by Thomas Donahue of the University of Michigan and Frank McDonald of NASA Headquarters. Their initial report, reprinted here, contained a series of recommendations on ways to put the relationship back on a productive footing.

The Universities and NASA Space Sciences

[1] **I.   Introduction**

From the beginning of the space program, university scientists have played a vital role in all phases of NASA's basic space research activity. It continues to be NASA's policy that a substantial portion of the basic research in space science should be conducted by university groups. The contributions from these groups have been an essential factor in the vitality of our nation's space program. Universities not only help generate new missions, design and build experiments, and interpret data, but most importantly, they are the essential conduit in transferring new knowledge and technology to other elements of society through the education and training of students.

A.   NASA/University Relations: The NASA space science program has evolved over 25 years from one with a high frequency of exploratory missions, to one based primarily on long-lived observatories and planetary orbiters. During the 60's, NASA's space science program involved an average of 4 or 5 flight missions per year. The scientific investigations for most of these missions were selected by a competitive process with the university groups historically supplying some 60% of the experiments. NASA further encouraged university participation through continuing multi-disciplinary research grants to more than 40 universities and through the construction of 37 space science buildings or additions. To increase the number of research workers, there was a nationwide program of NASA fellowships for graduate students. By the late 60's, more than 5,300 students had received 3 year graduate fellowship awards. The establishment of the Space Science Board, under the auspices of the National Academy of Sciences/National Research Council, provided university research scientists with a major role in advising NASA on science goals and policy for the U.S. space program.

By the early 80's the NASA launch rate of scientific satellites had declined to 1 or 2 per year. The graduate fellowship program had been dropped and the sustaining university grants were terminated. The Office of Space Science and Applications' [OSSA] budget, when measured in 1982 dollars, has decreased from a peak of 1.63 billion dollars in 1964 to .95 billion in 1984. Despite this decrease, the NASA science and applications budget remains one of the major funding sources for basic research in the United States. There have also been programmatic changes with a natural evolution toward larger and more complex missions as the exploratory phase of space studies has been completed. These new missions are taking the form of long-lived observatories such as Space Telescope and the Gamma Ray Observatory and planetary orbiters such as Galileo and the Venus Radar Mapper. A similar evolution has taken place with Explorers and the very exciting but technologically challenging missions such as IRAS [Infrared Astronomical Satellite] and COBE [Cosmic Background Explorer]. This sharp decrease in flight opportunities, accompanied by significant decreases in supporting research and data analysis funding have had the most deleterious effect, forcing drastic reductions in many university space research groups. Furthermore, contrary to original expectations, frequent opportunities to carry out scientific investigations on the Space Shuttle have not yet developed.

[2] **B. The Role of the University:** The elements of space sciences are a part of broader scientific disciplines, such as astronomy and astrophysics, earth and planetary sciences, and solar and space plasma physics. In their complete form, these provide both the rationale for the NASA programs and a framework for interpreting, incorporating, and communicating the results of those programs. It is through the continuing development and evolution of this disciplinary framework and the education of new scientists and engineers, that the universities play their unique and essential role in NASA's space program.

There are many facets to the universities' role in the space sciences which result naturally from its place in this broader spectrum of science and engineering research. In the universities, the space sciences maintain contact with related disciplines, benefiting from and contributing to the cross-fertilization of creative activity that stimulates innovation. Contact between disciplines in the universities also leads to the development of new areas of research. For example, high energy astrophysics emerged from such contact between physics and astronomy. Similarly, space plasma physics grew out of physics, geophysics, and solar physics. Within the universities there is the flexibility to respond to the opportunities offered by new developments in related fields of study, and there are young, innovative students anxious to develop and exploit new approaches to scientific endeavors.

Perhaps the most obvious role of the universities is in the education and training of graduate students. Students are an integral part of university research programs which are directed toward the increase of fundamental knowledge in the various scientific and engineering disciplines. The infusion of new talent, ideas, and innovation through the education of young people in the relevant disciplines is essential in maintaining the long-term vitality of space sciences. Equally important is the transfer of knowledge and technology that occurs when students trained in these disciplines move to industry and the national laboratories, taking with them research skills and familiarity with advanced technology characteristic of the space sciences.

The universities educate more than just space scientists. As the results from the space sciences are distilled and incorporated into coherent bodies of knowledge, they become part of the general education of all students and are eventually woven into the fabric of society.

C. The Requirements of University Space Science: The ways in which universities participate in space science can be broadly characterized as the formulation of new concepts and ideas, the development of new observational and experimental techniques, data analysis and interpretation, laboratory studies and theory. The relative importance of these various modes tends to differ among disciplines and to change with time. Organizationally, university participation has taken many different forms, from the creation of large research centers on some campuses to the involvement of small investigator groups at other institutions. The vitality of NASA's space science program is dependent on establishing an adequate research base at universities as well as at the NASA centers. It requires adequate research and analysis funding, a proper level of support for mission operation, data analysis and theoretical research, and continuing opportunity for participation in flight experiments.

[3] *Continuity of support is a very key factor in sustaining the vitality of university research groups.* To be effective, a typical university research activity must include professional faculty, key senior research faculty, postdoctoral fellows, and graduate students. Many activities also require a small core staff of engineers, programmers, and technicians in order to carry out the technical and managerial tasks characteristic of space science programs. Continuity is also important in the many cases where the innovations of subsequent programs often arise from the experience gained in previous programs.

Continuity of support for graduate students and postdoctoral fellows is also essential. The key element of graduate education is learning to be a researcher, a process that takes on average 6 years in space sciences, culminating in a doctoral dissertation. Undertaking such a lengthy educational process is feasible only if there is dependable continuity of support not only for the student, but for the university group's research program.

As a postdoctoral fellow, for a period of 2 to 3 years, the scientist continues to develop as a researcher, seeking to establish a solid research program and gaining recognition as an effective and independent scientist through the publication of research results. Since these objectives can be accomplished only with a sustained effort over several years, continuity of support is required during this important phase of an emerging scientist's research career.

Effective university programs also depend on the availability of modern instrumentation and computing facilities. Modern laboratory test equipment is critical not only in the development of new observational techniques, but also in training graduate students in the technology which is current in industrial and federal laboratories. Effective progress in space research depends on the existence of appropriately staffed and maintained major laboratory facilities, which must be periodically upgraded in order to address the scientific questions of greatest interest with the most modern techniques.

All of these attributes, including the provision of adequate flight opportunities, continuity of support, and the need for modern instrumentation and computing facilities, are necessary if the space sciences are to be sufficiently challenging to retain the interest of senior researchers, to offer realistic career opportunities to the most innovative younger researchers, and to attract capable, motivated graduate students.

D. NASA/University Study: Both NASA and the university scientific community have recognized for some time that a significant and undesirable erosion has occurred in the funding level of many university space research groups. (See Appendix 2 for a brief discussion of long-term funding trends in both NASA and in NASA funding to educational institutions.) After consultation with the Space Science Board, NASA felt that the best approach to defining the problem was to conduct a study with broad representation from NASA and the university community. The terms of reference and list of participants are included as Appendix 1. This group had meetings in March and April 1983. The strategy developed was to first explore short-term problems and issues and then spend the next year examining longer-term policy considerations and changes that might be made to reaffirm and/or redefine the NASA-university space science program.

[4] A letter describing the study and requesting comments from the community was sent to some 120 scientists. Thirty percent sent written responses. A representative sample of these letters is included in *Appendix 3*. Members of the group also had lengthy discussions with many university and NASA scientists. The Headquarters science discipline chiefs were also consulted in a series of meetings on the principal needs of their research areas.

In summary, the most important areas identified by the community and the discipline chiefs were:

1.   Increase the availability of low to moderate cost flight opportunities on the Space Shuttle, Explorers and the sub-orbital programs.
2.   Improve and modernize the university space science laboratories.
3.   Provide additional data analysis funding.
4.   Examine the adequacy of the research and technology base for the space science and applications program.

Three of these areas were identified where immediate steps could be taken which would have a positive impact on the health of university research groups. These were:

1.   University Equipment Grants to provide standard laboratory equipment, as well as larger facility instruments, to university groups actively engaged in NASA research.
2.   Graduate Student Fellowships to provide financial support to graduate students working on NASA related programs.
3.   Increased Funding for Data Analysis which many programs including Voyager, IUE [International Ultraviolet Explorer], Landsat and others, could greatly benefit from.

The specific recommendations and their rationale are given in the next three sections. Looking beyond these immediate steps the following longer range studies are planned for the coming year:

–   The Space Shuttle offers great promise for creating new experiment opportunities on a timely basis which can be exploited at a reasonable cost. A group will be established under the auspices of the NASA-university relations study   group to examine how these objectives can be met.
–   It is recommended that the NASA Space and Earth Science Advisory Committee examine the research and analysis program to ensure that the various discipline areas are being properly supported both in the development of new detectors, advanced analysis systems, and theoretical research.
–   The NASA-Center-university relations should be examined to consider means by which this partnership could be made more effective.

[5] **II. Laboratory Tools for the Space Sciences**
University scientists, with their students and staff, have made major contributions during the past quarter century in establishing the position of leadership in the space sciences and exploration that the United States enjoys. This achievement was made possible by the unique collaboration, or partnership, between the university community and NASA, that provided the resources, planning, and long-range objectives for our national space program. Central to this success was the recognition by NASA at the beginning of the Space Age (e.g., 1958-64) that university scientists needed the "tools" and equipment to design and develop innovative instrumentation for space flight, and to process and analyze the data returned from space missions. Thus, through the purchase of equipment and facilities with funds provided by NASA, and pooling of laboratory equipment existing in the university laboratories at that time, there came into existence well equipped facilities that generated a program of imaginative scientific research in space and permitted the training of a new generation of investigators, engineers and managers. However, during the past decade the equipment and special facilities acquired in the early 1960's and

70's became obsolete and increasingly difficult to maintain. Dramatic technological advances in space flight instrumentation made it very difficult, or impossible, to develop "state of the art" space flight instruments with the laboratory equipment of the 1960's and early 70's. As aging instruments have fallen into disrepair, it has become all but impossible to obtain replacement parts from industry. Consequently, technicians in the laboratories are preoccupied increasingly with repair of instruments, side-tracking them from more important tasks.

The development of experiments for future space missions requires full access to modern technology. All too often we are now training our next generation of scientists and engineers and designing new experiments with equipment from a past generation— instead of equipment that will keep them and our technology on the forefront of the engineering and experimental sciences. Indeed, many European and Japanese laboratories, with which we compete, are equipped with much more modern tools than those possessed by our own laboratories.

How did NASA and the universities fall so far behind in equipping university laboratories for space research and teaching of the next generation—after such an auspicious beginning? Two factors, both based on financial support, appear to be at the root of the problem:

1. NASA support for space experiments by university investigators is in the form of a contract which provides for the design, fabrication and testing of the instrument, followed by funds for data analysis. However, the contracts exclude funding for the acquisition of new capital equipment or facilities to carry out the commitment;

[6] 2. Since the principal support is through mission contracts, university scientists look to supporting Research and Technology (now Research and Analysis) grants for research and equipment funds. However, over the past decade the real level of support in these areas has steadily declined. Consequently, as the support declines, an ever-increasing fraction of the funds must be used to keep students and staff—with the result that funds for equipment become non-existent, especially for state-of-the-art equipment.

The time has come to take extraordinary steps to rectify this situation and again bring university laboratories into the same competitive position as laboratories in other countries, or even laboratories in our own NASA centers. Clearly, this will require an infusion of funds over a period of a few years devoted to this objective, but an early beginning is urgently needed.

In working out a program, we may define three general classes of equipment and facilities needed as follows:

1. Commercial laboratory equipment (e.g., oscilloscopes, test equipment, spectral analyzers, micro-processor development systems, etc.);

2. Small and medium sized computational equipment of the micro and minicomputer class (e.g., computers and peripheral hardware, tape and disc drives, terminals, couplers to national networks, commercial software, etc.), and interactive hardware which becomes especially important with the evolution of "observer class" space missions (e.g., Space Telescope, IUE, etc.);

3. Major facilities for use by several investigators collaborating at an institution (e.g., vibration and shock testing equipment), or for establishing interdisciplinary research (e.g., micro-ion probes, gas analyzers, etc.) which would be used by different groups of investigators within an institution. Another example would be large, fast computer facilities of the Cray class, which would be used by several investigators and jointly by investigators at several institutions. Major facilities of

this type require periodic technical support; and it is recognized that the necessary funding support for these facilities should be provided by NASA as part of this program.

Clearly, the NASA Discipline Chiefs within OSSA are closest to the needs of the investigators and their institutions, and are in the optimum position to make judgments on which institutions, investigators, and researchers would benefit the most from equipment funds. Therefore, we recommend that a line item be identified in each Discipline Chief's budget which would be available only for this purpose and would be funded through augmentation of current budgets by the appropriate amount.

[7] A preliminary survey by NASA staff indicates that the following annual budget levels for this purpose would be:

Astrophysics	$3.000M
Planetary Sciences	2.000M
Environmental Observations including Space Plasma Physics and Solar Terrestrial Theory	4.000M
Life Sciences	1.000M
Communications, etc.	1.000M
TOTAL Annual Commitment:	$11.000M

In view of the urgency, it is recommended that this program be instituted in the FY 1985 budget and continued at this, or higher level, for at least five years—with a somewhat lower level in future fiscal years.

How should the NASA Discipline Chiefs decide on which institutions to focus their attention? Suitable criteria might include:

    a.   The proven record of the investigators at institutions with regard to innovative instrumentation, discovery, and exploration in their disciplines;

    b.   The proven record of their training graduate students;

    c.   Evidence that the institution has demonstrated a commitment to the space sciences as an integral part of teaching, research, etc., in the departments of the institution;

    d.   New institutional support where a novel and important direction of research of interest to NASA has been identified.

A program of this type is essential for revitalizing—indeed retaining—those institutions and individuals and groups within institutions concerned with the space sciences, if they are to continue their vital role in space research and training for the 1980's and 1990's.

[9] **III. Graduate Research Fellowships**

The education and training of graduate students is one of the vital roles of the University. Training these students in space science is important to NASA and to the technology base of the country. They bring dedication and new insight to the ongoing research program and will design and build tomorrow's new generations of spacecraft, instrumentation, telescopes and rockets.

To attract the best students into the challenging areas of NASA activities and to reaffirm its commitment to graduate education, it is proposed that NASA re-establish a program of graduate research fellowships on a smaller and more focussed [sic] scale. Such a program would initially have 50 fellowships and would build to an annual steady state program level of 200 students.

The existence of such a fellowship program would constitute an announcement by NASA that the Agency is interested once again in seeing students of the highest quality involved in its programs and, we believe, would be a mechanism for attracting the best students. The competitive nature of the program we propose (as well as an attractive stipend) would help ensure that these fellowships would be regarded as prestigious awards. Such a fellowship program would permit awardees the freedom and stability to concentrate on their studies and research and allow them to progress through their graduate studies without being dependent on a particular NASA grant or flight program.

The fellowship program would be designed to attract students at two stages in their careers. The first of these is at the transition from undergraduate to graduate school when the student is selecting a field of study and a university department in which he wishes to pursue those studies. The objective of the fellowship program is to influence the best graduating seniors to select some field of space science. The second group of students is that which is at the stage of selecting and being accepted by a faculty research advisor with a view to choosing a thesis research topic. The purpose of the fellowship award, is to induce the best students already in a department or a university that has a space science program to do his thesis research in that program.

The preliminary prospectus for such a program is outlined in the following paragraphs:

Graduate Research Fellowships

Eligibility—The first class of students eligible consists of those entering graduate school who are accepted for study in a university department with a recognized program in some phase of space science. A list of such departments will be prepared by NASA. Students with outstanding undergraduate records and a[n] aptitude for success in some field of space science will be sought. Their continuing eligibility would be dependent on their selecting a space related thesis topic. The second class are students with proven ability in graduate study.

[10] **Duration**—In no case will a student be eligible to maintain a NASA graduate fellowship beyond the sixth year of graduate study. For students in category (a), the initial award shall be for three years. Renewal for a second three year period will be contingent on the student's admission to candidacy for the PhD degree and acceptance by a faculty research advisor for a thesis research project in space science. For students in category (b), the award should be for three years, subject to annual certification that the student is making normal progress toward a degree and is continuing to work in space science.

Stipend—The stipend should cover the full calendar year (not only the 9-month academic year) and be comparable to the best graduate research assistantships. The stipend should provide full tuition at whatever university the student attends (so as not to prevent students from attending private universities having higher tuition) and a living allowance of $13,000 per year that would be increased by $1,000 per year after each additional year after the first, up to a maximum of $16,000.

Application—Selection of candidates entering graduate school should be based on a one-page statement by the student describing their career goals, a transcript, G.R.E. results, and three letters of recommendation. For advanced students, the statement should describe the proposed research topic and one of the letters of recommendation must be from the proposed research advisor.

Selection—Applications should be submitted to NASA Headquarters and fellows should be selected by a board consisting of 3 members of the academic community and 3 NASA scientists, all appointed by the NASA Administrator. NASA discipline chiefs will be asked to review and grade the proposals in the appropriate disciplines.

Number—Approximately 50 new awards per year, of which at least 25 should be to entering graduate students, leading to a steady state number of about 250.

Publicity—We would urge wide publicity for the selected students including, if possible, articles in *Science News, Science,* etc.

[11] **IV. Data Analysis and Mission Operations**

It is the interpretation and analysis of the data from space missions that frequently offers the greatest intellectual challenge to researchers, post-docs and graduate students. The scientist takes the information from these remote laboratories, analyzes the data, and uses the results to extend our knowledge of the universe. This seeking of "new knowledge" is the primary reason for undertaking these new missions. However, as the experiments and spacecraft have become more complex and the costs of mission operations and data analysis have increased, the available funding has not always reflected this change.

There has also been a very positive development over the past ten years as the average lifetime of the NASA science missions has significantly increased. This enhanced longevity is due in large measure to the increase in space engineering experience and the development of a sophisticated technology base. Despite the decrease in launch rate, there are now some 14 active satellites returning valuable new data to a large number of space experimenters and guest investigators.

This increase in spacecraft lifetime frequently offers a very cost effective means of achieving new, high priority scientific objectives—objectives which were not part of the spacecraft's original intended mission. For example, ISEE-3 has been moved from the Lagrangian point 8,000,000 km in front of the Earth to a close lunar fly-by with repeated passes through the distant geomagnetic tail region. It will be the first detailed survey of this very dynamic portion of our magnetosphere. Later this year, ISEE-3 will be redirected toward the first encounter with a comet—Giacobini-Zinner in 1985. After completing its fly-by of Jupiter and Saturn, Voyager 2 has now been targeted for a Uranus encounter in 1986 and Neptune in 1989—thus making it possible to accomplish most of the objectives of the original "Grand Tour." Pioneer 10, now in its 12th year, is exploring the distant heliosphere beyond 30 AU and discovering a number of unexpected phenomena. ISEE I and II, Pioneer-Venus and IMP-8 [Interplanetary Monitoring Platform] are studying the changes that occur in the Earth's magnetosphere, the atmosphere of Venus and the interplanetary medium as the level of solar activity begins to significantly decrease. IUE has an almost unlimited number of classes of astronomical objects that are being observed for the first time in the ultra-violet portion of the electromagnetic spectrum. Nimbus-7 is in its fifth year of operation and its data set of earth observations is now long enough to permit study of long-term trends, interannual variations, and questions of climate. This fleet of active satellites is currently one of space science's greatest assets.

The Landsat programs produced vast quantities of high resolution imagery of the [E]arth's surface. As with the operational meteorological satellites, there is a significant need to analyze this data as part of scientific research into the functioning of the Earth. Extended analysis of this type of data must be established in the university community as an integral part of space science research or else this available wealth of information will not be adequately used for furthering the understanding of our own planet.

[12] The costs associated with the extended phases of these missions are generally modest, since most production and analysis programs have already been developed and only require updating. However, the funding levels for data analysis and mission operations have not been adequate to realize the full scientific return from this sustained and available flow of scientific data.

In fiscal year 1983, the data analysis and mission operation budget is about $155M or approximately 15% of the OSSA total budget. $14M of this was added by Congressional appropriations committees to ensure the continuation of such key programs as Pioneer-Venus and Pioneer 10. It is proposed that this be increased by $20M per year with most of this increased allocation going to University groups. This increase in data analysis funding will have an enormous impact on the vitality of the space science program.

A survey of the data analysis requirements was made by the NASA headquarter's [sic] discipline chiefs and the following augmentations were proposed:

Data Analysis Requirements

Solar System Exploration
    Inner Planets Data Analysis $4M/yr
        (Mariner 10, Lunar Orbiter, PVO)
    Outer Planets Data Analysis $6M/yr
        (Viking, Voyager, Pioneer 10/11)

Astrophysics
    Solar & Heliospheric Physics $1M/yr
        (SMM Guest Investigators, OSO-7, OSO-8)
    Astronomy & Relativity $1M/yr
        (OAO 2 & 3, Increase IUE)
    High Energy Astrophysics $2M/yr
        (HEAO-1, 2, 3, SAS-3)

Environmental Observations
    Space Plasma Physics $1M/yr
        (IMP-8, AE, ISEE-1, 2, 3)
    Climate Research $1M/yr
        (Nimbus 6 & 7, Sage 1)
    Upper Atmosphere Research $1M/yr
        (Nimbus 4, 6 & 7, Sage, SME)
    Global Weather, Tropical Air Quality $3M/yr
        (GOES, GMS, TIROS-N, NOAA-D, E, F)

[13] V. Future Studies

The most important area identified by the community was increasing the availability of low-cost flight opportunities via the Space Shuttle, Explorers, and the sub-orbital programs. This is a challenging task that requires more detailed study. The implementation of a more effective Explorer program should be pursued by the Space Science Board, NASA Headquarters and the NASA field centers. The current sub-orbital program is a good way for graduate students to conduct small but scientifically significant experiments which complement larger missions. The Space Shuttle offers great promise for creating new experimental opportunities. With the current flight plans, there should be frequent flight opportunities which can be exploited at a reasonable cost. The time scale from project approval to launch should be on the order of 18 months. It is not surprising that the great promise of the Shuttle for science has not been fully realized. The Shuttle itself has just reached operational status. Scientists and the manned program both need to learn how to use this new transportation system to greater advantage for science. The Study Group strongly urges that NASA establish a panel to study the Spacelab experience and make recommendations on new approaches. It would be highly desirable to have this task completed over the next 8-10 months.

The research and analysis funding provides the research base for the NASA science program both at the universities and at the NASA centers. Over the years there have been substantial changes in the NASA program. The Study Group recommends that a re-examination of the R&A [Research and Applications] program be made to ensure that the various discipline areas are being properly supported both in the development of new detectors, advanced analysis systems and theoretical research. It is recommended that this study be undertaken by the Space and Earth Science Advisory Committee.

The NASA centers in managing the space missions and sub-orbital programs play a crucial role in the space science program. The centers also maintain strong in-house research programs. It is important to re-evaluate the University-NASA Center relationship, both in the management of space missions and experiments as well as their relative roles in the conduct of space sciences. In the longer term activity of the Study Group, a com-mittee will be formed including representation from all the NASA centers strongly involved in space science and university representatives to consider ways in which the NASA-University partnership may be made more effective. Specific questions to be considered include the interchange of NASA and university personnel at several levels, methods of making center facilities, including large computer systems, more accessible to university users, and methods of providing center management and technical expertise to university principal investigators.

[1]

Appendix 1

Study of NASA-University Relations in Space Sciences
Terms of Reference

I. Motivation for Study

The agency recognizes that the benefits to the nation from a vital university space sci-ence program are large and diverse, and extend beyond the areas of scientific inquiry; that university-based space science research is a national resource which cannot be dupli-cated or obtained elsewhere; and that a healthy space science program at U.S. universities is essential to the agency space science program.

The nation's space science program has evolved so there is now greater emphasis on long-lived space observatories. These programmatic changes combined with a decline in the funding of the space science program has led to a marked decrease in new opportu-nities for flight experiments and to a decline in the viability of many long established research groups.

While there may have been early expectations that the university-based program could adjust itself at an appropriate level to support the agency space science program, there is now a growing body of evidence which strongly indicates that university relations and the resource represented in university space science will soon be insufficient to sup-port current levels of the space science program in the agency.

II. General Approach to the Study

The agency, after consultation with the Space Science Board and other outside groups, feels that the best approach to the problem is a study with broad representation from NASA and the university community. The information resources (statistics, manage-ment personnel, and contract network to the universities) are on hand within the agency.

However, the programmatic changes related to both the evolution of the flight program and the decrease in the number of flight activities depends on a combined NASA/University discussion. Of greater importance is the understanding of conditions in the university laboratories that can only be supplied by representatives from all levels of university-based space science (researchers, program managers, university administrators), and they should possess knowledge of the problems adequate at their respective levels to serve as representatives of their communities. The agency will assume responsibility for the management of the study and the study report.

[2] **III. Tasks for Study Group**

A study group consisting of approximately 12 people drawn from NASA and the university community will:

- Assess the health of university space science research groups and identify the problems.
- Examine, and redefine if necessary, the role of university groups in NASA future space science program.
- Identify the essential steps which must be taken in the agency and universities to restore university space science to a viable level.
- Present conclusions and recommendations to appropriate outside groups for comment and assessment and to the Space Science Board and to NASA management for review and action.

IV. Methodology

- Organizational meeting to pose the problem and scope of the study (2 1/2 days— early March 1983).
- Period of information gathering (statistics, funding trends, student trends, program trends) (April-June 1983).
  - Individual visits
  - Regional meetings/workshops
  - Study groups, representatives meeting with appropriate groups (NASA centers, university research groups, research administrators)
  - Collection of statistics
- Synthesis of data and information (mid-July 1983, one week).
- Determine follow-on as necessary.

[1]

Appendix 2

Funding Trends in NASA's Space Science and Applications Program

In this appendix, the long term funding trends in both the total NASA appropriation and the Office of Space Science and Application's [sic] (OSSA) portion of that budget are briefly examined. All of these budget numbers have been converted to 1982 dollars in order to compare the variation of equivalent real purchasing power. In Figure 1, the total NASA funding is shown for the 1960-1984 period. The corresponding OSSA numbers are also shown, but have been multiplied by a factor of 10 to emphasize the relative variation of the OSSA budget to the total NASA appropriation. The OSSA data has been compiled by the Administration and Resources Management Division of NASA's Office of Space Science and Applications. They have taken into account the reorganization and changing program office responsibilities that have occurred during this time.

Over a period of 3-5 years, there can be large variations in the annual OSSA funding level. . . . The most disturbing trend is the decrease from 1.55 billion dollars in 1973 to 0.9 billion in 1982. Most of this decrease occurred in the planetary program. . . . These changes reflect the launch of the Viking and Voyager spacecraft and the stretch-out of the Galileo program. However, there are reductions in other areas that also have a large impact on the science program. There has been a steady decrease in the annual appropriations for research and analysis in the space science area . . . , while the level in space applications has remained relatively constant. As emphasized in the main body of the text, this support is of great importance in maintaining the vitality of research groups. . . . [2]. . . [3] NASA total funding to universities has been almost exactly 3% of the total appropriations from 1973 to 1984 (Table 1). However, in 1982 dollars, there is a decrease from 261.3M in 1973 to 177.6 in 1983. It is this very large decrease in the real funding level that is the key problem in maintaining NASA research programs at the universities.

The marked decrease in the number of flight opportunities, (see Table 2), has been an additional factor that has greatly affected space science research groups. This change is more complex, since it represents both a decrease in the available funding, as well as a move to larger, more expensive missions.

Table 1
Total NASA Funding to Educational Institutions

Year	Total NASA R&D Obligations 1982 $(millions)	Real Year $(millions)	in Constant FY 1982 Dollars	% of Univ. Funding to Total NASA Funding
1973	7,710	114.9	261.3	3.0%
1974	6,420	100.9	214.0	3.0%
1975	6,160	112.4	215.1	3.0%
1976	6,210	122.5	215.1	3.0%
1977	6,030	124.9	198.0	3.0%
1978	5,950	135.3	199.0	3.0%
1979	6,100	147.8	198.5	2.8%
1980	6,330	177.3	215.1	3.0%
1981	6,010	191.1	209.1	3.0%
1982	6,020	185.6	185.6	3.0%
1983	6,210	197.2	177.6	2.8%

[4]

Table 2
NASA Space Science and Applications Launches

Year	# of Launches	5 Year Average/Year
1958	1	
1959	4	—
1960	2	
1961	3	
1962	4	3.8
1963	3	
1964	7	—
1965	7	
1966	5	
1967	8	6.2
1968	5	
1969	6	—
1970	2	
1971	4	
1972	4	4.2
1973	6	
1974	5	—
1975	7	
1976	1	
1977	5	4.0
1978	6	
1979	1	—
1980	1	
1981	2	
1982	0	1.0
1983	1	
1984	1	—

Document III-26

Document title: Section 201 of Title II of Public Law 100–147, "National Space Grant College and Fellowship Program," October 30, 1987.

Source: NASA Historical Reference Collection, NASA History Office, NASA Headquarters, Washington, D.C.

During the administration of Ronald Reagan, 1981 to 1989, Congress passed the "National Space Grant and Fellowship Act" as a means of making funding available to institutions of higher learning for the revitalization of the scientific and engineering disciplines. The act was deliberately modeled on the Morrill Land Grant College Act of the 1860s, which provided land for public sale with the proceeds going to public universities. The 1987 act created "space grant" universities and consortia eligible for public funds to foster aerospace research and development and education.

[no pagination]

Title II—National Space Grant College and Fellowship Program

Sec. 201. This title may be cited at the "National Space Grant College and Fellowship Act."

Sec. 202. The Congress finds that—

(1) the vitality of the Nation and the quality of life of the citizens of the Nation depend increasingly on the understanding, assessment, development, and utilization of space resources;

(2) research and development of space science, space technology, and space commercialization will contribute to the quality of life, national security, and the enhancement of commerce;

(3) the understanding and development of the space frontiers require a broad commitment and an intense involvement on the part of the Federal Government in partnership with State and local governments, private industry, universities, organizations, and individuals concerned with the exploration and utilization of space;

(4) the National Aeronautics and Space Administration, through the national space grant college and fellowship program, offers the most suitable means for such commitment and involvement through the promotion of activities that will result in greater understanding, assessment, development, and utilization; and

(5) Federal support of the establishment, development, and operation of programs and projects by space grant colleges, space grant regional consortia, institutions of higher education, institutes, laboratories, and other appropriate public and private entities is the most cost-effective way to promote such activities.

Sec. 203. The purposes of this title are to—

(1) increase the understanding, assessment, development, and utilization of space resources by promoting a strong educational base, responsive research and training activities, and broad and prompt dissemination of knowledge and techniques;

(2) utilize the abilities and talents of the universities of the Nation to support and contribute to the exploration and development of the resources and opportunities afforded by the space environment;

(3) encourage and support the existence of interdisciplinary and multidisciplinary programs of space research within the university community of the Nation, to engage in integrated activities of training, research and public service, to have cooperative programs with industry, and to be coordinated with the overall program of the National Aeronautics and Space Administration;

(4) encourage and support the existence of consortia, made up of university and industry members, to advance the exploration and development of space resources in cases in which national objectives can be better fulfilled than through the programs of single universities;

(5) encourage and support Federal funding for graduate fellowships in fields related to space; and

(6) support activities in colleges and universities generally for the purpose of creating and operating a network of institutional programs that will enhance achievements resulting from efforts under this title.

Sec. 204. As used in this chapter, the term—

(1) "Administration" means the National Aeronautics and Space Administration;

(2) "Administrator" means the Administrator of the National Aeronautics and Space Administration;

(3) "aeronautical and space activities" has the meaning given to such term in section 2452(1) of this title;

(4) "field related to space" means any academic discipline or field of study (including the physical, natural, and biological sciences, and engineering, space technology, education, economics, sociology, communications, planning, law, international affairs, and public administration) which is concerned with or likely to improve the understanding, assessment, development, and utilization of space;

(5) "panel" means the space grant review panel established pursuant to section 2486h of this title;

(6) "person" means any individual, any public or private corporation, partnership, or other association or entity (including any space grant college, space grant regional consortium, institution of higher education, institute, or laboratory), or any State, political subdivision of a State, or agency or officer of a State or political subdivision of a State;

(7) "space environment" means the environment beyond the sensible atmosphere of the Earth;

(8) "space grant college" means any public or private institution of higher education which is designated as such by the Administrator pursuant to section 2486f of this title;

(9) "space grant program" means any program which—

(A) is administered by any space grant college, space grant regional consortium, institution of higher education, institute, laboratory, or State or local agency; and

(B) includes two or more projects involving education and one or more of the following activities in the fields related to space—

(i) research,

(ii) training, or

(iii) advisory services;

(10) "space grant regional consortium" means any association or other alliance which is designated as such by the Administrator pursuant to section 2486f of this title;

(11) "space resource" means any tangible or intangible benefit which can only be realized from—

(A) aeronautical and space activities; or

(B) advancements in any field related to space; and

(12) "State" means any State of the United States, the District of Columbia, the Commonwealth of Puerto Rico, the Virgin Islands, Guam, American Samoa, the Commonwealth of the Northern Mariana Islands, or any other territory or possession of the United States.

Sec. 205. (a) The Administrator shall establish and maintain, within the Administration, a program to be known as the national space grant college and fellowship program. The national space grant college and fellowship program shall consist of the financial assistance and other activities provided for in this chapter. The Administrator shall establish long-range planning guidelines and priorities, and adequately evaluate the program.

(b) Within the Administration, the program shall—

(1) apply the long-range planning guidelines and the priorities established by the Administrator under subsection (a) of this section;

(2) advise the Administrator with respect to the expertise and capabilities which are available through the national space grant college and fellowship program, and make such expertise available to the Administration as directed by the Administrator;

(3) evaluate activities conducted under grants and contracts awarded pursuant to sections 2486d and 2486e of this title to assure that the purposes set forth in section 2486a of this title are implemented;

(4) encourage other Federal departments, agencies, and instrumentalities to use and take advantage of the expertise and capabilities which are available through the national space grant college and fellowship program, on a cooperative or other basis;

(5) encourage cooperation and coordination with other Federal programs concerned with the development of space resources and fields related to space;

(6) advise the Administrator on the designation of recipients supported by the national space grant college and fellowship program and, in appropriate cases, on the termination or suspension of any such designation; and

(7) encourage the formation and growth of space grant and fellowship programs.

(c) To carry out the provisions of this chapter, the Administrator may—

(1) accept conditional or unconditional gifts or donations of services, money, or property, real, personal or mixed, tangible or intangible;

(2) accept and use funds from other Federal departments, agencies, and instrumentalities to pay for fellowships, grants, contracts, and other transactions; and

(3) issue such rules and regulations as may be necessary and appropriate.

Sec. 206. (a) The Administrator may make grants and enter into contracts or other transactions under this subsection to assist any space grant and fellowship program or project if the Administrator finds that such program or project will carry out the purposes set forth in section 2486a of this title. The total amount paid pursuant to any such grant or contract may equal 66 percent, or any lesser percent, of the total cost of the space grant and fellowship program or project involved, except that this limitation shall not apply in the case of grants or contracts paid for with funds accepted by the Administrator pursuant to section 2486c(c)(2) of this title.

(b) The Administrator may make special grants under this subsection to carry out the purposes set forth in section 2486a of this title. The amount of any such grant may equal 100 percent, or any lesser percent, of the total cost of the project involved. No grant may be made under this subsection, unless the Administrator finds that—

(1) no reasonable means is available through which the applicant can meet the matching requirement for a grant under subsection (a) of this section;

(2) the probable benefit of such project outweighs the public interest in such matching requirement; and

(3) the same or equivalent benefit cannot be obtained through the award of a contract or grant under subsection (a) of this section or section 2486e of this title.

(c) Any person may apply to the Administrator for a grant or contract under this section. Application shall be made in such form and manner, and with such content and other submissions, as the Administrator shall by regulation prescribe.

(d) (1) Any grant made, or contract entered into, under this section shall be subject to the limitations and provisions set forth in paragraphs (2) and (3) of this subsection and to such other terms, conditions and requirements as the Administrator considers necessary or appropriate.

(2) No payment under any grant or contract under this section may be applied to—

(A) the purchase of any land;

(B) the purchase, construction, preservation, or repair of any building; or

(C) the purchase or construction of any launch facility or launch vehicle.

(3) Notwithstanding paragraph (2) of this subsection, the items in subparagraphs (A), (B), and (C) of such paragraph may be leased upon written approval of the Administrator.

(4) Any person who receives or utilizes any proceeds of any grant or contract under this section shall keep such records as the Administrator shall by regulation prescribe as being necessary and appropriate to facilitate effective audit and evaluation, including records which fully disclose the amount and disposition by such recipient of such proceeds, the total cost of the program or project in connection with which such proceeds were used, and the amount, if any, of such cost which was provided through other sources. Such records shall be maintained for three years after the completion of such a program or project. The Administrator and the Comptroller General of the United States, or any of their duly authorized representatives, shall have access, for the purpose of audit and evaluation, to any books, documents, papers and records of receipts which, in the opinion of the Administrator or the Comptroller General, may be related or pertinent to such grants and contracts.

Sec. 207. (a) The Administrator shall identify specific national needs and problems relating to space. The Administrator may make grants or enter into contracts under this section with respect to such needs or problems. The amount of any such grant or contract may equal 100 percent, or any lesser percent, of the total cost of the project involved.

(b) Any person may apply to the Administrator for a grant or contract under this section. In addition, the Administrator may invite applications with respect to specific national needs or problems identified under subsection (a) of this section. Application shall be made in such form and manner, and with such content and other submissions, as the Administrator shall by regulation prescribe. Any grant made, or contract entered into, under this section shall be subject to the limitations and provisions set forth in section 2486d(d)(2) and (4) of this title and to such other terms, conditions, and requirements as the Administrator considers necessary or appropriate.

Sec. 208. (a) (1) The Administrator may designate—

(A) any institution of higher education as a space grant college; and

(B) any association or other alliance of two or more persons, other than individuals, as a space grant regional consortium.

(2) No institution of higher education may be designated as a space grant college, unless the Administrator finds that such institution—

(A) is maintaining a balanced program of research, education, training, and advisory services in fields related to space;

(B) will act in accordance with such guidelines as are prescribed under subsection (b)(2) of this section; and

(C) meets such other qualifications as the Administrator considers necessary or appropriate.

(3) No association or other alliance of two or more persons may be designated as a space grant regional consortium, unless the Administrator finds that such association or alliance—

(A) is established for the purpose of sharing expertise, research, educational facilities or training facilities, and other capabilities in order to facilitate research, education, training, and advisory services, in any field related to space;

(B) will encourage and follow a regional approach to solving problems or meeting needs relating to space, in cooperation with appropriate space grant colleges, space grant programs, and other persons in the region;

(C) will act in accordance with such guidelines as are prescribed under subsection (b)(2) of this section; and

(D) meets such other qualifications as the Administrator considers necessary or appropriate.

(b) The Administrator shall by regulation prescribe—

(1) the qualifications required to be met under subsection (a)(2)(C) and (3)(D) of this section; and

(2) guidelines relating to the activities and responsibilities of space grant colleges and space grant regional consortia.

(c) The Administrator may, for cause and after an opportunity for hearing, suspend or terminate any designation under subsection (a) of this section.

Sec. 209. (a) The Administrator shall support a space grant fellowship program to provide educational and training assistance to qualified individuals at the graduate level of education in fields related to space. Such fellowships shall be awarded pursuant to guidelines established by the Administrator. Space grant fellowships shall be awarded to individuals at space grant colleges, space grant regional consortia, other colleges and institutions of higher education, professional associations, and institutes in such a manner as to assure wide geographic and institutional diversity in the pursuit of research under the fellowship program.

(b) The total amount which may be provided for grants under the space grant fellowship program during any fiscal year shall not exceed an amount equal to 50 percent of the total funds appropriated for such year pursuant to this chapter.

(c) Nothing in this section shall be construed to prohibit the Administrator from sponsoring any research fellowship program, including any special emphasis program, which is established under an authority other than this chapter.

Sec. 210. (a) The Administrator shall establish an independent committee known as the space grant review panel, which shall not be subject to the provis[i]ons of the Federal Advisory Committee Act (5 U.S.C. App.; Public Law 92-463).

(b) The panel shall take such steps as may be necessary to review, and shall advise the Administrator with respect to—

(1) applications or proposals for, and performance under, grants and contracts awarded pursuant to sections 2486d and 2486e of this title;

(2) the space grant fellowship program;

(3) the designation and operation of space grant colleges and space grant regional consortia, and the operation of space grant and fellowship programs;

(4) the formulation and application of the planning guidelines and priorities pursuant to section 2486c(a) and (b)(1) of this title; and

(5) such other matters as the Administrator refers to the panel for review and advice.

(c) The Administrator shall make available to the panel any information, personnel and administrative services and assistance which is reasonable to carry out the duties of the panel.

(d) (1) The Administrator shall appoint the voting members of the panel. A majority of the voting members shall be individuals who, by reason of knowledge, experience, or training, are especially qualified in one or more of the disciplines and fields related to space. The other voting members shall be individuals who, by reason of knowledge, experience or training, are especially qualified in, or representative of, education, extension services, State government, industry, economics, planning, or any other activity related to efforts to enhance the understanding, assessment, development, or utilization of space resources. The Administrator shall consider the potential conflict of interest of any individual in making appointments to the panel.

(2) The Administrator shall select one voting member to serve as the Chairman and another voting member to serve as the Vice Chairman. The Vice Chairman shall act as Chairman in the absence or incapacity of the Chairman.

(3) Voting members of the panel who are not Federal employees shall be reimbursed for actual and reasonable expenses incurred in the performance of such duties.

(4) The panel shall meet on a biannual basis and, at any other time, at the call of the Chairman or upon the request of a majority of the voting members or of the Administrator.

(5) The panel may exercise such powers as are reasonably necessary in order to carry out the duties enumerated in subsection (b) of this section.

Sec. 211. Each department, agency or other instrumentality of the Federal Government which is engaged in or concerned with, or which has authority over, matters relating to space—

(1) may, upon a written request from the Administrator, make available, on a reimbursable basis or otherwise, any personnel (with their consent and without prejudice to their position and rating), service, or facility which the Administrator considers necessary to carry out any provision of this chapter;

(2) may, upon a written request from the Administrator, furnish any available data or other information which the Administrator considers necessary to carry out any provision of this chapter; and

(3) may cooperate with the Administration.

Sec. 212. (a) The Administrator shall submit to the Congress and the President, not later than January 1, 1989, and not later than February 15 of every odd-numbered year thereafter, a report on the activities of the national space grant and fellowship program.

(b) The Director of the Office of Management and Budget and the Director of the Office of Science and Technology Policy in the Executive Office of the President shall have the opportunity to review each report prepared pursuant to subsection (a) of this section. Such Directors may submit, for inclusion in such report, comments and recommendations and an independent evaluation of the national space grant college and fellowship program. Such comments and recommendations shall be submitted to the Administrator not later than 90 days before such a report is submitted pursuant to subsection (a) of this section and the Administrator shall include such comments and recommendations as a separate section in such report.

Sec. 213. The Administrator shall not under this chapter designate any space grant college or space grant regional consortium or award any fellowship, grant, or contract unless such designation or award is made in accordance with the competitive, merit-based review process employed by the Administration on the date of enactment of this Act.

Sec. 214. (a) There are authorized to be appropriated for the purposes of carrying out the provisions of this chapter sums not to exceed—

(1) $10,000,000 for each of fiscal years 1988 and 1989; and

(2) $15,000,000 for each of fiscal years 1990 and 1991.

(b) Such sums as may be appropriated under this section shall remain available until expended.

Document III-27

Document title: NASA Commercial Space Policy, October 1984.

Source: NASA Historical Reference Collection, NASA History Office, NASA Headquarters, Washington, D.C.

The Reagan administration placed a high value on privatizing many government services and activities. This perspective extended to space and was coupled with the optimistic viewpoint that space commerce held the potential of growing into a multibillion dollar annual enterprise. In 1984, the administration released a National Commercial Space Policy, which outlined its views in this area. NASA responded soon after with its own commercial space policy, which attempted to redefine the agency's role, including being a partner with U.S. industry in exploring various areas of space commercialization.

[ii]

NASA Commercial Space Policy

Encouraging Private Enterprise in Space

The purpose of this Policy is to prepare NASA for expanding its mission in a new direction—the fostering of commercial enterprises in space.

This Policy, and accompanying Implementation Plan were drawn up by representatives from NASA headquarters and field centers. These representatives looked at the commercial possibilities in space and how NASA can encourage more private industrial ventures in orbit.

To supplement their perspective, the NASA representatives sought and received advice from experts in industry and universities as well as other outside specialists.

[iii]   *The United States Government will provide a climate conducive to expanded private sector investment and involvement in civil space activities. . . .*

President Ronald Reagan
in his National Space Policy, July 4, 1982

[iv]   *We should establish a policy which would encourage commercialization of space technology to the maximum extent feasible.*

Committee on Science and Technology,
U.S. House of Representatives,
Report, April 15, 1983

* * * * *

The Committee is fully supportive of efforts by the private sector to invest and seek commercial opportunities in space.

Committee on Commerce, Science and
Transportation, U.S. Senate,
Report, May 15, 1983

* * * * *

The extent to which past investment in space technology contributes to our future economic well-being and national growth will depend in large measure on policies and actions taken in a spirit of collaboration by the Federal Government and industry.

Unless the public and private sector join to develop the opportunities presented by new space technologies and unless entrepreneurial forces are engaged more fully, the United States will fall behind in the contest for leadership in space and the economic rewards associated with that position.

May 1983 Report,
National Academy of Public Administration

[v]

Preamble

The new chapter in the U. S. space program that opened early in this decade with the first flights of the Shuttle is now reaching a new phase: space technology is ripe for its transition from exploration to major exploitation, from experimentation to expanded profitable commercial uses.

To examine the opportunities for and impediments to expanded commercial activities in space, NASA formed a Task Force in mid-1983.

The Task Force's conclusions are straightforward:

- Commercial activities in space by private enterprise should be expanded now if our nation is to retain and improve its leadership in science and technology, its high living standards, and its advantage in international trade.
- Natural and bureaucratic barriers inhibiting the commercialization of space need to be and can be lessened or removed through joint actions by the Government and private enterprises.
- With firm resolve and the commitment of reasonable resources over a number of years, Government and private enterprise working together can turn space into a realm of immense benefit for our nation.
- A positive NASA Commercial Space Policy should be implemented to expedite the expansion of self-sustaining, profit-earning, tax-paying, jobs-providing commercial space activities.

[vi] The *NASA Commercial Space Policy* supports commercial space activities through:

- Reducing the risks of doing business in space to levels competitive with conventional investments.
  - To reduce technical risks, NASA will conduct and stimulate additional research relevant to commercial developments in space.
  - To reduce financial risks, NASA will provide easy and inexpensive access to orbit as well as to experimental ground facilities.

- To reduce institutional risks, NASA will help remove procedural impediments, offer organizational support, and maintain consistent policies regarding its relationship with commercial space ventures.
- Reaching out and establishing new links with the private sector to stimulate the development of private businesses in space.
  - NASA will expand its traditional links with the aerospace industry and academia to also embrace other industries such as new high-technology entrepreneurial ventures and the financial and non-aerospace industrial and academic communities.
  - NASA will expand and target dissemination of scientific information to stimulate domestic space commerce projects.
  - NASA will use public announcements, interviews, speeches, press releases, and articles in technical and business journals to provide information about commercialization opportunities and its commercialization activities to industry, academia, and the American public. . . .

[1]

NASA Commercial Space Policy

Introduction

NASA's thrust into the future is taking a new turn: NASA is encouraging free enterprise to participate in space by inviting industries and other private entities to finance and conduct business in space.

Private investment in space is called "space commercialization." Commercial projects would aim at developing profitable products and services in space for sale to consumers on Earth and for other space activities.

The NASA Policy to stimulate the commercialization of space will give new impetus and importance to traditional space efforts. At the same time, the Policy will give private enterprises the extra push they need to get started with permanent self-sustaining, tax revenue-producing establishments that will generate unique goods, well-paying jobs, and new outlets for innovation and ingenuity in space and on the Earth. The rewards can be immense for our nation.

The Policy calls for new ways of thinking about space. It requires revision and innovation in the traditional approaches and outlook. It calls for new roles by and alterations in relationships between NASA and private enterprises.

NASA has accumulated a long and proud history of working closely and productively with private enterprises. NASA-space programs have been based on participation and contributions by a trio of segments in our society—Government, industry, and academic institutions.

Since its earliest days, NASA has employed industries and universities as contractors. Since 1962, NASA has provided launch services for privately-owned commercial communications satellites. Beginning in 1972, NASA has entered into "partnership" arrangements with private firms for the commercial use of space. Now, the nature and character of NASA's relationship with private enterprise is changing still more. To persuade private investors to become involved in new space endeavors, NASA must be responsive to the needs and wants of these investors.

NASA must assure these investors of reliable and dependable roundtrip transportation for their projects between Earth and orbit. NASA must also help assure the availability of suitable work places for industries in orbit.

NASA will need to expand its basic research—with the advice of these customers and partners—to make sure it is relevant and helpful to private space ventures.

NASA must share its experience and know-how, including research information and NASA patents. NASA will need to establish space commercialization offices [2] at its headquarters and field centers to coordinate the new relationships with private investors. There may be opportunity for specialized companies to serve as, intermediaries—bridges—between NASA and private investors in space endeavors.

These and other approaches are outlined in the new NASA Commercial Space Policy on the following pages.

Space commercialization can have profound impact on the future of our Nation. We already know from our experiences with highly profitable, privately-owned communications satellites that free enterprise in space can work well. New leaps in technology which are likely to emerge from private initiatives in space could have major implications for the national economy, individual living standards and life styles, industrial activities and jobs and international trade.

The NASA Commercial Space Policy is designed to provide a foundation and framework for facilitating the realization of these opportunities.

[3]
NASA Commercial Space Policy

Executive Summary

Introduction
President Reagan, in his National Space Policy of July 4, 1982, made the expansion of private investment and involvement in space, a major objective of the United States Government. Committee reports from both Houses of Congress strongly endorsed this thrust in 1983. Supporting statements also have come from studies by non-government groups.

Opportunities for benefiting the nation are significant. Commercial space endeavors offer the potential for new industries, new jobs, lower product costs and an improved balance of trade. Technological advances from commercial use of space could help conquer diseases, produce computers faster and smarter than presently exist, develop metals lighter and stronger than any presently known, increase communications and information availability around the world and enhance our understanding of our environment and its resources.

NASA's Commercial Space Policy is designed to encourage private involvement in commercial endeavors in space to help take advantage of these opportunities. The Policy introduces approaches and incentives to reduce the risks inherent in commercial space ventures to levels competitive with conventional investments.

This "Executive Summary" presents an overview of the goals and principles of the NASA Commercial Space Policy, as well as a summary of major new initiatives NASA will implement to stimulate private investment in commercial space ventures.

Goals and Principles
The primary goal of NASA's Commercial Space Policy is to encourage and stimulate free enterprise in space.

Private investments in space, in turn, are expected to (a) yield important economic advantages; (b) advance science and technology; (c) help maintain in U.S. space leadership; and (d) enhance the nation's competitive position in international trade, thereby improving the in[-]U.S. balance of payments.

Implementation of the NASA Commercial Space Policy is to be guided by these five principles:

1. *The Government should reach out to and establish new links with the private sector.*

NASA will broaden its traditional links with the aerospace industry and the science community to include relationships with major non-aerospace [4] firms, new entrepreneurial ventures, as well as the financial and academic communities.

2. *Regardless of the Government's view of a project's feasibility, it should not impede private efforts to undertake commercial space ventures.*

If the private sector is willing to make the necessary investment, the project's feasibility should allowed to be determined by the marketplace and the creativity of the entrepreneur rather than the Government's opinion of its viability.

3. *If the private sector can operate a space venture more efficiently than the Government, then such commercialization should be encouraged.*

When developing new public space programs, the Government should actively consider the view of, and the potential effect on, private ventures.

4. *The Government should invest in high-leverage research, and space facilities which encourage private investment. However, the Government should not expend tax dollars for endeavors the private sector is willing to underwrite.*

This will provide at least two benefits. First it will enable NASA to concentrate a greater percentage of its resources on advancing the technological state-of-the-art in areas where the investment is too great for the private sector. Second, it will engage the private sector's applications and marketing skills for getting space benefits to the people.

5. *When a significant Government contribution to a commercial endeavor is requested, two requirements must be met. First, the private sector must have significant capital at risk, and second, there must be significant potential benefits for the nation.*

In appraising the potential benefits from and determining appropriate Government contributions to commercial space proposals, NASA will use an equitable, consistent review process.

A possible exception to these principles would be a commercial venture intended to replace a service or displace a NASA R&D program and/or technology development program of paramount public importance now provided by the Government. In that case, the Government might require additional prerequisites before commercialization.

Implementation

In implementing this Policy, NASA will take an active role in supporting commercial space ventures in the following categories, listed in order of importance:

• New commercial high-technology ventures.
• New commercial applications of existing space technology.
[5] • Commercial ventures resulting from the transfer of existing space programs to the private sector.

NASA will implement initiatives to reduce the technical, financial and institutional risks associated with doing business in space.

To reduce *technical risks*, NASA will:

Support research aimed at commercial applications; ease access to NASA experimental facilities; establish scheduled flight opportunities for commercial payloads; expand the availability of space technology information of commercial interest; and support the development of facilities necessary for commercial uses of space.

To reduce *financial risks*, NASA will:

Continue to offer reduced-rate space transportation for high-technology space endeavors; assist in integrating commercial equipment with the Shuttle; provide seed-funding to stimulate commercial space ventures; and, under certain circumstances, purchase commercial space products and services and offer some exclusivity.

To reduce *institutional risks,* NASA will:

Speed integration of commercial payloads into the Orbiter; shorten proposal evaluation time for NASA/private sector Joint Endeavor proposals; establish procedures to encourage development of space hardware and services with private capital instead of Government funds; and introduce new institutional approaches for strengthening NASA's support of private investment in space.

A high-level Commercial Space Office will be formed within NASA as a focal point for commercial space matters. This Office will be responsible for implementing the NASA Policy to stimulate space commerce. It will have sufficient authority and resources to fully carry out this assignment.

Document III-28

Document title: Office of the Press Secretary, "The President's Space Policy and Commercial Space Initiative to Begin the Next Century," February 11, 1988.

Source: Ronald Reagan Presidential Files, NASA Historical Reference Collection, NASA History Office, NASA Headquarters, Washington, D.C.

During the second Reagan administration, an alternative space policy making body concentrating on commercial spaceflight emerged to complement the National Security Council's Senior Interagency Group (Space), known as SIG (Space). Chaired by the Commerce Department, the Space Working Group of the White House Economic Policy Council worked on a new set of commercial space initiatives during 1987, at the same time that SIG (Space) was examining overall national space policy. SIG (Space) finished its review first, and its directive on national space policy was approved by President Reagan on January 5, 1988 (published in 1995 as Document III-42 in Volume I of Exploring the Unknown: Selected Documents in the History of the U.S. Civil Space Program). *However, its release was delayed until the space commerce review was completed. Both reviews were made public on February 11, 1988.*

---

[1]

The President's Space Policy and Commercial Space Initiative to Begin the Next Century

Fact Sheet

The President today announced a comprehensive "Space Policy and Commercial Space Initiative to Begin the Next Century" intended to ensure United States space leadership.

The President's program has three major components:

- Establishing a long-range goal to expand human presence and activity beyond Earth orbit into the Solar System;
- Creating opportunities for U.S. commerce in space; and
- Continuing our national commitment to a permanently manned Space Station.

The new policy and programs are contained in a National Security Decision Directive (NSDD) signed by the President January 5, 1988, the FY 1989 Budget the President will submit shortly to Congress, and a fifteen point Commercial Space Initiative.

I.   Expanding Human Presence Beyond Earth Orbit

In the recent NSDD, the President committed to a goal of expanding human presence and activity in the Solar System. To lay the foundation for this goal, the President will be requesting $100 million in his FY 1989 Budget for a major new technology development program "Project Pathfinder" that will enable a broad range of manned or unmanned missions beyond the Earth's orbit.

Project Pathfinder will be organized around four major focuses:
- Exploration technology;
- Operations technology;
- Humans-in-space technology; and
- Transfer vehicle technology.

This research effort will give the United States know-how in critical areas, such as human in space environment, closed loop life support, aero braking, orbital transfer and maneuvering, cryogenic storage and handling, and large scale space operations, and provide a base for wise decisions on long term goals and missions.

Additional highlight[s] of the NSDD are outlined in Section IV of this fact sheet.

[2] II.   Creating Opportunities for U.S. Commerce in Space

The President is announcing a fifteen point commercial space initiative to seize the opportunities for a vigorous U.S. commercial presence in Earth orbit and beyond—in research and manufacturing. This initiative has three goals:
- Promoting a strong U.S. commercial presence in space;
- Assuring a highway to space; and
- Building a solid technology and talent base.

Promoting a Strong U.S. Commercial Presence in Space

1. *Private Sector Space Facility:* The President is announcing an intent for the Federal Government to lease space as an "anchor Tenant" in an orbiting space facility satiable for research and commercial manufacturing that is financed, constructed, and operated by the private sector. The Administration will solicit proposals from the U.S. private sector for such a facility. Space in this facility will be used and/or subleased by various Federal agencies with interest in microgravity research.

   The Administration's intent is to award a contract during mid-summer of this year for such space and related services to be available to the Government no later than the end of FY 1993.

2. *Spacehab:* The Administration is committing to make best efforts to launch within the Shuttle payload bay, in the early 1990s, the commercially developed, owned and managed Shuttle middeck module: Spacehab. Manifesting requirements will depend on customer demand.

   Spacehab is a pressurized metal cylinder that fits in the Shuttle payload by and connect to the crew compartment through the orbiter airlock. Spacehab takes up approximately one-quarter of the payload bay and increases the pressurized living and working space of the orbiter by approximately 1,000 cubic feet or 400 percent in usable research volume. The facility is intended to be ready for commercial use in mid-1991.

3. *Microgravity Research Board:* The President will establish, through Executive Order, a National Microgravity Research Board to assure and coordinate a broader range of opportunities for research in microgravity conditions.

NASA will chair this board, which will include senior-level representatives from Departments of Commerce, Transportation, Energy, and Defense, NIH [National Institutes of Health], and NSF [National Science Foundation]; and will consult with the university and commercial sectors. The board will have the following responsibilities:

- To stimulate research in microgravity environments and its applications to commercial uses by advising Federal agencies, including NASA, on microgravity priorities, and consulting with private industry and academia on microgravity research opportunities;
- To develop policy recommendations to the Federal Government on matters relating to microgravity research, including tapes of research, government/industry/and academic cooperation, and access to space, including a potential launch voucher program;

[3] • To coordinate the microgravity programs of Federal agencies by:
- reviewing agency plans for microgravity research and recommending priorities for the use of Federally-owned or leased space on microgravity facilities; and
- ensuring that agencies established merit review processes for evaluating microgravity research proposals; and
- To promote transfer of Federally funded microgravity research to the commercial sector in furtherance of Executive Order 12591.

NASA will continue to be responsible for making adjustments on the safety of experiments and for making manifesting decisions for manned space flight systems.

4. *External Tanks:* The Administration is making available for five years the expended external tanks of the Shuttle fleet at no cost to all feasible U.S. commercial nonprofit endeavors, for use such as research, storage, or manufacturing in space.

NASA will provide any necessary technical other assistance to these endeavors on a direct cost basis. If private sector demand exceeds supply, NASA may auction the external tanks.

5. *Privatizing Space Station:* NASA, in coordination with the Office of Management and Budget, will revise its guide lines [sic] on commercialization of the U.S. Space Station to clarify and strengthen the Federal commitment to private sector investment in this program.

6. *Future Privatization:* NASA will seek to rely to the greatest extent feasible on private sector design, financing, construction, and operation of future Space Station requirements, including those currently under study.

7. *Remote Sensing:* The Administration is encouraging the development of commercial remote sensing systems. As part of this effort, the Department of Commerce, in consultation with other agencies, is examining potential opportunities for future Federal procurement of remote sensing data from the U.S. commercial sector.

Assuring a Highway to Space

8. *Reliance on Private Launch Services:* Federal agencies will be required to purchase expendable launch services directly from the private sector to the fullest extent feasible.

9. *Insurance Relief for Launch Providers:* The Administration will take administrative steps to address the insurance concerns of the U.S. commercial launch industry, which currently uses Federal launch ranges. These steps include:
- *Limits on Third Party Liability:* Consistent with the Administration's tort policy, the Administration will propose to Congress a $200,000 cap on noneconomic damage awards to individual third parties resulting from commercial launch accidents;

[4] • *Limits on Property Damage Liability:* The liability of commercial launch operators for damage to Government property resulting from a commercial launch accident will be administratively limited to the level of insurance required by the Department of Transportation.

If losses to the Government exceed this level, the Government will waive its right to recover for damages. If losses are less than this level, the Government will waive its right to recover for those damages caused by Government willful misconduct or reckless disregard.

10. *Private Launch Ranges:* The Administration will consult with the private sector on the potential construction of commercial launch range facilities separate from Federal facilities and the use of such facilities by the Federal Government.

11. *Vouchers for Research Payloads:* NASA and the Department of Transportation will explore providing to research payload owners manifested on the Shuttle a one time launch voucher that can be used to purchase an alternative U.S. commercial launch service.

Building a Solid Technology and Talent Base

12. *Space Technology Spin-Offs:* The President is directing that the new Pathfinder program, the Civil Space Technology Initiative [CSTI], and other technology programs be conducted in accordance with the following policies:
   • Federally funded contractors, universities, and Federal laboratories will retain the rights to any patents and technical data, including copyright, th[at] result from these programs. The Federal Government will have the authority to use this intellectual property royalty free;
   • Proposed technologies and patents available for licensing will be housed in a Pathfinder/CSTI library within NASA; and
   • When contracting for commercial development of Pathfinder, CSTI and other technology work products, NASA will specify its requirements in a manner that provides contractors with maximum flexibility to pursue innovative and creative approaches.

13. *Federal Expertise on Loan to American Schools:* The President is encouraging Federal Scientists, engineers, and technicians in aerospace and space related careers to take a sabbatical year to teach in any level of education in the United States.

14. *Education Opportunities:* The President is requesting in his FY 1989 Budget expanding five-fold opportunities for U.S. Teachers to visit NASA field centers and related aerospace and university facilities.

In addition, NASA, NSF, and DoD [Department of Defense] will contribute materials and classroom experiments through the Department of Education to U.S. schools developing "tech shop" programs. NASA will encourage corporate participation in this program.

15. *Protecting U.S. Critical Technologies:* The Administration is requesting that Congress extend to NASA the authority it has given the Department of Defense to protect the whole-sale release under the Freedom of Information Act those critical national technologies and systems that are prohibited from export.

[5] **III. Continuing the National Commitment to the Space Station**

In 1984, the President directed NASA to develop a permanently manned Space Station. The President remains committed to achieving this end and this requesting $1 billion in his FY 1989 Budget for continued development and a three year appropriation commitment from Congress for $6.1 billion. The Space Station, planned for development in cooperation with U.S. friends and allies, is intended to be a multi-purpose facility for the nation's science and applications programs. It will permit such things in space as: research, observation of the solar system, assembly of vehicles for facilities, storage, servicing of satellites, and basing for future space missions and commercial and entrepreneurial endeavors in space.

The help ensure a Space Station that is cost effective, the President is proposing as part of this Commercial Space Initiative actions to encourage private sector investment in the Space Station, including directing NASA to rely to the greatest extent feasible on private sector design, financing, construction, and operation of future Space Station requirements.

IV. Additional Highlights of the January 5, 1988 NSDD

- *U.S. Space Leadership:* Leadership is reiterated as a fundamental national objective in areas of space activity critical to achieving U.S. national security, scientific, economic and foreign policy goals.
- *Defining Federal Roles and Responsibilities:* Government activities are specified in three separate and distinct sectors: civil, national security, and nongovernmental. Agency roles and responsibilities are codified and specific goals are established for the civil space sector; those for other sectors are updated.
- *Encouraging a Commercial Sector:* A separate, nongovernmental or commercial space sector is recognized and encouraged by the policy that Federal Government actions shall not preclude or deter the continuing development of this sector. New Guidelines are established to limit unnecessary Government competition with the private sector and ensure that Federal agencies are reliable customers for commercial space goods and services.
- The President's launch policy prohibiting NASA from maintaining an expendable launch vehicle adjunct to the Shuttle, as well as limiting commercial and foreign payloads on the Shuttle to those that are Shuttle-unique or serve national security for foreign policy purpose, is reaffirmed. In addition, policies endorsing the purchase of commercial launch services by Federal agencies are further strengthened.
- *National Security Space Sector:* An assured compatibility for national security missions is clearly enunciated, and the survivability and endurance of critical national security space functions is [sic] stressed.
- *Assuring Access to Space:* Assured access to space is recognized as a key element of national space policy. U.S. space transportation systems that provide sufficient resiliency to allow continued operation, despite failures in any single system, are emphasized. The mix of space transportation vehicles will be defined to support mission needs in the most cost effective manner.
- *Remote Sensing:* Policies for Federal "remote sensing" or observation of the Earth are established to encourage the development of U.S. commercial systems competitive with or superior to foreign-operated civil or commercial systems.

Document III-29

Document title: Office of the Press Secretary, "Commercial Space Launch Policy," NSPD-2, September 5, 1990.

Source: NASA Historical Reference Collection, NASA History Office, NASA Headquarters, Washington, D.C.

During the administration of George Bush, 1989 to 1993, several commercial space policy documents emerged that affected the manner in which NASA conducted its relations. National Space Policy Directive-2 established a "Commercial Space Launch Policy" that reflected the administration's commitment to develop a thriving commercial space sector by establishing "the long-term goal of a free and fair [space launch] market in which the U.S. industry can compete" internationally.

[no pagination]

Statement by the Press Secretary

The President has approved a new National Space Policy Directive providing important guidance which will further encourage the growth of U.S. private sector space activities. This policy, developed by the Vice President and the National Space Council, is completely consistent with, and provided the policy framework for, the President's August 22, 1990, decision regarding participation by a U.S. firm in Australia's Cape York space launch project. The policy supplements the National Space Policy which the President approved on November 2, 1989.

The commercial space launch policy recognizes the many benefits which a commercial space launch industry provides to the United States. It balances launch industry needs with those of other industries and with important national security interests, and establishes the long-term goal of a free and fair market in which U.S. industry can compete. The policy specifies a coordinated set of actions for the next ten years aimed at achieving this goal.

Fact Sheet on Commercial Space Launch Policy

Policy Findings

A commercial space launch industry can provide many benefits to the U.S. including indirect benefits to U.S. national security. The long-term goal of the United States is a free and fair market in which U.S. industry can compete. To achieve this, a set of coordinated actions is needed for dealing with international competition in launch goods and services in a manner that is consistent with our nonproliferation and technology transfer objectives. These actions must address both the short-term (actions which will affect competitiveness over approximately the next ten years) and those which will have their principal effect in the longer term (i.e. after approximately the year 2000).

In the near term, this includes trade agreements and enforcement of those agreements to limit unfair competition. It also includes the continued use of U.S.-manufactured launch vehicles for launching U.S. Government satellites.

For the longer term, the United States should take actions to encourage technical improvements to reduce the cost and increase the reliability of U.S. space launch vehicles.

Implementing Actions

U.S. government satellites will be launched on U.S.-manufactured launch vehicles unless specifically exempted by the President.

Consistent with guidelines to be developed by the National Space Council, U.S. Government Agencies will actively consider commercial space launch needs and factor them into their decisions on improvements in launch infrastructure and launch vehicles aimed at reducing cost, and increasing responsiveness and reliability of space launch vehicles.

The U.S. Government will enter into negotiations to achieve agreement with the European Space Agency (ESA), ESA member states, and others as appropriate, which defines principles of free and fair trade.

Nonmarket launch providers of space launch goods and services create a special case because of the absence of market[-]oriented pricing and cost structures. To deal with their entry into the market there needs to be a transition period during which special conditions may be required.

There also must be an effective means of enforcing international agreements related to space launch goods and services.

Statement by the Press Secretary

The United States seeks a free and fair international commercial space launch market to further the use of outer space for the betterment of mankind. At the same time, because space launch technologies have significant military applications, important U.S. national security considerations must be addressed by our commercial space launch policy.

Over the past several weeks, the President has had detailed discussions with the Vice President and other senior advisors on U.S. commercial space launch policy developed by the National Space Council. The President has authorized the Secretary of State to approve a license application for participation by a U.S. firm in Australia's Cape York space launch project, provided certain agreements necessary to ensure U.S. national security interests are reached.

Specifically, the U.S. will seek agreements to ensure that:

(1) The USSR will provide launch services (boosters, equipment, technology, or training) only from Cape York or any other single location;

(2) The USSR and Australia will observe the Missile Technology Control Regime; and

(3) U.S. regulations on technology transfer to the Soviet Union will be observed.

The United States hopes and expects that these agreements can be concluded quickly so that the license can be granted.

To permit continued U.S. participation, the United States in the coming months will also be seeking agreements to ensure free and fair trade in the international commercial space launch market.

Details of the U.S. commercial space launch policy will be announced in the near future.

Document III-30

Document title: Executive Office of the President, "U.S. Commercial Space Policy Guidelines," NSPD-3, February 11, 1991.

Source: NASA Historical Reference Collection, NASA History Office, NASA Headquarters, Washington, D.C.

In 1991, the Bush administration refined its commercial space policy by issuing NSPD-3, which articulated in specific terms a commercial space policy "aimed at expanding private sector investment in space by the market-driven Commercial Space Sector." The intent was to move more of the onus for investment in space technology to the private sector, where it was assumed that market forces would drive down costs.

[1]
U.S. Commercial Space Policy Guidelines

A fundamental objective guiding United States space activities has been space leadership, which requires preeminence in key areas of space activity. In an increasingly competitive international environment, the U.S. Government encourages the commercial use and exploitation of space technologies and systems for national economic benefit. These efforts to encourage commercial activities must be consistent with national security and foreign policy interests, international and domestic legal obligations, including U.S. commitments to stem missile proliferation, and agency mission requirements.

United States space activities are conducted by three separate and distinct sectors: two U.S. Government sectors[—]the civil and national security[—]and a non-governmental commercial space sector. The commercial space sector includes a broad cross section of potential providers and users, including both established and new market participants. There also has been a recent emergence of State government initiatives related to encouraging commercial space activities. The commercial space sector is comprised of at least five market areas, each encompassing both earth and spacebased activities, with varying degrees of market maturity or potential:

[2] *Satellite Communications:* the private development, manufacture, and operation of communications satellites and marketing of satellite telecommunications services, including position location and navigation;

Launch and Vehicle Services: the private development, manufacture, and operation of launch and reentry vehicles, and the marketing of space transportation services;

Remote Sensing: the private development, manufacture, and operation of remote sensing satellites and the processing and marketing of remote sensing data;

Materials Processing: the experimentation with, and production of, organic and inorganic materials and products utilizing the space environment; and

Commercial Infrastructure: the private development and provision of space[-]related support facilities, capabilities and services.

In addition, other market-driven commercial space sector opportunities are emerging.

The U.S. Government encourages private investment in, and broader responsibility for, space-related activities that can result in products and services that meet the needs of government and other customers in a competitive market. As a matter of policy, the U.S. Government pursues its commercial space objectives without the use of direct federal subsidies. A robust commercial space sector has the potential to generate new technologies, products, markets, jobs, and other economic benefits for the nation, as well as indirect benefits for national security.

Commercial space sector activities are characterized by the provision of products and services such that:

- private capital is at risk;
- there are existing, or potential, nongovernmental customers for the activity;
- the commercial market ultimately determines the viability of the activity; and
- primary responsibility and management initiative for the activity resides with the private sector.

[3]
Implementing Guidelines

The following implementing guidelines shall serve to provide the U.S. private sector with a level of stability and predictability in its dealings with agencies of the U.S.

Government. The agencies will work separately but cooperatively, as appropriate, to develop specific measures to implement this strategy. U.S. Government agencies shall, consistent with national security and foreign policy interests, international and domestic legal obligation and agency mission requirements, encourage the growth of the U.S. commercial space sector in accordance with the following guidelines:

- U.S. Government agencies shall utilize commercially available space products and services to the fullest extent feasible. This policy of encouraging U.S. Government agencies to purchase, and the private sector to sell, commercial space products and services has potentially large economic benefits.
    - A space product or service is "commercially available" if it is currently offered commercially, or if it could be supplied commercially in response to a government procurement request.
    - "Feasible" means that products and services meet mission requirements in a cost-effective manner.
    - "Cost-effective" generally means that the commercial product or service costs no more than governmental development or directed procurement where such government costs include applicable government labor and overhead costs, as well as contractor charges and operations costs.
    - However, the acquisition of commercial space products and services shall generally be considered cost-effective if they are procured competitively using performance-based contracting techniques. Such contracting techniques give contractors the freedom and financial incentive to achieve economies of scale by combining their government and commercial work as well as increased productivity through innovation.
    - U.S. Government agencies shall actively consider, at the earliest appropriate time, the feasibility of their using commercially available products and services in agency programs and activities.
[4] - U.S. Government agencies shall continue to take appropriate measures to protect from disclosure any proprietary data which is shared with the U.S. Government in the acquisition of commercial space products and services.
- U.S. Government agencies shall promote the transfer of U.S. Government-developed technology to the private sector.
    - U.S. Government-developed unclassified space technology will be transferred to the U.S. commercial space sector in as timely a manner as possible and in ways that protect its commercial value.
    - U.S. Government agencies may undertake cooperative research and development activities with the private sector, as well as State and local governments, consistent with policies and funding, in order to fulfill mission requirements in a manner which encourages the creation of commercial opportunities.
    - With respect to technologies generated in the performance of government contracts, U.S. Government agencies shall obtain only those rights necessary to meet government needs and mission requirements, as directed by Executive Order 12591.
- U.S. Government agencies may make unused capacity of space assets, services and infrastructure available for commercial space sector use.
    - Private sector use of U.S. Government agency space assets, services, and infrastructure shall be made available on a reimbursable basis consistent with OMB [Office of Management and Budget] circular A25 or appropriate legislation.
- U.S. Government agencies may make available to the private sector those assets which have been determined to be excess to the requirements of the U.S. Government in accordance with U.S. law and applicable international treaty obligations. Due regard

shall be given to the economic impact such transfer may have on the commercial space sector, promoting competition, and the long-term public interest.

[5] •   The U.S. Government shall avoid regulating domestic space activities in a manner that precludes or deters commercial space sector activities, except to the extent necessary to meet international and domestic legal obligations, including those of the Missile Technology Control Regime. Accordingly, agencies shall identify, and propose for revision or elimination, applicable portions of U.S. laws and regulations that unnecessarily impede commercial space sector activities.

•   U.S. Government agencies shall work with the commercial space sector to promote the establishment of technical standards for commercial space products and services.

•   U.S. Government agencies shall enter into appropriate cooperative agreements to encourage and advance private sector basic research, development, and operations. Agencies may reduce initial private sector risk by agreeing to future use of privately supplied space products and services where appropriate.

    –   "Anchor tenancy" is an example of such an arrangement whereby U.S. Government agencies can provide initial support to a venture by contracting for enough of the future product or service to make the venture viable in the short term. Long[-]term viability and growth must come primarily from the sale of the product or service to customers outside the U.S. Government.

    –   There must be demonstrable U.S. Government mission or program requirements for the proposed commercial space good or service. In assessing the U.S. Government's mission or program requirements for these purposes, the procuring agency may consider consolidating all anticipated U.S. Government needs for the particular product or service, to the maximum extent feasible.

    –   U.S. Government agencies entering into such arrangements may take action, consistent with current policies and funding availability, to provide compensation to commercial space providers for future termination of missions for which the products or services were required.

[6] •   The United States will work toward establishment of an international trading environment that encourages market[-]oriented competition by working with its trading partners to:

    –   Establish clear principles for international space markets that provide an atmosphere favorable to stimulating greater private investment and market development;

    –   Eliminate direct government subsidies and other unfair practices that undermine normal market competition among commercial firms;

    –   Eliminate unfair competition by governments for business in space markets consistent with domestic policies that preclude or deter U.S. Government competition with commercial space sector activities.

The U.S. Commercial Space Policy Guidelines are consistent with the National Space Policy and the U.S. Commercial Space Launch Policy which remain fully applicable to activities of the governmental space sectors and the commercial space sector.

Reporting Requirements

U.S. Government agencies affected by these guidelines are directed to report by October 1, 1991, to the National Space Council on their activities related to the implementation of these policy guidelines.

George Bush

Biographical Appendix

Spiro T. Agnew (1918-1996) was elected vice president of the United States in November 1968, serving under Richard M. Nixon. He served as chair of the 1969 Space Task Group that developed a long-range plan for a post-Apollo space effort. *The Post-Apollo Space Program: Directions for the Future* (Washington, DC: President's Science Advisory Council, September 1969) developed an expansive program, including the building of a space station, a space shuttle, and a lunar base, as well as a mission to Mars (the last goal had been endorsed by the vice president at the time of the Apollo 11 launch in July 1969). President Nixon did not accept this plan, and only the Space Shuttle was approved for development. See Roger D. Launius, "NASA and the Decision to Build the Space Shuttle, 1969-72," *The Historian* 57 (Autumn 1994): 17-34.

Edward C. Aldridge, Jr. (1938-), spent his entire career in the aerospace community as a corporate and governmental official. He served as under secretary and then secretary of the Air Force during the Reagan administration. Before then, he was educated at Texas A&M University and the Georgia Institute of Technology, entering the Department of Defense (DOD) as assistant secretary for systems analysis from 1967 through 1972. He then went to LTV Aerospace Corporation for a year. In 1973 he was named as a senior management associate in the Office of Management and Budget (OMB) in Washington. Returning to DOD in 1974, he served as assistant secretary for strategic programs until 1976. He then moved back to private industry until reentering government service with the Air Force in 1981. See "Aldridge, Edward C.," biographical file, NASA Historical Reference Collection, NASA History Office, NASA Headquarters, Washington, D.C.

Anatoliy P. Aleksandrov (1903-) was a senior member of the of the Soviet Union's Academy of Sciences throughout much of the 1950s and 1960s and served as its president from 1980 to 1986. A physicist, Aleksandrov was born in the Ukraine and educated at Kiev State University. He was heavily involved in research on the physics of dielectrics and studies of the properties of compounds having high molecular weight. See "Aleksandrov, Anatoliy, P.," biographical file, NASA Historical Reference Collection.

Robert F. Allnutt (1935-) was a longtime NASA employee throughout the 1960s and 1970s. Born in Richmond, Virginia, and educated at Virginia Polytechnic Institute (now known as Virginia Tech) and the George Washington University Law School, Allnutt joined NASA in 1960 as a patent attorney. He then worked as a attorney with the Communications Satellite Corporation and as NASA's assistant general counsel (patents). In 1967 he was named as assistant administrator for legislative affairs; later, he was a member of the Apollo 13 Accident Review Board. He left NASA in 1983 to become legal counsel to the U.S. Committee for Energy Awareness. He became executive vice president for the Pharmaceutical Manufacturers Association. See "Allnutt, Robert F.," biographical file, NASA Historical Reference Collection.

William A. Anders (1933-) was a career U.S. Air Force officer, although a graduate of the U.S. Naval Academy. Chosen with the third group of astronauts in 1963, he was the backup pilot for Gemini XI and lunar module pilot for Apollo 8. Having resigned from NASA and the Air Force (active duty) in September 1969, he became executive secretary of the National Aeronautics and Space Council. He joined the Atomic Energy Commission in 1973 and became the chair of the Nuclear Regulatory Commission in 1974. He was named U.S. ambassador to Norway in 1976. Later, he worked as a vice president of General Electric and then as senior executive vice president of operations for Textron, Inc. Anders retired as chief executive officer of General Dynamics in 1993, but he remained chairman of the board. See "Anders, W.A.," biographical file, NASA Historical Reference Collection.

Clinton P. Anderson (1895-1975) (D-NM) was elected to the House of Representatives in 1940 and served through 1945, when he was appointed secretary of agriculture. He resigned from that position in 1948 and was elected to the Senate, where he served until 1973. See *Biographical Directory of the United States Congress, 1774-1989* (Washington, DC: U.S. Government Printing Office, 1989).

Neil A. Armstrong (1930-) was the first human to set foot on the Moon on July 20, 1969, as commander of Apollo 11. He had become an astronaut in 1962, after having served as a test pilot with the National Advisory Committee for Aeronautics (1955-1958) and NASA (1958-1962). He flew as command pilot on Gemini VIII in March 1966. In 1970 and 1971, he was deputy associate administrator for the Office of Advanced Research and Technology at NASA Headquarters. In 1971 he left NASA to become a professor of aerospace engineering at the University of Cincinnati and to undertake private consulting. See Neil A. Armstrong, *et al., First on the Moon: A Voyage with Neil Armstrong, Michael Collins and Edwin E. Aldrin, Jr.* (Boston: Little, Brown, 1970); Neil A. Armstrong, *et al., The First Lunar Landing: 20th Anniversary/as Told by the Astronauts, Neil Armstrong, Edwin Aldrin, Michael Collins* (Washington, DC: NASA EP-73, 1989)

Henry H. (Hap) Arnold (1886-1950) was commander of the Army Air Forces in World War II and the only air commander ever to attain the five-star rank of general of the armies. He was especially interested in the development of sophisticated aerospace technology to give the United States an edge in achieving air superiority. He fostered the development of such innovations as jet aircraft, rocketry, rocket-assisted takeoff, and supersonic flight. After a lengthy career as an Army aviator and commander that spanned the two world wars, he retired from active service in 1945. See Henry H. Arnold, *Global Mission* (New York: Harper & Brothers, 1949); Flint O. DuPre, *Hap Arnold: Architect of American Air Power* (New York: Macmillan, 1972); Thomas M. Coffey, *Hap: The Story of the U.S. Air Force and the Man Who Built It* (New York: Viking, 1982).

J. Leland Attwood (1904-) was a long-standing official of North American Rockwell, Inc. He began work as an aeronautical engineer for the Douglas Aircraft Corporation in 1930, and he moved to North American in 1934. He became assistant general manager in 1938 and was named North American's first vice president in 1941. He became president in 1948 and served continually until 1970, when he retired. (The company eventually became known as North American Aviation.) See "J.L. Attwood," biographical file, NASA Historical Reference Collection.

Norman R. Augustine (1935-) was born in Denver, Colorado, and has been longtime a key person in the aerospace industry. He became chairman and chief executive officer of the Martin Marietta Corporation in the 1980s. Previously, he had served as under secretary of the Army, assistant secretary of the Army for research and development, and assistant director of Defense Research and Engineering in the Office of the Secretary of Defense. In 1990 he was appointed to head the Advisory Committee on the Future of the U.S. Space Program for the Bush administration. This panel produced the *Report of the Advisory Committee on the Future of the U.S. Space Program* (Washington, DC: Government Printing Office, December 1990). The study was enormously important in charting the course of the space program in the first half of the 1990s. See Norman R. Augustine, Augustine's Laws (Washington, DC: American Institute for Aeronautics and Astronautics, 1984); "Norman R. Augustine," biographical file, NASA Historical Reference Collection.

B

George Ball (1909-1994) served as under secretary of state from 1961 to 1966. See *Who's Who in America, 1978-1979* (Chicago: Marquis Who's Who, 1978); NASA Headquarters Library, Washington, DC.

Richard J.H. Barnes was director of the International Affairs Division of the Office of External Relations at NASA throughout much of the 1980s. He had been a longtime NASA official, first coming to the agency in 1961 to work on international programs. See "Barnes, Richard J.H.," biographical file, NASA Historical Reference Collection.

Arnold O. Beckman (1900-) received his Ph.D. from the California Institute of Technology (Caltech) in 1928 and became an inventor and manufacturer of various analytical instruments. He became chairman emeritus of Caltech's Board of Trustees in 1981. See *Who's Who in America, 1996* (New Providence, NJ: Marquis Who's Who, 1995).

James E. Beggs (1926-) served as NASA administrator between July 10, 1981, and December 4, 1985, when he took an indefinite leave of absence pending disposition of an indictment from the Justice Department for activities taking place prior to his tenure at NASA. This indictment was later dismissed, and the U.S. attorney general apologized to Beggs for any embarrassment. His resignation from NASA was effective on February 25, 1986. Prior to NASA, Beggs had been executive vice president and a director of General Dynamics Corporation in St. Louis. Previously, he had served with NASA in 1968-1969 as associate administrator for the Office of Advanced Research and Technology. From 1969 to 1973, he was under secretary of transportation. He went to Summa Corporation in Los Angeles as managing director of operations and joined General Dynamics in January 1974. Before joining NASA, he had been with Westinghouse Electric Corporation, in Sharon, Pennsylvania, and Baltimore, Maryland, for thirteen years. A 1947 graduate of the U.S. Naval Academy, he served with the Navy until 1954. In 1955, he received a master's degree from the Harvard Graduate School of Business Administration. See "Beggs, James E.," biographical file, NASA Historical Reference Collection.

David E. Bell (1919-) was budget director for President Kennedy, 1961-1962. A Harvard University-trained economist, Bell had previously been a member of the staff of the Bureau of the Budget and special assistant to the president during the Truman administration before returning to the Harvard faculty during the late 1950s. Between 1962 and 1966, he served as head of the U.S. Agency for International Development and thereafter as vice president of the Ford Foundation. While budget director, Bell was responsible for working with NASA in establishing a realistic financial outlook for Project Apollo. See "Bell, David," biographical file, NASA Historical Reference Collection.

Lloyd V. Berkner (1905-1967) was involved in most of the early spaceflight activities of the United States in some capacity. Trained as an electrical engineer, he was at first interested in atmospheric propagation of radio waves, but after World War II he became a scientific entrepreneur of the first magnitude. He was heavily involved in the planning for and execution of the International Geophysical Year in 1957-1958, and he served in a variety of positions in Washington where he could influence the course of science policy. See "Berkner, Lloyd V.," biographical file, NASA Historical Reference Collection.

Henry E. Billingsley (1906-) was appointed NASA's director of the Office of International Cooperation in January 1959. Previously, he had served in the Navy in World War II; he later joined the Department of State. See "Henry E. Billingsley," biographical file, NASA Historical Reference Collection.

Richard M. Bissell (1909-1994) was a Central Intelligence Agency (CIA) official who was the deputy director for plans during the Bay of Pigs incident. He also was involved in various reconnaissance programs such as the U-2 airplane. See Evan Thomas, *The Very Best Men* (New York: Simon and Schuster, 1995); Richard M. Bissell, *Reflections of a Cold Warrior* (New Haven, CT: Yale University Press, 1996); CIA History Office, Washington, DC.

Anatoli A. Blagonravov (1895-1975) was head of an engineering research institute in the Soviet Union. As Soviet representative to the United Nations Committee on the Peaceful Uses of Outer Space (COPUOS) in the early 1960s, he served as a senior negotiator, along with NASA's Hugh L. Dryden, for cooperative space projects at the height of the Cold War in the early 1960s. He worked in developing infantry and artillery weapons in World War II and on rockets afterward. See "Blagonravov, A.A.," biographical file, NASA Historical Reference Collection.

Nancy W. Boggess was a scientist at the Goddard Space Flight Center working on the Cosmic Background Explorer (COBE) spacecraft in the latter 1980s and early 1990s. See "Miscellaneous NASA," biographical file, NASA Historical Reference Collection.

Herman Bondi was director general of the European Space Research Organization from 1967 through the early 1970s and the organization's transformation into the European Space Agency. A British citizen, Bondi later served as science advisor to the minister for energy. See "Biography, Foreign Miscellaneous, A-D," file, NASA Historical Reference Collection.

Roger M. Bonnet (1938-) of France became director of scientific programs for the European Space Agency (ESA) in 1983. Previously, he had been director of the Stellar and Planetary Laboratory of the French National Scientific Research Center and chair of ESA's Space Science Advisory Committee from 1978 to 1980. See "ESA Names New Scientific Chief," *Defense Daily*, January 27, 1983, p. 144.

Walter F. Boone was an admiral, who, after retiring from the Navy, became the NASA's deputy associate administrator for defense affairs. He held this post until retiring from NASA in 1968. See "Boone, Walter F.," biographical file, NASA Historical Reference Collection.

Frank Borman (1928-) was the commander of the December 1968 Apollo 8 circumlunar flight. He had been chosen as a NASA astronaut in the early 1960s and had been on the Gemini VII mission in 1965. After leaving the astronaut corps, he became president of Eastern Airlines. See Andrew Chaiken, *A Man on the Moon: The Voyages of the Apollo Astronauts* (New York: Viking, 1994); Frank Borman, with Robert J. Serling, *Countdown: An Autobiography* (New York: William Morrow, 1988).

Robert R. Bowie (1909-) was the deputy director and then director of the Policy Planning Staff at the Department of State from 1953 to 1958. Afterward, he became a consultant to the Department of State. From 1966 to 1968, he returned to the State Department to serve as a counselor. Biographical information from the *Biographic Register of the Department of State, 1957*, Department of State History Office, Washington, DC.

Ernest W. Brackett joined NASA in 1959 as director of procurement, after a lengthy career as an attorney (1925-1942) in Utica, New York, an Army Air Forces officer (1942-1946), and a civilian in the Department of the Air Force (1946-1959). He served as director of NASA procurement until 1968, during the Apollo era, and was appointed chair of the Board of Contract Appeals in 1968. Later, he served as chair of the Inventions and Contributions Board, before retiring from NASA in 1972. See "Brackett, Ernest W.," biographical file, NASA Historical Reference Collection.

Willy Brandt (1913-1992) was chancellor of the Federal Republic of Germany from 1969 to 1974. See "Brandt, Willy," obituary section, *Current Biography Yearbook 1992*, p. 628, from obituary, *New York Times*, October 9, 1992, p. A23.

Wernher von Braun (1912-1977) was the leader of the so-called "rocket team" that had developed the German V-2 ballistic missile in World War II. At the conclusion of the war, von Braun and some of his chief assistants—as part of a military operation called Project Paperclip—came to America and were installed at Fort Bliss in El Paso, Texas, to work on rocket development and use the V-2 for high-altitude research. They used launch facilities at the nearby White Sands Proving Ground in New Mexico. In 1950 von Braun's team moved to the Redstone Arsenal near Huntsville, Alabama, to concentrate on the development of a new missile for the Army. They built the Army's Jupiter ballistic missile, and before that the Redstone, used by NASA to launch the first Mercury capsules. The story of von Braun and the "rocket team" has been told many times. See, as examples, David H. DeVorkin, *Science With a Vengeance: How the Military Created the US Space Sciences After World War II* (New York: Springer-Verlag, 1992); Frederick I. Ordway III and Mitchell R. Sharpe, *The Rocket Team* (New York: Thomas Y. Crowell, 1979); Erik Bergaust, *Wernher von Braun* (Washington, DC: National Space Institute, 1976).

Leonid I. Brezhnev (1906-1982) was first secretary of the Communist Party of the Soviet Union between 1964 and 1982 and the Soviet leader during the entire official lunar program. He was responsible for the development of a succession of Soviet space stations built in the 1970s. See "Brezhnev, L.I.," biographical file, NASA Historical Reference Collection.

Geoffrey A. Briggs was director of the Solar System Exploration Division at NASA Headquarters throughout the 1980s. Educated in high-energy physics at the University of Virginia, Briggs became involved in the space program in 1967, working at Bellcomm, Inc., and at the Jet Propulsion Laboratory, where he was principal investigator on the Mariner Mars 1971 imaging team. He also worked on the Viking Orbiter imaging team and was leader of the Voyager imaging team. See "Briggs, Geoffrey A.," biographical file, NASA Historical Reference Collection.

Detlev W. Bronk (1897-1975), a scientist, was president of the National Academy of Sciences, 1950-1962, and a member of the National Aeronautics and Space Council. He also was president of Johns Hopkins University, 1949-1953, and Rockefeller University, 1953-1968. See "Bronk, Detlev," biographical file, NASA Historical Reference Collection.

Overton Brooks (1897-1961) (D-LA) was elected to represent Louisiana in the House of Representatives for twelve successive terms since 1937. He became chair of the House Committee on Science and Astronautics in January 1959 and was reappointed to this position in 1961. See "Brooks, Overton," biographical file, NASA Historical Reference Collection.

Wilber M. Brucker (1894-1968) was secretary of the Army between 1955 and 1961. An attorney, he had also held a number of important government positions, including governor of Michigan (1930-1932), prior to becoming secretary. Brucker had served with the Army in World War I. After leaving federal service, Brucker returned to his law practice in Detroit. See William Gardner Bell, *Secretaries of War and Secretaries of the Army: Portraits & Biographical Sketches* (Washington, DC: Center of Military History, 1982), p. 140; *New York Times,* October 29, 1968, p. 41.

Percival Brundage (1892-1981) was deputy director and then director of the Bureau of the Budget, 1954-1958. Thereafter, he worked in a series of business and financial positions.

McGeorge Bundy (1919-1996) was a professor of government before serving as the national security advisor to Presidents Kennedy and Johnson, 1961-1966. See *Who's Who in America, 1996* (New Providence, NJ: Marquis Who's Who, 1995).

George H.W. Bush (1924-) was president of the United States between 1989 and 1993. Before that, he had been a diplomat, director of the CIA, and vice president under Ronald Reagan (1981-1989).

C

James (Jimmy) Carter (1924-) was president of the United States between 1977 and 1981. Previously, he had been a naval officer and businessman before entering politics. He entered politics in the Georgia State Legislature (1962-1966) and served as the governor of Georgia (1971-1975).

Eugene A. Cernan (1934-), a career naval aviator, was chosen by NASA to enter the astronaut corps in the third group, in 1963. He served as the pilot of Gemini IX upon the death of a prime crew member. He was also back-up pilot for Gemini XII, backup lunar module pilot for Apollo 7, lunar module pilot for Apollo 10, backup commander for Apollo 14, and commander for Apollo 17 (becoming the eleventh American to walk on the Moon). Thereafter, he served as deputy director of the Apollo-Soyuz Test Project before resigning from NASA and the Navy on July 1, 1976, to become executive vice president–international at Coral Petroleum, Inc., in Houston. Later, he headed the Cernan Corporation in Houston. See "Eugene A. Cernan," biographical file, NASA Historical Reference Collection.

Robert H. Charles (1914-) became a special assistant to the NASA administrator in 1963, with responsibility for working with industry to accomplish Project Apollo. He was especially involved in the creation of incentive contracting mechanisms at the agency to reward exceptional performance by contractors. Previously, he had been an executive with the McDonnell Aircraft Corporation. After remaining with NASA for a short time, Charles became assistant secretary of the Air Force, where he was involved in the development of the C-5A total procurement package contract of the mid-1960s. He left that position in 1968 to return to industry. See "Biography, NASA Miscellaneous, Ch-Ci," file, NASA Historical Reference Collection.

William P. Clements, Jr., served as deputy secretary of defense from 1973-1977. He also was governor of Texas from 1979 to 1983 and from 1987 to 1991. See *Department of Defense Key Officials* (Washington, DC: Historical Office, Office of the Secretary of Defense, 1995).

William J. (Bill) Clinton (1946-) became president of the United States in 1993. Previously, he served as governor and attorney general of Arkansas.

Charles W. Cook served during the 1970s and 1980s as deputy under secretary and deputy assistant secretary of the U.S. Air Force in the Office of Plans, Policy and Operations of Space Systems. He also worked in the Office of the Secretary of Defense as director for defensive systems and served in positions in the Advanced Research

Projects Agency, the Central Intelligence Agency, and various aerospace companies. Since retiring from formal government service in 1988, Cook has worked as a consultant to the Institute for Defense Analyses, ANSER, the Defense Science Board, and several other aerospace organizations in areas related to U.S. and foreign space activities.

John J. Corson (1905-1990) had been a management consultant with McKinsey and Company, Inc. since 1951, remaining there until 1966. T. Keith Glennan contracted with McKinsey for a series of studies, including: "Organizing Headquarters Functions," two volumes, December 1958; "Financial Management—NASA-JPL Relationships," February 1959; "Security and Safety—NASA-JPL Relationships," February 1959; "Facilities Construction—NASA-JPL Relationships," February 1959; "Procurement and Subcontracting—NASA-JPL Relationships," February 1959; "NASA-JPL Relationships and the Role of the Western Coordination Office," March 1959; "Providing Supporting Services for the Development Operations Division," January 1960, on the transfer of the Army Ballistic Missile Agency to NASA; "Report of the Advisory Committee on Organization," October 1960; and "An Evaluation of NASA's Contracting Politics, Organization, and Performance," October 1960. All are in T. Keith Glennan, Correspondence Files, NASA Historical Reference Collection.

Edgar M. Cortright (1923-) earned an M.S. in aeronautical engineering from Rensselaer Polytechnic Institute in 1949, the year after he joined the staff of Lewis Flight Propulsion Laboratory. He conducted research at Lewis on the aerodynamics of high-speed air induction systems and jet exit nozzles. In 1958 he joined a small task group to lay the foundation for a national space agency. As soon as NASA was created, he became chief of advanced technology at NASA Headquarters, directing the initial formulation of the agency's meteorological satellite program, including the TIROS and Nimbus projects. After becoming assistant director for lunar and planetary programs in 1960, Cortright directed the planning and implementation of such projects as Mariner, Ranger, and Surveyor. He became deputy director and then deputy associate administrator for space science and applications in the next few years. In 1967 he was deputy associate administrator for manned space flight. In 1968 he became director of the Langley Research Center, a position he held until 1975, when he went to work for private industry, becoming president of the Lockheed-California Company in 1979. See "Cortright, Edgar M.," biographical file, NASA Historical Reference Collection.

Laurence C. Craigie (1902-1994) was a career Air Force officer and the first U.S. military jet pilot in 1942 when he flew the Bell XP-59. A graduate of the U.S. Military Academy at West Point, in 1923 he went into the Army Air Corps and became a pilot. In World War II, he served in a variety of weapons development programs, as well as in a combat role in North Africa and Corsica. After the war, he directed the Air Force's research and development programs, serving as deputy chief of staff for development, 1951-1954, and commander of Allied Air Force in southern Europe before his retirement following a heart attack in 1955. See "Lieut. Gen. Laurence Craigie, 92; First Military Jet Pilot for the U.S.," *New York Times*, March 1, 1994.

Malcolm R. Currie (1927-) was trained in physics and electrical engineering at the University of California at Berkeley and served in the U.S. Navy from 1944 to 1947. After military service, he returned to school to complete his Ph.D. In 1954 he joined Hughes Research Laboratories, eventually serving as director, before becoming vice president of Hughes Aircraft from 1964 to 1969. He then worked for Beckman Instruments, Inc., but in 1973 President Nixon appointed him director of Defense Research and Engineering in the Department of Defense, where he served until returning to Hughes in 1977. See "Currie, Dr. Malcolm R.," biographical file, NASA Historical Reference Collection.

D

Edward E. David, Jr. (1925-), served as science advisor to President Richard Nixon in 1970 and then as director of the Office of Science and Technology. Previously, he had served between 1950 and 1970 as executive director of research at Bell Telephone Laboratories. For a discussion of the President's Science Advisory Committee, see Gregg Herken, *Cardinal Choices: Science Advice to the President from Hiroshima to SDI* (New York: Oxford University Press, 1992).

James H. Douglas, Jr. (1899-1988), was secretary of the Air Force, 1957-1959, and deputy secretary of defense, 1959-1961. Trained as an attorney, Douglas practiced most of his career in Chicago but served as fiscal assistant secretary of the treasury, 1932-1933, and under secretary of the Air Force, 1953-1957, before serving as Air Force secretary. At the conclusion of the Eisenhower administration, Douglas rejoined his old law firm, Gardner, Carton, Douglas, Chilgren & Waud. See "Miscellaneous Department of Defense (DOD)," biographical file, NASA Historical Reference Collection.

Charles Stark (Doc) Draper (1901-1987) earned his Ph.D. in physics at the Massachusetts Institute of Technology (MIT) in 1938 and became a full professor there the following year, when he founded the Instrumentation Laboratory. Its first major achievement was the Mark 14 gyroscopic gunsight for Navy antiaircraft guns. Draper and the laboratory applied gyroscopic principles to the development of inertial guidance systems for airplanes, missiles, submarines, ships, satellites, and space vehicles—notably those used in the Apollo Moon landings. See John Noble Wilford, "Charles S. Draper, Engineer, Guided Astronauts to the Moon," *New York Times,* July 27, 1987, p. 2; Donald MacKenzie, *Inventing Accuracy: A Historical Sociology of Nuclear Missile Guidance* (Cambridge, MA: MIT Press, 1990), especially pp. 64-94; C. Stark Draper, "The Evolution of Aerospace Guidance Technology at Massachusetts Institute of Technology, 1935-1951: A Memoir," in R. Cargill Hall, ed., *Essays on the History of Rocketry and Astronautics,* Vol. II (Washington, DC: NASA Conf. Pub. 2014, 1977), pp. 219-252.

Hugh L. Dryden (1898-1965) was a career civil servant and an aerodynamicist by discipline who had also begun life as a child prodigy. He graduated at age 14 from high school and went on to earn an A.B. in three years from Johns Hopkins University (1916). Three years later, he earned his Ph.D. in physics and mathematics from the same institution, even though he had been employed full time by the National Bureau of Standards since June 1918. His career at the Bureau of Standards, which lasted until 1947, was devoted to studying airflow, turbulence, and particularly the problems of the boundary layer—the thin layer of air next to an airfoil that causes drag. In 1920 he became chief of the bureau's aerodynamics section. His work in the 1920s on measuring turbulence in wind tunnels facilitated research in the NACA that produced the laminar flow wings used in the P-51 Mustang and other World War II aircraft. From the mid-1920s to 1947, his publications became essential reading for aerodynamicists around the world. During World War II, his work on a glide bomb named the Bat won him a Presidential Certificate of Merit. He capped his career at the Bureau of Standards by becoming its assistant director and then associate director during his final two years there. He then served as director of the NACA from 1947 to 1958, after which he became deputy administrator of NASA under T. Keith Glennan and James E. Webb. See Richard K. Smith, *The Hugh L. Dryden Papers, 1898-1965* (Baltimore: The Johns Hopkins University Library, 1974); Michael H. Gorn, *Hugh L. Dryden's Career in Aviation and Space,* Monographs in Aerospace History #5 (Washington, DC: NASA, 1996).

Lee A. DuBridge (1901-1994), a physicist with a Ph.D. from the University of Wisconsin (1926), became director of the radiation laboratory at the Massachusetts Institute of Technology after an academic career capped by a deanship at the University of Rochester, 1938-1941. He was president of the California Institute of Technology between 1946 and 1969, when he resigned to serve as presidential science advisor to Richard Nixon. He had been involved in several governmental science advisory organizations before taking up his formal White House duties in 1969 and serving in that capacity until 1970. See "Lee A. DuBridge," biographical file, NASA Historical Reference Collection.

Allen W. Dulles (1893-1969), brother of President Eisenhower's more famous secretary of state, served as director of the Central Intelligence Agency (CIA) from 1953 to 1961. See "Miscellaneous Other Agencies," biographical file, NASA Historical Reference Collection.

John Foster Dulles (1888-1959) served as secretary of state under President Eisenhower, 1953-1959. See "Miscellaneous Other Agencies," biographical file, NASA Historical Reference Collection.

Frederick C. Durant III (1916-) was heavily involved in rocketry in the United States during the period between the end of World War II and the mid-1960s. He worked for several different aerospace organizations, including Bell Aircraft Corporation, Everett Research Laboratory, the Naval Air Rocket Test Station, and the Maynard Ordnance Test Station. He later became the director of astronautics for the National Air and Space Museum, Smithsonian Institution. In addition, he was an officer in several spaceflight organizations, including the American Rocket Society (president in 1953), the International Astronautical Federation (president from 1953 to 1956), and the National Space Club (governor in 1961).

E

Burton I. Edelson (1926-) was NASA's associate administrator for space science and applications between 1982 and 1988. He earned his B.S. from the U.S. Naval Academy in 1947 and served for 20 years in the service. He then returned to school and received a Ph.D. from the University of California at San Diego in 1969. Thereafter, he worked with the Communications Satellite Corporation for 14 years before arriving at NASA. See "Edelson, Burt I.," biographical file, NASA Historical Reference Collection.

Raymond Einhorn was a former General Accounting Office auditor who joined NASA in 1960 as its director of audits. He served in this position throughout the 1960s. See "Assorted NASA Officials," biographical file, NASA Historical Reference Collection.

Dwight D. Eisenhower (1890-1969) was president of the United States between 1953 and 1961. Previously, he had been a career U.S. Army officer and was supreme allied commander in Europe during World War II. As president he was deeply interested in the use of space technology for national security purposes and directed that ballistic missiles and reconnaissance satellites be developed on a crash basis. On Eisenhower's space efforts, see Rip Bulkeley, *The Sputniks Crisis and Early United States Space Policy* (Bloomington: Indiana University Press, 1991); R. Cargill Hall, "The Eisenhower Administration and the Cold War: Framing American Astronautics to Serve National Security," *Prologue: Quarterly of the National Archives* 27 (Spring 1995): 59-72; Robert A. Divine, *The Sputnik Challenge: Eisenhower's Response to the Soviet Satellite* (New York: Oxford University Press, 1993).

John D. Erlichman was a senior assistant to the president during the Nixon administration. See John Erlichman, *Witness to Power: The Nixon Years* (New York: Simon and Schuster, 1982).

F

Konstantin Petrovich Feoktistov (1926-) worked as a spacecraft engineer and cosmonaut. As a cosmonaut, he flew on the Voskhod 1 mission in 1964 and was also flight director on the Soyuz 18/Salyut mission in 1975. See *Who's Who in Russia and the New States* (London: J.B. Tauris and Co., 1993).

Peter M. Flanigan (1923-) was an assistant to President Nixon on the White House staff, 1969-1974. Previously, he had been involved in investment banking with Dillon, Read, and Company. He returned to business when he left government service. His position in the White House involved him in efforts to gain approval to build the Space Shuttle during the 1969-1972 period. See "Miscellaneous Other Agencies," biographical file, NASA Historical Reference Collection.

James C. Fletcher (1919-1991) was born on June 5, 1919, in Millburn, New Jersey. He received an undergraduate degree in physics from Columbia University and a doctorate in physics from the California Institute of Technology. After holding research and teaching positions at Harvard and Princeton Universities, he joined Hughes Aircraft in 1948 and later worked for the Guided Missile Division of the Ramo-Wooldridge Corporation. In 1958 Fletcher co-founded the Space Electronics Corporation in Glendale, California, which after a merger became the Space General Corporation. He was later named systems vice president of the Aerojet General Corporation in Sacramento, California. In 1964 he became president of the University of Utah, a position he held until he was named NASA administrator in 1971, serving until 1977. He also served as NASA administrator a second time, for nearly three years following the loss of the Space Shuttle *Challenger* in 1986 until 1989. During his first administration at NASA, Dr. Fletcher was responsible for beginning the shuttle effort. During his second tenure, he presided over the effort to recover from the *Challenger* accident. See Roger D. Launius, "A Western Mormon in Washington, DC: James C. Fletcher, NASA, and the Final Frontier," *Pacific Historical Review* 64 (May 1995): 217-41.

Gerald R. Ford (1913-) (R-MI) was elected to the House of Representatives in 1948 and served there until he became vice president in 1973 following the resignation of Spiro T. Agnew. He then became president, 1974-1977, following Richard M. Nixon's resignation in the wake of the Watergate scandal.

John S. Foster, Jr. (1922-), is a physicist who served as director of Defense Research and Engineering from 1965 to 1973, when he moved to the private sector. He has served on a number of scientific and technical government advisory boards. In 1995, he was the chair of a NASA federal laboratory review team. In 1992, he served on the Vice President's Space Policy Advisory Board that reviewed U.S. space policy after the cold war. See "Foster, John S., Jr.," biographical file, NASA Historical Reference Collection.

Robert A. Frosch (1928-) was NASA administrator throughout the administration of President Jimmy Carter, 1977-1981. He earned undergraduate and graduate degrees in theoretical physics at Columbia University, and between September 1951 and August 1963, he worked as a research scientist and director of research programs for Hudson Laboratories of Columbia University. Until 1953 he worked on problems in underwater sound, sonar, oceanography, marine geology, and marine geophysics. Thereafter, Frosch was first associate and then director of the laboratories. In September 1963, he came to Washington to work with the Advanced Research Projects Agency, serving as director for nuclear test detection (Project VELA), and then as deputy director of the agency. In July 1966 he became assistant secretary of the Navy for research and development, responsible for all Navy programs of research, development, engineering, test, and evaluation. From January 1973 to July 1975, he served as assistant executive director of the United Nations Environmental Program. While at NASA, Frosch was responsible for overseeing the continuation of the development effort on the Space Shuttle. During his tenure, the project underwent testing of the first orbiter, *Enterprise*, at NASA's Dryden Flight Research Facility in southern California. The orbiter made its first free flight in the atmosphere on August 12, 1977. He left NASA with the change of administrations in January 1981 to become vice president for research at the General Motors Research Laboratories. See "Frosch, Robert A., Administrators Files," NASA Historical Reference Collection.

Arnold W. Frutkin (1918-) was deputy director of the U.S. National Committee for the International Geophysical Year in the National Academy of Sciences when NASA hired him in 1959 as director of international programs, a title that changed in 1963 to assistant administrator for international affairs. In 1978 he became associate administrator for external relations, a post he relinquished in 1979 when he retired from federal service. During his career, he had been NASA's senior negotiator for almost all of the important international space agreements. See "Arnold W. Frutkin," biographical file, NASA Historical Reference Collection.

G

Charles A. Gabriel served as U.S. Air Force chief of staff between 1983 and 1986 and was the highest ranking uniformed official in the service. See "Miscellaneous DOD," biographical file, NASA Historical Reference Collection.

Yuri Gagarin (1934-1968) was the Soviet cosmonaut who became the first human in space with a one-orbit mission aboard the spacecraft Vostok 1 on April 12, 1961. The great success of that feat made the gregarious Gagarin a global hero, and he was an effective spokesman for the Soviet Union until his death in an unfortunate aircraft accident. See "Gagarin, Yuri," biographical file, NASA Historical Reference Collection.

Thomas S. Gates (1906-1983) was a businessman who served as secretary of the Navy and then, from 1959 to 1960, as secretary of defense. See Office of the Secretary of Defense Historical Branch, Department of Defense, Washington, DC; *Who's Who in America, 1972-1973* (Chicago: Marquis Who's Who, 1972).

Roswell L. Gilpatric (1906-) is a retired attorney who served as deputy secretary of defense from 1961 to 1964. See *Who's Who in America, 1996* (New Providence, NJ: Marquis Who's Who, 1995).

Robert R. Gilruth (1913-) was a longtime NACA engineer working at the Langley Aeronautical Laboratory, 1937-1946, then chief of the Pilotless Aircraft Research Division at Wallops Island, 1946-1952, who had been exploring the possibility of human spaceflight before the creation of NASA. He served as assistant director at Langley, 1952-1959, and as assistant director (crewed satellites) and head of Project Mercury, 1959-1961—technically assigned to the Goddard Space Flight Center but physically located at Langley. In early 1961, T. Keith Glennan established an independent Space Task Group (already the group's name as an independent subdivision of Goddard) under Gilruth at Langley to supervise the Mercury program. This group moved to the Manned Spacecraft Center in Houston in 1962. Gilruth was then director of the Houston operation from 1962 to 1972.

See Henry C. Dethloff, *"Suddenly Tomorrow Came . . .": A History of the Johnson Space Center* (Washington, DC: NASA SP-4307, 1993); James R. Hansen, *Engineer in Charge: A History of the Langley Aeronautical Laboratory, 1917-1958* (Washington, DC: NASA SP-4305, 1987), pp. 386-88.

John H. Glenn, Jr. (1921-), was chosen with the first group of astronauts in 1959. He was the pilot for the February 20, 1962, Mercury-Atlas 6 (*Friendship 7*) mission, the first American orbital flight, making three orbits. He left the NASA astronaut corps in 1964 and later entered politics as a senator from Ohio. See Lloyd S. Swenson, Jr., James M. Grimwood, and Charles C. Alexander, *This New Ocean: A History of Project Mercury* (Washington, DC: NASA SP-4201, 1966).

T. Keith Glennan (1905-1995) was the first administrator of the NASA. Born in Enderlin, North Dakota, in 1905, Glennan was educated at Yale University and worked in the sound motion picture industry with the Electrical Research Products Company. He was also studio manager of Paramount Pictures, Inc., and Samuel Goldwyn Studios in the 1930s. Glennan joined the Columbia University Division of War Research in 1942, serving through the war, first as administrator and then as director of the U.S. Navy's Underwater Sound Laboratories at New London, Connecticut. In 1947 he became president of the Case Institute of Technology. During his administration, Case rose from a primarily local institution to rank with the top engineering schools in the nation. From October 1950 to November 1952, Glennan served as a member of the Atomic Energy Commission. He then served as administrator of NASA while on leave from Case, between August 7, 1958, and January 20, 1961. Upon leaving NASA, Glennan returned to the Case, where he was continued to serve as president until 1966. See J.D. Hunley, ed., *The Birth of NASA: The Diary of T. Keith Glennan* (Washington, DC: NASA SP-4105, 1993).

Daniel S. Goldin (1940-) became the ninth NASA administrator in April 1992 and immediately began to earn a reputation as an "agent of change" by bringing reform to America's space agency. In addition to implementing many management changes, Goldin negotiated with his Russian counterpart, Yuri Koptev, the head of the Russian Space Agency, to construct an international space station with a partnership involving fourteen nations. Before coming to NASA, Goldin was vice president and general manager of the TRW Space & Technology Group in Redondo Beach, California. During a twenty-five-year career at TRW, he managed the development and production of advanced spacecraft, technologies, and space science instruments. Goldin began his career as a research scientist at NASA's Lewis Research Center in Cleveland in 1962, where he worked on electric propulsion systems for human interplanetary travel. See "Daniel S. Goldin," biographical file, NASA Historical Reference Collection.

Nicholas E. Golovin (1912-1969), born in Odessa, Russia, but educated in this country (Ph.D. in physics at George Washington University in 1955), worked in various capacities for the government during and after World War II, including the Naval Research Laboratory, 1946-1948. He held several administrative positions with the National Bureau of Standards from 1949 to 1958. In 1958 he was chief scientist for the White Sands Missile Range and then worked for the Advanced Research Projects Agency in 1959 as director of technical operations. He became deputy associate administrator at NASA in 1960. He joined private industry before becoming, in 1961, the director of the NASA-DOD Large Launch Vehicle Planning Group. He joined the Office of Science and Technology at the White House in 1962 as a technical advisor for aviation and space and remained there until 1968, when he took a leave of absence as a research associate at Harvard and as a fellow at the Brookings Institution. See his obituaries, *Washington Star,* April 30, 1969, p. B-6, and *Washington Post,* April 30, 1969, p. B14.

Mikhail S. Gorbachev (1931-) became leader of the Soviet Union in 1985 and restructured the nation, presiding over the demise of the communist state and the end of the cold war in 1989. In the process, he opened negotiations with the United States for significant international cooperation in space exploration. See Thomas G. Butson, *Gorbachev: A Biography* (New York: Stein and Day, 1985); "Gorbachev, Mikhail Sergeyevich," biographical file, NASA Historical Reference Collection.

Aristid V. Grosse (1905-) was born in Riga, Russia, and trained in engineering at the Technische Hochschule in Berlin. He came to the United States in 1930 and was on the chemistry faculty at the University of Chicago, 1931-1940. He then went to Columbia University briefly before working on the Manhattan Project during the war years. In 1948 he became a faculty member at Temple University, presiding over the Research Institute (now Franklin Institute) through 1969. See "Grosse, Aristid," biographical file, NASA Historical Reference Collection.

Richard W. Gutman (1921-) is a retired auditor and accountant who worked at the General Accounting Office (GAO) on defense and international programs. From 1968 to 1972, he was deputy director of the Defense Division, and when he retired from GAO in 1981, he was the director for the Defense Programs Planning and Analysis Staff. See *The GAO Review* (Fall 1968, Summer 1978, and Spring 1982) GAO's *Office of Personnel Management: Professional Staff Register,* entry for Gutman dated September 30, 1970. This information was obtained from the GAO Law Library, Washington, DC.

H

George H. Hage (1925-) was associated with Project Apollo in the 1960s. After completing his B.S. in electrical engineering from the University of Washington, he went to work for the Boeing Company in 1947. He was involved in the development of the Bomarc and Minuteman missile systems, and in 1962 he went to the Minuteman assembly and test complex in Florida in 1962. From there he took charge of Boeing's reconnaissance efforts, and in 1968 he came to NASA Headquarters as deputy director of the Apollo program. Soon afterward Hage returned to Boeing, and in 1973 he was appointed president of the Aerojet Solid Propulsion Company. See "Hage, George H.," biographical file, NASA Historical Reference Collection.

James C. Hagerty (1909-1981) had been on the staff of the *New York Times* from 1934 to 1942, the last four years as legislative correspondent in the newspaper's Albany bureau. He served as executive assistant to New York Governor Thomas Dewey from 1943 to 1950 and then as Dewey's secretary for the next two years before becoming press secretary for President Eisenhower from 1953 to 1961. See "Miscellaneous Other Agencies," biographical file, NASA Historical Reference Collection.

Grant L. Hansen (1921-) was an engineer in the aerospace industry before serving as assistant secretary of the Air Force for research and development from 1969 to 1973. See *Who's Who in America, 1996* (New Providence, NJ: Marquis Who's Who, 1995).

George Haskell (1940-) is a British physicist who has worked for the European Space Agency (ESA) since 1972. From 1972 to 1987, he worked in ESA's space science planning office, and from 1987 to 1992, he served as the liaison officer for scientific use of the space station. He has also served as associate dean and vice president for academic affairs of the International Space University. See "Miscellaneous Foreign," biographical file, NASA Historical Reference Collection.

Walter Hedrick (1921-) was an Air Force brigadier general who was involved in space systems throughout the 1960s. In 1967, he became the Air Force's director of space, deputy chief of staff, research and development. See U.S. Air Force biography, June 15, 1969, for Brigadier General Walter R. Hedrick, Jr., History Office, Air Force Materiel Command, Wright-Patterson Air Force Base, OH.

Richard C. Henry was a career U.S. Air Force officer involved in the development of space systems during the last part of his service. He was commander of Air Force Space Division in Los Angeles between 1978 and 1982 and vice commander of Air Force Space Command for almost a year, 1982-1983, retiring as a lieutenant general. See *Aerospace Daily,* February 9, 1983, p. 232.

Earl D. Hilburn (1920-) was trained in physics and mathematics at the University of Wisconsin and worked for more than twenty years in the electronics and aerospace industry before accepting a position at NASA in 1963 as deputy associate administrator. In that post, he was responsible for industry affairs, helping maintain liaison with the far-flung corporations involved in the production of NASA space hardware. In 1966 he left NASA and became president of Western Union. See "Hilburn, Earl D.," biographical file, NASA Historical Reference Collection.

Noel W. Hinners (1935-) was trained in geochemistry and geology at Rutgers University, the California Institute of Technology, and Princeton University. He began his career in 1963 with Bellcomm, Inc., working on the Apollo program, and he arrived at NASA Headquarters in 1972 as the deputy director of lunar programs in the Office of Space Science. From 1974 to 1979, he was NASA's associate administrator for space science. He also served as director of the Smithsonian Institution's National Air and Space Museum, 1979-1982, and as director of the Goddard Space Flight Center in Greenbelt, Maryland, 1982-1987. He then became associate deputy administrator of NASA before leaving the agency in 1989 to join the Martin Marietta Corporation as vice president of strategic planning. See "Hinners, Noel W.," biographical file, NASA Historical Reference Collection.

John Hodge (1929-) began a distinguished career at NASA in 1959. He worked in the area of flight control at the Langley Research Center and the Johnson Space Center until 1970. In 1982 he became director of the Space Station Task Force at NASA Headquarters. He then took on a series of increasingly responsible positions dealing with the Space Station, culminating with him being named associate administrator for operations, space station, in 1986. See "Hodge, John," biographical file, NASA Historical Reference Collection.

Valerie Hood (1945-) is a lawyer specializing in space law who has worked for the International Affairs Branch of the European Space Agency since 1976. See "Miscellaneous Foreign," biographical file, NASA Historical Reference Collection.

Donald F. Hornig (1920-), a chemist, was a research associate at the Woods Hole Oceanographic Laboratory, 1943-1944, and a scientist and group leader at the Los Alamos Scientific Laboratory, 1944-1946. He taught chemistry at Brown University starting in 1946, rising to the directorship of the Metcalf Research Laboratory, 1949-1957, and also serving as associate dean and acting dean of the graduate school from 1952 to 1954. He was Donner professor of science at Princeton from 1957 to 1964 as well as chairman of the chemistry department from 1958 to 1964. He was President Lyndon Johnson's special assistant on science and technology from 1964 to 1969 and president of Brown University from 1970 to 1976. See Gregg Herken, *Cardinal Choices: Science Advice to the President from Hiroshima to SDI* (New York: Oxford University Press, 1992).

H.C. van de Hulst of the Netherlands served as president of the Committee on Space Research (COSPAR).

J

Lee B. James (1920-) was a career Army officer, trained at the U.S. Military Academy at West Point and the California Institute of Technology, who was assigned to the Army Ballistic Missile Agency at Huntsville, Alabama, in 1956. In 1960 he became deputy director of the Army's newly formed Research and Development Division. In 1962 he was assigned to the Marshall Space Flight Center and the next year became deputy director of the Apollo program at NASA Headquarters. In 1968 he returned to Marshall to head the Saturn Program Office and retired from the Army as a colonel. Only a year later, he was elevated as the director of the overall program office at Marshall. James retired from NASA in 1971 and accepted a faculty position at the University of Tennessee Space Institute in Tullahoma. See "James, Lee. B.," biographical file, NASA Historical Reference Collection.

Robert Jastrow (1925-) earned a Ph.D. in theoretical physics from Columbia in 1948 and pursued postdoctoral studies at Leiden, Princeton (Institute for Advanced Studies), and the University of California at Berkeley before becoming an assistant professor at Yale, 1953-1954. He then served on the staff at the Naval Research Laboratory from 1954 to 1958. In 1958, he was appointed chief of the theoretical division at the Goddard Space Flight Center. He became director of the Goddard Institute of Space Studies in 1961 and stayed at its helm for twenty years before becoming professor of earth sciences at Dartmouth. He specialized in nuclear physics, plasma physics, geophysics, and the physics of the Moon and terrestrial planets. See "Jastrow, Robert," biographical file, NASA Historical Reference Collection.

Caldwell C. Johnson was a longtime NASA official who held a number of positions in the Apollo program at the Manned Spacecraft Center in Houston in the 1960s. He started work at the Langley Memorial Aeronautical Laboratory in Hampton, Virginia, in 1938 and worked in a variety of aeronautical engineering activities. He moved to Houston with the Space Task Group in 1962. He retired from NASA and became chief of design for Space Industries, Inc., in Texas. See "Johnson, Caldwell C.," biographical file, NASA Historical Reference Collection.

John A. Johnson (1915-), after completing law school at the University of Chicago in 1940, practiced in Chicago until 1943, when he entered military service with the Navy. From 1946 to 1948, he was an assistant for international security affairs in the Department of State. He joined the office of the general counsel of the Department of the Air Force in 1949 and served until October 7, 1958 (for the last six years as the general counsel), when he accepted the general counsel position at NASA. In 1963 he left NASA to become director of international arrangements at the Communications Satellite Corporation. The next year, he became a vice president of COMSAT, and, in 1973, senior vice president and then chief executive officer, retiring in 1980. See "Johnson, John A.," biographical file, NASA Historical Reference Collection.

Lyndon B. Johnson (1908-1973) (D-TX) was elected to the House of Representatives in 1937 and served until 1949. He was a senator from 1949 to 1961, vice president under John F. Kennedy from 1960 to 1963, and president from the time of Kennedy's assassination in November 1963 until 1969. Best known for the social legislation he passed during his presidency and for his escalation of the war in Vietnam, he was also highly instrumental in revising and passing the legislation that created NASA and in supporting the U.S. space program as chair of the Committee on Aeronautical and Space Sciences and of the preparedness subcommittee of the Senate Armed Services Committee. While he was vice president, he chaired the National Aeronautics and Space Council. On his role in support of the space program, see Robert A. Divine, "Lyndon B. Johnson and the Politics of Space," in Robert A. Divine, ed., *The Johnson Years: Vietnam, the Environment, and Science* (Lawrence: University of Kansas Press, 1987), pp. 217-53; Robert Dallek, "Johnson, Project Apollo, and the Politics of Space Program Planning," unpublished paper delivered at a symposium on "Presidential Leadership, Congress, and the U.S. Space Program," sponsored by NASA and American University, March 25, 1993.

Roy W. Johnson (1906-1965) was named director of the Advanced Research Projects Agency for the Department of Defense in 1958, serving until 1961. Previously, he had been with the General Electric Company. He was a strong proponent of exploiting space for national security objectives. See "Roy W. Johnson Dead; First U.S. Space Chief," *Washington Post*, July 23, 1965.

U. Alexis Johnson (1908-) was a longtime member of the U.S. Foreign Service and served in a number of embassies around the world. A specialist in Asian affairs, he was attached to the embassy in Tokyo, 1935-1938, served as consul general to Japan, 1947-1949, and served as ambassador to Japan, 1966-1969. He also served on several international commissions and in numerous senior positions with the Department of State in Washington, D.C., most significantly as under secretary of state for political affairs, beginning in 1969 until his retirement. See "Miscellaneous Other Agencies," biographical file, NASA Historical Reference Collection.

Vincent L. Johnson (1918-) was a longtime NASA official, joining the agency in 1960 after working as an aerospace engineer with the Navy since 1947. He managed the Launch Vehicle and Propulsion Programs Division at NASA Headquarters and had primary responsibility for the program management of Scout, Delta, and Centaur launch vehicle development. He retired from NASA in 1974, after having served as deputy associate administrator for space science. See "Johnson, V.L.," biographical file, NASA Historical Reference Collection.

Charlie Jonas (1904-) (R-NC) served in the U.S. House of Representatives from 1953 to 1973. See *Biographical Directory of the United States Congress, 1774-1989* (Washington, DC: U.S. Government Printing Office, 1989).

John Erik Jonsson (1901-1995) was an engineer and businessman who chaired the board of Texas Instruments, Inc., from 1958 to 1966. He later became the mayor of Dallas. Biographical information from the Corporate Archives Office of Texas Instruments, Inc., Dallas, TX.

K

William C. Keathley arrived at NASA's Marshall Space Flight Center in 1966. He served as the project manager for the Apollo Telescope Mount experiments that were flown on Skylab and as chief of the Skylab Optical Telescope Assembly project. In 1977 he was named manager of the Space Telescope Project (later named the Hubble Space Telescope). See *Marshall Star*, March 16, 1977, p.4, from the Marshall Space Flight Center History Office, Huntsville, AL.

Estes Kefauver (1903-1963) (D-TN) served in the U.S. House of Representatives from 1939 to 1949 and in the U.S. Senate from 1949 to 1963. He ran unsuccessfully as Adlai Stevenson's vice presidential choice in 1956. See *Biographical Directory of the United States Congress, 1774-1989* (Washington, DC: U.S. Government Printing Office, 1989).

M.V. Keldysh (1911-1978) was trained in physics and mathematics at Moscow University (where he received a Ph.D. in 1938) and became the chief theoretician of Soviet cosmonautics in the 1960s. He had previously served many years in a variety of positions at the Central Institute of Aerohydrodynamics, Moscow University, and the Steklov Mathematical Institute. He was vice president (1960-1961) and then president (until 1975) of the Soviet Academy of Sciences. See "M.V. Keldysh, Soviet Scientist, Dies," *Washington Post,* June 27, 1978.

Edward M. (Ted) Kennedy (1932-) (D-MA) has been a longtime Democratic member of the Senate from Massachusetts who was first elected in 1962.

John F. Kennedy (1916-1963) was president of the United States from 1961 to 1963. In 1960, as a senator from Massachusetts between 1953 and 1960, he ran for president as the Democratic candidate, with party wheelhorse Lyndon B. Johnson as his running mate. Using the slogan, "Let's get this country moving again," Kennedy charged the Republican Eisenhower administration with doing nothing about the myriad social, economic, and international problems that festered in the 1950s. He was especially hard on Eisenhower's record in international relations, taking a cold warrior position on a supposed "missile gap" (which turned out not to be the case) wherein the United States lagged far behind the Soviet Union in intercontinental ballistic missile (ICBM) technology. On May 25, 1961, President Kennedy announced to the nation the goal of sending an American to the Moon before the end of the decade. The human spaceflight imperative was a direct outgrowth of it; Projects Mercury (at least in its latter stages), Gemini, and Apollo were each designed to execute it. On this subject, see Walter A. McDougall, . . . *The Heavens and the Earth: A Political History of the Space Age* (New York: Basic Books, 1985); John M. Logsdon, *The Decision to Go to the Moon: Project Apollo and the National Interest* (Cambridge, MA: MIT Press, 1970).

Robert F. Kennedy (1925-1968) was attorney general during the administration of his brother, John F. Kennedy, and a candidate for the Democratic nomination for the presidency in 1968 at the time of his assassination. He was involved in the 1961 decision to go to the Moon as a senior advisor in the Kennedy administration. On his career, see Arthur M. Schlesinger, Jr., *Robert Kennedy and His Times* (Boston: Houghton Mifflin, 1978).

Robert S. Kerr (1896-1963) (D-OK) had been governor of Oklahoma from 1943 to 1947 and was elected to the Senate the following year. From 1961 until 1963, he chaired the Committee on Aeronautical and Space Sciences. See Anne Hodges Morgan, *Robert S. Kerr: The Senate Years* (Norman: University of Oklahoma Press, 1977).

Nikita S. Khrushchev (1894-1971) was premier of the Soviet Union from 1958 to 1964 and first secretary of the Communist Party from 1953 to 1964. He was noted for an astonishing speech in 1956 denouncing the crimes and blunders of Joseph Stalin and for gestures of reconciliation with the West in 1959-1960, ending with the breakdown of a Paris summit with President Eisenhower and the leaders of France and Great Britain in the wake of Khrushchev's announcement that the Soviets had shot down an American U-2 reconnaissance aircraft over the Ural Mountains on May 1, 1960. Then in 1962, Khrushchev attempted to place Soviet medium-range missiles in Cuba. This led to an intense crisis in October, following which Khrushchev agreed to remove the missiles if the United States promised to make no more attempts to overthrow Cuba's Communist government. Although he could be charming at times, Khrushchev was known for boisterous threats (extending even to shoe-pounding at the United Nations) and was a tough negotiator, although he believed, unlike his predecessors, in the possibility of Communist victory over the West without war. For further information about him, see his *Khrushchev Remembers: The Last Testament* (Boston: Little, Brown, 1974); Edward Crankshaw, *Khrushchev: A Career* (New York: Viking, 1966); Michael R. Beschloss, *Mayday: Eisenhower, Khrushchev and the U-2 Affair* (New York: Harper and Row, 1986); Robert A. Divine, *Eisenhower and the Cold War* (New York: Oxford University Press, 1981).

James R. Killian, Jr. (1904-1988), was president of the Massachusetts Institute of Technology (MIT) between 1949 and 1959, but he was on leave between November 1957 and July 1959 to serve as the first presidential science advisor. President Dwight D. Eisenhower established the President's Science Advisory Committee (PSAC), which Killian chaired, following the Sputnik crisis. After leaving the White House staff in 1959, Killian continued his work at MIT. In 1965 he began working with the Corporation for Public Broadcasting to develop public television. Killian described his experiences as a presidential advisor in *Sputnik, Scientists, and Eisenhower: A Memoir of the First Special Assistant to the President for Science and Technology* (Cambridge, MA: MIT Press, 1977). For a discussion of PSAC, see Gregg Herken, *Cardinal Choices: Science Advice to the President from Hiroshima to SDI* (New York: Oxford University Press, 1992).

V.A. Kirillin (1913-) was educated as a physicist and worked in thermodynamics. He was deputy chair of the Council of Ministers (in the early 1960s) and chair of the State Committee for Science and Technology (1965-1980) for the Soviet Union. He was stripped of his position in 1980 after the ascension of Leonid Brezhnev as head of the Soviet Union. See "Biography, Soviet, Miscellaneous (K-O)," NASA Historical Reference Collection.

Henry A. Kissinger (1923-) was presidential advisor for national security affairs from 1969 to 1973 and secretary of state (under Presidents Richard Nixon and Gerald Ford) thereafter until 1977. In these positions, he was especially involved in international aspects of spaceflight, particularly the joint Soviet-American flight, the Apollo-Soyuz Test Project, in 1975. See "Kissinger, Henry," biographical file, NASA Historical Reference Collection.

George B. Kistiakowsky (1900-1982) was a pioneering chemist at Harvard University, associated with the development of the atomic bomb. He later became an advocate of banning nuclear weapons. He served as science advisor to President Eisenhower from July 1959 to the end of the administration. He later served on the advisory board to the U.S. Arms Control and Disarmament Agency from 1962 to 1969. See *New York Times*, December 9, 1982, p. B21; "George B. Kistiakowsky," biographical file, NASA Historical Reference Collection.

Yuri N. Koptev (1940-) became general director of the Russian Space Agency. Trained as an engineer, he began work in 1965 at NPO S.A. Lavochkina, as head of the organization for spacecraft design. Beginning in 1969, he served in administration and eventually was appointed as senior engineer to the deputy minister at the design bureau. See "Koptev, Yuri N.," biographical file, NASA Historical Reference Collection.

Christopher C. Kraft, Jr. (1924-), was a long-standing official with NASA throughout the Apollo program. He received a B.S. in aeronautical engineering from Virginia Polytechnic University in 1944 and joined the Langley Aeronautical Laboratory of the National Advisory Committee for Aeronautics (NACA) the next year. In 1958, still at Langley, he became a member of the Space Task Group developing Project Mercury and moved with the group to Houston in 1962. He was flight director for all of the Mercury and many of the Gemini missions and directed the design of Mission Control at the Manned Spacecraft Center (MSC), redesignated the Johnson Space Center in 1973. He was named the MSC deputy director in 1970 and its director two years later, a position he held until his retirement in 1982. Since then, he has remained active as an aerospace consultant. See "Kraft, Christopher C., Jr.," biographical file, NASA Historical Reference Collection.

L

Edwin (Din) Land was president of the Polaroid Corporation and a member of the Purcell Panel that assessed spaceflight capabilities for the U.S. government in 1957-1958.

Harold R. Lawrence was assistant director of NASA's Office of International Programs. He resigned in 1960 to take a job at the Jet Propulsion Laboratory. See T. Keith Glennan, Correspondence Files, NASA Historical Reference Collection.

Theo Lefevre was the Belgian minister who served as chair of the European Space Conference, which was a policy-level organization created to coordinate European responses to U.S. positions on space issues. Lefevre led the European delegation in negotiations with the United States concerning launch assurance and post-Apollo cooperation during the early 1970s.

Curtis E. LeMay (1906-1990) was a career Air Force officer who entered the Army Air Corps in the 1920s and rose through a series of increasingly responsible Army Air Forces commands during World War II. After the war, LeMay built the Strategic Air Command into the premier nuclear deterrent force in the early 1950s. He also served as deputy chief of staff, 1957-1961, and chief of staff, 1961-1965, of the U.S. Air Force. He retired as a four-star general in 1965 and ran for vice president with independent candidate George C. Wallace in 1968. See Thomas M. Coffey, *Iron Eagle: The Turbulent Life of General Curtis LeMay* (New York: Crown Pub., 1986).

Reimar Leust (1923-) is a German theoretical physicist who held a variety of prestigious academic and advisory council posts before serving as director general of European Space Agency from 1984 to 1990. See "Miscellaneous Foreign," biographical file, NASA Historical Reference Collection.

Robert B. Lewis was a longtime government official who joined NASA in 1961 as director of financial management. He served until 1965, when he left the agency to return to the Office of the Secretary of Defense. See "Miscellaneous NASA," biographical file, NASA Historical Reference Collection.

Russell Long (1918-) served as a U.S. senator from Louisiana from 1948 to 1987. See *Biographical Directory of the United States Congress, 1774-1989* (Washington, DC: U.S. Government Printing Office).

Alan M. Lovelace (1929-) was born in St. Petersburg, Florida, and was educated at the University of Florida, receiving a B.S. in chemistry in 1951, an M.S. in organic chemistry in 1952, and a Ph.D. in organic chemistry in 1954. Shortly after the end of the Korean conflict, he served in the U.S. Air Force from 1954 to 1956. Thereafter, Dr. Lovelace began work as a government scientist at the Air Force Materials Laboratory, Wright-Patterson Air Force Base, Dayton, Ohio. In January 1964, he was named as chief scientist of the Air Force Materials Laboratory, and in 1967 he was named director of the laboratory. In October 1972, he was named director of science and technology for the Air Force Systems Command at Headquarters, Andrews Air Force Base, Maryland. In September 1973, he became the principal deputy to the assistant secretary of the Air Force for research and development. One year late, Dr. Lovelace left the Department of Defense to become the associate administrator of the NASA Office of Aeronautics and Space Technology. With the departure of George Low as NASA deputy administrator in June 1976, Dr. Lovelace became deputy administrator, serving until July 1981. He retired from NASA to accept a position as corporate vice president of science and engineering at the General Dynamics Corporation in St. Louis. See "Lovelace, Alan M.," Deputy Administrator files, NASA Historical Reference Collection.

George M. Low (1926-1984), a native of Vienna, Austria, moved to the United States in 1940 and received an aeronautical engineering degree from Rensselaer Polytechnic Institute (RPI) in 1948 and an M.S. in the same field from that school in 1950. He joined the National Advisory Committee for Aeronautics (NACA) in 1949; at the Lewis Flight Propulsion Laboratory he specialized in experimental and theoretical research in several fields. He became chief of manned spaceflight at NASA Headquarters in 1958. In 1960, he chaired a special committee that formulated the original plans for the Apollo lunar landings. In 1964 he became deputy director of the Manned Spacecraft Center in Houston, the forerunner of the Johnson Space Center. He became deputy administrator of NASA in 1969 and served as acting administrator in 1970-1971. He retired from NASA in 1976 to become president of RPI, a position he still held until his death. In 1990 NASA renamed its quality and excellence award after him. See "Low, George M.," Deputy Administrator files, NASA Historical Reference Collection.

Glynn S. Lunney (1936-) was a longtime NASA official. Trained as an aeronautical engineer, he came to the Lewis Research Center near the time of the creation of NASA in 1958 and became a member of the Space Task Group developing Project Mercury the next year. He worked on the Apollo program in a series of positions, including manager of the Apollo Spacecraft Program in 1973 and manager of the Apollo-Soyuz Test Project at the Johnson Space Center in Houston. Thereafter, he managed the development of the Space Shuttle and served in several other NASA positions. Lunney retired from NASA in 1985 and became vice president and general manager, Houston Operations, for Rockwell International's Space Systems Division. See "Lunney, Glenn S.," biographical file, NASA Historical Reference Collection.

M

Harold Macmillan (1894-1986) became a British member of Parliament in 1924, foreign secretary in 1955, and then prime minister from 1957 to 1963. See "Macmillan, (Maurice) Harold," *Current Biography Yearbook 1987*, p. 637.

Frank J. Malina (1912-1981) was a young California Institute of Technology Ph.D. student in the mid-1930s when he began an aggressive rocket research program to design a high-altitude sounding rocket. Beginning in late 1936, Malina and his colleagues started the static testing of rocket engines in the canyons above the Rose Bowl, with mixed results, but a series of tests eventually led to the development of the WAC-Corporal rocket during

World War II. After the war, Malina worked with the United Nations and eventually retired to Paris to pursue a career as an artist. See "Malina, Frank J.," biographical file, NASA Historical Reference Collection.

Vittorio Manno (1938-) is an Italian physicist who was a senior scientist at European Space Agency's Science Directorate from 1972 to 1989. From 1989 to 1995, Manno served as the scientific attaché at the Italian Embassy in Vienna. See "Miscellaneous Foreign," biographical file, NASA Historical Reference Collection.

Hans Mark (1929-) became NASA's deputy administrator in July 1981. He had previously served as secretary of the Air Force from July 1979 until February 1981 and as under secretary of the Air Force since 1977. In February 1969, Mark became director of NASA's Ames Research Center in Mountain View, California, where he managed the center's research and applications efforts in aeronautics, space science, life science, and space technology. Born in Mannheim, Germany, he came to the United States in 1940 and became a citizen in 1945. He received a Ph.D. in physics from the Massachusetts Institute of Technology in 1954. Upon leaving NASA, he became chancellor of the University of Texas at Austin. See "Mark, Hans," Deputy Administrator files, NASA Historical Reference Collection.

Robert T. Marsh, a general in the Air Force, was commander of the Air Force Systems Command from 1982 to 1984. See "Miscellaneous DOD," biographical file, NASA Historical Reference Collection.

John J. Martin was educated as a mechanical engineer, receiving a Ph.D. from Purdue University in 1951. He joined North American Aviation in 1951 and moved to the Bendix Corporation in 1953. In 1960 he joined the Institute for Defense Analyses and in 1969 moved to the staff of the President's science advisor at the White House. During 1973-1974, he served as the associate deputy to the director of the Central Intelligence Agency. He then was deputy assistant secretary of the Air Force for research and development, 1974-1976, before returning to Bendix. He became a NASA official in 1984, as associate administrator for aeronautics and space technology at NASA Headquarters, before returning to industry in 1985. See "Martin, Dr. John J.," biographical file, NASA Historical Reference Collection.

Sir Harrie S.W. Massey (1908-) was Quain professor of physics at University College in London and chaired the British National Space Research Committee in the early 1960s. He was the leader of a team of British scientists responsible for the selection of the experiments and instruments for the S.51 satellite project, a British-American cooperative effort begun in 1959 to launch individual instruments into space for scientific purposes. See "Biography, Foreign Miscellaneous, I-M," biographical file, NASA Historical Reference Collection.

James A. McDivitt (1929-) was a career Air Force officer, retiring as a brigadier general, who was chosen as a NASA astronaut in the second group selected, in 1962. He served as command pilot of the Gemini IV and commander of the Apollo 9 missions. He also managed the Apollo Spacecraft Program at Johnson Space Center from September 1969 to August 1972; he then resigned from NASA and the Air Force. Starting in 1975, he joined Pullman, Inc., in Chicago and then served as vice president, president of Pullman Standard, and executive vice president, in that order. He resigned from Pullman on January 31, 1981, to become vice president of strategic management for Rockwell International in Pittsburgh. He then became senior vice president of government and international operations for Rockwell International in Washington, D.C. See "McDivitt, James A.," biographical file, NASA Historical Reference Collection.

Frank B. McDonald (1925-) began a career with NASA in 1959 as head of the Energetic Particles Branch in the Space Science Division at the Goddard Space Flight Center in Greenbelt, Maryland. Thereafter, he served as project scientist on nine NASA satellite programs. In 1982 he became NASA's chief scientist, serving until 1987 when he returned to Goddard as associate director/chief scientist. See "McDonald, Dr. Frank B. (Chief Scientist)," biographical file, NASA Historical Reference Collection.

Neil H. McElroy (1904-1972) became secretary of defense in 1957 and served through 1959. He had previously been president of Procter & Gamble and returned there in December 1959 to become chair of the board. He served in that position until October 1972, a month before his death. See "McElroy, Neil," biographical file, NASA Historical Reference Collection.

Robert S. McNamara (1916-   ) was secretary of defense during the Kennedy and Johnson administrations, 1961-1968. Thereafter, he served as president of the World Bank, where he remained until retirement in 1981. As secretary of defense in 1961, McNamara was intimately involved in the process of approving Project Apollo by the Kennedy administration. See "McNamara, Robert S(trange)," *Current Biography Yearbook 1987,* pp. 408-13; John M. Logsdon, *The Decision to Go to the Moon: Project Apollo and the National Interest* (Cambridge, MA: MIT Press, 1970).

John B. Medaris (1902-1990) was a major general commanding the Army Ballistic Missile Agency when T. Keith Glennan tried to incorporate it into NASA in the late 1950s. He attempted to retain the organization as part of the Army, but with a series of Department of Defense agreements, the Air Force obtained primacy in space activities. Therefore, Medaris could not succeed in his effort. Medaris also worked with Wernher von Braun to launch Explorer I in early 1958. He retired from the Army in 1960 and became an Episcopal priest, later joining an even more conservative Anglican-Catholic church. See "John Bruce Medaris," biographical file, NASA Historical Reference Collection; John B. Medaris, with Arthur Gordon, *Countdown for Decision* (New York: Putnam, 1960)).

W.J. Mellors headed the Washington, D.C., office of the European Space Agency.

Clark B. Millikan (1903-1966) was a pioneer researcher in aerodynamics and guided missiles. With a Ph.D. in physics from the California Institute of Technology (Caltech), he was the son of Nobel Prize-winning Robert A. Millikan. He was appointed to the faculty of Caltech in 1928 and later became director of the Guggenheim Aeronautical Laboratory at the institute. He was enormously important in fostering rocket technology, both at Caltech and elsewhere, and he served as chair of the Guided Missile Committee for the Department of Defense during the late 1940s and early 1950s. See "Clark B. Millikan of Cal Tech Dead," *New York Times,* January 3, 1966.

Erwin Mitchell (1924-   ) (D-GA) served as a congressman from 1958 to 1961. He chaired the House Subcommittee on Patents and Scientific Inventions, which was under the Committee on Science and Astronautics. See *Biographical Directory of the United States Congress, 1774-1989* (Washington, DC: U.S. Government Printing Office, 1989).

Brooks Morris (1913-   ) was an aerospace engineer who worked as a manager of quality assurance and reliability at the Jet Propulsion Laboratory from 1961 to 1981. See *Who's Who in Aviation and Aerospace,* U.S. edition (Boston and New York: National Aeronautical Institute and Jane's Publishing Company, Ltd., 1983).

Donald Morris was a former Foreign Service official who joined NASA in 1967. Morris served as deputy assistant administrator for international affairs and then became deputy associate administrator for applications–management in 1976. In 1977 he was detailed to the President's Committee on Science and Technology. See "Assorted NASA Officials," biographical file, NASA Historical Reference Collection.

George E. Mueller (1918-   ) was associate administrator for the Office of Manned Space Flight at NASA Headquarters, 1963-1969, where he responsible for overseeing the completion of Project Apollo and for beginning the development of the Space Shuttle. He moved to the General Dynamics Corporation, as senior vice president in 1969, and remained there until 1971. He then became president of the Systems Development Corporation, 1971-1980, eventually becoming its chairman and corporate executive officer, 1981-1983. See "Mueller, George E.," biographical file, NASA Historical Reference Collection.

Robert Murphy (1894-1978) was a career Foreign Service and State Department official. He served as deputy under secretary of state for political affairs and then as under secretary in the 1950s. Biographical information from the *Biographic Register of the Department of State, 1959,* Department of State History Office, Washington, DC.

Dale D. Myers (1922-   ) served as NASA's deputy administrator from October 1986 until 1989. He had previously been under secretary of the Department of Energy from 1977 to 1979. From 1974 to 1977, he was vice president at Rockwell International and president at North American Aircraft Group in El Segundo, California. He also was the associate administrator for manned spaceflight at NASA from 1970 to 1974. From 1969 to 1970, Myers served as vice president/program manager of the Space Shuttle Program at Rockwell International. He was vice president and program manager of the Apollo Command/Service Module Program at North American-Rockwell from 1964 to 1969. After leaving NASA in 1989, Myers returned to private industry. See "Myers, Dale D.," Deputy Administrators files, NASA Historical Reference Collection.

John E. Naugle (1923-) was trained as a physicist at the University of Minnesota and began his career studying cosmic rays by launching balloons to high altitudes. In 1959 he joined NASA's Goddard Space Flight Center in Greenbelt, Maryland, where he developed projects to study the magnetosphere. In 1960 he took charge of NASA's fields and particles research program. He also served as NASA's associate administrator for the Office of Space Science and as the agency's chief scientist before his retirement in 1981. See John E. Naugle, *First Among Equals: The Selection of NASA Space Science Experiments* (Washington, DC: NASA SP-4215, 1991).

Homer E. Newell (1915-1983) earned his Ph.D. in mathematics at the University of Wisconsin in 1940 and served as a theoretical physicist and mathematician at the Naval Research Laboratory from 1944 to 1958. During part of that period, he was science program coordinator for Project Vanguard and was acting superintendent of the Atmosphere and Astrophysics Division. In 1958 he transferred to NASA to assume responsibility for planning and developing the new agency's space science program. He soon became deputy director of spaceflight programs. In 1961 he assumed directorship of the Office of Space Sciences, and in 1963, he became associate administrator for space science and applications. Over the course of his career, he became an internationally known authority in the field of atmospheric and space sciences, as well as the author of numerous scientific articles and seven books, including *Beyond the Atmosphere: Early Years of Space Science* (Washington, DC: NASA SP-4211, 1980). He retired from NASA at the end of 1973. See "Newell, Homer," biographical file, NASA Historical Reference Collection.

Richard M. Nixon (1913-1994) was president of the United States between January 1969 and August 1974. Early in his presidency, Nixon appointed a Space Task Group under the direction of Vice President Spiro T. Agnew to assess the future of spaceflight for the nation. Its report recommended a vigorous post-Apollo exploration program, culminating in a human expedition to Mars. Nixon did not approve this plan, but he did decide in favor of building one element of it, the Space Shuttle, which was approved on January 5, 1972. See Roger D. Launius, "NASA and the Decision to Build the Space Shuttle, 1969-72," *The Historian* 57 (Autumn 1994): 17-34.

O

Gerald D. O'Brien was assistant general counsel for patent matters at NASA between 1958 and 1965, when he was appointed an assistant commissioner of patents by President Lyndon B. Johnson. Previously, he had received a B.S. in electrical engineering at the U.S. Naval Academy and a law degree in 1940 from American University's Washington College of Law. He then served in the Navy as patent advisor to the National Defense Research Council during World War II. After the war, he became patent counsel of the Bureau of Ordnance, Department of the Navy, from 1946 to 1958. See "O'Brien, Gerald D.," biographical file, NASA Historical Reference Collection.

Henk Olthof (1944-) is a Dutch physicist who has worked at the European Space Agency since 1977. From 1977 to 1986, he was responsible for the secretariat of the Astronomy Working Group. Since 1986, Olthof has served as the head of space station and platforms for scientific users at the European Space Research and Technology Centre in the Netherlands. See "Miscellaneous Foreign," biographical file, NASA Historical Reference Collection.

P

Edgar Page (1935-) is an Irish physicist who specialized in cosmic ray research while at the European Space Research Organization from 1965 to 1975. He then became head of the European Space Agency's Space Science Department. Beginning in 1986, he has served as the science coordinator for the Ulysses spacecraft mission. See "Miscellaneous Foreign," biographical file, NASA Historical Reference Collection.

Thomas O. Paine (1921-1992) was appointed deputy administrator of NASA on January 31, 1968. Upon the retirement of James E. Webb on October 8, 1968, he was named acting administrator. He was nominated as NASA's third administrator on March 5, 1969, and confirmed by the Senate on March 20, 1969. During his leadership, the first seven Apollo crewed missions were flown, in which twenty astronauts orbited the Earth, fourteen traveled to the Moon, and four walked on its surface. Paine resigned from NASA on September 15, 1970, to return to the General Electric Company in New York City as vice president and group executive, Power Generation Group, where he remained until 1976. In 1985 the White House chose Paine as chair of the National Commission on Space to prepare a report on the future of space exploration. Since leaving NASA fifteen years earlier, Paine had been a tireless spokesperson for an expansive view of what should be done in space. The Paine Commission took most of a year to prepare its report, largely because it solicited public input in hearings throughout the United States. The report, *Pioneering the Space Frontier,* was published in a lavishly illustrated, glossy format in May 1986. It espoused a "pioneering mission for 21st-century America"—"to lead the exploration and development of the space frontier, advancing science, technology, and enterprise, and building institutions and systems that make accessible vast new resources and support human settlements beyond Earth orbit, from the highlands of the Moon to the plains of Mars." The report also contained a "Declaration for Space," which included a rationale for exploring and settling the solar system and outlined a long-range space program for the United States. See Roger D. Launius, "NASA and the Decision to Build the Space Shuttle, 1969-72," *The Historian* 57 (Autumn 1994): 17-34.

Frank Parker (1916-) was assistant director of Defense Research and Engineering at the Department of Defense from 1959 to 1961. This information is from the Office of the Secretary of Defense Historical Branch, Department of Defense, Washington, DC.

Robert J. Parks (1922-) was a longtime employee at the Jet Propulsion Laboratory (JPL), arriving there in 1947 after completing his education at the nearby California Institute of Technology. Closely associated with robotic planetary exploration, he worked on the Mariner, Ranger, and Surveyor programs. He served as JPL's planetary program director in the 1960s and then became JPL's associate and finally deputy director. See "Parks, Robert J.," biographical file, NASA Historical Reference Collection.

Kenneth S. Pedersen (1939-) served in numerous government agencies—the Office of Equal Opportunity, the Department of Commerce, the Atomic Energy Commission, and the Nuclear Regulatory Commission—prior to coming to NASA in 1982 as director of international affairs. In 1988 Pedersen was appointed as NASA's associate administrator for external relations, serving until 1990, when he left NASA to accept an academic appointment at Georgetown University. See "Pedersen, Kenneth S.," biographical file, NASA Historical Reference Collection.

Charles J. Pellerin, Jr., was a longtime NASA official who began his career at the Goddard Space Flight Center as he was completing his Ph.D. in physics from the Catholic University of America in 1974. The next year he moved to NASA Headquarters, where he managed the development and integration of scientific instrumentation for flight on the Space Shuttle. In 1983 he was named director of astrophysics in NASA's Office of Space Science and Applications, and in 1992, he was appointed as deputy associate administrator for safety and mission quality, serving until 1994. See "Pellerin, Charles J., Jr.," biographical file, NASA Historical Reference Collection.

Boris N. Petrov (1913-1980) was a leading Soviet scientist whose later years were devoted to space exploration. As a senior academician for the Soviet Academy of Sciences, Petrov chaired the Inter-Cosmos Council, which promoted cooperation in space among eastern European nations during the height of the cold war, 1966-1980. See "Boris Petrov, 67, Soviet Expert on Automation, Space Research," *Washington Post,* August 27, 1980; Kenneth W. Gatland, "Boris Petrov," *Spaceflight* 23 (January 1981): 29.

Franklyn W. Phillips (1917-) graduated from the Massachusetts Institute of Technology in 1941 with a degree in mechanical engineering. He then worked at the Langley Aeronautical Laboratory, later moving to Lewis Flight Propulsion Laboratory, where he conducted research on aircraft engine materials and stresses. In 1945 he became a member of the NACA director's staff and served as administrator for a variety of NACA research programs in aircraft engines and aircraft and missile structures and loads. In October 1958, he became special assistant to T. Keith Glennan, NASA's first administrator. He relinquished that position in January 1959 to become acting secretary of the National Aeronautics and Space Council, but in February 1960, he returned to his position as Glennan's assistant. He continued in that job under James E. Webb until 1962, when he became director of NASA's new northeastern office. In 1964 he became assistant director for administrative operations at the new

NASA Electronics Research Center in Cambridge, Massachusetts. This information is from background summaries of top NASA staff, NASA Historical Reference Collection.

Samuel C. Phillips (1921-1990) was trained as an electrical engineer at the University of Wyoming, but he also participated in the Civilian Pilot Training Program during World War II. Upon his graduation in 1942, Phillips entered the Army infantry but soon transferred to the air component. As a young pilot, he served with distinction in the Eighth Air Force in England—earning two distinguished flying crosses, eight air medals, and the French croix de guerre—but he quickly became interested in aeronautical research and development. He was involved in the development of the successful B-52 bomber in the early 1950s and headed the Minuteman intercontinental ballistic missile program in the latter part of the decade. In 1964 Phillips, by this time an Air Force general, was lent to NASA to head the Apollo lunar landing program, which, of course, was unique in its technological accomplishment. He returned to the Air Force in the 1970s and commanded the Air Force Systems Command prior to his retirement in 1975. See "Gen. Samuel C. Phillips of Wyoming," *Congressional Record*, August 3, 1973, S-15689; Rep. John Wold, "Sam Phillips: One Who Led Us to the Moon," *NASA Activities*, May/June 1990, pp. 18-19; obituary in *New York Times*, February 1, 1990, p. D1.

William H. Pickering (1910-) obtained his bachelor's and master's degrees in electrical engineering and then a Ph.D. in physics from the California Institute of Technology before becoming a professor of electrical engineering there in 1946. In 1944 he organized the electronics efforts at the Jet Propulsion Laboratory (JPL) to support guided missile research and development, becoming project manager for Corporal, the first operational missile JPL developed. From 1954 to 1976, he was director of JPL, which developed the first U.S. satellite (Explorer I), the first successful U.S. circumlunar space probe (Pioneer IV), the Mariner flights to Venus and Mars in the early to mid-1960s, the Ranger photographic missions to the Moon in 1964 and 1965, and the Surveyor lunar landings of 1966 and 1967. See "Pickering, William H.," biographical file, NASA Historical Reference Collection.

Kenneth S. Pitzer (1914-) was a chemist who served as director of the Atomic Energy Commission from 1949 to 1951. From 1961 to 1968, he served as president of Rice University. From 1964 to 1965, Dr. Pitzer also served on NASA's Science and Technology Advisory Committee, and in 1965, President Lyndon Johnson appointed him a member of the President's Science Advisory Committee. Biographical information from University Relations Office of Rice University, Houston, TX.

Herman Pollack (1920-1993) was a State Department official for 28 years before retiring in 1974. He served as the department's director of international scientific and technological affairs for ten years before retiring. See obituary, *Washington Post*, April 14, 1993, p. C6, in "Biography, Other Agency Miscellaneous, N-Z," file, NASA Historical Reference Collection.

Richard W. Porter was an electrical engineer who worked on missile programs with the General Electric Company before working on Earth sciences programs at the National Academy of Sciences. In 1964 he was the academy's delegate to the Committee on Space Research (COSPAR). See "Assorted Government Officials," biographical file, NASA Historical Reference Collection.

Thomas Power (1905-1970) was an accomplished pilot who served as a general during World War II. As chief of staff to General Curtis LeMay, he was one of several top planners of the atomic bombing of Hiroshima and Nagasaki. After World War II, Power served as the commander of the Air Research and Development Command, which developed early missiles. He served as commander of the Strategic Air Command from 1957 to 1964, when he retired. See "Power, Thomas," biographical file, NASA Historical Reference Collection.

Donald L. Putt (1905-1988) was a career U.S. Air Force officer who specialized in the management of aerospace research and development activities. Trained as an engineer, he entered the Army Air Corps in 1928 and served in a series of increasingly responsible posts at the Air Materiel Command and Air Force headquarters. From 1948 to 1952, he was director of research and development for the Air Force, and between 1952 and 1954, he was first vice commander and then commander of the Air Research and Development Command. Thereafter, until his retirement in 1958, he served as deputy chief of the development staff at Air Force headquarters. See "Putt, Donald," biographical file, NASA Historical Reference Collection.

Donald A. Quarles (1894-1959) was deputy secretary of defense between 1957 and 1959. Just after World War II, he had been a vice president first at the Western Electric Company and later at Sandia National Laboratories, but in 1953, he accepted the position of assistant secretary of defense for research and development. He also was secretary of the Air Force between 1955 and 1957. See "Quarles, Donald," biographical file, NASA Historical Reference Collection.

J. Danforth (Dan) Quayle (R-IN) served as a senator before becoming George Bush's vice president from 1989 to 1993. As vice president, he chaired the National Space Council and had significant involvement with the development of the space station, Space Shuttle replacement options, the Space Exploration Initiative, and NASA management.

Erik Quistgaard was the director general of the European Space Agency from 1980 to 1984, overseeing the Ariane rocket's development and Spacelab's many contributions to space science. See "Quistgaard, Erik," biographical file, NASA Historical Reference Collection.

R

Ronald Reagan (1911-) served as president of the United States from January 1981 until 1989. During his presidency, the maiden flight of the Space Shuttle took place. In 1984 he mandated the construction of an orbital space station. Reagan declared: "America has always been greatest when we dared to be great. We can reach for greatness again. We can follow our dreams to distant stars, living and working in space for peaceful, economic, and scientific gain. Tonight I am directing NASA to develop a permanently manned space station and to do it within a decade." See Sylvia D. Fries, "2001 to 1994: Political Environment and the Design of NASA's Space Station System," *Technology and Culture* 29 (July 1988): 568-93.

Felix Michael Rogers (1921-) was an ace fighter pilot who became an Air Force general. He was deputy chief of staff for development plans at the Air Force Systems Command and also served with the United Nations Military Armistice Commission in Korea. After working as the commander of Air University at Maxwell Air Force Base, he became commander of the Air Force Logistics Command. See U.S. Air Force biography, November 1977, for General Felix Michael Rogers, History Office, Air Force Logistics Command, Wright-Patterson Air Force Base, Dayton, OH.

William P. Rogers (1913-) was chair of the presidentially mandated blue ribbon commission investigating the *Challenger* accident of January 1986. It found that the failure had resulted from a poor engineering decision— an O-ring used to seal joints in the solid rocket booster that was susceptible to failure at low temperatures, introduced innocently enough years earlier. Rogers kept the commission's analysis on that technical level and documented the problems in exceptional detail. The commission, after some prodding by Nobel Prize-winning scientist Richard P. Feynman, did a credible job of grappling with the technologically difficult issues associated with the accident. See *Report of the Presidential Commission on the Space Shuttle Challenger Accident, Vol. I* (Washington, DC: U.S. Government Printing Office, June 6, 1986).

Dean Rusk (1909-1994) was a Rhodes scholar who studied philosophy, politics, economics, and law. After teaching government and international relations and serving in the military in World War II, Rusk joined the State Department in 1946. He held increasingly responsible positions, culminating in his appointment as secretary of state in 1961. He served as secretary for eight years, through the entire Kennedy and Johnson administrations. He was a strong supporter of U.S. involvement in Vietnam and also presided over U.S. foreign policy during the Bay of Pigs incident and the Cuban missile crisis. See "Rusk, Dean," biographical file, NASA Historical Reference Collection.

S

Roald Z. Sagdeyev (1932-) was one of the leaders of Soviet space science from the 1960s through the 1980s. He was involved in virtually every lunar and planetary probe of the Soviet Union during this era, including the highly successful Venera and Vega missions. He also advised Soviet leader Mikhail Gorbachev on space and arms control at the 1986 Geneva, 1987 Washington, and 1988 Moscow summits. In he late 1980s, he left the Soviet Union and settled in the United States, where he headed the East-West Science and Technology Center at the University of Maryland at College Park. See Roald Z. Sagdeyev, *The Making of a Soviet Scientist: My Adventures in Nuclear Fusion and Space From Stalin to Star Wars* (New York: John Wiley, 1995).

James R. Schlesinger (1929-) served in numerous governmental positions during the 1960s and 1970s. After a career at the University of Virginia, 1955-1963, and the RAND Corporation, 1963-1969, he worked for the Bureau of the Budget/Office of Management and Budget, 1969-1971. He also served as chair of the Atomic Energy Commission, 1971-1973, and secretary of defense, 1973-1975. In 1977 he was appointed head of the newly created Department of Energy. See "Schlesinger, James," biographical file, NASA Historical Reference Collection.

Bernard A. Schriever (1910-) earned a B.S. in architectural engineering from Texas A&M University in 1931 and was commissioned in the Army Air Corps Reserve in 1933 after completing pilot training. Following broken service, he received a regular commission in 1938. He earned an M.A. in aeronautical engineering from Stanford in 1942 and then flew 63 combat missions on B-17s with the 19th Bombardment Group in the Pacific Theater during World War II. In 1954, he became commander of the Western Development Division (soon renamed the Air Force Ballistic Missile Division), and from 1959 to 1966, he was commander of its parent organization, the Air Research and Development Command, renamed Air Force Systems Command in 1961. As such, he presided over the development of the Atlas, Thor, and Titan missiles, which served not only as military weapon systems but also as boosters for NASA's space missions. In developing these missiles, Schriever instituted a systems approach, whereby the various components of the Atlas and succeeding missiles underwent simultaneous design and test as part of an overall "weapons system." Schriever also introduced the notion of concurrency, which has been given various interpretations but essentially allowed the components of the missiles to enter production while still in the test phase, thereby speeding up development. He retired as a general in 1966. See Jacob Neufeld, "Bernard A. Schriever: Challenging the Unknown," *Makers of the United States Air Force* (Washington, DC: Office of Air Force History, 1986), pp. 281-306; Robert L. Perry, "Atlas, Thor . . .," in Eugene M. Emme, ed., *A History of Rocket Technology* (Detroit: Wayne State University Press, 1964), pp. 144-160; Robert A. Divine, *The Sputnik Challenge: Eisenhower's Response to the Soviet Satellite* (New York: Oxford University Press, 1993), p. 25.

Glenn T. Seaborg (1912-) earned a Ph.D. in physics from the University of California at Berkeley in 1937 and worked on the Manhattan Project in Chicago during World War II. Afterward, he became associate director of Berkeley's Lawrence Radiation Laboratory, where he and associates isolated several transuranic elements. For this work, Seaborg received the Nobel Prize in 1951. He also served as chair of the Atomic Energy Commission between 1961 and 1971; thereafter, he returned to the University of California at Berkeley as a member of the faculty. See David Petechuk, "Glenn T. Seaborg," in Emily J. McMurray, ed., *Notable Twentieth-Century Scientists* (New York: Gale Research Inc., 1995), pp. 1803-1806.

Robert C. Seamans, Jr. (1918-), had been involved in aerospace issues since he completed his Sc.D. degree at the Massachusetts Institute of Technology (MIT) in 1951. He was on the faculty at MIT's Department of Aeronautical Engineering between 1949 and 1955, when he joined the Radio Corporation of America as manager of the Airborne Systems Laboratory. In 1958 he became the chief engineer of the Missile Electronics and Control Division. He then joined NASA in 1960 as associate administrator. In December 1965, he became NASA's deputy administrator. He left NASA in 1968, and in 1969, he became secretary of the Air Force, serving until 1973. Seamans was president of the National Academy of Engineering from May 1973 to December 1974, when he became the first administrator of the new Energy Research and Development Administration. He returned to MIT in 1977, becoming dean of its School of Engineering in 1978. In 1981 he was elected chair of the board of trustees of the Aerospace Corporation. See "Seamans, Robert C., Jr.," biographical file, NASA Historical Reference Collection; Robert C. Seamans, Jr., *Aiming at Targets* (Washington, DC: NASA SP-4106, 1996).

Frederick Seitz (1911-) was trained in mathematics and physics at Stanford and Princeton Universities and worked at a variety of corporations, laboratories, and government organizations throughout his career. He served on the National Defense Research Committee from 1941 to 1945, was a consultant to the secretary of war in 1945, served as director of the atomic energy training program at Oak Ridge from 1946 to 1947, was a science advisor to the North American Treaty Organization (NATO) from 1959 to 1960, and was a faculty member of several universities during his career. In 1962 he was elected president of the National Academy of Sciences, and he was reelected to a six-year term in 1965. In 1968 he left the academy to become president of Rockefeller University in New York City and served until his retirement. See "Seitz, Frederick," biographical file, NASA Historical Reference Collection.

Eduard A. Shevardnadze (1927-) was a reform leader of the Soviet Union along with Mikhail Gorbachev in the late 1980s. He was heavily involved in the transformation of the nation from a Communist state to one built on capitalism. Serving in a variety of senior positions, he negotiated with the United States for international cooperation in space, including the building of a space station in the 1990s. See Eduard Shevardnadze, *The Future Belongs to Freedom* (New York: Free Press, 1991).

George P. Shultz (1920-) served as director of the Office of Management and Budget after 1970, during the Nixon administration. Before that time, he had been Nixon's secretary of labor. During the Reagan administration, 1981-1989, Shultz served as secretary of state. See "Shultz, George P.," *Current Biography Yearbook 1988*, pp. 525-30.

S. Fred Singer (1924-), a physicist at the University of Maryland, proposed a Minimum Orbital Unmanned Satellite of the Earth (MOUSE) at the fourth Congress of the International Astronautics Federation in Zurich, Switzerland, in the summer of 1953. It had been based on two years of previous study conducted under the auspices of the British Interplanetary Society, which had built on the post-war research of the V-2 rocket. The Upper Atmosphere Rocket Research Panel at White Sands discussed Singer's plan in April 1954, and a month later, Singer presented his MOUSE proposal at the Hayden Planetarium's fourth Space Travel Symposium. MOUSE was the first satellite proposal widely discussed in nongovernmental engineering and scientific circles, although it never was adopted. See "Singer, S. Fred," biographical file, NASA Historical Reference Collection.

Walter D. Sohier (1924-), a graduate of Columbia Law School, had worked for the Central Intelligence Agency (CIA) and the Air Force before joining NASA in 1958 as assistant general counsel. He became deputy general counsel in 1961 and general counsel in 1963. He left NASA in 1966 to become a partner in a New York law firm. See "Sohier, Walter," biographical file, NASA Historical Reference Collection.

Thomas P. Stafford (1930-), a career military officer who retired as a lieutenant general in the U.S. Air Force, was chosen by NASA in the second group of astronauts, in 1962. He served as the backup pilot for Gemini III and the pilot for Gemini VI. He became command pilot for Gemini IX upon the death of a prime crew member and was the backup commander for Apollo 7, the commander of Apollo 10, and the commander of the Apollo-Soyuz Test Project. He resigned from NASA on November 1, 1975, to become commander of the Air Force Flight Test Center, at Edwards Air Force Base in California. He was promoted to Air Force deputy chief of staff for research and development in March 1978. He then retired from the Air Force in November 1979 and became executive vice president of commercial sales and finance for American Farm Line in Oklahoma City. He also worked as a consultant with Defense Technology in Oklahoma City and thereafter as vice chairman of Stafford, Burke and Hecker, Inc., in Alexandria, Virginia. He joined the Spectrum Information Technologies Technical Advisory Board in 1993. See "Stafford, Thomas P.," biographical file, NASA Historical Reference Collection.

Homer J. Stewart (1915-) earned his doctorate in aeronautics from the California Institute of Technology (Caltech) in 1940, joining the faculty there two years before that. In 1939 he participated in pioneering rocket research with other Caltech engineers and scientists, including Frank Malina, in the foothills of Pasadena. Out of their efforts, the Jet Propulsion Laboratory (JPL) arose, and Stewart maintained his interest in rocketry at that institution. He was involved in developing the first American satellite, Explorer I, in 1958. In that year, on leave from Caltech, he became director of NASA's Office of Program Planning and Evaluation, returning to Caltech in 1960 in a variety of positions, including chief of the Advanced Studies Office at JPL from 1963 to 1967 and professor of aeronautics at Caltech itself. See "Stewart, Homer," biographical file, NASA Historical Reference Collection; Clayton R. Koppes, *JPL and the American Space Program: A History of the Jet Propulsion Laboratory* (New Haven, CT: Yale University Press, 1982), pp. 23, 32, 44, 47, 79-80, 82.

Vladimir S. Syromiatnikov (1934-) was educated at Bauman Technical University in Moscow and went to work for RKK Energia of Kalingrad after graduating in 1956. He was the designer of one of the most successful pieces of space hardware used by the Soviet Union, the docking collar used to link two spacecraft together. It was adapted for use in the Apollo-Soyuz Test Project in 1975 and has been successful in more than 200 dockings of Soviet/Russian missions. It will be used aboard the International Space Station being constructed at the end of the twentieth century. See "Vladimir S. Syromiatnikov, Russian Docking System Engineer," *Space News*, February 12-18, 1996, p. 22.

T

Brian Taylor (1940-) joined the European Space Agency (then the European Space Research Organization) in 1967 as a staff scientist. In 1971, he became the head of the High Energy Astrophysics Division and then, in 1984, the head of the Astrophysics Division. See "Miscellaneous Foreign," biographical file, NASA Historical Reference Collection.

Albert Thomas (1898-1966) (D-TX), a lawyer and World War I veteran, had first been elected to the House of Representatives in 1936 and served successively until 1962. In 1960 he was chair of the independent offices subcommittee of the House Appropriations Committee and thus exercised considerable congressional power over NASA's funding. See "Thomas, Albert," biographical file, NASA Historical Reference Collection.

Shelby G. Tilford was a NASA scientist in the late 1980s and early 1990s in the Office of Space Science, for which he was director of Earth sciences. In 1992 he was appointed acting associate administrator for Mission to Planet Earth and served until 1994. See "Tilford, Shelby," biographical file, NASA Historical Reference Collection.

Holger N. Toftoy (1903-1967) was a career U.S. Army officer and an expert in ordnance who was responsible for bringing the German "rocket team" under the leadership of Wernher von Braun to the United States in 1945. He became commander of the Redstone Arsenal in Huntsville, Alabama, in 1954 and worked closely with von Braun's team in the development of the Redstone and Jupiter missiles. In the aftermath of the first successful Sputnik launch in 1957, he persuaded the Department of Defense to allow the launch of the first U.S. Earth-orbiting satellite aboard the Jupiter missile; the result was the orbiting of Explorer I on January 31, 1958. He also held a number of other positions in the Army; he was the head of the Rocket Research Branch of the Chief of Ordnance in Washington, D.C., and the commander of the Aberdeen Proving Ground in Maryland. He retired from the Army in 1960 with the rank of major general. See "Maj. Gen. Holger Toftoy Dies; Leader in U.S. Rocket Program," *New York Times*, April 20, 1967, p. 41.

H.S. Tsien (1909-) was a Chinese national who received a Ph.D. in aeronautics in 1939 from the California Institute of Technology (Caltech) and worked on the development of rocket technology at his alma mater through World War II. He was on the faculty of the Massachusetts Institute of Technology from 1946 to 1949, when he returned to Caltech. In the 1950s, his loyalty to democratic institutions was questioned, and he was deported from the United States to the People's Republic of China. There, he was largely responsible for the development of intercontinental ballistic missile rocket technology, especially the "Long March" launch vehicle. See Iris Chang, *Thread of the Silkworm* (New York: Free Press, 1996).

Nathan F. Twining (1897-1982) was a career pilot in the Army and the Air Force, commanding the 13th Air Force in the Pacific, the 15th Air Force in Europe, and then the 20th Air Force in the Pacific during World War II. He became chief of staff of the Air Force in 1953 and chaired the Joint Chiefs of Staff from 1957 to 1960. See Donald J. Mrozek, "Nathan F. Twining: New Dimensions, a New Look," in John L. Frisbee, ed., *Makers of the United States Air Force* (Washington, DC: Office of Air Force History, 1987), pp. 257-80.

V

James A. Van Allen (1914-) was a pathbreaking astrophysicist best known for his work in magnetospheric physics. Van Allen's January 1958 Explorer I experiment established the existence of radiation belts—later named for the scientist—that encircled the Earth, representing the opening of a broad research field. Extending outward in the direction of the Sun approximately 40,000 miles, as well as stretching out with a trail away from the Sun to approximately 370,000 miles, the magnetosphere is the area dominated by Earth's strong magnetic field. See James A. Van Allen, *Origins of Magnetospheric Physics* (Washington, DC: Smithsonian Institution Press, 1983); David E. Newton, "James A. Van Allen," in Emily J. McMurray, ed., *Notable Twentieth-Century Scientists* (New York: Gale Research Inc., 1995), pp. 2070-72.

Hoyt S. Vandenberg (1899-1954) was a career military aviator who served as chief of staff of the U.S. Air Force between 1948 and 1953. He was educated the U.S. Military Academy at West Point and entered the Army Air Corps after graduation, becoming a pilot and air commander. After numerous command positions during World War II, most significantly as commander of Ninth Air Force, which provided fighter support in Europe during the invasion and march to Berlin, he returned to Washington and helped with the formation of the Department of Defense (DOD) in 1947. As Air Force chief of staff, he was a senior official at DOD during the formative period of rocketry development and the work on intercontinental ballistic missiles. See Phillip S. Meilinger, *Hoyt S. Vandenberg: The Life of a General* (Bloomington: Indiana University Press, 1989).

W

Alan T. Waterman (1892-1967) was the first director of the National Science Foundation (NSF), from its founding in 1951 until 1963. He received his Ph.D. in physics from Princeton University in 1916. He then served with the Army's Science and Research Division during World War I, on the faculty of Yale University in the interwar years, with the War Department's Office of Scientific Research and Development during World War II, and then with the Office of Naval Research between 1946 and 1951. He and NASA leaders contended over control of the scientific projects to be undertaken by the space agency, with Waterman's NSF being used as an advisory body in the selection of space experiments. See "Waterman, First NSF Head, Dies at 75," *Science* 158 (December 8, 1967): 1293; Norriss S. Hetherington, "Winning the Initiative: NASA and the U.S. Space Science Program," *Prologue: The Journal of the National Archives* 7 (Summer 1975): 99-108; John E. Naugle, *First Among Equals: The Selection of NASA Space Science Experiments* (Washington, DC: NASA SP-4215, 1991).

James E. Webb (1906-1992) was NASA administrator between 1961 and 1968. Previously, he had been an aide to a congressman during the New Deal era in Washington, an aide to Washington lawyer Max O. Gardner, and a business executive with the Sperry Corporation and the Kerr-McGee Oil Company. He also had been director of the Bureau of the Budget between 1946 and 1950 and under secretary of state from 1950 to 1952. See W. Henry Lambright, *Powering Apollo: James E. Webb of NASA* (Baltimore: Johns Hopkins University Press, 1995).

Caspar W. Weinberger (1917-), a longtime Republican government official, was a senior member of the Nixon, Ford, and Reagan administrations. For Nixon, he was deputy director (1970-1972) and director (1972-1976) of the Office of Management and Budget. In this capacity, he had a leading role in shaping the direction of NASA's major effort of the 1970s, the development of the reusable Space Shuttle. For Reagan, he served as secretary of defense, where he also oversaw the use of the Space Shuttle in the early 1980s for the launching of classified Department of Defense payloads into orbit. See "Weinberger, Caspar W(illard)," *Current Biography Yearbook 1973*, pp. 428-30.

Edward C. Welsh (1909-1990) had a long career in various private and public enterprises. He had served as legislative assistant to Senator Stuart Symington (D-MO), 1953-1961, and was the executive secretary of the National Aeronautics and Space Council through the 1960s. See "Welsh, Edward," biographical file, NASA Historical Reference Collection.

Fred L. Whipple (1906-) received a Ph.D. in astronomy from the University of California at Berkeley. He then served on the faculty of Harvard University. He was involved in efforts during the early 1950s to expand public interest in the possibility of spaceflight through a series of symposia at the Hayden Planetarium in New York City and articles in *Collier's* magazine. He was also heavily involved in planning for the International Geophysical Year, 1957-1958. As a pathbreaking astronomer, he pioneered research on comets. See Raymond E. Bullock, "Fred Lawrence Whipple," in Emily J. McMurray, ed., *Notable Twentieth-Century Scientists* (New York: Gale Research Inc., 1995), pp. 2167-70.

James F. Whisenand (1911-) was trained as an aeronautical engineer at the University of Illinois and entered the Army Air Corps in 1934. Serving in a variety of command and staff positions, including in combat in World War II and Korea, he served as special assistant to the chair of the Joint Chiefs, General Nathan F. Twining, beginning in 1957 as a major general. See "Biography, DOD Miscellaneous, N-Z," biographical file, NASA Historical Reference Collection.

Gordon P. Whitcomb (1940-) is a British engineer who began his career working on automatic landing systems for civilian aircraft. In 1974 he joined the European Space Research Organization to work on spacecraft system design. Currently, he is the head of the European Space Agency's Future Science Projects Office. See "Miscellaneous Foreign," biographical file, NASA Historical Reference Collection.

Thomas D. White (1901-1965) was a career Air Force officer who served in a succession of increasingly responsible positions until his retirement in 1961. He was director of legislation for the secretary of the Air Force between 1948 and 1951, deputy chief of staff for operations from 1951 to 1953; vice chief of staff from 1953 to 1957, and chief of staff from 1957 to 1961. See "White, T.D.," biographical file, NASA Historical Reference Collection.

Clay T. Whitehead was a White House staff assistant during the Nixon administration between 1969 and 1972. He was heavily involved in space policy associated with the decision to build the Space Shuttle and post-Apollo planning for NASA. See Roger D. Launius, "NASA and the Decision to Build the Space Shuttle, 1969-72," *The Historian* 57 (Autumn 1994): 17-34; Roger D. Launius, "A Western Mormon in Washington, D.C.: James C. Fletcher, NASA, and the Final Frontier," *Pacific Historical Review* 64 (May 1995): 217-41.

Jerome B. Wiesner (1915-1994) was science advisor to President John F. Kennedy. He had been a faculty member of the Massachusetts Institute of Technology and had served on President Eisenhower's Science Advisory Committee. During the presidential campaign of 1960, Wiesner had advised Kennedy on science and technology issues and prepared a transition team report on the subject that questioned the value of human spaceflight. As Kennedy's science advisor, he tussled with NASA over the lunar landing commitment and the method of conducting it. See Gregg Herken, *Cardinal Choices: Science Advice to the President from Hiroshima to SDI* (New York: Oxford University Press, 1992).

Lynette (Lyn) Wigbels is the assistant director for international programs on the Global Learning and Observations to Benefit the Environment (GLOBE) program. She joined NASA's International Affairs Division in 1979 and developed the space station agreements covering cooperation with Europe, Japan, and Canada. She has also held several other policy and internationally related positions at NASA. Biographical sketch from Lyn Wigbels and "Wigbels, Lyn," "Miscellaneous NASA Officials," biographical file, NASA Historical Reference Collection.

Y

John F. Yardley (1925-) was an aerospace engineer who worked with the McDonnell Aircraft Corporation on several NASA human spaceflight projects between the 1950s and the 1970s. He also served as NASA associate administrator for spaceflight between 1974 and 1981. Thereafter, he returned to McDonnell Douglas as president, 1981-1988. See "Yardley, John F.," biographical file, NASA Historical Reference Collection.

Boris N. Yeltsin (1930-) became leader of Russia in the immediate post–Cold War era in the early 1990s and carried even further democratic reforms than had his predecessor, Mikhail Gorbachev. One of his principle objectives was closer ties to the West, and under his leadership, the international partnership to build a space station came much closer to reality. See "Yeltsin, Boris N.," biographical file, NASA Historical Reference Collection.

Index

A

Adams, Mac C., 363

Administrative Services Act of 1949, 440

Advanced Research Projects Agency (ARPA), 249-51, 285-86, 293, 296, 298-301, 311, 314, 449, 480; and creation of NASA, 248-53; and loses program, 254-55

Aerojet General Corporation, 371, 510

Aeronautics and Astronautics journal, 213

Aeronautics and Astronautics Coordinating Board (AACB), 254, 296-97, 346, 349, 352, 357, 359, 360, 372, 383, 389

AEROSAT cooperative satellite system, 85, 87

Aerospace Corporation, 320, 322

Afghanistan, 215

Agena rocket, 254, 257, 373, 482

Agnew, Spiro, 40, 95

Air Force, U.S., 164-65, 181, 233, 238, 299, 301, 303-04, 309, 311, 313-14, 350, 355, 389-410, 423, 444-45, 469, 476, 481-82, 509; and Air Force Space Center (AFSC), 369; and ARPA, 249-51, 254-55, 285-86, 298, 311, 339, 366-79, 383, 384, 449; and Blue Gemini, 258-63; and Civil-Military Liaison Committee, 253-54; and creation of NASA, 248-53; and defense reorganization, 247-48; and Defense Special Projects Agency (DSPA), 246-50; and Dyna-Soar program, 166, 234, 246, 252, 259, 260-62, 268, 280-85, 310; and Explorer I, 244-45; and *Feasibility of Weather Reconnaissance from a Satellite Vehicle*, 237; and "Freedom of Space," 239, 244-45, 273; and ICBM development, 242-44; and IGY, 237-38, 275-77; and Manned Orbital Laboratory (MOL), 262-63; and "Meeting the Threat of Surprise Attack," 238-39; and Minimum Orbital Unmanned Satellite of the Earth (MOUSE), 237-38; and NASA roles and missions, 290-92; and "Preliminary Design for an Experimental World Circling Spaceship," 236; and Project Feed Back, 237; and Project RAND, 236, 271-72; and relations with NASA, 233-410; and Robert McNamara, 257-63, 312; and scientific satellite program, 240-42; and Space Shuttle, 26-69; and space studies, 236-38; and Sputnik, 244-46, 251; and "Statement of Policy for a Satellite Vehicle," 236; and *Utility of a Satellite Vehicle for Reconnaissance*, 237; and Vanguard Program, 242-45, 277-80

Air Force Ballistic Missile Division (AFBMD), 254, 255-56

Air Materiel Command, 236, 271-72

Air Research and Development Command (ARDC), 243, 257, 310

Albert, John, 370

Aldridge, Edward C., 267, 410

Aleksandrov, A. P., 215, 217

Allnutt, Robert J., 26-27

American Astronomical Society (AAS), 541, 548

American Bar Association, 461

American Institute of Aeronautics and Astronautics (AIAA), 94, 182

American Rocket Society, 238, 240

American Society of Electrical Engineers (ASEE), 525

Ames Research Center, California, 264

Anders, William A., 8, 63-65, 68-69, 370

Anderson, Clinton P., 182, 415

Antarctica, 71

Apollo, Project, 5, 7-9, 13, 34, 48-51, 53-61, 64, 66-74, 77-84, 91-92, 95-97, 100, 165, 176, 81-82, 189-211, 214-15, 322-23, 330, 343, 357, 365, 383, 411, 417-21, 427-29, 485, 490, 495, 497, 520; and Apollo 8 mission, 62; and Apollo 11 mission, 6, 34-35, 40, 62, 72; and Apollo 204 capsule fire, 34, 426-27, 527-37; and Apollo-Soyuz Test Project (ASTP), 14, 182, 189-211, 214-15, 379; and Robert McNamara, 257-63

Applications Technology Satellite (ATS) program, 24

Ariane launch vehicle, 85, 137-38

Ariel 1, 3, 10, 19

Armed Services Procurement Act of 1947, 413-14, 438, 445, 457, 480

Arms Control and Disarmament Agency (ACDA), 96, 103
Army, U.S., 235, 236, 299, 301, 303, 314, 476, 481; and Army Ballistic Missile Agency (ABMA), 288-90; and ARPA, 249-51, 254-55, 285-86, 298, 444; and Civil-Military Liaison Committee, 253-54; and creation of NASA, 248-53; and defense reorganization, 247-48; and Defense Special Projects Agency (DSPA), 246-50; and Explorer I, 244-45; and Janus Report, 243 246; and Jet Propulsion Laboratory (JPL), 287-88, 503; and "Meeting the Threat of Surprise Attack," 238-39; and *Minimum Satellite Vehicle: Based on Components available from missile development of the Army Ordnance Corps, A,* 238; and NASA roles and missions, 290-92; and Robert McNamara, 257-63; and scientific satellite program, 240-42; and Vanguard Program, 242-45, 277-80
Army Air Forces, see Army, United States
Army Ballistic Missile Agency (ABMA), 240-45, 246, 416, 480-81; and ARPA, 249-51, 254-55; and Civil-Military Liaison Committee, 253-54; and creation of NASA, 248-53, 412; and Defense Special Projects Agency (DSPA), 246-50; and NASA roles and missions, 290-92; and transfer to NASA, 288-90
Army-Navy Aeronautical Board, 236
Army Ordnance Enemy Equipment Intelligence Section, 235
Army Signal Corps, 481
Arnold Engineering Development Center, 372
Arnold, Gen. Henry H. (Hap), 236
Asia, 182
AT&T Corporation, 419; and Bellcomm, 419
Atkinson, Paul, 370
Atlantic Missile Range, 18
Atlas launch vehicle, 38, 243, 254, 267, 327, 332, 476; and Atlas-Agena B launch system, 321; and Atlas-Centaur launch system, 384, 399
Atomic Energy Commission (AEC), 51, 167, 274, 299-302, 414-15, 439-40, 443, 451-52, 458-60, 471-72, 476, 481, 487, 492, 509, 514-15
Attached Pressurized Module (APM), 113, 117-23, 126, 129, 133
Atwood, J. Leland, 429, 527, 536
Australia, 582-83; and Cape York launch program, 582-83
Aviation Week and Space Technology, 246, 428
Austin, 46
Australia, 6, 42, 44, 157
Austria, 97

B

Bacher, Robert F., 511-12
Baikonaur launch complex, 203
Baker, Bobby, 428
Baker, Kenneth, 107-108
Ball Brothers Corporation, 482
Ball, George, 162-64, 166
Baltimore, 430
Banks, Peter, 109
Barinov, Y. A., 153, 155, 158
Barnes, Richard
Bean, Alan, 201
Beckman, Arnold O., 503, 510, 512
Beckman Instruments, Inc., 510
Beggs, James M., 10, 90, 100, 105, 107, 110, 410, 432
Belgium, 38, 59
Bell, David, 421; and "Bell Report," 421-22
Bellcomm, see AT&T Corporation
Bell Telephone Company, 482
Berkner, Lloyd V., 240, 418, 493, 495-96
Berlin Wall, 11

Billingsley, Henry E., 17
Bissell, Richard, 245
Black, Fred, 428
Blagonravov, Anatoli, 12, 152-67, 172, 177-81
Blue Gemini, Project, 258-63
Boeing Company, 212, 329, 427-28
Bogart, Frank, 370
Boggess, Nancy, 32
Boise Cascade Corporation, 490
Bolshoi Theater, 184
Bombay, 24
Bondi, Herman, 40-41
Bonn, Germany, 105-106
Bonnet, Roger M., 31-34
Boone, Adm. W. Fred, 259, 347-48
Borman, Frank, 35, 37, 63
Boston, 423
Boston University, 237
Bowden, Lewis, 153, 155, 158
Bowie, Robert B., 273-74
Bowman, Isaiah, 498
Brackett, Ernest, 414
Brandt, Willy, 43
Braun & Co., C. F., 510
Braun, John G., 510
Braun, Wernher von, 233, 236, 238, 243, 286, 288-90; and ARPA, 249-51, 254-55; and "arsenal" system, 412; and
    Civil-Military Liaison Committee, 253-54; and creation of NASA, 248-53; and defense reorganization, 247-48;
    and Defense Special Projects Agency (DSPA), 246-50; and Robert McNamara, 257-63; and scientific satellite
    program, 240-45; and Sputnik, 244-46, 251
Brazil, 4, 24
Brezhnev, Leonid A., 14
Briggs, Geff, 220
British National Space Committee, 3
Bronk, Detlov, 240
Brooks, Overton, 256, 257-28, 311-12, 317
Brown, George, 492-93
Brown, Harold, 338, 340
Brucker, Wilber M., 287-90
Brundage, Percival, 244-45, 277-80
Brussels, Belgium, 59, 62
Bundy, McGeorge, 162, 164-67
Bureau of the Budget, 167, 340, 346, 421, 423, 443, 492; also see Office of Management and Budget
Burke, Arleigh, 298
Bush, George H. W., 15, 220, 433, 582-83, 586
Business Week, 486

C

California Institute of Technology (Caltech or CIT), 235, 244, 287-88, 423-24, 496, 503-20
Canada, 4-6, 10, 42-44, 49, 87, 95, 98-100, 107-108, 110-111, 181, 439; and Remote Manipulator System
    (Canadarm), 7, 48-49, 98; also see Ministry of State for Science and Technology (MOSST) of Canada
Cape Canaveral, Florida, 146, 264, 378, 419, 491
Carter, James (Jimmy), 14, 26, 265
Case Institute of Technology, 412
Cassini spacecraft program, 33

D

Dallas, 493

Daniels, Chester S., 551

David, Edward E., Jr., 52, 54, 58

Dayton, Ohio, 243

Deep Space Network, 509

Defense Astronautical Agency, 298

Defense Early Warning (DEW) Line, 275-76

Defense, Department of (DOD), 6-7, 46, 56, 70, 74, 76, 96-98, 103, 107, 109, 163-64, 167-68, 413-15, 439-40, 443-58, 462-63, 470, 480-81, 492-93, 518-19, 579-80; and ARPA, 249-51, 254-55, 285-86, 293,; and Blue Gemini, 258-63; and Civil-Military Liaison Committee, 253-54; and creation of NASA, 248-53; and defense reorganization, 247-48; and Defense Space Operations Committee, 403-09; and Defense Special Projects Agency (DSPA), 246-50; and "Draft Statement of Policy on U.S. Scientific Satellite Program" (NSC 5520), 241, 245, 273, 277; and Dyna-Soar program, 166, 234, 246, 252, 259, 260-62, 268, 280-85; and Explorer I, 244-45; and "Freedom of Space," 239, 244-45, 273; and Janus Report, 243, 246; and "Meeting the Threat of Surprise Attack," 238-39; and NASA roles and missions, 290-92; and Office of Defense Mobilization, 238-39, 274; and relations with NASA, 233-410, 417; and Robert McNamara, 257-63; and scientific satellite program, 240-42; and Space Shuttle, 26-69; and Technological Capabilities Panel, 238-39, 240, 273-76; and Vanguard Program, 242-45, 277-80

Defense Meteorological Satellite Program (DMSP), 399

Defense Production Act, 446

Defense Special Projects Agency (DSPA), 246-50

Defense Mobilization, Office of, 238-39, 274

Delta launch vehicle, 46, 384, 482

DePauw University, 469

Donahue, Thomas, 553

Dornier Corporation, 375

Douglas Aircraft Company, 482

Douglas, James H., 296, 298

"Draft Statement of Policy on U.S. Scientific Satellite Program" (NSC 5520), 241, 245, 273, 277

Draper, Charles Stark, 420

Drew, Russell, 63

Dryden, Hugh L., 2, 12, 17-18, 21-23, 152-64, 167, 172, 177-80, 280-85, 338, 340, 347-48, 417-19, 423, 427, 495-97

DuBridge, Lee A., 35, 424, 496, 505, 510, 514

Dulles, Allen, 240

Durant, Frederick C., III, 238

Dyna-Soar program, 166, 234, 246, 252, 259, 260-62, 268, 280-85, 310, 322

E

Eastern Test Range, 367

Echo satellite program, 154, 159, 163, 168, 481-82

Economic Policy Council, see White House

Edelson, Burton I., 31, 33-34

Education, Department of, 580

Eggers, Alfred J., Jr., 363, 367

Einhorn, Raymond, 512-13

Eisenhower, Dwight D., 1, 11, 16, 298, 412, 417; and ABMA, 288-90; and ARPA, 249-51, 254-55; and creation of NASA, 248-53; and defense reorganization, 247-48; and DSPA, 246-50; and "Draft Statement of Policy on U.S. Scientific Satellite Program" (NSC 5520), 241, 245, 273, 277; and Dyna-Soar program, 285-86; and Explorer I, 244-45; and "Freedom of Space," 239, 244-45, 273; and Jet Propulsion Laboratory (JPL), 286-88; and NASA roles and missions, 290-92; and scientific satellite program, 240-42; and Sputnik, 2, 234, 245-47, 251; and Vanguard Program, 242-45, 277-80

Electronics Research Center, 422-23

Elms, James C., 363
Erlichman, John, 50
Emery, Roger, 34
Energiya, see NPO Energiya
Energy, Department of, 579
Essmeier, Charles T., 371
Europa III launch vehicle program, 53, 57, 71, 74
Europe, 3-7, 10, 21-24, 28, 36-39, 41, 43-45, 48, 50-100, 106, 108, 197, 545
European Launcher Development Organization (ELDO), 36-41, 43, 45, 372
European Prepatory Commission for Space Research, 21
European Science Foundation (ESF), 541
European Space Agency (ESA), 4-5, 9-10, 26-33, 36, 81, 87-88, 95, 97, 99-100, 106, 108, 110-12, 115-42, 583; and
    European Space Technology Education Centre, 32; and Hubble Space Telescope, 545
European Space Conference (ESC), 8, 43, 45, 56-57, 60-62, 79, 81-84
European Space Research Organization (ESRO), 3-4, 21-22, 35, 37-42, 45
Evans, Harry L., 363
Explorer spacecraft program, 244-45, 252, 561

F

Feasibility of Weather Reconnaissance from a Satellite Vehicle, 237
Federal Aviation Administration (FAA), 41
Federal Civil Defense Authority (FCDA), 274
Feed Back, Project, 237, 239, 240, 242
Feibleman, James A., 371
Feoktistov, K. P., 187
Finarelli, Margaret, 26
Fink, Daniel J.
Flanigan, Peter, 8, 49-50, 52-53, 62-66, 68
Fletcher, James C., 48, 50, 52-53, 59, 64, 66-69, 79, 198, 264-65, 378, 380-82, 385, 390
Flight Telerobotic System (FTS), 113
Florence, 29, 81
Ford, Henry, 488
Ford, Ralph, 370
Forman, Howard L., 470-71
Fortune magazine, 487
Foster, John S., Jr., 361, 364, 378
France, 21, 26, 37-39, 42-43, 85-86, 88, 97, 100, 107
Freedom of Information Act, 580
French Guiana, 38
Frosch, Robert A., 430, 550, 552
Frutkin, Arnold W., 2, 9, 22-23, 48-49, 58-59, 63, 65, 85-87, 152, 155, 158, 162, 187, 190-91, 194-95, 197, 201-203,
    213
Fuqua, Don, 379

G

Gabriel, Gen. Charles A., 266, 390-91
Gagarin, Yuri, 11, 143, 185; and Yuri Gagarin Cosmonaut Training Facility, 224-25; also see Star City
Galileo mission, 5, 554, 564
Gamma Ray Observatory (GRO) program, 26, 554
Gates, Thomas S., Jr., 292
Gemini program, 165, 181, 210, 330-31, 337-38, 340-44, 350, 357, 426
General Accounting Office (GAO), 550-51
General Dynamics Corporation, 212

General Electric Company, 419
General Mills Corporation, 482, 490
General Services Administration (GSA), 440, 445
Geneva, Switzerland, 161-64
George Washington University, 389
Geostationary Operational Environmental Satellite (GOES) program, 561
Germany, 36-37, 39, 41-43, 55, 88, 106-107, 375
Gilbert, M. Raymond, 369-70
Gilmore, Robert B., 510-11
Gilpatric, Roswell L., 165, 338, 343, 492-93
Gilruth, Robert, 186-87, 419, 494
Glenn, John H., Jr., 12, 147, 155
Glennan, T. Keith, 3, 17, 19-20, 175, 252, 292, 296, 298, 411-18, 429, 436-37, 442, 472-74; and ABMA, 288-90; and Dyna-Soar program, 285-86; and Jet Propulsion Laboratory (JPL), 286-88; and NASA roles and missions, 290-92
Global Atmospheric Research program, 41
Global Positioning System (GPS) program, 399, 408-09
Goddard Space Flight Center (GSFC), Maryland, 48-49, 209, 429-30, 475-76, 482, 547
Goldberg, Leo, 543
Goldin, Daniel S., 220, 228, 232
Golovin, Nicholas, 258, 318, 320
Goodpaster, Gen. Andrew, 250
Gorbachev, Mikhail S., 15, 217
Gore, Albert, Jr., 222, 230
Government Patent Board, 454
Graham, Brig. Gen. Wallace, 238
Green, Cecil, 492-93
Griffin, Gerry, 379
Grosse, Aristide V., 238
Grumman Corporation, 375, 427
Guggenheim Aeronautical Laboratory, California Institute of Technology (GALCIT), 235
Guhin, Michael, 58
Gutman, Richard W., 550-51

H

Hage, George, 34
Hagerty, James C., 16
Hahn, Herbert L., 510
Haig, Alexander, 63
Halley's Comet, 4-5
Hansen, Grant, 268, 366, 369-70
Harbridge House, 510
Hardy, George K., 186-88
Harriman, W. Averell, 166
Harris, Ray M., 469-72
Harvard University, 418, 423, 469, 485, 493
Haskell, George, 34
Heaton, D. H., 474
Hedrick, Walter, 369
Henry, Lt. Gen. Richard C., 267, 398, 400
Hermes program, 137-38
High Energy Astronomy Observatory (HEAO), 561
Hilburn, Earl, 422, 511, 513
Hinners, Noel W., 430, 541, 544

Hodge, John, 9, 90
Holmes, D. Brainerd, 348
Hood, Valerie, 34
Hornig, Donald F., 153-55, 158, 546, 548
Horowitz, Norman, 511
Horwitz, Abraham, 502
Houston, 43, 209, 223, 225, 371, 418-19, 428-30, 491-92, 538-39
Houston Ship Canal, 493
Hubble Space Telescope, 5, 544; also see Space Telescope
Hulst, H. C. van de, 2, 17, 18
Huntsville, Alabama, 240, 244, 371, 475, 477

I

India, 4, 44, 225
Indian Ocean, 37
Inertial (or Interim) Upper Stage (IUS), 28, 382, 393
Infrared Astronomical Satellite (IRAS) program, 32, 554
Infrared Space Observatory (ISO) program, 32
Institute of Geochemistry (of Soviet Union), 211
Institute of Space Research (of Soviet Union), 211
Institute for Space and Astronautical Science (of Japan) (ISAS), 4, 33
Institute for Space Studies, 429-30
Inter-Agency Consultative Group, 5
International Academy of Astronautics, 174
International Astronautical Federation, 237-38
International Council of Scientific Unions (ICSU), 2
International Geophysical Year (IGY), 2, 5, 11, 17, 237, 240, 275-76
International Solar Polar Mission (ISPM), 5, 26-28, 31
International Space Station program, 15, 87-90, 223-32
International Mobile (formerly Maritime) Satellite Organization (INMARSAT), 221-22
International Telecommunications Satellite Consortium (INTELSAT), 8, 24-25, 36, 38-39, 46-47, 51, 55, 60, 62, 71, 105
International Telecommunications Union (ITU), 154, 157
International Ultraviolet Explorer (IUE), 545, 556-57, 561
International Union of Geodesy and Geophysics, 163
Italy, 29, 37-38, 43, 81, 88, 106-107

J

J-2 engine, 415
James, Lee, 369-70
Janus Report, 243, 246
Japan, 4-7, 10, 22-24, 42, 44-47, 66, 87, 95, 98-100, 106-08, 110-11, 157, 160, 202
Japanese Experiment Module (JEM), 114, 126, 129, 133
Jastrow, Robert, 429
Jet-assisted takeoff (JATO), 235
Jet Propulsion Laboratory (JPL), 160, 197-98, 235, 244, 245, 252-53, 286-88, 320, 329, 423-24, 474-77, 482, 503-20
Jodrell Bank Observatory, 168
Johns Hopkins University, 430, 498
Johnson, Caldwell C., 187-88
Johnson, Charles E., 166-67
Johnson, John, 414, 442-472
Johnson, Lyndon B., 13, 22, 37, 46, 168, 170-71, 179, 260, 342, 417-19, 423, 425, 428, 492

Johnson, Roy W., 250, 252, 285-86

Johnson Space Center (JSC), Texas, 183, 203, 223, 225, 384, 418-19, 434, 495; also see Manned Spacecraft Center and Houston

Johnson, U. Alexis, 23-24, 46, 53, 58-62, 64, 78, 82-83

Johnson, Vincent, 369-70

Joint Chiefs of Staff, 254-55

Jonas, Charlie, 418, 492-93

Jones, Thomas V., 510

Jonsson, Jon Erik, 492-93

Journal of Medical Education, 502

Juno rocket, 252

Jupiter, 5, 560

Jupiter launch vehicle program, 252, 332, 476

Justice, Department of, 472

K

Kaiser, Robert, 191

Kaliningrad, 208

Kavanau, Lawrence L., 320

Keathley, William C., 544

Kefauver, Estes, 415

Keldysh, M. V., 12-14, 44, 182-83, 189-95, 197, 200, 211, 213-14

Kelly, William P., 474

Kennedy, Edward, 423

Kennedy, John F., 5, 11-13, 15, 143, 147, 149, 153, 156, 159, 161-62, 165-66, 168-71, 175, 178, 234, 257-62, 311-12, 317-18, 342, 417, 423, 492, 494, 496

Kennedy Space Center (KSC), Florida, 369, 371, 378, 381, 383-85, 398, 491, 532-33

Kenya, 37

Kerr, Robert S., 317, 418, 492-93

Khrushchev, Nikita, 11-12, 147, 149, 152-53, 156, 159, 161, 164, 170, 175, 178, 180

Killian, James R., 238-39, 273, 274, 298; and ARPA, 249-51, 254-55, 285-86; and creation of NASA, 248-53; and defense reorganization, 247-48; and DSPA, 246-50; and "Freedom of Space," 239, 244-45, 273; and scientific satellite program, 240-42

Kinzel, Augustus B., 510

Kirillin, V. A., 202, 212

Kirk, John E., 363

Kissinger, Henry, 42, 52-53, 57-58, 63, 65-69, 79, 81, 190, 196-98

Kistiakowsky, George B., 298

Kitt Peak National Observatory, 430, 542

Knolle, Frank, 370

Kohl, Helmut, 106

Koptev, Yuri, 220, 228, 232

Kosygin, Alexei, 14

Kraft, Christopher C., Jr., 203, 369-70; and *Report of the Space Shuttle Management Independent Review Team,* 434

L

L3S launch vehicle, 85-86; also see Ariane launch vehicle

Laird, Melvin, 367

Lambright, W. Henry, 411-435

Land, Edwin (Din), 239, 250

Landsat program, 432, 556, 560

Langley Research Center, Virginia, 355, 476-77, 481-82

Large Launch Vehicle Planning Group (LLVPG), 318-37

Large Space Telescope (LST), 48-49, 548; also see Space Telescope and Hubble Space Telescope
Latin America, 182
Launius, Roger D., 412-13
Lavrow, V. A., 187
Lawrence, Harold R., 17
Lee, Chet, 200-01
Lefevre, Theo, 53, 56, 59, 61-62, 82
Leighton, Robert B., 511
LeMay, Gen. Curtis E., 236, 312
Leust, Reimar, 110, 142
Lewis, Robert B., 512-13
Lincoln Center, New York City, 487
Lindvall, Frederick C., 511
Life magazine, 176
Limited Test Ban Treaty, 13
Lincoln Laboratory, 476
Lockheed Aircraft, Inc., 239, 243, 246, 254, 482
Lockheed Martin Corporation, 434
London, 19, 105, 108, 203
London Economic Summit, 107-108
Long, Russell, 415
Louisiana State University, 522
Lovelace, Alan M., 26-27, 215-17, 386-88
Low, George M., 14, 34-35, 39, 49-50, 52, 58, 63, 68, 189-90, 197-99, 201, 204, 213-15, 366-67, 378-81, 389-90
Loweth, Hugh, 541
Lucas, William
Luedecke, A. R., 509, 516
Luna program, 211
Lunar Excursion Module (LEM), 176-77, 427
Lunar Module (LM), 35
Lunar and Planetary Missions Board, 539
Lunar Science Institute, 429-30, 538-39
Lunney, Glynn S., 183-84, 187, 189, 190-95, 198, 200-05, 208, 210, 212, 215

M

MacDonald, Frank, 32, 34
MacGregor, Clark, 198
Magruder, Jeb, 67
Main Battle Tank program, 55
Manned Orbital Laboratory (MOL), 262-65, 322
Manned Orbital Research Laboratory (MORL), 360
Man-Tended Free Flyer (MTFF), 113, 116-42
Manned Spacecraft Center (MSC), 43, 183, 190-91, 369-71, 418-20, 427, 428-30, 491, 495, 538-39; also see Johnson Space Center and Houston
Manned Space Flight Experiments Board (MSFEB), 361-64
Manno, Vittorio, 34
Mark, Hans, 90, 264-66, 268
Mariner, Project, 504-05, 561
Mars, 11, 14, 42, 44, 113, 145-52, 180, 211, 214, 220
Mars Geoscience Climatology Orbiter (MGCO) program, 33
Mars Observer program, 220
Marsh, Gen. Robert T., 266, 390, 398
Marshall Space Flight Center (MSFC), Alabama, 320, 322, 369-71, 419, 427, 475, 477, 482, 537, 542, 544-45
Martin, John J., 264, 382, 385-87

Newell, Homer E., Jr., 2, 17-18, 363, 430, 511
New York City, 149, 158, 184
New York Times, 165, 191
Nimbus satellite program, 153, 560-61
Nixon, Richard M., 6-8, 14, 34-35, 40, 48-50, 62-63, 71-72, 91, 95-96, 100, 198, 250, 262, 263, 364, 425
Noordwijk, 38
Norman, Memphis, 541
North American Air Defense (NORAD) system, 98
North American Aviation, Inc., 237, 419, 426-29, 527-37
North American Rockwell, 371, 375
North Atlantic Air Traffic Control and Navigation Satellite program, 41
North Atlantic Treaty Organization (NATO), 2, 38, 97, 275-76
Northrop Corporation, 510
Nova launch vehicle, 306-08, 325, 328-30
Novosty news agency, 212
NPO Energiya, 221, 224

O

O'Brien, Gerald D., 438, 442-72
Odishaw, Hugh, 496
Office of Management and Budget (OMB), 26, 51, 96, 198-99, 380, 400, 541, 571, 579, 585
Office of Science and Technology Policy (OSTP), 63, 96, 103, 167-68, 346, 571
Ohio University Research Foundation, 237
Olthof, Henk, 32, 34
Orbiter, Project, 240
Orbiting Astronomical Observatory (OAO), 545, 561
Orbiting Geophysical Observatory, 523
Ottawa, 105
Outer Space Treaty, 109

P

Pacific Missile Range, 18
Page, Edgar, 34
Paine, Thomas O., 6-7, 13, 35, 39-40, 42, 44-45, 49, 51, 62-64, 66, 72, 78, 95, 182-83, 263, 364, 367-68
Pakistan, 25
Palomar Observatory, 512
Paris, 21, 26, 31, 43, 81, 105, 123, 191
Paris Air Show, 207
Parker, Robert N., 386-88
Parks, R. J., 509
Parsons Company, see Ralph M. Parsons Company
Pasteur, Louis, 499
Pathfinder, Project, 578, 580
Patrick Air Force Base, 371
Pedersen, Kenneth S., 9, 27, 34, 90, 100, 109
Pellerin, Charles, 34
Peterson, Pete, 67
Petrov, Boris N., 187, 191, 195, 200-06, 212-13, 216
Phleger, Herman, 272-74
Phillips, Franklyn W., 17
Phillips, Samuel C., 428, 527-28, 538; and "Phillips Report," 428-29, 527
Phobos program, 220
Pickering, William H., 244, 246, 496, 504-05, 508, 510-11

Pioneer spacecraft, 560-61
Pitzer, Lyman, 496, 540
Platt, Mark, 107
Pollack, Herman, 22, 58, 63-64, 82-85
Polar Platform, 111-42
Polaris Missile, 332, 476-77, 491
Polaroid, Inc., 239
Porter, Richard, 17, 19
Power, Gen. Thomas, 243
Pratt and Whitney, 371
"Preliminary Design for an Experimental World Circling Spaceship," 236
President's Board of Consultants on Foreign Intelligence Activities, 250
President's Science Advisory Committee (PSAC), 247, 301, 541; and defense reorganization, 247-48, 298
Presidium (of Soviet Academy of Sciences), 194-95, 201, 203; also see Soviet Academy of Sciences
Price Waterhouse, 508
"Priorda" module, 229, 231
Project Moonbase, 233
Prospector program, 504
Prussia, 490
Putt, Lt. Gen. Donald L., 280-85

Q

Quarles, Donald A., 238, 240-42, 243, 246, 268; and creation of NASA, 248-53; and defense reorganization, 247-48; and DSPA, 246-50
Quayle, J. Danforth (Dan), 433, 582-83
Quistgaard, Erik, 26

R

Radio Corporation of America (RCA), 237, 243
Ralph M. Parsons Company, 371
Ramsey, Norman, 548
RAND, Project, 236, 237, 271-72
Ranger, Project, 423-24, 482, 490, 503-06, 511, 515
Reagan, Ronald, 5, 9-10, 14, 26-27, 96, 105, 108, 217, 266, 432-33, 565, 572, 575, 577
Redstone Arsenal, Huntsville, Alabama, 238, 412
Redstone Missile, 252-53
Remote Control (Recon) system, 41
Remote Manipulator System (RMS), 7, 98; also see Canada
Rensselaer Polytechnic Institute, 63, 68, 199, 214
Report of the Space Shuttle Management Independent Review Team, see Space Shuttle program
Republic Aviation Company, 176
Research Application Module (RAM), 48, 56, 69-70, 72-79, 81-83, 91; also see Spacelab
Research Triangle, North Carolina, 418, 493
Rice, David, 64
Rice, Donald, 66
Rice University, 418, 493, 495, 539-40
Rocketdyne Company, 371, 415-16, 482
Rockwell International Corporation, 434
Rodgers, F. M., 369
Rogers, William P., 8, 52, 54, 57, 65-66, 68, 79-80, 197
Rome, 36-37, 105
Rose, John, 64
Ross, Frank, 370

Rover project, 452
Royal Society (United Kingdom academy of science), 3; and British National Space Committee, 19
Rubel, John H., 320, 338
Rumsfield, Donald, 390
Rusk, Dean, 22, 47
Russia, 15, 220-32
Russian Space Agency (RSA), 220-232

S

Sagdeyev, Roald S., 202, 211
Salt Lake City, 52, 59
Salyut program, 14-15, 190, 193, 197, 206, 214, 216; and *Salyut*-Shuttle program, 216
SAMOS reconnaissance program, 254, 262
San Clemente (California), 190
Sanders, Newell, 474
San Diego, 197
San Marco satellite, 38
Satellite Instructional Television Experiment (SITE), 24
Saturn, 33, 560
Saturn Launch Vehicle program; 146, 263, 286, 307-08, 323-24, 326-27, 330-31, 334, 415, 419, 427-28, 482, 527-37
Schlesinger, James, 264-65, 378
Schriever, Gen. Bernard A., 237, 242, 243, 257, 268, 312, 363
Science Adviser, Office of, see Office of Science and Technology Policy (OSTP)
Science, 560
Science News, 560
Science and Technology Agency (STA) of Japan, 111-42
Scout launch vehicle, 18, 20, 482
Scull, John R., 474
Seaborg, Glen, 492-93
Seamans, Robert C., Jr., 258, 263, 318, 320, 337-39, 347-48, 364, 366-69, 417-19, 422-23, 427, 488-90, 511
Seitz, Frederick, 538, 541
Senate, U.S., 1, 16, 316-17, 415, 429, 442, 458, 460, 493, 573, 575
Senior Interagency Group (SIG) (Space), see National Security Council
Sharp, Robert P., 511
Shepard, Alan, 494
Shevardnadze, Eduard, 217, 219
Shultz, George P., 10, 105-106, 217, 219
Singer, S. Fred, 237-37
"Skunkworks," Lockheed, 239
Skybolt program, 55
Skylab program, 201
Sohier, Walter D., 438, 474
Solar and Heliospheric Observatory Satellite (SOHO) program, 32-33
Solaris space station, 100
South Africa, 157
Southern Interstate Nuclear and Space Board, 487
Soviet Academy of Sciences, 12, 44, 88, 152, 168, 170, 177-78, 182, 189-90, 192-96, 199, 202, 213-15
Soviet Ministry of General Machine Building, 220
Soviet Union, 1, 5-6, 11-15, 42, 47, 87-88, 143-220, 328, 350, 367, 417, 583
Soyuz program, 185, 188-90, 193-211, 214-15, 221, 224
Space Astronomy Institute (SAI), 541
Spacehab, 578
Space Infrared Telescope Facility (SIRTF) program, 32

Space and Missile Systems Organization (SAMSO), 370-71

Spacelab, 5, 9, 81, 86-89, 91-92, 95, 97, 99

Space Museum of the Soviet Union, 203, 208

Space Science Board, see National Academy of Sciences

Space Shuttle program, 7-8, 14, 26, 32, 43, 45, 48-49, 54-55, 58, 63, 65-67, 69-70, 74-88, 91, 94, 98, 109, 214-16, 221-32, 366, 431, 433-34, 554, 556, 561, 573, 577-81; and Hubble Space Telescope, 544-45, 548; and military, 263-69, 367-410; and *Report of the Space Shuttle Management Independent Review Team,* 434; and Shuttle Carrier Aircraft, 406; and Shuttle-Mir program, 223-32; also see Space Transportation System

Space Station program, 9-10, 32, 45, 69-70, 87-100, 105-42, 356, 398, 579-81; also see International Space Station program

Space Station Task Force, 9, 90-96

Space Task Group, 6-7, 40, 43-44, 263, 364-65, 383, 494

Space Telescope, 542-49, 554, 557; also see Hubble Space Telescope

Space Telescope Science Institute, 430, 541-49

Space Transportation System (STS), 53, 56, 78-79, 83-84, 95, 97, 99, 116-17, 119-20, 125, 127-28, 136, 139; and military, 364-65, 367-410; and Space Transportation System Committee, 263-64, 369-77, 383, 390; also see Space Shuttle program

Space Tug, 55-56, 58, 65, 67, 69-70, 72-78, 80-82, 85, 91-92, 372-73, 378-79

Sparks, Brian, 504-05, 508

Spaso House, 192

"Spektr" module, 229, 231

Spinning Solid Upper Stage (SSUS), 382

Sputnik, 2, 234, 245-47, 251, 411

Stafford, Thomas, 201, 205

Stuhlinger, Ernst, 244

Star City, 202-03, 205-13, 225; also see Yuri Gagarin Cosmonaut Training Center

State, Department of, 7-9, 11, 22, 24, 30, 35, 39, 41, 45-46, 49, 58-59, 63-66, 68-69, 72, 76, 78, 81, 96, 103, 105, 162, 166-68, 172, 177, 191, 197, 240, 244, 494; and Office of United Nations Affairs, 469

"Statement of Policy for a Satellite Vehicle," 236

Steelman, Donald L., 369

Stevenson, Adlai E., 179

Stewart, Homer J., 509

Strashevsky, G. S., 153

Surveyor, Project, 504-05

Suslennikov, V. V., 187-88

Sweden, 38, 43, 97

Switzerland, 38, 97

Symphonie communications satellite program, 39

Synchronous Meteorological Satellite program, 41

Syromyatnikov, V. S., 187, 201

T

Taylor, Brian, 34

Teague, Olin E. (Tiger), 163

Tech, Jack, 191, 200

Technological Capabilities Panel, 238-39, 273-76

Tennessee Valley Authority (TVA), 471-72, 487

Test Ban Treaty, see Limited Test Ban Treaty

Tether program, 99

Texas Instruments, Inc., 492

Thacher, Peter, 153, 155-56, 158

Thomas, Albert, 418, 492-93

Thor Delta launch vehicle, 18, 252; also see Delta launch vehicle

Thor launch vehicle, 38, 332

Thurer, Byron, 371
Tilford, Shelby, 34
Timerbaev, Roland H., 153
Television and Infrared Operational Satellite (TIROS) weather satellite program, 153, 481, 561
Titan (moon of Saturn), 33
Titan launch vehicle program, 166, 263, 267, 323-24, 326-27, 330-32, 343, 373, 390-92, 395, 399, 400, 410
Toftoy, Holger, 235
Tokyo, 23, 46, 105
Townsend, John W., 153, 155, 157-58
Tracking and Data Relay Satellite System (TDRSS) space network, 117-21, 130
Transportation, Department of, 41, 367, 433, 579-80; and Office of Commercial Space Transportation, 433
Treasury, Department of, 519
Truman, Harry S, 238
Turner, Thomas, 176-77
Twining, Gen. Nathan F., 254

U

Uhuru, 211
Union Carbide Corporation, 490, 510
United Kingdom (U.K.), 3, 19-20, 37-38, 42-43, 88, 107-108; and U.K. Steering Group on Space Research, 20
United Nations (UN), 12-13, 35, 43, 147-53, 156, 158, 161, 165-66, 170-71, 178-79; and Outer Space Affairs Group, 43; and Outer Space Committee, 43, 149, 161
United Space Alliance (USA), 434
University of Arizona, 522
University of California at San Diego, 418, 493
University of California at Los Angeles, 521
University of Chicago, 493
University of Japan, 23
University of Michigan, 43, 521, 525, 553
University of Tokyo, 22
University of Utah, 52, 59
University Space Research Association (USRA), 430, 541
Uranus, 560
U.S. Information Agency, 167
Utah, 52, 59
Utility of a Satellite Vehicle for Reconnaissance, 237

V

V-2 rocket, 235, 238, 240
Van Allen, James A., 157, 237-38, 244; and creation of NASA, 248-53
Vandenberg, Gen. Hoyt S., 236, 271-72
Vandenberg Air Force Base, California, 264, 265, 378, 380-81, 384-85, 391, 394, 398, 405
Vanguard Program, 242-45, 277-80
Vapor Magnetometer Project, 480
Vega rocket, 254, 482, 504-05
Venus, 11, 145-52, 220, 560-61
Venus 8 spacecraft, 212
Venus Orbiting Imaging Radar (VOIR) program, 26
Venus Radar Mapper, 554
Vesta program, 220
Vienna, 11, 23
Vietnam, 422, 425
Viking spacecraft program, 42, 44, 211, 561, 564

Vostok program, 173
Voyager program, 504, 556, 560-61, 564

W

Walsh, John, 49, 63, 66, 69, 79
Washington Post, 191
Waterman, Alan T., 240
Weapons System (WS) 117L, 243, 245, 246, 250, 262, 268
Weapons System (WS) 464L, 249
Webb, James E., 12-13, 22-23, 25, 35, 165, 168, 170, 182, 347-48, 411, 416-29, 485, 492, 494-97, 503, 508, 527, 538, 541; and Robert McNamara, 257-63, 337-47, 356-60
Weinberger, Caspar, 267, 400-01
Welsh, Edward C., 164, 317, 342, 492
Western Electric Company, 482
Western Development Division, 242-44
WESTFORD project, 156, 160
Whisenand, Brig. Gen. James F., 254
Whitcomb, Gordan, 34
White, Gen. Thomas D., 256, 257, 280-85, 311-12
Whitehead, Clay T., 49-50, 52, 58, 63, 66
White House, 5, 7-8, 14, 35, 40, 42, 44, 49, 51-52, 63-64, 66, 72, 79, 91, 96, 162, 165, 168, 189-91, 196-97, 220, 298, 315, 417, 432; and White House Economic Policy Council, 577; and White House Space Task Force Group, 95
White Sands Proving Grounds, New Mexico, 240
Wiesner, Jerome B., 11, 143, 164, 317, 338
Wigbels, Lynnette A., 34
Wilford, John Noble, 191
Wilson, Jim, 379
Wilson, Robert, 370
Woods Hole, Massachusetts, 542
World Meteorological Organization (WMO), 148, 153-54, 157, 159, 163
Wright Field (later Wright-Patterson Air Force Base), Ohio, 243, 355

X

X-15, 249
XB-70, 355

Y

Yardley, John F., 264, 382, 385-87
Yeltsin, Boris N., 15, 220

Z

Zaitzev, Valentin A., 153, 155, 158
Zisch, William E., 510
Zurich, Switzerland, 237

The NASA History Series

Reference Works, NASA SP-4000

Grimwood, James M. *Project Mercury: A Chronology* (NASA SP-4001, 1963)

Grimwood, James M., and Hacker, Barton C., with Vorzimmer, Peter J. *Project Gemini Technology and Operations: A Chronology* (NASA SP-4002, 1969)

Link, Mae Mills. *Space Medicine in Project Mercury* (NASA SP-4003, 1965)

Astronautics and Aeronautics, 1963: Chronology of Science, Technology, and Policy (NASA SP-4004, 1964)

Astronautics and Aeronautics, 1964: Chronology of Science, Technology, and Policy (NASA SP-4005, 1965)

Astronautics and Aeronautics, 1965: Chronology of Science, Technology, and Policy (NASA SP-4006, 1966)

Astronautics and Aeronautics, 1966: Chronology of Science, Technology, and Policy (NASA SP-4007, 1967)

Astronautics and Aeronautics, 1967: Chronology of Science, Technology, and Policy (NASA SP-4008, 1968)

Ertel, Ivan D., and Morse, Mary Louise. *The Apollo Spacecraft: A Chronology, Volume I, Through November 7, 1962* (NASA SP-4009, 1969)

Morse, Mary Louise, and Bays, Jean Kernahan. *The Apollo Spacecraft: A Chronology, Volume II, November 8, 1962–September 30, 1964* (NASA SP-4009, 1973)

Brooks, Courtney G., and Ertel, Ivan D. *The Apollo Spacecraft: A Chronology, Volume III, October 1, 1964–January 20, 1966* (NASA SP-4009, 1973)

Ertel, Ivan D., and Newkirk, Roland W., with Brooks, Courtney G. *The Apollo Spacecraft: A Chronology, Volume IV, January 21, 1966–July 13, 1974* (NASA SP-4009, 1978)

Astronautics and Aeronautics, 1968: Chronology of Science, Technology, and Policy (NASA SP-4010, 1969)

Newkirk, Roland W., and Ertel, Ivan D., with Brooks, Courtney G. *Skylab: A Chronology* (NASA SP-4011, 1977)

Van Nimmen, Jane, and Bruno, Leonard C., with Rosholt, Robert L. *NASA Historical Data Book, Vol. I: NASA Resources, 1958–1968* (NASA SP-4012, 1976, rep. ed. 1988)

Ezell, Linda Neuman. *NASA Historical Data Book, Vol II: Programs and Projects, 1958–1968* (NASA SP-4012, 1988)

Ezell, Linda Neuman. *NASA Historical Data Book, Vol. III: Programs and Projects, 1969–1978* (NASA SP-4012, 1988)

Astronautics and Aeronautics, 1969: Chronology of Science, Technology, and Policy (NASA SP-4014, 1970)

Astronautics and Aeronautics, 1970: Chronology of Science, Technology, and Policy (NASA SP-4015, 1972)

Astronautics and Aeronautics, 1971: Chronology of Science, Technology, and Policy (NASA SP-4016, 1972)

Astronautics and Aeronautics, 1972: Chronology of Science, Technology, and Policy (NASA SP-4017, 1974)

Astronautics and Aeronautics, 1973: Chronology of Science, Technology, and Policy (NASA SP-4018, 1975)

Astronautics and Aeronautics, 1974: Chronology of Science, Technology, and Policy (NASA SP-4019, 1977)

Astronautics and Aeronautics, 1975: Chronology of Science, Technology, and Policy (NASA SP-4020, 1979)

Astronautics and Aeronautics, 1976: Chronology of Science, Technology, and Policy (NASA SP-4021, 1984)

Astronautics and Aeronautics, 1977: Chronology of Science, Technology, and Policy (NASA SP-4022, 1986)

Astronautics and Aeronautics, 1978: Chronology of Science, Technology, and Policy (NASA SP-4023, 1986)

Astronautics and Aeronautics, 1979–1984: Chronology of Science, Technology, and Policy (NASA SP-4024, 1988)

Astronautics and Aeronautics, 1985: Chronology of Science, Technology, and Policy (NASA SP-4025, 1990)

Gawdiak, Ihor Y. Compiler. *NASA Historical Data Book, Vol. IV: NASA Resources, 1969–1978* (NASA SP-4012, 1994)

Noordung, Hermann. *The Problem of Space Travel: The Rocket Motor.* In Ernst Stuhlinger and J.D. Hunley, with Jennifer Garland, editors (NASA SP-4026, 1995)

Management Histories, NASA SP-4100

Rosholt, Robert L. *An Administrative History of NASA, 1958-1963* (NASA SP-4101, 1966)

Levine, Arnold S. *Managing NASA in the Apollo Era* (NASA SP-4102, 1982)

Roland, Alex. *Model Research: The National Advisory Committee for Aeronautics, 1915-1958* (NASA SP-4103, 1985)

Fries, Sylvia D. *NASA Engineers and the Age of Apollo* (NASA SP-4104, 1992)

Glennan, T. Keith. *The Birth of NASA: The Diary of T. Keith Glennan,* edited by J.D. Hunley (NASA SP-4105, 1993)

Seamans, Robert C., Jr. *Aiming at Targets: The Autobiography of Robert C. Seamans, Jr.* (NASA SP-4106, 1996)

Project Histories, NASA SP-4200

Swenson, Loyd S., Jr., Grimwood, James M., and Alexander, Charles C. *This New Ocean: A History of Project Mercury* (NASA SP-4201, 1966)

Green, Constance McL., and Lomask, Milton. *Vanguard: A History* (NASA SP-4202, 1970; rep. ed. Smithsonian Institution Press, 1971)

Hacker, Barton C., and Grimwood, James M. *On Shoulders of Titans: A History of Project Gemini* (NASA SP-4203, 1977)

Benson, Charles D. and Faherty, William Barnaby. *Moonport: A History of Apollo Launch Facilities and Operations* (NASA SP-4204, 1978)

Brooks, Courtney G., Grimwood, James M., and Swenson, Loyd S., Jr. *Chariots for Apollo: A History of Manned Lunar Spacecraft* (NASA SP-4205, 1979)

Bilstein, Roger E. *Stages to Saturn: A Technological History of the Apollo/Saturn Launch Vehicles* (NASA SP-4206, 1980)

Compton, W. David, and Benson, Charles D. *Living and Working in Space: A History of Skylab* (NASA SP-4208, 1983)

Ezell, Edward Clinton, and Ezell, Linda Neuman. *The Partnership: A History of the Apollo-Soyuz Test Project* (NASA SP-4209, 1978)

Hall, R. Cargill. *Lunar Impact: A History of Project Ranger* (NASA SP-4210, 1977)

Newell, Homer E. *Beyond the Atmosphere: Early Years of Space Science* (NASA SP-4211, 1980)

Ezell, Edward Clinton, and Ezell, Linda Neuman. *On Mars: Exploration of the Red Planet, 1958-1978* (NASA SP-4212, 1984)

Pitts, John A. *The Human Factor: Biomedicine in the Manned Space Program to 1980* (NASA SP-4213, 1985)

Compton, W. David. *Where No Man Has Gone Before: A History of Apollo Lunar Exploration Missions* (NASA SP-4214, 1989)

Naugle, John E. *First Among Equals: The Selection of NASA Space Science Experiments* (NASA SP-4215, 1991)

Wallace, Lane E. *Airborne Trailblazer: Two Decades with NASA Langley's Boeing 737 Flying Laboratory* (NASA SP-4216, 1994)

Butrica, Andrews J. *To See the Unseen: A History of Planetary Radar Astronomy* (NASA SP-4218, 1996)

Center Histories, NASA SP-4300

Rosenthal, Alfred. *Venture into Space: Early Years of Goddard Space Flight Center* (NASA SP-4301, 1985)

Hartman, Edwin, P. *Adventures in Research: A History of Ames Research Center, 1940-1965* (NASA SP-4302, 1970)

Hallion, Richard P. *On the Frontier: Flight Research at Dryden, 1946–1981* (NASA SP-4303, 1984)

Muenger, Elizabeth A. *Searching the Horizon: A History of Ames Research Center, 1940-1976* (NASA SP-4304, 1985)

Hansen, James R. *Engineer in Charge: A History of the Langley Aeronautical Laboratory, 1917-1958* (NASA SP-4305, 1987)

Dawson, Virginia P. *Engines and Innovation: Lewis Laboratory and American Propulsion Technology* (NASA SP-4306, 1991)

Dethloff, Henry C. *"Suddenly Tomorrow Came . . .": A History of the Johnson Space Center, 1957–1990* (NASA SP-4307, 1993)

Hansen, James R. *Spaceflight Revolution: NASA Langley Research Center From Sputnik to Apollo* (NASA SP-4308, 1995)

General Histories, NASA SP-4400

Corliss, William R. NASA *Sounding Rockets, 1958–1968: A Historical Summary* (NASA SP-4401, 1971)

Wells, Helen T., Whiteley, Susan H., and Karegeannes, Carrie. *Origins of NASA Names* (NASA SP-4402, 1976)

Anderson, Frank W., Jr. *Orders of Magnitude: A History of NACA and NASA, 1915–1980* (NASA SP-4403, 1981)

Sloop, John L. *Liquid Hydrogen as a Propulsion Fuel, 1945–1959* (NASA SP-4404, 1978)

Roland, Alex. A *Spacefaring People: Perspectives on Early Spaceflight* (NASA SP-4405, 1985)

Bilstein, Roger E. *Orders of Magnitude: A History of the NACA and NASA, 1915–1990* (NASA SP-4406, 1989)

Logsdon, John M., with Lear, Linda J., Warren-Findley, Jannelle, Williamson, Ray A., and Day, Dwayne A. *Exploring the Unknown: Selected Documents in the History of the U.S. Civil Space Program, Volume I: Organizing for Exploration* (NASA SP-4407, 1995)

Logsdon, John M., with Day, Dwayne A., and Launius, Roger D., eds. *Exploring the Unknown: Selected Documents in the History of the U.S. Civil Space Program, Volume II: External Relationships* (NASA SP-4407, 1996)